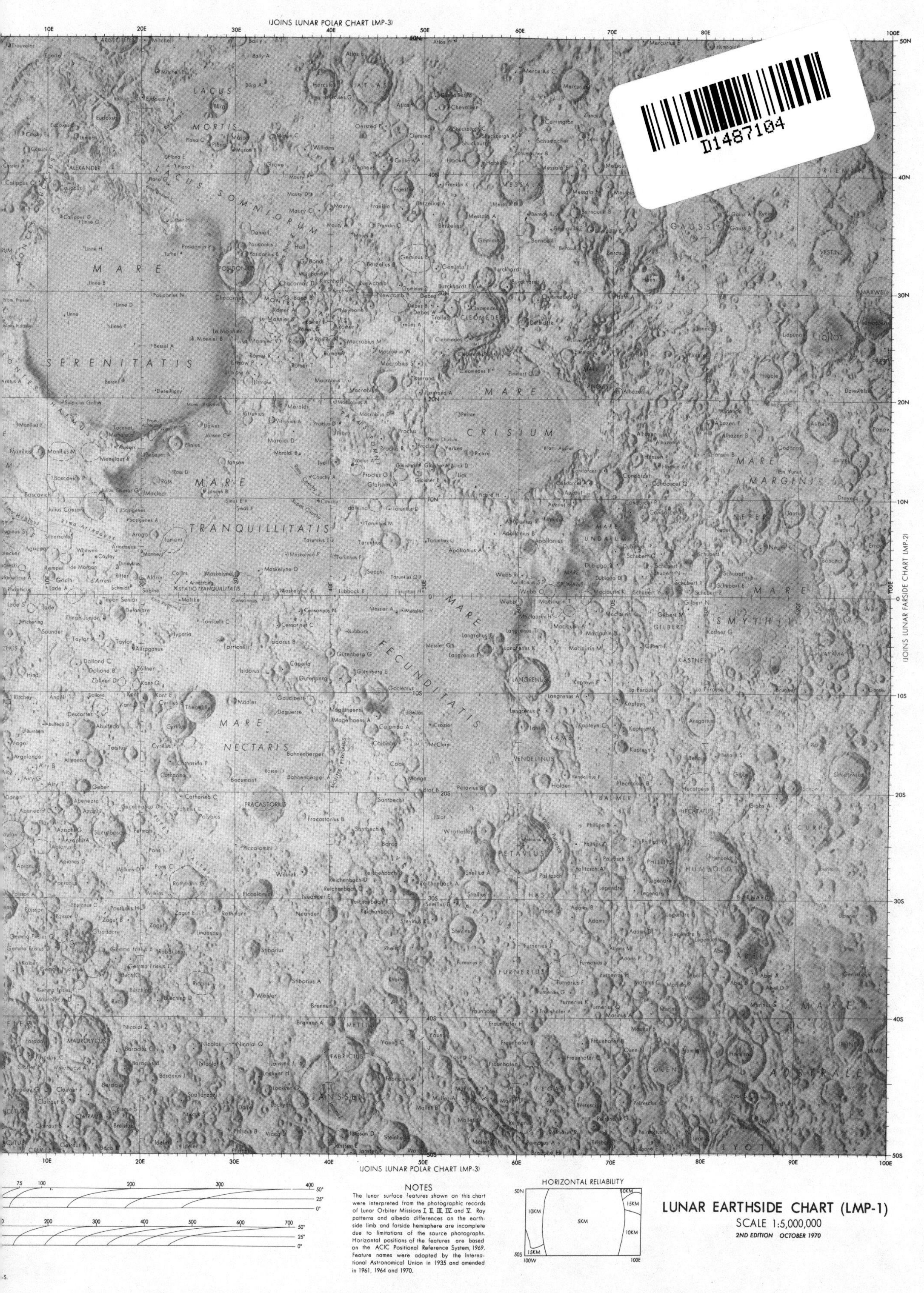

(JOINS LUNAR POLAR CHART LMP-3)
(JOINS LUNAR FARSIDE CHART LMP-2)
LACUS MORTIS
LACUS SOMNIORUM
MARE SERENITATIS
MARE CRISIUM
MARE TRANQUILLITATIS
MARE FECUNDITATIS
MARE NECTARIS
MARE UNDARUM
MARE SPUMANS
MARE MARGINIS
MARE SMYTHII
MARE AUSTRALE
HORIZONTAL RELIABILITY
10KM
15KM
5KM
10KM
15KM
50N
50S
100W
100E
NOTES
The lunar surface features shown on this chart were interpreted from the photographic records of Lunar Orbiter Missions I, II, III, IV and V. Ray patterns and albedo differences on the earth-side limb and farside hemisphere are incomplete due to limitations of the source photographs. Horizontal positions of the features are based on the ACIC Positional Reference System, 1969. Feature names were adopted by the International Astronomical Union in 1935 and amended in 1961, 1964 and 1970.
LUNAR EARTHSIDE CHART (LMP-1)
SCALE 1:5,000,000
2ND EDITION OCTOBER 1970

NATIONAL AERONAUTICS AND SPACE ADMINISTRATION

LUNAR CHART

NORTH POLAR REGION

NOTES

The Lunar Surface Features shown on this chart are interpreted from the photographic records of Lunar Orbiter Missions I, II, III, IV, and V. Ray patterns and Albedo differences on the earthside limb and farside hemisphere are incomplete due to limitations of the source photographs. Horizontal position of the features is based on the ACIC Positional Reference System, 1969. Feature names are adopted from the 1935 International Astronomical Union Nomenclature System, as amended by I.A.U. in 1961 and 1964. The application of names is restricted to only those associated features that have been positively identified.

LUNAR DATA

Earth/Moon Mass Ratio	M_e/M_m 81.3015
Density (mean)	3.34 g/(cm)^3
Synodic Month (new Moon to new Moon)	29.530 588d
Sidereal Month (fixed star to fixed star)	27.321 661d
Inclination of Lunar orbit to ecliptic	5°8′43″
Inclination of equator to ecliptic	1°32′40″
Inclination of Lunar orbit to Earth's equator	18°.5 to 28°.5
Distance of Moon from Earth (mean)	238,328M (384,400km)
Optical libration in longitude	±7°.6
Optical libration in latitude	±6°.7
Albedo (average)	0.07
Magnitude (mean of full Moon)	−12.7
Temperature	−244°F to +273°F (120°K to 407°K)
Escape velocity	1.48 mi/sec (2.38km/sec)
Diameter of Moon	2160 mi (3476km)
Surface gravity	162.2 cm/sec^2
Orbital velocity	0.64 mi/sec (Moon) 18.5 mi/sec (Earth)

PLANETARY GEOLOGY

PLANETARY GEOLOGY

Nicholas M. Short
Earth Survey Applications Division
National Aeronautics and Space Administration
Goddard Space Flight Center
Greenbelt, Maryland

Prentice-Hall, Inc.
Englewood Cliffs, New Jersey

Library of Congress Cataloging in Publication Data

Short, Nicholas M.
Planetary geology.

Bibliography: p. 335
Includes index.
1. Planets. 2. Lunar geology. I. Title.
QB601.S55 559.9 74-31303
ISBN 0-13-679290-1

Printed in the United States of America

10 9 8 7 6 5 4 3 2

Prentice-Hall International, Inc., *London*
Prentice-Hall of Australia, Pty. Ltd., *Sydney*
Prentice-Hall of Canada, Ltd., *Toronto*
Prentice-Hall of India Private Limited, *New Delhi*
Prentice-Hall of Japan, Inc., *Tokyo*

This book is dedicated to my wife, Eleanor, and son, Nick Jr.—in partial payment of the deep debt of gratitude owed them for their patience, understanding, and cooperation throughout the many weekends and nights over the last four years during which I was lost to them while this book dominated my every spare moment.

Contents

Foreword

This book is one of a series of single-topic publications intended primarily for use in undergraduate geology and earth science courses. It was prepared under the direction of the Instructional Materials Program (IMP) panel of the Council on Education in the Geological Sciences (CEGS), a National Science Foundation supported project of the American Geological Institute, and is presented in cooperation with Prentice-Hall, Inc., Englewood Cliffs, New Jersey, one of the publishers for products in the series.

The collective purpose of these works is to serve as models from which modern aspects of concepts basic to geology can be introduced. Typically, they are inquiry- and problem-oriented and deal with interdisciplinary, contemporary, and pragmatic aspects of subject matter that is difficult to treat in a more ordinary way. It is intended that the materials be open-ended so that ideas from them can be incorporated into upper division studies or extracted from them for introductory classes. It is hoped that they will inspire teachers to develop similar materials in areas of their own interest and competence.

This publication is the most extensive of the series and can be used as the core of a one-semester course. In addition, parts of units may be chosen for inclusion in an introductory geology course.

The IMP series was conceived by J. W. Harbaugh, former Director of CEGS, and work on *Planetary Geology* was started while Peter Fenner was the CEGS Director. An earlier version of this publication benefited from the critical testing and evaluation of numerous students, teachers, and other scientists. Space limitations preclude acknowledging the efforts of all those individuals by name, but we recognize with deep appreciation and gratitude that their suggestions and comments have contributed significantly to the evolution of the book.

The preliminary version was tested by 174 geology and non-science majors in 11 colleges and universities during the spring of 1972 under the direction of the following people: Brewster Baldwin, Middlebury College; James N. Gundersen, Wichita State University; René A. De Hon, Northwestern State University of Louisiana; Gene C. Ulmer, Temple University; Richard D. Dietz, University of Northern Colorado; James V. O'Connor, University of Maryland; Luciano Ronca, Wayne State University; J. Hatten Howard, University of Georgia;

W. C. Shellenberger, Northern Montana College; Richard C. Peterson, Adams State College; Dean Eppler, St. Lawrence University. We extend our gracious thanks to these teachers and their students for their detailed criticism.

Dr. Nicholas M. Short, the author, took these suggestions, corrections, and student reports into account during the extensive revision in which he also continuously updated the data by incorporating scientific results through July 1974. His cooperation with the CEGS staff and his efforts in this fast-changing field are greatly appreciated.

We would also like to thank NASA and William H. Grimshaw and Bruce Williams of the Prentice-Hall staff. Additional credit is due the CEGS support staff, Dorothy Kelty, Cheri Ford, Becky Smith, Sheldon Iskow, Jackson E. Lewis, F. D. Holland, Jr., and William H. Matthews III, who administered the testing program and handled the many details prior to publication.

Peter Fenner
George R. Rapp, Jr.

Preface

This book is an outgrowth of the realization in recent years that geology has an important role to play in the exploration of outer space. In the last few years the successful unmanned space probes to the Moon, Mars, Venus, Jupiter, and Mercury and the astronaut landings on the Moon have returned vast amounts of scientific data about planetary bodies relative to their composition, structure, surface-forming processes, and history. This new information is itself a "feedback" leading to some fundamental insights into the geologic nature of our Earth and its environs. The coupling of the wealth of knowledge and the methodology of terrestrial geology with the highly sophisticated technological systems by which planets can now be directly examined has fostered the creation of the new field of *planetary geology.* This discipline, alternatively called *astrogeology,* represents in the stricter sense an extension of the study of the Earth to embrace all those members of the solar system in which rock materials are thought to be a major component. In a more general sense the science of *planetology* considers not only the Moon, Mars, Mercury, and Venus and the asteroids and meteorites, but is concerned also with the gas-rich outer planets as other samples of the evolutionary products from a cosmogony operating in our part of the universe.

Heretofore, planetary geology has seemed much too esoteric and specialized to fit into the general program of studies through which geologists are trained. Any mention of the fruits of the space program as they apply to the earth sciences was once restricted to outside readings or brief surveys incorporated in certain graduate courses. However after results from Apollo 11 and later missions were spotlighted before the public, it became obvious that the concepts and implications arising from our first on-site inspection of another planetary body must be carried right through the undergraduate curriculum even to the most basic levels in the introductory courses.

Planetary Geology was originally conceived and prepared as a contribution of about 150 pages to the Concepts in Introductory Geology module series sponsored by the Council on Education in the Geological Sciences (CEGS) of the American Geological Institute. In the early stages of preparation, the range of topics and level of presentation were carefully attuned to the non-major seeking additional knowledge to round out his experience while meeting the science requirements in a traditional liberal arts program. As the book evolved, however, the scope and depth of development appearing in the present version were continually expanded in response to a growing awareness of the diversity and complexity of the subject matter which must be treated to properly encompass planetary geology. After work on the book first began in 1970, each subsequent major conference or summary report on results of space exploration opened up fresh topical areas demanding coverage and added volumes of data deserving to be documented. Then too, as soon as sections of the book were recast to accommodate this, NASA or the Russians would launch another probe to the Moon or a planet, generating quantities of exciting and provocative information that simply had to be fitted into the text. This indeed has been the mark of the space age since its inception: fast-moving events leading to one discovery or new concept after another. The literature of space, once dominated by science fiction, has burgeoned into shelf after shelf of reports, reviews, technical documents, symposia proceedings, and popular summaries for the public.

Yet even today there remains a glaring lack of textbooks directed toward presenting a comprehensive overview of the accomplishments in space exploration to the serious student in search of a professional career in science, a supplemental background in space technology, or perhaps only a better understanding of the planets within the context of his educational growth. This is particularly true where geology and the space program find common ground. Only a very few books that touch upon planetary geology are available as references, and none of these is uniquely oriented to teaching the subject. Thus, it was decided to organize this book at a level that makes it suitable both as a survey text for undergraduate geology majors and as a sourcebook for graduate students and practicing professionals alike. In this respect, I have attempted to achieve what may well prove impractical, if not impossible; that is, to structure the book in such a way that it can still be used, with judicious selection and appropriate guidance, in those introductory courses in which teachers choose to give extraterrestrial geology a prominent role. At the least, the introductory student who merely reads critically through each chapter, peruses the illustrations, and answers the review questions can learn enough to justify the effort. Regardless of how it is used, *Planetary Geology* provides a practical test of how well one has learned to apply the understanding obtained from other geological studies, inasmuch as many principles of mineral and rock formation, stratigraphy, landform development, and the like must be extrapolated from the more familiar terrestrial examples to the exotic, somewhat abstract realm of the Moon and planets.

The prime objective of this book is to survey the important facts and principles that together comprise the subject matter of planetary geology. In many instances such knowledge can be redirected toward a better appreciation of the geology of the Earth itself. Some tie-ins with our awakening concern for the total environment in which we live may even be extracted from the perspectives being gained through man's penetration of outer space. Wherever appropriate, data and practical applications relevant to terrestrial investigations in the earth sciences that result from the study of the other planets are examined in varying detail within the book. However, as conceived and developed, *Planetary Geology* is mainly concerned with utilizing geologic principles as guides and techniques for the scientific exploration of the planets. The emphasis, therefore, is on the results pertinent to the planets themselves rather than on the continuing reappraisal of the significance of these new discoveries to Earth-oriented geology alone.

Although "planetary" is the key word in the book's title, the critical reader will note that the subject matter appears heavily weighted toward the Moon and its exploration. This reflects a simple fact: Most knowledge of other planetary bodies of direct significance to geologists is still confined to what has been learned from the Moon. This is the obvious consequence of our satellite's proximity and hence its accessibility to effective examination by telescope and space probe and, since Apollo 11, by men with some understanding of geological principles and methods. Thus, most of the book after Chapter 4 concentrates on the results of lunar studies and their implications for our concepts of the planets. However, the approaches and techniques developed to explore our nearest neighbor, as discussed in the chapters dealing with lunar observations by telescope and unmanned probes that lead to descriptive maps, are essentially the same as those now being applied to Mercury, Venus, Mars, and Jupiter. This becomes evident in Chapter 12 in which the tremendous advances brought about by the recent Mariner, Pioneer, and Venera space probes are compared with earlier knowledge of the planets secured almost exclusively from the telescope. Currently, Mars and Mercury are being mapped in much the same manner and detail as the Moon. As Viking and other unmanned landers or fly-by satellites return visual or instrumental data for other planets, it should be possible within a decade to recast a book such as this to achieve a new balance between the Moon and the other members of the solar system insofar as their geological characteristics are concerned. Meanwhile, the reader who wishes to learn more of what is known today about the planets from the viewpoint of physicists and astronomers should consult such excellent references as W. K. Hartmann's *Moons and Planets* and W. M. Kaula's *Introduction to Planetary Physics* (see References). Hartmann's book, in particular, provides an exceptional supplement in scope and treatment to *Planetary Geology* as it delves extensively into many nongeological topics that are afforded only limited consideration in the present book. Other sources of information and materials useful in teaching or self-study of planetary geology are listed under Sources of Materials at the back of the book.

Insofar as practical, the book is constructed from professed facts as documented in the open literature. I have tried to be objective in the many instances where several competing hypotheses or explanations developed from the facts are still seriously considered. Thus, alternate ideas are presented impartially with coequal weight despite my own knowledge that some are less favored than others in the current rankings. But numerical and conceptual misstatements are inevitable—especially in sections of the book dealing with the most recent results in the Apollo and Mariner programs. The written literature has not yet caught up with the preliminary reports, papers at meetings, and press conferences which, for some topics, are still the only available data sources. The reader should be fully cognizant of the tentative, rather than dogmatic, label that must be placed on those parts of the book which attempt to chronicle the implications and interpretations growing out of accomplishments in space in the last few years.

It has proved impossible, however, to record the reference or data sources for the majority of the facts and interpretations discussed throughout the book without cluttering up almost every sentence with one or more citations. Literally hundreds of papers, books, personal communications, newspaper clippings—even tape recordings made while attending meetings—were consulted while I prepared the book. The references given at the end of the book serve as a comprehensive list of the publications used as the data base. Specific references are cited in the text or captions wherever a particular statement requires support from or credit to the precise source.

A select group of professional workers within the planetary sciences, contacted at various stages as the book progressed, were another indispensable source of assistance in checking the facts, assessing the interpretations, correcting for omissions, and evaluating the overall effectiveness of the text. I am grateful to the following for their constructive comments and advice prior to the first revision of the manuscript: B. Glass, University of Delaware; M. H. Hait, U.S. Geological Survey; W. Kaula, University of California at Los Angeles; P. D. Lowman, NASA, Goddard Space Flight Center; G. McGill, University of Massachusetts; T. Mutch, Brown University; V. R. Oberbeck, NASA, Ames Research Center; J. A. O'Keefe and C. C. Schnetzler, NASA, Goddard Space Flight Center; and D. E. Wilhelms, U.S. Geological Survey.

After the test version was distributed, these individuals kindly offered helpful suggestions for improvements: E. Anders, University of Chicago; R. B. Baldwin, Oliver Machinery Co.; R. Brett, NASA, Johnson Spacecraft Center; S. G. Brush, University of Maryland; W. S. Cameron, NASA, Goddard Space Flight Center; G. W. Colton, U.S. Geological Survey; W. Elston, University of New Mexico; G. Fielder, University of Lancaster (England); B. M. French, National Science Foundation; V. Gornitz, NASA, Institute for Space Studies, New York; B. Hapke, University of Pittsburgh; W. K. Hartmann, Tucson; J. Hartung, NASA, Johnson Spacecraft Center; A. Marcus, Johns Hopkins University; C. C. Mason, NASA, Johnson Spacecraft Center; J. F. McCauley, U.S. Geological Survey; J. Pearl and J. Philpotts, NASA, Goddard Space Flight Center; W. Phinney, NASA, Johnson Spacecraft Center; E. Roedder and L. C. Rowan, U.S. Geological Survey; and G. J. Wasserburg, California Institute of Technology.

To the many contributors of photographs and line drawings, acknowledged where appropriated in the figure captions, I offer my sincere gratitude for their generosity. Mrs. M. Ware of NASA Headquarters; R. Wright of NASA, Johnson Spacecraft Center; D. Chiddix, NASA, Goddard Space Flight Center; and G. W. Colton, U.S. Geological Survey merit my personal thanks for their responses to my requests for illustrations.

Special thanks are given to P. Hewitt of the State University of New York at Brockport and to P. Fenner of Governors State University (Illinois) for their generous assistance in editing the original and revised manuscripts. The aid and counsel provided first by F. D. Holland and then W. H. Matthews III, successive Directors of Education at the American Geological Institute, along with the support of R. Smith and others of their staff, were essential to the successful completion of all major phases of my work on the book during the last year and a half.

Greenbelt, Maryland NICHOLAS M. SHORT

CHAPTER 1

Prologue: A History of Man's Quest to the Stars

Imagine the excitement, the challenge, and the burst of activity generated among the world's scientific community if overnight, in some incredible way, an entirely new continent were to emerge from the sea—ready for inspection. Experts from every discipline would be mobilized and committees appointed to organize a comprehensive and methodical program for exploration and research. Expeditions with teams of specialists carrying the latest in equipment would be dispatched to every accessible point to map and sample on a grand scale. All this would culminate in symposia and seminars followed by volumes of learned papers. In short, we can be sure in this twentieth-century age of science that discovery of such a vast area of unknown lands would bring about the complete marshaling of the best available talent and resources to conduct one of the most thorough and rewarding intellectual enterprises ever attempted.

Is this just fantasy, or a touch of science fiction? Not when we reflect on the greatest adventure in exploration since Marco Polo, Christopher Columbus, Captain Cook, and others expanded our horizons to Asia and the New World. In less than two decades, we have left the confines of our own planet, penetrated outer space, and reached the Moon, Mars, Venus, Mercury, and Jupiter with instruments capable of observing and carefully measuring many of their physical and chemical features. Finally, in a triumph of dedication and cooperation perhaps never before matched by civilization, we have placed man himself on the surface of a new "continent" beyond the Earth.

The Moon is just that—a new continent now available for direct study. Its surface area on the side visible from Earth is over 19 million square kilometers, comparable to all of South America in extent. Its farside, accessible to us for the first time through photographic documentation, is similar in size to Antarctica. But, unlike the terrestrial continents whose surfaces may be partially hidden by vegetation or ice, the Moon is completely free of any masking cover other than its own rock debris. What a joy for geologists to contemplate—a pristine surface of rock materials still preserved to allow us to reconstruct a record of planetary history extending to the dawn of the solar system. And, as if this new "continent" were not enough, in 1972 the entire surface of Mars was photographed in preparation for eventual landings by unmanned exploratory probes.

This present "golden era" of direct exploration of the

nearby planets stands in strong contrast to the past. Not long ago, solar system astronomers—particularly selenologists (those who study the Moon)—were a lonely lot. Most astronomers looked far out to the stars rather than to our planetary neighbors for information about the universe. This has not changed but, as rockets become a reality and satellites proved feasible, a new emphasis was placed on the solar system as a practical object for study.

The Beginnings of Space Exploration

The modern space age is a culmination of five thousand years of a persistent, and often enlightened, search for an understanding and meaning to the visible celestial bodies. Astronomy, the most ancient of the sciences, provided the observational basis for cosmology—originally a branch of philosophy concerned with the nature and origin of the universe. The Babylonians first chartered the constellations, followed in Egypt and China by studies of planetary motions and in Greece by a series of measurements of the Earth and its neighbors that placed astronomy on a quantitative level. The Moon received particularly reverent attention because of its beauty. Anaxagoras in the fourth century B.C. correctly associated the phases of the Moon with the progressive rise of the Sun across its face. Aristarchus, and then Erathosthenes, hit upon clever geometric methods of estimating distances to the Sun and Moon. Hipparchus devised a scheme for the universe which allowed for the movement of stars and planets around the Earth. This geocentric view was extended by Ptolemy to an elaborate system that permitted predictions of planetary motions and positions well into the future.

Figure 1-1 One of the early maps of the Earthside hemisphere of the Moon, made in 1651 by the Italian Riccioli. This observer began the practice of naming the larger craters after scientists and philosophers of the times. (Taken from *Moon, Man's Greatest Adventure,* Davis Thomas, ed., Harry N. Abrams, Inc., New York, 1972)

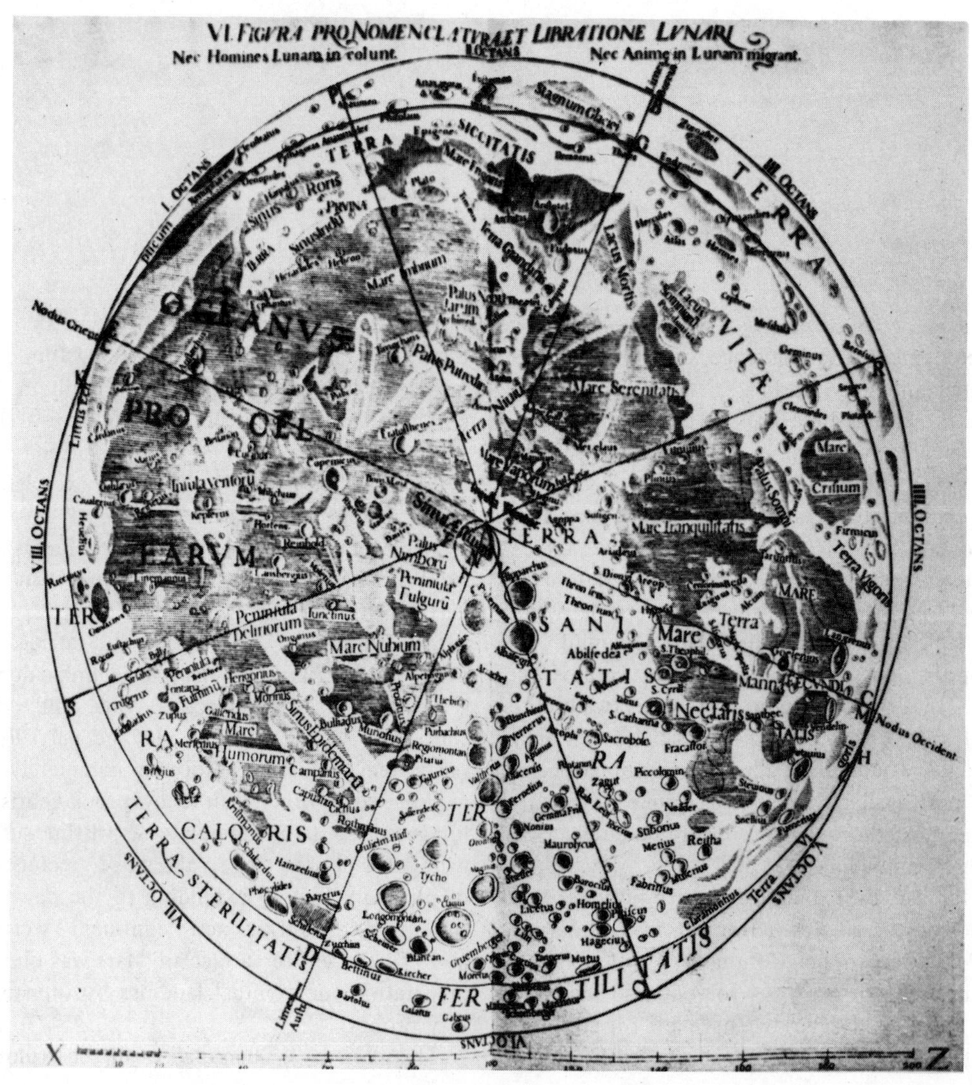

The Ptolemaic model, in which the planets seemed to move along epicyclic paths while the stars followed circular arcs, stood as the foundation of astronomy for nearly fifteen hundred years until its overthrow by Copernicus. The latter's heliocentric theory, stating that the Earth rotates on its own axis and revolves around the sun. enabled Kepler to discover the fundamental laws of planetary motion which lie at the heart of modern celestial mechanics. These concepts, in turn, were the basis on which Newton formulated his celebrated laws of gravitation and force.

Astronomy advanced rapidly after Galileo applied the newly invented telescope to a survey of the heavens. With this instrument he discovered sunspots and the moons of Jupiter. Herschel, in 1781, used the telescope to discover Uranus. Neptune's existence was first inferred from mathematical calculations of orbit perturbations, made in the mid-1800s and then quickly proved by visual sighting. The outermost planet, Pluto, was likewise postulated to exist in order to explain Neptune's orbital irregularities. Many years went by, however, before Pluto could be confirmed as a member of the solar system in 1930 by interpretation of anomalous shifts of a starlike object on photographs taken through the telescope.

Arguments about Life Elsewhere in the Solar System

They arose from observations on the other planets. Venus, appearing cloud-covered in the telescope, was once thought to support living matter at least in a primitive way. Life on Mars, the red planet, became the subject of intense speculation owing to the suggestion in 1877 by Schiaparelli that regular long and narrow markings, which he named *canali* (literally "channels"), occasionally were visible under ideal viewing conditions. His hypothesis that these might be natural channels of some sort was quickly twisted by others into a concept citing them as evidence of intelligent beings. Even as that view abated, apparent signs of seasonal changes during the Martian year were seized upon as indicators of some forms of living matter.

The First Studies of the Moon

The Moon, by virtue of its proximity, was repeatedly examined by telescope. Galileo, more than any other, must be heralded as the founder of selenology. He noted the topographic irregularities of the Moon's surface and recognized that circular depressions or craters were the cause of many of its peculiar markings. He deduced that the darker areas were level enough to be oceans—hence the name *mare* (Latin for "sea"; plural, *maria*). The lighter, non-mare areas appeared to rise above these "water" bodies and were termed *terrae* (plural of Latin *terra*, "land") or, more generally, uplands or highlands. Galileo recorded his findings on a sketch map showing some of the major basins and craters in their correct positions. Soon thereafter others produced their own versions of Moon maps; some of these, such as that published in 1651 by Riccioli (Figure 1-1), contain most of the landmark features which we use today in setting up a general lunar cartography. Descriptive observations, documented on maps, continued throughout the nineteenth century.

The development of photography in the mid-1800s provided the next great impetus to the study of the planets and, especially, the Moon. However, visual observations continued well into our own century because of the superiority of the human eye in singling out details well below the resolving power of long-exposure telescopic photography, which is subject to blurring by atmospheric fluctuations. These observations, coupled with photography, led to production of outstanding lunar maps, culminating in the mid-1900s with the magnificent photographic atlases of Kuiper, Alter, and Miyamoto. The atlases include selected photographs from the Lowell, Lick, Mount Wilson, Catalina, and Pic du Midi Observatories. One of the finest examples of a photograph showing the entire front face of the Moon is reproduced as Figure 1-2. A widely used reference map of the Moon, showing the maria, major craters, and some well-known terrain features, appears in Figure 1-3. The landing areas for U.S. and Soviet manned and unmanned lunar spacecraft are also plotted on this map.

Controversy over the nature and origin of many of the lunar surface features began almost with the first reports by Galileo. In 1665, the Englishman Robert Hooke argued from simple model experiments that the Moon's large craters are bursted bubbles that resulted from escaping vapors. Herschel in 1779 cited volcanism as a cause of many of the Moon's attributes. The debates diversified in the nineteenth and early twentieth centuries as four opposing schools set about to interpret the circular structures that, by then, were recognized as the dominant characteristic of the Moon's topography. Thus, the craters have been variously explained as (1) vapor-induced bubbles, (2) collapse features related to tidal flexing, (3) explosive volcanoes and calderas, and (4) impact scars made by meteorites. The last process was first advocated in 1802 by von Bieberstein and then supported in 1815 by von Moll. Its chief protagonist in the last century was the famed geologist G. K. Gilbert, who ironically failed to appreciate the same origin for Meteor Crater in Arizona despite years of careful field study. Well-known proponents of the impact theory just before the opening of the space age include Urey, Daly, Baldwin, Dietz, and Öpik. Their views were strongly opposed by the geologist J. E. Spurr, who drew upon his field experience in structural geology and volcanology to propose a completely internal, volcanic origin for the bulk of the lunar features—an up-

Figure 1-2 The frontside of the Moon as seen through the Lick Observatory 36-inch telescope. In order to obtain the optimum lighting effects, this photograph is a composite constructed from two half-moon views (note the different orientations of shadows within the craters). Contrast this photo with a "true" full moon picture (Figure 5-19), in which crater relief is subdued at high sun angles. The major surface features can be identified by comparison with Figure 1-3. (Photo by Moore and Chappell, 1937, courtesy Lick Observatory.)

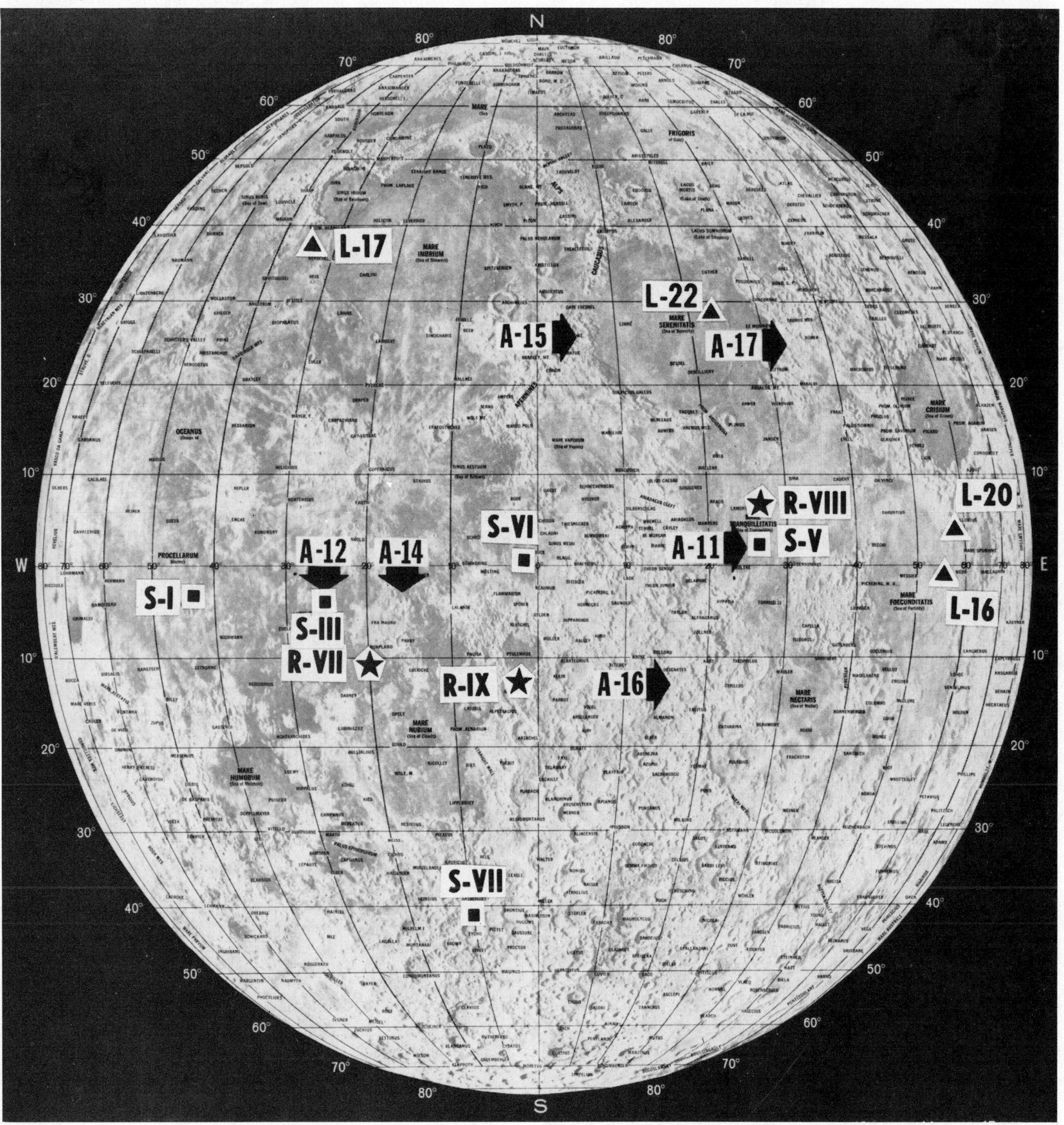

Figure 1-3 A map of the frontside of the Moon, prepared by the U.S. Air Force's Aeronautical Chart and Information Center, showing the principal landmarks on the lunar surface and the landing sites of the *Ranger* (R), *Surveyor* (S), and *Luna* (L) unmanned probes and the *Apollo* (A) manned lunar modules. The spacing of latitudes and longitudes varies in this projection, but the distance between any two latitude or longitude lines (10° intervals) has a ground equivalent of approximately 300 km near the equator. (NASA photo; sites added by the author.)

TABLE 1-1

Fundamental Facts about the Moon

Average Distance (Center to Center) from Earth to Moon	384,400 km
Diameter of Moon	3476 km (0.2723 of Earth diam.)
Mass of Moon	7.35×10^{22} kg (0.0123 of Earth mass)
Density of Moon	3.34 g/cm^3
Escape Velocity	2.38 km/sec
Orbital Eccentricity	0.0549
Inclination of Orbital Plane to Ecliptic	5°9′
Inclination of Lunar Equator to Ecliptic	1°32′
Temperature Range	-189°C to +117°C

date of earlier ideas brought forth by Dana in 1846. Spurr's 1945 book, *Geology Applied to Selenology,* is a classic milestone in the scientific study of the Moon.

In the first half of the twentieth century, despite addition of spectral analysis to telescope observations, acquisition of new data bearing on the characteristics of nearby planets began to level off. The diversity of views on the nature and origin of the Moon remained unresolved because of the sparsity of new, critical evidence. Some fundamental facts about the Moon, known prior to the space age, are summarized in Table 1-1.

Figure 1-4 Part of the farside of the Moon, photographed by the Russian spacecraft *Zond 3* on July 22, 1965, from a distance of about 10,000 km. The dark area in the right center is the Orientale Basin (see also Figure 1-8b). The first pictures of the farside, televised by *Lunik 3* on October 6, 1959, led to a generalized map showing major craters, basins, and maria. Many of the features never before seen from Earth were named after Russian cities, scientists, and space pioneers. (TASS News Agency photograph, reproduced by NASA.)

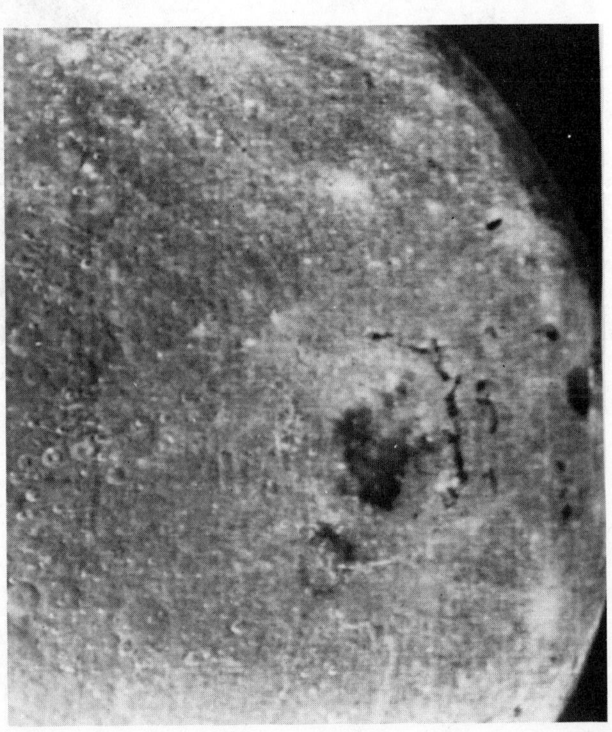

Early Attempts to Enter Space

What was now sought was the means to reach and measure the Moon and other planets through direct contact. The need to press into space had become urgent from a scientific standpoint.

The notion of space travel is not new. Kepler in 1634, and shortly thereafter Bishop Godwin, conceived of voyages to the Moon. When lack of water and air on that body was later demonstrated, writers set their sights further out to the planets. Then, in 1865, two classic novels—*Journey to the Moon* by Alexandre Dumas and *From the Earth to the Moon* by Jules Verne—actually touched upon some of the basic problems and approaches which had to be mastered if man were ever to venture beyond his planet. The foundations of modern rocketry were laid by the German Ganswindt in 1891 and the Russian Tsiolkovsky in 1898 in publications describing their experiments. In the United States, the efforts of R. H. Goddard between 1908 and 1944 are now recognized as pioneering steps which greatly influenced von Karman in America and a German group led by Wernher Von Braun to begin practical development of space rocketry. The first rockets to leave the lower atmosphere were the guided missiles or V-2s used as weapons by the Germans near the close of World War II. Both the Soviets and Americans put some of the German technical experts to work in building their rocket systems for military purposes in the cold war period between 1945 and 1957.

The Start of the Space Age

The invasion of outer space as a scientific endeavor dates from October 4, 1957, when a 72-kg (kilogram) satellite the *Sputnik 1**—was placed in orbit around the Earth by the Soviet Union. Life entered space on November 3 of that year with the launching of a dog aboard *Sputnik 2.* On January 31, 1958, the United States joined this "space race" with its *Explorer 1* satellite (12 kg), propelled by a Redstone rocket to an elliptical orbit whose perigee (closest approach) was 350 km (kilometers) above the Earth. Most such probes in the early years of satellite launches were designed to gather data about the Earth's atmosphere and radiation fields, but some biological experiments were directed toward the eventual entry of men into outer space. Again the Soviets achieved this first with the orbiting of Yuri Gagarin once around the world on April 12, 1961. This was matched by the United States

*Although roman numerals are used in the official names of these spacecraft—Sputnik I, Apollo XII, etc.—this text uses arabic numerals throughout for the sake of readability.

on February 20, 1962, by John Glenn's three orbits in a *Mercury* capsule. This nation's commitment to a superior space program reached for new horizons with the proposal by President John F. Kennedy on May 25, 1961, of a goal to place men on the Moon by 1970.

Events in space exploration moved swiftly after the first *Sputnik*. Because of the United States' decision to land men on the Moon, the need to obtain detailed information about the nature and characteristics of the lunar surface assumed paramount importance in planning space exploration. Accordingly, a series of experimental programs was developed in the 1960s to extend our view of the Moon beyond the limits of telescopic resolution ultimately to actual contact with the surface.

Unmanned Exploration of the Moon and Planets

The first successful spacecraft to the Moon dates back to the early *Sputnik* era. The Soviet *Luna 3* in 1959 provided scientists with their first peek at the farside (Figure 1-4). The United States scored its first major triumph in lunar exploration when, on July 28, 1964, after a series of failures, a rocket-launched probe crash-landed on the Moon (see Figure 1-3) after transmitting a series of television pictures to Earth. This *Ranger 7* craft was followed by two more that provided views of the surface superior to any achieved by telescope (Figure 1-5).

A giant stride in our knowledge of the lunar surface resulted from the first successful soft-landing of an instrumented spacecraft, the Soviet *Luna 9*, in January 1966. The TV photos transmitted from this probe showed a rough, pitted surface (Figure 1-6) capable of supporting both men and machines. Later that year the first United States *Surveyor* duplicated this feat and sent back even higher-quality pictures of a small area within the equatorial belt on the Moon's western edge. Over the next two years three more Surveyors landed in the maria along this belt (Figure 1-3) while the last in this series put down in the southern highlands. These sophisticated soft-landers carried instruments that dug into the loose surficial rubble and determined the chemical composition of materials at their touchdown sites. Close-up views of individual rocks disclosed a range of textures (Figure 1-7) in the blocks scattered about the surface, including large crystals, pits and vugs, fragments, and dark coatings. The *Surveyors* also revealed that the lunar surface is dotted with many craters whose diameters are less than 1 meter.

From 1966 to late 1967 five *Lunar Orbiter* spacecraft were sent on photographic missions around the Moon to obtain vertical and oblique pictures, sometimes at resolutions as high as 1 meter, of proposed landing sites and other areas of scientific interest. Several of the most spectacular scenes returned from these *Orbiter* probes are shown in Figure 1-8. *Orbiter 4* provided complete

(a)

Figure 1-5 (a) The large craters Ptolemaeus (top), Alphonsus, and Arzachel, with the 100-km wide Albategnius (right) and Purback (below), as seen through the Lick Observatory 36-inch refractor. (b) The area around Alphonsus, with its central peak, ridges, and rilles, from a television picture taken by *Ranger 9* on March 24, 1965, about three minutes before impact at a distance of 415 km from the surface. Alpetragius appears in the lower left. Compare these views with the *Orbiter 4* photograph of the same region (Figure 9-4). (NASA photo.)

(b)

Figure 1-6 One of the first pictures returned by the Russian *Lunik 9* spacecraft which soft-landed on February 3, 1966 (prior to *Surveyor 1*), in the western edge of Oceanus Procellarum. This television image shows what appears to be a pitted and cratered surface covered with scattered rocks. Some initial interpretations considered this surface to be vesicular basalt but later *Surveyor* views demonstrated the apparent "roughness" to be related to small craters and clods of "lunar soil." (TASS News Agency photograph, reproduced by NASA.)

Figure 1-7 One of the first close-up views of a rock in place on the surface of another planet. This *Surveyor 1* photograph shows a block some 50 cm in length with rounded edges, erosion pits and/or vesicles, a mottled exterior, and several distinct fractures. (NASA photo.)

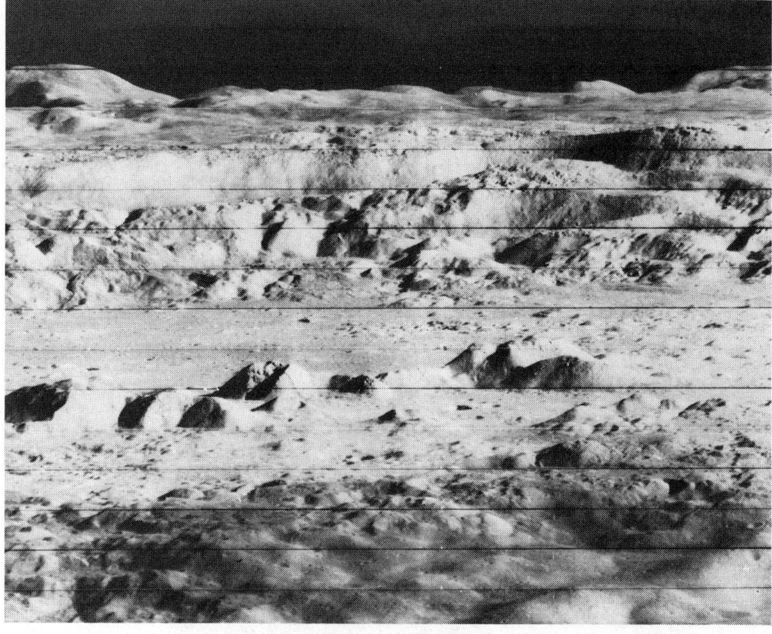

(a)

Figure 1-8 Three of the most famous and instructive photos taken by the *Lunar Orbiter* spacecraft. (a) An *Orbiter 2* oblique view looking north into the crater Copernicus from a point about 240 km above the surface. The central peaks on the floor of the crater are 300–400 m or more in height and the top of the rim is several kilometers higher. Note the several terraces fronted by steep escarpments, the lack of obvious bedding in exposed walls, the channeling of lower slopes (left center), and the hummocky terrain in the foreground consisting of ejecta deposits beyond the rim nearest the viewer. (b) The Orientale Basin near the west limb of the Moon's frontside, as seen by *Orbiter 4* in a near vertical view from about 2700 km above the surface. Three nearly circular ring scarps can be discerned: the more subdued inner one encloses the dark lavas of Mare Orientale; the middle ring, some 750 km wide, comprises the Rook Mountains; the outer ring, named the Cordillera Mountains, has an average diameter greater than 900 km and rises more than 6 km above the basin. Note especially the two larger craters within the "bulls-eye" (see Figure 8-15) and the radial valleys which resemble immense grooves. (c) A view of the Earth (south pole in darkness near the left edge of the crescent) seen from *Orbiter 1,* which was then nearly 389,000 km away in an orbit 1200 km above a cratered portion of the terrae or highlands on the farside of the Moon. At this time the *Orbiter* spacecraft was moving in a direction toward the bottom of the picture so that the Earth would appear in any continuing photo sequence as though it were setting.

(b)

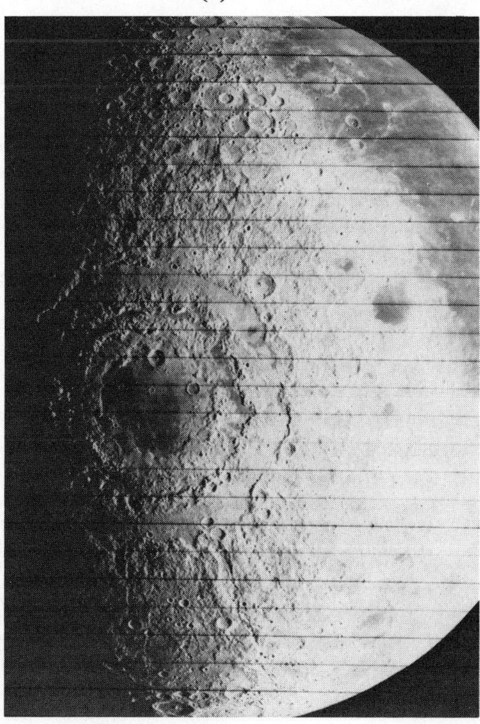

(c)

Figure 1-9 Much of the farside of the Moon, never visible from Earth, as photographed by astronaut Thomas Mattingly from the orbiting *Apollo 16* Command Module in April, 1972. The region seen here lies in the Moon's northern hemisphere along and beyond the eastern limb of the frontside. The large dark areas in the upper left are Mare Crisium (near edge), Mare Marginis, and Mare Smythii. The craters Pasteur, Borman, Lovell, and Anders appear in the lower left. (NASA photo.)

coverage of the Moon's frontside for NASA (the National Aeronautics and Space Administration), while nearly all of the farside was also photographed by the several *Orbiters* (see Figure 1-9 for a recent view of much of this farside taken during the *Apollo 16* mission). By combining photos from each *Orbiter*, a map describing the cratered surface of the entire moon (see end papers) has been compiled by the U.S. Air Force.

Although the Moon was destined for most attention because of its nearness, exploration by space probes has now included Venus, Mars, Mercury, and Jupiter. On August 27, 1962, the United States' *Mariner 2* was launched toward Venus and, after traveling 56 million km, on December 14, 1965, the spacecraft began to send back data on the particle and magnetic fields, surface and atmospheric temperatures, and atmospheric composition as it approached to within 41,000 km of the planet. Other American and Soviet probes have since penetrated closer to Venus and in 1970 a Soviet spacecraft successfully landed on the surface. The *Mariner 4*, similar to the Venusian probes, left on November 28, 1964, for a Mars encounter some 210 million km distant. The first TV close-ups (Figure 1-10) of a planet outside the Earth-Moon system were obtained on July 15, 1965, as the craft sped past Mars at distances as near as 9100 km. Two more *Mariners* (6 and 7) launched in early 1969 returned even better pictures and other data between July 31 and August 2 of that year. In late 1971, *Mariner 9* began to orbit Mars and to obtain high-resolution photographs suited to detailed mapping of the surface. Two Soviet space probes, launched earlier, attempted a soft landing on the Martian surface; one actually landed but ceased to function after less than one minute. The first journey to an outer planet started in 1972 with the launch of *Pioneer 10*, scheduled to fly past Jupiter in late 1973.

The USSR scored a series of firsts by placing fully automated unmanned spacecraft on the Moon to carry out specialized missions. On September 19, 1970, they landed and relaunched a lunar sampler (*Luna 16*) that had collected materials from a site in Mare Fecunditatis for

Figure 1-10 The surface of Mars near Atlantis between Mare Sirenum and Mare Cimmerium, as viewed by a TV camera onboard the *Mariner 4* spacecraft as it passed to within 10,000 km of the planet on July 14, 1965. This photo is a reconstructed image made from digitally-coded data radioed to the Goldstone Receiving Station in California. This and other views obtained during the mission disclosed Mars to be heavily cratered, although less so than the lunar highlands. The large crater defined by an incomplete rim exceeds 115 km in diameter. (Jet Propulsion Laboratory photo, released by NASA.)

Figure 1-11 Photograph taken during the 1973 Paris Air Show of a mate to the Russian robot rover, *Lunokhod 1*, in operation on the Moon's surface. Using batteries powered through solar cells (dish displayed in front of the rover), the vehicle moves along eight flexible wheels and receives guidance information from a TV camera. *Lunokhod* can survey the lunar surface with this camera system and contains sampling instruments for physical and chemical analysis. At the end of a lunar operating day (about two weeks) the vehicle becomes "dormant" for the long lunar night. Each time after being successfully reactivated, *Lunokhod* has carried out further excursions over distances of a kilometer or more beyond the *Luna 17* landing stage on which it arrived, until it ceased to operate nearly a year later. (Photo courtesy of *Aviation Week and Space Technology*.)

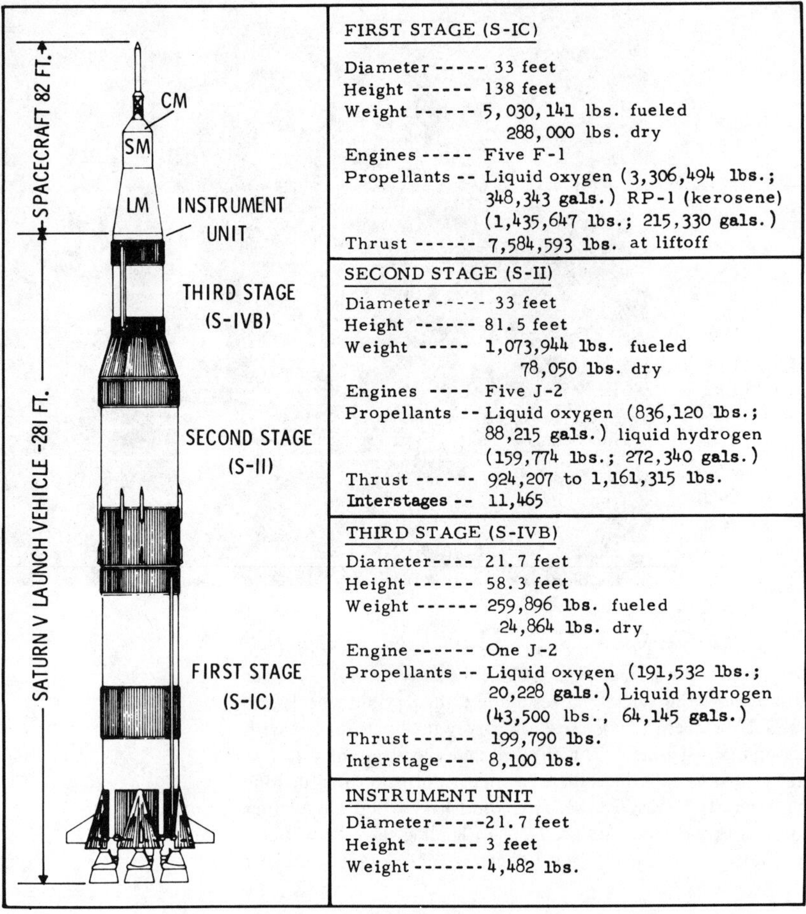

NOTE: Weights and measures given above are for the nominal vehicle configuration for Apollo 12. The figures may vary slightly due to changes before launch to meet changing conditions. Weights of dry stages and propellants do not equal total weight because frost and miscellaneous smaller items are not included in chart.

Figure 1-12 Vital statistics of the *Saturn 5* launch vehicle used to transport U.S. astronauts into orbit or to the Moon in the *Apollo* and *Skylab* programs. Within the spacecraft unit, LM refers to the Lunar Module (for placing two astronauts on the lunar surface); SM and CM (CSM) are the Service and Command Modules respectively. (NASA illustration.)

analysis after return to Earth. The USSR followed this in late 1970 by detaching an unmanned roving vehicle (*Lunokhod 1*) from the *Luna 17* lander (Figure 1-11) and continuing to operate it during each long lunar day well into 1971. A second unmanned sampler (*Luna 20*) brought back soil materials from Mare Fecunditatis in 1971. A second *Lunokhod* was deployed from *Luna 22*, an orbiter-lander that began operation in January 1973 in the eastern sector of Mare Serenitatis.

The Apollo Program

The 1960s were an era of greatly heightened activities among a rapidly increasing number of workers dedicated to lunar research. Professional meetings and symposia soon became arenas for renewed controversies. As more information built up, two opposing camps–the Impacters and the Volcanologists–squared off to press their respective positions on lunar problems. New suggestions concerning the origin and structure of the Moon–and, particularly, whether it was always *cold* (unmelted) or had been or is now *hot* (at least partially melted and differentiated)–stimulated these debates. Only a series of on-the-spot investigations and samplings by trained human observers could resolve some of these fundamental questions about the Moon's nature and history. This, then, was a prime objective of *Project Apollo,* the culmination of the Kennedy mandate to go to the Moon.

The transport of men to other planets had to await development of powerful and reliable multistage rockets capable of launching heavy payloads into an Earth-parking orbit and then carrying a spacecraft out of the Earth's gravity field on an accurate trajectory to their destination (see Figure 10-1). The mighty *Saturn 5* rocket, brainchild of Wernher Von Braun and his staff of designers, gave the United States a vehicle that could accomplish these objectives (Figure 1-12).

The payoff for the years of effort to "reach for the stars" began on December 24, 1968, when U.S. astronauts Borman, Lovell, and Anders became the first humans to circle the Moon in their *Apollo 8* spacecraft. This feat was duplicated on March 6, 1969, by astronauts McDivitt,

Scott, and Schweickart during the *Apollo 10* mission. Then, on July 20, 1969—a memorable day in the history of civilization—the world thrilled to the message that "the Eagle has landed," marking the moment when men were joined to another planet. (The *Apollo* landing sites discussed here are shown in Figure 1-3.) Just before midnight (EDT), Neil Armstrong touched that surface and a unique scientific expedition was underway (Figure 1-13). On July 24, the *Apollo 11* mission became legend when Armstrong, Aldrin, and Collins returned to Earth with a priceless collection of rocks and soil soon to be distributed to 143 scientists and their coinvestigators. The *Apollo 12* lunar module, crewed by Conrad and Bean, set down on the Moon on November 17, 1969, and later joined with astronaut Gordon's command module to return another cargo of lunar materials for analysis. On February 5, 1971, astronauts Shepard and Mitchell descended in a lunar module to the Moon's hilly upland region north of Fra Mauro crater while astronaut Roosa piloted the *Apollo 14* command module in lunar orbit. Three excursions around the Hadley Rille-Apennine Mountains front were carried out by *Apollo 15* astronauts Scott and Irwin from July 30 to August 3, 1971, while astronaut Worden conducted experiments from the orbiting command mod-

Figure 1-13 Man's first venture on the Moon. Astronaut Edwin Aldrin, Jr. is deploying two scientific experiment packages during the "moon walk" made on the night of July 20, 1969, after the landing of the *Apollo 11* LM. His footprints leave deep impressions on the powdery "soil" of weakly cohesive rock debris. A shallow depression, perhaps a small secondary crater formed by a piece of material ejected from a larger crater elsewhere, lies in the left foreground. Other craters and numerous rock fragments dot the generally level surface of the southwestern part of Mare Tranquillitatis. (NASA photo.)

ule during an ambitious and productive lunar exploration. The study of the ancient lunar highlands began on April 20, 1972, as the *Apollo 16* lunar module placed astronauts Young and Duke on the surface near the crater Descartes even as astronaut Mattingly carried out orbital observations. The final U.S. trip to the Moon–and perhaps the last manned landing for many years–involved *Apollo 17* astronauts Cernan and Schmitt on the surface and Evans overhead. The landing site, northeast of Mare Serenitatis in the Taurus-Littrow region of the surrounding highlands, was reached on December 11, 1972. The mission was the longest stay in the series at a single site and included the first scientist-geologist ("Jack" Schmitt) to fly in the program.

Outlook

Thus, with these successful landings on the Moon by manned and unmanned explorers from both the United States and the USSR and with the prospects for eventual landings on Mars, the study of space has entered a new phase that promises many benefits for geology. Clearly, one major objective of planetary geology–a field which has matured in the space age–will be an expanded understanding of the Earth and its place in the hierarchy of the solar system.

REVIEW QUESTIONS

1. State the contributions made by Galileo to planetology.
2. Distinguish between the Ptolemaic and Copernican models of the solar system.
3. What are maria and terrae? What terms are synonymous with these?
4. List some of the competing ideas for the origin of lunar craters.
5. Identify the following: *Sputnik; Explorer; Luna; Ranger; Orbiter.*
6. Who was the first Soviet and who the first American to orbit the Earth?
7. What was the principal information returned from the *Surveyor* probes?
8. Mention one contribution to space exploration made by each of these individuals: Jules Verne; Wernher Von Braun; R. H. Goddard; John F. Kennedy; J. E. Spurr.
9. Name the first two men to set foot on the Moon; give the date of this achievement.
10. What are the major objectives of planetary geology?

CHAPTER 2

The Solar System

Our solar system is a collection of nine planets, their satellites, many thousands of asteroids, millions of comets, innumerable rock and dust fragments, and vast quantities of gases and charged particles—all associated with a central star (the Sun). Most scientists today believe that the material in this system became organized into large bodies, such as the Sun and the major planets, by condensation from a vast dispersed nebular gas-solid mixture. Motions acquired during this process have allowed most of the larger bodies to remain in about the same relative orbital positions since the time the Sun attained its present dimensions.

Planetary Systems within Galaxies

For centuries men considered the solar system to be unique within the universe. It is now recognized that our Sun is only an average star in the Milky Way galaxy—one of an estimated 100 billion stars, a typical number for each of the 100 billion or more galaxies spread throughout the observable universe. The Sun occupies a "point" about two-thirds out from the assumed center of the galactic disc. This disc (actually a series of spiral arms; see Figure 4-1a) has a diameter of approximately 100,000 *light-years* and a thickness of about 10,000 light-years. Each light-year, equal to 9.461×10^{12} km (9,461,000,000,000 km), measures the distance traveled in one year by a light photon moving at a velocity of 299,800 km/sec (i.e., the speed of light). This galaxy—and probably most if not all others—developed from clots of nebular material which organized into countless separate star-planet systems whose dimensions, characteristics, and properties must be similar to our own solar system. Thus, the term *solar system* more properly should be replaced by the general term *planetary system* in reference to the high probability of repeated occurrence of this phenomenon throughout the universe, although there is no direct evidence of the existence of planets other than those we see within the system to which the Earth belongs. Certain perturbations in the motions of other stars can theoretically be attributed to the influence of satellitic masses. The observed

---

For a lucid introduction to astronomy that provides a general background for this chapter, the reader is referred to Donald H. Menzel, Fred L. Whipple, and Gerard deVaucouleurs, *Survey of the Universe* (Englewood Cliffs, N.J.: Prentice-Hall, 1970) and Paul Hodge, *Concepts of the Universe* (New York: McGraw-Hill, 1971). A more detailed treatment of the solar system and the planets can be found in S. Glasstone, *Sourcebook on the Space Sciences* (Princeton, N.J.: Van Nostrand, 1965) and W. K. Hartmann, *Moons and Planets* (New York: Bogden and Quigley, 1972).

movements of Barnard's Star are best explained by assuming that two massive planets orbit around it. Many cosmologists now infer that perhaps millions of these planetary objects are present within our galaxy and elsewhere in the universe.

Definition of a Planet

The meaning of a *planet* is best defined in terms of the features of those in our solar system. A planet in this context must show these critical characteristics:

(1) It moves in a fixed orbit around a more massive luminous body or *star* that exerts a controlling influence through gravitational forces.
(2) It is configured as a more or less spherical body.
(3) It is large enough to reflect sunlight detectable from Earth.
(4) It consists of agglomerations of solidified materials, including silicates, metals, "ices" (frozen H_2O, NH_3, etc.), and compressed gases (H, He, N, etc.) in varying proportions, enveloped by minute to extensive quantities of gases and possibly liquids in its outer regions.
(5) It does not experience high-temperature nuclear reactions (e.g., hydrogen-to-helium fusion) because any fusionable matter, if present, is below a critical density.

These defining characteristics do not completely exclude planetary satellites, asteroids and meteoroids, and comets unless a limit to either size or mass is specified for a planet (Figure 2-1). No arbitrary values have been proposed, but most satellites are normally very much smaller than any of the planets; only the planet Mercury is as small as the three largest satellites in the solar system. Asteroids are sometimes considered minor planets, but none of these bodies has a diameter larger than 800 km. Comets can fit all of the above criteria; however, the size of the solidified central body is less than 100 km—far smaller than any of the principal planets.

The Sun as a Star

Our Sun is a relatively average, young star, existing less than half of the 12 billion years estimated to be an upper limit for the age of the universe. The Sun belongs to the GO spectral class of stars and falls on the main sequence of star distributions as plotted on the Hertzsprung-Russell diagram (Figure 2-2) which relates luminosity (brightness) to temperature. Stars near and below this luminosity are more likely to have planetary systems than the hotter ones to the left of the Sun's plotted position. The Sun's diameter is 1,392,000 km (about 109 times greater than the Earth's diameter) and its mass is 1.99×10^{30} kg (330,000 times greater than Earth's); its density is 1.41 g/cm^3 (grams per cubic centimeter). The rotational period at its equator is 25.4 days; despite the mass involved, this slow rotation leads to a very small amount of angular momentum relative to the planets.

A general diagram describing the principal features of the Sun, including its internal structure, appears in Figure 2-3. The Sun's white corona is beautifully displayed during total solar eclipses (Figure 2-4). Most of the energy radiating to space emanates for a 100-200-km thick layer below the *photospheric* surface. This surface has an emission temperature of ~5800°K (degrees Kelvin: °K = °C + 273) or 10,700°F. All matter is in the gaseous state at the temperatures measured for the photosphere, chromosphere, and corona. At such temperatures, atomic and

Figure 2-1 The distribution of planets, satellites, asteroids, comets, and meteorites in the solar system as a function of their relative masses (expressed as log M). (From NASA Publication TM X-64677.)

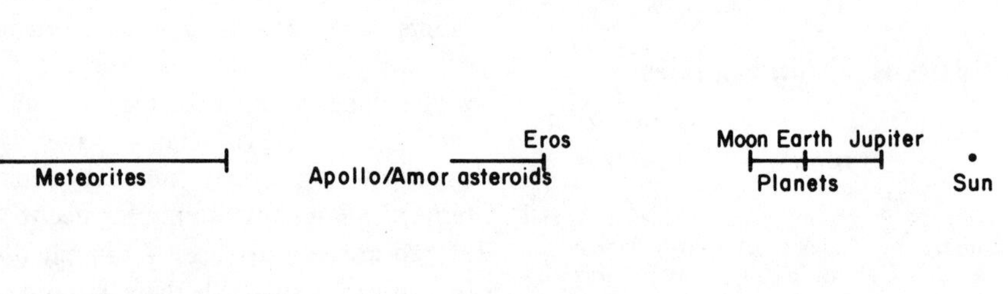

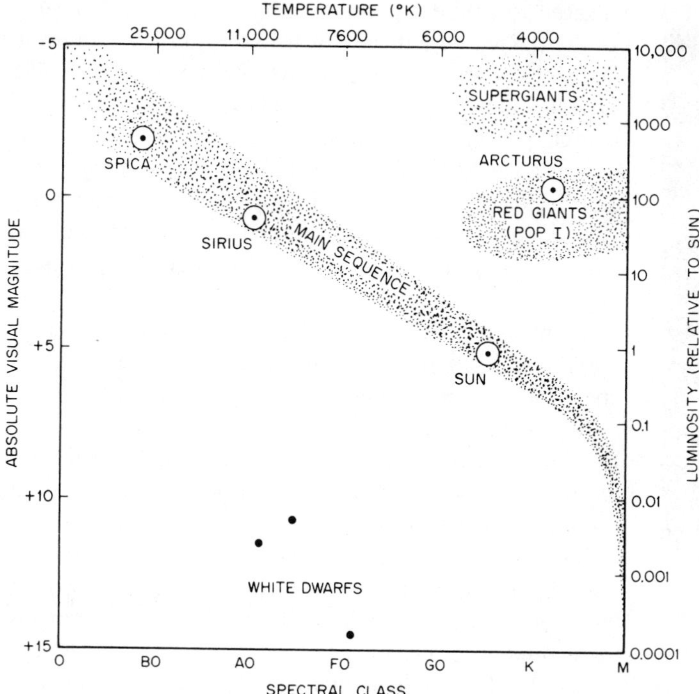

Figure 2-2 The Hertzsprung–Russell (H–R) diagram on which stars have been plotted according to their absolute magnitude or luminosity (ordinate on log scale) versus a parameter such as spectral class (abscissa, shown by symbols from M through O), temperature (in degrees Kelvin), or color (O class stars are bluish-white; GO stars are yellow; M stars are reddish). Luminosity refers to the total radiant power (amount of energy radiated from the source in unit time) within the visible range; small, massive stars such as white dwarfs (hot) are less luminous (low end of ordinate) than large, less dense red (cooler) supergiants (upper ordinate). Most stars plot along the main sequence. In spiral galaxies, population I stars are older and occur mostly near the central region of the galactic disc, whereas younger population II stars tend to form in the outer regions. Various genetic schemes for the life history of a star can be plotted relative to the H–R diagram. For example, a star somewhat more massive than the Sun could start its condensation to the right of the main sequence, then reach that sequence as it contracts and grows hotter. Later, as mass is consumed, the star moves into a red giant stage and becomes more luminous. Finally, as additional mass is lost, it begins a pulsation phase or blows off even more material as a nova or supernova (moving to left of sequence). The end product in this evolution is usually a small, dense, hot dwarf which cools and darkens with loss of luminosity. (From *Sourcebook on the Space Sciences* by S. Glasstone, copyright © 1965 by Litton Educational Publishing Inc. Reprinted by permission of Van Nostrand Reinhold Company.)

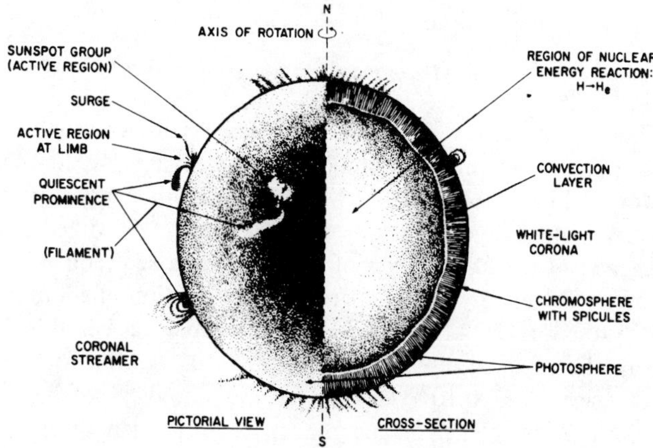

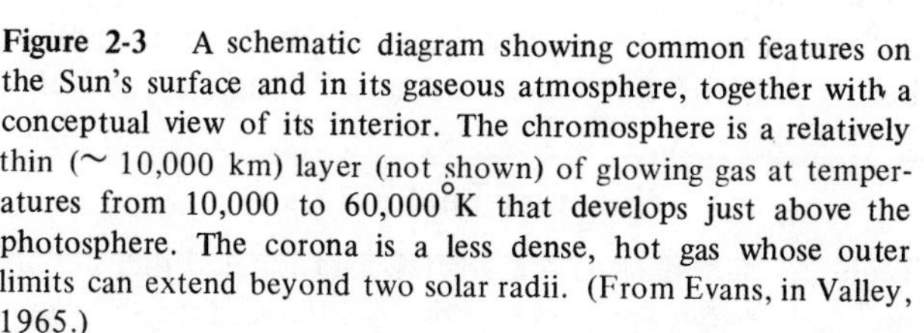
Figure 2-3 A schematic diagram showing common features on the Sun's surface and in its gaseous atmosphere, together with a conceptual view of its interior. The chromosphere is a relatively thin (~ 10,000 km) layer (not shown) of glowing gas at temperatures from 10,000 to 60,000°K that develops just above the photosphere. The corona is a less dense, hot gas whose outer limits can extend beyond two solar radii. (From Evans, in Valley, 1965.)

Figure 2-4 The white light corona of the Sun observed during the total eclipse of March 7, 1970. The luminosity of this gaseous envelope results from electron excitations (ionization) owing to the high coronal temperatures. When an eclipse is total, the blotting out of light from the photosphere allows this coronal glow to be observed. (NASA photo by Sheldon Smith and Leonard Weinstein using a reflecting telescope with a 41-cm aperture.)

molecular gases normally are dissociated into *plasmas* of positively charged particles and electrons.

Because of the much lower gas densities in the outer solar atmosphere, the temperatures recorded for that region can locally reach as high as 5,000,000°K, even though this thermal activity is associated with only a small fraction of the total energy being emitted. In the Sun's interior, temperatures rise rapidly to a maximum computed to be about 15,000,000°K in a core composed mainly of hydrogen (H) and helium (He) nuclei and free electrons. Most of the energy released from the interior results from the consumption of its main constituent, hydrogen. One principal fusion reaction follows the *proton-proton chain,* involving a series of steps which includes generation of deuterium nuclei, as summarized by the overall equation:

$$4\,{}_1H^1 = {}_2He^4 + \text{energy}$$

The energy released in this fusion of four hydrogen atoms to a single helium atom amounts to approximately 26.4 Mev (million electron volts). Expressed another way, for every gram of hydrogen converted to helium, 1.50×10^{11} *calories* or 6.2×10^{18} *ergs* of energy are released (equivalent to the energy obtained from the detonation of about 150 tons of TNT). The Sun is using up hydrogen in this way at an annual rate approaching 2×10^{22} g. But for a solar age of 5×10^9 years, so far only about 5% of the total mass of solar hydrogen of nearly 2×10^{33} g has been changed to helium. Carbon present in the Sun at temperatures attained near its core will also be burned, by proton and alpha particle capture, to intermediate stages involving nitrogen and oxygen isotopes and eventually a product of carbon plus helium. In the carbon cycle, that element acts as a catalyst and is regenerated during energy conversion. Most heavier elements (with atomic masses greater than oxygen) are not being synthesized in the present Sun inasmuch as its evolution has not yet progressed to a stage (e.g., supernova activity) in which these elements form by those nuclear processes, which require very high temperatures and pressures. The presence of these heavier elements in the Sun indicates their production during an earlier stage of cosmological evolution extending back perhaps 12 billion years—one current estimate for the age of the universe—followed by incorporation in the Sun from its parent nebular materials.

Energy released within the Sun's interior is transferred by radiation until it encounters the cooler, more opaque outer zones below the photosphere. Thereafter, energy is redistributed mainly by mass transfer through convection, although relatively little hydrogen from the deep interior ever reaches the surface. Innumerable convection cells are visible as granular "blotches" on the photosphere. Larger, darker (hence cooler) marks, called *sunspots*, appear periodically (reaching their maximum in number and intensity over an eleven-year cycle) in groups that migrate across the Sun's face. They are associated with very strong magnetic fields, typically as much as 3000 times greater than the general magnetic field of 1 gauss at the Sun's surface, that cause local reductions in convection and turbulence.

These sunspots, as well as great masses of incandescent gases that are propelled through the solar atmosphere as *flares* (typically out to 2000-10,000 km beyond the photosphere) or even larger *prominences* (up to 1,000,000 km), are associated with bursts of high-energy particles, X-rays, and UV (ultraviolet) radiation. The stream of protons (hydrogen nuclei), heavier nuclei, and electrons that moves as a plasma at velocities from 350-750 km/sec from the Sun to all parts of the solar system and beyond is called the *solar wind.* These particles normally have an interplanetary density of ~5 protons/cm³ (compared with an intergalactic density of ~1 atom/cm³) except where focused by strong magnetic fields around such planets as Earth and Jupiter to produce "radiation belts" within the magnetosphere (Figure 2-5). The two Van Allen belts (one of the first major discoveries by artificial satellites circling our planet) reach maximum proton densities (and radiation intensities) at distances about 1.5-1.8 and 3-4 Earth radii for the inner and outer belts, respectively.

Planetary Parameters

Table 2-1 summarizes the orbital, rotational, dimensional, and mass parameters and Table 2-2 the surface, interior,

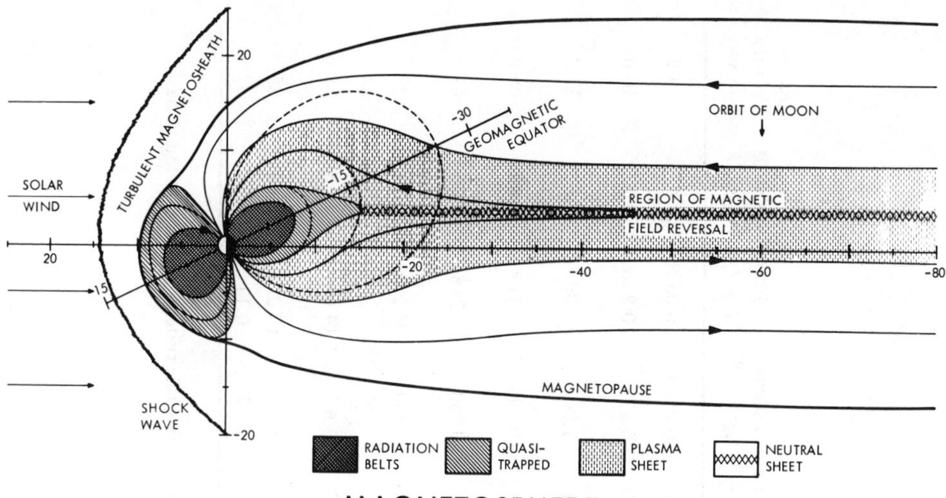

Figure 2-5 Development of the magnetosphere by interactions between the Earth's magnetic field and ionized particles making up the solar wind. This results in a distortion of the geomagnetic lines of force emanating from the Earth's magnetic poles. Close to the Earth, particles are trapped in radiation belts (inner and outer Van Allen belts) by magnetic forces. The magnetosphere consists of extensions of the force field into a long tail (more or less cylindrical) that has been traced for millions of kilometers in directions away from the Sun (depending on the position of the Earth in its orbit). At these distances, the geomagnetic field gradually merges with the interplanetary magnetic field associated with the Sun itself. The outer boundary of the magnetosphere is the magnetopause beyond which the geomagnetic field is absent. The plasma sheet contains a flux of low-energy electrons derived from the solar wind. Around the axis of the tail, the neutral sheet indicates a reversal in the polarity or direction of field flow. As particles in the supersonic solar wind encounter the geomagnetic field towards the Sun, a disturbed region (magnetosheath) of magnetic turbulence (unstable and shifting field lines) is produced beyond the magnetopause preceding the Earth. The outer limit of the magnetosheath is marked by a hydromagnetic shock wave a few Earth radii upstream. (Illustration supplied by N. F. Ness, NASA Goddard Space Flight Center.)

Figure 2-6 Sizes of the planets and asteroids relative to the Sun (only a portion of which is shown) at the same scale. (Jupiter has a diameter of approximately 70,000 km.) (NASA photograph.)

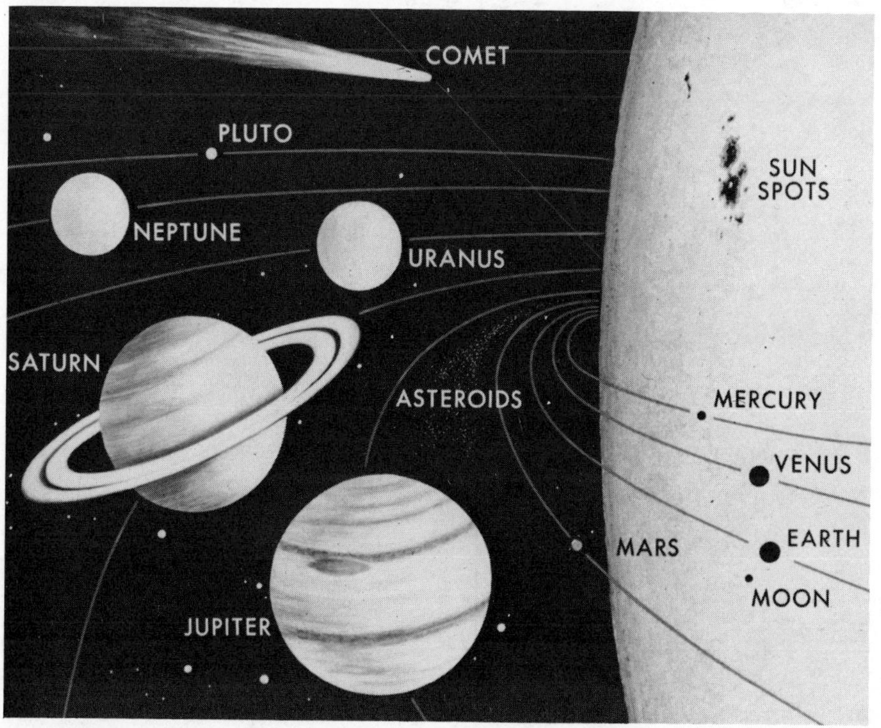

TABLE 2-1††

Orbital, Rotational, Dimensional, and Mass Parameters of the Planets

	Av Distance from Sun													
	*A.U.**	*10^6 km*	*Orbital Eccentricity*	*Sidereal Period*	*Mean Velocity (km/sec)*	*Inclination (°)*	*Rotational Period*	*Angle between Ecliptic and Rotational Axis (°)*	*Average Diameter (km)*	*Radius (E = 1)†*	*Volume (E = 1)†*	*Mass (E = 1)†*	*Density (g/cm^3)*	*Escape Velocity (km/sec)*
Mercury	0.390	58.3	0.206	88 days	48	7.00	58.6 days	7	4,864	0.380	0.0549	0.0536	5.4	4.2
Venus	0.723	108.2	0.0068	224.7 days	35	3.39	243 days retrograde	6	12,112	0.958	0.879	0.816	5.1	10.0
Earth	1.000	149.2	0.0167	365.3 days	29.8	0	23 hr 56.07 min	23.45	12,740	1.00	1.00	1.00	5.52	11.2
Mars	1.52	228	0.0933	687 days	24.1	1.85	24 hr 37.4 min	25.0	6,740	0.53	0.149	0.108	4.05	5.0
Jupiter	5.20	778	0.0484	11.86 years	13.1	1.03	9 hr 55 min	3.01	139,800	10.8	1260	318	1.35	61
Saturn	9.53	1426	0.0557	29.46 years	9.65	2.50	10 hr 30 min	26.7	117,000	9.02	734	95.2	0.71	37
Uranus	19.2	2870	0.0472	84.02 years	6.85	0.77	10 hr 50 min retrograde	98.0	46,600	3.77	53.6	14.6	1.56	22
Neptune	30.0	4496	0.0086	164.8 years	5.47	1.79	16 hr	22.8	44,200	3.38	39.1	17.2	2.47	25
Pluto	39.5	5900	0.249	248.4 years	4.84	17.2	6.4 days	?	6,000	0.5	0.09	0.11	4.9	?

*A.U. = astronomical unit. An A.U. is defined as the mean (average) distance between the centers of the Sun and Earth, i.e., 1.496×10^8 km; using this system, Saturn is found to be $9.530 \times 1.496 \times 10^8$ or 1.426×10^9 km from the Sun.

†Ratioed to Earth as unity: Earth radius = 6371 km; volume = 1.082×10^{21} m^3; mass = 5.975×10^{24} kg.

††Most data in Tables 2-1 and 2-2 extracted from Fairbridge (1967), Glasstone (1965), and Valley (1965). See General References.

TABLE 2-2††
Surface, Interior, and Atmospheric Properties of the Planets

	Number of Satellites	*Diameter of Largest Satellite (km)*	*Magnetic Field*	*Interior Composition*	*Atmospheric Composition*	*Atmospheric or Interior Pressure (atm)*	*Temperature of Illuminated Side (°K)*	*Albedo: Light-Reflecting Capacity*
Mercury	0	–	?	Silicates: Ni-Fe core*	?: A, K, Xe	<0.003	600 - 700°K Surface	0.07
Venus	0	–	weak to absent	Silicates; Carbonates; Ni-Fe Core*	CO_2 >90% N_2 3% H_2O, A	~100	650 - 700°K Surface	0.7 – 0.8
Earth	1	3,476	strong 0.3–0.7 gauss	Silicates; Ni-Fe Core	N_2 78% O_2 21% A 1%	1	240 - 320°K Surface	var. av ~0.39
Mars	2	~16	very weak	Silicates; Core (?)	CO_2 >90% H_2O	0.005 - 0.007	203 - 295°K Surface	0.15
Jupiter	12	3,000	1000 gauss	H_2 78%† He 22% Silicate Core (?)	NH_3, CH_4, H_2, He, Ne (banded)	100 million at center	130°K Atmosphere	0.51
Saturn	9	25,800	Present	H_2 63%† He 37%	NH_3, CH_4, minor H_2, He	50 million at center	125°K Atmosphere	~0.50
Uranus	5	2,600	?	H_2, NH_3 or H_2, He, metal	CH_4; some NH_3, H_2	?	~70°K† Atmosphere	~0.66
Neptune	2	4,200	?	Similar to Uranus	CH_4; minor NH_3, H_2	?	~56°K†	~0.62
Pluto	0	–	?	Metal (?)	?	?	?	~0.16

*Inferred state.

†Estimated value.

††Most data in Tables 2-1 and 2-2 extracted from Fairbridge (1967), Glasstone (1965), and Valley (1965). See General References.

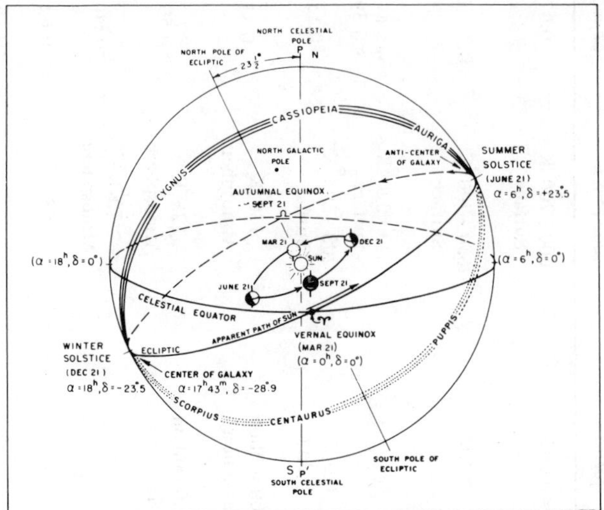

Figure 2-7 The celestial sphere defined in relation to the Earth's rotational poles. The North celestial pole in the 20th Century lies within 1° of Polaris (North Star) in the constellation Ursae Minoris. The reference star changes with time owing to the precession (circular wobbling) of the Earth's axis over a period of 26,000 years. All stars are fixed in definite positions on the sphere with respect to the Sun, but their apparent positions as viewed from Earth vary with the seasons. Some of the common constellations are plotted on the celestial sphere. The projection of the Earth's orbital path on the sphere defines the ecliptic (shown in the diagram as the apparent path of the Sun when the Earth is the observing station). Owing to the $23\frac{1}{2}°$ tilt of the Earth's axis relative to the ecliptic, that plane also makes an angle of $23\frac{1}{2}°$ with the celestial equator. (From Evans, in Valley, 1965.)

and atmospheric properties of the nine principal planets. The distance of each planet from the Sun is conveniently expressed in terms of the *Astronomical Unit* (A.U.), which is defined as the mean (average) distance between the Earth and the Sun—1.496×10^8 km or 93 million miles. Mean distances are measured as the length of the semi-major axis of the ellipse of revolution of an orbiting planet. When the actual distance of any planet is divided by the Earth-Sun distance, the resulting quotient is a decimal fraction or multiple of 1.00; the A.U. thus determines the number of times a planet is closer to or farther from the Sun than is the reference planet Earth. Pluto lies almost 40 A.U. from the Sun. The solar system itself has a radius of ~1000 A.U., although some comets influenced by the Sun orbit out to 50,000 A.U. or more.

It is customary to group the four planets closest to the Sun into the *inner* or *terrestrial* planets, of which the Earth is the largest in size. All have interiors composed primarily of relatively dense silicates, with or without nickel-iron cores. The remaining *outer* planets are called the *major, giant,* or *Jovian* planets because of their much larger diameters (excepting Pluto; see p. 29); their lower densities, however, result from predominance of the lighter elements and compounds in their makeup. The relative sizes of these planets with respect to the Sun are depicted in Figure 2-6.

Motions of the Planets

Although most planetary orbits are nearly circular (except for Mercury and Pluto), mathematically they are still best described as *elliptical.* The orbital planes for the planets nearly coincide with that described by the circumsolar path of the Earth which defines the *ecliptic* (Figure 2-7). As seen by a viewer looking down on the ecliptic from the celestial north pole, the planets revolve around the Sun in a counterclockwise (ccw) or *prograde* direction. Most planets undergo counterclockwise rotation as well (the Earth, as an example, spins on its axis from west to east so that the Sun appears to rise in the east as a western reference point moves toward it). Venus and Uranus move in a clockwise (cw) or *retrograde* manner.

The orbital motions of the planets, as they appear to observers on Earth, were accurately charted but poorly understood for several millenia. The first major breakthrough in deducing the physical basis for *orbital mechanics* was achieved by Kepler between 1609 and 1618 as stated by his three laws of planetary motion:

(1) Each planet moves along an elliptical (eccentric) orbit with the Sun as one focus.
(2) An (imaginary) line between the Sun's center and a planet's center sweeps through equal areas in equal time periods.

(3) For any two planets X and Y, their distances (a_x and a_y) with respect to the Sun and their periods of revolution (T_x and T_y) are related according to the equation

$$\frac{(\text{Period of } X)^2}{(\text{Period of } Y)^2} = \frac{(\text{Mean distance of } X \text{ from Sun})^3}{(\text{Mean distance of } Y \text{ from Sun})^3},$$

or

$$\frac{(T_x)^2}{(T_y)^2} = \frac{(a_x)^3}{(a_y)^3},$$

where a is the semimajor axis of an ellipse, taken as equivalent to the mean distance from the Sun for planetary bodies of low eccentricities.

When Kepler's laws are coupled with Newton's laws of *universal gravitation* ($F = Gm_1m_2/r^2$), *inertia* (the tendency of bodies to remain at rest or in uniform motion in a straight line until acted upon by an external force) and *interaction* (every force action is accompanied by an equal opposing reaction), the behavior of stars, planets, comets, and natural and artificial satellites can be grasped in general terms. Consider Figure 2-8, which describes the motion of any satellite S (planet or moon) in elliptical orbit around its more massive primary P (Sun or planet). It is assumed that, at any initial instant (as at S), the orbiting body has an inherited momentum (mass times velocity) to carry it along a path within the system. According to Newton's inertial law, this path will follow a straight line (in Euclidian geometry) nearly tangent to the ellipse of reference. However, the primary exerts some force at that point which translates by the gravitational law to an acceleration inversely proportional to the square of the distance between the bodies. If the satellite were not in forward motion (e.g., if it remained at rest), this force would cause it to fall in toward the primary. But, being in motion, the satellite moves neither along a straight line gradually away from the primary nor directly toward it. Instead, the path of the satellite is the resultant vectorial average for each of these possible motion components.

The satellite is thus continually deflected toward its primary. At each point along the path there is some critical velocity below which the satellite would gradually move inward and above which it would tend to move gradually away from the primary in nonlinear paths that depart from the ellipse. The appropriate critical velocity at each point must be maintained if the body is to stay in the same orbit. In order to launch an artificial satellite into a circular orbit, it must be accelerated to just that constant terminal velocity at which its motion tangential to Earth is *balanced* against the pull of gravity. It can then move forward and down in equal increments at all points in its path. The speed of a satellite in circular orbit is inversely proportional to the distance from its primary in response to the decreasing pull of gravity.

For any planet moving in an elliptical orbit, the gravitational acceleration varies with distance, and the velocity required to maintain the planet in this orbit changes proportionately. Thus, at a point (2) in Figure 2-8 further from the primary, the planet's speed, in terms of distance traveled per unit time, will decrease (if it remains the same, the planet would move beyond its orbit). At a point (3) close in, this speed will increase (relative to 1) in order for the satellite to overcome (maintain balance with) the stronger gravitational force from the primary. The net effect of this variation is that a planet's angular momentum (product of its mass m, its changing straight-line velocity v_1, and distance r to the primary, or mv_1r) remains constant throughout its orbit. As long as no ex-

Figure 2-8 A diagram illustrating Kepler's second law of planetary motion. A secondary body S (e.g., a planet) orbits along an elliptical path around its primary P (e.g., the Sun) located at one of the foci of the ellipse (P′ marks the other focus). Three different path segments A, B, and C (subdivided arbitrarily into unit lengths) are chosen such that the areas they subtend (shaded) are equal in extent. According to Kepler's equal area law, the body S will pass through each of these segments in the same amount of time. Owing to varying path lengths, however, the average orbital velocity will be different in each segment. The letters ap and p refer to the *aphelion* and *perihelion* positions, i.e., the farthest and closest points reached by an orbiting planet relative to the Sun. The letter a labels the semi-major axis of the ellipse—the distance between the + and p or ap.

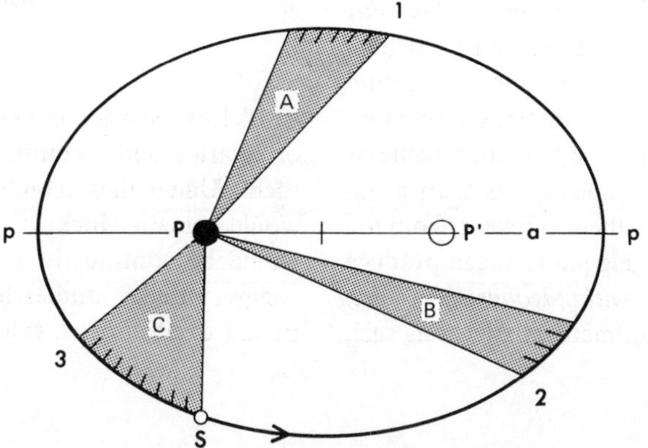

ternal forces perturb this mutual balance of forces, the planet will revolve indefinitely without change. Planets in the near-vacuum of space move without frictional drag (from an atmosphere), but over long time spans the *tidal forces* developed between gravitationally-interacting bodies will cause changes in the orbital distance between primary and satellite and changes in the *rotational* periods of each. These same principles apply, of course, to any moon or artificial satellite in motion around its primary (such as a planet).

In actuality, the orbital motions of a simple planet-satellite system are complicated by the presence of other bodies in space. Thus, both Earth and Moon are affected gravitationally by the Sun and, to a much lesser extent, by the remaining planets and even the stars. Whenever three or more orbiting bodies mutually interact, solution of relative motions cannot be accomplished by the mathematics of the Keplerian and Newtonian laws alone, but requires an exacting numerical treatment (utilizing a computer) to calculate the effects of the many forces acting at varying distances. The motions of an artificial satellite moving between Earth and Moon exemplify a typical three-body problem.

Motions of Small Bodies

Large natural bodies in a planetary system tend to remain interconnected through mutual orbits having circular to elliptical shapes. However, it is possible for small bodies located on the surface of a primary or orbiting close to it to acquire a velocity sufficiently high to break loose from the gravitational control of the primary. This *escape velocity* is a characteristic parameter of each planetary body in our solar system and varies for the different planets in direct proportion to their masses. The path assumed by a small object, such as a spacecraft, which is launched from the primary's surface at precisely the escape velocity will be *parabolic* in shape. Any velocity that exceeds this critical value will cause the object to follow a *hyperbolic* path. In either case the small body ultimately will leave the solar system to travel indefinitely through or beyond the galaxy.

Although motions of all objects in orbit within the system are influenced by mutual gravitational actions, other forces can substantially alter the positions of very small bodies such as dust particles and gas molecules. For example, the solar wind—a plasma of charged particles—can drive off the outer fringes of atmospheres around planets or the interplanetary gases in directions away from the Sun. Photons emanating from the Sun are capable of exerting a *radiation pressure* on bodies less than a micrometer in diameter that impels them outward from the light source. However, the impinging photons can produce a counteraction, known as the *Poynting-Robertson effect,* on particles of millimeter to centimeter dimensions such that these particles move progressively along spiral paths inward toward (and eventually into) the Sun. The effect depends on generation of a resisting drag as the photon is absorbed, leading to a net loss in kinetic energy and a resulting decrease in orbital radius. In this way, much of the finer-sized debris in outer space is continually being carried into the Sun ever since the solar system became organized.

The Spacing of Planets

The average distance of each planet from the Sun appears to be governed by a regular numerical relationship called *Bode's law*: If a series of terms starting with 0.3 are added to zero as a geometric progression such that each successive term is twice the previous one, the following sequence develops:

0.0 0.3 0.6 1.2 2.4 4.8 9.6 19.2 38.4 76.8

A constant value of 0.4 is then added to each term producing the new sequence

0.4 0.7 1.0 1.6 2.8 5.2 10.0 19.6 38.8 77.2

This sequence is very close to the actual distance in *Astronomical Units* (A.U.) of each planet out from the sun, namely:

0.39	0.72	1.00	1.5	5.2
Mercury	Venus	Earth	Mars	Jupiter

9.5	19.2	30.0	39.5
Saturn	Uranus	Neptune	Pluto

The correspondence between calculated and observed A.U. values is remarkably close except for Neptune and (especially) Pluto, whose orbit may be derived from that of Neptune which in turn may have been perturbed by Pluto. The gap where 2.8 A.U. should occur is precisely that now assigned as the average distance of the asteroids from the Sun; indeed, this prediction by Bode's law of a missing "planet" aided in discovery of those bodies. A theoretical basis for Bode's law has not been established, but the regularity of spacing it describes must relate in some way to the dynamics of formation of the solar system.

Characteristics of the Planets and Other Solar Bodies

Mercury

Until 1965, visual observations indicated Mercury's periods of rotation and revolution to coincide exactly at 88 days each. Under these conditions the same side of the planet would remain "locked" on the Sun so that this surface would be continually exposed to a large dosage of solar energy. Radar studies have now confirmed a rotational period of 58.6 days, exactly two-thirds that of its revolu-

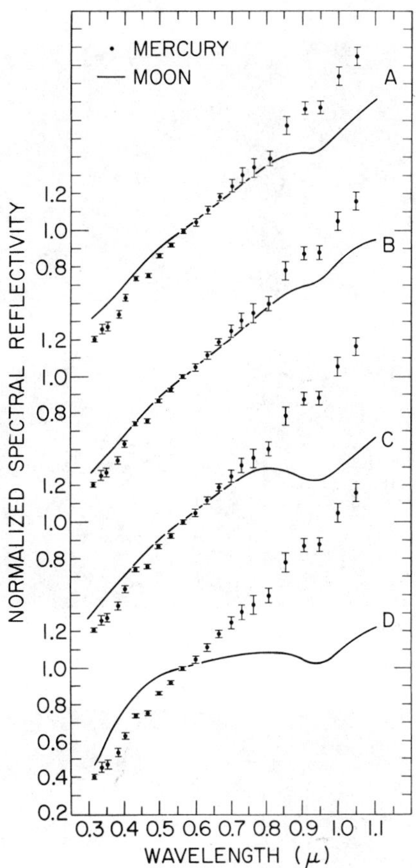

Figure 2-9 Spectral reflectivity curves (solid lines) for four different lunar terrains (A. uplands; B. maria; C. bright upland craters; D. mare bright craters) over which has been superimposed the measured reflectivities (with error bars) of the integral disk of Mercury scaled to unity at 0.564 micrometers. The best fit is that of the lunar maria, suggesting that Mercury may be largely covered by dark, basic lavas. (From McCord and Adams, 1972.)

tion. The surface of Mercury was not seen until Mariner 10—partly because of its small size, partly because of its closeness to the bright Sun, and partly because it attains maximum illumination when in *superior conjunction* (i.e., position in orbit when farthest from the Earth, along a line in which the Sun is between it and Earth). Light and dark markings reminiscent of the lunar surface are observed visually through the telescope. Spectrophotometric measurements made by telescope (see Chapter 5) have produced a reflectivity curve which closely matches those obtained for the lunar terra and mare surfaces, leading to the conclusion (Figure 2-9) that a lunarlike soil of basaltic composition has developed on Mercury (presumably by impact). Because of the absence of a shielding atmosphere, a heavily cratered surface has been predicted. Mercury's high density (5.5 g/cm^3) has been cited as evidence for a metallic core (comprising up to 70% by weight of the planet) which, if verified, implies complete planetary melting and subsequent differentiation.

Venus

Venus is the brightest starlike object in the sky. The brightness results largely from the high reflectivity of sunlight from a cloud deck above the surface. Being on an inner orbit relative to Earth, Venus experiences a notable decrease in its apparent diameter as it moves toward its superior conjunction (260 million km). At this point Venus is nearly fully illuminated by the Sun. As Venus nears its *inferior conjunction* (point of closest approach as Venus crosses the line between Earth and Sun), only a small fraction of its illuminated surface is visible from Earth, so that it appears as a thin crescent sliver through a telescope. These phase changes (Figure 2-10a) are similar to those described for the Moon (Chapter 5). It was once thought that Venus' similarity in size to Earth favors the existence of life in some form on its surface or in its atmosphere; measurements of high surface and atmospheric temperatures have now greatly diminished this possibility (see Chapter 12).

Earth

Viewed as a full orb from outer space by the astronauts or by imaging satellites, the terrestrial surface has a distinct bluish tone over its oceans and greenish to brownish tones where large land masses are visible. Great swirls and banks of highly reflective white clouds attest to operation of a circulating atmosphere containing condensible gases. From distances beyond those of most Earth-orbiting satellites, high-resolution imaging devices (such as cameras) have failed to detect any direct evidence of living organisms—including human construction and other activities of man—that mark our planet as possibly unique in that respect within the solar system. Only the cyclic variations of green tones, which surface dwellers know to be seasonal changes in vegetation, would afford any clues to an extraterrestrial observer as to the presence of life forms on Earth.

Mars

Mars, the "red" planet (Figure 2-10b), is another possible candidate for life in the solar system because its distance from the Sun gives it a temperature range not too alien for development and survival of certain simple organisms. However, its smaller size relative to Earth reduces the gravitational pull on some of the lighter gaseous molecules, including water, so that the bulk of any gases below molecular weights of ~30 are likely to have escaped over time unless these are being replenished from the planet's interior. A thin atmosphere of mainly CO_2 now remains. Mars is obscured periodically by fast-moving yellowish clouds consisting of silicate particles of dustlike nature (see Chapter 12).

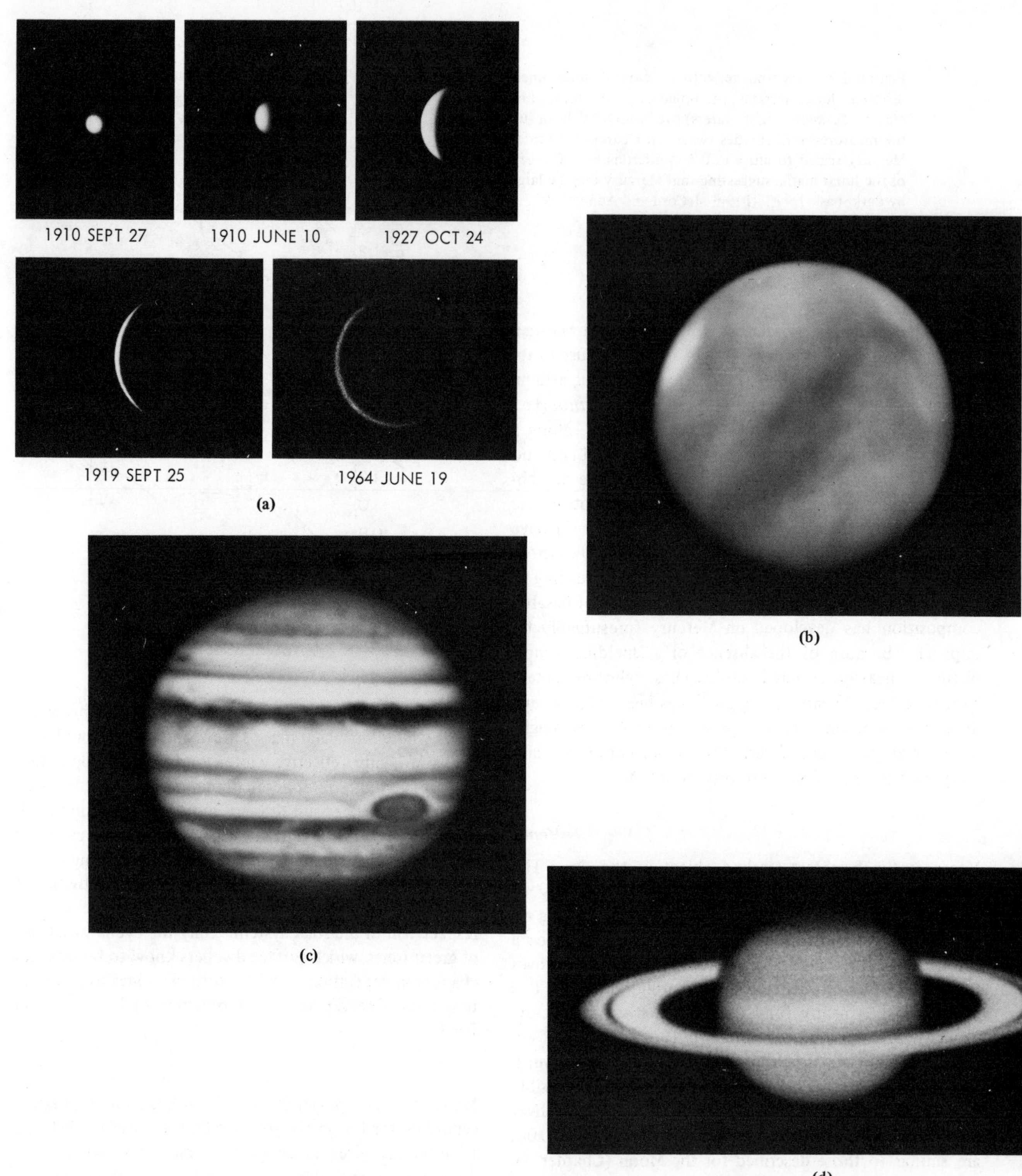

Figure 2-10 (a) Photographs of the phases of Venus, taken at the Lowell Observatory. At nearly full illumination, the planet lies in superior conjunction (farthest away from Earth). When at inferior conjunction (closest to Earth), Venus reaches its largest angular diameter but the area lighted in this crescent phase is minimal (NASA photos). (b) Mars, photographed by R. Leighton through the 60-inch reflector at the Mount Wilson Observatory. The north polar cap appears at the top left. The broad dark band in the equatorial region includes Mare Cimmerium and Mare Tyrrhenum, with Syrtis Major on the left (see Figure 12-1). (c) Jupiter, photographed through a blue filter by the Mount Palomar 200-inch telescope. The Great Red Spot appears near the limb in the lower right quadrant. (d) Saturn, with its three main rings tilted slightly towards the Earth, as seen in the 100-inch reflector telescope of the Mount Wilson Observatory. The outer ring begins beyond the dark band.

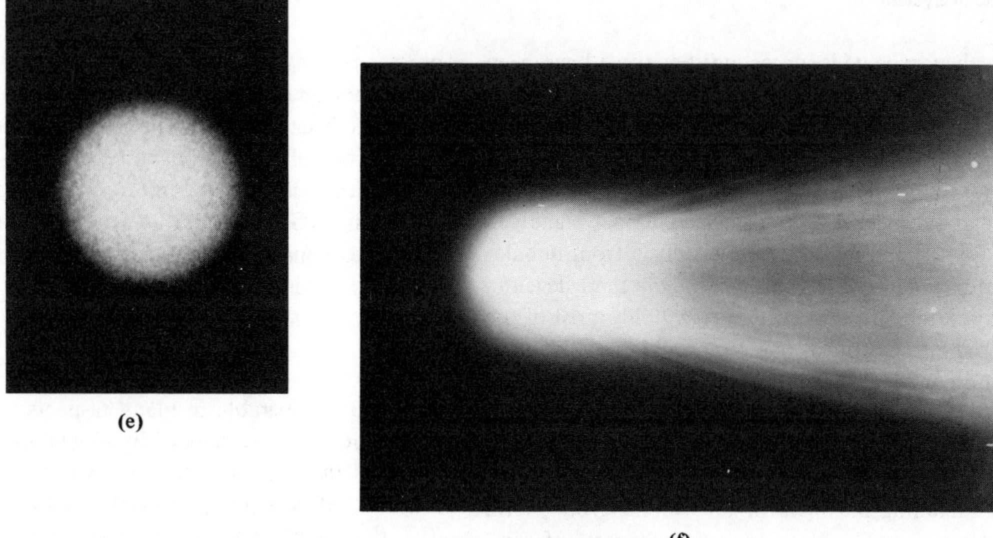

Figure 2-10 (continued). (e) Uranus, photographed by a 36-inch telescope carried aloft in an unmanned balloon, *Stratiscope 2,* to an altitude of 25 km. (f) Close-up view of the coma and inner tail of Halley's Comet, as photographed through the 60-inch reflector telescope, Mount Wilson Observatory, on May 8, 1910. (Photographs in b, c, d, and f reproduced by permission of the Hale Observatories, California Institute of Technology.)

Visual sightings and photographs have shown a pattern of light and dark patches and bands ranging from reddish-orange to brownish in tone. By following fine details, those patterns are seen to shift during the Martian year, along with bright, whitish areas around the poles consisting mainly of solid CO_2 ("dry ice") and probably some frozen water. Some of the narrower surface markings coincide with the linear features called *canali* by Schiaparelli (see Chapter 12).

Asteroids

At an average distance of 2.8 A.U. the "asteroids," a series of small *planetoids*—planetary bodies whose shapes are subspherical to ellipsoidal to irregular—move in groups and swarms in orbits near the ecliptic. Among the largest are Ceres (maximum dimension = 800 km), Pallas (390 km), Vesta (490 km), and Juno (250 km) (Figure 2-11). Most of the others among the estimated fifty thousand or more detected by telescope and radar are less than 50 km in diameter (including Eros, 35 X 16 km; Apollo, ~3 km; and Icarus, 2 km). Although most asteroids lie in a broad belt whose distance from the Sun varies between 2.1 and 3.5 A.U.—i.e., between Mars and Jupiter—a few have been perturbed into very eccentric orbits which can carry them near Jupiter (the Trojan group) or close to Earth (the Apollo group). Thus, Apollo itself has approached to within 3.2 million kilometers of Earth; Adonis comes as close as 1.6 million km; and Hermes swung to within 800,000 km in 1973.

Asteroids are believed to be the sources of most of the meteorites in the solar system. If so, those bodies at least are composed of basaltic and metallic phases (see Chapter 3). Large asteroids may have collided with the Earth and the Moon in the past and, along with comets, could account for formation of the larger impact craters. Although the present estimated mass of all asteroids combined is only about a thousandth that of the Earth (hence about a tenth of the Moon), a single planet recombined from all asteroids would still be about an eighth the diameter of the Earth. Based on data gathered during passage of the *Pioneer 10* space probe through the asteroidal belt, the number and density of smaller particles in the belt was

Figure 2-11 Artist's conceptual illustration of the external appearance of several of the better known planetoidal bodies in the asteroid belt. Compare with Phobos, the Martian moon shown in Figure 12-40. (NASA illustration.)

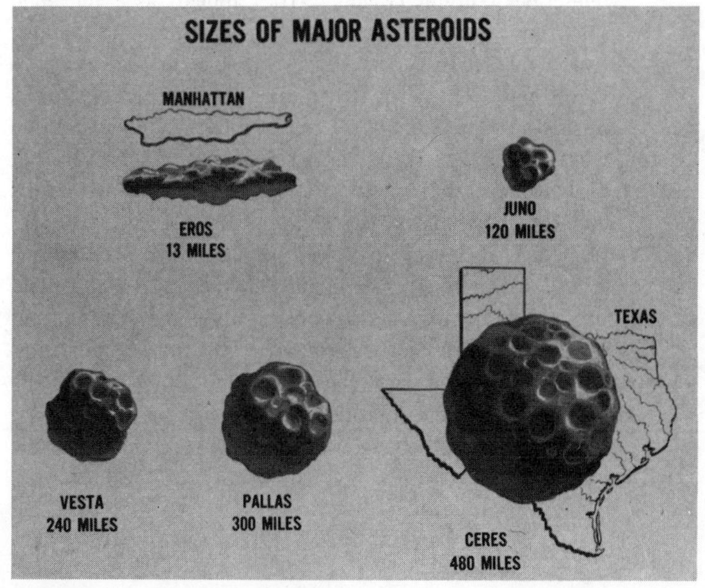

much less than expected, suggesting that these have been effectively swept up into the larger bodies.

One prevalent view considers the asteroids to be small primary accreted bodies which, because of gravitational perturbation by Jupiter or other major masses, never grew large enough to organize into a single spherical planet; in this concept the asteroids became residuals from nebular condensation that were unable to complete their development as discrete planets. Another view holds most of the asteroids to be fragments of one or several larger planetesimals disrupted by collisions or by strong tidal (gravitational) forces during passage near a large planet such as Jupiter. Ceres, Pallas, and Vesta may be original survivors that were not fragmented in this way. Some asteroids, including members of the Apollo group, have orbital velocities and eccentricities characteristic of cometary bodies.

Jupiter

Jupiter, the largest planet in the solar system (about 0.001 the mass of the Sun), is unique in that it is giving off energy (in part, as strong electromagnetic waves in the radio frequencies) from its sphere at a rate far in excess of that which could come from reradiation of the Sun's heat or from an internal heat source such as radioactivity. Jupiter's strong magnetic field is probably related to a process deep in its interior involving compression of its principal constituent hydrogen to an electrically conductive metallic (solid and/or liquid) state. Pressures exerted by the great overlying mass of dense gaseous substances exceed 100 million atmospheres near Jupiter's center, but temperatures there are still far too low (about 2000°K) to initiate nuclear reactions of the types operating in the stars.

Jupiter's atmosphere is probably continuous with its main interior mass, i.e., there is no sharp pressure discontinuity that defines a discrete surface boundary, although density discontinuities may exist at various levels. Through telescopes, Jupiter (Figure 2-10c) appears as a flattened spheroid (owing to outward bulging around its equator from centrifugal forces) whose outline is diffuse (owing to "atmospheric haze"). By far its most conspicuous visual property is the series of alternating light- and dark-colored bands (yellowish zones associated with grayish, bluish, brownish, or reddish belts) oriented subparallel to Jupiter's equator. These bands shift in position and thickness over time and also vary in color. A large oval area (about 45,000 km in length), termed the Red Spot, is a conspicuous and apparently permanent marking. Some bands and the Red Spot rise vertically well up into the Jovian atmosphere as though participating in convection cells. The origin of the bands and the spot remains unknown. The bands are most likely gaseous (e.g., NH_3) impurities in the hydrogen-rich atmosphere which are stabilized in their present regions by electrical or magnetic fields. The rotational motion of the planet could have smeared these mixtures into broad streaks. The Red Spot may be a localized convection cell of swirling gases moving up through the Jovian atmosphere. Of Jupiter's twelve satellites, the four larger ones (Galilean satellites), of moonlike dimensions, appear to consist of frozen gases (including water) and a dark component that may be rock.

Saturn

Saturn is comparable in many respects to Jupiter except that its lower total mass of hydrogen and helium is much less than the critical limit above which strong thermal and magnetic effects are produced. Saturn's lower density (0.7, the lowest of all the planets) suggests a more tenuous compaction of its gases and a much smaller liquid or solid metallic hydrogen core.

Saturn has three prominent luminous rings (Figure 2-10d) that are visible by eye in the telescope. The inner ring (~179,000 km in diameter, 21,000 km wide, and probably no more than 3 to 10 km thick) is not easily seen. The middle ring (234,000 km in diameter and 26,000 km wide) is the brightest. It is separated by a dark zone or gap from the outer ring (274,000 km in diameter and 16,000 km wide). A fourth, innermost ring, faint and tenuous, was discovered in 1969 by long-exposure photography. For years these rings were thought to be composed of minute dispersed objects (reflected light from Saturn's surface can pass through them) which may be frozen gases or ice-covered solids of dust size that have been trapped in orbit without ever coalescing to form other satellites. Alternatively they may have resulted from the disintegration of a satellite that moved within the Roche limit (see p. 55) for Saturn. This latter view is now supported by the first radar observations of Saturn in 1973, which discovered the rings to contain numerous solid bodies a meter or more in size having irregular, rough surfaces typical of rocky fragments.

Uranus and Neptune

Both Uranus and Neptune are mixtures of frozen water, methane, and ammonia—or possibly hydrogen and helium (perhaps with rocky cores). Both appear pale greenish through the telescope. Uranus has a somewhat granular surface (Figure 2-10e). The rotational axis of Uranus is inclined at 98° from the north ecliptic pole; this near-parallelism to the ecliptic is a far greater tilting than noted in any other planet and accounts for the retrograde rotation of Uranus. Neptune may be more completely condensed than Uranus owing to its lower internal temperatures, at which much of the methane and most of the ammonia will have solidified; its atmosphere is thus relatively enriched in CH_4 and impoverished in NH_3. Triton, one of Neptune's two satellites, is almost as large as the largest Jovian satellites.

Pluto

The small "planet" Pluto is anomalous in its present state. Because its density (of uncertain value—estimates have ranged between 2 and 8 g/cm^3, with an average of 3.5 g/cm^3) is appropriate to a terrestrial-type body of silicate composition, Pluto is considered by many planetologists to be a large satellite that escaped from Neptune.

Comets

Literally millions of bodies composed mainly of frozen gases travel through the solar system along a wide variety of orbits. At any given time only a very few of these comets are close enough to the Sun to display a luminosity visible from Earth (Figure 2-10f). As comets approach to within 1 to 2 A.U. from the Sun, interactions between their constituents and pressure exerted by solar radiation produce a glowing *tail* of diffuse minute particles that stream more or less radially away from the Sun over measurable lengths of 75 to 150 million kilometers (Figure 2-10f). Toward the Sun the tail connects to a *head* whose diameter ranges between 15,000 and 1,000,000 km or more. This head is composed of gases and dispersed particles that form a very-low-density "cloud" or *coma* with a diffuse outer boundary.

The core of the coma consists of a small *nucleus*, commonly only a fraction of a kilometer to a few tens of kilometers in diameter, a total mass less than 10^{20} grams. This solid body is made up possibly of silicate rock materials and complex hydrocarbons (similar to carbonaceous chondrites—see Chapter 3) embedded in a water ice and frozen gas matrix—ammonia (NH_3), carbon dioxide (CO_2), cyanogen (C_2N_2), and others. Sublimation by solar heating and photodissociation release such gaseous species as H, OH, CN, C_2, NH, NH_2, N_2, and metastable oxygen from the nucleus into the expansive coma. When near the Sun, cometary heads and tails obtain their luminosity by reflecting and scattering sunlight, by emitting visible radiation produced through molecular excitation (mainly of C_2), through ionization of gases to CH^+, CO^+, $N_2{}^+$, etc., and by resonance fluorescence of atomic sodium and other metallic elements.

The orbits of most comets are typically inclined at high angles to the ecliptic and tend to be highly eccentric. The majority of comets are concentrated in swarms (the "Oort cloud") that extend to the outer limits of the influence of solar gravitation well beyond the farthest major planet. The *long-period* comets sweep into near-parabolic orbits out to regions of the solar system as distant as 50,000 A.U. and pass infrequently near the Sun only at intervals from a hundred thousand to millions of years. The *short-period* comets, which comprise less than 20% of the nearly six hundred catalogued comets, have smaller eccentricities and thus make more frequent approaches to the Sun at regularly recurring intervals. Earth itself encounters particulate matter from some of these inner comets as it intersects remnants of their tails during its annual revolution about the Sun. Entry of this cometary material into the atmosphere gives rise to meteor showers (e.g., the Perseids and the Leonids) that recur each year at specific dates.

The origin of comets remains unsolved. Most workers contend that comets represent planetesimals of primeval matter formed from the original gas-dust cloud during the early history of the solar system. Far from the Sun, the nebular gases of higher molecular weights froze and admixed with silicates. As such, these cold bodies are the best surviving examples of the average composition of this cloud, exclusive of hydrogen. Some cosmologists believe that comets are probably being continually created from gases and dust picked up from the solar system and from interstellar space as the Sun and its planets move through the galaxy.* Only a few comets each century are perturbed into orbits that bring them close enough to the Sun to undergo gradual destruction. The lifetime of those individuals that repeatedly pass near the Sun is restricted to a few thousand years at most, owing to the rapidity with which the nuclear materials evaporate and disperse.

Having surveyed the planets in their broad aspects, in the next chapter we turn to the best evidence available—the meteorites—of the materials that comprised the dust which condensed into the silicate fraction of these bodies.

REVIEW QUESTIONS

1. Distinguish between rotation and revolution of a planet.
2. What is retrograde motion as applied to a spinning planet?
3. Which gases are most common in the outer planets?
4. Express in a simple summary the essential ideas of Kepler's laws.
5. How do the short-period comets differ from the long-period comets? Why are comets important as clues in understanding the origin of the solar system?
6. What change in Saturn's rings would you expect to note if you observed it occasionally through a telescope during one revolution of this planet?
7. Why does Venus display large variations in size

*Discovery of hydrogen cyanide and methyl cyanide in Comet Kohoutek, a large (6 km nucleus) comet that passed around the Sun in late 1973, and early 1974, suggests that these bodies probably form far beyond the confines of the solar system inasmuch as those compounds are observed as common components in interstellar dust.

when viewed over an extended period through the telescope?

8. In what sense can Jupiter be compared with the Sun? How are its markings explained?

9. What causes the magnetosphere about the Earth?

10. Briefly discuss the nature and origin of the asteroids.

11. Explain the conditions under which an artificial satellite must be launched in order to maintain itself in orbit around its primary (e.g., the Earth).

12. Describe the main characteristics of the Sun: What is its principal source of energy?

13. Review the expression and implications of Bode's law.

14. What is the essential difference in the behavior of small particles subjected either to the Poynting-Robertson effect or to radiation pressure from the Sun?

15. How might the Earth appear if viewed through binoculars from the Moon?

CHAPTER 3

Meteorites: Samples of the Universe

The popular notion of *meteorites* treats them as rocks from outer space that are found by chance on the land surface. They are associated also with many of the bright streaks of light or "shooting stars"—technically, *meteors**—frequently seen in the night sky. For thousands of years, these "heaven stones" have been held in awe and superstition. The present scientific view, which is that meteorites represent samples of planetesimals which populate other regions of the solar system, can be traced back only to the first part of the nineteenth century. Until rock materials were returned from the Apollo lunar landings, meteorites were the only specimens of other planet-forming bodies available as guides to either the primitive or the differentiated states of terrestrial-type planets beyond the Earth. Meteorites are particularly important as indicators of the composition of planetary interiors, from which information about the Earth's inaccessible interior can be deduced. Modern interest in meteorites as vital clues to understanding the physical and chemical makeup of the Earth and its planetary neighbors has led to a dramatic upswing in research in the field of meteoritics. Many of our present concepts about the origin and early history of the solar system are now based on the results of this research since the 1950s.

*These meteors result when bits of interplanetary material called *meteoroids,* mostly low-density, fragile cometary debris, burn to incandescence after entering the Earth's air shield. The vast majority of meteoroids never reach the surface as discrete meteorites.

Occurrence

The lure of finding exotic pieces of extraterrestrial matter seemingly should foster many discoveries all over the world. It is estimated that some 500 tons of rock debris, ranging from mostly cosmic dust to rarely large meteoroids, enter the Earth's atmosphere each day. Most of this, however, ends up as micrometeorites, which become highly diluted within the Earth's soil and sediments. Arrival at the Earth's surface of large (meters to possibly kilometers) fragments capable of producing impact craters on landing appears to be exceedingly rare. This is partly because of the scarcity of large fragments as free bodies in outer space and partly also because these meteoroids tend to break up into smaller pieces as they encounter the denser atmosphere. No wonder, then, that only about 500 meteorites larger than 10 cm in diameter survive passage through the air each year. Of these, more than two-thirds drop into the oceans, leaving perhaps 150 to strike the continents. Only rarely are more than 5 meteorites recovered annually. The number of known meteorites currently is given as just over 1800; more no doubt have been handled by individuals who failed to appreciate their true identity.

We can distinguish between *falls* and *finds* depending on the circumstances by which the meteorites are located. A fall refers to recovery of a meteorite seen or heard while moving through the atmosphere. Scientifically, this permits the date of entry to be fixed, sometimes the orbit or trajectory path to be calculated, and the distribution of different types of meteorites to be evaluated statistically. Finds, by contrast, depend on chance collection after an unobserved event and on the curiosity of the discoverer who (hopefully) submits the object to an expert for identification. The proportion of the two main classes of meteorites—*irons* and *stones*—obtained this way is biased, since the irons tend to survive much longer on the land surface (especially in dry regions) and are unusual attention-getters by virtue of their weight, metallic appearance, and surface markings.

Physical Appearance

Many meteorites are indeed hard to recognize unless one knows what to look for. Nevertheless, the exterior of a meteorite often shows certain characteristics that serve to separate it from ordinary rocks. Most iron meteorites are dark reddish-brown (sometimes rusted), and the surfaces of many are pockmarked with scalloped depressions or "thumbprints" (Figure 3-1a). Stones frequently are covered with a glazed (shiny to dull) blackish to yellowish fusion crust and elongated furrows (Figure 3-1b). Both

Figure 3-1 (a) A small iron meteorite, with "thumbprints"; one of many fragments recovered at the Henbury craters in Australia (photo courtesy Floyd R. Getsinger). (b) The fusion crust, including flow ridges, formed on the Johnstown (Colorado) stony meteorite during passage through the Earth's atmosphere (length of meteorite about 10 cm). (c) Widmanstätten structure (criss-crossing bands) developed in the Edmonton (Kentucky) meteorite; the narrow white bands are kamacite that enclose gray angular areas of plessite; note the iron sulphide (troilite) inclusion on the right. (d) Texture typical of an ordinary chondrite; the largest chondrule shown in this slab from the Clovis (New Mexico) meteorite is 2 mm in diameter.

(e)

(f)

Figure 3-1 (continued). (e) Neumann bands in a single crystal of kamacite in the Edmonton (Canada) hexahedrite; these narrow lines are actually the effects of crystal twinning caused by mechanical stress such as that resulting from shock pressures at impact. (f) Polished surface of the Cumberland Falls (Kentucky) brecciated enstatite achondrite (whitish fragments) containing large inclusions of a black chondrite; specimen approximately 10 cm in length. (Photos b through f courtesy Dr. B. Mason, U.S. National Museum.)

crust and markings are skin-deep effects of aerodynamic ablation that can remove most of a meteoroid's mass by flaking, vaporization, and melting as it passes through the atmosphere. The best immediate test for recognizing a meteorite involves cutting into the interior (the amateur should saw off no more than a small piece at a corner to avoid degrading the specimen before examination by specialists). Many irons, after etching in a suitable solvent, reveal characteristic internal textural figures (the *Widmanstätten pattern*) that readily confirm their identity (Figure 3-1c). Most stones, when properly sectioned and polished, disclose the presence of *chondrules* (Figure 3-1d), rounded silicate bodies, unlike any found in terrestrial rocks, that appear to be quenched droplets of a melt.

Classification

Meteorites are classified in several ways, but the simplest approach relies on composition (mineralogical and chemical) and texture as the defining parameters. Iron meteorites are subdivided according to the proportions of several metallic phases and their textural interrelations, which in turn depend on the amounts of alloyed nickel and the cooling history of the materials when they were still within their parent bodies. Stones can be grouped into two major categories, those with chondrules (*chondrites*) and those without (*achondrites*). Some meteorites are termed *stony-irons* if metal and silicates are present in proportions roughly between 1:2 and 2:1 respectively. These have been subdivided on the basis of silicate mineralogy into pallasites (olivine-rich) and mesosiderites (pyroxene with minor olivine). Various classes of meteorites are defined by their principal mineral constitutents in Table 3-1. In addition, each meteorite discovery is catalogued by the name of a geographic locality at or near the site from which it was first collected.

Iron Meteorites

The iron meteorites consist of one or more phases of metallic iron each of which is an alloy containing somewhat variable amounts of nickel (Ni) and phosphorus (P) as important components. The conditions of temperature, pressure, and composition which govern the formation of each phase are interrelated by the phase (equilibrium) diagram for the system Fe-Ni-P (Figure 3-2). The high-temperature nickel-iron phase is known as *taenite*, which crystallizes with octahedral faces in the cubic system. Taenite contains varying amounts of nickel (between 25 and 65%) distributed randomly throughout the crystal structure. A low-temperature phase, called *kamacite*, is characterized by hexahedral (cube) faces and a nickel content below 6%.

When metallic iron containing between 6 and 15% nickel cools below some transition temperature (from 900°C to less than 600°C, depending on the amount of nickel present), it undergoes a transformation in which some of the nickel will redistribute by diffusion (exsolve) into a banded or lamellar intergrowth of nickel-rich taenite and nickel-poor kamacite that forms the distinctive Widmanstätten pattern. The ratio of taenite to kamacite and the size of the bands depends on the initial percentage

TABLE 3-1

Mineralogy and Classification of Meteorites†

The Common Minerals of Meteorites

Kamacite	α-(Fe, Ni)	(4-7% Ni)
Taenite	γ-(Fe, Ni)	(30-60% Ni)
Troilite	FeS	
Olivine	$(Mg, Fe)_2SiO_4$	
Orthopyroxene*	$(Mg, Fe)SiO_3$	
Pigeonite	$(Ca, Mg, Fe)SiO_3$	(About 10 mole% $CaSiO_3$)
Diopside	$Ca(Mg, Fe)Si_2O_6$	
Plagioclase	$(Na, Ca)(Al, Si)_4O_8$	

*Divided into enstatite, with 0-10 mole % $FeSiO_3$, bronzite, 10-20%, and hypersthene, >20%; these minerals are orthorhombic, and have monoclinic polymorphs known as clinoenstatite, clinobronzite, and clinohypersthene.

The Classification of Meteorites

Group	*Class*	*Principal Minerals*
Chondrites	Enstatite(11)	Enstatite, nickel-iron
	Bronzite(227)	Olivine, bronzite, nickel-iron
	Hypersthene(303)	Olivine, hypersthene, nickel-iron
	Carbonaceous(31)	Serpentine, olivine
Achondrites*	Aubrites(8)	Enstatite
	Diogenites(8)	Hypersthene
	Chassignite(1)	Olivine
	Ureilites(3)	Olivine, clinobronzite, nickel-iron
	Angrite(1)	Augite
	Nakhlite(1)	Diopside, olivine
	Howardites(14)	Hypersthene, plagioclase
	Eucrites(26)	Pigeonite, plagioclase
Stony-Irons	Pallasites(2)	Olivine, nickel-iron
	Siderophyre(1) (find)	Orthopyroxene, nickel-iron
	Lodranite(1)	Orthopyroxene, olivine, nickel-iron
	Mesosiderites(6)	Pyroxene, plagioclase, nickel-iron
Irons	Hexahedrites(7)	Kamacite
	Octahedrites(32)	Kamacite, taenite
	Ni-rich ataxites(1)	Taenite

Figures in parentheses are the numbers of observed falls in each class.

*Sometimes subdivided into calcium-poor achondrites (aubrites, diogenites, chassignite, ureilites) and calcium-rich achondrites (angrite, nakhlite, howardites, eucrites).

†From B. Mason, 1971, "Meteorites" (Ch. 8). In *Understanding the Earth*, Artemis Press, pp. 115-121.

of dissolved nickel. If, however, the nickel content is originally low, Widmanstätten lamellae fail to develop and the cooled meteorite consists entirely of kamacite. Again, if the percentage of nickel is greater than about 12-15%, this rearrangement into bands does not happen and may be replaced by a microscopic mixture of taenite and kamacite grains called *plessite.*

Iron meteorites are divided into three classes according to their nickel content and structure. *Hexahedrites* (4-6% Ni) contain only kamacite, which occurs as crystals or granular aggregates. The hexahedrites frequently are crisscrossed by lamellae, called *Neumann bands* (Figure 3-1e), that are mechanically induced twins commonly caused by pressures (e.g., shock waves). *Octahedrites* (6-14% Ni) are characterized by Widmanstätten lamellae. Five subdivisions of that class are defined by the bandwidth of the kamacite, from coarse (>2.5 mm) through medium (0.5-1.5 mm) to finest (<0.2 mm). The nickel-rich *ataxites* (Ni >12-14%) consist mainly of plessite and are marked by a fine granular texture. Iron meteorites have also been classified by the ratios of their trace-element contents of gallium (Ga) and germanium (Ge); four distinct groups are thus defined in

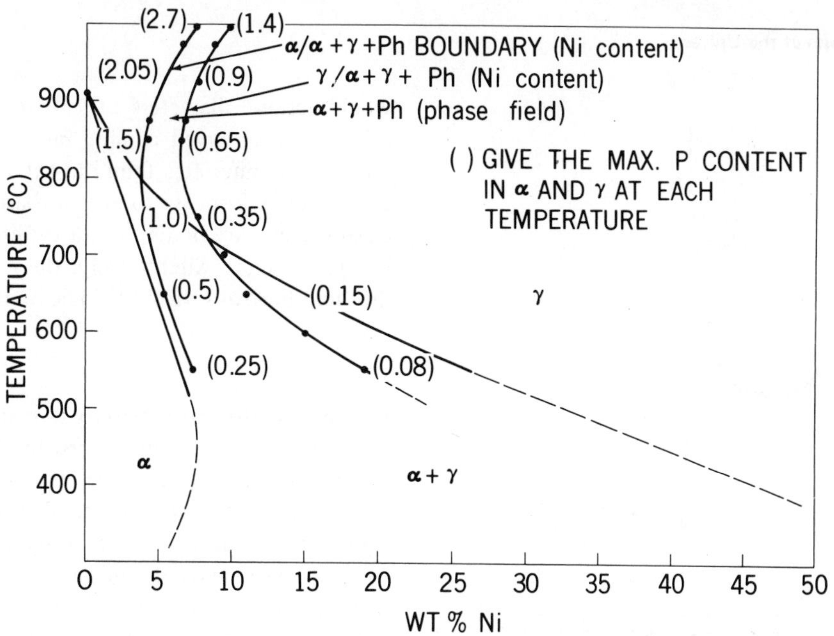

Figure 3-2 The standard phase diagram for the binary Fe–Ni system, also showing the effects of a third component, phosphorus, on the field boundaries. For low Ni contents, iron is present as the α-phase. The γ-phase occurs where Ni concentrations are high. Both phases co-exist for intermediate ranges of Ni. The field boundaries, however, shift with temperature. For example, at 600°C and a 5% Ni content, only α-iron (kamacite) would crystallize; for the same temperature and 25% Ni, γ-iron (taenite) is present alone. An iron melt with 12% Ni forms an intergrowth of kamacite and taenite (Widmanstätten structure) in some proportion at 600°C; a different proportion is produced if the initial Ni content is 9%. Other proportions are associated with specific temperatures and Ni contents. As any Fe-Ni melt with some given Ni content (e.g., 15%) cools through falling temperatures, the first phase to appear would be taenite containing that percentage of Ni. At some temperature, the $\alpha + \gamma$ field is entered and kamacite, with its Ni content defined by the field boundary at the entry temperature, begins to form. With further cooling, the alloy compositions (percent Ni) of both kamacite and taenite change systematically (by diffusion) with each new temperature until the melt has completely crystallized. If the initial Ni content is below ~6%, the final phase would be α-iron (see text for additional details). The presence of phosphides (Ph) in meteoritic iron substantially shifts the field boundaries. One such set of changed boundaries is indicated in the figure. Note that, at high temperatures, both the $\alpha / \alpha + \gamma +$ Ph and $\gamma / \alpha + \gamma +$ Ph boundaries reverse slope directions. A three-dimensional diagram is required to show all phase relationships for the ternary Fe–Ni–Ph system in which phosphorus contents vary up to their maximum solubilities. (From Goldstein and Doan, 1972.)

which compositional differences have been related to the sizes and cooling rates of differentiated parent bodies.

Cooling rates of bodies containing large continuous masses of nickel-iron have also been inferred from the width of kamacite bands in the octahedrites. With the advent of the electron microprobe,* it is now possible to obtain an even better estimate of cooling rates and to deduce the approximate sizes of the parent bodies from whence the iron meteorites were derived. Data needed for this determination include (1) the total nickel content of the octahedrite, (2) the distribution of Ni within adjacent taenite and kamacite crystals and across mutual boundaries, and (3) diffusion coefficients of nickel within each iron phase as a function of temperature over the cooling interval. Figure 3-3 provides an example typical of the nickel distribution or profile encountered in a microprobe traverse across Widmanstätten lamellae. Nickel content usually reaches its maximum near the outer edge of the taenite next to any adjoining kamacite. There is also a steady decrease in percent of nickel approaching the center of a taenite crystal as well as a slight decrease of nickel in kamacite immediately at the phase boundary. This characteristic distribution is explained as follows.

As the nickel-bearing metallic iron cools below initial crystallization temperatures, the initial taenite begins to undergo phase separation into taenite plus kamacite at a

*An analytical instrument that measures distributions of elements over small areas (~10 square micrometers). When atoms in a mineral are bombarded by a narrow electron beam, these give rise to secondary X-rays having wavelengths characteristic of the elemental species and intensities proportional to the amounts present. With this technique, variations in element concentrations in a chemical phase can be determined spatially in great detail.

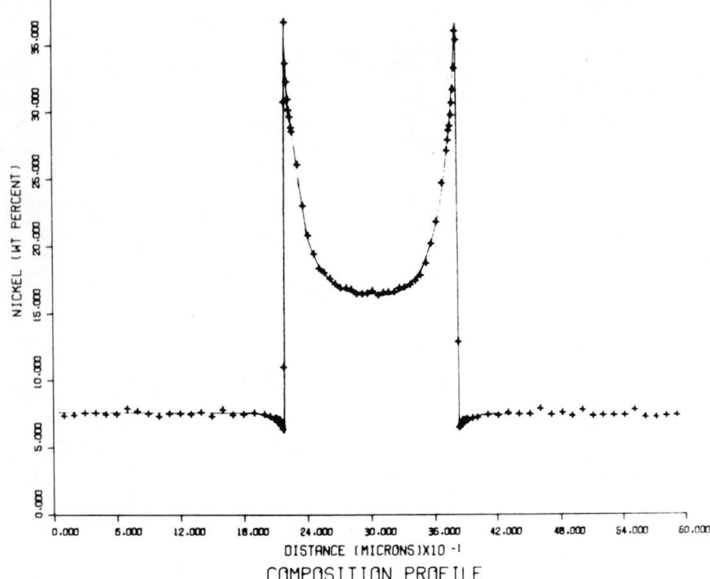

Figure 3-3 Concentration gradient of nickel across a kamacite–taenite–kamacite area in the Grant iron meteorite (octahedrite). Note the Ni dip in the kamacite (Agrell effect) at the α/γ boundary and the Ni buildup, above the bulk composition of 9.4 weight percent Ni, in the taenite. Maximum Ni concentrations occur within the taenite just at the α/γ boundary. (Courtesy J. Goldstein.)

transition temperature appropriate to the total nickel content. At any given temperature, there will be some specific amount of nickel in each phase. This depends both on the original nickel content and on the relative proportions of nickel incorporated in the taenite and kamacite at equilibrium conditions as defined by the phase diagram for the Ni-Fe alloy system. Both crystal phases grow in the space occupied by the high-temperature metal resulting in interleaved plates of kamacite and taenite. In effect, kamacite crystals are enlarged and taenite diminishes as cooling proceeds through a range of temperatures. The growth of the two phases takes place by diffusion of nickel from the high-Ni phase to the low-Ni phase. The rate of diffusion diminishes progressively from a maximum at the transition temperature to minimal values at much lower temperatures.

The measured change in nickel content on either side of a two-phase interface expresses the effect of this decreasing diffusion as each phase grows during cooling. The resulting gradients (variations of nickel with distance from the interface) are established by the cooling rates. Computer-generated profiles of nickel distribution as a function of cooling rates are matched to the measured profiles. Most octahedrites subjected to this analysis have computed cooling rates between 1° and 100°C per million years. Using the thermal conductivities characteristic of iron and stony meteorites, it can be shown that planetoids consisting of an iron core surrounded by a silicate mantle would require diameters of 150 to 400 km in order to achieve the observed cooling rates. To account for the range of cooling rates found in octahedrites, one explanation requires from six to eleven (based on distinct Ge-Ga classes) planetary bodies having different dimensions within those limits. Alternatively, the iron could come from different horizons within a single body.

Chondritic Meteorites

Chondrites, the largest group of meteorites (over 80% of the falls), are readily identified by the presence of spheroidal bodies, or *chondrules,* that display a diversity of textures (Figure 3-4). The chondrules are normally about 1 mm in diameter and contain grains, blades, and needles of the same minerals found in some basic igneous rocks—namely, *plagioclase*, *olivine*, and the pyroxene minerals *enstatite*, *bronzite*, and *hypersthene* (defined by the percentage of ferrous iron [Fe^{+2}] in each). This mineral assemblage is characteristic of higher temperatures of formation (above 1000°C). Rapid cooling or quenching is indicated in some chondrites by the unusual radiating crystals within chondrules and by glass filling the spaces between these crystals. Minute grains of Ni-Fe phases and of an iron sulphide called *troilite* (also found in iron meteorites) occur both in the chondrules and in the matrix.

The origin of chondrules is still open to debate. Their shapes, textures, and mineral composition all point to a high-temperature environment distinct from that in which the matrix of carbonaceous chondrites (see below) existed during accretion. The roundness of undeformed chondrules suggests they were once droplets of molten material—an idea originally proposed by Sorby in 1877. Some meteoriticists consider chondrules to have formed from fusion of part of the nebular "dust" by means of such proposed heating mechanisms as solar radiation, lightning discharges, or shock-wave compressions during the T Tauri phase (see p. 50) of the Sun's growth. Other, less favored genetic hypotheses include: melting of superficial debris on asteroids or planets by hypervelocity impacts during accretion; direct ejection of liquid silicates from the Sun; rapid condensation by supercooling of very hot silicate gases; expulsion by "volcanism" from already-formed planets; mechanical rounding in turbulent dust clouds; crystallization as immiscible phases in magmas; and growth in the solid state during metamorphism in primitive planets. Many scientists have surmised that, regardless of mode of origin, chondrules are a fundamental feature of primeval planetary matter.

Element distributions within the chondrites show discrete groupings of concentrations—particularly for the trace elements. A continuous series of chemical changes, such as characterizes many terrestrial differentiates, is therefore not developed in the chondrites. Iron, especially, displays significant clusterings of values in the different chondrites in terms of total content and oxidation state.

Figure 3-4 Types of chondrules as seen in thin sections. (a) Radiating fibers of pyroxene, Bjurbole. (b) Barred olivine in a single crystal filling chondrule in Mezo-Madaras. (c) Olivine phenocrysts in dark microcrystalline matrix from Bjurbole. (d) Granular olivine with native iron and troilite, Seres. (e) Olivine crystals and devitrification crystals in glass, from Mezo-Madaras. (f) Plates of olivine and troilite in plagioclase, Dhurmsala. (John Wood, 1963; photographs b, d, e, and f originally taken by G. Tschermak in 1885.)

Based on the ratio of metallic (reduced) Fe^0 to total Fe*, five chemical groups have been established, as shown in Table 3-2.

TABLE 3-2

Classification of Chondrites by Degree of Oxidation of Iron (After Urey and Craig)

Group Symbol	*Fe^0/Fe**	*Representative Class*
E	0.80	Enstatite chondrites
H (High)	0.63	Bronzite chondrites
L (Low)	0.33	Hypersthene chondrites
LL	0.08	Amphoteric chondrites
C	0.0-0.1	Carbonaceous chondrites

These various classes of chondrites are conveniently compared by plotting their iron contents in the reduced state (as metal and troilite) against the percentage of total iron in a more oxidized state within the ferromagnesian silicates (Figure 3-5). In addition, chondrites also generally follow two relationships stated as *Prior's rules:* Even as the amount of the nickel-iron phase scattered throughout a chondrite decreases, (1) the Ni/Fe ratio within that phase correspondingly increases, and (2) the ratio of FeO to MgO for the entire meteorite also increases.

Figure 3-5 A plot of the distribution of iron in the principal subgroups of chondrites (after B. Mascon). The two straight lines drawn through the cluster of points call attention to two broad groups, the H- and L-types defined by *h*igh and *l*ow total iron contents respectively. In carbonaceous chondrites nearly all Fe is oxidized and resides largely in the ferromagnesian silicates. As more Fe is reduced, the plots trend up and to the left such that each chondritic group contains a distinctive pyroxene associated with characteristic ratios of oxidized to reduced iron. The amphoteric chondrites are transitional to the Ca-poor achondrites but contain some hypersthene-bearing chondrules. (B. Mason, 1962.)

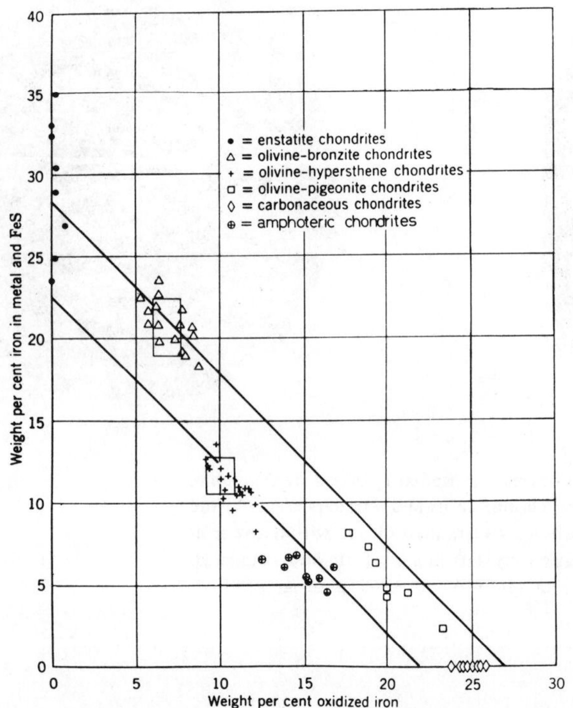

The chemical composition of all classes of chondrites considered together shows only small variations compared, for example, with rocks in the Earth's crust. This suggests that the parent bodies for chondrites initially formed from similar materials of nearly uniform composition. These parents, furthermore, remained as closed chemical systems unmodified by extensive element transfer. The overall element abundances in chondrites are more easily correlated with cosmic abundances determined from analysis of the Sun and stars than with abundances associated with the Earth and Moon. This adds support to the concept that chondrites represent relatively undifferentiated materials which were never subjected to the extreme alterations accompanying formation of the terrestrial planets.

The aforementioned chemical variations are paralleled by mineralogical and textural changes generally ascribed to thermal metamorphism and recrystallization at low pressures (Figure 3-6). Van Schmus and Wood (1967) have recognized six petrologic types based on the effects of metamorphism on various characteristic properties (Table 3-3). Each type represents an equilibration of the mineral phases under conditions of similar temperatures and pressures. The trend, overall, points to a general coarsening of crystal sizes in the matrix, conversion of glass to plagioclase and other minerals, and a gradual destruction of the crystalline structures of the constituent minerals and eventually of the outlines of the chondrules themselves.

One or more petrologic types are associated with each of the five chemical groups of Table 3-2. Thus, petrologic types 4 through 6 are observed in each of the *ordinary* (bronzite, hypersthene, and amphoteric—or H, L, and LL) chondrites, and types 1 through 3 correspond to differences in the three types of carbonaceous chondrites. However, this dual association of chemistry with mineralogy does not imply that the different chondrite classes are derived from one another. Thus, bronzite and hypersthene chondrites are not necessarily interrelated genetically and may have stemmed from separate parent bodies or from disconnected regions of the same body. The carbonaceous chondrites lack a direct link between any two of their three types; they probably formed independent of one another in different parts of the solar nebula in which thermodynamic and physicochemical conditions varied significantly. There is also no evidence that the ordinary chondrites are simply more highly metamorphosed and chemically modified variants of the carbonaceous chondrites; development of chondrules from matrix by a metamorphic process is very unlikely. The enstatite chondrites formed or were metamorphically altered as a distinct group under strongly reducing conditions which excluded FeO from the silicates.

A widely held view today accepts the carbonaceous chondrites, divided into types I, II, and III (or, C1, C2, and C3) in order of decreasing carbon content and increasing numbers of chondrules, as representing the most primitive (i.e., essentially unfractionated or undifferentiated) of the condensed silicate bodies that formed in the cooler regions of the solar nebula. Because of this primordial nature, analysis of a type I carbonaceous chondrite constitutes perhaps the best means of sampling the solar (cosmic) abundances of elements associated with a star and its gas-dust envelope. Since they weather rapidly, carbonaceous chondrites are not known as finds and are comparatively scarce (about 5%) among chondrite falls. Despite this scarcity, this chondrite group is believed to be common to prevalent among meteoroids in space.

A C1 chondrite contains up to 5% (by weight) of carbon combined with hydrogen and other elements in hydrocarbons and other compounds associated with abiogenic organic matter, and up to 20% water, along with other volatiles. This type is characterized by a homogeneous, fine-grained matrix (nearly devoid of chondrules) containing inorganic minerals (e.g., serpentines, clays) stable only

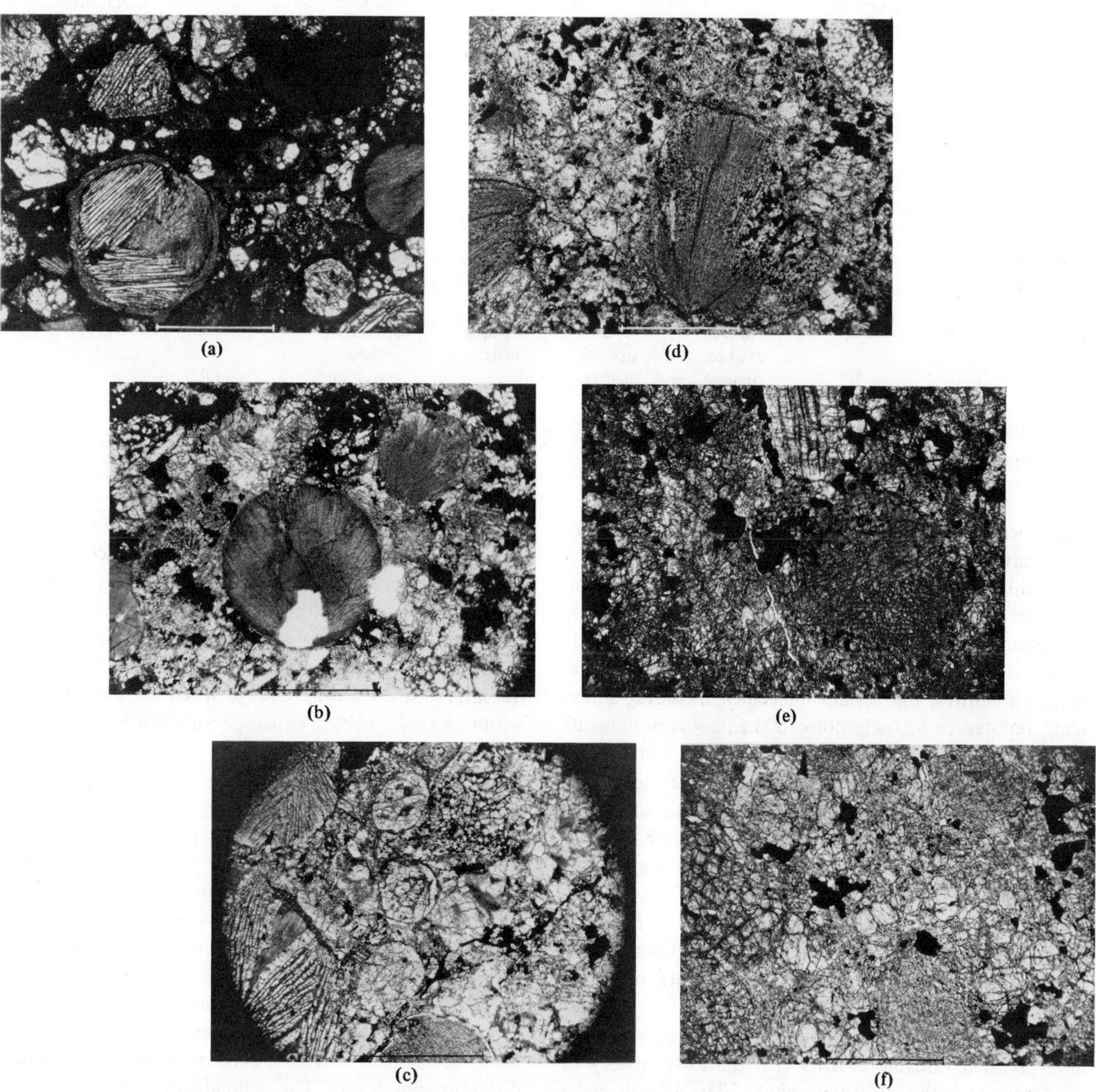

Figure 3-6 Examples of changes in texture within various hypersthene chondrites owing to effects of increasing thermal metamorphism in which distinctness of chondrules decreases from a through e: a, Tieschitz; b, Yatoor; c, Aldsworth; d, Lumpkin; e, Milena; f, Lissa. Scale bars represent 1 mm. (Figure from J. Wood, 1963.)

TABLE 3-3

Summary of Petrologic Types for the Chondrites

	Petrologic Types					
	1	*2*	*3*	*4*	*5*	*6*
(i) Homogeneity of olivine and pyroxene compositions	–	Greater than 5% mean deviations		Less than 5% mean deviations to uniform	Uniform	
(ii) Structural state of low-Ca pyroxene	–	Predominately monoclinic		Abundant monoclinic crystals	Orthorhombic	
(iii) Degree of development of secondary feldspar	–	Absent		Predominately as microcrystalline aggregates		Clear, interstitial grains
(iv) Igneous glass	–	Clear and isotropic primary glass; variable abundance		Turbid if present	Absent	
(v) Metallic minerals (maximum Ni content)	–	(<20%) Taenite absent or very minor		kamacite and taenite present (>20%)		
(vi) Sulfide minerals (average Ni content)	–	>0.5%		<0.5%		
(vii) Overall texture	No chondrules	Very sharply defined chondrules		Well-defined chondrules	Chondrules readily delineated	Poorly defined chondrules
(viii) Texture of matrix	All fine-grained, opaque	Much opaque matrix	Opaque matrix	Transparent microcrystalline matrix		
(ix) Bulk carbon content	$\sim$2.8%	0.6-2.8%	0.2-1.0%		<0.2%	
(x) Bulk water content	$\sim$20%	4-18%		<2%		

at temperatures less than 500-600°C. The matrix may have accreted through gravitational and electrostatic attraction, collisions, cementing by organic matter, or chemical reactions in the cool and diffuse outer regions of the collapsing solar nebula. No indisputable evidence exists for the presence of any living matter within the carbonaceous chondrites, but certain molecules, including amino acids, reputed to be indigenous to (i.e., not contaminants of) the meteorites, represent abiotic forms from which simple organisms could, in principle, be generated under appropriate conditions on a planet. If a planet like Earth accreted initially from meteoritic material, carbon could have been driven off as inorganic gaseous compounds whenever heating brought about melting of its interior. This carbon could then have accumulated in favorable environments (such as the oceans) that ultimately promoted its combination with other elements into true organic compounds.

Type C3 carbonaceous chondrites, although very rare, have revealed valuable insights into the early history of condensation within the solar nebula. The Allende meteorite, which fell in northern Mexico in 1969, provided over two tons of material for extensive analysis. Besides an olivine-rich matrix low in water and carbon, which is typical of C3 meteorites, Allende contained unusually large (up to 1 cm) chondrules that were enriched in such refractory elements as Al, Ca, and Ti. Some chondrules were composed almost entirely of calcic plagioclase. These chondrules must have formed at high temperatures and may represent examples of the first accumulations of condensates formed during a stage when the dust and gases within the solar nebula were especially hot (see p. 51).

Achondrites

It is tempting to carry the concept of increasing metamorphism of the chondrites through a complete melting of some chondritic parent bodies and development of differentiated products. This may lead to the *achondrites,* derived possibly as fragments of a crust or outer shell composed of lighter-density materials that concentrate near the surface of a once-molten planetary body. Some achondrites are notably brecciated, as though they had been reconsolidated from a zone of rubble derived from a differentiated layer. The achondrites follow a trend of increasing Ca and Fe and decreasing Mg within the silicates (Figure 3-7) that may have their counterpart in the Earth's interior. Thus, the *eucrites* particularly resemble the subcrustal basalts that differentiate from ultrabasic rocks residing in the Earth's upper mantle.* If a large enough planetary body underwent general melting, a core of metallic nickel-iron (with possibly some silicon as a reduced metal) and a lower mantle of mixed iron and silicates (pallasites and mesosiderites) would result. In this extreme case, the small quantities of Na, K, and Al in the initial chondrites would tend to concentrate in an outer shell through "flotation" after melting, giving rise to a thin crust of alkaline and silicic composition.

silicates condensed and aggregated within an interval of 20 to 100 million years after formation of these elements in the Sun.

The uranium-lead and rubidium-strontium isotope methods have fixed the time of accretion of the chondrites and of first melting and differentiation of the parent bodies for the achondrites and some irons at approximately 4.4 to 4.6 billion years before the present. Potassium-argon dating, which determines the time when gaseous argon-40 began to be retained, indicates that cooling of the solids

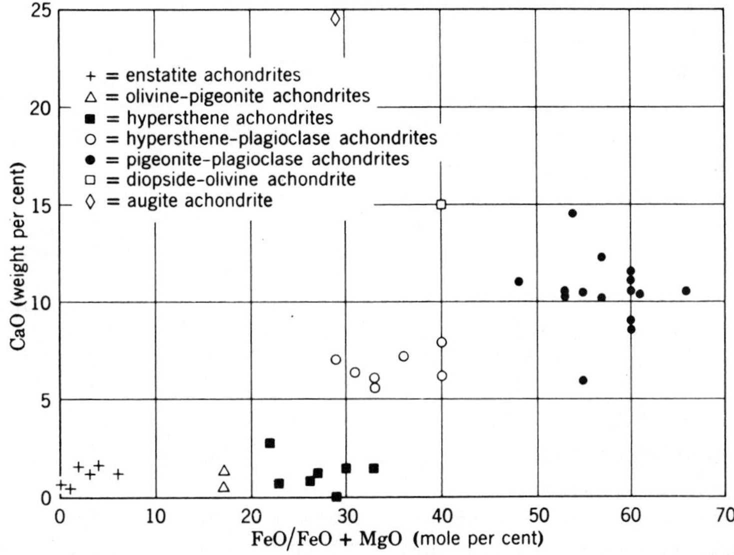

Figure 3-7 Variation of CaO with respect to the ratio of FeO to total FeO + MgO (relative proportions of Fe^{++} to Mg^{++}) in the achondrites (see Table 3-1; the hypersthene-plagioclase and pigeonite-plagioclase types correspond to the howardites and eucrites respectively). Several possible interrelationships among the achondrites and sequences of differentiation by which they were derived have been proposed; the single specimens of chassignite and angrite, and perhaps the aubrites, may be fragments of large single crystals from other achondrite types. Crystallization trends relating diogenites, eucrites, and howardites are similar to those established for some terrestrial basic igneous rocks. (From B. Mason, 1962.)

Meteorite Ages

Meteorites comprise the oldest samples of solar system materials available to us for direct age-dating by isotope analysis. The chondrites contain small amounts of the gaseous element xenon (Xe), often enriched in the isotope Xe^{129}. This isotope is produced by decay of the extinct radioactive isotope iodine-129 (I^{129}), whose relatively short half-life (16 million years) allows it only brief existence after element formation by galactic nucleosynthesis. Presence of daughter isotopes from I^{129} and other extinct radionuclides–e.g., plutonium-244 (Pu^{244}) with an 80-million-year half-life–supports the deduction that the

to temperatures below ~400°C was completed over an interval between 3.5 and 4.5 billion years ago. Argon-40 itself may subsequently be lost by shock-induced heating during collisional breakup of the parent body, so that the new K^{40}/A^{40} ratio can be used to date that event. Another indicator of collision that "resets the clock" is associated with loss of nearly all the helium gas accumulated from uranium and thorium isotope decay. New helium produced from the remaining uranium thus dates the rupture itself. K-A, and U, Th-He gas retention ages of about 500 to 1000 million years are characteristic of certain chondrites and octahedrites. Evidently, large chunks of the broken planetary bodies–perhaps equivalent to the asteroids–were subjected to further comminution to fragments comparable in size to present meteorites. As fresh surfaces are produced by this fragmentation, they

*Spectral reflectivity studies (see Chapter 5) of the surfaces of the Martian satellite Phobos and the asteroid Vesta indicate compositions similar to the eucrites.

will be exposed to cosmic-ray irradiations that generate characteristic radioisotopes such as He^3, Ne^{21}, Al^{26}, and Ar^{38}. Exposure ages of these meteorites mark a final size reduction at times ranging between 2000 years and 80 million years for stones and as much as 2 billion years for some irons.

Genetic Implications

There is general agreement that meteorites record some aspects of each of four stages crucial to the history of the solar system prior to organization of solids into planetary bodies: (1) condensation of high-temperature refractory silicates, oxides, and metals; (2) separation of silicates from metals as granular particles in the nebula; (3) condensation of lower-temperature or volatile phases; and (4) varying degrees of remelting of the early condensates. Opinion is still much divided on whether the different classes of meteorites represent different stages of a single genetic process or instead originated by multiple, and possibly unrelated, processes.

Unless a large planet underwent erratic melting in which large rock masses at depth experienced only moderate thermal metamorphism, it is difficult to see how a single parent body could be the source of all the varieties of meteorites in the general classification. It is more likely that meteorites are derived from numerous (from few to perhaps many) parent bodies of different sizes. Each body will have changed to some degree from a primitive state comprised of generally homogeneous carbon-bearing chondrites either to a state involving one or more of the different chondrites or to more advanced states marked by a plagioclase-pyroxene outer layer that grades into inner layers constructed from varying proportions of pyroxene, olivine, and nickel-iron metal. Judging from the very small sizes of most meteorites, we can also conclude that their parent bodies were broken apart most probably by collisions with one another. Fragmentation caused by collisions is supported by signs of shock-wave damage both in iron (Neumann bands) and stone (brecciation) meteorites (Figures 3-1e and 3-1f)

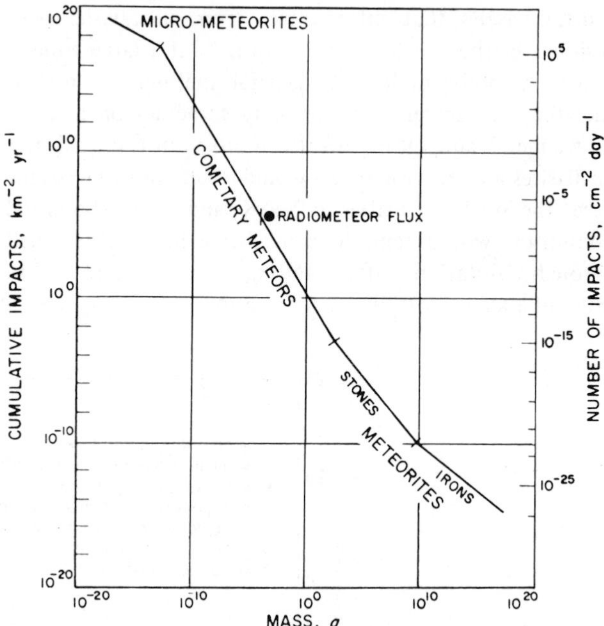

Figure 3-8 A general meteoroid flux curve for objects of different mass in interplanetary space, based on radio, visual, and photographic observations and rocket and satellite soundings. The divisions on both abscissa and ordinate (in log scale) are the exponents to which the base 10 must be raised to obtain the number represented (thus 10^{-10} equals 0.0000000001 and 10^5 refers to 100,000). The ordinate scale relates the number of specific impacts (size of resulting craters unspecified) expected to occur anywhere on the Earth's surface within the stated areas per unit time; sometimes this is given as 10^6 km^2 per 10^6 years. As an example of the use of this curve: there will be only 10^{-10} impacts possible in an area of 1 square kilometer each year caused by meteoritic particles whose masses are equal to or greater than 10^{10} grams; each day, over an area of 1 square centimeter, a flux composed of particles 1 gram (10^0) or larger will produce, on average, about 10^{-12} impacts. Note that the curve slopes upward to the left (negative slope); this means that smaller and smaller sized (by weight) particles strike a target of fixed area with increasing frequencies. (After Hawkins, 1964.)

Regions of Origin

In what parts of the solar system did the parent bodies of meteorites originate? It is likely that meteorites come from several different regions of outer space. Two lines of evidence suggest that most, however, were once orbiting with the present-day asteroids in the belt between Mars and Jupiter. First, the fact that the incidence of falls reaches a maximum in the local afternoon and a minimum near dawn is explained by assuming that meteoroids will encounter the Earth on its sunset side if they move in direct (counterclockwise) orbits and have aphelia lying beyond the Earth's orbit. Second, the paths of two nighttime meteorite falls (Pribram, Czechoslovakia, 1959; Lost City, Oklahoma, 1970) have been documented by recording cameras at separate locations, allowing their final trajectories to be accurately calculated. These trajectories can be related to orbits similar to some short-period comets that became perturbed from source regions beyond Mars, most probably from within the asteroidal belt. This shift of normal orbits of smaller fragments in the belt to an Earth-crossing orbit requires perturbations by increased gravitational attraction during favorable conjunctions with one or more planets or stray cosmic bodies. The possibility that some meteoritic materials—particularly low-density carbonaceous chondrites—may derive from cometary nuclei cannot be ruled out. Some achondrites, such as the brecciated howardites, may be surficial materials spashed off the Moon by impacts.

An inventory of particulate matter of varying sizes that reaches or can be observed from Earth permits estimation of a *meteorite flux rate curve* (number of particles in various size groups that pass through or strike a reference area during some unit time period) applicable to at least the inner planets (Figure 3-8). Most of this matter is dust-sized (micrometer range). This dust, which concentrates in the plane of the ecliptic, is responsible for the *zodaical light*—a bright luminous band seen in a dark sky around sunrise or sunset as sunlight reflects from tiny particles in space. Cosmic dust and meteorite infall data obtained from deep sea cores and aerial meteor counts, respectively, supplemented by rocket and satellite surveys above the atmosphere, are used to derive the flux rates for the smaller bodies. The flux curve can be extended to larger-sized bodies by calculating the masses needed to excavate impact craters in a given time span over selected areas up to continental sizes. Asteroids above a few kilometers in diameter can be located and counted by telescope and radar techniques. This modern flux curve has been extrapolated with caution back well into geologic time to approximate the flux densities in space and the corresponding frequencies of impact to be expected on planetary surfaces during their early histories.

Tektites

There is still another group of unusual rocks which appear to have originated by impacts with meteoroids. These are the *tektites* (Figure 3-9) that have puzzled many investigators for decades. Tektites are fragments of a glassy material which superficially resembles obsidian but contains no crystals (other than a few incompletely fused mineral grains in the Muong Nong type). Their compositions are generally silicic (70-80% SiO_2) although the smallest sizes (microtektites) tend to be more basic (silica-

Figure 3-9 Representative samples of tektites from various localities. Top row shows some sculptures and fluted examples (possibly further etched out by natural corrosion); elongate forms appear in the second row; and tear-drop and button-shaped tektites are placed in the third row. Types: a. Moldavites (from Czechoslovakia); b. Australites (from Australia); c. Bediasites (from Texas); d. Billitonites (from the Isle of Billiton). Scale: tektite in bottom center is 3 cm in diameter. (Photo from H. H. Nininger; published originally in *Sky and Telescope*, Feb., 1943.)

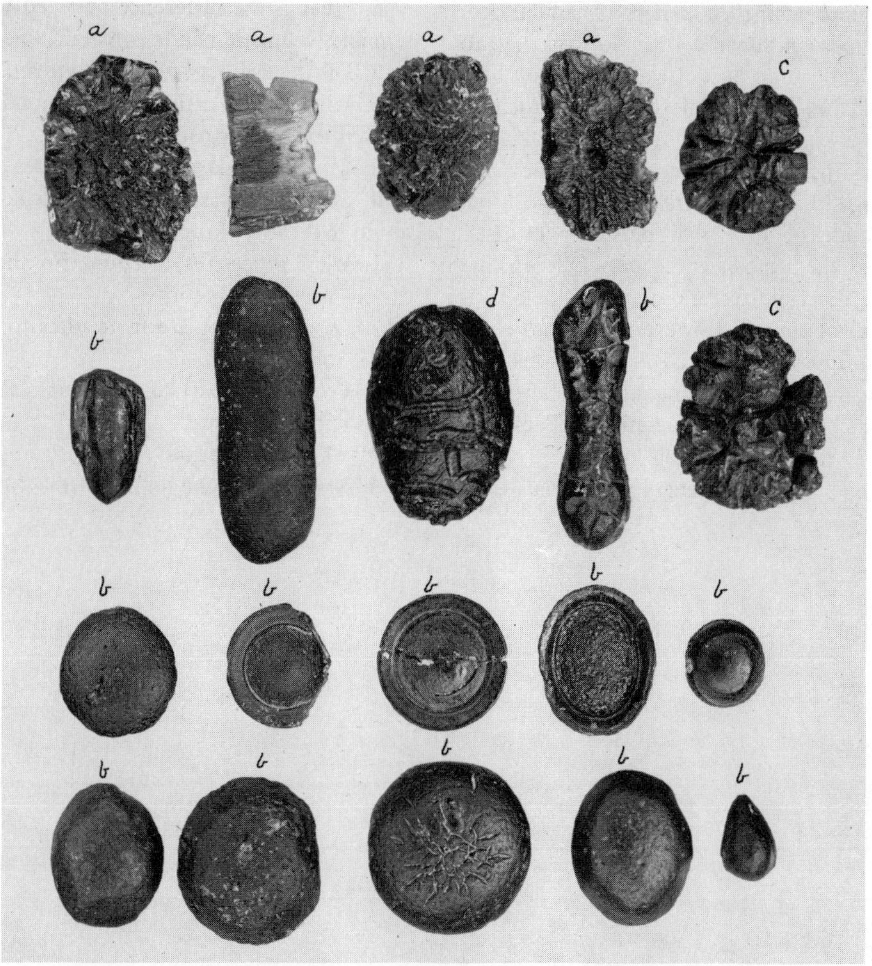

poor). Most tektites are button- or teardrop-shaped, and many are grooved and pitted. These features are interpreted to indicate a later fusion and aerodynamic sculpturing of their outer parts that most likely occurred during high-speed flight through the Earth's atmosphere. Some tektites contain evidence of having been completely melted earlier by intense heating (above 1700°C), and a few also reveal signs of having been shocked at high pressures. Presence of a few Ni-Fe particles in some tektites suggests that iron meteorites were involved in the events that melted them.

Tektites are distributed throughout *strewnfields* in Southeast Asia, Australia, West Africa (Ivory Coast), and central Czechoslovakia and at isolated localities in Texas and Georgia. Microtektites–mostly globular glass beads less than a millimeter in diameter–have been found in cores penetrating younger marine sediments off West Africa and in the Indian and Pacific oceans off Australia. The youngest strewnfield was formed just 700,000 years ago; the oldest was deposited about 32 million years ago (middle Oligocene).

The genesis of tektites remains a mystery. Their composition, together with relict minerals frequently found in sedimentary sandstones, suggests a terrestrial derivation. The African and European tektites may be associated with known impact craters, but the American and Asian ones have not been tied to identified craters of similar age. Some workers propose a lunar source for tektites to account for their nearly total lack of water and their high degree of homogeneity. Furthermore, infall velocities, estimated to reach 6 km/sec or greater, would be difficult to attain if tektites originated from terrestrial impacts and then passed through the atmosphere before deposition, unless the magnitude of the infall plus impact itself momentarily forced the atmosphere aside. O'Keefe has recently suggested that tektites are silica-enriched lavas ejected from lunar volcanoes. However, so far no rocks on the Moon with appropriate initial compositions have been returned from any Apollo or Luna missions, although some highlands and/or subcrustal materials collected at the *Apollo 12* and *14* sites show possible kinships. Tektites are not likely to be samples of planetary materials from outside the Earth-Moon system because they show no evidence of cosmic-ray effects that would indicate long residencies in orbit.

We are left, then, with a persisting dilemma. Dynamically, it is hard to posit that tektites are derived from the Earth; chemically, it is very unlikely that they came from the Moon; agewise, it is clear that tektites do not originate beyond the Earth-Moon system. The problem of the source of tektites remains unsolved, even though most meteoriticists today favor a terrestrial meteorite impact as the mechanism by which the tektite-forming melt is produced.

REVIEW QUESTIONS

1. How do the terms *meteor*, *meteoroid*, *meteorite*, and *meteoritics* differ in meaning?
2. Which are more abundant in meteorite collections, falls or finds?
3. Why do most meteoroids fail to reach the Earth's surface?
4. Why are falls more valuable than finds as indicators of the relative proportions of different types of meteorites in space?
5. What is the best method for identifying a rock as a meteorite?
6. What is the difference between Widmanstätten pattern and Neumann bands in metallic meteorites?
7. Distinguish between kamacite and taenite.
8. Why are the carbonaceous chondrites especially important to cosmologists?
9. On what basis are the chondrites classified?
10. Suggest a reason why chondrules have never been found in terrestrial rocks.
11. What process(es) account for the derivation of the various types of meteorites?
12. Of what value are meteorites to our understanding of the solar system?
13. From what part of the solar system do most meteorites appear to come?
14. Where are the glassy tektites found on Earth?
15. State briefly the nature of the "tektite dilemma."

CHAPTER 4

The Origin of Planets

Two modern views of the origin of the solar system are maintained by cosmologists. The first assumes the Sun to be one of only a few stars among the millions within the Milky Way, a typical spiral galaxy (Figure 4-1a), that because of favorable circumstances succeeded in capturing a denser concentration of interstellar materials. A variant of this idea considers an exploding supernova (Figure 4-1b) to be responsible for this dense cloud–consisting of particles of silicates and metals mixed with hydrogen, helium, and other gases. The other, more widely accepted view holds the Sun and planets to be one of many such systems derived universally from condensations within vast rotating nebulas or "clouds" (Figure 4-1c and 1d).

Constraints on the Origin of the Planets

Any attempt to account for the origin of the solar system should consider at least the following outstanding facts:

(1) The planetary orbits are regular and nearly circular; the planets all revolve in the same prograde (counter-clockwise) direction within a narrow zone about the ecliptic plane; most planets also rotate in a prograde manner.

(2) There is a notable similarity between Sun-planet and planet-satellite mass ratios–i.e., most planets are about a thousandth to a hundred-thousandth of the Sun's mass, and most satellites (excepting the Moon) have nearly the same fractional range of mass relative to their primaries.

(3) The planets lie in stable positions at regular intervals from the Sun (following Bode's rule, except for Neptune and Pluto).

(4) The terrestrial planets are small, dense, slow in rotation, and have few satellites; the giant planets are large, less dense, faster rotating, and have many satellites.

(5) The Sun possesses less than 2% of the angular momentum of the system even though it contains nearly all the mass.

Early Explanations for Planetary Origin

The foundation for a scientific explanation of the formation of the planets was laid by Kepler in his laws of

---

The contents of this chapter are based mainly on ter Haar and Cameron (1963), Wood (1968), and Hartmann (1972). (See the references at the end of this book.)

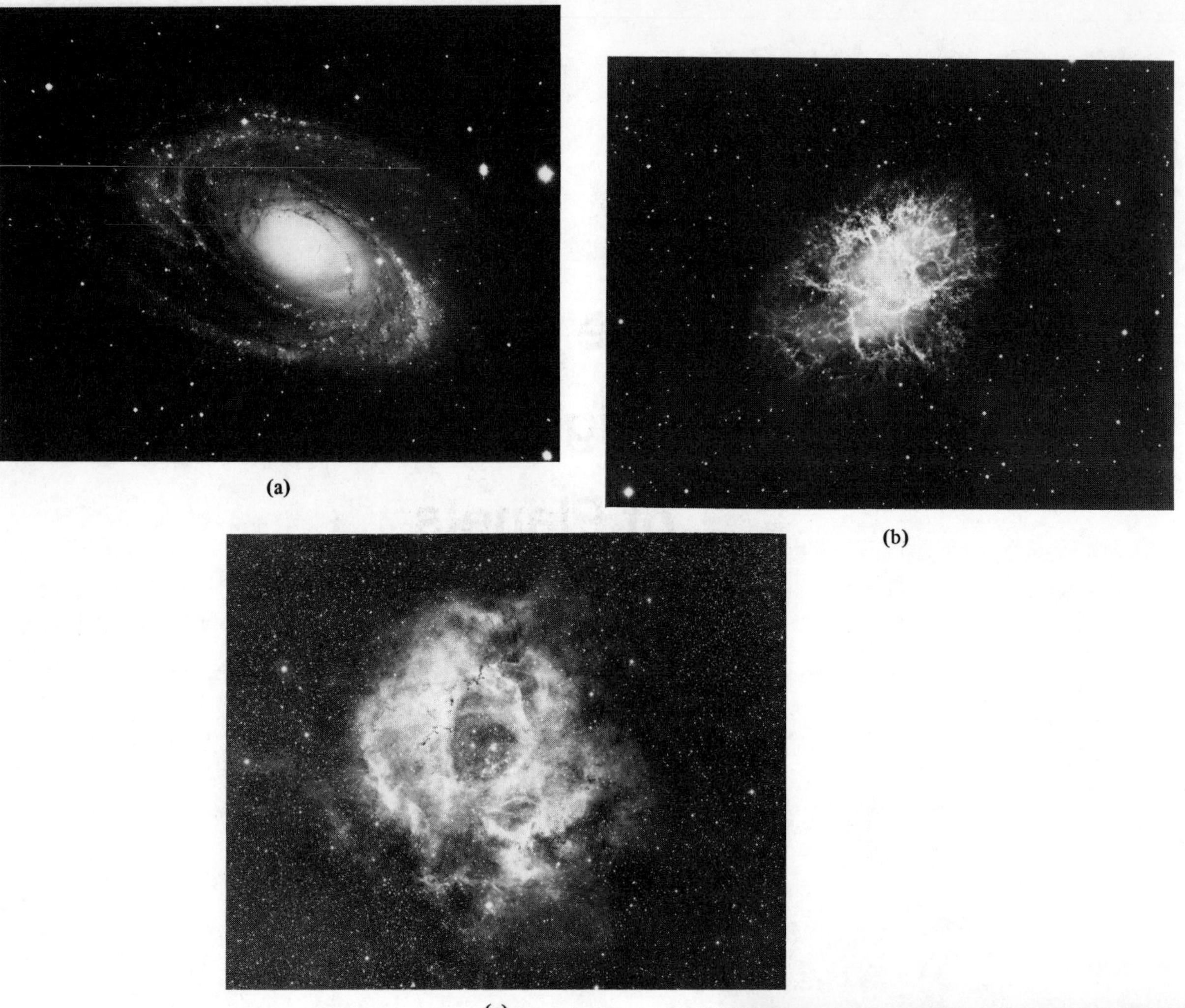

Figure 4-1 Four examples of stellar clusters in the universe that display some of the cosmological phenomena associated with evolution of planetary systems. (a) Messier 81, a spiral galactic nebula (type Sb) in *Ursa Major.* (b) Remains of the supernova event, first witnessed on Earth in 1054 A.D., within the "Crab" Nebula in *Taurus,* photographed in red light. (c) A gaseous nebula without well-defined rotational organization–the "Rosette" Nebula in *Monoceras,* photographed in red light. (d) A planetary nebula in *Aquarius,* photographed in red light. (Illustrations a, b, and d were photographed through the 200-inch reflector telescope and c through the 48-inch reflector telescope of the Hale Observatory, Mount Palomar, in California. Reproduced by permission of the Hale Observatories, California Institute of Technology.)

planetary motion (see Chapter 2). The first serious "modern" hypothesis was the vortex theory proposed in 1644 by the philosopher-mathematician Descartes. He postulated planetary development to be the outgrowth of condensation of matter from swirling eddies. Since then more than thirty theoretical models have been advocated, but most can be rejected when asked to explain one or more of the above main facts. These models fit broadly into two categories: *monistic* or *evolutionary*–formation of the solar system as a single unit–and *dualistic* or *catastrophic*–formation through interaction of the Sun with a foreign body.

Buffon, in 1745, proposed the first dualistic model; it held that the planets originated from materials ejected from the Sun after collision with a comet. This concept was fundamentally erroneous, for later studies of comets show them to have far too little mass to affect a star after infall. The competing monistic models of Kant (1755) and Laplace (1796) each assumed contraction of a gaseous nebula in which most of the condensed matter accumulated into a central protosun. Kant considered the contraction to have been initiated by gravitational forces, while Laplace held it to have resulted from cooling. According to both, the cloud gradually flattened into a disc which increased its rotational velocity. This caused centrifugal forces to exceed gravitational forces, so that rings of matter separated from the region of the Sun and combined into the planets.

Foundations of Modern Cosmologies

In 1901 Chamberlin and Moulton proposed a dualistic model according to which a passing encounter between two stars drew a gaseous protuberance or filament from one (the Sun) that condensed to form planetesimals (small aggregations of condensed matter), which were presumed to grow into larger planets by collision. The motions imparted during extraction from the Sun were held to account for the orbital characteristics now observed. Jeffreys and Jeans in 1916 modified and expanded this concept; in their hypothesis, the Sun underwent a major distortion in its shape as a second star passed within grazing distance, which led to separation of part of the Sun's mass. Variations in these views include collisions between nebulae (See, in 1910) or stars (Arrhenius, in 1913) or total breakup of the smaller member of a binary star system (Russell, in 1935), possibly as a supernova explosion (Hoyle, in 1942).

The role of magnetic forces in redistributing ions of different masses to different distances out from the Sun prior to condensation was first cited by Birkeland in 1912 and then refined by Alfven in 1942. This interaction between magnetic fields emanating from the Sun and ionized gases was capable of producing strong motions in the solar nebula—giving rise to *magnetohydrodynamic* effects that dispersed and separated ions according to their mass and charge. Rotation of the field source resulted in complex twisting of the field lines. The field motions caused accelerations of nebular materials, but the Sun itself was slowed by magnetic drag.

Berlage, between 1934 and 1940, developed a scheme by which concentric rings formed through dissipation of energy by viscous interactions with a nonturbulent gaseous envelope. Each ring was dominated by one or more elements, the species depending on the charge-to-mass ratios of the various nebular constituents. However, according to von Weizsäcker (1946), turbulence was important in producing small vortices within a swirling gaseous cloud. These vortices rotated in prograde directions like "roller bearings" between the larger clouds that were moving in a retrograde (clockwise) sense. By this means gravitational forces could have overcome disrupting internal pressures to aid condensation and transfer momentum. Turbulence rather than gravitational instability was considered by ter Haar (1950) as the prime factor in promoting collisional accretion. This accretion was aided by condensation of materials according to compositional differences at varying distances (and temperatures) from the Sun.

Whipple (in 1947), returning to the nebular view, allowed the Sun to form through contraction of a dust cloud of gas and solid particles, which then captured a second cloud with its own independent momentum. Excess gases were driven off the inner planets by heat from the Sun and from accretion. Schmidt, in the 1950s, pursued a similar view of particle capture from a previously formed sun in which nonvolatile elements condensed nearer the Sun and spiraled inward while low-temperature volatiles solidified further away and moved outward. Both Kuiper (in 1949) and Urey (in 1950) resorted to condensation of a nebula through self-gravitation; stabilized clots of dust and gas combined to form protoplanets that acted as nuclei, which eventually enlarged into planets. Kuiper proposed that noncondensable gases were driven away from the inner planets by the solar wind. Urey countered that these gases were removed by heat developed from accretionary, impact, melting, and radiogenic processes that affected small, moon-sized planets as they recombined into larger bodies.

Recent Views on the Formation of Planets

The trend was clear by the dawn of the space age. Ideas of planetary formation based on wrenching of materials from the Sun by passing bodies had fallen into disfavor. The notion of simultaneous organization of the Sun and its surrounding planets from contracting nebular material was now almost fully accepted. Most of the major concepts needed to derive a plausible model for the origin of planets had already been formulated. Specific data about the physical and chemical characteristics of planets and the interplanetary environment recorded since *Sputnik* by the space probes, the study of meteorites, and the acquisition of lunar samples have been influential, but not instrumental, in the development of currently popular models of the solar system's origin.

In recent years the origin of the Sun and its family of planets has been tied to a more general stellar cosmogony. This recognizes that the formation of our solar system was not a unique and special event in the history of the universe. There is a high statistical probability that similar planetary systems are generated repeatedly throughout the Milky Way galaxy and, by inference, within the millions of other galaxies as well. It does not follow, however, that each such system contains the essential conditions for life itself to develop spontaneously and diversify to the degree that characterizes Earth.

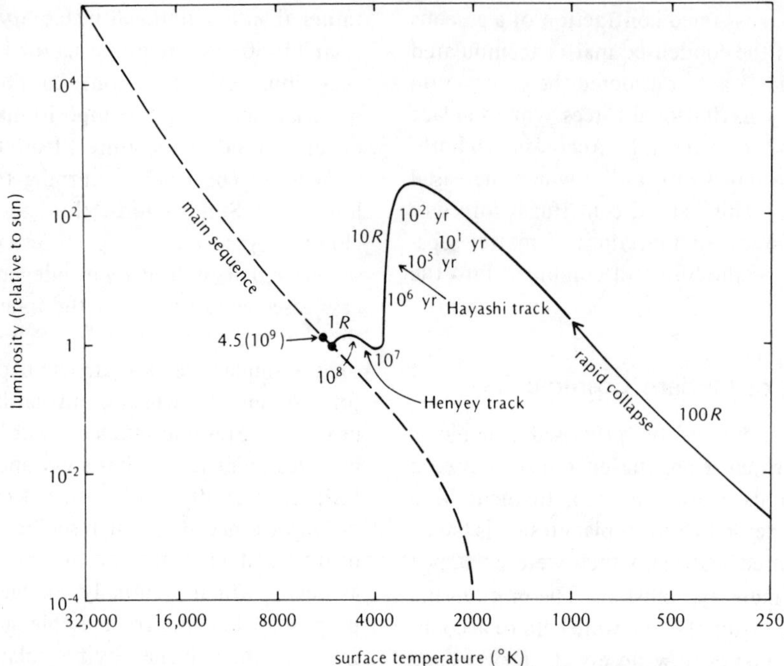

Figure 4-2 A possible evolutionary history of the early Sun and its solar system as plotted on the Hertzprung–Russell diagram. The process of organization of the nebula begins when a gas-particulates cloud (initially at temperatures of a few hundred degrees Kelvin), whose radius is about 1000 A.U., begins to collapse rapidly (in months to years). The central star that begins to form is about 100 R (where R is the present radius of the Sun). Temperatures rise continually during this phase. As the protosun shrinks to about 10 A.U. (in a few thousand years), it enters the Hayashi track stage, in which heat liberated internally is transported by convection to its outer zone. For a time, this protosun is 100 to 1000 times brighter than the present Sun. After about 1 million years, slower contraction brings the protosun to about twice its present size, at which stage (Henyey track) convection is superseded by radiative energy transfer processes. This continues for some 10 million years, during which nuclear reactions are initiated in the higher temperature regions of the interior. The Sun continues some further contraction over the next 100 million years or so until its present dimensions are reached. The planets probably formed near the close of this entire process. (From Hartmann, 1972, p. 111.)

The Evolution of a Star

Various stages in the birth and growth of stars have been monitored by optical and radio telescopes capable of observing literally millions of stars in different evolutionary states. The reconstructed history of a single star whose initial mass and sequential development are typified by the Sun can be traced on the H-R (Hertzsprung-Russell) diagram (Figure 4-2; see also Figure 2-2). The starting condition for this stellar process assumes that a region of nebular material within a galaxy is approaching some critical density or concentration of its principal constituents (gaseous or frozen hydrogen, helium, methane, ammonia, and inert gases; solid iron and silicate particles). Densities of hydrogen, the chief constituent, are as low as 10^{-9} g/cm^3 in intergalactic space, but increase to $\sim 10^0$ g/cm^3 in the dispersed parts of a galaxy. In a nebular region undergoing active star formation, these densities rise to 10^3-10^4 g/cm^3. Such a region of concentrated gas and dust commonly contains up to 10,000 solar masses of hydrogen. Matter within this cloud has been set into motion by radiation pressure (mainly from neighboring massive blue stars that are already burning nuclear fuel).

Densification of Nebular Materials

Dispersed particles (gas molecules, solids, liquids) within a nebula can be collected into denser units by three processes: (1) gravitational attraction, (2) accretion (juncture by collision between moving particles), and (3) condensation (adhesion of atoms and molecules to a larger particle). Wherever above-average densities occur, there is a high probability that occasional molecule-forming collisions of gas atoms with one another and with dust will take place. Possibly, also, the solid particles are made to stick together by "ices" (from H_2O, CO_2, and other gases below their freezing points) or by electrostatic attraction. Regardless of the cohesion mechanism, the end result is the formation of tiny growing clusters of particulates that in turn contribute to still greater gravitational instability.

One necessary condition for effective redistribution of nebular materials is establishment of local gravitational instabilities. As soon as higher densities are attained, the nebula begins to collapse mainly through gravitational attraction among gas molecules and particulates. Energy released from this gravitational action must exceed the thermal energy within the nebula; some of the thermal energy is derived from heating associated with the collapse process itself. Because the collapsing cloud becomes rotationally unstable, the nebular materials ultimately contract into numerous clots of dense hydrogen and minor amounts of the heavier elements. Each such entity–a protostar–is on the order of a few solar masses. These growing stars begin to collect concentrations of localized nebular materials (typically of dimensions comparable to the solar system). Each protostar inherits some of the angular momentum possessed by the parent nebula and begins to rotate ever faster as hydrogen and other gases fall into it. These protostars congregate in clusters that gradually reorganize by fragmentation or assimilation into individual stars that survive during the aging process.

The final distribution consists both of single stars and binary or multiple stars orbiting about each other. Binary stars tend to develop wherever large segments of this cloud material remain dispersed in the early stages. But if the mass is able to assemble into a single central star, the tendency to rotate faster causes its associated cloud to flatten into a discoid shape–a condition that may lead to planetary by-products from some of the material still left in the cloud.

Contraction Processes

This initial contraction into distinct entities occurs very rapidly over a time span as short as a few hundred years. The protostars are still not massive enough to undergo nuclear reactions that transfer radiative energy. Thereafter, the evolution to Sun-like stars that have reached an equilibrium state on the main sequence of the H-R diagram is one of further contraction as dust and gases are partitioned among individual stars. Typically, a cloud of matter is controlled by a protostar out to about 1000 A.U. In just a few thousand years much of this matter is swept into the star as it contracts to about 10 A.U. During the next million years, further contraction instigates convective motions within the star in response to heat transfer. The star now flares up brightly as the shrinking continues. Hydrogen burning begins in the next 10 million years when the star's mass is compacted to a critical density. Interior temperatures have then risen above 800,000°K–the level at which fusion through proton-proton reactions is initiated. This stage occurs as the protostar has contracted to about 2 A.U.

The final contraction to solar dimensions proceeds over the next 100 million years. Visible light emitted at this stage is difficult to detect because of a "shroud" of unorganized dust mixed with gas that extends outward for many Astronomical Units. This cloud of remaining nebular material (sometimes referred to as a cocoon nebula) is only a fraction of the total mass involved in generating the star, but it provides the main constituents from which any planetary or cometary bodies can form. Because the opacity of this smaller cloud has now increased, the heat released from the star raises the temperatures within the cloud as the star itself becomes hotter. The higher temperatures cause much of the star's outer layers and the surrounding nebular materials to become ionized. The inward transfer of matter to the growing star is countered by pressures that stop the collapse of the nebula. As the planets, satellites, and comets are generated, the obscuring dust is removed to expose the star to view.

While the final contraction was proceeding, most of the material components that eventually combine into a planetary system had moved toward the same relative positions they now occupy (even later perturbations are needed to shift the planets into Keplerian orbits). Around each central star, the nebular residue has assumed the shape of a thin disc. Particles in the disc rotate rapidly around the star. The balance between (1) centrifugal forces from this rotation, (2) outward-directed gas and radiation pressures, and (3) inward gravitational forces is responsible for the flattening of the cloud into a discoid.

Factors in Material Transfer

As deduced specifically for the solar system, most of the nebular material remaining outside the Sun undergoes redistributions of its element constituents according to their masses so that the lighter elements concentrate preferentially in the outer regions, and metallic elements with higher vaporization temperatures (e.g., Fe and Si) condense closer to the Sun. The strong magnetic field built up as the Sun contracts is a major factor in the transfer. This field propels ionized gases outward along lines of action; elements with lower atomic masses are carried farther from the Sun. Because of the motions of the Sun and its external cloud these lines become "twisted' so that the ionized particles follow spiral paths in the rotating disc. This allows the Sun to become magnetically coupled to its nebula, giving rise to a magnetic torque by which the Sun's angular momentum is transferred to these outer accumulations. At the same time, the Sun is forced to slow its rotation until only a small fraction of the total momentum remains within it.

The motions imposed on swirling materials within the cloud also influence the location of its constituents.

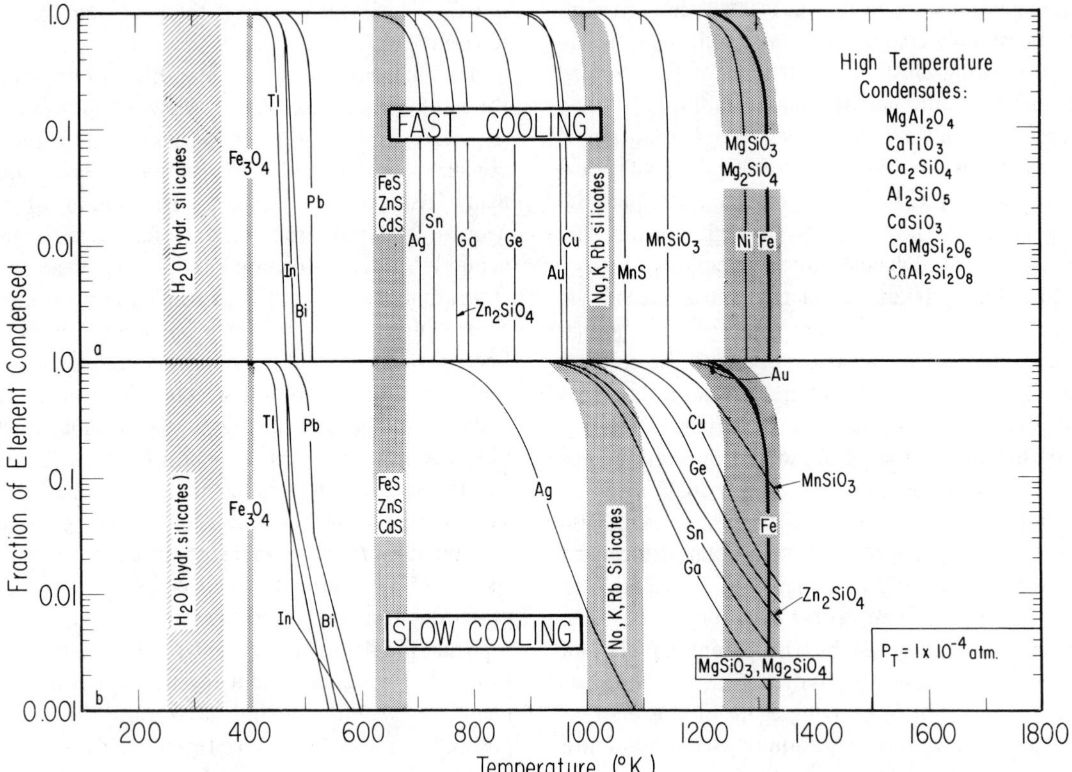

Figure 4-3 Sequence in which elemental constituents in a gaseous nebula of solar composition condense to produce meteoritic matter. The initial total pressure is assumed to be 6.6×10^{-3} atm. Nickel-iron condenses first, then magnesium and the alkali silicates. More than 90% of all condensable matter has condensed when the cooling nebula has reached temperatures of 1100° K; sulphur is the main constituent still uncondensed. Sequence a: "Fast cooling" involves deposition of condensates on surfaces of earlier-formed grains, without diffusion of added materials to the grain interior; Sequence b: "Slow cooling" assumes some diffusion of condensates into interiors, forming solid solution phases. Cooling times are on the order of 10^2–10^4 years. (From Larimer and Anders, 1967.)

Faster-moving gases and solids spiral inward and slower-moving materials extend outward. Greater turbulence in the inner regions increases viscous drag that slows materials revolving around the Sun and causes them to adopt smaller orbits. Materials beyond Uranus (where most of the original mass was concentrated) are further accelerated away from the Sun by magnetic forces; in that region whatever mass aggregates into planets follows larger orbits, while other substances are eventually lost from the solar system where gravitational restraints cannot overcome the effects of centrifugal forces.

During the final collapse phase, but prior to organization of the planets, temperatures in the solar nebula may have risen to 2000°K or higher. Thereafter, as the nebula cools (in part due to the effect of shielding from sunlight by dust), various chemical constituents in a gas-vaporized dust mixture (containing H, He, N, O, C, Si, Mg, Fe, Na, K, and other elements in solar proportion) begin to condense, each at a characteristic (pressure-dependent) temperature, to form a regular sequence of compounds (Figure 4-3). Condensation is probably the most important process at this stage by which the elements become fractionated within the contracting nebula.

The condensation sequence forms the basis of the equilibrium-condensation model for planetary genesis of J. S. Lewis, and others.* Around 1600°K, refractory oxides of calcium, aluminum, and titanium are the first to condense from the gas. Metallic iron and nickel appear next at ~ 1300°K, followed by magnesium silicate (enstatite) at 1200°K, and sodium-rich silicates (feldspars) at 1000° to 800°K. Near 680°K hydrogen sulphide reacts with available iron to form troilite, a dense phase of FeS. Between 1200° and 500°K, some iron is oxidized to FeO which enters into pyroxene and olivine produced by reaction with enstatite. Between 550° and 175°K, water becomes an essential part of the silicates reacting in equilibrium with it to form serpentine and other hydrous minerals. Below 175°K, the H_2O will solidify into water ice and at 150° and 120°K gaseous ammonia and methane,

*See J. S. Lewis, The Chemistry of the Solar System, *Scientific American*, March, 1974, pp. 51-65.

respectively, condense as hydrates. Finally, from 65° to 25°K the now impoverished gas loses argon, neon, and other inert elements leaving only helium and possibly hydrogen as uncondensed phases.

As applied to the solar system, the Lewis model assumes a decreasing temperature-pressure gradient outward from the Sun to the edge of the nebula. The proportions of silicates and metals to condensed gases vary with distance from the Sun. Close in, reactions over a narrow temperature range around 1200°K produce refractory minerals, metallic iron, and magnesium pyroxene that organize into Mercury. The Moon, as a minor, possibly displaced planet, may be even more enriched in refractory oxides and Mg-rich silicates than Mercury. Farther out, Venus forms at temperatures near 800°K so that the magnesium silicates now holding $\sim$ 0.5% FeO and some alkali-bearing silicates are produced. Earth develops in the 650-600°K range, within which some hydrous silicates, magnesium silicates with up to 10% FeO, and troilite are the chief condensed phases. For Mars, at temperatures near 400°K, metallic iron is used up in FeS or in FeO that replaces 25-50% of the MgO in the silicates.

In the model, each inner planet undergoes melting and differentiation so as to rearrange the mineral mixtures into an orderly crust, mantle, and core. Mercury ends up with a refractory crust, enstatite mantle, and Fe-Ni core; Venus has a more alkaline crust, enstatite mantle, and metallic core; Earth forms an alkaline crust, olivine-pyroxene mantle, and a sulphur-bearing metal core, and retains some water (locked up in minerals or released); and Mars gets an alkaline crust, pyroxene mantle, more water (now mostly lost), and a troilite core. The asteroids, farther from the Sun, contain mainly Fe-Mg silicates, serpentine, some troilite, and some carbon compounds; lack of melting preserves the original state of asteroidal material–seen to us as carbonaceous chondrites. Jupiter and Saturn are special cases of very large clots of primordial solar gases (H and He) but their satellites probably are differentiated mixtures of water ice and methane or ammonia, the lower temperature end numbers expected from the condensation series. Uranus and Neptune, whose compositions are inexactly known, also contain considerable methane and ammonia.

The Chondritic Nature of the Terrestrial Planets

Implicit in this discussion is the assumption that the original silicate materials making up the planets were chondritic in nature. However, the chondrules themselves may not have formed directly from condensing silicate vapors. At the time when planetary bodies were actively accreting, the dust in the cloud probably consisted of cool, fine powder. If this dust were to aggregate without change, only a carbon-rich product like the matrix of a type I carbonaceous chondrite would have resulted. Unless the chondrules first obtained their shapes and compositions after incorporation within the planets (considered unlikely), we must postulate that an efficient heating process acted to fuse some of the particles in the cool cloud into droplets, which then mixed with remaining dust to produce the carbonaceous chondrites. One suggested energy source for melting the dust involves sporadic discharges of electricity ("lightning flashes") within the dust cloud as electrons stripped from atoms built up to high voltages. Another would be shock waves associated with pulsating energy release within the nebula. Still a third possibility requires bursts of variable amounts of thermal energy during the Sun's T Tauri stage (in this case, the chondrites we now observe may have come into existence after the main planets had formed). Each of these impulsive mechanisms seemingly accounts for some of the variations in numbers of chondrules within the primitive carbonaceous chondrites. The ratio of hydrogen to oxygen (or degree of oxidation) played an important role in determining the particular composition, indicated mainly by the ratio Fe^{+2}/Fe^{+3}, of the local materials that aggregated to form a given planet.

Regardless of origin or location within the nebula–which only partly explains differences in ferrous iron, volatile elements, and carbon contents in the matrix–the subsequent histories of chondritic planetary bodies greatly influenced the final condition of the meteoritic materials prior to their arrival on Earth. It is likely that the larger terrestrial planets all underwent partial to complete melting soon after their formation owing to heat generated by short-lived isotopes (Al^{26}, I^{129}, Pu^{244}), gravitational compression, impacts accompanying aggregation, or a combination of these. A large planet like Earth presumably melted in toto to produce a metal core and perhaps to destroy all vestiges of the chondrules themselves in its interior. In contrast, a small protoplanet elsewhere may have stayed cool enough so that only metamorphism occurred, altering the chondrule textures and causing volatilization of carbonaceous compounds, accompanied by local reduction of iron, diffusion of Fe^{+2} through the matrix, and formation of free metal grains. Intermediate degrees of modification of the primitive carbonaceous chondrites can be postulated for each inner planet, although there is no direct evidence that any of these (including the Moon) consisted of chondritic material at its outset.

Growth of Planets

Details of the processes by which the planets grew from the solar nebula are still inferential. Clots of matter probably built up into tiny planetesimals held together by frozen ammonia or water, organic matter, and magnetic or electric charges. The initial aggregation may have resulted from random collisions in turbulent vortices in the cloud or from faster-moving particles overtaking slower ones in the path-crossing elliptical orbits that prevailed in the rotating disc. At first these planetesimals were probably only meters in size but as individuals reached critical masses they began to have significant gravitational effects on their neighbors and thus attracted smaller particles into themselves. Growth rates increased with time–the initial clots formed in less than 100,000 years; asteroidal-sized bodies took no more than 10 million years; the entire planetary system reached its present dimensions in less than 100 million years. Eventually, the larger bodies, growing at the expense of the smaller ones, swept up so much of the remaining materials that only a few planet-sized aggregates were left. Urey holds that many smaller bodies developed at this stage and later broke up into fragments which then combined to form the present planets. During the terminal growth stage, the planets were peppered by intermediate and small bodies responsible for the cratered surfaces that still survive on Mercury, the Moon, and Mars. However, at the onset each inner planet presumably had a thick coating or crust of frozen volatiles (mainly water) and a dense atmosphere of hydrogen and other gases much like that of Jupiter today. Some process then operated to drive off most of the residual light gases from Mercury, Venus, Earth, and Mars (and probably the Moon). A likely explanation for this loss assumes the Sun had passed through its T Tauri stage of evolution after these inner planets had grown to their maximum sizes. The original gaseous envelopes around the terrestrial planets were "blown away" by this explosive process leaving behind central cores of condensed silicates. After removal of the heat-absorbing atmosphere, exposure to sunlight would soon vaporize the water-ice crust, with its eventual loss to space from planets other than Earth.

Satellites

According to Alfven and Arrhenius (1972), satellites of the planets with regular (prograde) orbits most probably formed from leftover materials either by a coagulation process similar to their primary (producing one or more small bodies near the parent) or by accretion in several bands or rings where the gravitational potential reached specific values (producing many very small bodies extending 10 to 30 Earth radii from the primary). Satellites with unusual orbits (initially retrograde) are more likely to have formed away from their primaries but were captured when perturbed into crossing orbits.

Having acquired a background on the nature, composition, and origin of planets in the last 3 chapters, we can now proceed, in the remainder of the book, to study in detail the geologic aspects of the Moon, Mars, Mercury, and Venus–those planetary bodies whose surfaces have already been carefully studied by telescope or by orbiting or landing space probes.

REVIEW QUESTIONS

1. Review the main facts that must be accounted for in any model of the origin of the solar system.

2. Mention the contributions made by Descartes, Kant, Chamberlin and Moulton, Alfven, and Kuiper to concepts of the formation of plants.

3. List several ways in which dispersed particles in the nebula can be aggregated into denser bodies.

4. Describe the process(es) by which a nebula begins to collapse.

5. Identify: protostar; cocoon nebula; dualistic model; retrograde rotation; magnetohydrodynamic.

6. Discuss the processes or factors that may have caused redistribution of elements according to their masses into different regions of the solar system.

7. Review the temperature history of the nebula from its inception to reorganization into planets.

8. Mention the energy sources capable of melting a primitive planet.

9. Examine the growth history of accreting planetesimals.

10. What is a T Tauri event? Mention several problems in planetary development which might be explained by this process.

CHAPTER

5

An Earth's-Eye View of the Moon

For more than two hundred years, almost all information about our lunar neighbor was obtained primarily from telescopic observations. In this century, these observations have been supplemented by recording on photographic plates. More recently, spectroscopic and electronic devices have measured radiation emanating from the Moon's surface in the ultraviolet, visible, infrared, microwave, and radar regions of the electromagnetic spectrum (Figure 5-1). Thus, by 1965 extensive knowledge had accumulated from Earth-based studies relating to the Moon's figure, topography, surface properties, and gross morphological features.

The Moon's Motions

To the casual observer even in ancient times, the two most obvious characteristics of the Moon are its steady reappearance every 29.5 days (varying from month to month by as much as 6.5 hours from this average) and the regular growth cycle in which the frontside disc is illuminated through a series of definable phases. As astronomical observations became more precise, certain aspects of the lunar orbit were accurately determined. The Moon was found to revolve around the Earth along an elliptical path with an average *perigee* (point of closest approach to Earth) of 356,410 km and an average *apogee* (farthest distance) of 406,700 km. This orbit is inclined to the ecliptic by an average of $5^{\circ}9'$. The total orbital period relative to some fixed reference point (a distant star) was established to be just 27.3 days (the *sidereal* month). The longer *synodic* or lunar month, set by the time needed for the Moon to return precisely to the same phase (e.g., full moon), extends to 29.5 days. This increase in time by approximately 2.2 days results from the combined relative motions of both Earth and Moon forward in orbit around the Sun during which the illumination angle with respect to the Sun will be shifted; the Moon thus requires this extra time for the first sliver of reflected sunlight at new moon to be seen from a reference point on Earth to compensate for the angular advance of the Earth-Moon system since the previous lunar month began.

The Moon's *phases* (defined in Figure 5-2) can be explained simply as the proportion of the observable lunar surface which has begun to reflect sunlight continuously to a viewer on Earth. The *phase angle* (α) is at 90° or $+90^{\circ}$ (first and third quarters, respectively) when the Sun

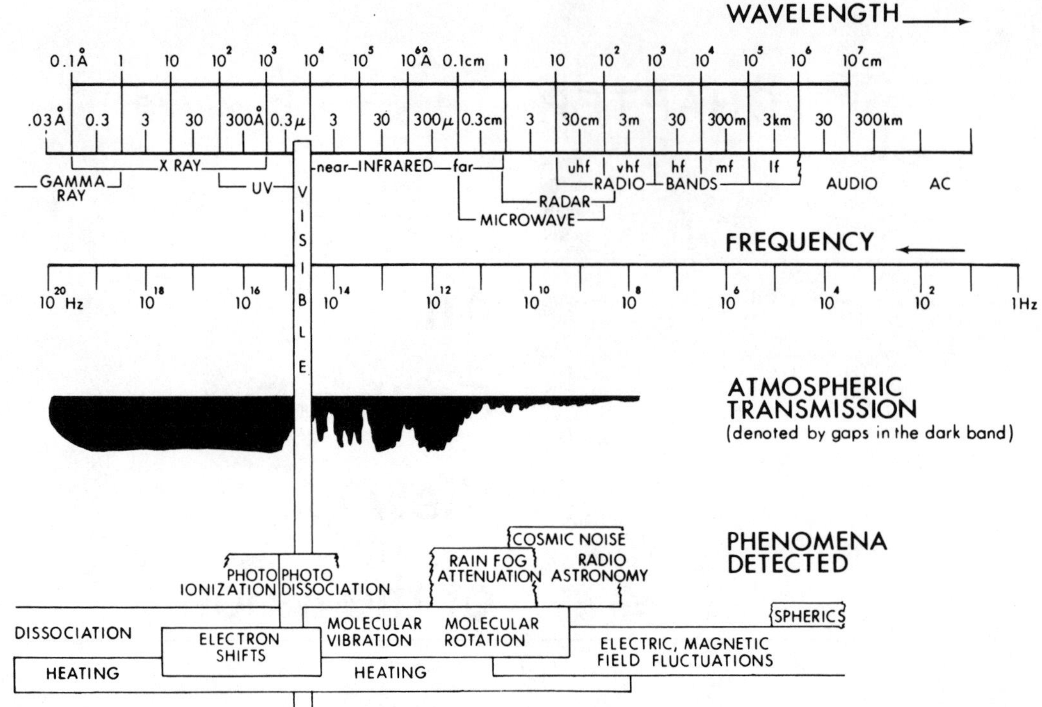

Figure 5-1 The electromagnetic spectrum subdivided into different spectral regions or working ranges (e.g., X-ray, visible, radar) within which data can be obtained by various remote sensing techniques. Electromagnetic radiation can be treated in terms of energy transfer within a continuum by means of wave motion. Any radiation level is uniquely specified by its *wavelength* (from angstroms, A, to kilometers), its *frequency* (number of repeated vibrations per second, in units of hertz, Hz), or its *energy* state (in units of ergs, joules, calories, electron volts, etc.). Radiation characteristic of some spectral region is the result of atomic (nuclear or electronic) or molecular motions within the source material that are usually activated by interaction with some external generating process (electron bombardment, heating, magnetic fields, etc.). The radiation can be detected by such remote sensing devices as geiger counters, spectrometers, photomultiplier tubes, photographic films, photometers (which measure the quantity of photons received), infrared scanners, radio-frequency receivers, and magnetometers. If radiation must pass through a gaseous atmosphere before detection (e.g., when an orbiting satellite looks at the Earth, or when light beamed from the Sun or stars or reflected from a planetary surface is gathered by a telescope), certain frequencies will be absorbed (dark bands in the transmission diagram) owing to atomic and molecular interactions. Thus, most UV and some IR radiation from the Moon fails to reach an Earth-based telescope because it is screened out by the atmosphere; a lunar-orbiting sensor system can detect almost all UV and IR radiation emitted and reflected from the Moon's surface. (Courtesy G. Rabchevsky.)

is positioned at right angles to the Earth-Moon join (the straight line between their centers) and at 0° (full moon) when the Sun, Earth, and Moon lie on this join with the Earth in the middle. The changes of phase from each new moon to the next (one full cycle from -180° to +180°), as observed from Earth, comprise a *lunation.* Telescopic views of several lunar phases are shown in Figure 5-3. The arcuate boundary between light and dark portions of the lunar disc is called the *terminator.* This junction actually appears as a broad, somewhat irregular wavy band when looked at under magnification because of the effects on the light scattered from an irregular surface having natural relief.

Careful inspection of these pictures discloses a vital fact about our satellite. By selecting certain major features on the Moon as reference points, one can note that the same face of the lunar sphere is seen during the entire period of its revolution around the Earth. This curiosity is the result of the Moon's sidereal revolution period of 27.3 days being equal to its axial rotation period. In effect, a center point on the lunar frontside is seemingly *locked* on the Earth—i.e., an imaginary line perpendicular to the lunar surface at that point would extend toward the Earth's center at all times during the Moon's orbital transit. This remarkable condition describes a state of *captured* or *synchronous* rotation.

Although the Moon is nearly a sphere, one side of which now always faces Earth, inspection of photographs of its visible surface taken on different dates discloses an apparent shift of major landmarks relative to reference points at the *limbs* or disc outline (see Figure 5-3). Mapping demonstrates that almost 60% of the surface of the lunar sphere has been seen from Earth, although no more than 50% is visible at any one time. Because this seems to imply a "wobbling" or "rocking back and forth," the term *libration* (from the Latin word for "balance') is applied to this phenomenon. The Moon's librations result in part from *apparent* or optical displacements that allow us to look at equatorial regions normally beyond the limbs (east-west or longitudinal librations) as well as more distant parts of the polar regions (north-south or latitudinal librations). The motions responsible for exposing some of the farside include variations in angular velocity of the Moon's revolution in an elliptical orbit, tilt of its rotational axis relative to its orbital plane, and, to a lesser extent, parallax effects owing to differences in an observer's initial location or to the shift in his viewing angle as the Earth rotates. But there is also a *real* or dynamic libration related to varying gravitational pull on the Moon, whose geometric shape and mass distribution depart from perfect sphericity.

Variations in Earth-Moon Distance

In the early history of the Earth-Moon system (at a time inferred to be at least 3 billion years ago), the Moon's periods of revolution and rotation were different from today's value and also from each other. The Moon once may have been much closer to Earth, although no closer than 2.9 Earth radii, or about 18,400 km–a critical distance known as the *Roche limit* for the Moon.* Inside this limit the satellite would disintegrate as strong gravitational forces exerted by the Earth exceeded the tensile strength of the Moon. There is no direct evidence on just how close the Moon has ever been to Earth, but the rotational history of each body sheds important light on the changes in distance between the two.

The mutual attraction between the Earth and Moon gives rise to tidal forces within both their solid bodies and also in the Earth's oceans and atmosphere that lead to dissipation of energy from frictional effects. This causes a dragging force that slows down the rotation of each body. According to the physics of celestial bodies, a slowing down in Earth, the primary, is accompanied by a transfer of angular momentum to the Moon that requires an increase in its average distance from Earth. This in turn brings about an expansion of the Moon's orbit and a corresponding increase in its period of revolution. Thus, over time the Moon is presumed to be gradually receding from the Earth. Eventually, it is possible that the rotational period of the Earth (i.e., one day) will coincide with the period of the Moon's revolution so that the lunar month and terrestrial day will be equivalent.

Modern evidence for deceleration of the Earth's rotation comes from changes in location of total eclipses over the last few thousand years. Striking evidence for a shorter day (about 21.5 hours long) in the Devonian period, some 400 million years ago, is recorded in corals that add a ridge marking each day to their outer surface during an annual growth cycle: these corals have about 400 such ridges compared with 360 or so on similar recent corals. Extrapolation back in time indicates even faster rotation rates early in the Earth's history. If the Moon had been much closer to Earth then, tidal effects on Earth should leave some record in the rocks themselves. No convincing evidence of this has been found, but growth habits of ancient, reef-forming stromatolites some 2 billion years ago has been cited as a possible response to very large ocean tides.

Figure 5-2 The phases of the Moon as seen from Earth; the observed phases depend on the relative positions of the Sun, Earth, and Moon. The phase angle α is at 0° at full moon, -90° at first quarter, $\pm 180^\circ$ at new moon, and $+90^\circ$ at last quarter (see also Figure 5-3). (From S. Glasstone, *Sourcebook on the Space Sciences*, copyright © 1965 by Litton Educational Publishing Inc. Reprinted by permission of Van Nostrand Reinhold Company.)

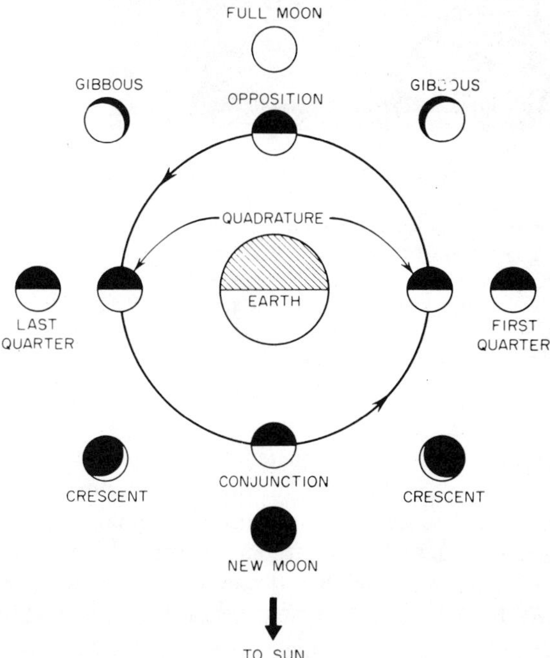

The Figure of the Moon

Like the Earth and other larger planetary bodies, the figure or shape of the Moon is described with reference to its departure from a perfect sphere. Both Earth and Moon

*Any planet-satellite system has a specific Roche limit, or distance of closest approach of each secondary body without rupture; the distance is controlled by the relative masses of primary and secondaries, which influence the magnitude of tidal forces between the bodies. If both the primary and satellite have the *same* density, the Roche limit will be 2.44 greater than the primary's radius.

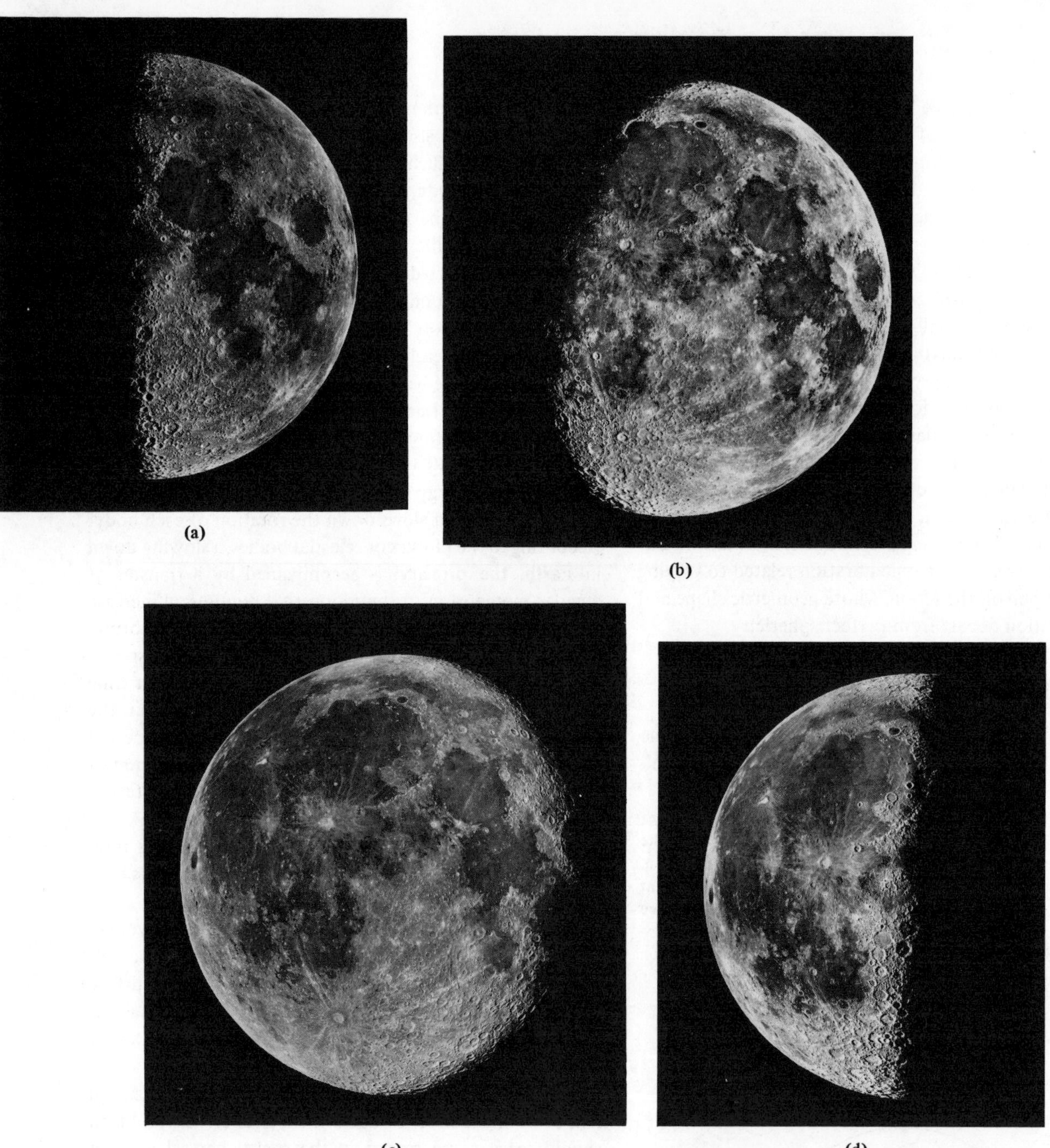

Figure 5-3 Four phases or stages of illumination of the frontside of the Moon, photographed through the Lick Observatory 36-inch telescope. (a) First quarter, age 7 (Earth) days. (b) Waxing gibbous phase, age 10.1 days. (c) Waning gibbous phase, age 17.8 days. (d) Last quarter, age 22 days. Note the effects of the heavy shadows on topographic relief in the region of the terminator. The effects of lunar librations can be detected by noting different areas on the lunar surface that appear to shift in the photo views taken from the same sighting point (Lick Observatory) on the Earth.

show variations in radii measured from the geometric centers of their respective bodies. The Earth, in fact, undergoes a significant flattening of approximately 31 km in its polar radius relative to the equator. Radii to different points on the equator are themselves not equal. The Earth's shape is thus approximated by a triaxial ellipsoid.

Because of the great distance between the Earth and Moon and a lack of an adequate reference surface (such as mean sea level on Earth) from which to measure absolute elevation differences, the problems in using telescope observations to determine the geometric shape of the Moon would seem insurmountable. However, Baldwin, Mills, and others, through painstaking mathematical analysis of shifts in control points on the lunar surface as seen in photographs of different librations, have succeeded in producing generalized contour maps of the lunar frontside (Figure 5-4). Mill's map indicates differences in first-order relief that may reach up to 4 km. Other workers have shown that the highlands average about 1 to 2 km higher than the maria. These variations in elevation can then be translated into estimates of lunar radii at different points on the surface.

Based on these optical observations, the radii obtained for the Moon are non-equal. Assuming an ellipsoidal form, the length of the polar semiaxis appears to be shorter by about a kilometer than that of the semiaxis directed toward the Earth. Occultation measurements along the limbs provide further data that suggest an intermediate value for the equatorial semiaxis normal to the Earth. The apparent elongation of the Earth-pointing semiaxis has led to postulation of a "bulge" in the geometric figure generally toward the Earth. This bulge coincides roughly with the elevation maximum of Mill's contour plot.

Additional information on the Moon's figure comes from determination of its moments of inertia, I. The moments for a rotating sphere interrelate the distribution of densities as a function of radial distance from the center. Nonuniform moments can result from asymmetric differences in density (i.e., irregular mass distribution) or from variations in radial dimensions or both. According to the moments derived for the Moon, the shape must be that of a triaxial ellipsoid if the density distribution were radially uniform within the interior (it is permissable for density to vary systematically with depth as a series of concentric layers or shells).

Further insight into lunar density distributions results from calculating the value of I/MR^2 (M = lunar mass and R is the mean radius of the Moon)—a dimensionless quantity. For any homogeneous sphere, this quantity has a value of 0.4. The corresponding value for the Moon is about 1% less than 0.4. This implies a very slight increase in density with depth, but does not support a model of a Moon with well-defined layers. A metallic core of any considerable size would seem to be ruled out.

The Moon's shape must also be considered in a dynamic sense. Over long time periods, the Moon's interior might be expected to respond to internal stresses much like a fluid body. If rotation is neglected, such a body would assume a hydrostatic figure that takes the form of a perfect sphere. The correspondence between this figure and the equipotential gravitational surface would be exact.

However, the Moon's rotation distorts the ideal hydrostatic figure into that of an oblate spheroid. A second distorting force results from the tidal attraction exerted by the Earth. For a fluid Moon in nonsynchronous rotation, this would give rise to a tidal bulge that travels across the lunar surface much like the ocean tides move on Earth. This bulge, however, would remain fixed in one position for a Moon whose rotational period keeps one face always pointed toward Earth.

Neither the effect of rotation nor the influence of gravity (which should further distort the Earthward equatorial region by about 40 meters) can account for the magnitude of the bulge and other radial differences. Thus, the observed figure of the Moon, supplemented by data on its shape obtained from spacecraft in orbit, shows a considerable departure from a body in hydrostatic equilibrium. It is concluded, therefore, that the Moon presently is not in hydrostatic equilibrium, although it may have been so at some time in its past.

Despite the seeming triviality of the slight differences in lunar radii relative to the hydrostatic model, this departure is actually a significant clue to the nature and history of the lunar interior. If the bulge represents a tidal-force distortion developed about the time the Moon "locked" on to the Earth, then Urey explains its present persistence

Figure 5-4 Two hypsometric maps (showing relative heights or relief with reference to an arbitrary datum plane) derived from the frontside of the Moon. (a) A map generalized from Baldwin (1963) in which higher (+) areas are distinguished from lower (-) areas. (b) A contoured map, made by Mills (1968), showing elevation differences at a 1 km interval. (From Mutch, 1970.)

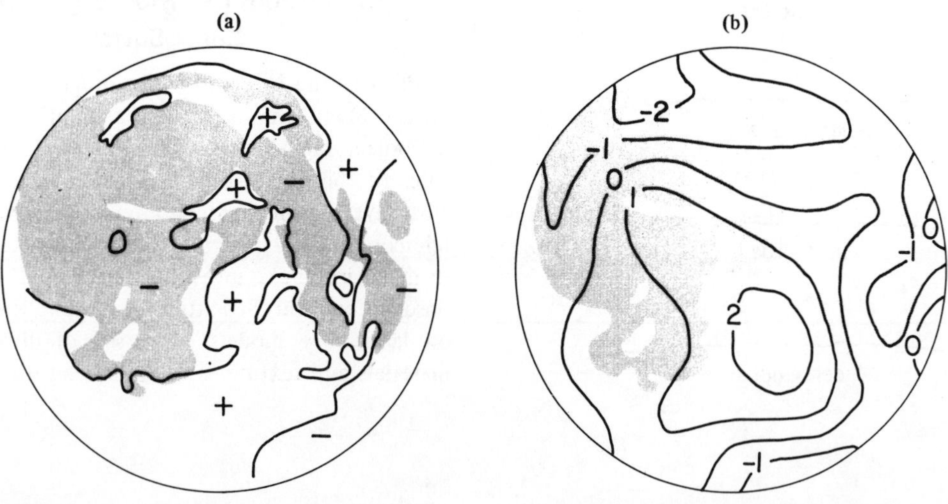

by considering it to be a "fossil" upwarp. This upbulging was a response to forces that deformed a plastic or possibly partially melted (hot) Moon when it was much closer to Earth. The bulge was "frozen in" only after development of a rigid outer layer that resulted from cooling down to a depth of perhaps several hundred kilometers. To preserve the bulge, this layer or crust requires sufficient strength to withstand the stress differences ($\sim 2 \times 10^7$ dynes/cm^2) at the base of the crustal bulge segment acting to remove the deformity. Runcorn once treated the bulge as a localized concentration of materials whose densities depart from their surroundings owing to fortuitous accumulations during accretion or cooling; failure to adjust isostatically would account for the surviving bulge. He now considers that the bulge more likely reflects the continuing influence of thermal convection currents that sustain the distortions.

Recent results from an altimeter profiling experiment on several Apollo missions (p. 186) show the farside highlands to be higher relative to the Moon's center of figure than the corresponding highlands on the frontside. Gravitational calculations, furthermore, lead to an indication of the distribution of mass within the Moon. The center of mass is displaced about 2 km Earthward with respect to the center of figure. A variable crustal thickness is a suggested factor in explaining this noncoincidence of centers.

The Topography of the Moon

Various lines of evidence all point to a topographically rugged lunar surface. Overall, the Moon's surface appears essentially flat at full moon. Locally, relief becomes increasingly difficult to recognize at higher Sun angles. Thus, topography is best revealed by shadows cast at low Sun angles (in which solar light rays meet the lunar surface at angles less than 10-20°). Examination of some telescope and space probe photographs taken at low Sun angles allows one to use shadows to calculate with fair accuracy the elevations of individual hills or depressions. This requires knowledge of a good value for local ground scale (to permit calculation of distances between shadow and source points) and either the specific Sun angle or the shortest distance to the nearby terminator. By solving geometrically for heights of many individual surface features, a rough indication of relative relief between crater rims and floors, between adjacent craters, and between prominences and mare plains can be established in selected areas of the Moon. The inner wall heights of certain large (50 km and wider) craters are estimated to lie between 1 and 4 km (depending on the extent of central infilling, degree of degradation, and other factors). Seen in cross section at a 1-to-1 scale, however, on a base representing the curvature of the lunar surface, these craters appear in their true context as slight depressions and ridges (Figure 5-5). A more elaborate surveying method, called *photoclinometry*, results in elevation profiles across "traverse lines." This method utilizes variations in surface brightness (caused by slope changes) over an area whose surface light-reflecting characteristics are determined independently and can be related to Sun angle (phase). Applying combinations of the above techniques to telescope observations and photographs, cartographers have produced a series of topographic maps for most of the frontside lunar surface (Figure 5-6). Photographs taken at oblique angles from orbiting spacecraft (*Lunar Orbiter*, *Apollo*) from known altitudes also permit calculation of surface relief within the viewed scenes; the *Surveyors* can estimate the heights of local irregularities from a ground perspective but uncertainities as to distances to the measured objects lead to an inherent inaccuracy. More accurate measurements of topographic relief have now been made with a laser-beam altimeter that scanned the surface below the orbiting command modules of Apollo 15 and 16. Measurements of relief over broad regions have also been made with Earth-based radar (see p. 63).

Figure 5-5 Profiles through some of the larger impact craters drawn to true scale in relation to the curvature of a segment of the lunar surface. (From Kopal, 1966.)

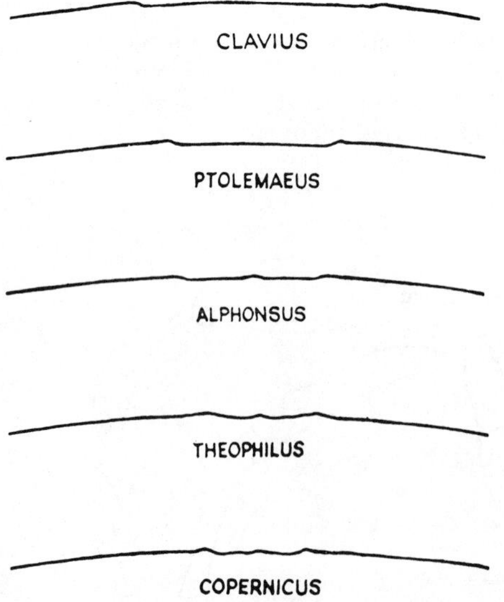

Interactions of Visible Light with the Lunar Surface

Prior to the U.S. and Soviet programs for soft-landings on the Moon, the main sources of information—much of it conjectural—about the fine-scale (close-up) characteristics of the lunar surface—or more specifically, its topmost layers—came through inferences drawn from optical telescope and radar surveys of different areas on the Moon. In many instances, investigators gathered data from lunar observations and then attempted to duplicate these results on laboratory models composed of different types of materials and textures by illuminating these artificial sur-

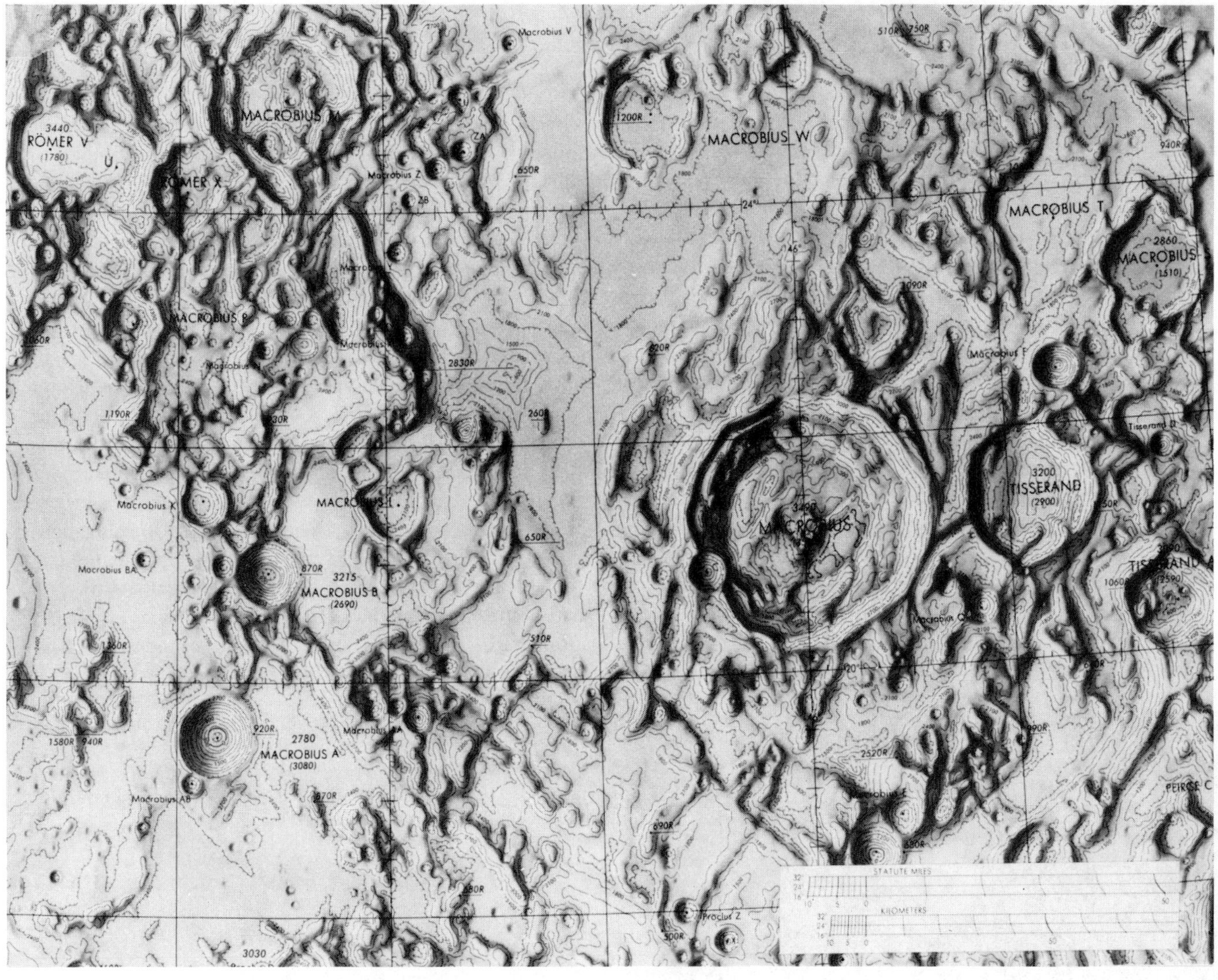

Figure 5-6 A portion of a topographic map of the Macrobius quadrangle (LAC 43) prepared in 1965, at an original scale of 1:1,000,000, by the Aeronautical Chart and Information Center (ACIC), U.S. Air Force. The contour interval is 300 meters.

faces under controlled conditions. In this procedure, response curves (e.g., percentage of scattered, emergent light versus illumination angle) for each material are compared with analogous curves obtained from lunar surface measurements; those which most closely fit the lunar curves are considered to identify or match the compositional and/or textural states of the Moon's outer layers. Often, several diverse model materials produce good matches to the lunar case, but experiments involving measurements of other parameters sometimes resolve the ambiguities. A few examples in the following paragraphs will demonstrate this approach.

Albedo is defined as the ratio of the total light reflected from a fully illuminated surface to the light incident on that surface. Maximum variations in albedo or reflecting ability of the lunar surface are evident during full moon, giving rise to distinctive lighter and darker patterns. The Moon's average visible light albedo is low (about 0.07, or 7%) compared to that of Earth (average 0.39; continents 0.10-0.45; clouds 0.60-0.90) or Venus (0.70-0.90, high values resulting from the great reflectivity of the clouds in the Venusian atmosphere). The maria typically have very low albedo values between 0.05 and 0.08, while those of the terrae lie between 0.09 and 0.12. Presuming that albedo bears a direct relation to composition, a clue to the nature of the lunar surface materials can be disclosed by comparisons to albedos of some terrestrial rocks. Candidates for lunar rock types would seem to be confined to scoria, dunite, and possibly gabbro, basalt, or andesite. However, the albedo is also influenced by the physical state of the reflecting materials; different values will be obtained from dense, porous, powdered or crushed, and reaggregated samples of the same substance.

The first detailed map of albedo variations over the entire front face of the Moon has been published by the U.S. Geological Survey,* using as a base a carefully pro-

*In U.S.G.S. Professional Paper 599-E (1970).

cessed telescope photograph supplemented by meticulous observations through the telescope and by measurements with electronic photometers. Under full moon lighting conditions (see Figure 5-19), certain craters such as Tycho, Copernicus, and Kepler are revealed to be surrounded by bright aprons (presumably ejecta deposits) and to lie at the intersections of a series of broad streaks or *rays* that have some of the highest albedo values (0.15 to 0.18) known for the Moon. However, when Tycho, for example, is illuminated at lower Sun angles, these features all but disappear (Figure 5-7). One early interpretation held these rays to consist of powdered, more highly reflective (hence brighter) materials tossed out of younger craters in linear deposits not yet destroyed by bombardments or burial. But, examination of *Apollo* samples now reveals most lunar fines (comminuted, granular debris) to be darker than most larger lunar rocks. A more likely cause of the higher albedos was deduced from careful inspection of some rays seen close-up in *Orbiter* photos. These reveal the occurrence of numerous small depressions (mainly secondary craters—see p. 90) that effectively lighten the terrain as seen from Earth. Photos taken by *Apollo* astronauts during lunar orbits provide striking examples of albedo contrasts (Figure 5-8) in which craters and rubble-covered hills appear much lighter than mare plains in the same scene. Inner walls of craters have especially high albedos (0.18 to 0.20), possibly because creep and landslides expose lighter materials underneath. Contrasts in apparent brightness also vary with the viewing angle.

A *specular* or smooth (polished) surface reflects light such that the surface appears brighter when the angle of reflection equals the angle of incidence. A *Lambert* or diffuse (unpolished to rough) surface of finite area reflects light equally in all directions independent of the angle of incidence. A fully illuminated specular surface of a sphere at zero phase would show a central bright spot or highlight (reflected image of light source) with rapid darkening toward the limbs; this spot would appear to migrate across the sphere as the phase angle approaches or departs from zero. A diffuse spherical surface at zero phase

Figure 5-7 Tycho (T) and its neighboring craters, including Clavius (C), as viewed through the 36-inch Lick Observatory telescope during a waning lunar phase in which the Suns' angle was less than 30°. Under these conditions, the ejecta apron and rays associated with this young crater all but disappear.

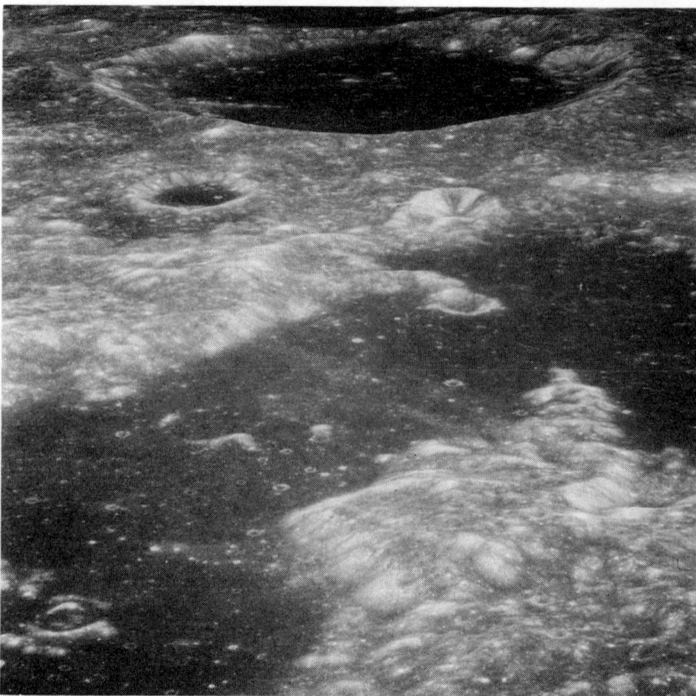

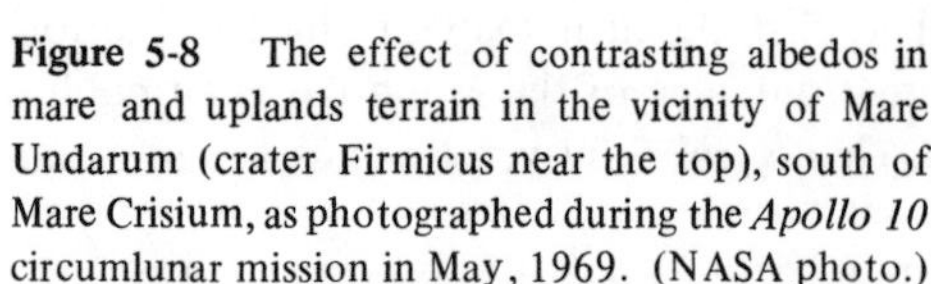

Figure 5-8 The effect of contrasting albedos in mare and uplands terrain in the vicinity of Mare Undarum (crater Firmicus near the top), south of Mare Crisium, as photographed during the *Apollo 10* circumlunar mission in May, 1969. (NASA photo.)

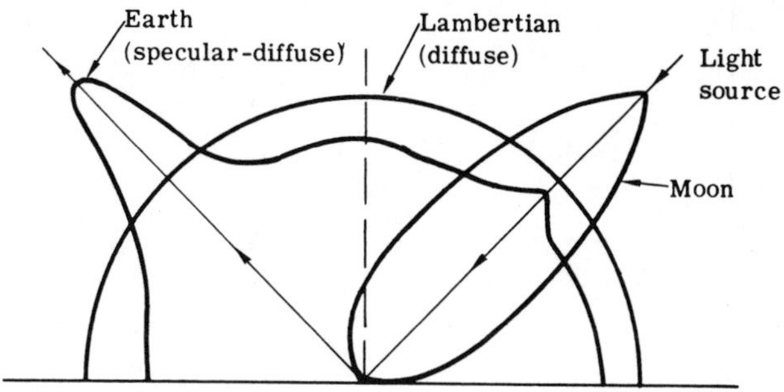

Figure 5-9 Comparison of the reflection of light from the illuminated hemispheres of the Earth and the Moon (see text). (From an unpublished Itek brochure prepared by W. C. Kinsey.)

would be most strongly lit in the central region while gradually darkening toward the limbs—that is, the brightest area is directly under the source of light. However, the surface would show some brightness when viewed from any direction. Relative to a fixed observer, surface brightness changes in proportion to the cosine of the direction of incident light (i.e., as phase angle shifts). The Moon actually behaves differently either from an ideal diffuse spherical reflector or from the Earth itself (which shows combined specular-diffuse effects) in that it remains uniformly bright at full moon with no distinctive central highlight or limb darkening (Figure 5-9). Most light reflects from the Moon back in the direction of incidence and displays a pronounced reduction in intensity where reflectance angles differ appreciably from the angle of incidence.

The variation of relative brightness with phase angle is termed the general *lunar photometric function* when determined for the Moon as a whole (Figure 5-10). Measurements made over individual sectors of the Moon's surface indicate that different areas, such as maria and large craters, each have a characteristic relative brightness level as a function of phase angle (Figure 5-11). Although differences exist from place to place, the overall similarity of these curves at locations where albedos are not the same suggests that there is some common property present over essentially all the Moon. A further insight into the nature of this property comes from experiments in which the surface brightness of various materials, each spread as a flat surface, is measured at different illuminating (phase) angles and from one constant observing angle in the plane of incidence. Reflectance measurements obtained from diverse materials such as powdered metal, furnace slag, volcanic ash, corals, and sponges—all of which cause pronounced backscattering of light—indicate the lunar surface to be highly porous and rough on a small scale. Thus, a wide variety of porous to fluffy or dendritic surface structures produce curves similar to those from the Moon after adjustment to the same observing angle. Prior to the *Apollo* landings, this evidence supported the argument that the textural characteristics of the lunar surface are comparable in importance to intrinsic albedos in determining the photometric function defined for the Moon. Studies of samples from the lunar soil demonstrate the importance of a dark, blackish coating (glass from vapor condensation) on the surfaces of many grains as another contributing factor.

Backscattered sunlight reaching Earth from the Moon is also weakly plane-polarized (i.e., the sinusoidal light waves vibrate in a single direction perpendicular to the direction in which a light beam travels). Different areas will show variations in the degree of maximum polarization as phase angles and albedos change (Figure 5-12). These differences are small, however, and are consistent with the polarizing properties of a dark, fine dust or ash that may cover the entire lunar surface (Dollfus, 1962).

The *Apollo* astronauts on the Moon observed that the lunar surface appears to be more or less uniformly medium gray in color, with brownish to buff overtones depending on lighting conditions. Earlier measurements made through telescopes had established the Moon to have a narrow range of colors as well as albedos; both parameters show much less variation than does the Earth's surface (Figure 5-13). *Spectral reflectivity* (intensity of reflected light at different wavelengths), as determined through the telescope, shows steadily increasing brightness levels from the near ultraviolet (UV) (wavelength λ = 0.3-0.4 μm) to the near infrared (IR) (λ = 0.7-1.5 μm). For the visible region (λ = 0.4-0.7 μm), these measurements proved quite accurate when compared with those made on lunar samples returned from the Moon; in the near infrared region of the spectrum, telescope-acquired values are somewhat

Figure 5-10 The photometric function for the entire visible face of the Moon. The variations in total brightness (I, a relative intensity function) are shown with respect to changing phase angles (angle of incidence of rising or setting Sun) for the conditions in which the observer looks along the Earth–Moon join ($\epsilon = 0^\circ$). The curve demonstrates quantitatively that the Moon displays its maximum brightness when fully illuminated ($\alpha = 0^\circ$). (After Hapke, 1963, as reproduced in Kopal, 1966.)

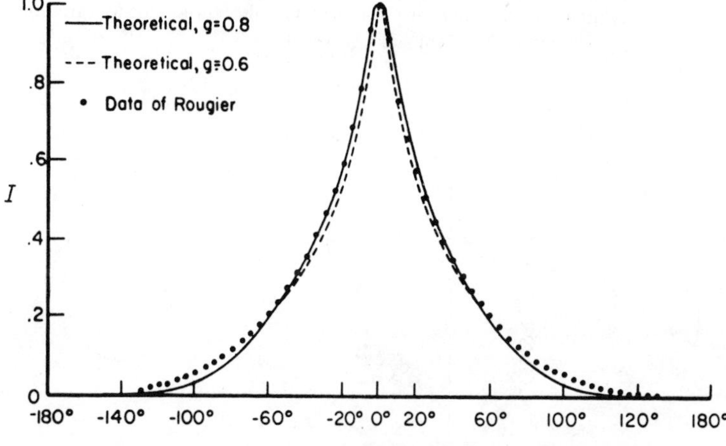

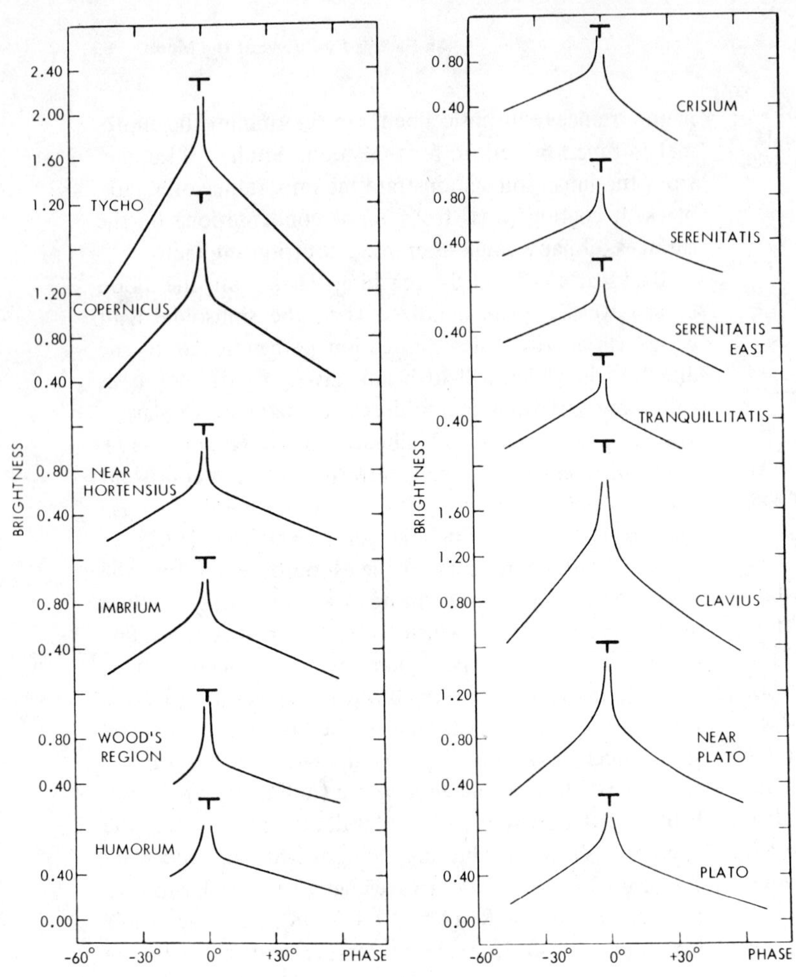

Figure 5-11 Brightness-phase angle curves for some selected mare and terra regions on the lunar surface. The T symbol refers to extrapolated maximum brightness at zero phase. (After Gehrels, 1964; from Kopal, 1966.)

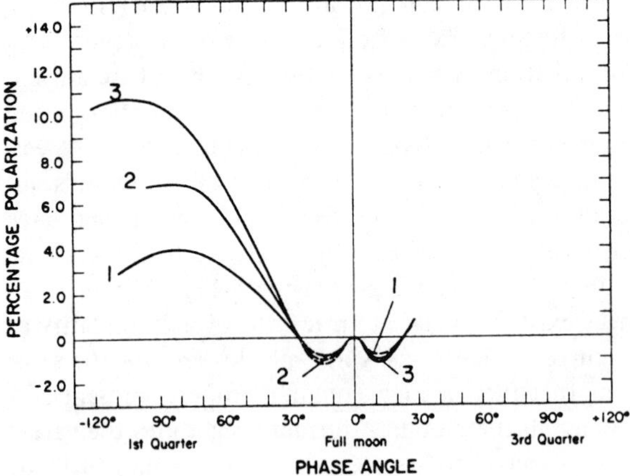

Figure 5-12 Changes in degree of polarization (in percent) of light reflected from the Moon as a function of phase angle. The plane of polarization (in terms of the electric vector of light) is either parallel or perpendicular to the equator of illumination. This orientation shifts approximately every 90° and is nearly coincident with the quadratures (the third quarter curves are symmetrical about the 0° axis). Polarization passes through zero at 0° and reaches its maxima at ± 90°. The three curves describe materials from different regions (1. Proclus ray; 2. Cayley Formation; 3. Procellarum Group) which also show albedo variations. (Courtesy U.S. Geological Survey.)

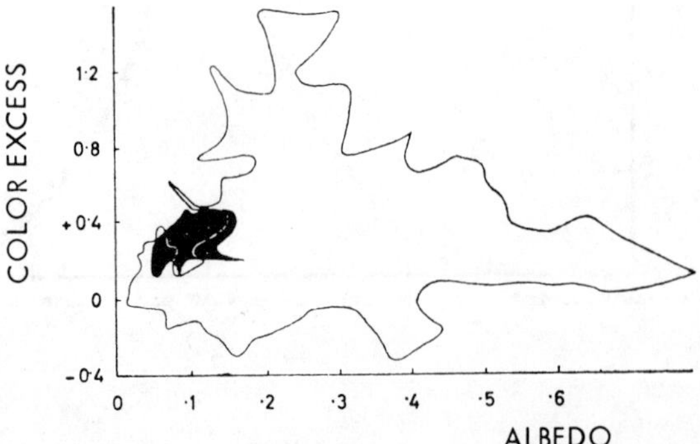

Figure 5-13 Variations in color and brightness measured telescopically for points on the Moon (black area) and in the laboratory for terrestrial rocks (white area). Color excess refers to the difference between the color index for the object and for the Sun (source of white light). Negative excess values describe bluish colors, those near zero are gray, and positive values denote yellow to brown to red colors. In this system, the Moon's average color would be identified as a dull brownish-gray. (Modified from data by N. N. Sytinskaya; extracted from Fielder, 1965.)

higher (Figure 5-14). Parts of the lunar surface reflect more brightly in one wavelength region than in others. Whitaker (1966) has demonstrated this by photographing parts of the Moon in UV and IR light. He then successively prints the UV negative and the IR positive (in register) through blue and red filters, respectively, to form a single red-and-blue picture. This technique brings out variations in color shades that appear to define different surface units (such as lava flows). Some maria tend to be blue while others are inclined toward the red. Differences between maria and highlands are also accentuated. These subtle color variations are related to the influence of certain elements, such as iron, titanium, and calcium, on the behavior of reflected light at different wavelengths.

Radar Studies of the Moon

Optical methods, then, yield direct information about surface-covering materials whose dimensions are not much larger than the wavelengths of visible light. When radar waves from manmade transmitters are reflected from ("bounced off") the Moon's surface, the longer wavelengths used (typically about 3.5 cm, but up to several meters) provide data pertaining to the relative smoothness of the surface at scales ranging from millimeters to meters. Analysis of backscattered radar-beam impulses returning to Earth indicates considerable surface roughness at the millimeter level, but a broadly undulating to smooth surface at the meter scale. Thus, over only a few square meters, gradients are generally low and do not depart significantly from uniform values. Perhaps 10 to 15% of these surface areas is occupied by objects as large as 3 to 4 cm; these could be rocks, small craters, or tiny protrusions. Radar waves, furthermore, can penetrate up to several meters below the surface, so that some idea of the three-dimensional structural and compositional makeup of the topmost layers can be inferred. Some of the first indications of variations in thickness of these layers (the regolith) were obtained in this way. Experiments that determine the radar response of substances with different dielectric constants (electric charge-storing capacities), particle sizes, porosities, and packing support the view that the bulk of the surface materials consists of fine-sized fragments.

Radar has also been used to generate surface facsimiles (Figure 5-15), similar to photographs, by programming the beam to scan successive linear strips—in much the same way as a television camera tube scans a scene. Very slight differences in the times in which the return signals are received result from variations in slope angles and elevations of the reflecting surface. In this way, a "relief," on the order of hundreds of meters, can be defined for areas covering many square kilometers even though each individual scan element represents the smoothed relief of much smaller areas.

Figure 5-14 Spectral reflectivity curve for an 18-km diameter field of view (in the telescope) containing the *Apollo 12* landing site, compared with (scaled) curves from measurements on finely divided grains sifted from surface and core tube "soil" samples recovered at the *Apollo 12* site. (From J. B. Adams and T. B. McCord, *Science*, 1971, copyright © 1971 by the American Association for the Advancement of Science.)

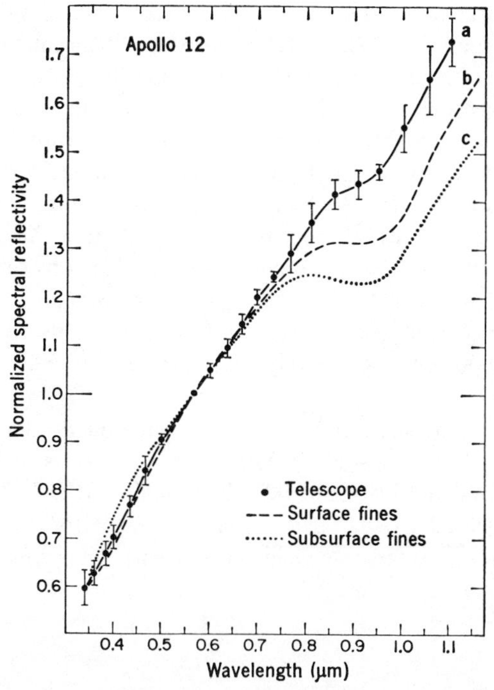

The Thermal State of the Moon

The Moon emits as well as reflects electromagnetic radiation. Solar and internal heat sources affect the temperatures recorded for its surface by sensors mounted in specially adapted telescopes. According to measurements taken with the assumption that the Moon behaves as an ideal black-body radiator, the lowest temperatures on the nonilluminated surface just before new moon are 102°K (-171°C) at the equator and 70°K (-203°C) near the poles. A maximum temperature of nearly 390°K (117°C) occurs at the subsolar point at full moon (Figure 5-16). The variations in temperature can be used to deduce the thermal inertia—a property depending on the combined effects of thermal conductivity, heat capacity, and bulk density—of various possible surface materials. Measured values of thermal inertia indicate a surface covering with high porosity (low density) and low thermal conductivity. This conclusion conforms to other Earth-acquired observations favoring loose, powdery materials almost completely blanketing any bare bedrock. These fine materials make up a highly insulating blanket such that, below a few tens of centimeters beneath the surface, the temperature has adjusted to a constant value (a conclusion supported by radio astronomy measurements).

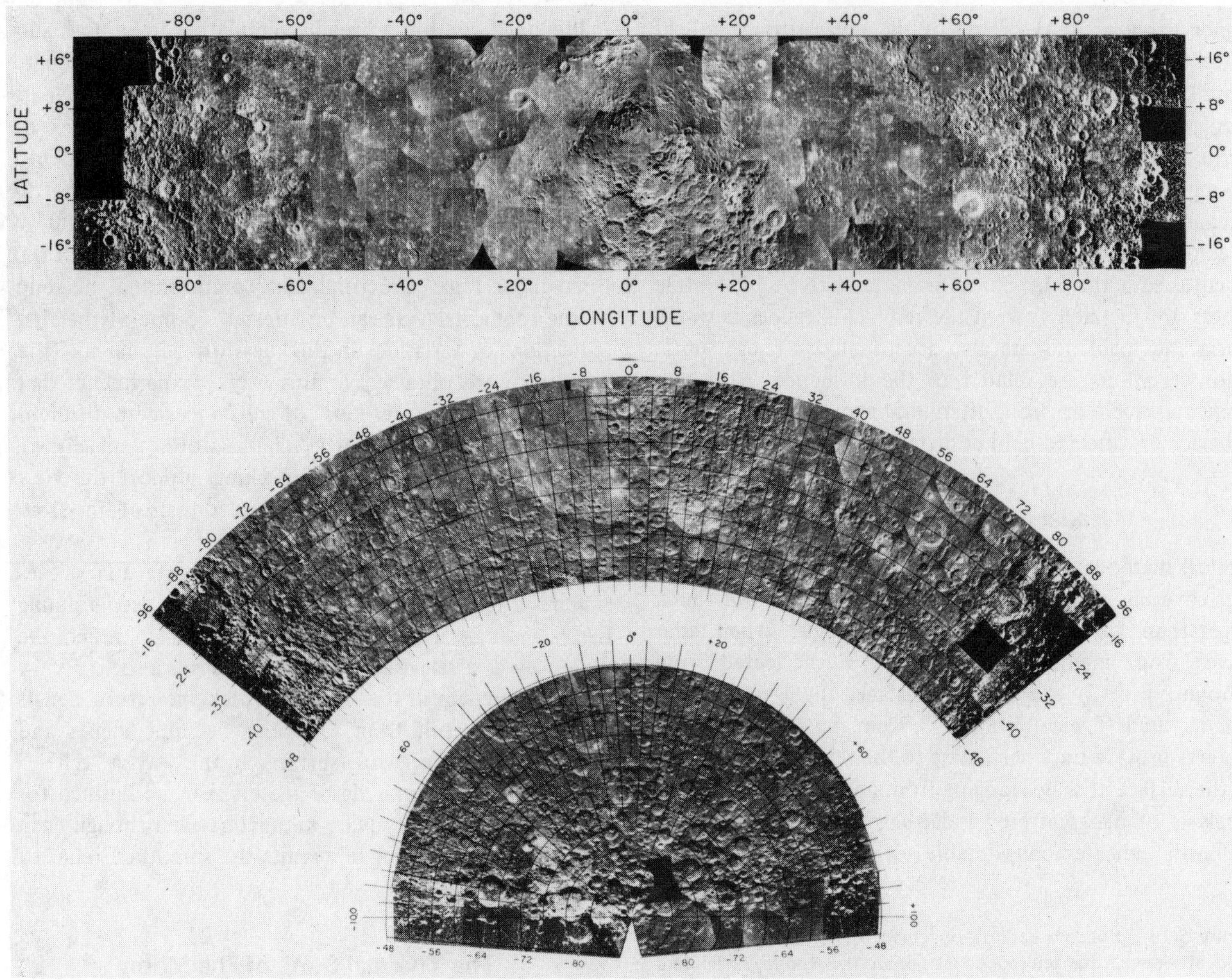

Figure 5-15 Images of the lunar surface (mounted as strip mosaics gridded in Mercator, Lambert conformal, and polar stereographic projections) formed by processing polarized radar backscatter waves received as reflections of 3.8-cm wavelength pulses sent to the Moon from M.I.T.'s Haystack planetary radar facility. (Courtesy J. V. Evans, M.I.T.)

During a total lunar eclipse, the front face begins to cool rapidly from its maximum temperature state. Under this special condition, the surface emits higher amounts of thermal radiation in the far infrared region (4-30 μm) than during its normal period of total darkness. When examined with an infrared scanning sensor, the Moon appears as an image (Figure 5-17) in which lighter (brighter) areas correspond to surfaces with apparently hotter temperatures. Most "hotspots" coincide with recent craters, around which are found boulders and other fragmental debris that possess lower thermal inertia, i.e., cool more slowly by radiation and hence remain warmer than their surroundings (Figure 5-18). Certain bright spots in the infrared image correlate with younger craters or regions in or adjacent to several maria where, in the view of some workers, volcanic activity may have taken place well after the main features were formed. A few bright areas coincide with the sites of reported *transient events*—visual telescopic sightings of color changes such as "red glows" around Alphonsus, Aristarchus, Schroeter's Valley, and elsewhere (Middlehurst, 1967). Over twelve hundred sightings of transients have been reported from more than a hundred locations (Figure 5-19); of these, more than three hundred are associated with Aristarchus, seventy-five with Plato, and twenty-five with Alphonsus (W. Cameron, personal communication). These lunar transients may be signs of gas discharges or of current volcanic activity. Kozyrev and others claim to have identified by spectroscopy certain gaseous molecules, including diatomic carbon or C_2, that could occur in emanations from volcanoes or fissures but these reports are now largely held to be discredited.

It will be interesting to note in subsequent chapters dealing with "on-site" investigations by probes and by the astronauts just how well these analytical deductions and predictions that have been developed from remote sensing measurements stand up when compared with actual surface studies.

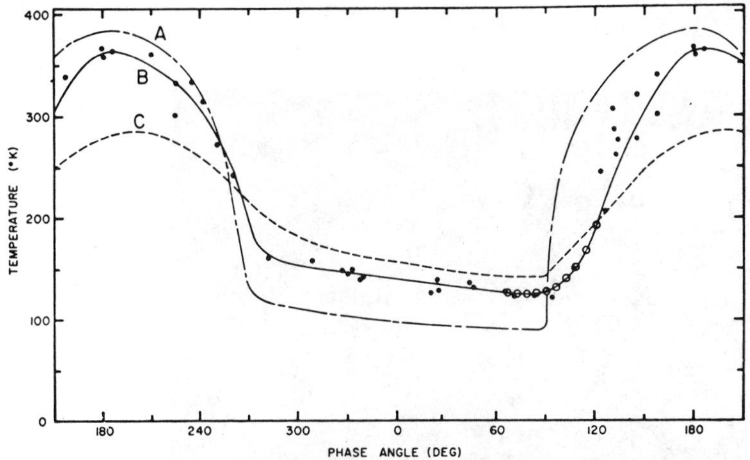

Figure 5-16 Variations of temperature at the central region of the frontside disc of the Moon during a lunation. (After Davidson and Low, 1965; extracted from Kopal, 1966.)

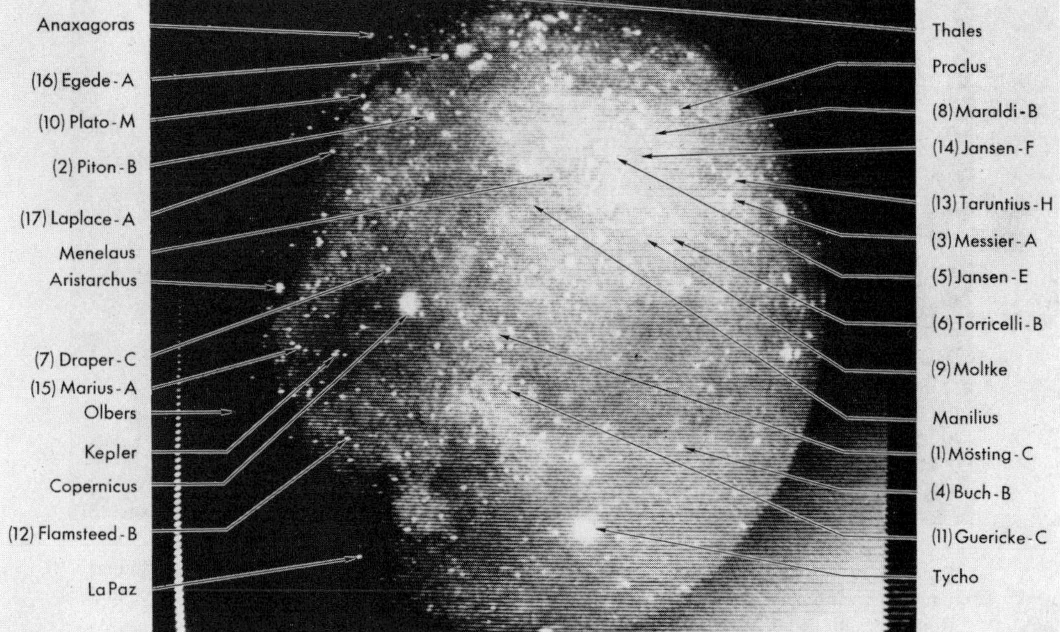

Figure 5-17 Results of an infrared (IR) scan (of emitted thermal radiation in the 10–12μm wavelength region) carried out during the total eclipse of December 19, 1964, by Saari and Shorthill (1965). Increased or enhanced thermal emission is recorded on this composite photograph as white spots. Craters and other lunar features apparently associated with the principal bright or "hot" spots are identified by arrows. (Photo courtesy R. W. Shorthill.)

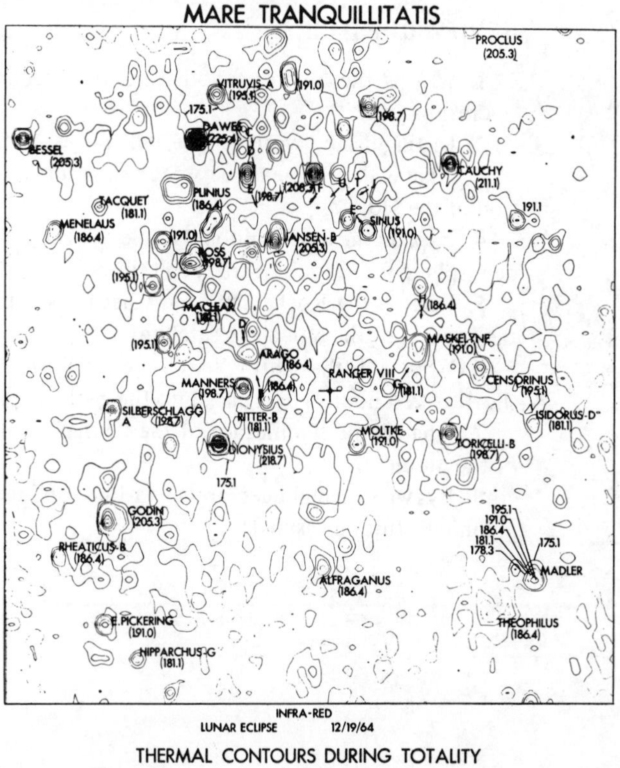

MARE TRANQUILLITATIS

INFRA-RED
LUNAR ECLIPSE 12/19/64

THERMAL CONTOURS DURING TOTALITY

Figure 5-18 Thermal contours (numbers in parentheses are temperatures in degrees Kelvin) constructed for the surface of most of Mare Tranquillitatis and part of the adjacent highlands from IR data obtained during the total lunar eclipse of December 19, 1964. Most "hot spots" coincide with large craters; Dawes and Dionysius are less degraded and presumably younger. (Z. Kopal, 1966.)

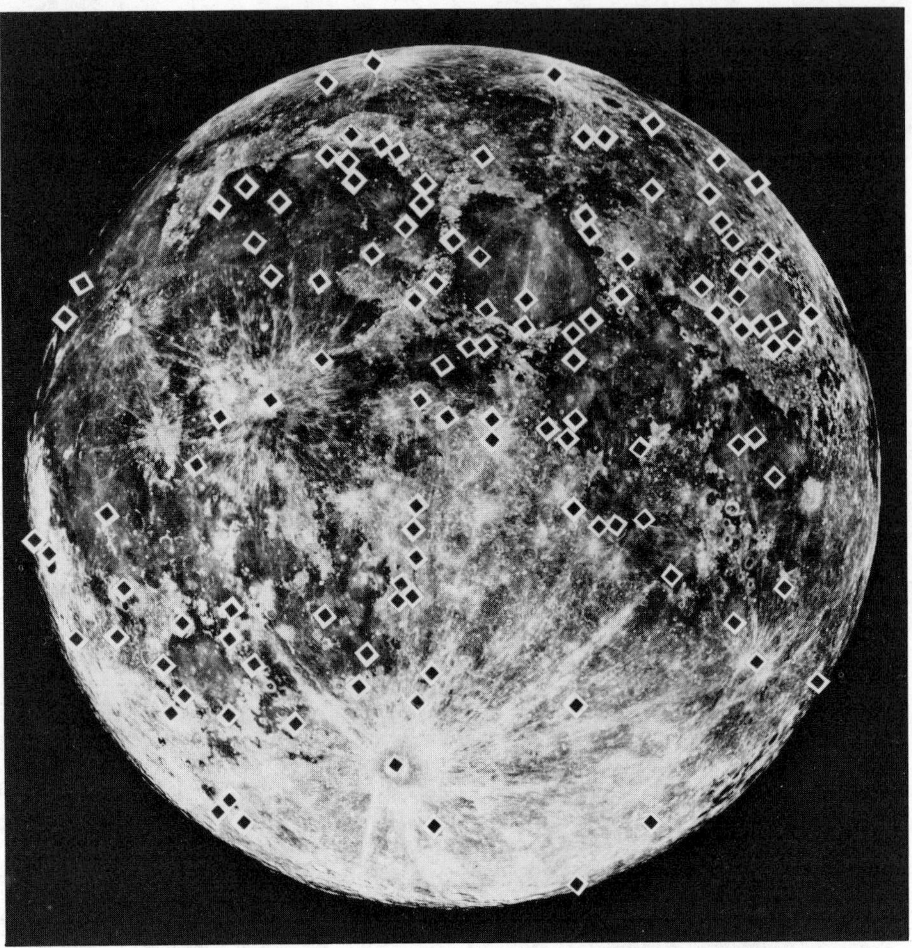

Figure 5-19 Full moon photo on which the locations of many of the sightings of lunar transient phenomena (LTP) have been plotted as black diamonds on a white base. (Prepared from data supplied courtesy of W. S. Cameron, NASA Goddard Space Flight Center.)

REVIEW QUESTIONS

1. Why is the Moon's synodic month longer than its sidereal month?

2. Distinguish between perigee and apogee.

3. What is the difference between a lunation and a libration of the Moon?

4. Explain how the Moon is able to keep only one hemisphere faced to the Earth at all times.

5. What is the Roche limit and what does it imply for the survival of the Moon as a satellite?

6. Why did the Earth "day" become progressively shorter (<24 hours) going back into the geologic past?

7. What is the significance of the "bulge" in the shape of the Moon?

8. By what techniques are elevations of lunar topographic features measured?

9. What is albedo? How is it measured?

10. What lunar surface features have the lowest and the highest albedos?

11. What is the difference between a specular and a diffuse reflecting surface? Compare the Moon relative to these types of reflectors.

12. Describe the experimental method by which surface brightness and phase angle studies aid in identifying the nature of the materials on the Moon.

13. What information does radar provide concerning the characteristics of the lunar surface?

14. What causes the thermal "hotspots" observed on a darkened Moon by infrared techniques?

15. What are lunar transients, and how might they be produced?

CHAPTER 6

Previews of the Lunar Surface

Prior to the mid-1960s, knowledge of the lunar surface had been acquired largely by such remote sensing techniques as telescopic viewing, optical measurements, and infrared and radar surveys. As noted in Chapter 5, certain inferences can be made by comparisons with terrestrial analogs. Still, competing hypotheses had been advanced to explain the same observations: for example, the materials filling the mare basins have been variously identified with lava flows, volcanic ash deposits, impact-ejecta fallback, dust from the highlands, and even water-laid "aqueous" deposits—each of which was seriously considered from the evidence on hand. Interpretation of data from albedo, photometric function, light polarization, and radar-wave-scattering determinations seemed to indicate that most of the Moon's topmost layers (down to about 10 cm) are made up of a fine powder—harkening back to the once-popular notion of "moondust." Clearly, however, what was needed to settle (or perhaps rekindle) the debate were closer looks at the surface, topped ultimately with actual on-the-spot examination by robot probes. The risks and uncertainties in placing men on a surface whose fundamental characteristics remained swathed in mystery could only be removed by designing and completing a series of interrelated experiments around and on the Moon.

The Ranger Flights to the Moon

American entry into effective lunar surface exploration began with the Project Ranger series of probes which sent back "live" pictures by TV using telephoto and wide-angle lenses to view selected areas of the Moon at ever-closer distances up to final collision. (See Figure 1-3 for the location of the landing sites of the *Ranger* and *Surveyor* probes.) Between mid-1964 and early 1965, three *Ranger*s returned altogether more than seventeen thousand pictures taken at lunar altitudes from beyond 1600 kilometers to lower than 500 meters. Figures 1-5 and 6-1 show typical scenes monitored by the TV sensors.

The *Ranger* views confirmed the opinion of most lunar scientists that craters would be found to extend over the entire size range between the resolution limits of telescopes (approximately 500 meters) and the last TV scans before crash (about 1 meter). This meant that the surface remains rough and uneven at dimensions measured at the meter level—a result suspected from remote sensing studies. How-

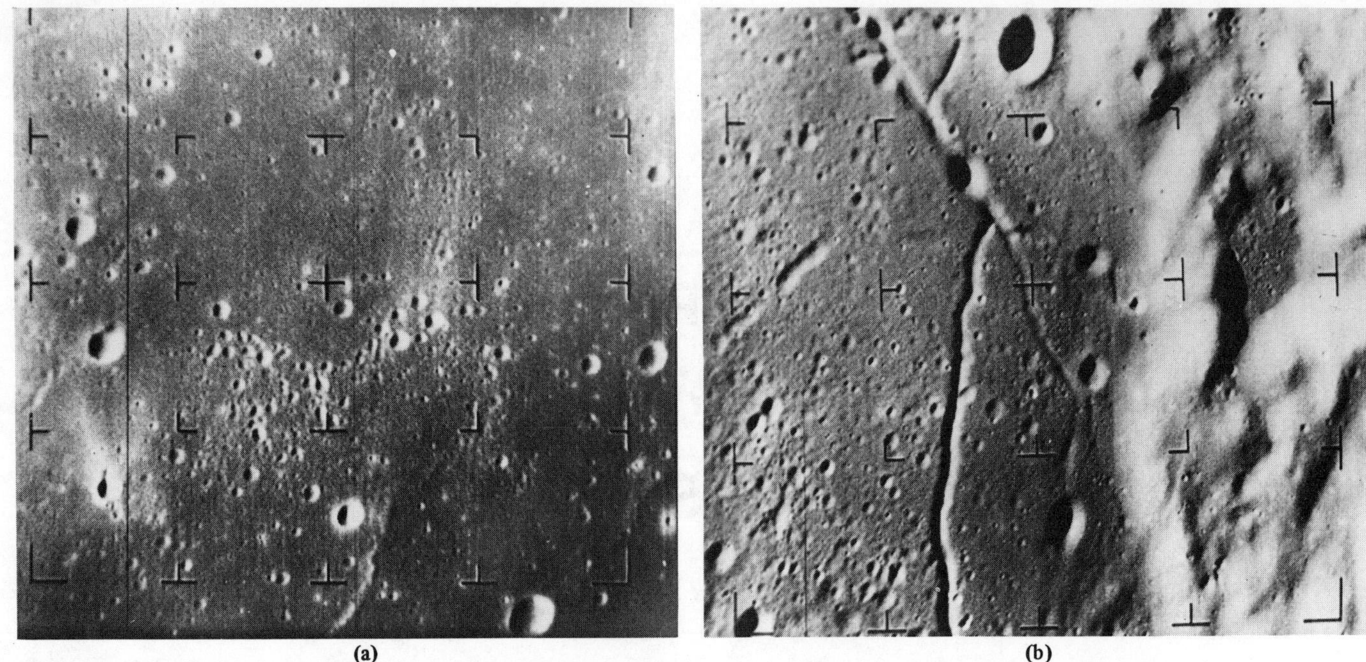
(a) (b)

Figure 6-1 (a) Part of A-camera photograph 190 from *Ranger 7* showing a portion of Mare Cognitum near the lunar equator south of Copernicus. Distance between tick marks is approximately 5 km. The larger rimmed craters are primaries; the one near the lower left tick mark shows distinct rays. The cluster of smaller craters near the center is part of a ray traceable to Tycho. Such rays are commonly composed of secondary impact craters which increase the brightness level of the enclosing area. (b) Frame B76 from *Ranger 9* taken at an altitude of 185 km; distance between tick marks (reseaus) is 6 km. The scene covers the northeast corner of Alphonsus. The lava-filled floor contains several irregular, branching rilles and several dark-haloed craters (lower and mid-center). (Photos: Jet Propulsion Laboratory.)

ever, areas of apparent smoothness between separated craters of 5 to 10 meters and larger in diameter offered promise for selection of suitable sites for manned landings. The presence of occasional large (meters) boulders (distinguishable from small craters by reversal of shadow directions) supported the contention that the outer layers had been fragmented by impact. This likewise suggested that these layers were *not* composed of cohesionless, loose and porous powder into which large, heavy objects would sink. The *Ranger* pictures also disclosed the existence of several types of small craters, including those with rims and bowl-shaped interiors (impact), elongate forms often occurring in clusters (secondary swarms), and rimless ones sometimes surrounded by dark deposits or "halos" (volcanic cinders?) or characterized by "dimple-shapes" in which inner slopes steepen to a narrow central concavity (drainage?). The quality of the *Ranger* images allowed the first detailed topographic and geologic maps of the surface to be made at scales comparable to many terrestrial maps.

The Lunar Orbiter Missions

However, only a small fraction of the Moon's surface was covered by the *Ranger* series. The logical follow-up to these missions was some system capable of transmitting sequences of pictures that could cover the entire Moon at resolutions permitting large-scale mapping suited to location of landing sites relatively free of small craters. To meet the need, five unmanned "photolabs" were placed in elliptical orbits around the Moon between August 1966 and August 1967. The first three *Lunar Orbiters* moved in orbits at low angles to the Moon's equator, while the last two were put into near-polar orbits. Each spacecraft contained two cameras, one of which took photographs at medium resolutions (M-frames) to obtain panoramic views. At perilune (point of closest approach to the lunar surface during orbit), the smallest visible object is about 8-10 meters for *Orbiters* 1-3 (perilunes between 48 and 56 km), 120-150 meters for *Orbiter 4* (2700 km), and 16-25 meters for *Orbiter 5* (100 km). The other camera took "close-ups" at *high* resolutions (H-frames) which were better by a factor of 8–i.e., the limit of resolution for the smallest object is 1/8 that of the medium-resolution camera. As a broad rule of thumb for determining scales in the *Lunar Orbiter* pictures, the actual width of the ground scenes imaged in individual *photostrips* (see examples in Figure 1-8) of the high-resolution photos is ~200 m in *Orbiters* 1-3, ~3000 m in *Orbiter 4*, and ~800 m (but variable) in *Orbiter 5*. A strip covers 8 times

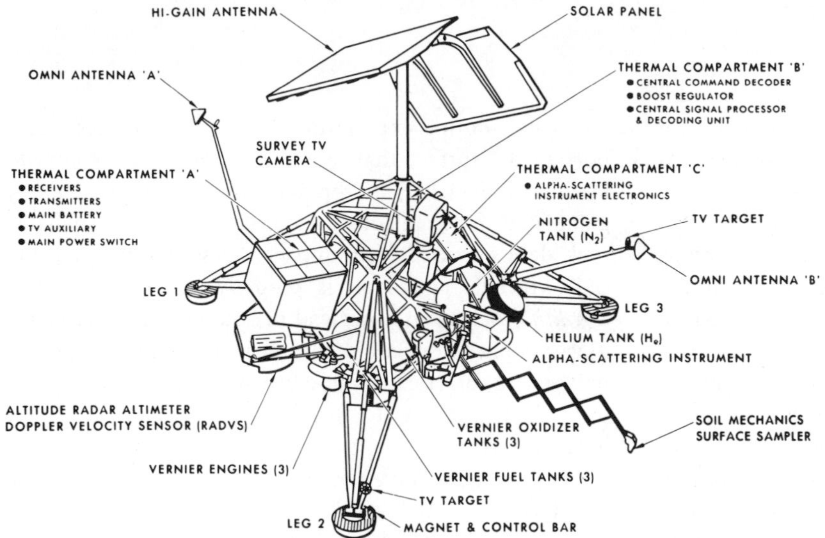

SURVEYOR SPACECRAFT LANDED MODE

(a)

(b)

(c)

(d)

Figure 6-2 (a) A drawing of the *Surveyor 7* spacecraft indicating the positions of the surface sampler, alpha-scattering device, and TV camera. (b) A mosaic of photos showing the surface sampler of *Surveyor 7* in extended mode and a series of trenches which it produced. (c) The magnet mounted on a footpad of *Surveyor 5* during a prelaunch test on Earth; basalt powder containing fine particles of magnetite cling to the magnet. These test photos of different rock types with varying magnetic iron contents were compared with those observed when the spacecraft reached the lunar surface to estimate the nature of the rock materials attracted to the magnet. (d) The alpha-scattering instrument in place on the lunar surface at the *Surveyor 5* site. (NASA photos.)

the ground width in the medium-resolution photos from each of the *Orbiters*. The spacing sequence for some photos permits overlap for stereo viewing.

The *Lunar Orbiter* program was a remarkable achievement in the U.S. space effort. All five scheduled missions were completed without major setbacks. Parts of the Moon were photographed at resolutions that represent a hundredfold improvement over previous telescope coverage. Nearly all the lunar farside, seen for the first time in detail, was imaged at mappable scales. The polar regions, hidden or poorly visible from Earth, were also fully photographed. Cartographers will long be busy in carefully revising and upgrading existing maps. For years to come selenologists will be guided in making their interpretations by nearly two thousand photos taken by the *Orbiters*. Few photos made in this century are more striking than some of those shot as "obliques" from these spacecraft (Figure 1-8).

The main scientific results of these missions are: (1) the precise visual description of crater morphology (terraces, inner floor features, ejecta deposits, etc.); (2) recognition of specific aspects of volcanic structures and other evidence of volcanism; (3) better definition of structural features and other signs of crustal deformation; (4) indications of surface modification by mass-wasting; (5) ability to discriminate between "stratigraphic" units; (6) development of methods for estimating thicknesses of unconsolidated deposits on the surface; and (7) discovery that the lunar farside lacks extensive maria and consists mostly of highland-type terrain having crater densities and albedos similar to the terrae on the frontside. Above all, success in meeting the prime objective to get detailed information about candidate or potential *Surveyor* and *Apollo* landing areas allowed the selection of final sites to conform to the exacting safety specifications and scientific criteria.

The Surveyor Program

Even before the *Orbiters* began to return their data, another U.S. program had scored an unexpected immediate triumph. Despite predictions that the first few attempts would fail, on the first try an unmanned landing vehicle, the *Surveyor 1*, set down on a mare surface in Oceanus Procellarum in May of 1966. The Soviets only 116 days earlier had achieved a similar success in landing their *Luna 9* further to the north (see Figure 1-6). A television camera on each probe witnessed an apparently rough, granular, boulder-strewn surface. At first the visible evidence was insufficient to foster any definitive interpretation as to whether this surface was a vesicular flow, a porous and pitted soil, or a consolidated pebbly sediment layer covered with microcraters. The higher-quality photos from later *Surveyors* provided strong assurances that soil-like deposits occupied most of the observable surface. These close-up pictures, as well as many taken by the astronauts during exploration of the *Apollo* sites, often portrayed a surface that resembled "sand" peppered with innumerable small "raindrop-like" impressions. This rubbly granulated material, consisting of layers of unconsolidated rock debris derived from impacts, mass-wasting, and possibly volcanic emissions, is properly referred to as the *regolith* (rock mantle) instead of soil which, in terrestrial usage, implies association with growing plants. However, it is general practice to interchange the terms *regolith* and *soil* when referring to unconsolidated materials at the lunar surface.

As four more *Surveyors* (of the seven launched) landed intact at different localities on the Moon, their sophisticated payload instruments (Figure 6-2a) gradually accumulated enough data for more convincing answers to be given to some of the pressing questions about the nature of the lunar surface. Thus, *Surveyor 3* was equipped with the "soil mechanics surface sampler" (Figure 6-2b), a small scoop on a pantograph arm which allowed certain mechanical tests to be made on the soil or regolith at the touchdown site. This remarkable gadget was also carried on *Surveyor 7*. The last three *Surveyors* (*5*, *6*, and *7*) were also outfitted with a simple magnet (Figure 6-2c) on which metallic particles could be seen by the TV camera. They also contained an alpha-scattering device for obtaining a semiquantitative analysis of the chemical elements present in the soil and rocks. This instrument (Figure 6-2d), operates on the principle of alpha-particle back-scattering. A radioactive source in the sensor head emits alpha particles (high-speed helium nuclei) over a narrow cone onto a small area of lunar surface beneath. Some particles penetrate through the network of atoms in the crystal structures within the thin (~1 mm) skin of material analyzed at the surface. Others, however, collide with nuclei of the different elements in this material and recoil or back-scatter with energies that diminish according to the masses of the nuclei with which the particles interact. A detector measures the return velocity of each particle and records the results as a spectrum of particle energies in which intensity amplitudes are proportional to amounts of each element present. Because the instrument is calibrated before launch by running a series of analyses on rock types having different compositions, the results obtained during lunar operation can be made more precise. Later improvements in the data after these are telemetered back to Earth are possible through use of computer techniques.

More than eighty-seven thousand photos of the areas around the five *Surveyor* landing sites were secured during these missions. Different scenes were obtained by moving a mounted mirror rather than by "panning" the TV cameras. Individual photos covering only small areas of view have been joined as mosaics into striking panoramic scenes. Some of the more spectacular and instructive of these photographs appear in Figure 6-3. The major scientific findings of the *Surveyor* program are as follows (p. 72).

(a)
(b)
(c)
(d)
(e)
(f)

Figure 6-3 View of a strewnfield of angular blocks (many approximately 1 meter wide) associated with a 12-meter crater within the flank of the larger crater in which *Surveyor 3* landed. (b) A mosaic panorama of the lunar surface southeast of *Surveyor 6* showing how the low angle (4°) at lunar sunrise produces shadows that emphasize the many small depressions. (c) *Surveyor 7* view of a large crater on the horizon and a small, rock-filled depression in the foreground, interpreted to be a secondary crater still containing the ejecta block that excavated it. (d) Clods of soil-like material exposed by footpad of *Surveyor 5*. (e) Impression of the footpad of *Surveyor 3* in the soft regolith; the preservation of the waffle pattern from the pad base and the steep walls indicate considerable cohesion. (f) A "spotted" rock about 25 cm in width lying near the *Surveyor 7* spacecraft. (NASA photographs.)

1. The upper layers of lunar regolith are readily compressible, yet have sufficient bearing strength to support spacecraft and astronauts on excursion. The mechanical behavior of this surface material can be compared to that of damp, sandy soil. The regolith appeared more or less uniform down to the 15-cm depth exposed by the scoop.

2. This soil has an average density of about 1.5 g/cm^3 (range 1.3-1.8), a variable porosity which can exceed 50%, and a very low compression-wave velocity of ~100 m/sec (well below that of typical Earth soil or alluvium). Most particles in the soil lie within a size range between 2 and 60 microns (average ~10 μm), which is similar to that of terrestrial clays.

3. By using a color calibration chart mounted on the spacecraft and three color filters on the camera, the colors of the surface materials were reconstituted from selected *Surveyor* photos. With few exceptions, the soil and most rocks appear to be uniformly medium-dark gray (comparable to blackboard slate) and have albedos between 0.07 and 0.08.

4. Larger particles, ranging in size from about a millimeter to slightly greater than a meter, occur over most of the surface at each *Surveyor* site. Some of these objects clearly are rocks, but most of the smaller particles are clumps or clods of loosely aggregated fine-sized materials.

5. Some of the surface roughness at the millimeter-to-centimeter scale is controlled by the presence of these clods, but much of the irregularity represents pits and undulations caused by countless small craters produced by micrometeorite bombardment.

6. Discernible craters on the surface were shown to extend in size down to millimeter diameters. Blocks of rock scattered about the surface near larger craters indicate the mare regolith to be relatively thin (1-20 meters); *Surveyor 7* in particular viewed a rocky terrain on the highlands where the regolithic cover may be only a few centimeters to a meter thick. Depressions enclosing clusters of rock fragments are probably secondary craters (see p. 148) made by ejecta tossed out from primary craters in which the impacting bodies broke up on landing but did not disperse. Certain craters are grouped in rows; some workers interpret this alignment as evidence that these craters originated by drainage of unconsolidated surface material into cracks in the underlying bedrock.

7. Various lines of evidence indicate the regolith to be constantly undergoing turnover ("tilling") and reworking, although at slow rates. Presumably this churning and comminution process is related to disturbances created during meteorite impacts. Sometimes rocks or soil clumps surmount small soil pedestals or have fillets of piled-up soil along one or more sides owing to erratic distribution of ejecta particles. Many rocks have distinctive rounded to knobby shapes that resulted from wearing away by micrometeorite bombardment and solar wind abrasion. That these shapes are derived from initially more angular fragments was demonstrated by digging up several rocks using the surface sampler; the portion of the fragment that lay below the surface was much less rounded owing to its protection by the lunar soil. One explanation for the observation that the soil's uppermost layer (1 mm or so thick) is lighter in albedo than the underlying material relies on a "bleaching" effect induced by charged particles (e.g., protons) in the solar wind; subsequent studies of *Apollo* materials indicate that particle-size differences may be a more important factor.

8. No actual rock outcrops or lava flow surfaces were evident at any of the sites. However, individual rocks show textural features which resemble the phenocrysts and vesicles (Figure 6-3f) characteristic of volcanic materials. Other rocks display crude laminations or well-defined fractures. A few rocks consist of consolidated fragments, as though derived by compaction of a part of the soil containing some larger particles. Only a small amount (0.25%) of the soil consists of free iron particles, some being equivalent to dispersed iron meteorites, as evidenced from the magnet coatings.

9. The element analysis by alpha back-scattering shows that sites V and VI are very similar in composition, but that site VII is somewhat different. By comparing the atomic percentages, recalculated as oxides, of the major constituents in the lunar materials to common rock types on Earth and to meteorites (Figure 6-4), the best match for sites V and VI is found to be certain terrestrial basalts and the basaltic achondrites or eucrites. Deficiencies in Mg and excesses in Ca and Al in the lunar material demonstrate these to be different from (but not necessarily unrelated to) chondrites or from rocks similar to those in the Earth's mantle. Other differences appear to rule out granitic rocks at the *Surveyor* sites, which also tends to eliminate such sites as sources of the common groups of tektites. The lower percentages of Fe and Ti and increased amounts of Na, Al, and Mg at the highlands site VII is interpreted to suggest a less basic rock type (one proposed possibility is anorthositic gabbro; see Chapter 11); lesser amounts of iron lead to higher spectral reflectivities and hence higher albedos.

10. Two unusual firsts associated with the *Surveyors* are: (1) an on-site survey of topographic features leading to the first remotely acquired surface map (Figure 6-5) and (2) the first robot-operated repair job on another planet, made when the surface sampler was used to force and guide a stuck alpha back-scattering device into position.

Mascons

We have become accustomed to serendipity in science. The fortuitous and unexpected are part of data gathering in research. Such was the case with the *Lunar Orbiters*. Thus, P. M. Muller, a scientist at the Jet Propulsion Laboratory in Pasadena, California, working on calculations of

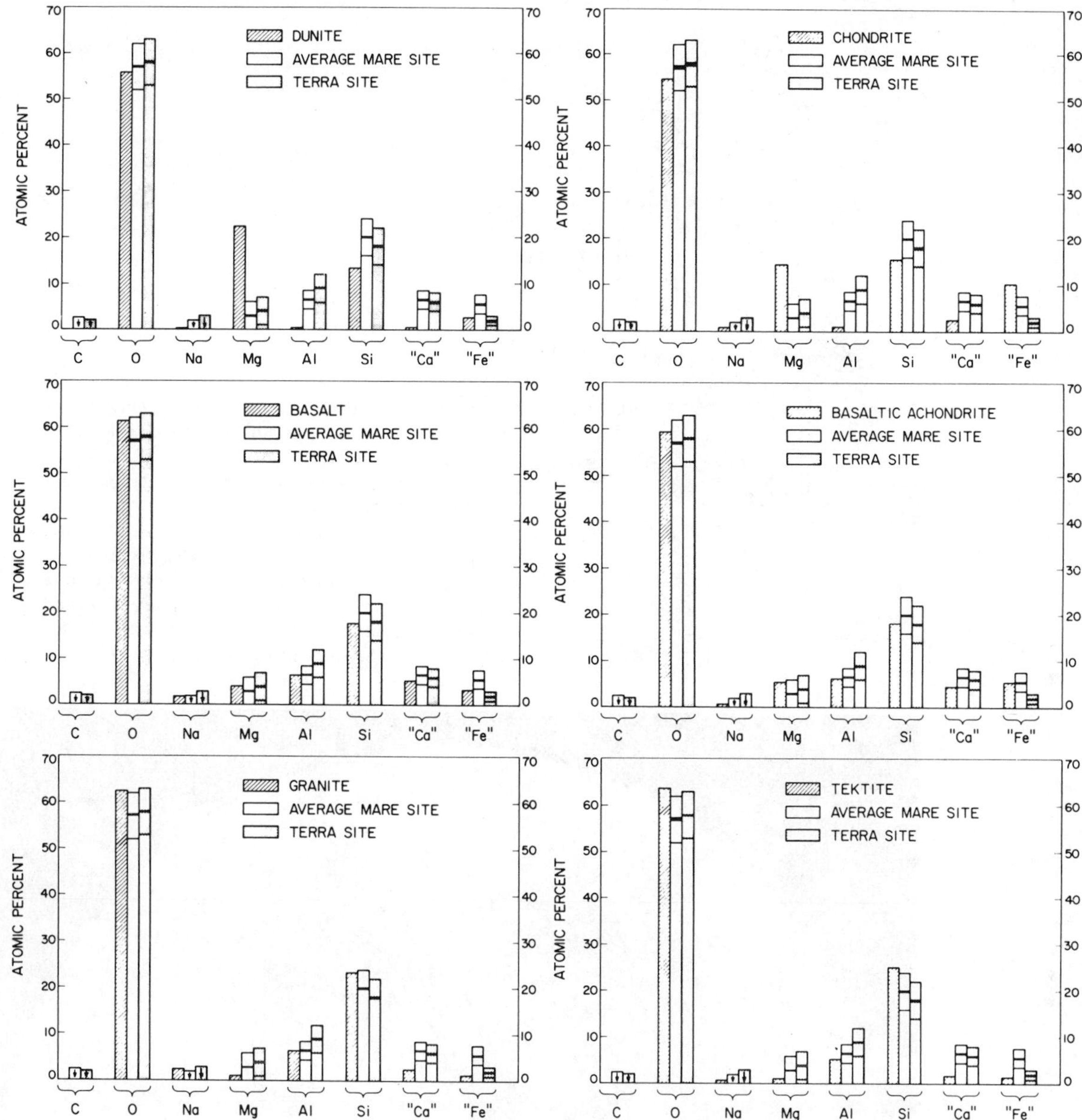

Figure 6-4 Plots of the chemical analyses made by the alpha-scattering instrument on materials at *Surveyor* sites V and VI (mare) and VII (terra) compared with average compositions of selected materials representing common terrestrial rock types, meteorites, and tektites. The histogram bars for each element are scaled to atomic percent values recalculated by a computer program which utilizes the "basic" data returned from each site. (Experimental results of A. Turkevich et al., reported in *Surveyor* Program Results, NASA Document SP-184.)

the actual orbits of the *Orbiters* around the Moon, noticed that there were significant departures in the accelerations of the spacecraft owing to differences in gravitational attraction. This effect has been likened to motions of a "rollercoaster," but the analogy is rather crude and exaggerated. Joining with W. Sjogren, Muller analyzed the tracking data in detail for clues to the cause of these perturbations. These workers concluded that gravity anomalies similar to those found on the Earth were responsible for the observed variations. The most characteristic anomalies were associated with positive (excess) *mass concentrations*, or *mascons*. Using geophysical methods to integrate the data over most of the lunar frontside, Muller and Sjogren constructed a gravimetric map (Figure 6-6) contoured to show the selenographic positions of gravity highs (positive anomalies or accelerations) and lows (negative anomalies or decelerations). The most pronounced highs coincided with some of the large circular basins,

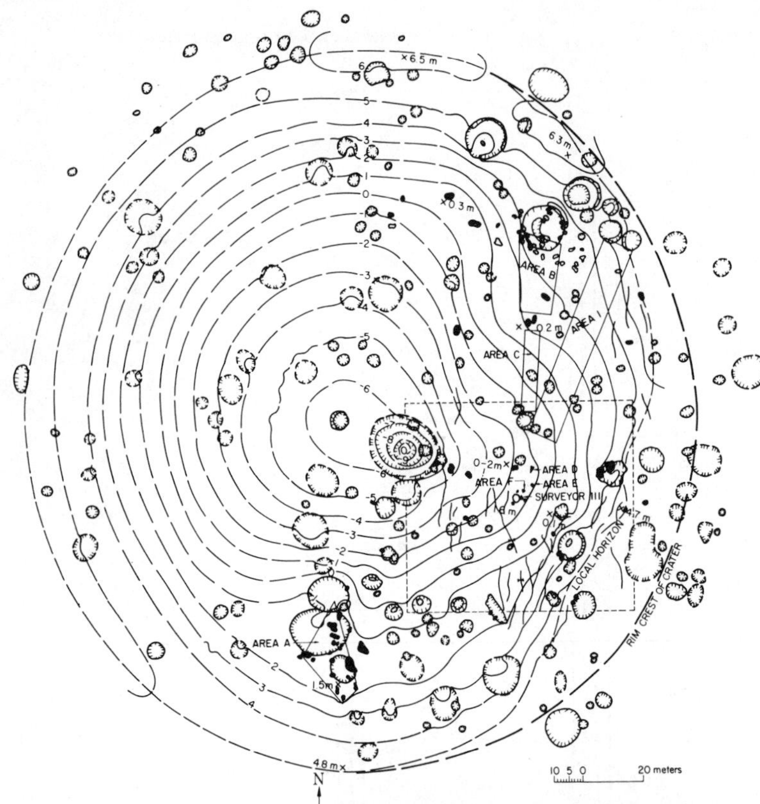

Figure 6-5 A contour map of the 200-meter-wide crater, on the flank of which *Surveyor 3* landed in 1967. The zero baseline is the horizontal projection of the *Surveyor* TV camera line of sight. Contour lines (1-meter interval) were drawn by interpolation between control points computed by the photographic trigonometry method. Heights were calculated from horizontal distances measured on a *Lunar Orbiter 3* high resolution photo and vertical angles taken from *Surveyor 3* pictures. (NASA photo of U.S.G.S. map.)

N

W

E

S

UNITS LEGEND

ABOVE × 0.1 = mm/sec^2

ABOVE × 10.0 = milligals

including Imbrium, Orientale, Serenitatis, Humorum, Nectaris, and Crisium. Thus, these large mascons tend to occur within mare-filled depressions, although other basins like Tranquillitatis and the irregular Oceanus Procellarum appear devoid of these features.

Association of the mascons with the lowlands has been puzzling inasmuch as terrestrial anomalies in physiographic or structural basins normally tend to be negative (e.g., island arc trenches). It is difficult to account for an excess of mass where there seems to be a deficiency of filled volume. It appears also that the mascon regions are not in isostatic equilibrium with the neighboring highlands.

Any new, unforeseen discovery invariably leads to a variety of hypotheses—often contradictory ones—to explain it. Urey accepts the large mascons as further proof of formation of impact basins by relatively slow-moving iron meteorites in which some of the metal becomes dispersed within the fallback materials. Baldwin holds the mascons to result from basin-filling lavas of different densities and/or thicknesses that are attempting to maintain isostatic equilibrium with the higher uplands through adjustments of the basin floors accompanied by transfer of (plastic) materials. Wise and Yates and also O'Keefe conclude that the mascons reflect the presence of plugs of denser mantle material shoved up into the basins after impact. Kaula considers the mascons to be supported by a rigid interior to depths of several hundred kilometers, implying an elastic crust and upper mantle of high strength. None of these hypotheses fully satisfies a fundamental observation—namely, several apparently circular basins, nearly all large craters, and many lava-filled noncircular lowlands do not contain mascons; however, these may represent regions in which early-formed mascons have been reduced by isostatic adjustments.

Additional information on the mascons has accumulated from the *Apollo* missions. Current explanations for these positive gravity anomalies are discussed in Chapters 10 and 13.

REVIEW QUESTIONS

1. How did the *Ranger* probes produce views of the Moon at successively larger scales?

2. What was the chief contribution of the *Ranger* series to our understanding of the lunar surface?

3. What were the principal objectives of the *Lunar Orbiter* missions?

4. Summarize the accomplishments of the *Orbiter* program.

5. What kinds of scientific instruments were mounted on the *Surveyor* probes? How did they operate?

6. Review the main achievements of the *Surveyor* series of missions.

7. Provide a simple description of the lunar surface as observed by the first four successful *Surveyor probes.*

8. How does the Tycho site (*Surveyor 7*) differ from those examined by the previous *Surveyors?*

9. Summarize the chemical information obtained by the alpha-particle back-scattering experiments on the *Surveyors.*

10. What are mascons? State several of the hypotheses proposed to explain them.

Figure 6-6 (facing page) A gravimetric map for the frontside of the Moon, prepared by Muller and Sjogren (1968) from calculations on changing accelerations in the orbital motions of *Orbiter* spacecraft. Gravity anomalies resulting from the effects of perturbations related to the lunar gravitational field are displayed here as contours in units of tens of milligals (a variation amounting to 1 milligal is equivalent to a change in acceleration of 0.01 mm/sec^2). Contoured areas showing high positive accelerations (e.g., the Imbrium, Serenitatis, Nectaris, Humorum, and Crisium Basins) are inferred to represent regions of increased mass concentrations (mascons) relative to their surroundings. (P. M. Muller and W. L. Sjogren, *Science,* copyright © 1968 by the American Association for the Advancement of Science.)

CHAPTER 7

Impact Cratering

From the evidence on the present land surface, collisions between large meteorites or comets and the Earth are rare events. Over the entire globe perhaps sixty structures whose diameters range from less than 100 meters to more than 60 km show some signs of formation by impact. Many of these are geologically young, others were produced in the Paleozoic, and a few appear to be more than a billion years old. Compared to volcanoes and mountain ranges, meteorite craters are still considered curiosities of little relevance to terrestrial geology.

Occurrence of Impact Craters

Nearly all of these impact structures are circular in outline, contain central depressions filled with fragmental materials, and possess well-defined to vestigial rims (Figure 7-1). However, erosion can greatly modify any original crater morphology so that only the remnants of the intense deformation resulting from these explosionlike events may remain. At least seventeen impact structures have been identified on the Precambrian Canadian Shield (Figure 7-2). Many are infilled with sediments, others are worn down to scars (called *astroblemes*), and several are structurally distorted. From somewhat uncertain evidence most of these appear to have originated less than 500 million years ago. Now, if this number is representative of the flux of meteorites striking the Earth over this period, and if allowance is made for the shielding of our planet by the atmosphere, for infall into the oceans, and for removal of impact structures by erosion, mountain-building processes, and covering by sediments, then the actual number of meteorite craters expected to have been impressed on the crustal surface should be in the thousands. Thus, impact cratering may be more common than we realize. The possibility of much higher meteorite fluxes early in the Earth's history implies that impact cratering was once capable of producing millions of circular structures on a primitive surface. This seems especially likely if accretion were a major factor in the Earth's origin. It follows that where a planet retains its degassing products (atmosphere and water) the effects of this intense bombardment will be initially inhibited and later erased by erosion.

The Moon is pockmarked with millions of craters ranging in diameter from millimeters in pitted rocks, to meters locally on typical surfaces, up to hundreds of kilometers for structures that approach the size of circular

(a)

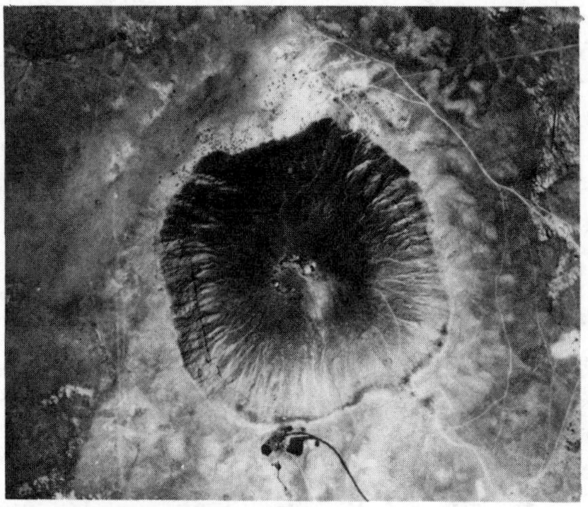

(b)

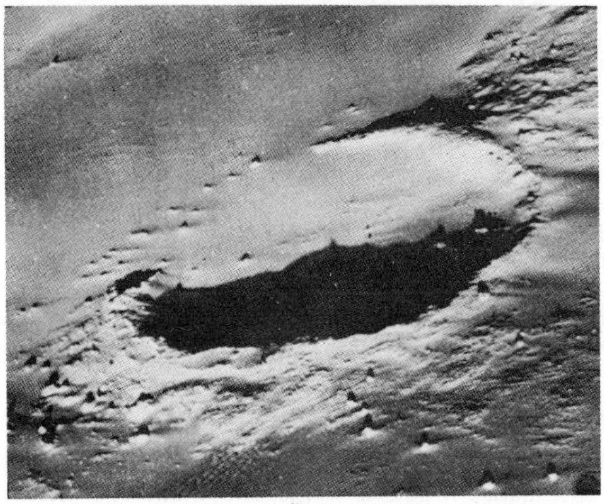

(c)

(d)

(e)

Figure 7-1 Photographs of some typical impact structures on Earth. (a) Oblique aerial view of Meteor Crater, east of Flagstaff, Arizona (diameter 1.2 km; depth to top of interior fill about 100 meters). (Courtesy John S. Shelton.) (b) Vertical aerial view of Meteor Crater; note the roughly square (polygonal) outline of the rim resulting from adaptation of rim to joints in the strata during crater formation and subsequent erosion. (Courtesy U.S. Geological Survey, Washington, D.C.) (c) Wabar Crater in Arabia (diameter about 100 meters); this crater apparently formed in loose desert sand, but shock-lithification produced numerous lumps and blocks of "instant rock". (Courtesy Arabian–American Oil Company.) (d) Roter Kamm (diameter 2 km), a partly eroded crater in Southwest Africa. (Courtesy R. S. Dietz, NOAA, Miami, Florida.) (e) Gosses Bluff in central Australia; the circular range of hills (several kilometers in diameter) is part of a central uplift within the larger (19 km wide) crater whose rim is now vestigial. (Courtesy R. S. Dietz.) (f) East (right) and West (left) Clearwater Lakes (diameter 30+ and 26 km, respectively) in northern Quebec, Canada, shown in a photo-mosaic. These astroblemes occur in glaciated Precambrian crystalline rocks. Remnants of a central ring or peak appear in West Clearwater Lake.) (Courtesy Michael Dence, Dominion Observatory.)

(f)

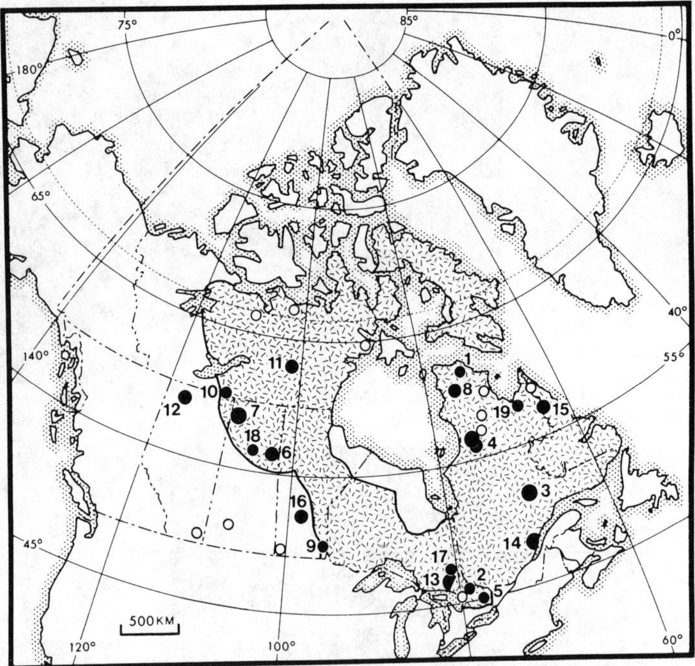

Figure 7-2 Location and distribution of proven (solid black dots) and possible (open circles) impact structures in Canada (dash-patterned area is the Canadian Shield, consisting primarily of Precambrian crystalline rocks). In the approximate order of discovery (no. 1 identified in 1852; no. 19 in 1972), these meteorite craters (diameters in kilometers in parentheses) are: 1. New Quebec (3); 2. Brent (4); 3. Manicouagan (65); 4. West Clearwater Lake (25); East Clearwater Lake (14.5); 5. Holleford (2); 6. Deep Bay (9); 7. Carswell (30); 8. Lac Couture (10); 9. West Hawk Lake (3); 10. Pilot Lake (5); 11. Nicholson Lake (12.5); 12. Steen River (13.5); 13. Sudbury (100); 14. Charlevoix (35); 15. Lake Mistastin (20); 16. Lake St. Martin (24); 17. Lake Wanapitei (8.5); 18. Gow Lake (5); 19. Lac La Moinerie (8). (Courtesy M. R. Dence, Department of Energy, Mines, and Resources, Ottawa, Canada.)

basins such as Imbrium and Crisium. Because the Moon apparently has lacked an atmosphere and hydrosphere over most (perhaps all) of its history, it is reasonable to infer that the majority of still-visible lunar craters record a complex and possibly continuing impact activity that was instrumental in shaping its surface. By analogy, a similar conclusion can be reached for Mars—as now confirmed by the *Mariner* results. Some scientists suspect that Venus and Mercury may have retained evidence of collisions with meteoroids. If this view is correct—namely, that impact craters have occurred in abundance on the surfaces of all the inner planets including, at one time, Earth—then presumably we have defined one of the universal and significant processes in planetary development. An in-depth study of the characteristics, mechanics of formation, and destruction of impact craters is thus in order.

Formation of Impact Craters

Impacts, like earthquakes and volcanic eruptions, are transient events that result in rapid (in the geologic time frame) transfer of energy. To appreciate how much energy is involved, assume a body 20 meters in diameter that travels in space toward a planet (e.g., Earth) at a velocity of 10 km/sec. If this is an iron meteorite with a density of 8 gm/cm^3, its kinetic energy (KE) at impact would be 1.6 X 10^{22} ergs (from the relation KE = ½ mv^2, where m is mass and v is velocity). Such energy is comparable to that released from moderate earthquakes of magnitude 5 on the Richter scale; a violent volcanic eruption such as that which destroyed Krakatoa in 1883 expends energies on the order of 10^{23} ergs. Using the relationship that 4.2 X 10^{19} ergs is equivalent to 1 kiloton of TNT, the above impact is calculated to generate energy comparable to detonating a nuclear bomb of 380 kilotons' yield. If this bomb were buried at a depth of 300 meters in alluvium, on detonation a crater some 500 meters wide and 125 meters deep (after fallback filling) similar to that shown in Figure 7-3 would be produced in less than 1 minute. Impacts responsible for some of the larger Canadian craters involve energies equivalent to thousands of megatons (1 megaton = 10^3 kilotons).

Both impact and underground-explosion craters result from energy released almost instantaneously from sources having a small initial volume. How, then, does a 20-meter object traveling at velocities from a few to tens of kilometers per second (bullets typically travel only hundreds of meters per second) produce a hole (crater) 25 times its diameter? The answer lies in the rapid transfer of the kinetic energy from the incoming projectile (meteorites or possibly comets) to the surrounding target (planetary crust). The energy is carried outward from the impact point by means of *shock waves*. These are pressure waves that move at supersonic velocities and subject the rock initially to extreme compression. Pressures within the shocked rock can rise to hundreds of kilobars (1kbar = 1000 atmospheres); within the Earth, static pressures of similar magnitudes are attained only at great depths (hundreds of kilometers) within the mantle and the core. Solids, such as rocks, respond in drastic ways to these pressures. Thus, granite will be permanently crushed at pressures around 250 kbars, will begin to melt (from heat obtained through compression) at 450-500 kbars, and can vaporize if shocked above 600 kbars.

In explosions and impacts, pressures spread out more or less radially from the point of energy release. This divergence along a spherical front dissipates energy so that peak pressures decrease with distance outward from the point. Compression alone, however, will not create large cavities. The key mechanism in cratering depends on interaction between compressive waves and *free* surfaces such as the ground itself beyond the impact point. This gives rise to a rarefraction—a tensional wave—which as it advances decompresses and fractures the rock and sets it into motion along planes of shear. Although the physics of this process is complicated, the net effect is to convert the rock target momentarily into a fluidlike material which moves both downward and laterally upward and out in a steady "flow" as excavation proceeds. Solid rock is fragmented and put into motion such that blocks and particles eject from the growing crater. These leave at generally low angles along an expanding cone of fast-moving material that carries most *ejecta* past the final crater edge. A hemispherical depression, related to the spherically divergent shock waves, represents the evacuated volume once occupied by materials that are given sufficient momentum to escape. Cratering ceases at distances where pressures decay below critical levels (usually about 2 to 5 kbars for rocks). The upper edge of the final crater is commonly uplifted to form a rim whose height may be further increased by deposition of falling ejecta. Most excavated material comes to rest as an *ejecta blanket* of dust and rubble extending out to 2-4 times the radius of the crater; however, a small fraction of the ejecta falls directly back into the crater. The above process is diagramed schematically in Figure 7-4. The main compression stage is depicted sequentially in Figure 7-5, in which a hint of the fate of the incoming projectile is given. Shock waves are also generated in that body as it strikes its target. These waves, initially compressional, will become rarefactions as they encounter the projectile surface. At these high pressures, the projectile in effect "explodes" much like a hand grenade, vaporizing, melting into droplets, and shattering into tiny pieces. Thus, the iron meteorite that produced Meteor Crater, Arizona was not lodged inside the rubble below the crater floor, as was once thought, but is found as "shrapnel" well outside the rim.

Experimental Cratering

Much of what is known about impacts has resulted from laboratory experiments. Unlike most geological processes,

Figure 7-3 The Sedan explosion crater (diameter ~ 350 meters) formed by detonating a 100-kiloton nuclear device in alluvium 200 meters below the surface at the A.E.C.'s Nevada Test Site north of Las Vegas. Three smaller craters, produced by chemical and nuclear explosives, appear in the background. The layers exposed below the crater rim have been tilted up and overturned in places. Note the irregular, hummocky deposits of ejecta beyond the rim. (Photo courtesy Lawrence Radiation Laboratory, Livermore, California.)

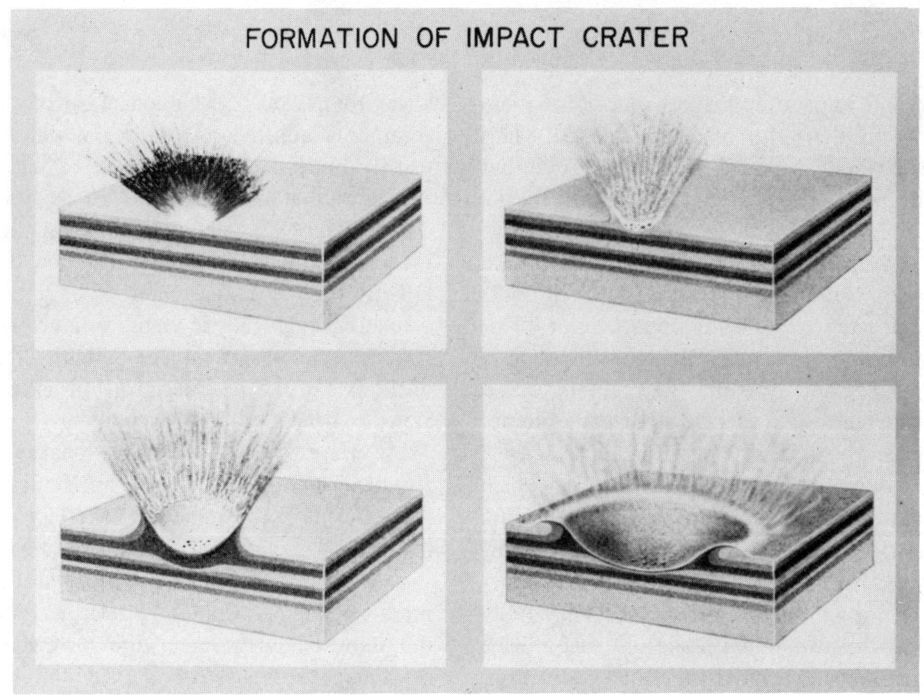

Figure 7-4 A schematic diagram showing the sequential development of an impact crater in a weak-layered medium. Upper left: The instant of impact accompanied by a bright flash of light (caused by electron excitations in vaporized material). Upper right: The crater begins to form as the dispersed meteorite mixes into the curtain of ejecta. Lower left: The crater continues to grow as rarefaction waves aid in deforming the layers along the expanding rim. Lower right: Cratering has ceased and most material from the excavation now lies beyond the rim, itself underlain by contorted and overturned layers. (D. E. Gault et al., 1968.)

Figure 7-5 (a) Schematic representation of the compression stage in the earlier moments of development of an impact crater. Terminal engulfment is accompanied by destruction of the impacting projectile, some of which is dispersed within the target. (b) Diagram, based on an experiment involving impact into a sand target, showing the directions and magnitudes (arrows) of displacement and flow of individual masses (points) at some instant during growth of the crater. The masses within the region of tangential flow will shortly leave the crater as a curtain of ejecta. V_i refers to the terminal velocity of the projectile at impact; ϕ is the angle of incidence—in this instance, normal to the surface. (D. E. Gault et al., 1968.)

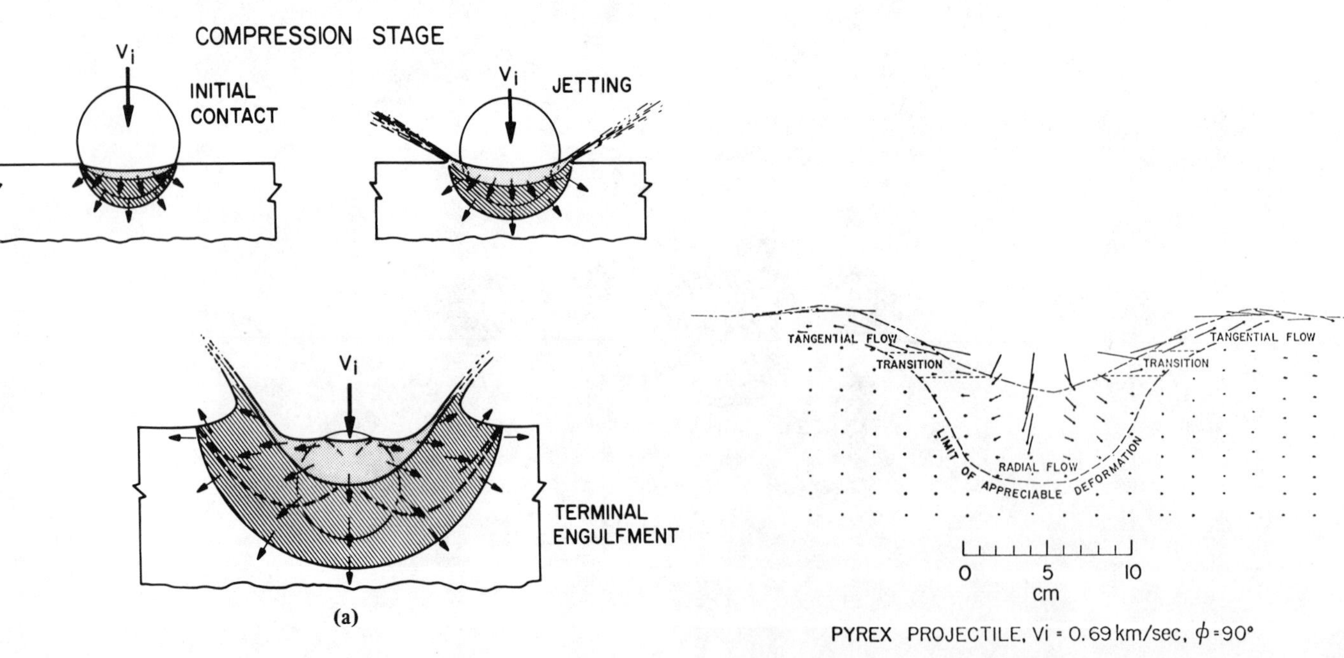

Figure 7-6 A series of frames from a motion picture sequence using a high-speed camera to show the progressive development of the ejecta curtain and surge cloud (lower right panels) from an experimental impact of a tiny projectile into a loose quartz sand target. This target had been ruled beforehand to simulate a set of gridded "faults," out of which the sand moves as the surface rises during cratering. (D. E. Gault et al., 1968.)

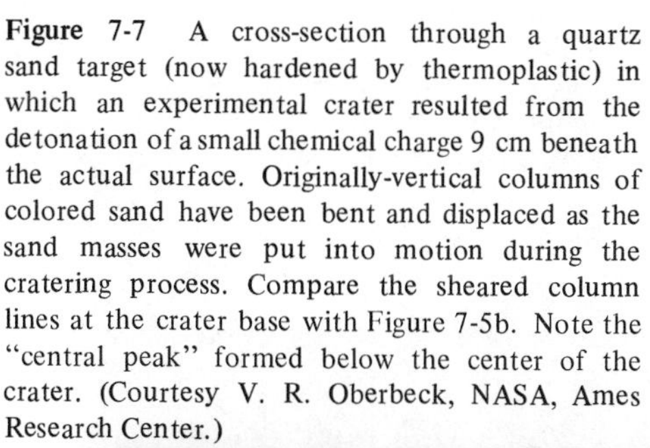

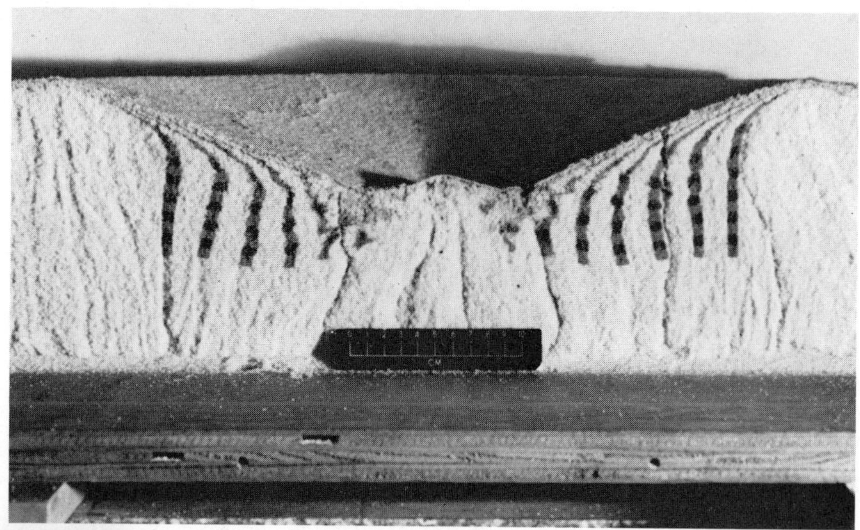

Figure 7-7 A cross-section through a quartz sand target (now hardened by thermoplastic) in which an experimental crater resulted from the detonation of a small chemical charge 9 cm beneath the actual surface. Originally-vertical columns of colored sand have been bent and displaced as the sand masses were put into motion during the cratering process. Compare the sheared column lines at the crater base with Figure 7-5b. Note the "central peak" formed below the center of the crater. (Courtesy V. R. Oberbeck, NASA, Ames Research Center.)

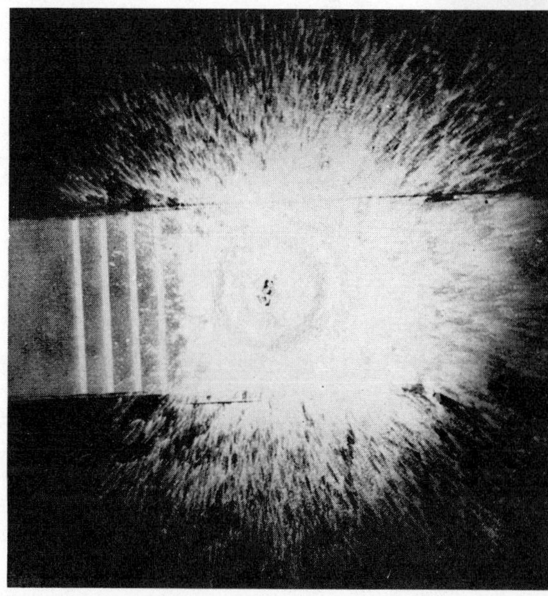

Figure 7-8 Vertical photograph of the ejecta blanket around the central crater produced by the high-speed impact of a gun-fired projectile into loose sand (covered at the surface with dark emory powder). Because the angle of incidence for the incoming projectile was less than 30°, the pattern of ejecta distribution is asymmetric around the crater. Note the tendency for rays to form along the outer fringe of the blanket. (D. E. Gault et al., 1968.)

impacts are completed almost instantaneously, so that they can be duplicated by scale modeling without concern for the time factor. A simple approach involves firing tiny projectiles from gas guns through evacuated barrels into rock or sand targets. Various physical measurements are made during the firing, and high-speed photography is used to record crater growth. Figure 7-6 shows the development of a miniature crater in loose sand as a plastic sphere about BB size strikes the surface at a terminal velocity of 6.5 km/sec. When such targets are sectioned, the distortion of once-vertical columns in the sand becomes evident (Figure 7-7). If the impacting body comes in at high angles to the target, the spray of materials that ends up in the ejecta blanket beyond distributes in a broad annular deposit which thins rapidly away from the rim. If the collisional angle is low, however, the deposits become dispersed, as indicated from overhead photography (Figure 7-8) and from studies of rocket missile impacts in the New Mexico desert.

Ejecta from Craters

During cratering by nuclear and chemical explosions and in certain types of volcanic eruptions that involve rapid escape of gases, a ring-shaped cloud of hot vapors, melted and fragmented rock debris, and dust moves rapidly outward (at velocities as high as 200 km/hr) as a *base surge*. Because the gas-particulate mixture is denser than the surrounding air, the advancing surge remains close to the surface. In volcanic events, deposition of material from the annular cloud produces *ignimbrites* (tuffs in which the particles are fused or "welded" together) similar to those associated with a *nuée ardente* or glowing avalanche (such as that which destroyed St. Pierre in Martinique during the 1902 eruption of Mount Pelée). Terrestrial impact structures sometimes have deposits preserved outside their rims that resemble ignimbrites. By inference, therefore, a base surge should develop in an impact even as vaporized rock, steam, and dust stirred up by falling ejecta are pushed away from the growing crater. Some of the suevites (breccias containing glass spatter, shocked rock fragments, and a granular matrix cemented by a glass condensed directly from the vapor state) found in the ejecta blanket near the outer fringes of the great Ries impact basin in Bavaria are thought to be base-surge deposits.

In both experimental and natural impacts the outer fringes of the ejecta blanket deposits are marked by irregular lobes and by radial streaks of jetted materials that form rays often extending many crater radii beyond the rim. Where coherent materials (such as blocks of rock) are included in the ejecta, these may form a swarm of secondary craters that dot the blanket at distances from 2 to 4 radii. Closer to the rim, the upper surface of the deposits is usually irregular and hummocky.

Crater Morphology

The shape of an impact crater depends in part on its size. M. R. Dence distinguishes between *simple* and *complex* craters (Figure 7-9) mainly on the basis of the presence of

Figure 7-9 Schematic cross sections through a simple crater (small without central peak) and a complex crater (large, with central uplift zone and melt in the depressed ring beyond) as defined from studies of Canadian meteorite impact structures by M. R. Dence, Dominion Observatory.

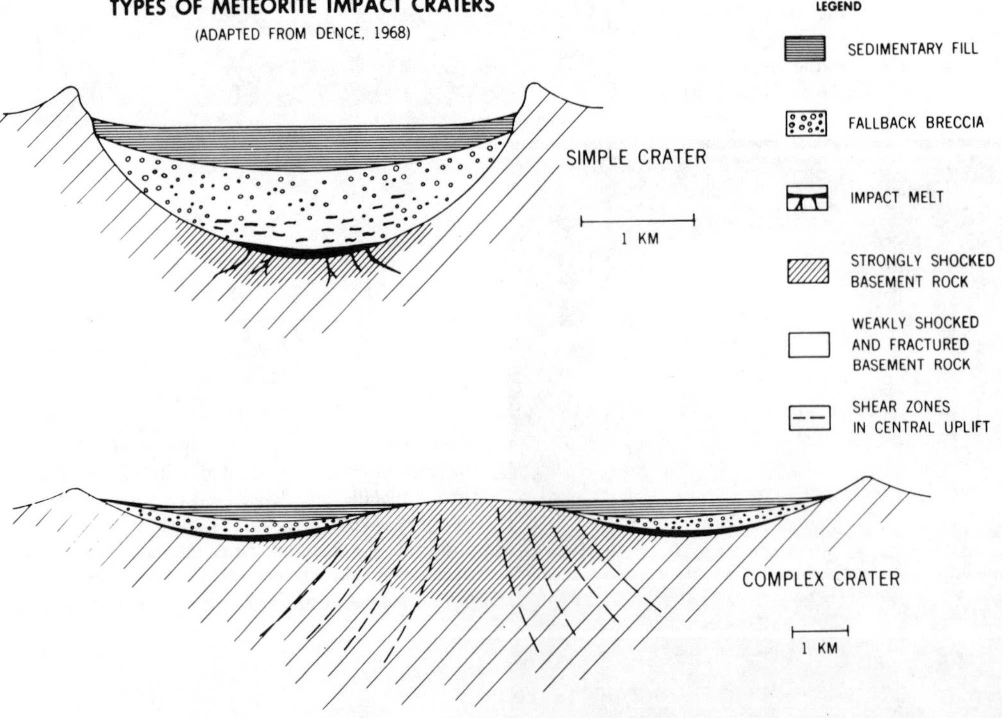

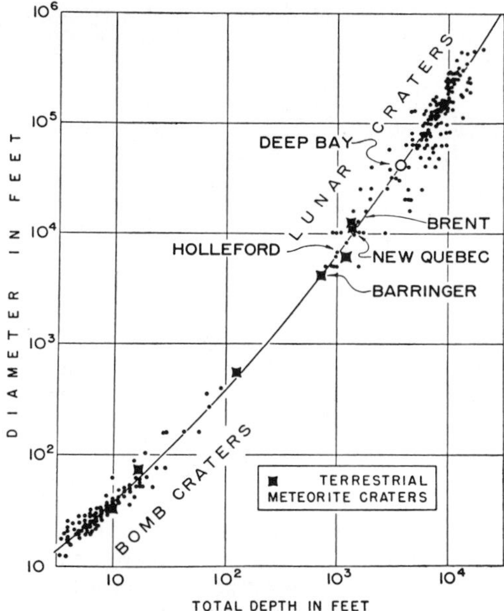

Figure 7-10 One of a series of classic diagrams based on studies by R. Baldwin showing the interrelationships between explosion crater parameters such as diameter, apparent depth (to top of fallback), real depth (to base of excavation), rim height, rim width, depth of brecciation, and energy. When plotted on log–log coordinates, pairs of these parameters tend to produce straight lines. These lines can also shift somewhat in slope and intercept as functions of the scaled depth of burst (distance below the surface of the effective point of energy release, normalized to some standard energy or yield value). The curve shown here, which relates crater diameter to apparent depth, also demonstrates that larger terrestrial and lunar craters tend to become more shallow as diameters increase. (From C. S. Beals et al., in *The Solar System*, edited by Middlehurst and Kuiper, copyright © 1961 by the University of Chicago Press.)

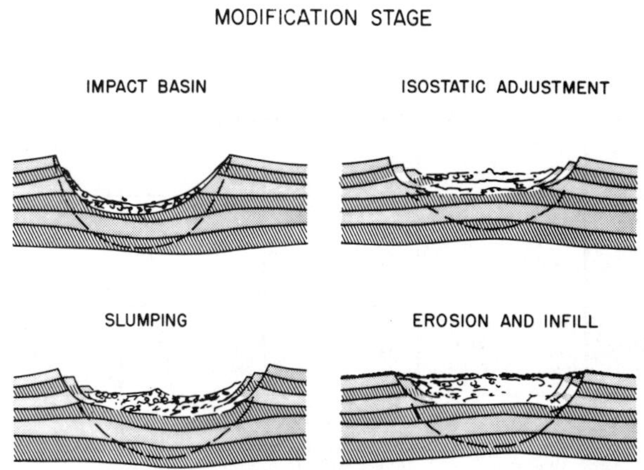

Figure 7-11 Examples of long-term modifications imposed on meteorite craters following initial impact (upper left). Slumping (lower left) can occur almost immediately after formation and continues intermittently until the rim walls stabilize. In effect, this process increases the observed crater diameter (typically by 10 to 20%). Isostatic adjustment (upper right) over a long period raises the crater base and lowers the rim so that the diameter-to-depth ratio gradually diminishes. Rim erosion and central infill (lower right) reduce a crater to an astrobleme; the rate of lowering of a regional surface can sometimes be estimated from the degree to which an impact structure has been eroded, provided the time of impact can be fixed from other evidence. (D. E. Gault et al., 1968).

central uplifts or peaks in the latter. Craters on Earth smaller than about 10 km have a relatively smooth bowl-shaped based. The bottom of a larger crater is marked by an upward bulge presumed to develop by combinations of decompression (rebound) and distortions from slumping during the terminal phases of excavation. Experiments with explosives placed close to the ground surface have produced such bulges. This indicates that central uplifts tend to form preferentially in shallow craters where the energy burst is near ground level. In general, the larger craters are also shallower, i.e., the ratio of their depths (either to crater base or to the top of the interior fill) to diameters becomes progressively less with size, and hence are more likely to contain central peaks. This holds true for essentially all craters of explosive origin, including bomb craters produced by chemical and nuclear detonations, some volcanic craters, and terrestrial and lunar impact craters (Figure 7-10).

Following its formation, a crater is subject to further modification (Figure 7-11). Almost immediately, steep unstable walls fail by slumping, enlarging the crater and partially refilling it with materials from the rim region. These slumps often involve concentric slices of wall rock that slip differentially to produce a set of nested terraces. In time, the gravitational imbalance arising from the expulsion of ejecta can bring about isostatic adjustments that raise the floor and lower the rim. This, together with gradual erosion and infill (and covering by ejecta deposits if other craters are nearby), will subdue and eventually destroy the initial depression.

Recognition of Impact Structures

The above descriptive properties of craters aid in identifying relatively "fresh" structures. However, as craters age, or where similar volcanic structures are possible, more definitive criteria are needed to prove an impact origin for a given structure. Based on terrestrial crater studies, recognition of impact structures depends on certain diagnostic features (Figure 7-12). Presence of associated meteorites obviously identifies impact structures. The style of deformation in the fractured zone beyond the limits of excavation can also be conclusive. Layered rocks commonly show pronounced folds (some overturned) and extensive faulting (especially thrusts). As shown in Figure 7-13, the degree of folding, character of faulting, and extent of fall-

CRITERIA FOR RECOGNITION OF METEORITE IMPACTS

TYPE	GENERAL ACCEPTED (METEORITES)	YOUNGER, GOOD EVIDENCE	OLDER STRUCTURES (CANADIAN)	CRYPTO-EXPLOSION (SEDIMENTS)	LARGER STRUCTURES
EXAMPLES	METEOR CRATER WABAR HENBURY	RIES L. BOSUMTWI L. MIEN	BRENT HOLLEFORD (H) CLEARWATER MANICUOAGAN	STEINHEIM WELLS CREEK FLYNN CREEK SIERRA MADERA GOSSES' BLUFF	SILJAN RING SUDBURY (S) VREDEFORT (V) BUSHVELD (B)
GENERAL	CIRCULARITY FOLDING FAULTING BRECCIATION "BRECCIA LENSES"	X	X	X	V, (S, B)? S, V S, V S, V S, V
PETROGRAPHIC HIGH PRESSURE	COESITE STISHOVITE	X X	(H)		
SHOCK	LAMELLAE (QUARTZ) ANOMALOUS OPTICS "GLASSY" PHASES (MASKELYNITE)	X X X X	X X X X	X	S, V
HIGH TEMPERATURE (>1500°C.)	FUSION GLASSES LECHATELIERITE OPAQUE DECOMP.	X X X	X X		S (?)
		SHATTER CONES	X	X	S, V
		CENTRAL UPLIFTS	X (larger)	X	V, B (?)
		IGNEOUS ACTIVITY			S, B

Figure 7-12 A simplified chart indicating principal criteria for recognizing impact structures. Different categories of structure (based on size and extent of erosion) are listed in the columns (a cryptoexplosion structure occurs in sedimentary rocks and resembles a volcanic pipe or maar but shows little other evidence of a volcanic origin; careful study demonstrates some of these to be impact and others to be volcanic in origin). Maskelynite is a glassy or amorphous form of feldspar in which the original crystal shape is retained. Lechetalierite is a form of silica glass derived usually from melting of quartz without complete mixing. (From B. M. French, 1968.)

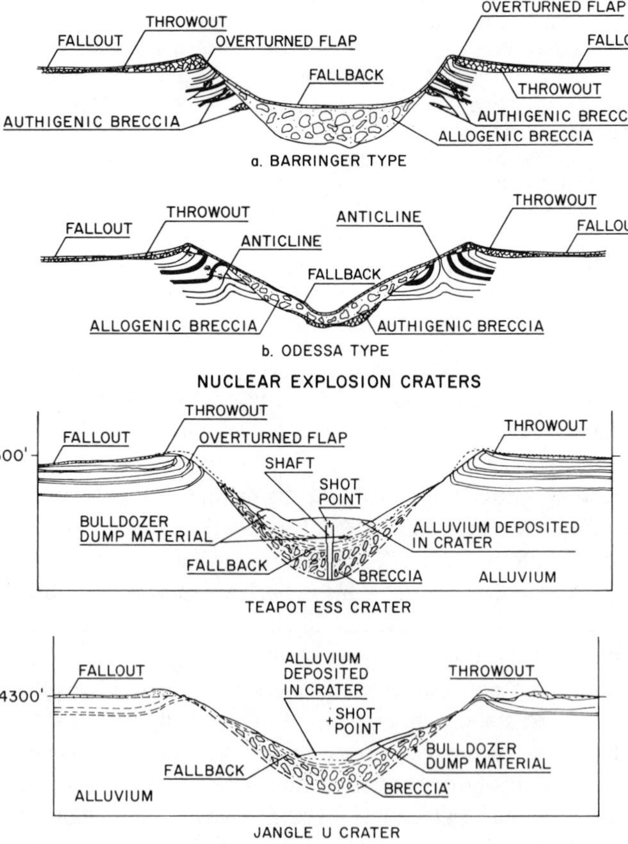

Figure 7-13 Generalized cross-section through Meteor (Barringer) Crater, Arizona (diameter 1.2 km) and Odessa Crater, Texas (diameter 150 meters) as examples of impact structures, and Teapot Ess (diameter 100 meters) and Jangle U (diameter 75 meters), Nevada Test Site (A. E. C.) as illustrations of explosion craters. Meteor Crater and Teapot Ess display similar extensive overturning of strata in their rims and a greater thickness of infilling breccia, whereas Odessa and Jangle U both are characterized by asymmetric rim folding and shallower breccia deposits. These differences are attributed to the greater scaled depth of energy release (represented by shot point for the explosion craters) at Meteor Crater and Teapot Ess. The term authigenic breccia refers to deposition of fragmental rock debris produced essentially in place (minimal transport), whereas allogenic breccia describes ejecta derived from various parts of the crater that becomes mixed and redistributed as fallback and throwout. (From Short, 1965, adapted from Shoemaker, 1963.)

back fill depend in part on the effective depth at which the crater-forming energy is released. The deformation abruptly dies out beyond the zone just below the crater base and rim. This confinement of structural disturbance to the close vicinity of the crater is in contrast to most other kinds of circular structures.

Shock Metamorphism

The best evidence for impact comes from shock effects within the rocks themsleves. The bulk of the shock damage is inflicted on the target material near the point of impact. Much of that material is tossed out of the crater

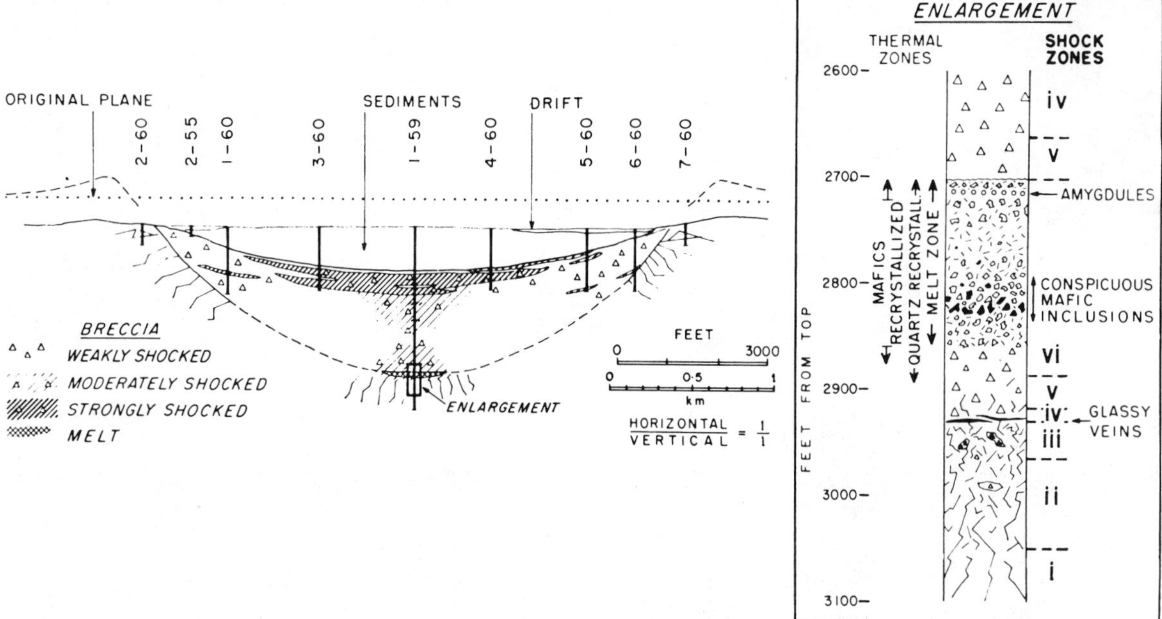

Figure 7-14 The distribution of shocked rock materials, based on data obtained from core drilling (numbers such as 1-59 refer to drill hole designations), as displayed in a cross-section of Brent Crater in Canada. Most of the breccias containing more strongly shocked minerals occur in the central part of the crater infill. Maximum shock effects, including melting, are observed near the crater base; zones of varying shock damage (right; VI is the most intensely shocked zone) can be defined from microscope studies of recovered core samples. The broad lens of strongly shocked breccia near the top of the crater fill originates partly from late-time fallback and partly from later redeposition of materials from beyond the rim as the crater enlarges by slumping. (M. R. Dence, 1968.)

so that these effects may be commonplace in the ejecta blanket. Only the central fill remains in older structures (astroblemes). Breccia in the fill will contain variable amounts of shocked rocks (Figure 7-14) depending on the relative proportions of direct fallback and later slump infill.

Quartz normally is an abundant mineral in most rocks. Under high shock pressures, this mineral can be compressed to *coesite* and *stishovite*, two higher-density forms of silica. Graphite may convert to diamond. These transformations would occur under regional metamorphic conditions only at great depths within the terrestrial crust or mantle. Thus, presence of coesite in quartz-bearing rocks now in a near-surface environment constitutes a unique indicator of pressure conditions attributable to shock waves.

As shock pressures increase, quartz and other minerals tend to shatter and granulate (Figure 7-15a). At pressures equivalent to mantle depths, diagnostic planes of deformation develop within the crystal structure of some minerals, including quartz. Multiple sets of close-spaced *planar features* (Figure 7-15b) are known to occur only in shocked rocks. At still higher pressures, grains of quartz, feldspars, and other phases may be completely changed into glasslike states *without loss of their original shapes* (thetomorphs) (Figure 7-15d and e). Thus, quartzites retaining their initial sandstone textures but consisting entirely of glassy grains have been found at Meteor Crater and elsewhere. At pressures near 500 kbars, most rock types begin to melt under shock-loading (Figure 7-15f).

Breccias containing rock and mineral fragments that show varying degrees of shock damage are one sign of an impact event (Figure 7-16a). Shatter cones (Figure 7-16b) are another indicator. Those fracture surfaces seemingly develop under tension around the crater base where shock pressures have decayed to low values. When deformed sedimentary units containing shatter cones are rotated back to their original (undisturbed) positions on a map, the apices (peaks) of the cones within the beds generally all point in directions that converge on a common center, roughly coincident with the spot at which the initial impact occurred.

Impact Craters on the Moon

Because they can be investigated by mapping and drilling, a great deal is now known about impact structures on Earth. Investigations of craters on the Moon and Mars are still restricted largely to examining their external appearance or morphology through a telescope, or orbital space-

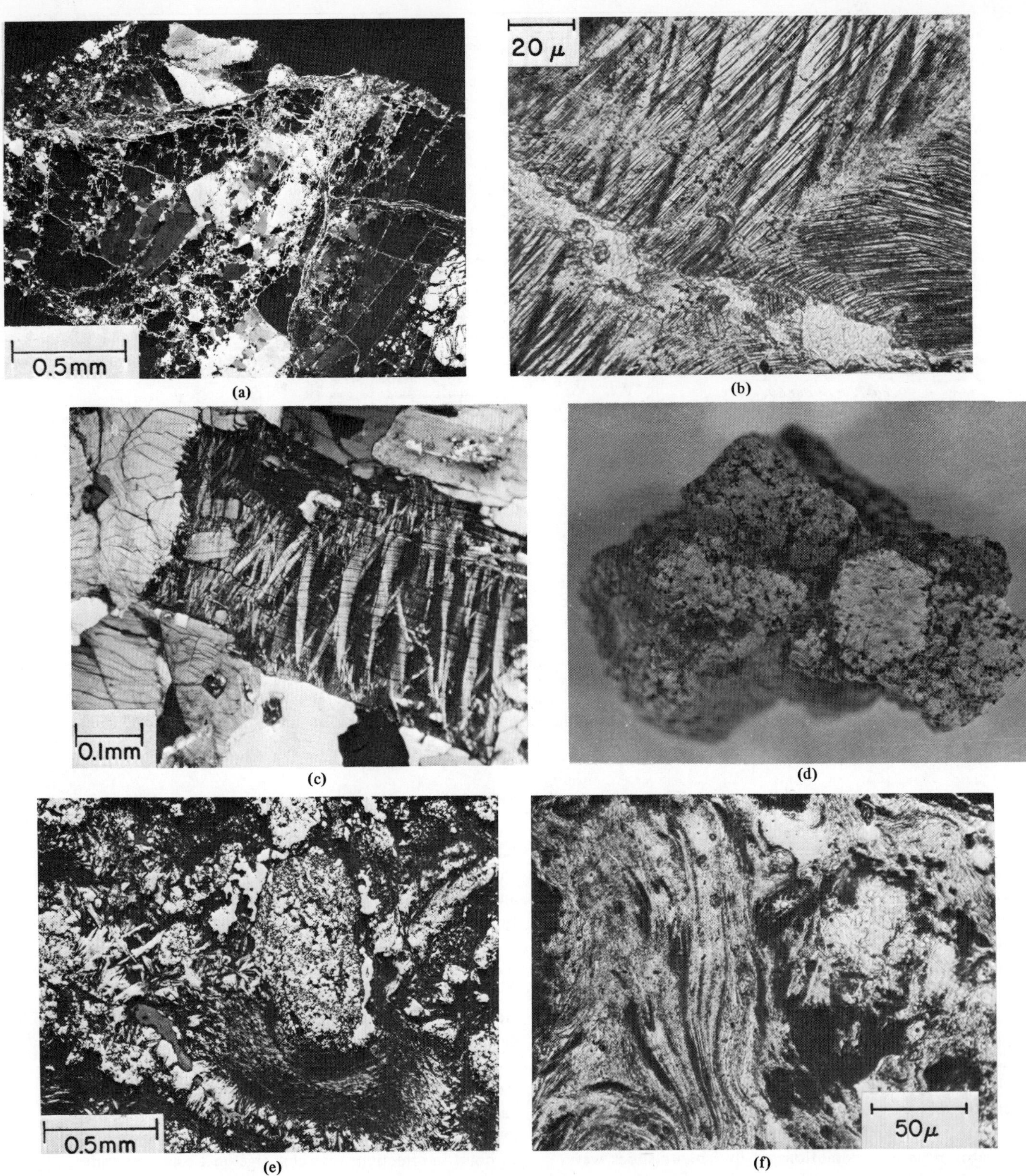

Figure 7-15 Examples of shock features in minerals recovered from impact and explosion craters, as observed under the microscope. (a) A shattered quartz grain, Brent Crater, Canada. (b) Planar features in quartz, Manson structure, Iowa. (c) Crumpled flakes of biotite mica, from a nuclear explosion in granite, Nevada Test Site. (d) A granite block in which component minerals have been converted to glass without loss of shape (note larger feldspar phenocryst), Sedan nuclear explosion crater, Nevada Test Site. (e) An intensely shocked granite specimen from Clearwater Lakes, Canada, in which both quartz and feldspar crystals, after conversion to a glassy state (and distorted plastically), have now completely recrystallized into fine granular aggregates. (f) Shock-melted alluvium, Sedan nuclear explosion crater, Nevada Test Site (note flow lines in quenched glass on left). (Courtesy N. M. Short, NASA, Goddard Space Flight Center.)

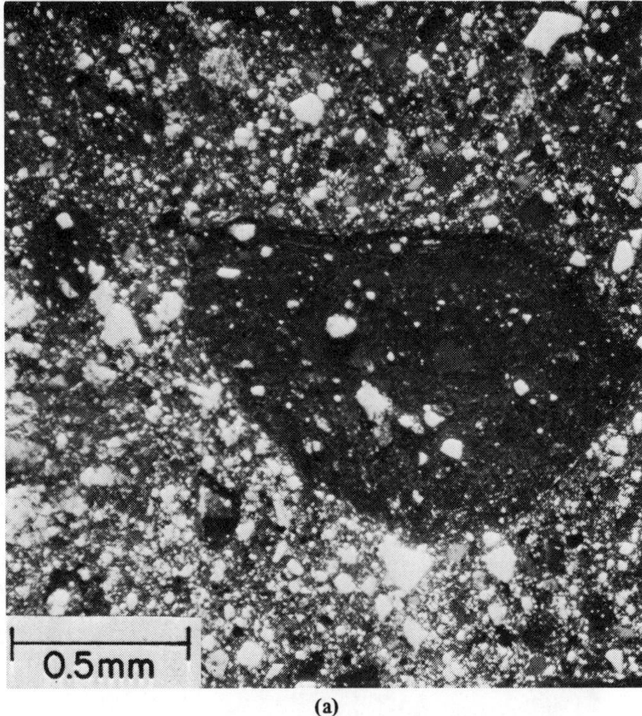

(a)

(b)

Figure 7-16 (a) Texture of breccia from the Brent Crater, Canada. Granulated pieces of minerals, showing varying degrees of shock damage, are mixed with rock fragments and blebs of glass, all bound together by rock powder and dispersed glass in the matrix. (Photo by N. M. Short.) (b) A cluster of shatter cones in dolomite from the Wells Creek cryptoexplosion structure in Tennessee. Note the horsetail-like patterns (striations) on individual cones and their common orientation towards one direction. (Photo by C. W. Wilson, courtesy R. S. Dietz, NOAA, Miami.)

craft photographs, or by limited sampling around *Apollo* landing sites. That impacts are commonplace on the Moon is now confirmed by discovery of numerous shock effects in rocks returned from lunar excursions (see Chapter 11).

Lunar craters come in a surprising diversity of sizes and shapes (Figure 7-17). Overall, the circularity of smaller craters tends to be nearly perfect. Craters larger than about 20 km are somewhat more polygonal. Craters larger than 50 km usually contain "telescoping" terraces arranged in concentric benches and slopes that extend downward from outer rim to floor.

Very large circular depressions, almost certainly of impact origin, are referred to as *basins*. The Orientale Basin is the type example on the Moon (see Figure 1-8b). Hartmann and Wood (1971) now recognize thirty-one impact basins, of which eleven occur on the lunar farside. Well-defined frontside basins include Imbrium, Crisium, Nectaris, Humorum, Smythii, Grimaldi, and Humboldtianum. (Short and Forman [1972] also list the Serenitatis and Tranquillitatis mare-covered areas as basins; McCauley and Wilhelms [1971] describe two other ancient, largely buried basins in the Oceanus Procellarum region.) Essential characteristics include (1) one or more concentric rings, each consisting of arcuate ridges and mountainous terrain that extend into lower valleys sloping downward toward the next outer ring; (2) radial lineaments and groovelike valleys; and (3) variable amounts of mare lavas in the central depression extending out to the low regions of the rings (basins without mare fillings have been termed *thassaloids*). Basins are typically larger than 100 km; some have multiple rings out to 500-1300 km from the center. These rings tend to develop at distances from the basin center such that the ratios of ring diameters relative to the innermost or most prominent ring vary systematically by multiples of $\sqrt{2}$ (Figure 7-18). One opinion holds these rings to be scarp-bounded blocks associated with inward-dipping circumferential faults. Another considers the rings as preserved "crests" of a standing shock wave moving outward through momentarily fluidized rock overstressed by a huge impact.

Many fresh craters have well-defined rims and regular, steep slopes that form simple bowl-shaped depressions; in some (the so-called dimple craters), the slopes appear to steepen toward the center as though material had been drained downward. Most larger craters have relatively flat, broad floors filled either with lavalike (dark-albedo) rock and/or landslide debris or other fragmental rubble. Individual or clustered promontories which *usually* lie near the center of these structures are conspicuous in craters wider than about 40 km. As crater diameters increase, the char-

(a)

(b)

(c)

(d)

Figure 7-17 (a) An oblique view of Aristarchus, a 42-kilometer-wide crater showing extensive terracing, a flat to hummocky floor, and interior peaks characteristic of many similar major craters on the Moon (*Apollo 15* pan camera 15-0333). (b) The large crater Aristoteles, northeast of the Imbrium basin at the south edge of Mare Frigoris. This young, terraced structure is surrounded by a broad ejecta blanket marked by hummocky topography, radial grooves, and numerous secondary craters in chains and bifurcating clusters (*Orbiter 4*-103_1). (c) The interior and rim region of Tycho, one of the youngest large craters on the Moon. The inner floor shows a rough, irregular surface attributed to volcanism (see Figure 8-5c), and the blanket deposits formed from the ejecta carried out of the central excavation during the cratering process (*Orbiter 5*-125M). (d) The craters Goclenius (foreground; 50-km-diameter), Magelhaens, Magelhaens-A, and Colombo-A (rear-right, center, and left, respectively) at the west edge of Mare Fecunditatis, seen looking southward from the orbiting *Apollo 8* spacecraft (Freme 2225). (NASA photos.)

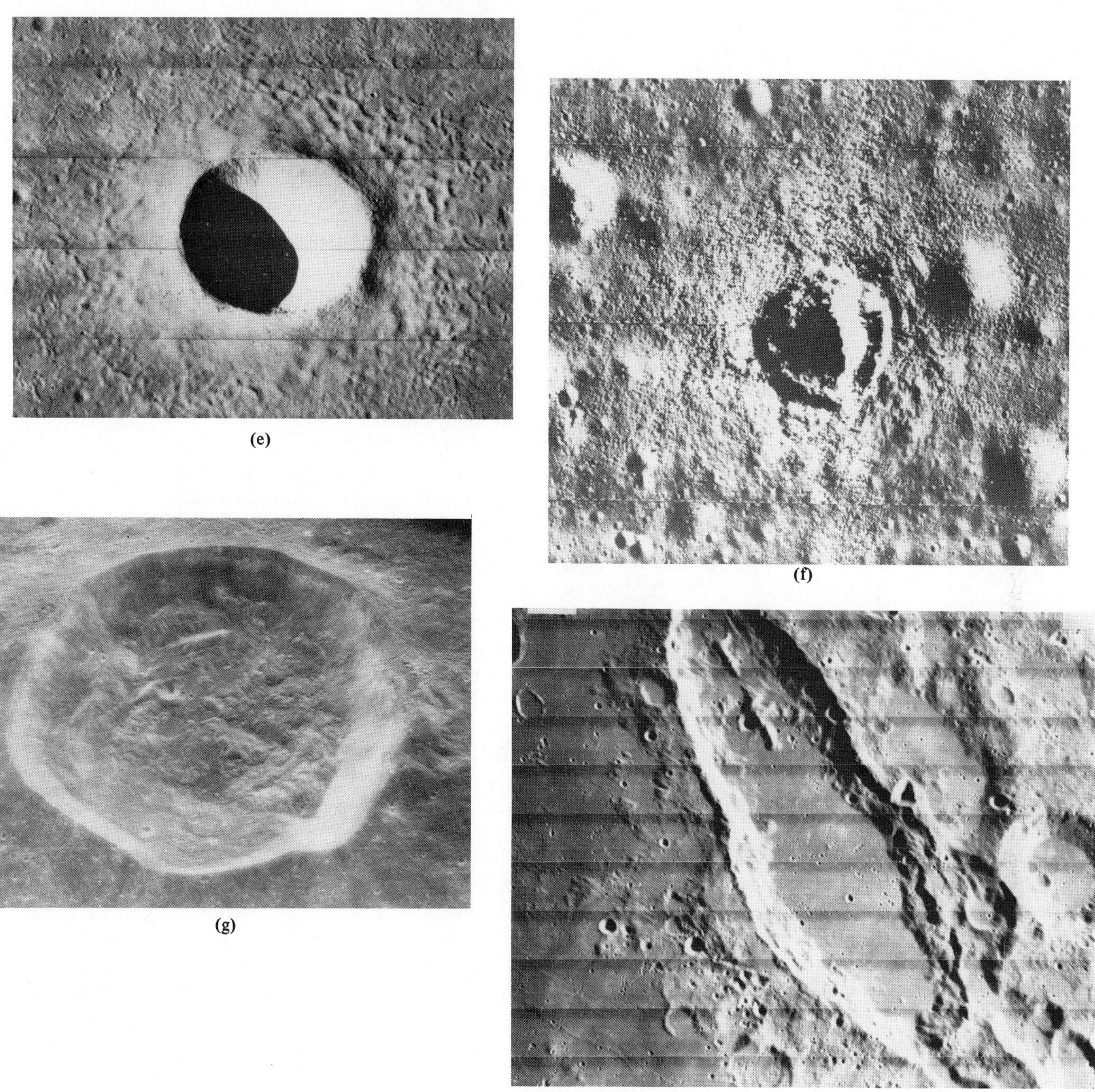

(e)

(f)

(g)

(h)

Figure 7-17 (continued) (e) Mosting C (4 km wide), a fresh, young crater in the edge of the highlands northwest of Ptolemaeus. The "recent" origin of this crater is suggested by the blocks along the rim, the presence of distinctive bifurcating ("bird's foot" impressions) gouges caused by secondary impacts of low-angle ejecta, and the braided, elongate dune-like ridges (probably produced as base surge deposits) more or less concentric to the rim. All of these features tend to disappear as a crater is eroded and/or buried by younger deposits. (f) A small (~ 200 meters), fresh crater in Oceanus Procellarum with a sharp outer rim and a nearly complete inner rim (possibly resulting from slumping or incipient terracing). The surface beyond this crater is covered with blocks of rock derived as ejecta from an assumed underlying lava flow (*Orbiter 3*-H164). (g) A moderately large, steep-walled crater lying along the equator (at 135°E) on the lunar farside north of Tsiolkovsky; the crater interior is characterized by partial terraces and bulbous to hummocky hills which may represent remnants of the original fallback surface, inasmuch as dark, mare-like lavas appear to be absent (*Apollo 10,* Frame 4198–Magazine P). (h) The crater Schiller, southwest of Tycho, an extremely elongate (~ 100 by 180 km) structure of undetermined origin (suggestions: volcanic caldera; tectonic graben; very low-angle impact; simultaneous impact of two or more bodies producing a compound crater) (*Orbiter 4*-154H_3). (NASA photos.)

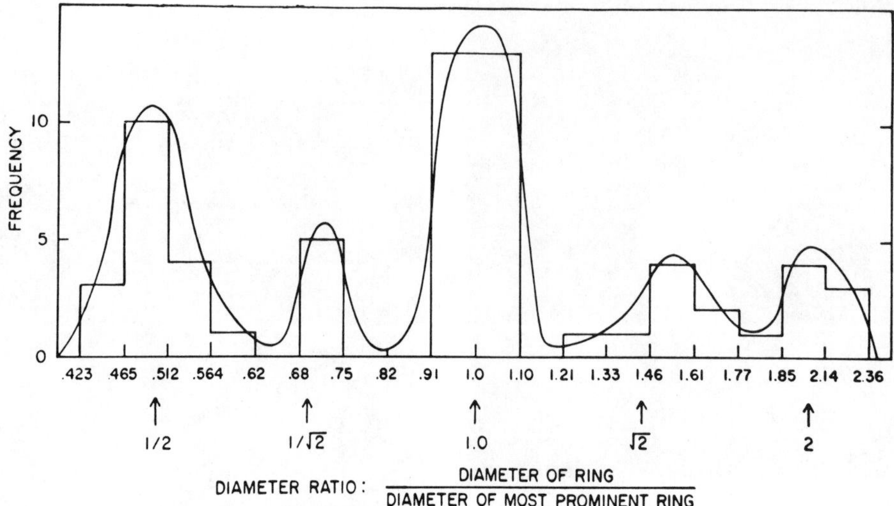

Figure 7-18 A plot of ring diameter ratios for one or more rings associated with 31 lunar basins. The most prominent ring is not necessarily the innermost ring. Histogram bars are used to depict the frequency distribution. These bars are then connected by smooth curves. The curves peak at intervals that increase by $\sqrt{2}$. (Reprinted from W. K. Hartmann and C. A. Wood, "Moon: Origin and Evolution of Multi-Ring Basins," *The Moon*, Vol. 3, No. 1, pp. 3-78. Copyright 1971 by D. Reidel Publishing Company.)

acter of these central peaks changes progressively from single promontories through multiple groups to rings in basin-sized depressions (Figure 7-19). Debate continues as to whether these central peaks originate through impact, rebound, volcanism, landslide, or isostasy. Other features related to the movements of material ejected during crater development, to secondary impacts, and to later adjustments by mass-wasting and gravity movements are depicted in Figure 7-20. Figure 7-21 shows the characteristic distribution of secondary craters around a large primary. These secondaries sometimes line up in chains or loops that tend to point toward the central primary crater.

Shoemaker (1962) has analyzed the development of secondary ejecta craters around Copernicus (Figure 7-21) in terms of cratering theory and ejection ballistics. He applies several equations which relate such variables as particle velocity at ejection, velocity of the incoming meteorite or comet, relative densities of cratered materials and of the meteorite, effective depth of origin of the shock waves, initial (pre-ejection) distance of a particle from the impact point, angle of ejection, and range or distance out that a particle travels before returning to the surface. Solution of these equations for Copernicus lead to some important relationships which apply to cratering in general. The eventual relocation of ejected fragments depends closely on the angle of ejection, as shown in Figure 7-22a. As derived for the impact conditions which formed Copernicus, the initial velocity at ejection is a

SEQUENCE OF BASIN INTERIOR PROFILES

PROFILE	EXAMPLE	DIAMETER
	CENSORINUS	4 Km
	LANSBERG	40
	BULLIALDUS	59
	GASSENDI	100
	COMPTON	175
	SCHRÖDINGER	300

Figure 7-19 Schematic size sequence of crater forms, from no central peak (top) through more complex and separated peaks, to arcuate peak or ridge-rings (bottom). (Reprinted from W. K. Hartmann and C. A. Wood, "Moon: Origin and Evolution of Multi-Ring Basins," *The Moon*, Vol. 3, No. 1, pp. 3-78. Copyright 1971 by D. Reidel Publishing Company.)

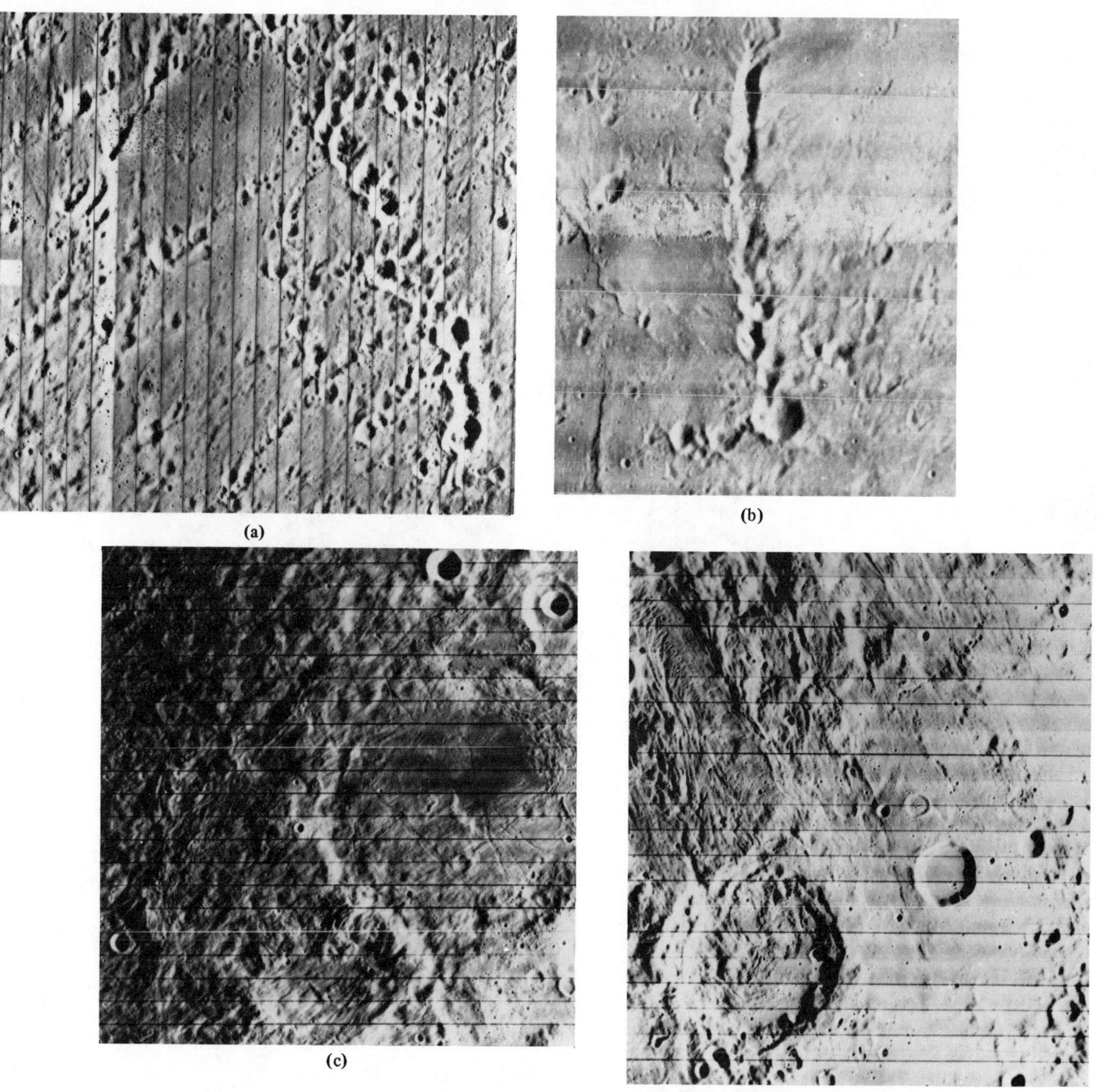

Figure 7-20 (a) A swarm of secondary craters northeast of Copernicus (the Stadius Rille complex) produced by the impact of ejecta tossed from that large crater during its formation. Note the tendency of these small craters to cluster, form chains, dig out gouges whose long axes are oriented toward Copernicus (located beyond the lower left corner), and develop straight to curved surface flow marks which point roughly in the direction of ejecta motion (*Orbiter 5*-142M). (b) Part of Rima Stadius IV, a trough consisting of connected craters that form part of the satellitic crater field lying just north of the area shown in (a). Some of the small gouges associated with V-shaped surface markings (producing a distinctive "herringbone" pattern) to the right of Stadius IV are also secondary craters. The width of the framelet is 42 km (*Orbiter 5*-121H). (c) A portion of the extensive ejecta blanket from the Orientale Basin around the crater Riccioli (center right; diameter–150 km) showing the hummocky nature of the blanket deposits. The surface is sculptured by a braided network of very long, subparallel grooves and ridges that are aligned in the general direction of movement from the southwest (lower left) of a base surge (high velocity gas and particle cloud that moves close to the surface) transporting debris expelled from the basin. The broader channels are oriented at high angles to these groove marks that are straight rilles (*Orbiter 4*-173H). (d) The approximate outer limit of the base surge deposits from the center of Mare Orientale, 1100 km to the northwest (upper left). These deposits appear to pass into the crater Inghirami (90 km diameter; lower left) and then thin out as braided and concentric ridges without leaving the crater. Similar ridges and grooves or depressions mantle the region north of this crater. Some of these flow features swirl and deflect from near radial trends to positions nearly normal to the principal direction of motion (*Orbiter 4*-172H_2).

(e)

(f)

Figure 7-20 (continued) (e) Track prints produced by rolling boulders (visible in the lower edge of the photo) on the surface of the inward sloping wall of the crater Vitello. The larger irregular track on the left is almost 25 meters wide (*Orbiter* 5-168H). (f) Close-spaced ridges and grooves (typically, 20 to 50 meters wide) located just outside the northern rim of Tycho. These markings are oriented concentric to (parallel with) the rim line. This feature, which makes up one type of patterned ground (see Figure 8-12d), is of uncertain origin. It may be related to an underlying fracture system, flow deposition from a base surge, or downslope mass movements. Note the large boulders (some larger than 50 meters across) that collected in a topographic low (*Orbiter* 5-127H). (NASA photo.)

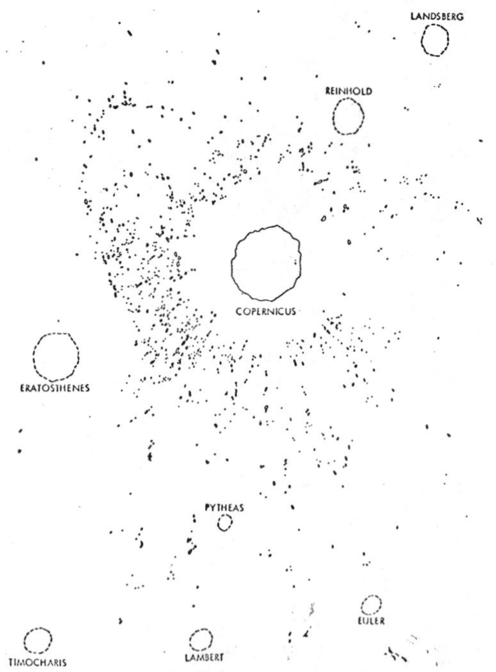

Figure 7-21 The distribution of unnamed secondary craters produced by impact of ejecta from Copernicus on the lunar surface. Most of the secondaries lie in a circumferential belt extending out to two to four times the radius of Copernicus, but a few can be traced outward up to eight to ten times this radius. (From E. M. Shoemaker, in Z. Kopal, 1962.)

function of this angle (Figure 7-22b). The distance to which any given particle located *at the surface* will travel along an arching trajectory depends on its original position in the volume of rock that is included within the eventual excavation (Figure 7-22c).

During the excavation phase of an impact crater as it forms on the Moon, particles are put into motion as ejecta at different times over different distances from the point of impact and move out of the crater at different velocities and angles of ejection to different final deposition sites in the resulting ejecta blanket. Furthermore, each particle comes from some specific position relative to the advancing shock wave, so that a wide variety of particles displaying different degrees of shock damage will result. The problem of determining the source point, ejection path, flight history, and intermediate and final resting places of a large number of particles in motion is complex; the calculations require a computer to keep track of the many interrelated factors and to insert changing values into the equations that solve this problem.

The final shape and the presence or absence of terraces and interior peaks are influenced by the nature and degree of consolidation of the materials making up the maria and terrae. Quaide and Oberbeck (1968) have defined three common interior morphologies found in small craters on the maria which seemingly depend on the ratio of the thickness of an assumed overlying layer of unconsolidated

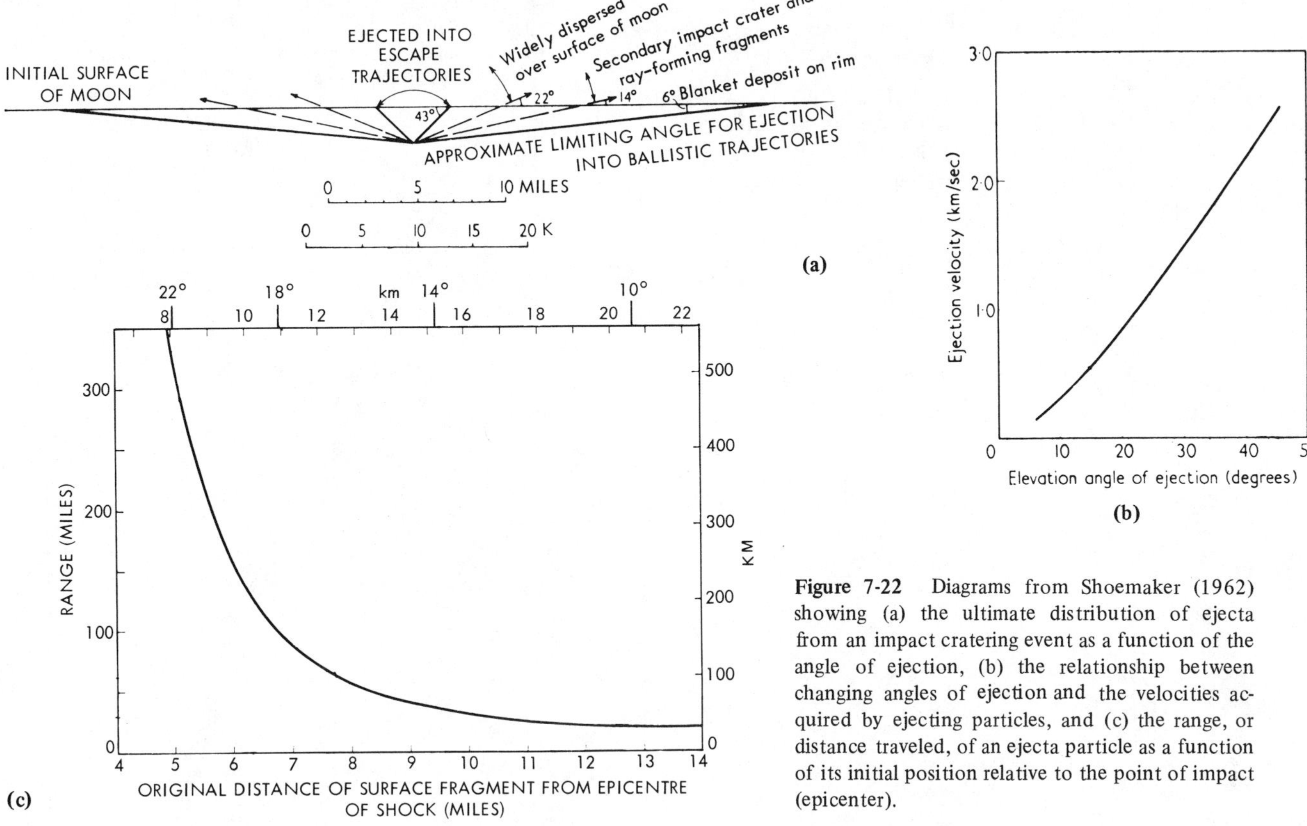

Figure 7-22 Diagrams from Shoemaker (1962) showing (a) the ultimate distribution of ejecta from an impact cratering event as a function of the angle of ejection, (b) the relationship between changing angles of ejection and the velocities acquired by ejecting particles, and (c) the range, or distance traveled, of an ejecta particle as a function of its initial position relative to the point of impact (epicenter).

rubble (regolith) resting on hard rock (lava?) to the size or diameter of a given crater (Figure 7-23). Thus, by measuring the ranges in widths of each crater type in some part of a mare surface, Quaide and Oberbeck were able to set limits on the thickness of the regolith in the area enclosing these craters. In this way, they have noted that the regolith at several proposed *Apollo* landing sites varies in thickness generally between 1 and 20 meters.

Crater Frequencies

Let us accept the hypothesis that most of the millions of circular structures on the Moon have originated by impact. By selecting some convenient region on the front face (such as any part of the surface with an area of 10^6 km^2), with time and patience one can count all visible craters and group these in size classes, i.e., sets of all craters whose diameters fall within specified limits as, for example, 2 to 4 km, or 16 to 32 km. If these classes are then arranged in order of decreasing size in a table that also lists the number of individuals in a class, it is evident that there are relatively few larger craters and an increasing abundance of smaller and smaller craters. This relationship is best expressed by graphs of the tabulated observations. In an *incremental graph* (Figure 7-24a) the number of craters of different discrete sizes or within specified size classes, as counted over a unit area (typically, 1 square kilometer), make up a series of frequency values which are then plotted for each corresponding diameter or interval. Alternatively, the crater count in each smaller

Figure 7-23 The relationship of crater geometry (best observed vertically from overhead) to thickness of regolith in the two-layer model of Quaide and Oberbeck (1967). This model has been applied to small craters developed in the rubble debris (derived mainly from repeated bombardments by meteorites) overlaying the hard rock lava units on the maria. R refers to the ratio of crater diameter to regolith thickness (R = d/t). In practice, this thickness is determined from the largest craters of a given geometry present in the area considered. Thus, if such an area contains mainly normal craters together with a few flat-bottomed craters (largest, 100 meters) and one concentric crater (diameter, 200 meters), then the thickness (t) is estimated to be: t = d/R = 100/11 = 9 meters and t = 200/18 = 11 meters for the flat-bottomed and concentric cases respectively. (W. L. Quaide and V. R. Oberbeck, *Journal of Geophysical Research,* vol. 72, p. 4700, 1967.)

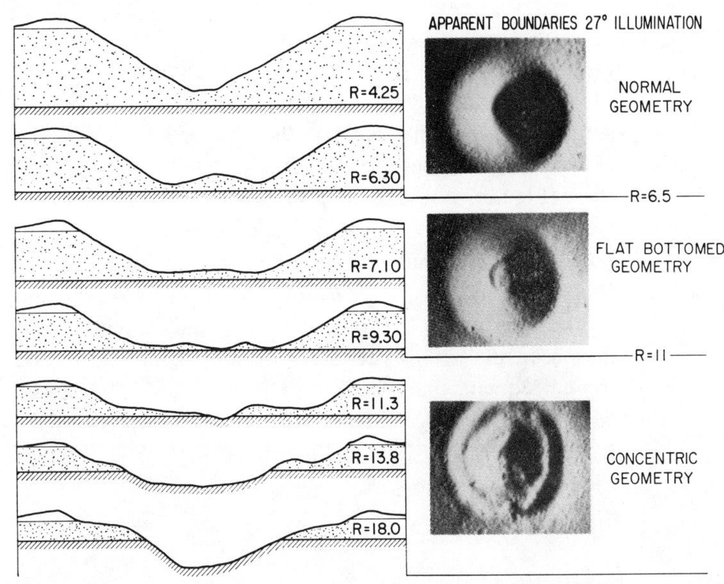

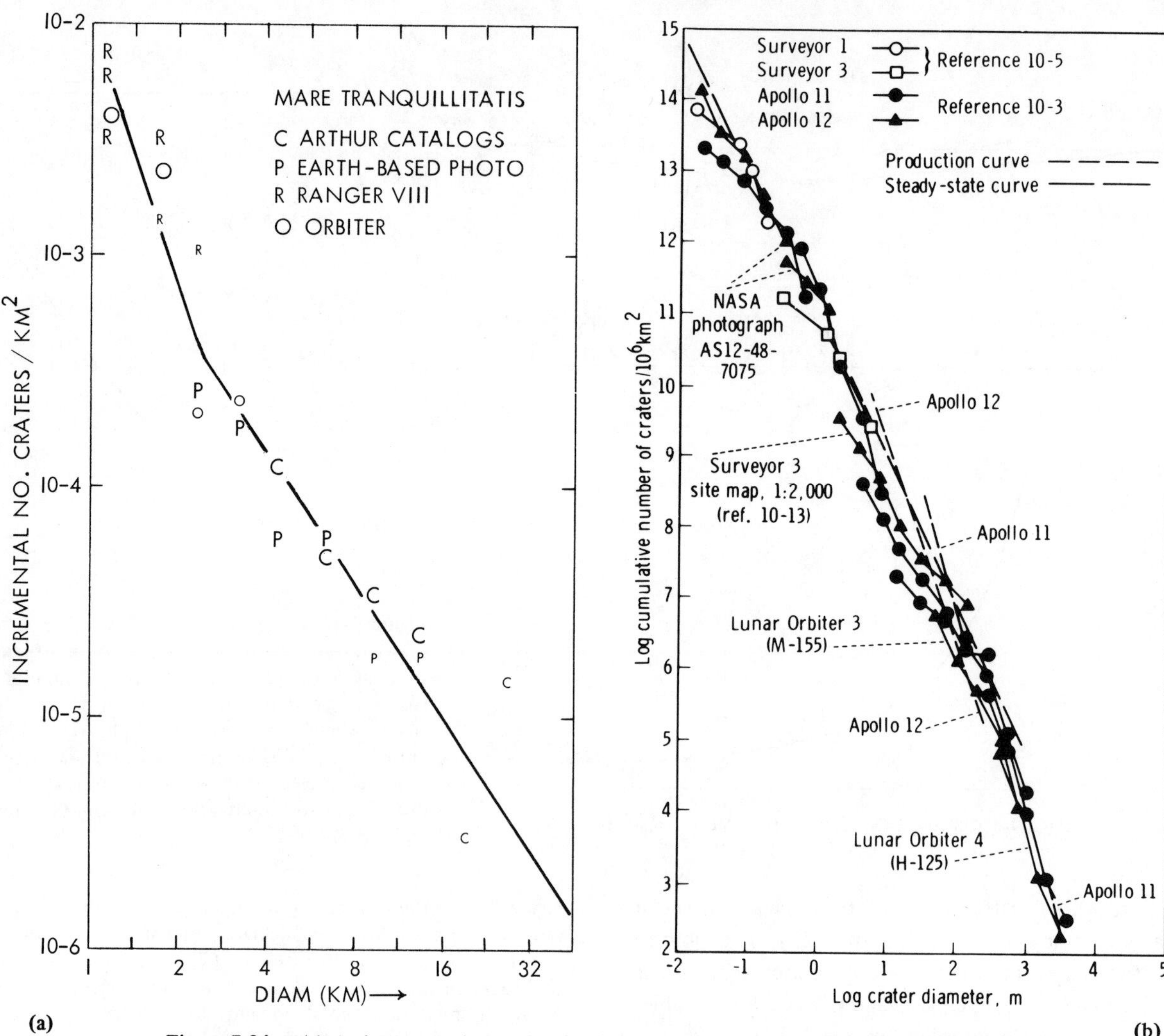

Figure 7-24 (a) An incremental curve in which the frequency or number of craters per square kilometer in region of Mare Tranquillitatis is plotted as a function of corresponding sizes. The ordinate points increase by powers of 10, whereas the abscissa is based on log $\sqrt{2}$ increments (courtesy W. K. Hartmann). (b) A cumulative size frequency distribution of craters in the vicinity of the *Apollo 12* site, using *Orbiter, Surveyor,* and *Apollo* photographs to measure numbers and diameters of craters. As examples of the use of this curve, consider point (1) and (2) (arrows). At (1) the curve indicates that there would be an average of 10,000 (log value 4 or 10^4) craters of size 1000 meters (log value 3 or 10^3) and larger to be found in a region of 10^6 km^2, which contains a crater distribution like that observed at the *Apollo 12* site. Likewise, at (2) there would be a total of 10^{13} craters of size 0.1 meters (log value −1 or 10^{-1}) and larger within this region. (Illustration from NASA Document SP-235.)

class can be added progressively to an accumulating total of all preceding (greater-diameter) classes. The running sum (in logarithmic units) is then plotted for each successive crater-diameter class. This results in a *cumulative curve* (Figure 7-24b) that represents another way to express frequency of occurrence as a function of size. When such a curve is compared with that constructed for meteorite fluxes (Figure 3-7), a clear correlation emerges: because there are presumably more small meteoroids moving about in space, small craters resulting from collisions should be more frequent than large ones produced from impacts with the fewer larger bodies. This relationship is evident from the negative slope inclination (extending upward to the left rather than to the right in Figure 7-24). The frequency decreases numerically along the ordinate as object size (crater diameter) increases along the abscissa.

Evolution of Cratered Surfaces

Now, assuming that the flux particles (from comets and asteroids to micro-meteorites) have an equal opportunity to strike anywhere on the surface of a planet, it seems evident that an older surface will have more craters impressed on it regardless of whether the flux has remained constant or varies with time. This effect on older surfaces is generally recognized by comparison of the cumulative curve plots of cratering in several areas in which the initial surfaces were formed at different times. As the number of craters in each class increases with age, a curve will be shifted toward the right (Figure 7-25), i.e., for any given diameter, the total numbers will be greater.

The Moon's surface, because it is little changed by disturbing processes (such as atmospheric erosion or tectonic deformation), should prove to be an almost *ideal impact counter.* Now that different surfaces have been age-dated by analysis of returned lunar samples, scientists are beginning to deduce the history of meteorite fluxes in the Earth-Moon capture region perhaps back in time to the earliest days of planetary formation.

There are, however, several complications. First, the meteorite flux has almost certainly not remained constant over time throughout the history of the solar system. In the first billion years or so—and especially during the brief periods when planets were forming by accretion—the number of large objects available to impact on growing planetary surfaces decreased rapidly. This change in flux density, therefore, increases exponentially back into time. Another difficulty arises from the likelihood that similar-appearing craters of diverse origins (impact, volcanic, collapse) can coexist in the same area. The different genetic types must be separated before information on impact and flux histories can be extracted. In particular, the many secondary craters around a large young crater must be subtracted from any count attributed to direct impacts.

Next, consider the changes appearing in Figure 7-26, which records the time-sequential effects of crater development in a modeling experiment that allows a succession of gun-fired projectiles of different sizes to hit a target of granulated carborundum. At first, an initially uncratered plain can accommodate newly formed craters of all sizes. But with continuous impacting, craters begin to interfere with one another and younger ones may partially obliterate those formed earlier. A stage is reached where little or no surface remains unaffected by cratering so that new impact structures invariably interact with previous ones. In effect, the terrain has become *saturated* with craters below a certain limiting size. This size tends to increase over time, and ever larger craters will both completely destroy smaller ones over which they superimpose and modify those of similar and larger sizes. Note that in the last panels (lower right), the more recent craters stand out as fresh and sharp, whereas many of the craters appearing in the first panels have all but disappeared. The surface is said to be in *equilibrium* when continuous cratering has progressed to the point that as many older craters are being wiped out as new craters are introduced, irrespective of size.

The erosional or destructive aspects of impacts are more or less balanced by the aggradational process of ejecta deposition. Given time, both processes tend to smooth out irregularities and subdue the youthful appearance of the initial cratered surfaces. On the Moon, estimates (Short and Forman, 1972) of the average thickness of impact-ejecta deposits range between 1.5 and 2.5 km—the upper value being more than twice that of unmeta-

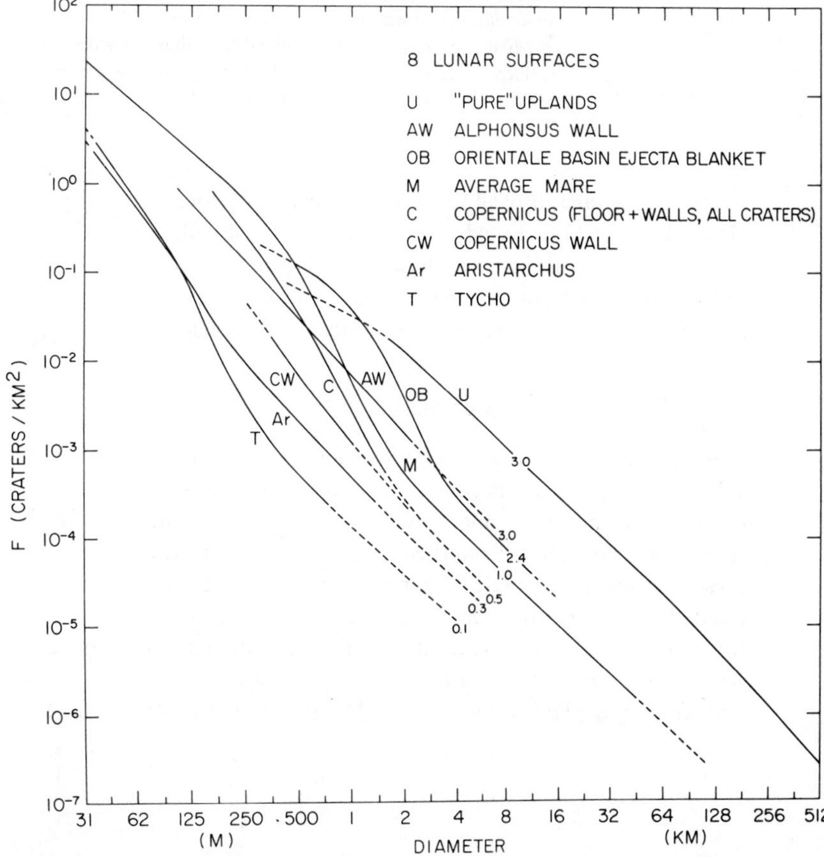

Figure 7-25 An incremental frequency plot of crater distributions in 8 regions of the Moon. Most of the curves are nearly parallel to one another, indicating the same relative proportions of craters at all size ranges. The "pure" uplands surface is probably the oldest, whereas those inside of Aristarchus and Tycho are youngest. This means that for any given size there will be more craters per unit area in the uplands than in the maria and, in turn, more mare craters than within Tycho. If there were relatively more smaller than larger craters at some site than is characteristic of the distributions shown here, that fact would be indicated in an incremental plot by a steeper slope. (W. K. Hartmann, 1970.)

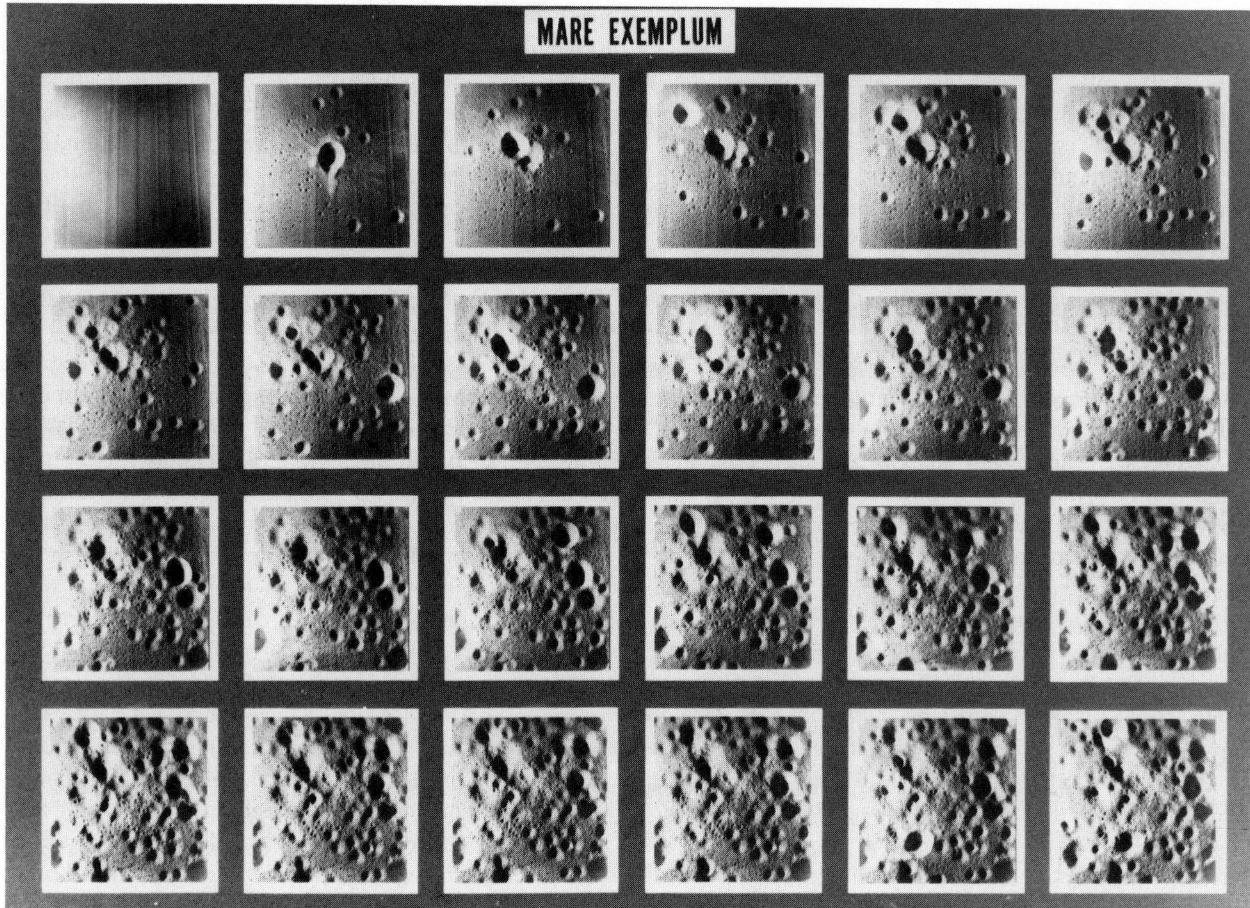

Figure 7-26 "Mare Exemplum," an artificial lunar surface made by firing small projectiles of different sizes from a gas-gun at a sand box filled with dark granular material. The sequence of time-lapse photographs reproduced here illustrates development of an immature cratered surface (relatively few large craters) first followed by eventual saturation with progressively larger craters. (From experiments by D. E. Gault, H. J. Moore et al.; photo courtesy U.S. Geological Survey.)

morphosed sediments of all kinds on the Earth's continents. Most of this rubble accumulated during the first billion years of the Moon's history. In addition, both lava flows and volcanic ash accumulations may have further modified terrain roughness and covered once heavily cratered regions.

Summary

Thus, in sum, there is a growing conviction that impact, along with volcanism, is a major and probably universal process that influences and modifies the surfaces of the terrestrial-type planets from their earliest days even to the present. However, the rate of impact seems to have diminished significantly throughout the solar system after the first billion years or so as most asteroidal and meteorite-sized bodies were swept up by gravitational interaction with the larger planets. Only on the Moon and, to a lesser degree, on Mars (and perhaps Mercury) has the initial record of the heaviest period of impacting been preserved as a memento of the juvenile stage in a planet's history.

REVIEW QUESTIONS

1. Why are so few meteorite craters found today on the Earth's surface?
2. What is a shock wave? What levels of pressure are reached at its peak?
3. What is the justification for considering impact cratering to be a major process in the development and history of the terrestrial planets?
4. Describe briefly the mechanism and sequence of development of a meteorite crater.
5. Why are meteorites only rarely found at impact craters?

6. What is the distinction between a meteorite crater and an astrobleme?

7. List at least two processes involved in producing ejecta blankets. How is an asymmetric blanket deposit formed?

8. How are craters modified over time?

9. Review the principal criteria by which meteorite craters can be recognized. What is the single most definitive proof of an impact origin for such structures?

10. What are the chief distinctions between craters and basins on the Moon?

11. How can crater morphology be used to estimate the thickness of the regolith in the cratered area?

12. What types of surface features are produced beyond the rim as a crater is formed?

13. What general relationship is evident from inspection of cumulative curves plotted for lunar crater distributions?

14. Discuss the conditions under which an impacted surface becomes saturated or reaches equilibrium.

15. What is the estimated average thickness of ejecta deposits on the lunar surface? Will thicker deposits tend to occur in the highlands or under the maria?

CHAPTER 8

Lunar Igneous Processes

Volcanic eruptions and lava outpourings are commonplace events at the Earth's surface. However, most active volcanism is selectively distributed over the globe in belts closely associated with earthquakes and mountain building. Thus, chains of volcanoes and recent lava flows are largely restricted to the circum-Pacific "Ring of Fire," the Mid-Atlantic Ridge, and tectonic zones in the Mediterranean and East Africa. At present the layers of basalt (and/or gabbro) that form the lower crust in places are being continuously added to or destroyed by upwellings and subductions of volcanic materials along the "conveyor belt" system that drives the drifting continental and oceanic plates (see Dewey, 1972). The oceanic crust has been generated almost entirely by outpourings of mafic lavas. The continents, on the whole, consist of ancient granitic and metamorphic rocks veneered with younger sedimentary rocks that later become metamorphosed and fused into more crystalline rocks. Continental volcanic rocks are interspersed with the sedimentary units mainly at the sites of ancient geosynclinal deformation. At the land surface, active volcanism is thus only a subordinate, although important, process in developing the present continental segments of the crust. However, at depth magmatic processes provide the major new ("juvenile") materials added to the substructure of the Earth's continents.

The Moon's Volcanism

The Moon also displays an igneous character and history that is fundamental in assessing the nature and development of the entire surface visible to us today. If we accept, with certain reservations, the view now held by many scientists that the Moon perhaps underwent complete interior melting—or, at least, general melting of its outer layers—soon after its formation, then the original lunar crust solidified from this melt must have shown many manifestations of volcanism. (Keep in mind that volcanism by definition refers to the surface expression of interior magmatism.) Volcanic activity during the cooling stage most likely occurred concurrently with extensive impact cratering that constantly modified and masked the flows and volcanic structures that were being formed in the outer crust. The maria, which postdate most of the highlands, are now known to consist of volcanic rocks whose ages are at least 1 billion years younger than the present estimates of the Moon's birth date. These lava bodies, whose cumulative thickness is not known, are much less cratered than the terrae. Thus, despite overall changes wrought by impact cratering, the maria still retain prolific signs of later, episodic volcanism. In contrast, the terrae, wherever not thoroughly degraded by postvolcanic impact grinding and deposition, show only a few undisturbed vestiges of the early volcanism that accompanied crustal building. The *important* point is that igneous processes have *produced* the materials that make up the Moon's outer layers, whereas impacts have mainly *reshaped* and modified these layers near its surface.

Diversity of Crater Origins

The task, then, is to define the characteristics of these igneous processes and to set forth criteria for recognizing those volcanic features that remain on the lunar surface. At first glance, removal of all visible craters from consideration would seem to simplify this effort if these are exclusively impact in origin. However, two complications prevent such an approach.

One rests on the claim by some geoscientists that the Moon's craters, like most of the Earth's craters, were caused by volcanism and that many associated characteristics outlined in the previous chapter as uniquely diagnostic of impact can also be produced by volcanic processes. Indeed, a few selenologists continue to maintain with conviction and often with tantalizing "proofs" (Figure 8-1) that an undetermined number (conceivably, the majority) of the Moon's craters are primary volcanic structures.* The problem of distinguishing between similar-looking impact and volcanic craters becomes especially complex in terrains in which large numbers of both types are mixed together in seemingly random fashion. Sometimes cumulative frequency curves will reveal the presence of volcanic craters through slope steepening that results from increases in populations of certain sizes. (Note, however, that secondary craters from either impact or volcanic processes will also steepen the curves.)

The other complication results from the possibility that impact itself can trigger some kind of volcanic activity. This may take the form of localization within craters of igneouslike products produced by direct (shock-induced) melting in response to the kinetic energy added from impacts. Alternatively, removal of some crustal material as ejecta from large impact craters and basins might reduce overburden pressures enough to allow very hot but still solid rock at depth to respond by melting and moving upward into the fractured zone leading to the depressions. Volcaniclike features on the floors of some craters younger than the maria may typify this mode of emplacement.

*The AGI/EBEC film ***Controversy Over the Moon*** might be viewed with profit in this regard. Order from Encyclopaedia Britannica Educational Corporation, Preview/Rental Library, 1822 Pickwick Avenue, Glenview, Ill. 60025.

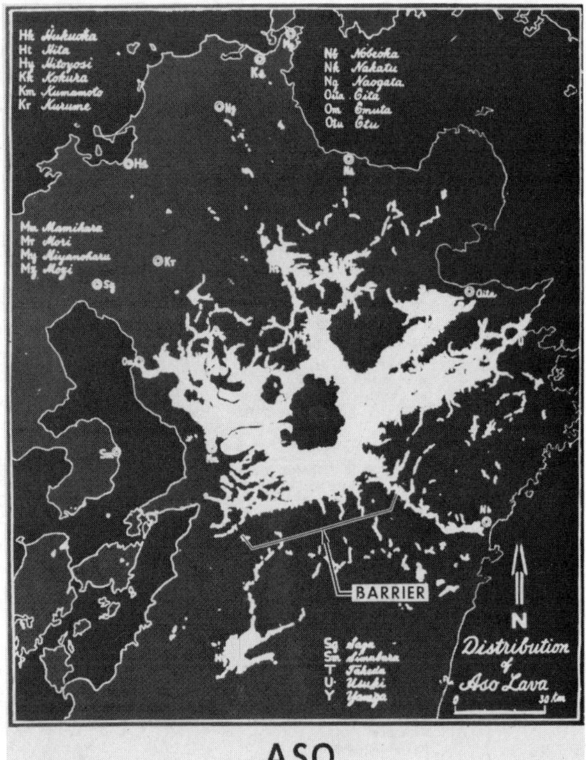

Figure 8-1 Top: The distribution of a tuff ignimbrite (an ash-flow deposited from a "nuée ardente") around the Aso caldera in Japan. The ray pattern and barrier line are controlled by low hills in the region. Bottom: Rays and possible barrier line around the young crater Kepler on the Moon. Most workers consider these deposits to be impact-derived ejecta. Green (1965) points to a strong similarity of distribution patterns between this crater and Aso caldera and concludes that most such craters on the Moon are volcanic structures surrounded by ash deposits. (Courtesy J. Green, California State University at Long Beach; from Green, 1965.)

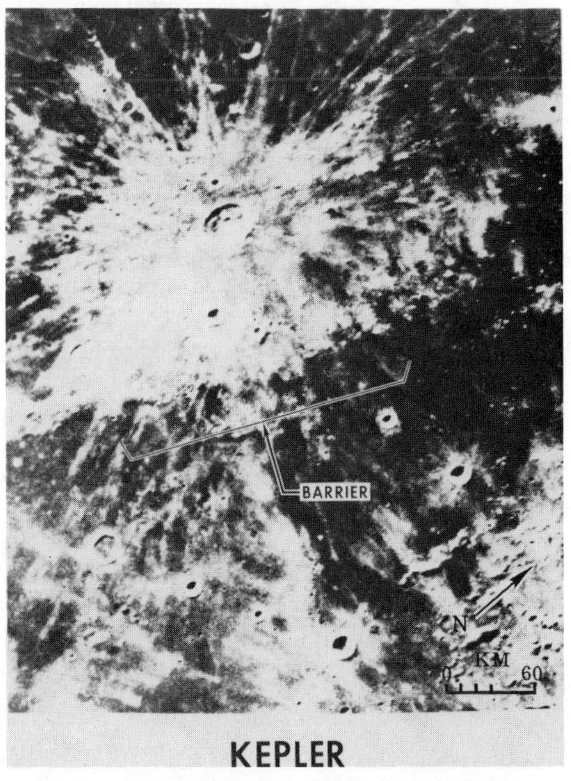

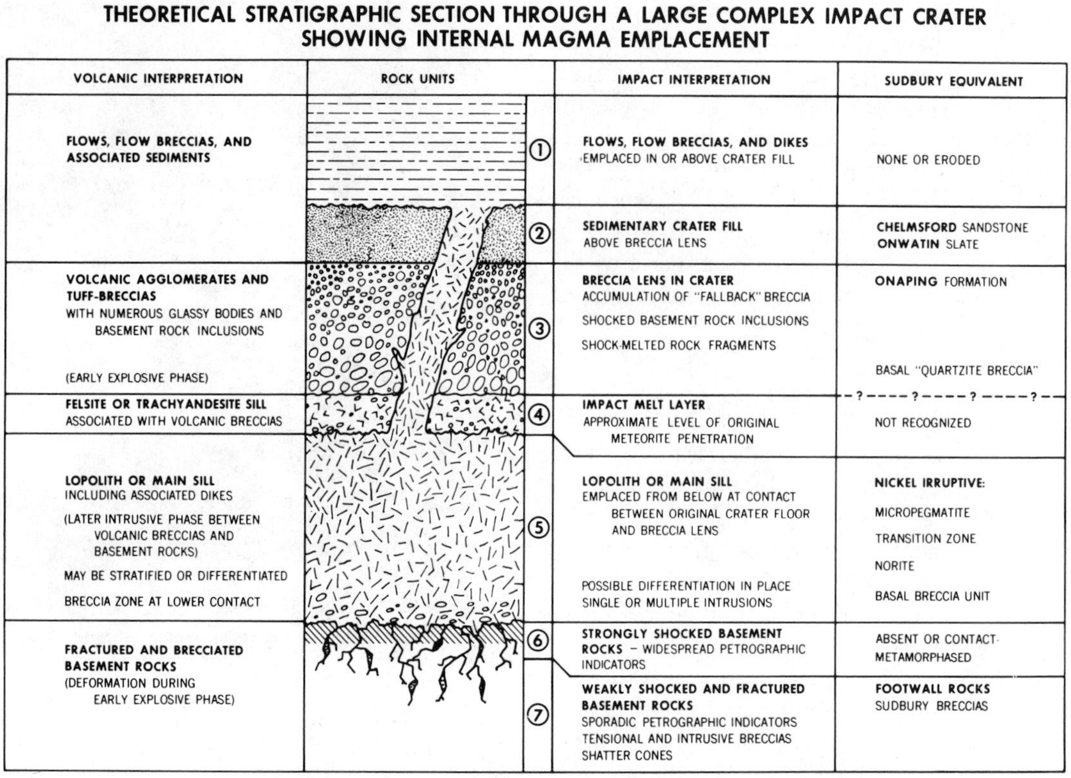

THEORETICAL STRATIGRAPHIC SECTION THROUGH A LARGE COMPLEX IMPACT CRATER SHOWING INTERNAL MAGMA EMPLACEMENT

VOLCANIC INTERPRETATION	ROCK UNITS	IMPACT INTERPRETATION	SUDBURY EQUIVALENT
FLOWS, FLOW BRECCIAS, AND ASSOCIATED SEDIMENTS	①	**FLOWS, FLOW BRECCIAS, AND DIKES** EMPLACED IN OR ABOVE CRATER FILL	NONE OR ERODED
	②	**SEDIMENTARY CRATER FILL** ABOVE BRECCIA LENS	**CHELMSFORD** SANDSTONE **ONWATIN** SLATE
VOLCANIC AGGLOMERATES AND TUFF-BRECCIAS WITH NUMEROUS GLASSY BODIES AND BASEMENT ROCK INCLUSIONS (EARLY EXPLOSIVE PHASE)	③	**BRECCIA LENS IN CRATER** ACCUMULATION OF "FALLBACK" BRECCIA SHOCKED BASEMENT ROCK INCLUSIONS SHOCK-MELTED ROCK FRAGMENTS	**ONAPING** FORMATION BASAL "QUARTZITE BRECCIA"
FELSITE OR TRACHYANDESITE SILL ASSOCIATED WITH VOLCANIC BRECCIAS	④	**IMPACT MELT LAYER** APPROXIMATE LEVEL OF ORIGINAL METEORITE PENETRATION	NOT RECOGNIZED
LOPOLITH OR MAIN SILL INCLUDING ASSOCIATED DIKES (LATER INTRUSIVE PHASE BETWEEN VOLCANIC BRECCIAS AND BASEMENT ROCKS) MAY BE STRATIFIED OR DIFFERENTIATED BRECCIA ZONE AT LOWER CONTACT	⑤	**LOPOLITH OR MAIN SILL** EMPLACED FROM BELOW AT CONTACT BETWEEN ORIGINAL CRATER FLOOR AND BRECCIA LENS POSSIBLE DIFFERENTIATION IN PLACE SINGLE OR MULTIPLE INTRUSIONS	**NICKEL IRRUPTIVE:** MICROPEGMATITE TRANSITION ZONE NORITE BASAL BRECCIA UNIT
FRACTURED AND BRECCIATED BASEMENT ROCKS (DEFORMATION DURING EARLY EXPLOSIVE PHASE)	⑥	**STRONGLY SHOCKED BASEMENT ROCKS** – WIDESPREAD PETROGRAPHIC INDICATORS	ABSENT OR CONTACT-METAMORPHASED
	⑦	**WEAKLY SHOCKED AND FRACTURED BASEMENT ROCKS** SPORADIC PETROGRAPHIC INDICATORS TENSIONAL AND INTRUSIVE BRECCIAS SHATTER CONES	**FOOTWALL ROCKS** SUDBURY BRECCIAS

Figure 8-2 Idealized stratigraphic section through a structure produced by a large meteorite impact followed by emplacement of internally-derived magma. Extrusive rocks have been deposited above the fallback breccia lens by penetration of dikes of magma through the lens. The series of rocks is complex and can be explained either as a conventional volcanic structure (left column) or as impact crater (right column). Only the breccia lens (unit 3) and the shocked basement rocks (units 6 and 7) would show distinctive shock effects if the structure had an impact history. Equivalent features in the Sudbury structure, in southern Ontario, Canada, are indicated on the far right. This structure, once thought to be a great lopolith of volcanic origin, now has been reclassified as an impact structure because of the presence of shocked rocks, shatter cones, and other signs of its having once been a meteorite crater comparable to many on the Moon. Following the impact, basic magmas from depth invaded the structure and deposited vast amounts of nickel and copper as a by-product. (From B. M. French, 1970.)

Figure 8-2 illustrates with a well-known terrestrial example how rock units of volcanic character found in a bowl-shaped depression have been interpreted either as the products of conventional volcanism or as evidence of internally derived magma emplaced within a large impact structure.

We need, therefore, to set about cataloging those definite and probable lunar features for which a direct igneous origin can be demonstrated. Only then can the nature, extent, and implications of volcanism on the Moon be evaluated as a contributing factor in lunar evolution.

Terrestrial Volcanism

Before looking at examples of lunar volcanism, some common terrestrial features will be considered first (Figure 8-3—read the captions for a synopsis of relevant characteristics).

These generalizations about the principal types of volcanic forms* on Earth that may have lunar counterparts are pertinent:

1. Most volcanoes are cone-shaped (stratovolcanoes) to mound-shaped (shield volcanoes) in profile.
2. These structures erupt from a central vent or from line(s) of fissures.
3. Nearly all stratovolcanoes possess at least one craterlike depression, typically no more than a kilometer wide, at or near the summit or center of the structure.
4. Larger craters (from 1 to 4 km, relative to the base area of the entire structure) grade into *calderas* (collapse or explosion-produced depressions that usually

*Most of the volcanological terms used here and elsewhere in this chapter are defined in the Glossary of the Volcanic Atlas by Green and Short (1971) (see References).

Figure 8-3 (a) The stratovolcano Farallol de Pajaros in the Marianas Islands of the Pacific, a steep-sloped cone volcano developed from magmas more silicic than those of basaltic nature. Conical structures similar to this are rare on the Moon (U.S. Navy Photo). (b) The summit caldera (maximum rim width 7 km) on Isla Fernandina, Galapagos Islands, Ecuador, developed gradually by explosion and collapse processes. Note the "central peak-like" cone within the caldera and the numerous lava flows along the flanks (U.S. Air Force Photo). (c) A cinder cone formed along the Laki fissure in Iceland from ash and cinders accompanying basaltic lava outpourings (Photo: R. Decker). (d) A linear pyroclastic cone whose elongation results from gas-lava emissions along an extended fissure; this is one of many volcanic vents (mostly cinder cones) in the San Francisco volcanic field near Flagstaff, Arizona (Photo by J. McCauley). (e) A radar image of a part of the San Francisco volcanic field in northern Arizona. Note the cinder cones with circular to elongate vents, the domical hills, and the irregular lava flow surfaces (Photo by J. McCauley). (f) Crater Elegante, a near circular maar (formed explosively by steam as volcanic lavas and gas encounter shallow ground water) in the Pinacate Field of northern Mexico. Note the subdued rim (and apparent lack of distortion of the underlying layers), the "smooth" ash blanket beyond, and the several small tuff rings (foreground and right) (Photo by D. Roddy).

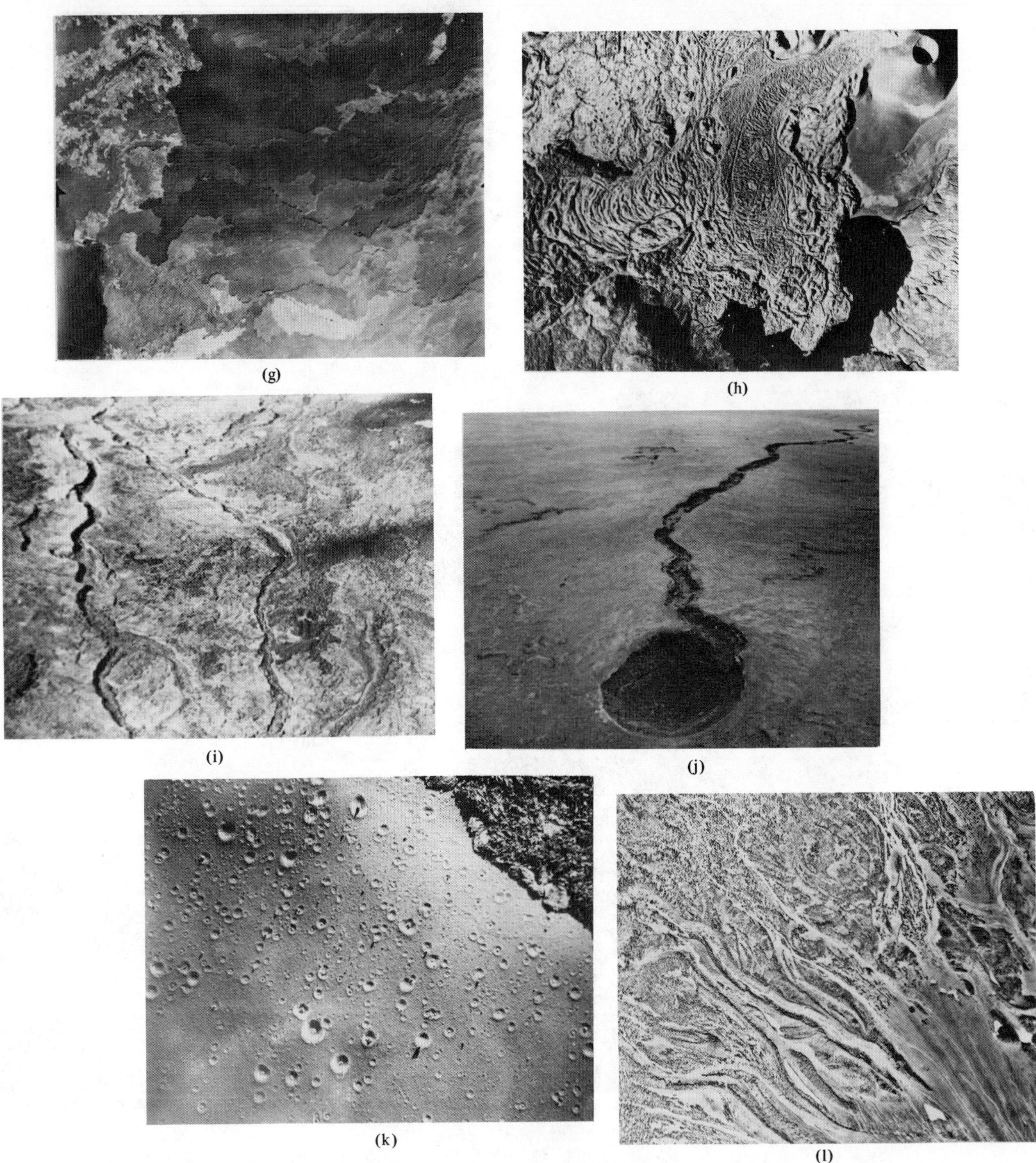

(g) (h) (i) (j) (k) (l)

Figure 8-3 (continued) (g) A series of overlapping basalt flows (darkest are most recent) on Isabella Island, Galapagos Islands, in the Pacific Ocean (U.S. Air Force Photo). (h) Recent rhyolite flows (note coarse crenulations or "corda" on their surfaces associated with pyroclastic cones) in southcentral Iceland (Photo by J. Green). (i) Lava channels (10 to 20 meters wide), some with pronounced sinuosity, within a lava flow field on Mount Hualalai, Island of Hawaii (U.S. Air Force Photo). (j) Butte Crater (diameter 100 meters), of maar-like character, and an associated collapsed lava tube that winds over 5 km to the south, Island Park Caldera region of Idaho (Photo by J.W. Dietrich). (k) A swarm of secondary craters in ash beds on the slope of Kapeho crater near Kilauea, Hawaii. These craters (some up to 3 meters in diameter) result from "impacts" caused by large blocks ejected from an erupting cinder cone (Photo: G. Kuiper). (l) Channels and levees on Mount Shasta, California, produced by mudflows of ash and volcanic rock debris mixed with rainwater that move down steep slopes during and after eruptions (U.S. Geological Survey Photo). (All photos in this figure are reproduced from *Volcanic Landforms and Surface Features: A Photographic Atlas and Glossary,* J. Green and N. Short, Ed., 1971: by permission of Springer-Verlag, Inc., New York.)

range from 4 to 20 km in diameter), but these generally still occupy the central parts of volcanic cones or low, broad shields.

5. The so-called *cryptovolcanic* (or cryptoexplosion) structures, containing breccias of country rock mixed with volcanic rocks, and many smaller circular features such as maars, lava sinks, or pit craters resemble some impact-formed structures in morphology; absence of shock features distinguishes these volcanic structures from those resulting from impacts.
6. Blanketed or layered deposits of volcanic debris that extend over wide areas are related to ash falls or ash flows (ignimbrites) derived usually from volcanoes with silicic lavas; other laterally extensive units include flood basalts (e.g., Columbia and Deccan Plateaus).
7. Many surface features developed outside the central vent are produced or affected by water vapor, running water, air mass movements (winds), and compaction of ash or collapse of drained lava tubes.

Thus, because of the presence of gases (including considerable water vapor derived from interaction of invading magma with sediments, the oceans, ground water, etc.) and the broad range of rock compositions in a lithosphere that is coupled with an oxygenating atmosphere, terrestrial volcanism results in a wide variety of characteristic features. The absence of a lunar atmosphere and hydrosphere (or, at least, of evidence supporting their one-time existence), together with an apparent scarcity of silicic lava rocks (e.g., rhyolites), suggests that most types of lunar volcanic landforms and associated phenomena should be less diversified than those on Earth.

Lunar Volcanic Features Associated with Lavas

Evidence for volcanism on the Moon is widespread. Opinion is divided, however, on the identification and interpretation of many specific features; on-site inspection and sampling in the *Apollo* program have helped to resolve some of these arguments. Most workers agree on the lunar features shown in Figure 8-4 as most probably volcanic in nature. The lobate flow front shown in Figure 8-4a is an obvious example. On a grander scale, the Rumker Hills (Figure 8-4b) in northern Oceanus Procellarum appear as a plateau built up of overlapping flows capped in places by low, gently sloping domes and protrusions. The Marius Hills (Figures 8-4c, d, and e), 1000 km to the south, are a collection of domes amidst other signs of volcanism. Many of the broad-based domes in this field resemble terrestrial shield volcanoes while others with steeper sides may represent extrusions of more viscous lavas. Flow channels (Figure 8-4f) along the outer slopes or inner walls of craters are indicators of fluid transport—either lava or gas clouds—resulting from volcanic activity.

Erratically-patterned fractures in floor-fillings of some craters (Figure 8-5a and b) are attributed to shrinkage cracking of a rapidly cooled lava outpouring which invaded these structures. The central interiors of several of the younger impact craters—Tycho, Copernicus, Aristarchus—large enough to have tapped a subcrustal magma reservoir are filled with lava whose surface shows numerous smaller-scale bulges and markings (Figures 8-5c, d, e, and f). Except for being much larger, these mounds resemble the hornitos, tumuli, and pressure ridges found in some terrestrial calderas and especially in Hawaiian-type lava lakes and flows.

Another group of lunar features, generally believed related to lava volcanism, are the *rilles* (rimae) and *mare ridges* (sometimes called wrinkle ridges) that are particularly prevalent in the basalt-filled lowlands (Figure 8-6). Although some rilles may be variants of one another, attempts to classify the varieties encountered call attention to considerable differences that may reflect a diversity of origins (Figure 8-7). Thus, those rilles having prominent zigzag patterns could be large cracks related to cooling. Those with pronounced sinuosities presumably result from flow of fluids or suspensions (lavas; gases; glowing ash materials; water [?]) that erode channels, much like those formed by meandering rivers, as the escaping substances disperse from vents (usually a flat-bottomed crater at the apparent head; sometimes missing). Some rilles may result from collapse of lava tubes which underlie broad warps in volcanic flows within the maria; similar features are common in terrestrial flows. However, at least twenty sinuous rilles are more than 100 km in length—much longer than any ash flow channel or lava tube known on Earth. Other rilles, by contrast, are straight to arcuate with steep walls and flat floors. Because of a notable resemblance to graben structures on Earth (e.g., the East Africa Rift Valley), they may have originated by down-dropping along faults in response to tension in the cooling mare lavas. Disconnected chains of craters possibly represent "blow-out" holes (analogous to maars) along a fissure from which gas was vented. But in some instances these chains could have formed merely as a succession of secondary impacts along a line of ejecta from a distant impact crater or an explosively active volcano. Mare ridges may be extrusions of underlying still-liquid lavas along tension cracks or they may be squeezed out as upwarps during a compression stage in the maria (Strom, 1971).

Gas-Related Volcanism

Volcanic materials other than the dark-albedo mare lavas are now recognized in both the maria and terrae. These materials generally have albedos of 0.09 to 0.12, usually form more or less level plains units that smooth out irregularities on the underlying surface, sometimes are "ponded" within depressions on crater terraces (Figure 8-8), and

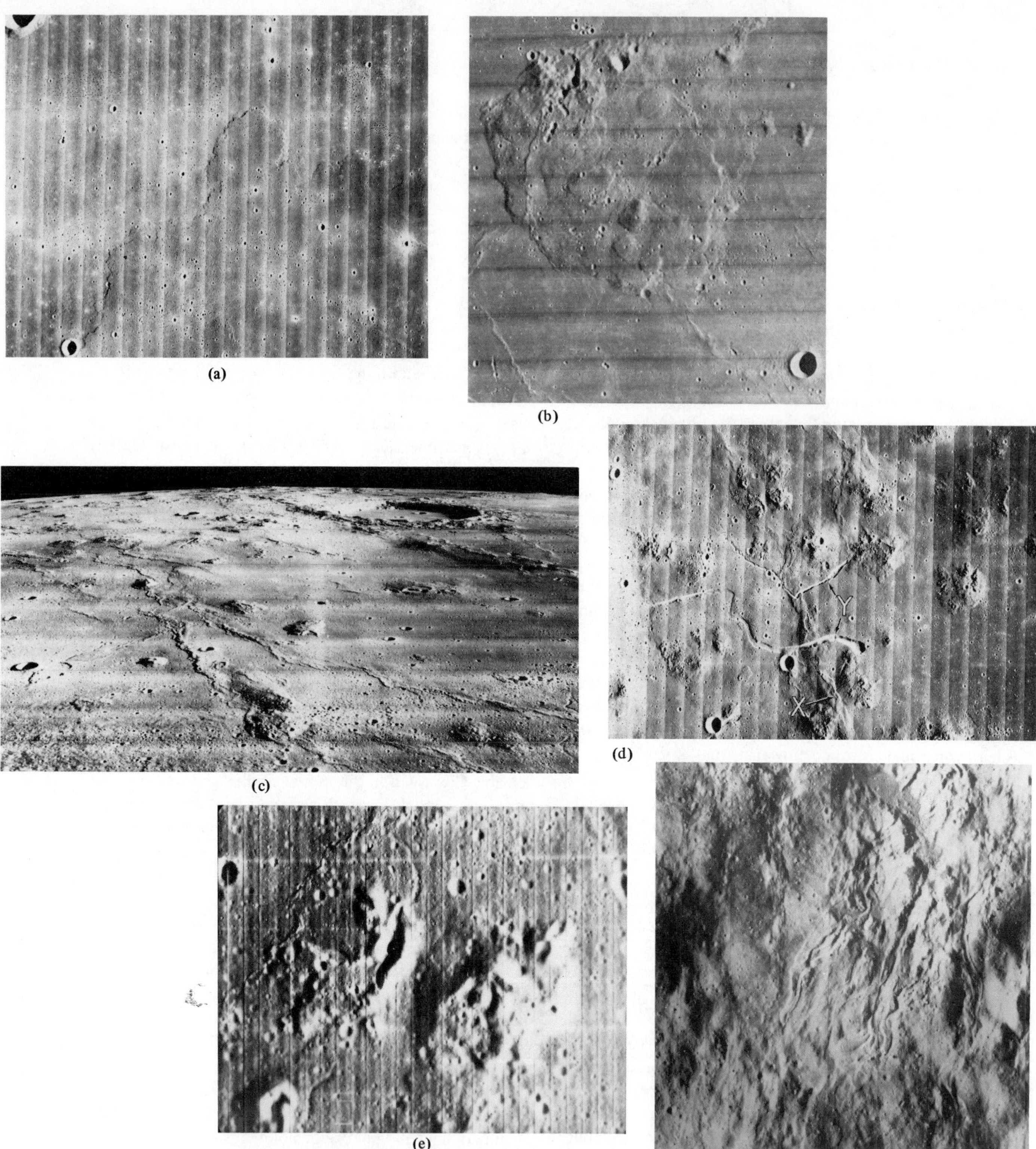

Figure 8-4 (a) The scarp (about 10 meters high) of an apparently younger lobate lava flow on the mare surface of Oceanus Procellarum (*Orbiter 5*-159M). (b) A broadly constructional landform (some 40 km wide) known as the Rumker Hills, east of Aristarchus. Note the three dome-like protrusions near the center of the picture (*Orbiter 4*-163H). (c) An oblique view from *Orbiter 2* of the Marius Hills in central Oceanus Procellarum, displaying domes, cone-like structures, wrinkle ridges, sinuous to straight-walled rilles, and possible collapsed lava tubes (*Orbiter 5*-213M). (d) Vertical view of the Marius Hills, the classic example of a volcanic field on the Moon. The Marius Hills are considered by J. McCauley and others of the U.S. Geological Survey to have a close terrestrial analog in the San Francisco volcanic field east of Flagstaff, Arizona (compare with the radar image of that field shown in Figure 8-3e). (e) Part of *Orbiter 5*-213H, showing several cone-like land masses with elongate central craters (vents?). Compare with Figure 8-3d. (f) Possible lava channels and levees just outside the north rim of Aristarchus, a probable impact crater with associated volcanism, where a number of lunar transient events have been sighted in modern times (*Orbiter 5*-201H). (NASA photos.)

(a) (b) (c) (d)

Figure 8-5 (a) A group of large craters in the highlands northeast of the Orientale Basin and west of Oceanus Procellarum. Mare lavas fill the crater interiors and spread beyond their rims. Within the craters these surfaces are broken by networks of concentric to irregular large "cracks" or rilles presumably developed during cooling of the lavas. The origin of the enclosing craters is uncertain–they may be lava-filled impact structures or volcanic calderas ("ring-structures" as termed by G. Fielder) (*Orbiter 4*-189H_2). (b) Details of the interior filling of two large craters on the farside located about 500 km northeast of the crater Tsiolkovsky. The smaller crater (about 23 km in diameter) has the typical "turtle-back" structure associated with shrinkage of presumed volcanic flows (*Orbiter 1*-115H_2). (c) The floor of Tycho near its western rim showing a wide variety of features attributed to volcanism, including numerous small domes (similar to terrestrial tumuli), crenulations or "corda," flow channels, and possible shrinkage fractures. Some workers note a ropy, pahoehoelike texture in parts of this floor, but the resolution is still too low for positive identification of such a small-scaled feature (*Orbiter 5*-125H_2). (d) Close-up (enlarged) of part of the previous picture (bottom center) showing details of some of the larger domes. Note the exfoliation-like structure associated with the dome in the lower right, which may result from flowage of melted material off a rising mound.

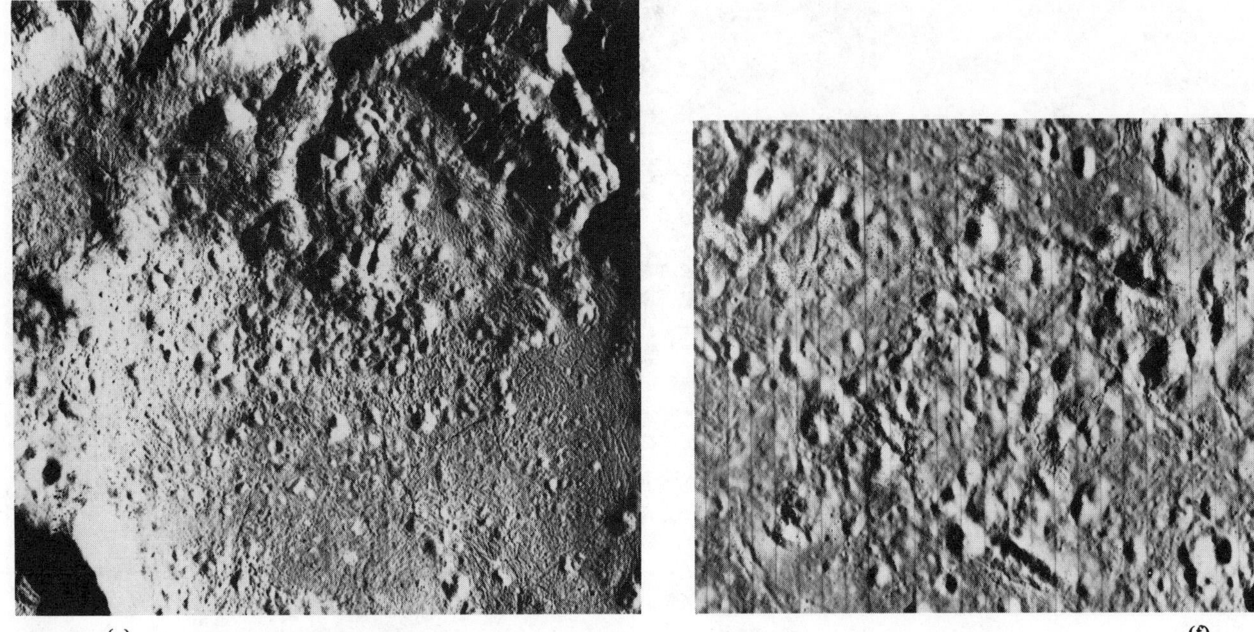

(e) (f)

Figure 8-5 (continued) (e) Mounds or protrusions on the floor of Aristarchus, together with numerous cracks and flow markings, all of probable volcanic nature (*Orbiter* 5-199H_2). (f) Dome-like hills, linear ridges and depressions, and flow markings characteristic of the southern part of the floor of Copernicus (*Orbiter* 5-155M). (NASA photos.)

Figure 8-6 (a) A group of sinuous rilles, several apparently emanating from source craters, just beyond the nearly completely inundated north rim of Prinz Crater east of Aristarchus in central Oceanus Procellarum. These rilles occur in the Harbinger Mountains complex. Note the unusual "tongue depressor" feature, possibly a small graben, on the right (*Orbiter* 5-191M). (b) Hadley rille, a meandering feature trending northward from the Apennine Mountains south of Autolycus. This rim seems to originate from a deep, elongate depression (bottom of picture). *Apollo 15* has landed next to this rille (*Orbiter* 5-105M).

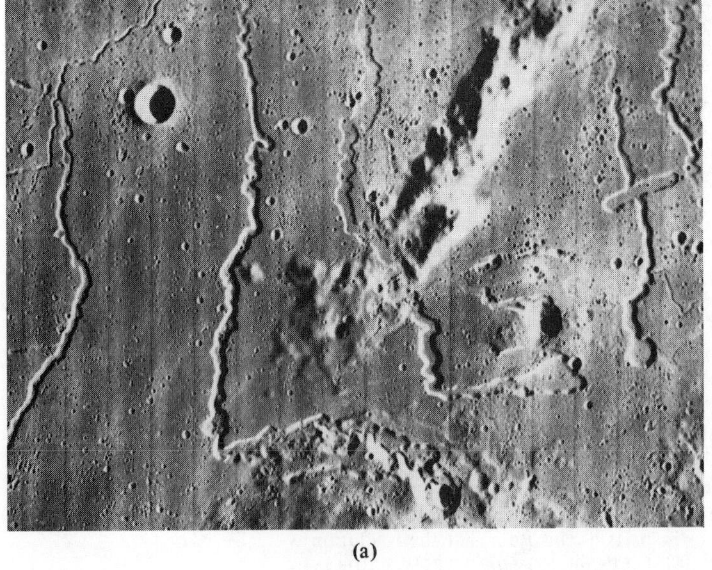

(a)

(b)

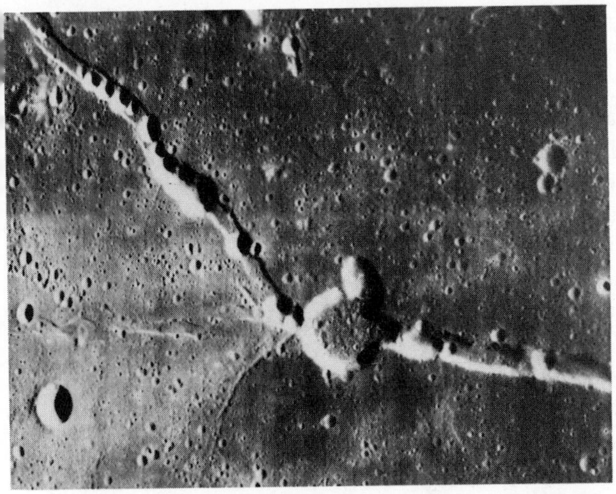

(c)

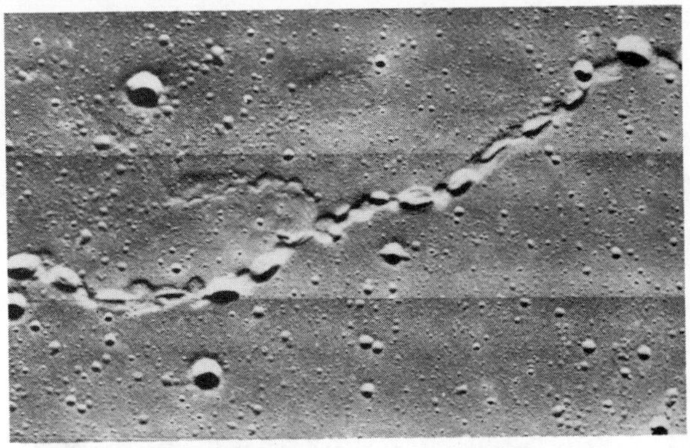

(e)

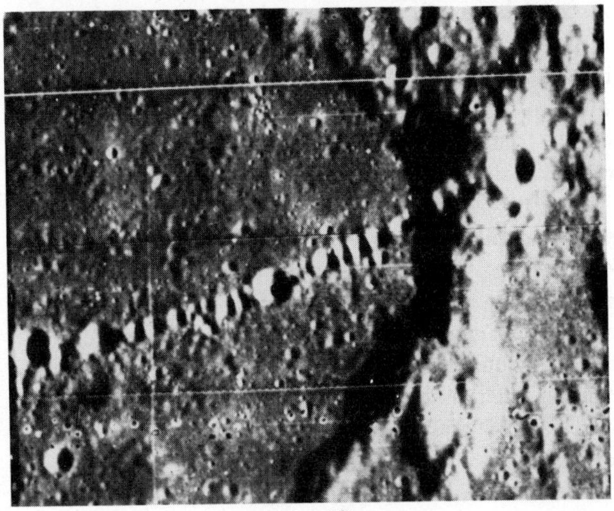

(d)

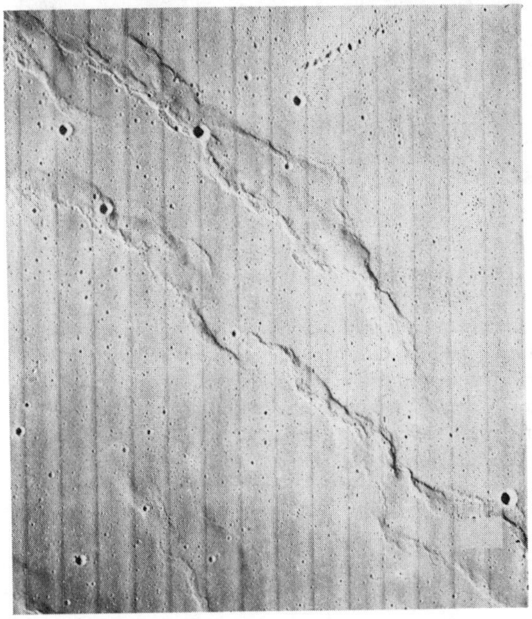

(f)

Figure 8-6 (continued) (c) Hyginus rille in the uplands south of Mare Vaporum, a generally straight depression which branches into two segments at the Crater Hyginus (center; 11 km diameter). Note the chain of craters within the left segment (*Orbiter* 5-95M). (d) A chain of craters (each about 2 km in diameter) extending from Davy G (not shown) into a mare region west of Ptolemaeus. These craters may be analogous to lines of volcanic maars, of explosive character, found on Earth (*Orbiter* 4-H108). (e) A chain of more or less connected, elongate craters (each about 0.5 – 1.0 km in length) located in the northern part of Oceanus Procellarum. This feature may be a series of connected vents, collapsed lava tubes, or some unknown volcanic phenomenon (*Orbiter* 5-183M). (f) Several wrinkle ridges in southern Oceanus Procellarum near Flamsteed crater (*Orbiter* 5-170M). (NASA photos.)

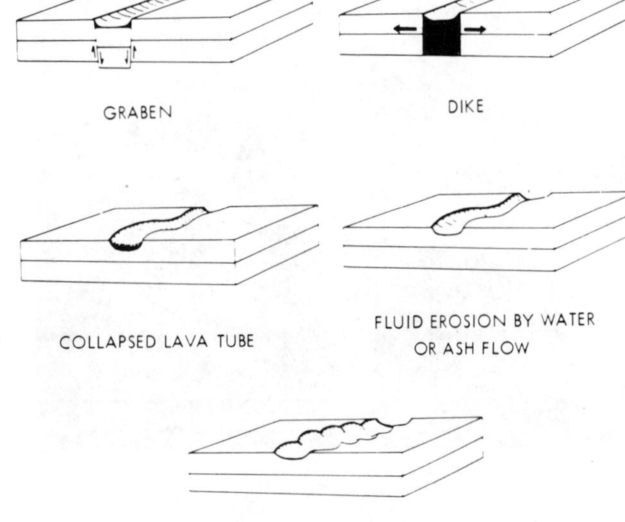

Figure 8-7 Schematic block diagrams illustrating most of the proposed mechanisms for development of lunar rilles. Straight rilles may be grabens caused by downfaulting or trenches resulting from erosion of softer dike fillings. Sinuous rilles are thought to result from lava tube collapse or possibly channel erosion by water or ash clouds. Chain craters, which can coalesce into rilles, may develop from gas escape along volcanic vents or explosive maars. (Courtesy J. W. Salisbury, Air Force Cambridge Research Laboratory.)

Figure 8-8 A group of large "ponds" or "playas" (arrows) (5 to 10 km in maximum extent) occupying depressions within the hummocky ejects deposits just beyond the northeast rim (upper left) of the crater Tycho. These features are generally flat, have notably darker albedos than surrounding terrain, and, in high resolution photos, usually show relatively few craters and considerable fracture and flow characteristics. Similar "ponds"—but commonly smaller in areal coverage—fill in low spots on the flatter surfaces of terraces in the inner walls of many larger craters. The characteristics of these "pond" fills suggest small pockets of lava (possibly impact-derived but more likely from later extrusions) whose mode of emplacement remains uncertain (*Orbiter* 5-125M). (NASA photos.)

often contain sinuous rilles, domes, and subdued cones. Older units commonly are heavily cratered, hummocky, and furrowed. The appearance and distribution of many of these plains-forming units led some lunar geologists to suggest that they were ash flows, pyroclastic deposits, and possibly bedded tuffs interstratified to some degree with lava flows and/or impact ejecta. Because the units are widely distributed over the Moon in many places and at different stratigraphic levels, early lunar volcanism was surmised to have been a continuing, although perhaps intermittent, process which reached its climax with emplacement of the mare lavas. Later *Apollo* missions, especially *Apollo 16*, are forcing a reassessment of some of these views in that certain units (e.g., the Cayley formation) have proved to be impact-generated breccias deposited from base surges.

Effusive deposits of volcanic gases, leading to surface coatings of sulphur and other sublimates, are rarely observed on the Moon because of the vigorous reworking of the regolith and other surficial materials by meteoroid bombardments, moonquakes, etc. In view of the widespread signs of volcanism over most of the Moon, such deposits ought to have been commonplace in the lunar past. Two examples of possibly "recent" gaseous precipitates on the lunar surface are shown in Figure 8-9a and b.

Some smaller, usually isolated craters are surrounded by a mantle or blanket of very dark materials. These dark-halo craters may be similar to explosive maars and diatremes found in volcanic regions on earth (see Figure 8-3f). The deposits forming the halo may be analogous to basaltic cinders or vesicular scoria.

Figure 8-9 (a) Reiner Gamma (left center), a broad (20-30 km), long (~ 150 km) surface marking recognized by its light color and high albedo. This feature has been interpreted as a coating of effusive materials, such as sulphur, of possible volcanic nature (*Orbiter* 2-215M). (b) The "Delta" feature, a series of "dome-like" protrusions, possibly covered with volcanic gas deposits, observed in dark mare lavas north of Mare Vaporum. This unique feature, about 3.2 km across, was photographed by astronaut Worden during the *Apollo 15* mission (Apollo Pan Camera: Roll 16, frame 0181). (NASA photo.)

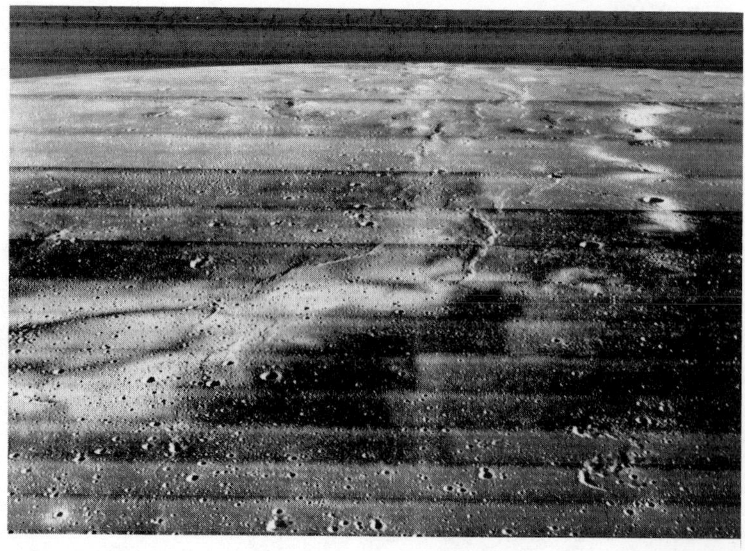

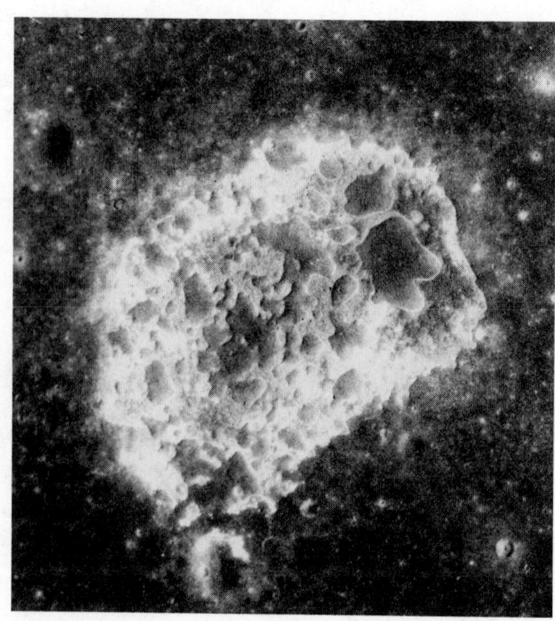

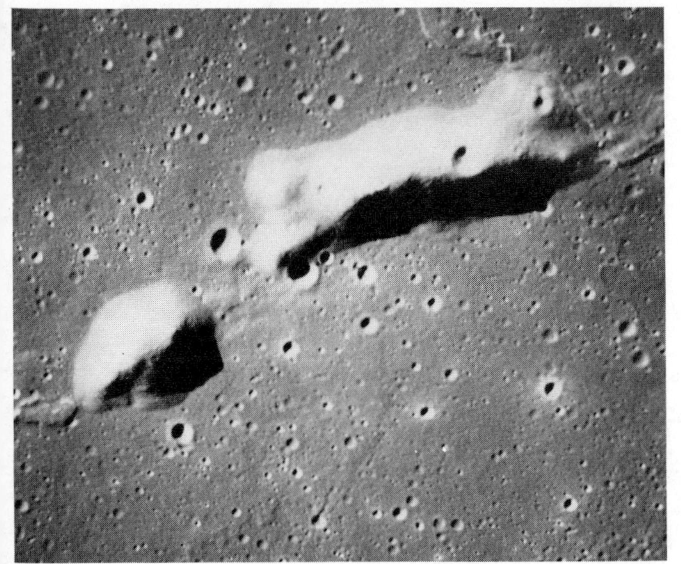

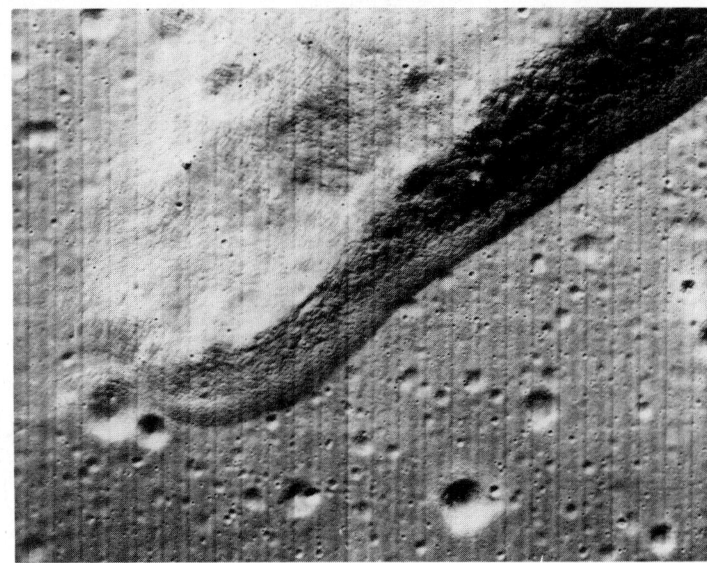

Figure 8-10 (a) A pair of isolated hills photographed on the nearside of the Moon during the *Apollo 16* missions. Although of unknown origin, these mound-like protuberances could be extrusive domes or surface expressions of deeper intrusive bodies. (b) One of the elongate hills that form the Flamsteed Ring (about 100 km in diameter) in southern Oceanus Procellarum. Note the sharp line of juncture between the mare surface and the toe of a bulge near the base of the inward-facing wall (towards ring center below the lower part of the photograph). Both bulge and wall show fewer craters than on the gentler slope (upper left), suggesting their destruction by the downward movement of erosional debris that piles up towards the bottom (*Orbiter 3*-199H). (NASA photos.)

Subsurface Igneous Activities

Since the lunar crust consists predominantly of igneous rocks (a reasonable inference from studies of the *Apollo* samples), there is a high likelihood that differentiation and multiple intrusions have produced a variety of plutonic bodies (Figure 8-10a) analogous to the layered mafic and ultramafic complexes and perhaps also to the batholiths, stocks, ring dikes, and feeder vents that make up much of the continental segments of the Earth's crust. However, no confirmed examples of intrusive plutons have been recognized on the lunar surface. Intrusions may not be readily visible at the lunar surface because of the absence of a counterpart to the combination of isostatic uplift and fluvial erosion processes by which the interiors or roots of mountains are eventually exposed on Earth. On the Moon, materials from deep within the crust can ultimately be exposed, however, as ejecta from large craters that penetrate many kilometers into the crust. Some rocks returned from the *Apollo* sites have textures which suggest slower cooling and deep burial consistent with a plutonic or intrusive origin.

O'Keefe et al. (1967), Fielder (1967), and Elston (1965), among others, consider many larger craters on the Moon to be built up as "ramparts" by extrusions along ring dikes (see Figure 8-4c for one example). Thus, the hills and ridges associated with the Flamsteed "Ghost" Ring near Flamsteed Crater (Figure 8-10b) may be surface expressions of shallow extrusions or of domical upwellings of viscous lavas emplaced along a system of steep conical fractures. The "hump" at the base of the steep face of the hill shown in Figure 8-10b is viewed as evidence of differential extrusion. Structures such as Flamsteed, however, are subject to an alternate interpretation as remnants of old crater rims or hummocky terrain now almost completely inundated by mare lava flows. Those who favor the inundation origin envision the hump to be the result of piling up of fine fragmental debris from mass-wasting of the rim face. Still a third view considers the Flamsteed Ring to be a caldera filled with lavas younger than those in the surrounding mare. The lower crater density within the ring favors this interpretation. Flamsteed is a prime example of how the same feature can receive notably contrasting explanations—each plausible in itself—that depend as much on the initial "bias" of the investigator as on the observable facts.

Distribution of Volcanic Features on the Moon

Most of the more than four hundred lunar domes, the dark-halo craters, rings, wrinkle ridges, straight and sinuous rilles, and other endogenetic features of volcanic origin are associated with the mare lavas, volcanic fillings in large craters, or, in some instances, the smooth plains (Figure 8-11). The majority of these features are concentrated on the nearside of the Moon, as would be expected from the prevalence of the maria on the face. Some domes and a few rilles appear on the highlands, but the bulk occur within circular basins near their margins, where they tend to be concentrically distributed. Domes are rare in the central

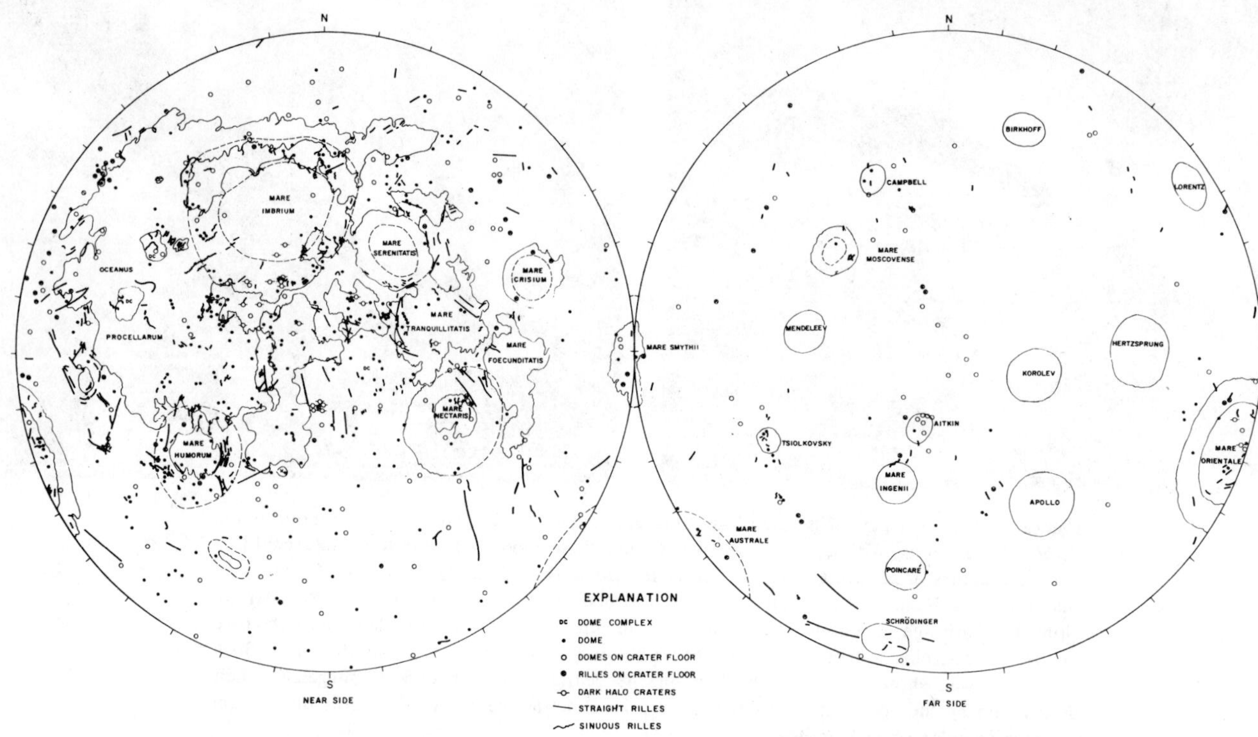

Figure 8-11 Distribution of domes, rilles, and dark halo craters—all presumably volcanic features of endogenetic origin—on the nearside and farside of the Moon. (From Smith, 1973.)

parts of lava-filled basins. Wrinkle ridges are almost exclusively located within the maria, again most frequently near the edges. Deep fractures, produced during basin formation, are a principal factor in localizing many of these volcanic features.

Lunar Tectonism

Certain linear features on the Moon are attributed to structural deformation (Figure 8-12), even though some interrelation to volcanism can be inferred in many instances. Persistent alignments of such lunar features as chain craters, rilles, wrinkle ridges, and possible fracture systems along three principal trend lines or directions (Figure 8-13) have been noted. On a small scale, aligned linear grooves (about 1 cm high, 2-4 cm wide, and up to 1 meter long) and ridges have been recognized in photographs from the *Apollo* sites; they follow trends and spacing similar to the *patterned ground* (Figure 8-12d) observed first in the *Lunar Orbiter* pictures.* All of these alignments constitute the so-called *lunar grid system*—a Moonwide linear pattern which may reflect an underlying structural control of the distribution of many surface features of nonimpact origin. In places, several large craters also tend to align along preferential directions; these may be volcanic calderas whose locations reflect lines of weakness in the lunar crust. Similar linear trends on Earth are often the consequence of fracture-joint systems in crystalline basement rocks or of oriented forces operating during mountain building.

On the lunar sphere it is postulated that rapid rotation should develop a Moonwide system of regular linear trends during cooling of the primordial crust. Being continuous, and cooled fairly uniformly throughout, the crust of the solidifying melt would be subjected to tensional stresses that produce multiple sets of fractures (joints) whose final positions are controlled (in part) by additional forces related to the higher rates of rotation of the early Moon. Tidal forces exerted by the Earth may also have acted over time to initiate or widen the cracks presumed to underlie the present superficial cover of rubble extending over the lunar surface. Recent discovery of a time coincidence between large increases in the number of moonquakes detected by Apollo seismometers (see p. 176) and periods in which the Moon is near either perigee or apogee has been interpreted to indicate that degassing (and possible accompanying volcanism) occurs when the crust is flexed or relaxes as the tidal forces vary.

Despite coassociation of volcanism and active mountain building on Earth, no similar relationship can be verified for the Moon. Thus, except possibly for mare ridges, signs

*There is growing opinion that many of these small-scale lineaments are actually "illusions" related to lighting effects on the surface (see p. 158).

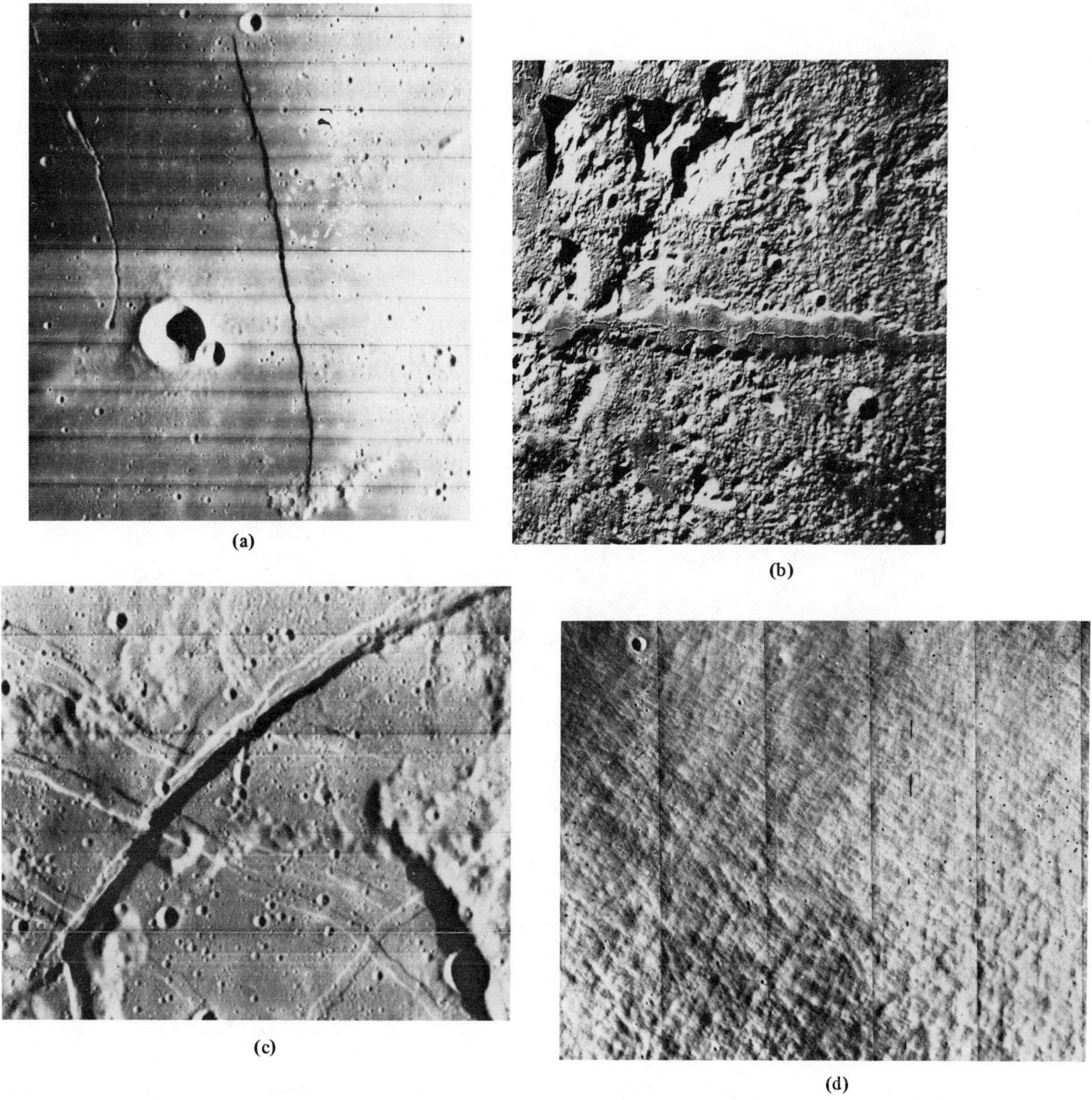

(a)

(b)

(c)

(d)

Figure 8-12 (a) The Straight Wall, exposed along the eastern edge of Mare Nubium, is one of the best examples of probable fault displacement observed on the Moon. The scarp shown here has about 250 meters of relief and extends continuously for nearly 150 km. Movement along this fault is presumed to be vertical inasmuch as none of the structures through which it passes show any apparent lateral offset (*Orbiter 4*-113H). (b) The Alpine Valley, a cigar-shaped depression some 120 km long and up to 10 km wide, located in the "Alps" highlands along the northeast edge of Mare Imbrium. This feature may be a large graben, but the scalloped walls (right) and the presence of a sinuous rille along its axis suggest other possible origins (*Orbiter 5*-102M). (c) Part of the Sirsalles rille system, beginning in uplands ejecta terrain east of the crater Darwin (near the southwest edge of Oceanus Procellarum) and extending northeastward to the crater Sirsalles. Sets of straight rilles, with steep (30°+) slopes and flat floors, intersect in this view. Many examples of this type of rille resemble terrestrial grabens and may result from tensional stresses of tectonic nature, but others appear related to stresses set up by shrinkage during cooling (*Orbiter 4*-161H_2). (d) An exceptional example of "patterned" ground, here represented by several intersecting sets of close-spaced, narrow ridges and grooves (whose spacing and heights are measured in meters). These linear features are possibly surface manifestation of joints in underlying bedrock. They develop gradually as depressions and rises through drainage into these cracks (aided by shaking from moonquakes or mass wasting). Alternatively, they may be base surge flow markings. The scene is the highlands near the south pole (*Orbiter 5*-179H_1). (NASA photo.)

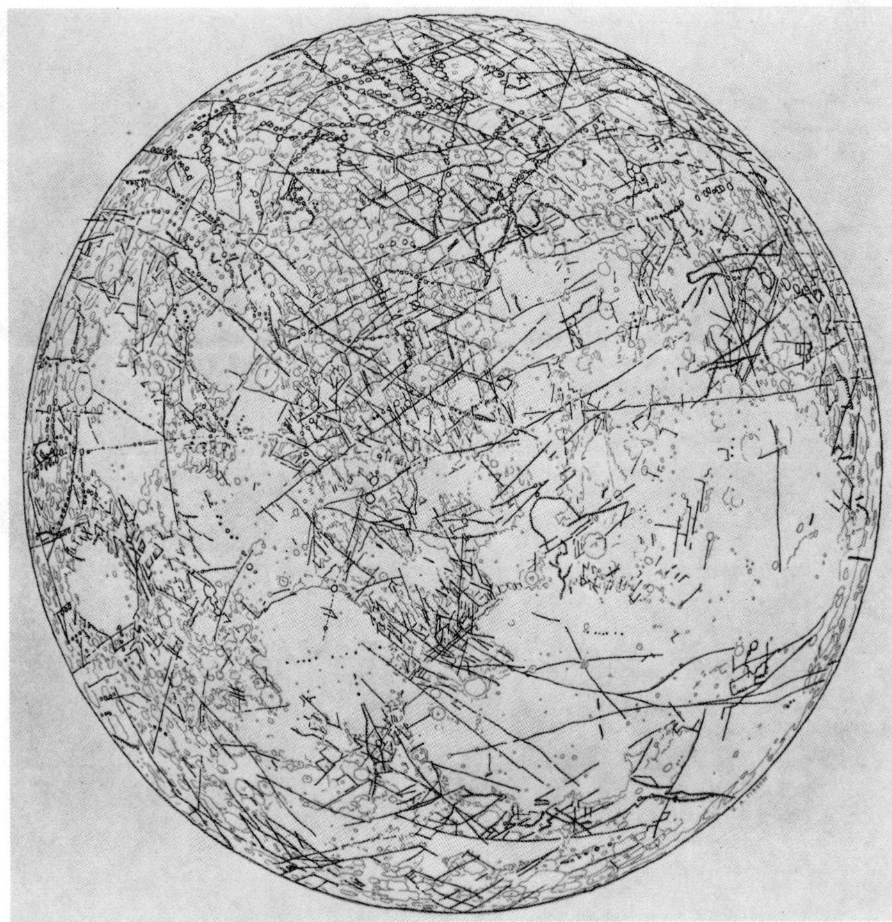

Figure 8-13 A map showing the Lunar Grid System, defined by linear structures (faults, straight rilles, grabens, chain craters, and unidentified lineaments) of possible tectonic and volcanic origin. When the trends or "strikes" of these features are plotted graphically relative to lunar north, they appear to follow three prominent directions (northeast, northwest, and north-south). (From V. A. Firsoff, *Surface of the Moon*, 1961.)

of intense folding and crumpling of lunar rock units or their offsetting by major thrusts or rifts have not been detected in photographs of the Moon's surface. Occasionally high-angle dip faults and grabenlike rilles are seen to intersect crater rims or lava flow units. This low degree of structural adjustment is not unexpected, of course, in view of absence of readily deformable sedimentary rocks, but it also indicates that compressional forces coincident with tectonism are uncommon to rare on the Moon. On Earth, the causes of these forces are generally correlated with drifting crustal plates, driven by sea-floor spreading and underlying convection currents, but conditions for the operation of these processes may be nonexistent or ineffective in the Moon. One possible exception may be within the central Oceanus Procellarum basin, where the Rumker Hills, Aristarchus Plateau, Schroeter's Valley, Marius Hills, and Flamsteed Ring all line up along a north-south trend that could mark an upwelling of lavas from spreading centers.

Effects of Gravity on Internal Processes

The lower gravity on the Moon may be a major factor in the operation of lunar igneous processes because of its effect in reducing pressure gradients. Inasmuch as the temperature of melting decreases with a lowering of pressure, the depths at which rocks should start to melt will be less on the Moon than on Earth—assuming identical thermal gradients. In order to activate melting, a smaller pressure release is required in lunar subcrustal zones where rocks are just below their melting points.

J. Green (1965) and others propose that most craters on the Moon stem from a defluidization process beginning in the subcrust as it is stressed by tidal flexing. Craters, in this view, may result from either violent eruptions or caldera collapse, or both. Heat supplied by radioactive decay provides the thermal input needed for melting and volatilization. Bubbles of gas (composition unspecified) will nucleate at greater depths, grow to larger sizes, and enhance the explosivity of the highly fluid (but water-free?) lavas that reach the surface. Indeed, the vesicular nature of many basalts returned from *Apollo* sites indicates that many lunar lavas are rapidly degassed. During degassing at the surface, fragments are explosively ejected to much greater distances outward—influenced as well by the lower gravity—than in similar volcanic events on Earth (Figure 8-14). Likewise, ash flows would travel farther in the lunar gravity field and hence spread more uniformly across the terrain.

Summary

It can be said with considerable confidence that the Moon has undergone a long, varied, but not particularly complex history of volcanic activities. Volcanism probably was first manifested during cooling of a melted crust, but evidence of specific volcanic landforms in the older highlands has been obscured by the later stages of the intense impact cratering which devastated that ancient terrain. Most still-recognizable volcanic features occur on the younger mare surfaces or the lava fillings in the bigger craters, where meteorite bombardment has not erased the larger structures developed in lava flows or ash fields. Because these flows appear to consist mostly of high-temperature mafic silicate rocks that were water-free and gas-poor when erupted, the resulting features are characteristic of very fluid, rapidly cooled lavas such as those found on the flanks and in the pit craters of volcanoes on the island of Hawaii or in parts of Iceland. Spectacular and graceful stratocones—comparable in form to Vesuvius, Rainier, and Fujiyama—that represent the piling up of andesitic lavas and ash are extremely uncommon on the Moon, unless subsequent investigations demonstrate that central peaks in craters such as Tycho and Copernicus are products of volcanism.

The vast majority of bonafide volcanic landforms on the Moon consist of low *protrusions*, whose terrestrial analogs may be shield volcanoes; *wrinkle ridges*, comparable perhaps to pressure ridges in terrestrial volcanic flows but much larger in scale; and *rilles*, analogous to flow levees and channels on Earth, but again of considerably greater dimensions. These features are particularly conspicuous in parts of the relatively young Oceanus Procellarum. The Marius Hills (see Figures 8-4c, 8-4d, and 8-4e) serve as the classic type locality for the kinds of volcanic activity characteristic of the lunar surface. The crater Marius itself is of ambiguous origin: it shows some signs of having been produced by impact but also resembles in some respects the typical lunar "caldera" proposed by Green (1972) as strictly volcanic in origin. Whereas terrestrial calderas generally lie at or near the summit of large, sloping cones or shields, lunar calderas lie at about the same levels as their surroundings except for their bounding rims.

Most astrogeologists now accept a duality of possible origins for the larger lunar craters. The bulk still appear to be impact-derived but some, with distinctly different morphology, are probably calderas in the sense of having formed in volcanic terrain by such endogenetic processes as explosive gas release and/or collapse. In the northern part of the inner "bullseye" of the Mare Orientale basin (Figure 1-8b) are two large craters (at 12 and 3 o'clock) considered by some astrogeologists to be type examples of impact and volcanic craters respectively (Figure 8-15).

Figure 8-14 Calculated range of ejection (distance in kilometers from source to landing point) of fragments from millimeter to 10 meter average diameters expelled from hypothetical or real volcanoes on the Moon, Mars, Venus, and the Earth. The calculations include the effects of atmospheric drag (from vacuum to tenuous, normal, and dense atmospheres) operative on the four planetary bodies in the above-named sequence. The normal case used in the calculation is a volcano with reservoir depth of 40 km, reservoir temperatures of 1000°C, and magma water content of 10% by weight. For the Moon, two additional water contents of 3% and 5% are considered. (T. R. McGetchen and G. W. Ulrich, *Journal of Geophysical Research*, Vol. 78, pp 1832-1852, 1973, copyright by American Geophysical Union.)

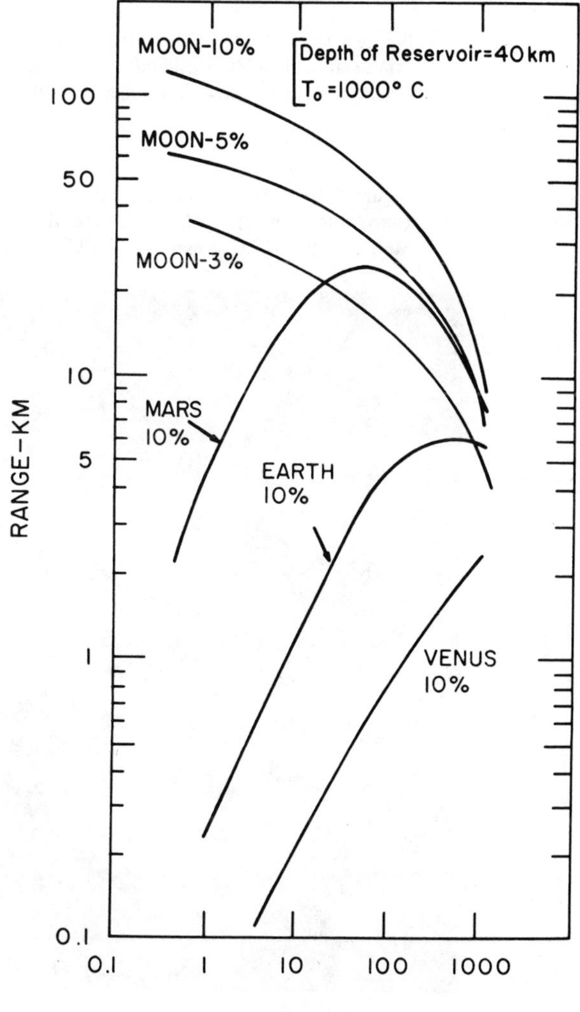

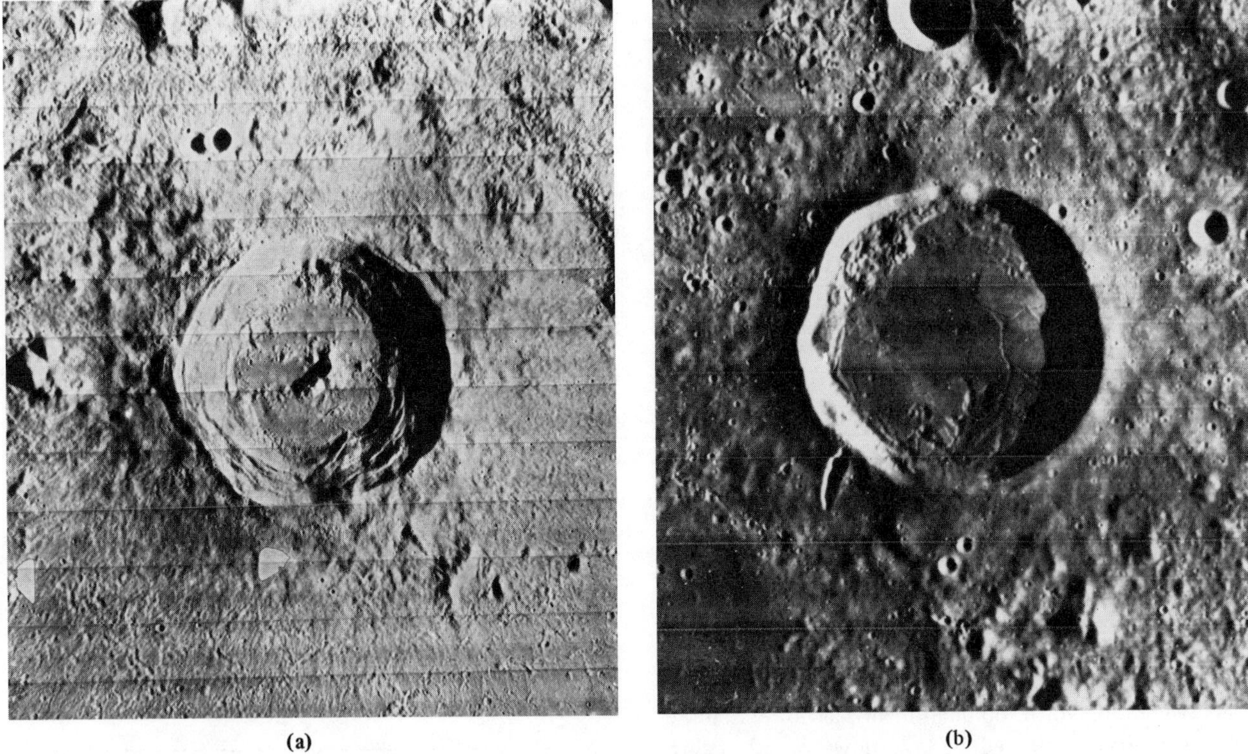

Figure 8-15 *Orbiter 4* high resolution photos of (a) a presumed impact crater and (b) a volcanic crater, possibly a caldera, ring located in the mare lava floor within the inner ring of the Orientale Basin. Both craters are about 40 km in diameter. (NASA photo.)

Figure 8-16 An *Apollo 15* oblique photograph showing the craters Aristarchus (left) and Herodotus (center), with the meandering Schroeter's Valley, one of the larger rilles on the Moon, in the right foreground. (NASA photo.)

Both show evidence of nearly contemporaneous emplacement following formation of the Orientale basin. Comparison of the various features associated with each crater leads to these descriptive or identifying criteria (implicitly, distinguishing features) for each type:

Impact	*Volcanic*
–Asymmetric rim	–Inverted V-shaped rim
–Central peak(s)	–No central peak group
–Interior terracing common	–Minimal terracing
–"Rough", hummocky floor (may or may not be "lava")	–"Smooth", lava-filled floor
–Extensive, irregular ejecta blanket beyond rim, marked by grooves, ridges, and hillocks	–Generally smooth, tapering exterior surface, often with few craters; sometimes channeled
–Clusters of secondary craters, often aligned in radial or concentrio groupings	–No apparent association of related secondary craters

The large juxtaposed craters Aristarchus and Heroditus (Figure 8-16) may represent another definitive example of the two genetic types.

Arguments over the nature and modes of development of the Moon's circular depressions continue to persist even after the close of the Apollo exploration program. It is indeed regrettable that no mission was ever scheduled to visit and sample one or more of the distinctive large craters of either category described above.

REVIEW QUESTIONS

1. What is the importance of volcanism relative to impact as applied to the lunar crust?
2. What rock and structural features associated with volcanoes can sometimes be confused with similar features in impact craters?
3. Which types of terrestrial volcanoes seem to be absent on the Moon? What factor(s) might account for this?
4. Distinguish between a summit crater, a caldera, and a maar.
5. Discuss possible origins for the lunar rilles and wrinkle ridges.
6. Cite the evidence for volcanic gases and ash deposits on the Moon.
7. What is the lunar grid system? What orientations are observed?
8. State several hypotheses for the origin of patterned ground.
9. Suggest several causes of fracturing of the lunar crust.
10. How much less is the pull of gravity on the lunar surface compared with Earth? In what ways can this influence volcanic activity?

CHAPTER

9

Mapping the Moon

The object of geologic mapping is to define distinguishable lithologic features on planetary surfaces and to represent these on scaled-down, two-dimensional diagrams. The resulting maps depict rock bodies according to their material composition, geographic location, areal extent, relative sequence of development, structural or geometric positions, and, wherever possible, ages or times of formation. On Earth, stratigraphy, with its concepts of uniformitarianism, superposition, crosscutting relationships, and faunal succession, forms the basis on which geologic maps are constructed. Structural geology and petrology also make substantial contributions.

Mapping from Telescope Observations

Close scrutiny through a telescope will soon convince an observer that the lunar surface records a history of events discernible as a succession of units and features having characteristic differences suited to mapping. Certain areas within the highlands, for example, are pockmarked with numerous craters showing varying degrees of degradation. In places, individual craters either overlap or are overlapped by other craters and their surrounding blankets and are obvious examples of transection (crosscutting) in which each superposing structure will usually also display a "fresher" or younger morphological appearance. Large craters that cut into the mare plains deposit their ejecta indiscriminantly across older mare and terra surfaces. Mare-covered lowlands seem to be lapping onto the more heavily cratered highlands and even "submerging" patches of preexisting basin terrain.

In the early 1960s, the U.S. Geological Survey undertook a comprehensive program to define, differentiate, and map the stratigraphy and structure of the lunar nearside surface as a major step toward deciphering the Moon's origin and history. This effort was aimed specifically at establishing the physical characteristics, spatial relations, sequence of formation, and, insofar as possible, the composition, processes of development, and origin of units that could be distinguished by direct observation of the lunar surface. A practical, time-dependent objective which both stimulated and accelerated the program was the necessity of setting up a geologic framework according to

---

Much of this unit has been adapted from D. E. Williams, "Summary of Lunar Stratigraphy–Telescopic Observations," U.S. Geological Survey Professional Paper 599-F (1970).

which sites for study in the *Ranger, Surveyor, Lunar Orbiter*, and *Apollo* missions could be selected well in advance of flight times. The Geological Survey's task at the outset was faced with such formidable difficulties as the need to carry out all mapping remotely by telescope at resolutions rarely better than 1 km, the lack of available time markers (fossils, beds datable by radioactivity, etc.), the uncertainties about types of rocks involved, the scarcity of exposed sequences in cross section, and the lack of opportunity to field-check correlations on-the-spot. However, mapping under these limitations bears many similarities to the approaches used in photogeologic studies of the Earth's surface, except for the inability (before *Apollo*) to acquire "ground truth" and control points on the Moon.

A curious crossbreed of astronomer-geologist evolved to meet these mapping requirements. Before the advent of *Orbiter* photographs, his field work consisted of hours of patient night-viewing through telescopes in order to locate boundaries more accurately, describe and measure features, and correct or improve interpretations already made from photographs taken under the best conditions at astronomical observatories. This visual technique takes advantage of the eye's ability to see more detail in those smaller objects which may be barely recognizable or "blurred out" in pictures. Using *objective* criteria as much as possible, the astrogeologist would test the decisions made from preliminary photogeologic analysis against the actual scene reviewed from night to night as a lunation proceeded and would sketch in his reinterpretation wherever problems were encountered.

Definition of Mapping Units

The chief difficulty in mapping in this fashion lies in selecting meaningful units that have both regional significance and temporal value. Stratigraphers have solved similar problems in selection and classification of terrestrial units by setting up three categories into which rocks in the crust are grouped for mapping purposes. *Geologic time units* (eras, periods, epochs, ages) divide all time since the beginning of the Earth into convenient intervals whose durations are measured in years; events, rather than materials, are considered in these units. *Rock-stratigraphic units* (groups, formations, and members) treat rocks as units defined by lithology (physical properties—e.g., composition, texture, color) rather than by time significance; however, the relative sequence of position, in a geometric sense, of units present at any location establishes a valid stratigraphic ordering that exists independently of the actual times during which the sequence was developed. A *time-stratigraphic unit* (system, series, stage) is composed of the rocks of a region which are formed as a sequence during a definite geologic time unit. Thus, rocks of a terrestrial system were deposited during a period, those of a series during an epoch, and those of a stage during an age. (There is no generally accepted time-stratigraphic equivalent of an era.)

Early in the lunar mapping program, consideration was given to use of time-stratigraphic units as the fundamental subdivisions for mapping. This of course requires some knowledge of actual ages for features present on the lunar surface. One can postulate a time scale for lunar history that might be comparable to that assumed for the duration of the solar system and, probably as well, to the total age of the Earth. However, there is as yet no a priori or direct basis for assuming that any subdivisions in such a scale for the Moon are contemporaneous with named time-stratigraphic units on Earth. In fact, age-dating of some *Apollo* samples indicates that most lunar events occurred before the time represented by the earliest Precambrian rocks now visible at the Earth's surface. There still is no sure way to determine whether these events were completed rapidly in succession over a short duration (from tens to a few hundred million years) or took place uniformly over most of this lunar geologic time or have proceeded intermittently at nonuniform rates within indeterminate intervals.

Most of the visible lunar surface is covered by deposits that superpose, overlap, interstratify, and thin out much as do sedimentary units on Earth. However, well-defined layering has been only rarely observed in *Lunar Orbiter* photography. The nature of most lunar units can be likened to poorly sorted continental-type sedimentary deposits such as those comprising glacial tills, talus accumulations, or alluvial fans. But the presumed absence of water on the Moon—a condition persisting over most if not all lunar history—rules out any sedimentation process that requires an aqueous environment. Most lunar units are inferred to have been emplaced by ballistic sedimentation (crater ejection), transport as volcanic ash or lava flows, or mass movements under gravity. All of these processes differ sharply from conventional sedimentation in that the materials involved move to their sites of deposition at much faster rates. Thus, both impact ejecta and landslide debris are dispersed from (usually single) source areas to broad depositional regions in a matter of minutes—incredibly rapid when compared to normal transport times and accumulation rates for similar thicknesses of terrestrial shales and limestones on a shelf or craton. Ash and lava flows likewise will travel to their terminal locations in only hours. These extremely fast rates imply that many thousands of events involving impacts or volcanic eruptions, each responsible for some specific and sometimes distinguishable deposit, can transpire on the Moon over intervals of time in which some individual marine stratigraphic units formed on Earth. It would be confusing, therefore, to attempt to map the deposits from each lunar event as a distinct unit. Instead, the deposits within some region (or even Moonwide) representing a continuum of sporadic events during an arbitrarily selected time interval have been grouped into convenient stratigraphic units that can exist independent of specific time boundaries or modes of origin.

Thus, because absolute time markers could not be established without direct access to the rocks for dating, lunar geologists rejected time-stratigraphic units as a basis for mapping and decided to rely on rock-stratigraphic units as the basis. This objective approach allows distinguishable materials to be arranged sequentially without regard for actual times of formation. Eventually, as materials returned from various *Apollo* sites underlain by different rock-stratigraphic units are accurately age-dated, at the least it may then be possible to establish boundaries for major time-stratigraphic units on the Moon. However, in ejecta blankets or regoliths, age determinations for individual fragments derived from many source localities and then mixed together would in themselves be of little value in assigning the host materials to specific time-stratigraphic units.

The fundamental rock-stratigraphic unit for the Moon is designated a *lunar materials* unit. Such a unit is defined by its distinguishing properties rather than by its lithologic character. All connected-to-discontinuous areas that show uniformity or similarity of topographic and surficial properties are outlined in a region as a single specific lunar materials unit. Superposition and transection relations, freshness of craters and topography, variations in crater density and in albedo, are among the guidelines used to work out the relative sequence of different materials units in the region (see pp. 121ff). Generally, the criteria for differentiating units depends on those two-dimensional features that can be seen through the telescope or on a photo, in contrast to terrestrial stratigraphic methods that rely on evidence from layered sequences as exposed in outcrop walls or from subsurface drill cores. For this reason, the type section (vertical stratigraphic column at the locality where the unit was first studied) used on Earth to designate lithologic characteristics of formations or members is replaced on the Moon by a (horizontal or lateral) *type area* within which the distinctive properties of a materials unit are well displayed.

Where a lunar materials unit is both widespread and shows signs of having developed from a single event or from events closely interrelated in time and space (e.g., excavation of a basin by impact), this unit is commonly identified as a *formation*, preceded by a geographic name that refers to its type area (e.g., Fra Mauro Formation, found around the Fra Mauro Crater). Distinguishable topographic variants within a formation can be singled out as *members*. Several formations within a region comprise a *group* if they appear to have been emplaced in succession over a limited time span. In many instances, however, formal rock-stratigraphic names are not appropriate for recognizable materials units that are limited in extent, have poorly defined boundaries or contacts, or lack definitive properties. These are then described by some distinctive characteristics of their terrains followed by the word *material(s)*; examples include "crater floor material," "plains-forming materials," "dark-haloed crater material," and "sinuous rille materials." Within the confines of a single crater, subunits (or members) of the more general "crater material" can be recognized, as indicated in Figure 9-1.

Many materials adjacent to the lowlands and extending on to parts of the uplands can be traced to the large circular basins now covered by much of the mare lavas. These multiringed basins and their associated features constitute first-order type areas for a wide variety of materials derived from them. The sequence of events accompanying formation of a basin and its deposits is depicted schematically in Figure 9-2. Some of the characteristics of these basins will disappear in time as they become progressively worn down by meteorite bombardment or further inundated by lavas. However, careful analysis of the surficial features around regions underlain by basins allows these to be ranked in relative age from oldest to youngest.

Figure 9-1 Diagram showing general stratigraphic relations of subunits among crater materials associated with a single crater. The symbol "c" alone refers to undivided crater materials. The various members are identified as follows: *cr*, rim material; *crh*, hummocky, and *crr*, radial, materials representing laterally gradational facies located beyond the rim of larger craters. *cw*, wall material; *cf*, floor material; *cfs*, and *cfh*, smooth and hummocky floor materials; *cp*, peak materials. Materials shown in the lower boxes form more or less contemporaneously with crater excavation. Wall and some floor materials are emplaced at later times. (From D. E. Wilhelms, 1970.)

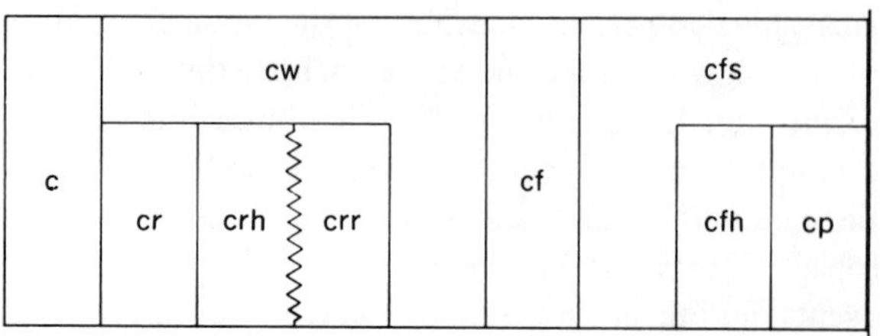

Development of a Lunar Stratigraphy

Gradually, as some region of the Moon is mapped in ever-greater detail, it becomes possible to build up local stratigraphic columns that list the order or succession of lunar materials according to estimates of relative age. Although such a column for each given region is defined by its sequence of materials units, it has proved expedient to group these into major time-stratigraphic units or subdivisions (systems and series) for which geologic time equivalents (periods and epochs) are also specified. This classification was first proposed by Shoemaker and Hackman in 1962 as an analogy to the terrestrial conventions by which stratigraphic nomenclature has evolved. Like the

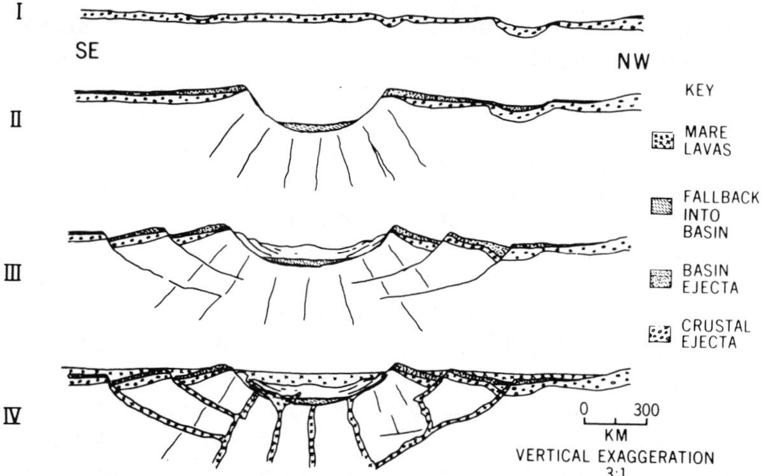

Figure 9-2 The inferred sequence of events (generalized) involved in production of a major impact basin (based in part on the study of the Mare Imbrium region by Hartmann and Kuiper, 1962). I: Formation of a continuous thick blanket of ejecta derived from an intense pre-basin episode of cratering over the entire crust. II: Essentially instantaneous development of a basin of excavation by impact of a large meteorite or comet. Most basin material is ejected on to surrounding crustal ejecta deposits forming a new (younger) ejecta blanket as much as 2 km thick near the initial crater rim. But a small fraction of the ejecta returns immediately into the basin as fallback. III: Formation of a series of roughly circular rings (3 or more) around the basin by a process similar to gravity sliding. The basin itself expands and partly infills by slumping of its unstable walls. The entire time span during which the ringed valleys and scarps and the slump terraces were produced may have been only minutes to days. IV: Invasion of lavas through the deep, impact-produced fractures penetrating into the sub-crust, filling both the central basin and most of the surrounding ringed valleys but leaving here and there some of the scarps to form ringed ramparts above the mare plains. (From N. M. Short and M. L. Forman, 1972, *Modern Geology*, Gordon and Breach Science Publishers, Ltd.)

situation in England in the early nineteenth century in which the Paleozoic was gradually subdivided into systems and series by establishing superposition relations without regard to absolute times of formation, lunar systems and series are defined solely on evidence of relative age. Each system is recognized as older than any systems that altered or otherwise affected those materials units assigned to it. The Shoemaker-Hackman classification originated from their study of materials found around the southern part of the Imbrium Basin near the craters Copernicus and Eratosthenes.

Thus, in this scheme, the time over which the Imbrium Basin was carved out by impact and modified before being invaded by lavas was designated as the *Imbrian Period* and corresponding deposits defined as the *Imbrian System.* Deposits identified as Imbrian in age are superimposed on parts of the older central highlands. However, other large impact basins also must have ejected materials on to the highlands and into any preexisting basins. Several basins near Imbrium (e.g., Serenitatis) have ejecta deposits now overlapped by Imbrian materials. Evidently, basins were being formed over an extended time period (but not necessarily long in relation to total lunar time) and were cut into still older surfaces. Because stratigraphic relations are obviously complex and once-sharp boundaries become obscured, all units which existed prior to the initial Imbrium Basin event (at its onset, a fleeting instant in lunar time) were classified as *pre-Imbrian*—a term broadly analogous to Precambrian as applied to all geologic time before the Paleozoic on Earth. If, however, the Imbrium Basin developed early in lunar time, as now confirmed by *Apollo* age determinations, pre-Imbrian deposits result from a myriad of fast-moving events executed over a comparatively short time span relative to the duration of the Precambrian. Pre-Imbrian materials are exposed mainly on the highlands wherever basin ejecta blankets, large craters, and lavas have not obscured events recorded on the older surfaces.

Events following those of the Imbrian Period produced deposits that fall into several definitive sequences amenable to further subdivisions as systems. In the original Shoemaker-Hackman concept, the filling of all basins with lavas occupied a more or less continuous time span in which volcanic deposits of the *Procellarian System* were emplaced. After the mare lavas solidified, craters cut into these units and continued to disrupt older highlands terrain. The *Eratosthenian System* was defined to include all mare craters and their terra counterparts which are old enough for superficial ray deposits to have been obliterated by subsequent erosion or deposition. Another class of fresher-looking craters, with associated rays and ejecta blankets, is considered to represent the youngest impact events on the lunar surface; all such craters and other features developed during this time interval belong to the *Copernican System.* However, some smaller rayless craters and certain dark-haloed craters are recognized as Copernican in age if they are superimposed on or otherwise related to larger rayed craters that formed during that period. No obviously younger events can be confirmed on the Moon; however, it is still uncertain whether Copernicus, Kepler, Tycho, and other more recent craters have been forming intermittently up to the present or were "frozen in" without significant change at the end of activity long ago on a now-dead Moon.

them to be younger than Apenninian units were included in the *Archimedian Series,* defined from exposures around the crater Archimedes.

As mapping continued to provide more detailed information on interrelationships among units, other modifications of the original Shoemaker-Hackman classification were introduced. The most recent version of the major time-stratigraphic divisions used by the U.S. Geological Survey is given in Table 9-1 along with appropriate subdivisions. In one change, mare units that were once the basis of the Procellarian System are now considered a group placed within the upper Imbrian System. They include the closing sequence of volcanic deposits which flooded the Imbrium and other basins. Some even younger volcanic deposits are recognized as Eratosthenian or Copernican in age. Certain units previously designated as series

TABLE 9-1

Time-Stratigraphic Classification for Lunar Map Units*

Period (System)	*Age of Boundary Between Periods† (billions of years before present)*	*Events and Associated Units*
Copernican		Deposits of Copernicus and other fresh-appearing rayed craters
	~2.0-2.5	
Eratosthenian		Deposits of Eratosthenes and similar slightly subdued craters whose rays are no longer visible or are very faint at high sun illumination angles
		Dark mare materials in the Imbrium Basin and Oceanus Procellarum
	~3.1-3.8	
Imbrian		Deposits of Archimedes and other mare-flooded craters superposed on circum-Imbrium deposits
		Circum-Imbrium deposits and structures
	~3.9-4.0	
pre-Imbrian		Deposits of Julius Caesar and other similar degraded craters cut by the Imbrium Basin structure and covered by its deposits
Birth of Moon	~4.6	

*Strictly applicable only to region of Moon around Imbrium Basin but generally applied to the Moon as a whole.

†Based on radiogenic age dating of Apollo mare and terra rocks and on photogeologic interpretations of relative crater densities in superposed materials units.

After the Shoemaker-Hackman classification first appeared, attempts were made to subdivide the systems still further as mapping focused on specific units. Those materials derived by excavation from the Imbrium Basin and then modified by further cratering prior to lava extrusion were placed in the *Apenninian Series,* named from the type area in the rugged Apennine Mountains that was apparently covered by a pile of ejecta from the Imbrium Basin. Because they were emplaced almost instantaneously, materials directly identified as Imbrium Basin ejecta constitute a major time plane or horizon marker in the lunar stratigraphic column to which all other units or sequences can be referenced as younger or older. Deposits in and around the Imbrium Basin whose characteristics indicate

were dropped in favor of formation names applied to ejecta deposits or presumed volcanic materials related to each basin where such materials are still recognizable. For example, the lower Apenninian Series in the Imbrium region becomes the Fra Mauro Formation and rocks of the Archimedian Series are called "crater materials" and the Cayley Formation. Possibly contemporaneous units surrounding different basins are individually named from newly specified type areas (usually, from nearby large craters). But because correlations are rather inexact, the relative positions within a generalized time-stratigraphic column for the whole Moon are not always determinable. Thus, ejecta deposits around the Orientale Basin are known as the Hevelius Formation, whereas those around the Nectaris and Humorum basins are included in the Janssen and Vitello Formations, respectively. These changes now appear on later geologic maps of the Moon. As higher-resolution photographs of the lunar surface from *Orbiter* and *Apollo* are examined in detail, more units are being recognized and traced to different regions.

Lunar materials units defined for specific regions around the major basins and the highlands have now been arranged by the U.S. Geological Survey in a composite stratigraphic columnar section applicable to the frontside of the Moon (Figure 9-3). Correlation from any one basin to other distant areas is difficult because of lack of widespread rock datum planes (areally extensive units) analogous to thin deposits on Earth laid down over much of a continent during a single invasion of an epeiric sea. The Fra Mauro Formation—an excellent datum plane owing to its extension well beyond the Imbrium Basin—can be traced over an area representing about a quarter of the lunar nearside, but it has no recognizable exposures in either the Orientale or Humorum basins. The Sulpicius Gallus Formation, on the other hand, is adjacent to both the Imbrium and Serenitatis basins. In the early stages of mapping, the most useful time marker or datum plane was thought to be the mare lavas that covered most of the basins and spilled over into the lowlands. While this assumption still generally holds, age dating of *Apollo* samples indicates that lava emplacement took place over an interval of no less than 500 million years.

Determination of Relative Ages

This section reviews the principal criteria and methods by which astrogeologists have identified lunar units, delineated their boundaries, and classified them according to relative age.

Terrain Types

Major units on the Moon can be separated to some extent by their characteristic topographic expressions. An earlier (1966) U.S. Geological Survey classification of the physiography of the lunar equatorial belt recognized some twenty-seven morphological types based on shapes, slope characteristics, terrain roughness, relative relief, and distribution patterns. Where age and/or mode of origin differences are large, adjacent units are often easily distinguished by their terrain contrasts. Thus, a hummocky surface surrounded or embayed by smooth plains material will display boundaries that can be clearly outlined. Again, deposits associated with a rille commonly can be recog-

TABLE 9-2

Regional Physiographic Characteristics of the Lunar Nearside Geological Provinces

Unit	*Regional Characteristics*
A. Mare Plateaus Younger Mare Older Mare	Dark fill of major multi-ring basins and depressed areas of irregular shape such as Oceanus Procellarum. Most recent examples of widespread flooding and crustal modification
B. Dark blanketed Terra Light blanketed Terra	Subdues topographically varied subjacent surface—albedo contrast marked. Terra modification is peripheral to basins
C. Light Terra Plains Hilly and Furrowed Terra Hilly and Pitted Terra	Smooth, light plains filling structural troughs, large old craters and irregular depressions. Furrowed and pitted units obscure older more cratered surfaces. Modification and hybridization of primitive terra materials.
D. Circum-basin Blankets and Structures	Structurally modified and/or blanketed terrain surrounding Mare Imbrium and Mare Orientale.
E. Degraded Terra, undivided Densely cratered Terra	Blocky hills and unevenly filled depressions without distinctive superposed landforms. Of mixed origin including debris from pre-Imbrium circular basins. Densely cratered regions are primitive and exhibit least modification by subsequent endogenetic processes.

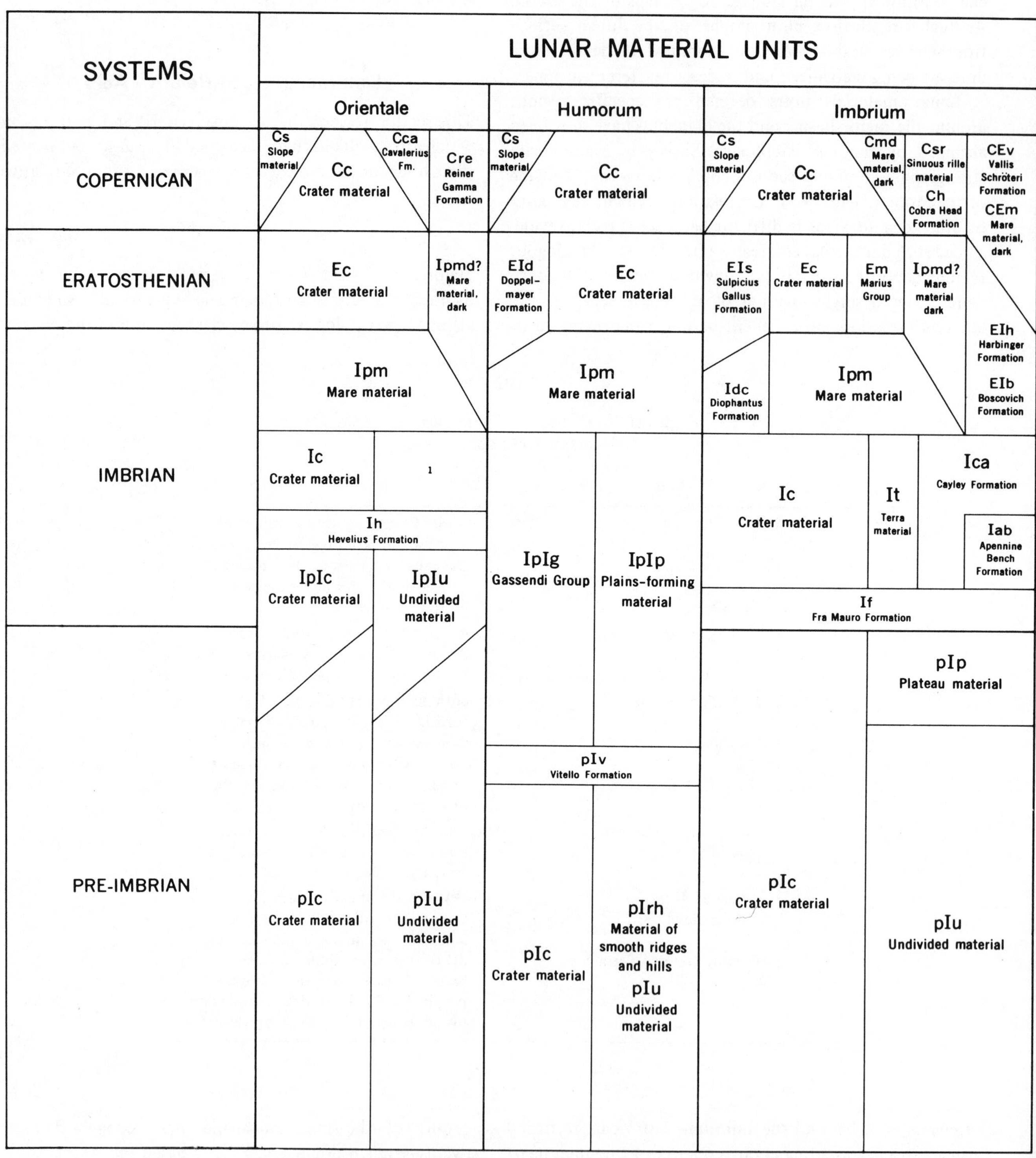

Figure 9-3 (Here, and on opposite page) Stratigraphic columns showing the relative sequence of major lunar materials units mapped around the indicated lunar basins and terrae. The estimated age of formation of the different units is given in terms of the time-stratigraphic classification as indicated in the column marked "Systems." Symbols used for individual materials units correspond to those appearing on maps published by the U.S. Geological Survey. (From D. E. Wilhelms, 1970.)

LUNAR MATERIAL UNITS — Continued

Serenitatis	Nectaris	Fecunditatis	Crisium	terra
Cs Slope material; Cc Crater material	Cs Slope material; Cc Crater material; Ct Theophilus Formation	Cs Slope material; Cc Crater material	Cs Slope material; Cc Crater material	Cs Slope material; Cc Crater material; Complex units
EIs Sulpicius Gallus Formation; Ec Crater material; Emd Mare material, dark; Et Tacquet Formation	Ec Crater material	Ec Crater material	Ec Crater material	EIc Crater material
Ipm Mare material	Ipm Mare material	Ipm Mare material	Ipm Mare material	
IpIc Crater material; IpIp Plains-forming material; IpIhf Hummocky material, fine	Ica Cayley Formation; Ik Material of Kant Plateau; It Irregular terra material; IpIc Crater material; IpIp Plains-forming material	IpIc Crater material; IpIp Plains-forming material	IpIc Crater material; IpIp Plains-forming material	IpIc Crater material
2	2	2	2	
pIc Crater material; pIu Undivided material	pIc Crater material; pIu Undivided material	pIc Crater material; pIu Undivided material	pIc Crater material; pIu Undivided material	pIc Crater material; pIhc Hummocky material, coarse; pIhf Hummocky material, fine

Notes on the chart layout:

- In the Serenitatis column, the IpIhf box (Hummocky material, fine) runs down through the IpIc/IpIp level and the pIc/pIu level.
- In the terra column, "Complex units" runs the full height of the chart.

nized by their spatial confinement to the vicinity of the resulting channel.

The U.S. Geological Survey has developed a general classification of major geologic provinces on the Moon's nearside which are defined, in large part, by their regional physiographic characteristics (Table 9-2). The arrangement of these morphological units does, however, have some general stratigraphic significance in that each principal subdivision is broadly associated timewise with events to which the lunar formations and groups are correlated. Thus, degraded terra units are largely pre-Imbrian, whereas most mare units are presumably late (upper) Imbrian in age.

Contacts

Major lunar units in juxtaposition, such as mare lavas that lap on to terra deposits, usually are clearly distinguishable.

Figure 9-4 The craters Arzachel (lower), Alpetragius (above and left), Alphonsus (center), and Ptolemaeus (top, in part) shown in *Lunar Orbiter 4* photograph 108H_2. These craters are superimposed on materials mapped as Fra Mauro Formation. The floors of Alphonsus and Ptolemaeus are partly covered with volcanic lavas and/or ash associated with the Cayley Formation. Younger mare lavas of the Procellarum Group lap on to the edge of the highlands along the left edge of the photo. (NASA photograph.)

More generally, it is frequently difficult to define and trace contacts between individual materials units on lunar photographs or to infer their positions from telescope observations. Crater deposits commonly merge with their surroundings. In addition, the lithologic variants exposed at the surface may be generally similar in composition and hence show little contrast in albedo to the eye or on photos. The range of reflectivity on the Moon is small and observed changes from place to place do not necessarily coincide with unit boundaries. Albedo is most useful in distinguishing mare from terra materials. Slight albedo variations within a mare plain may correspond to different flow units, as illustrated by Mare Serenitatis. In a few instances, a younger flow will possess a well-defined front (Figure 8-4a). Albedo is also effective in "spotlighting" some younger events such as Copernican craters and landslides that create surfaces with higher reflectivities (Figure 5-8).

Contacts between overlapping units are susceptible to destruction by meteorite bombardment, local "shakedown" during moonquakes, and burial under younger ejecta. Where turnover and mixing in the regolith have been extensive, any initial differences between contacting units formed from specific large events will be obliterated as these units are homogenized. The region through which a contact is drawn may vary imperceptibly in some subtle characteristic (perhaps seen only through the telescope); limits sometime have to be placed arbitrarily as some defining property reaches a critical value or visual contrast, much as boundaries between facies are decided in terrestrial stratigraphic problems.

Transection

The usual argument given for crosscutting of one unit by another—namely, that the unit which has been invaded, overlapped, or offset is older—applies to lunar as well as terrestrial features. However, owing to the absence of well-defined layers at the resolutions applicable to most lunar photographs, it is frequently difficult to recognize transection effects. Repeated or missing sections—marks of faulting in terrestrial terrains—are likewise not noticeable on the lunar surface, partly of course because of the apparent scarcity of normal and reverse faults in the tectonic regime attributed to the Moon. Clearcut signs of age differences deduced from crosscutting relations include craters, rilles, wrinkle ridges, domal extrusions, and escarpments, all of which are younger in time of formation than their host units.

Overlap

Where deposits from one crater encroach upon those of an adjacent crater, the overlapping one must be more recent. An excellent example, on a grand scale, is given by the relative sequence of deposits from the craters grouped around Ptolemaeus (see Figure 9-4). Volcanic ash units also will generally overlap all preexisting units over which they flow. Overlap and other superposition geometries can sometimes be detected in *Orbiter* and *Apollo* stereophotos by the slight differences in relief or elevation brought out in the three-dimensional image. With age, overlap relations tend to become diffuse as contacts are wiped away. Sometimes only slight differences in the terrain or variations in crater densities between units will remain to separate deposits.

Crater Frequencies

Regardless of whether a constant, steadily decreasing (in time), or erratically variable meteorite flux best represents the history of projectile bombardment on the lunar surface, the fact remains that a larger number of impact craters within a unit or over a reference area will develop in older terrain than in younger. This simple relationship—that of increasing frequency with age of a surface—holds until that surface either becomes saturated with craters below some size or is removed from view by erosion or burial. The saturation level depends on crater size; craters above a certain diameter may still be infrequent enough so that others in the upper size ranges can be superimposed on to the surface without interfering with each other.

Comparison of the cumulative frequency curves of crater distributions in different regions enables estimates to be made of the relative ages of the cratered surfaces. A curve lying parallel and to the right of another curve describes a surface with a greater apparent age in which the number of craters of all sizes has increased uniformly. If the curve for the older cratered surface shows a decrease in slope, this would imply saturation in the lower size classes (new craters in these classes in effect destroy an equivalent number of older ones) together with a continuing increase of larger, although as-yet-unsaturated, size classes.

In Figure 9-5, cumulative frequency curves determined from crater counts are shown for typical highlands and mare areas respectively. The higher crater population in the highlands (in places as much as thirty times more than maria) attests to the generally older ages of the surfaces now visible. The relative ages of several unconnected maria themselves are distinguishable as well, as suggested by the curves in Figure 9-6 for representative surfaces within three basins which, from other evidence, appear to have originated at different times. The less steep slope for crater frequency distributions on the Humorum Basin rim illustrates the effect of saturation with larger craters; alternatively, burial of smaller craters by volcanic or other deposits could explain this change.

To be valid, crater counts should cover large areas within a unit whose age is sought. Local variations in frequency

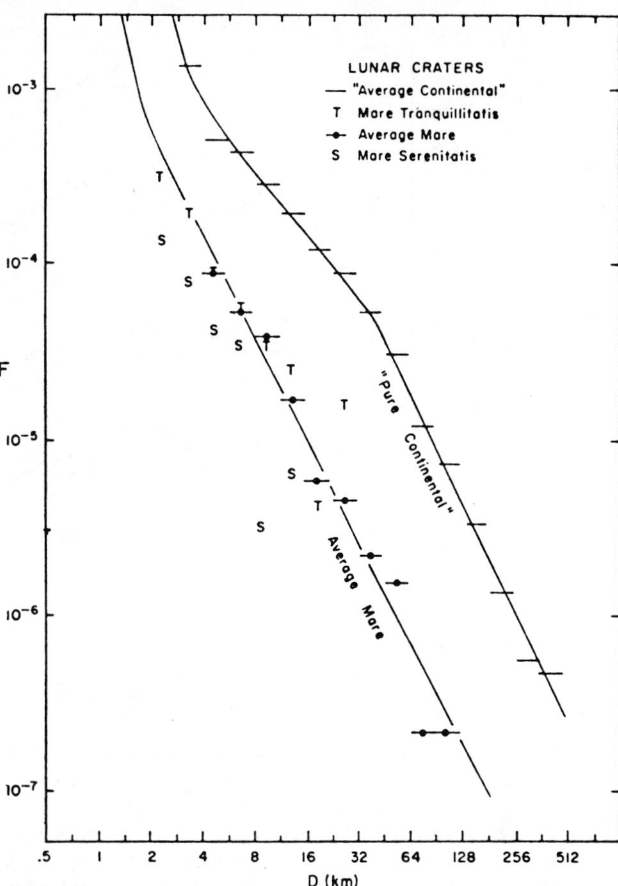

Figure 9-5 Crater densities on "average" continental and mare surfaces and on the highly and sparsely cratered maria of Tranquillitatis and Serenitatis, respectively, as indicated by an incremental crater frequency plot based on data from telescope and *Ranger* observations. (After W. K. Hartmann, 1966.)

Figure 9-6 Cumulative crater frequency curves derived from counts of craters located on the rims of the Orientale, Imbrium, and Humorum Basins. Because Humorum has the most and Orientale the least number of craters of a given size (e.g., 10 km), they are respectively the oldest and youngest of the basins in time of formation, provided the majority of these craters are related to impacts resulting from a steady or consistently decreasing influx of extralunar particles. (Curves prepared by A. R. Kelley; cited by McCauley, 1967.)

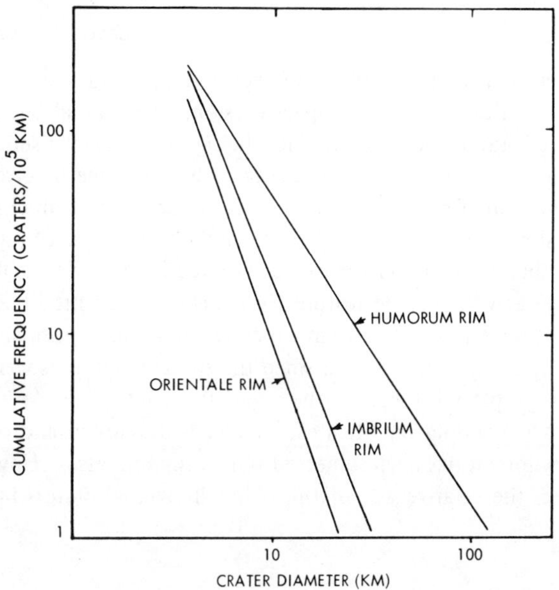

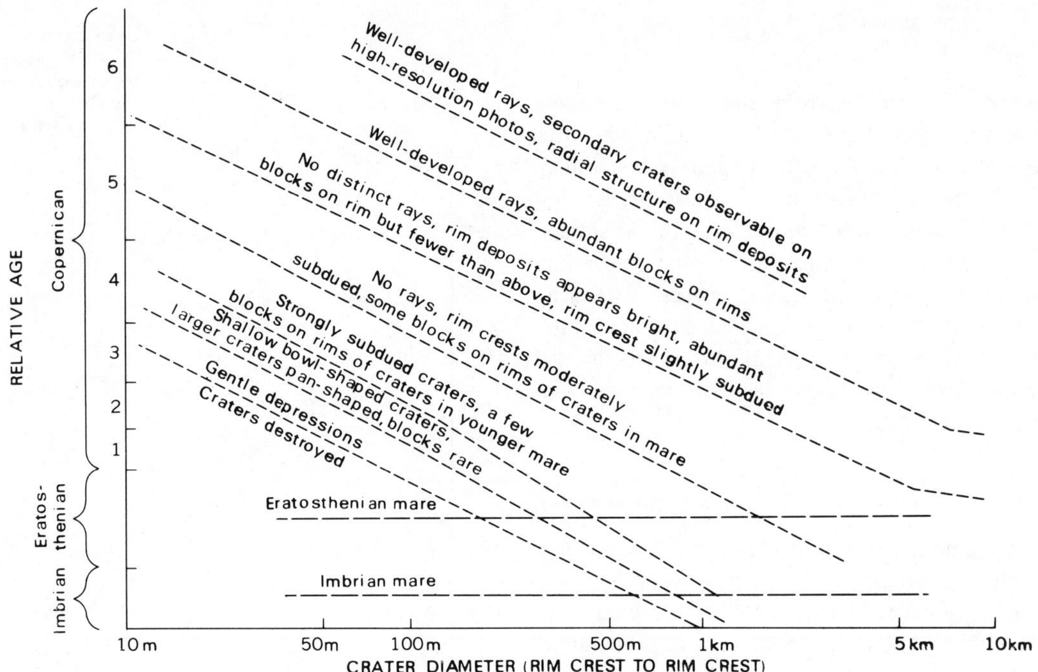

Figure 9-7 The Trask classification of the relative age of craters between about 10 meters and 10 kilometers in diameter occurring on the maria. Criteria for each stage in changing crater morphology (essentially a degradation process) are included in the areas bounded by dashed lines. The postulated ages are expressed on the ordinate (using time-stratigraphic terms). The apparently oldest and youngest Copernican craters are indicated by the numbers 1 and 6 respectively. As an example of the use of this diagram, consider the category *strongly subdued craters.* If such a crater were 50 meters in diameter, it would belong to the Copernican class falling between 2.5 and 3.5; but if it were 500 meters wide, it would be placed in the Eratosthenian age group. The upper limit in age of a surface containing different sizes of craters in varying stages of degradation is determined by the presence of the oldest individual craters in the total population considered to have been imposed on this surface. (Courtesy N. Trask, U.S. Geological Survey.)

(or crater density) are to be expected wherever secondary crater swarms, volcanic crater fields, or drainage craters become mixed with impact craters. The influence of these "interlopers" can sometimes be disclosed from cumulative frequency curves by slope changes at some critical size or by statistical analysis of crater distributions.

Crater "Ages"

As a lunar crater is subjected over time to isostatic adjustments, slump infilling, impact erosion, and gradual burial by ejecta, ash, or lava, its morphology undergoes a series of progressive changes. In general, the sequence involves reduction of rim height, decrease in crater depth, rounding of slopes, burial of blocky ejecta, and obliterations of rays and blankets. A relative age can be cited for each definable stage at which some feature has appeared, been modified, or vanished. In any area in which craters were continually being formed and eroded since the surface materials were first emplaced, there will be a variety or range of relative ages depending on when each crater is developed and on how long it has been subjected to altering processes. However, the relative age of the oldest individual craters impressed on the exposed surface materials will be approximately equivalent to the age that describes the initial emplacement in the underlying unit or first sculpturing of the terrain. The units present in a region can be tied into the lunar time scale by carefully assessing the ages of many craters in different sections of the surface, as well as on the proportion of craters in each age group. However, the relative ranking alone will not establish the age of each unit until one or more of these units is correlated by independent methods (e.g., radiometric dating) to appropriate type areas. Part of the difficulty lies in the likelihood of nonconstant meteorite flux rates, including the possibility that there was a time gap between the development of a crustal surface and the inception of intensive meteorite bombardment. In terrestrial sediments a similar problem arises from attempting to use variable rates of deposition to assign ages to formations within the stratigraphic column.

Trask has proposed a scheme for classifying relative ages (ranging from Imbrian to Copernican) of craters smaller than about 10 km. His criteria for aging are described in Figure 9-7 as categories bounded by dashed lines. A numerical age span is defined by intersection of

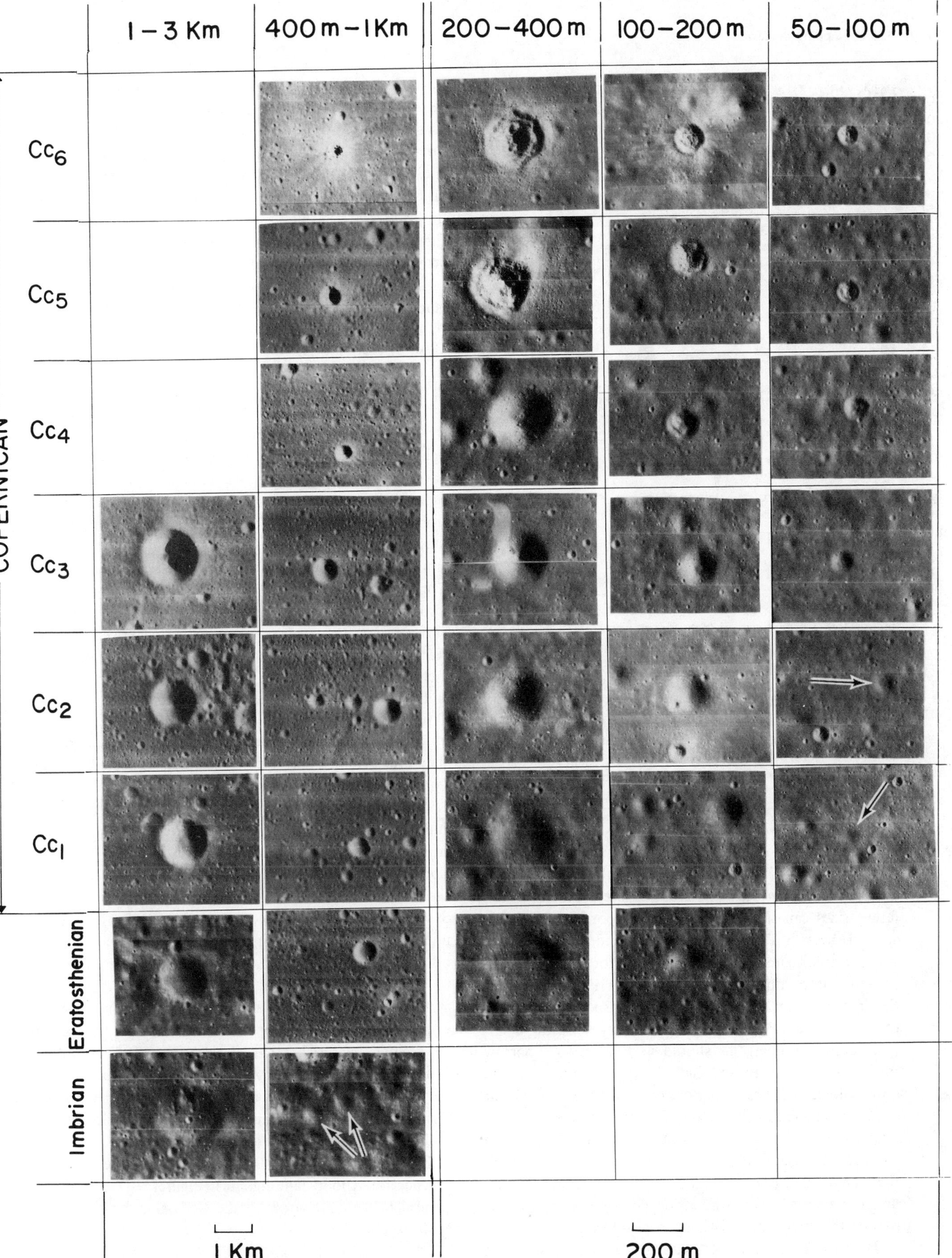

Figure 9-8 Examples taken from Lunar Orbiter photographs of different-sized craters displaying various stages of degradation as defined by the categories and relative age index described in Figure 9-7. (Courtesy N. Trask, U.S. Geological Survey.)

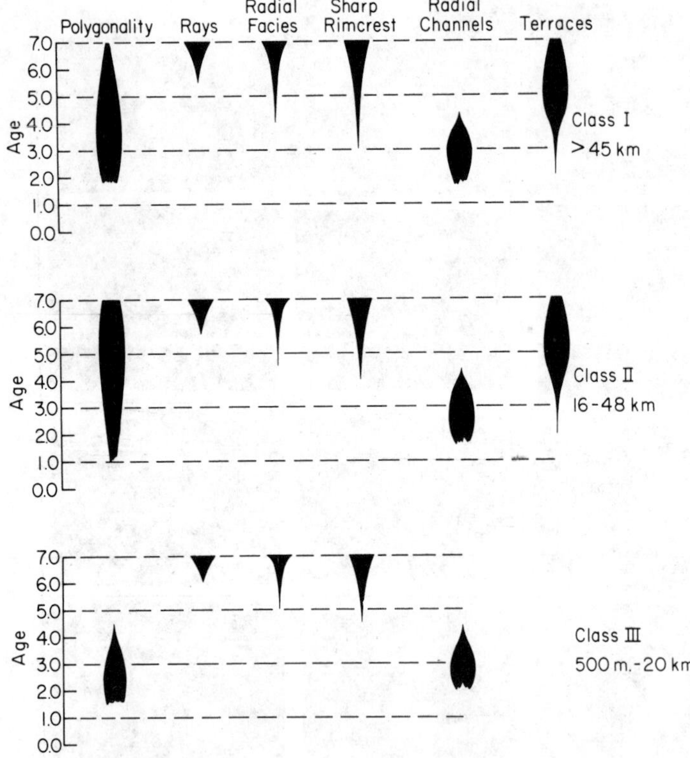

Figure 9-9 Progressive changes in morphological features of craters grouped into three size classes and arranged in order of inferred relative ages (youngest at top). The width of the black band under each feature category refers to the degree or extent of maximum development; tapering ends denote appearance or disappearance of a diagnostic feature. The age sequence (7 to 0) is assumed to reflect stages of increasing erosional modifications resulting from further impact bombardments, gradual burial by ejecta and volcanic products, and other processes. (From Pohn and Offield, 1969.)

these lines with the ordinate; the largest number (6+) represents the youngest age. The age boundaries slope downward with increasing crater diameter. Craters of different sizes can therefore have contemporaneous relative ages. However, if a large (e.g., 1-km) and a small (50-m) crater formed simultaneously at some locality, at a considerably later time (higher horizontal line in diagram) the smaller crater will have degraded ("aged") much more than the large. For effective application to dating a surface, crater comparisons should be limited to a narrow size range in order to find the "oldest" ones present. The morphologies of these aging craters as a function of their sizes are depicted in Figure 9-8. Trask constructs his classification on three basic assumptions: (1) craters when formed are fresh, (2) smaller craters degrade more rapidly than large, and (3) degradation rates are approximately uniform everywhere on the Moon. His age method works best on the mare surfaces where the regolith is usually thin.

Determination of ages for smaller craters is meaningless in highlands terrain because the surfaces have long since become saturated with these craters. Pohn and Offield (1969) have therefore devised an age classification based on somewhat different criteria (Figure 9-9). They grouped lunar craters into three principal size classes; furthermore, those in class III initially are circular, whereas those in class I tend to be strongly polygonal. Representatives of different ages for one of these classes (II) are illustrated in Figure 9-10. Using this classification, these workers have attempted to date some of the large lunar basins and certain of the cratered plains in both near and farside highlands (Figure 9-11). Examples of lunar surfaces amenable to the Trask and the Pohn and Offield classifications are given in Figure 9-12.

The Lunar Mapping Program

By 1970 the U. S. Geological Survey had prepared 44 quadrangle maps at a 1:1,000,000 scale (Figure 9-13) covering most of the lunar frontside except the polar and extreme limb regions. Of these, 32 were charted primarily from telescope observations while 4 more maps were drawn both from these observations and *Orbiter* photography. The remaining 8 maps have been prepared entirely from *Orbiter 4* M-frame photographs, whose optimum resolutions of 120-150 meters are as much as five times better than the corresponding telescope coverage. Currently, many of the original maps are being re-edited to include more of the details, both on smaller surface features and on boundaries, present in the higher resolution *Orbiter* and *Apollo* photos.

The Copernicus quadrangle is usually cited as the standard example of a lunar map. The region around Copernicus was first mapped by Shoemaker and Hackman (1962) using both telescope photographs (Figure 9-14) and direct viewing. They defined the Imbrium System from hummocky materials found north and south of Copernicus. Dark lavas to the east and west that overlap the Imbrium deposits were placed in the Procellarum System. The crater Eratosthenes penetrates these mare units northeast of Copernicus; deposits from this and other craters without conspicuous rays were grouped into the Eratosthenian System. All materials in these older systems are covered in places by the ejecta blanket from Copernicus itself. The Copernican System was thus set up to include materials that derive from fresher-looking craters with associated rays comparable to Copernicus as the prototype. The original Copernicus map was later revised by Schmitt, Trask, and Shoemaker in a color version published in 1967 (Figure 9-15).

Special purpose maps, such as those made to cover prospective Apollo landing sites (Figures 9-16 and 9-17), have also been prepared by the U. S. Geological Survey. Most vary in scale from 1:250,000 to 1:5,000 depending on the amount of detail needed. In addition, the Survey has now published a 1:5,000,000 scale geologic map of

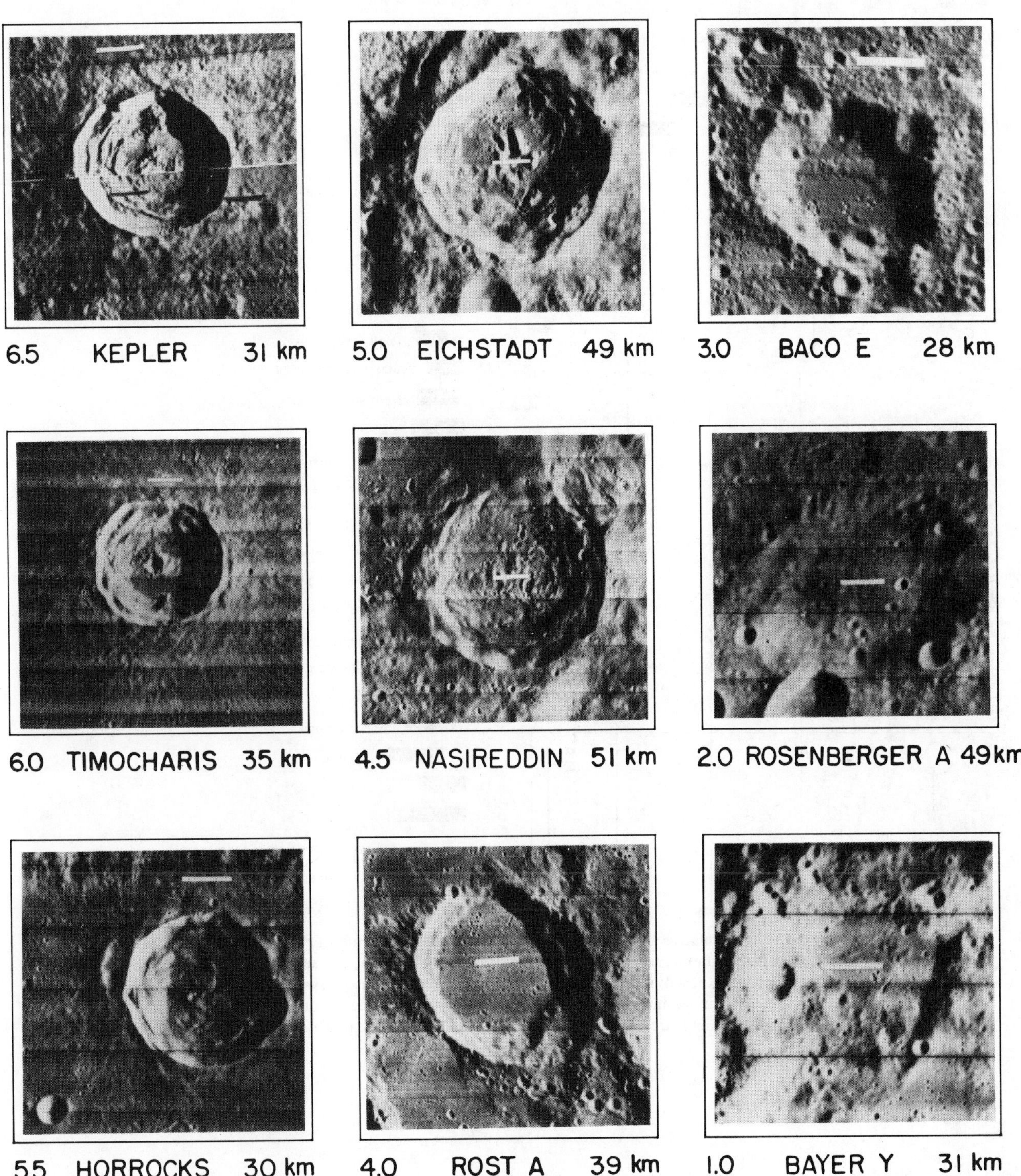

Figure 9-10 Examples of progressive modification of the morphology and exterior deposits associated with Class II craters, arranged according to the relative age sequence (index numbers to left of crater name) defined by Pohn and Offield (1969). (Courtesy H. A. Pohn, U.S. Geological Survey.)

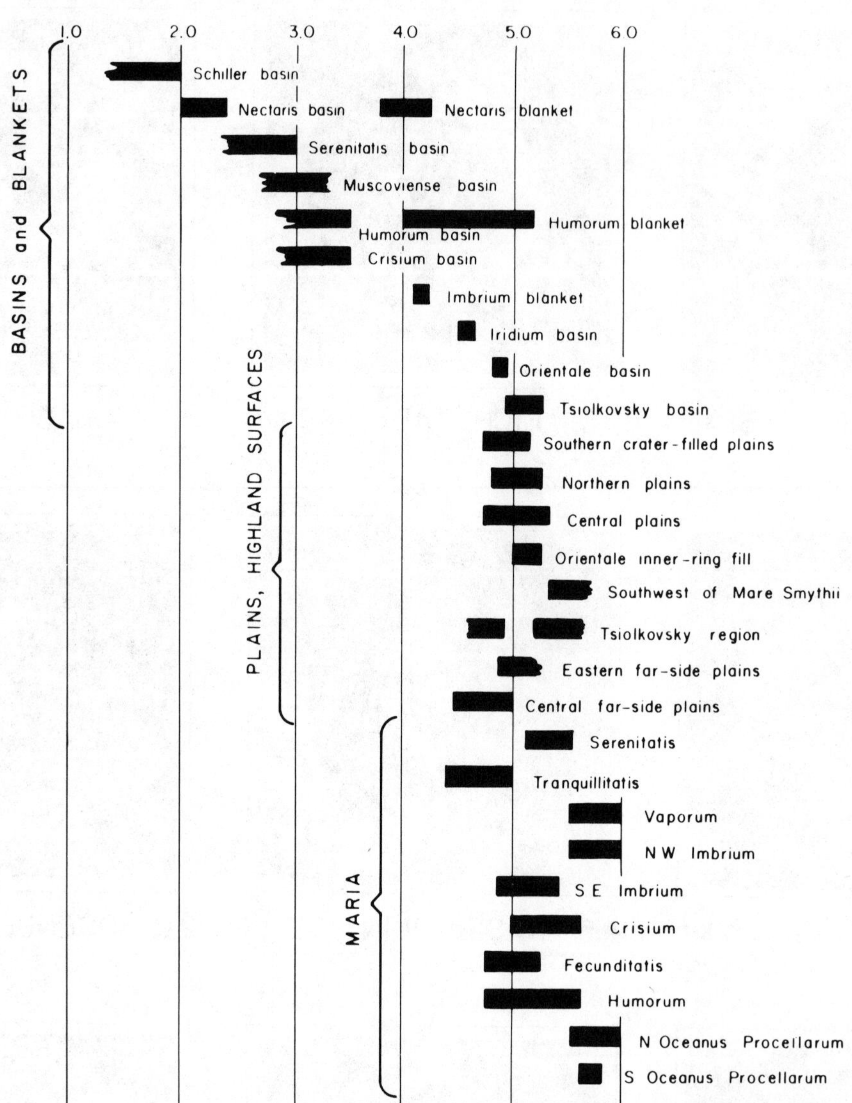

Figure 9-11 The relative ages, based on the crater morphology scale of Pohn and Offield, of some major physiographic regions of the Moon. Within each region, the age is defined by the oldest crater type exposed on the surface (width of the black bars expresses the uncertainty in assigning a precise crater morphology age). Absolute ages for a region can be approximated from data presented in Figure 13-9. (From Pohn and Offield, 1969; extracted from Mutch, 1970.)

(a)

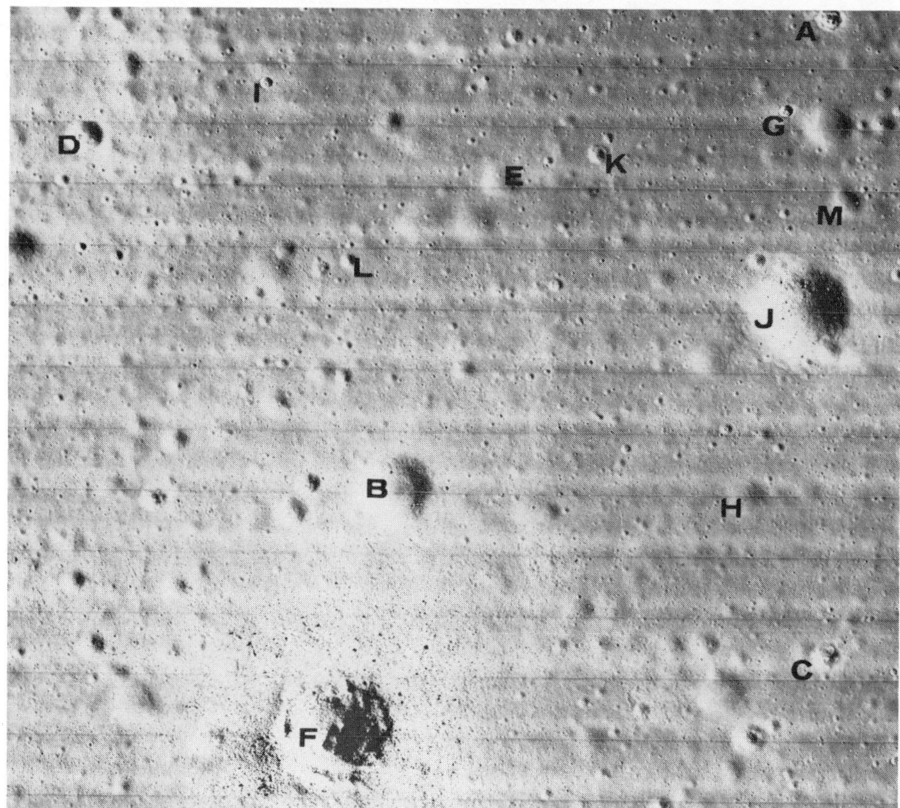

Figure 9-12 (a) Small craters on a mare surface, which can be ranked according to their relative ages using the Trask classification. (b) Large craters on part of the farside highlands whose differing morphologies allow their relative ages to be classed in terms of the criteria developed by Pohn and Offield. The large feature marked E is the 200-km wide crater Tsiolkovsky on the lunar farside (viewed looking toward the southern hemisphere), the only major structure in this part of the highlands invaded by large volumes of dark (basaltic?) lavas similar to those emplaced in the maria. The reader is encouraged to attempt a relative age classification of the above small and large crater groups as an exercise, using the letters to identify each and listing the oldest first in the sequence. (NASA *Orbiter* photos.)

(b)

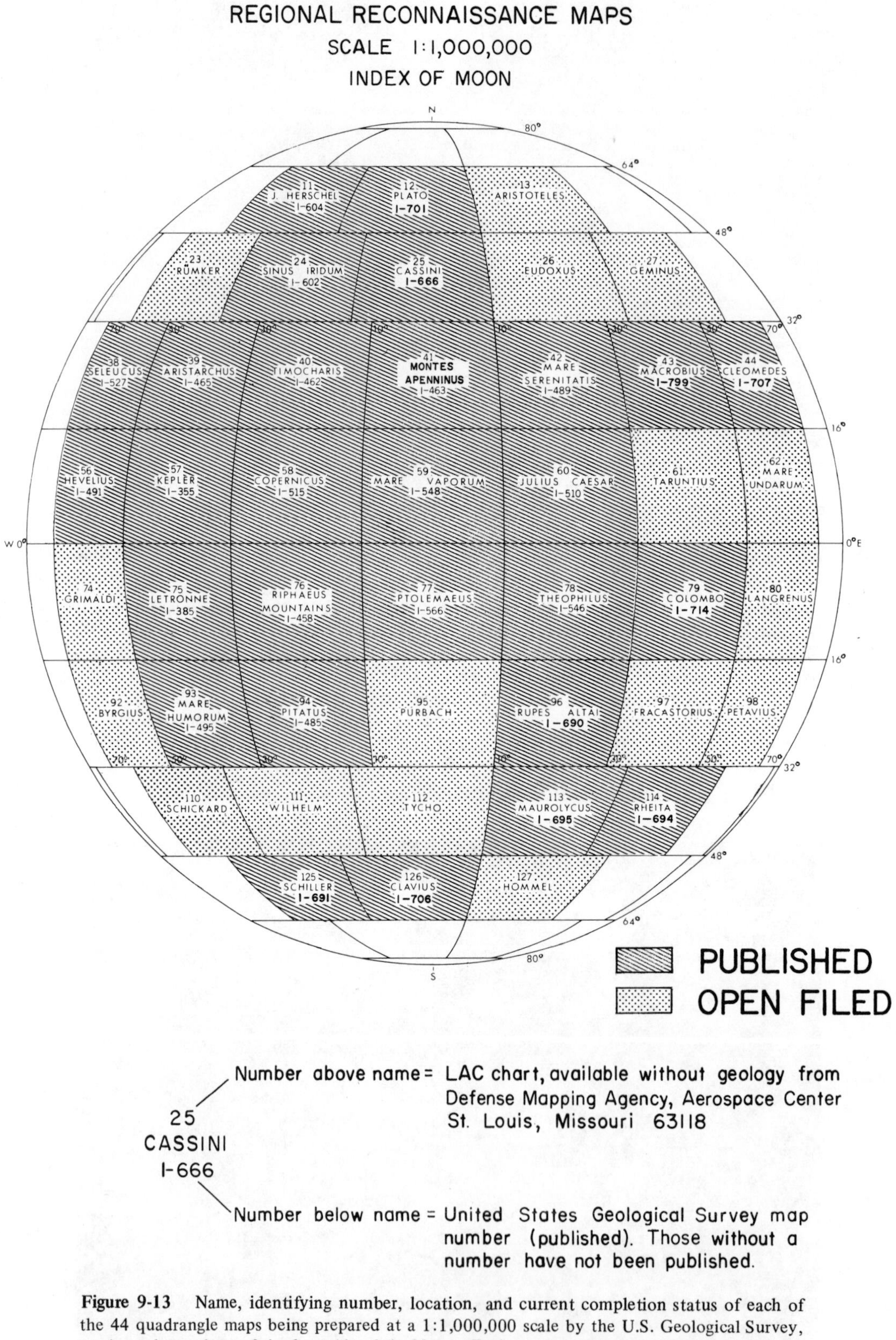

REGIONAL RECONNAISSANCE MAPS
SCALE 1:1,000,000
INDEX OF MOON

N
80°
64°
48°
32°
16°
0°E
16°
32°
48°
64°
80°
S
W 0°

11 J. HERSCHEL I-604
12 PLATO I-701
13 ARISTOTELES
23 RÜMKER
24 SINUS IRIDUM I-602
25 CASSINI I-666
26 EUDOXUS
27 GEMINUS
70° 50° 30° 10° 10° 30° 50° 70°
38 SELEUCUS I-527
39 ARISTARCHUS I-465
40 TIMOCHARIS I-462
41 MONTES APENNINUS I-463
42 MARE SERENITATIS I-489
43 MACROBIUS I-799
44 CLEOMEDES I-707
56 HEVELIUS I-491
57 KEPLER I-355
58 COPERNICUS I-515
59 MARE VAPORUM I-548
60 JULIUS CAESAR I-510
61 TARUNTIUS
62 MARE UNDARUM
74 GRIMALDI
75 LETRONNE I-385
76 RIPHAEUS MOUNTAINS I-458
77 PTOLEMAEUS I-566
78 THEOPHILUS I-546
79 COLOMBO I-714
80 LANGRENUS
92 BYRGIUS
93 MARE HUMORUM I-495
94 PITATUS I-485
95 PURBACH
96 RUPES ALTAI I-690
97 FRACASTORIUS
98 PETAVIUS
70° 50° 30° 10° 10° 30° 50° 70°
110 SCHICKARD
111 WILHELM
112 TYCHO
113 MAUROLYCUS I-695
114 RHEITA I-694
125 SCHILLER I-691
126 CLAVIUS I-706
127 HOMMEL

PUBLISHED
OPEN FILED

25
CASSINI
I-666

Number above name = LAC chart, available without geology from Defense Mapping Agency, Aerospace Center St. Louis, Missouri 63118

Number below name = United States Geological Survey map number (published). Those without a number have not been published.

Figure 9-13 Name, identifying number, location, and current completion status of each of the 44 quadrangle maps being prepared at a 1:1,000,000 scale by the U.S. Geological Survey, to show the geology of the frontside of the Moon. Each quadrangle is named from some major feature (usually a large crater) found within it. (Courtesy U.S. Geological Survey.)

the Moon's nearside compiled both from previous quadrangle maps and from interpretation of contacts using the *Orbiter 4* frontside imagery. The reader is encouraged to purchase this map as a basic reference to lunar geology and to read the accompanying explanatory text which serves to summarize and expand upon many of the concepts introduced in this chapter.

To be sure, the major geological processes which act to produce the different units depicted on a lunar map are reducible to only a few–impact, volcanism, mass-wasting–in contrast to a much wider range of interacting processes that shape and redistribute the Earth's outermost layers. Thus, the essentials of lunar stratigraphy–and the underlying causes that form the surface units–were well defined and broadly understood without ever having set foot on the Moon itself. By the time men could expect with confidence to reach the Moon's surface, the framework of a lunar geology had been erected. From a stratigraphic viewpoint, the remaining goals centered on establishing the identification of the principal rock types, the radiometric ages of the units sampled, and the genetic patterns in emplacing these materials. The *Apollo* missions have carried us to a realization of these goals.

Summary

Before closing, it is proper to emphasize once more that mapping the surfaces of the other inner planets–despite their distances and inaccessibility–finds firm roots in the conventional techniques which originated in the early nineteenth century with William Smith, Charles Lyell, Adam Sedgwick, and other great stratigraphers. When compared with classical geologic maps of parts of the Earth, the recent maps of both the Moon and Mars (see Chapter 12) show every bit as much complexity and diversity. The sequential relations portrayed on those maps, however, are determined and interpreted with the same procedures as applied to terrestrial mapping.

REVIEW QUESTIONS

1. Describe briefly the technique(s) by which a lunar geologist makes a map of part of the Moon's surface. Why is this task more difficult than preparing a terrestrial geologic map?

2. What are the three types of units with which stratigraphers classify rock bodies? Indicate how they differ in meaning and application.

3. Discuss the concept of a materials unit as it applies to lunar stratigraphy.

4. In what way does rapid sedimentation (from impact events, etc.) complicate the effort to establish time-significant mapping units for the lunar surface?

5. Outline the general sequence of events in the history of a large impact basin.

6. List the major systems (periods) in the lunar time-stratigraphic classification in the order of increasing age.

7. What type of unit is best suited as a time marker or datum plane in setting up a lunar stratigraphy? Why?

8. Discuss briefly how each of the following is used in distinguishing materials units and determining relative ages: terrain types; transection; overlap; crater frequencies.

9. How can changes in slope and position of cumulative frequency curves describing crater distributions in several regions be used to estimate relative ages of surfaces in each region?

10. Explain in some detail how the Trask classification uses crater morphology to estimate the relative ages of different surfaces on the Moon.

Figure 9-14 (page 134) Earth-based telescope photo of the crater Copernicus and its surroundings taken through the 120-inch reflector telescope at the Lick Observatory. This was one of several classic photos used by Shoemaker and Hackman (1962) and later by Schmitt, Trask, and Shoemaker (1967) to prepare a geologic map of the Copernicus region. In addition, these workers viewed the lunar surface directly through smaller telescopes to assist them in making interpretations of materials units relations.

Figure 9-15 (page 135) Part of the revised geologic map of the Copernicus quadrangle prepared by H. H. Schmitt, N. J. Trask, and E. M. Shoemaker and published in color in 1967 by the U.S. Geological Survey (U.S.G.S. Map I-515) at a scale of 1:1,000,000. Major map units (oldest first) are: If, Fra Mauro Formation (h and f are hummocky and smooth members) of Imbrian System; Ic, Imbrian crater material (w, r, and f refer to wall, rim, and floor); Ipm, Procellarum Group mare material; Ipd, Procellarum mare dome; Eh, Eratosthenian Sulpicius Gallus Formation; Ec, materials from rimless craters of Eratosthenian age; Ech, Eratosthenian chain crater material; Cc, crater materials (w, r, rr, and f refer to wall, rim, rim radial, and floor) formed during the Copernican period; Csc, Copernican satellite crater material; Ccd, dark-halo crater material; Stippled pattern, ray ejecta material.

Figure 9-14

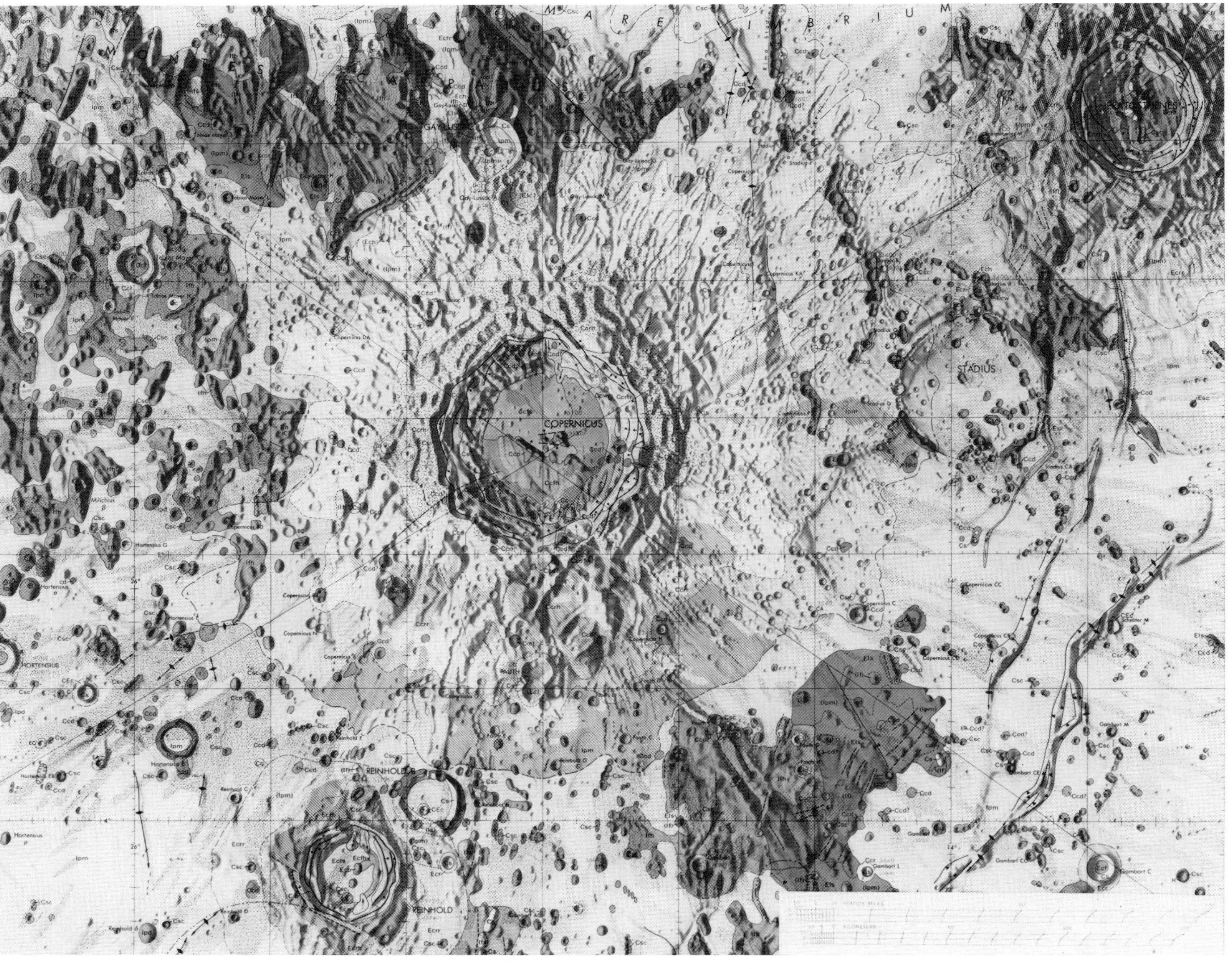

M A R E I M B R I U M
COPERNICUS
STADIUS
REINHOLD
REINHOLD B
HORTENSIUS

Figure 9-15

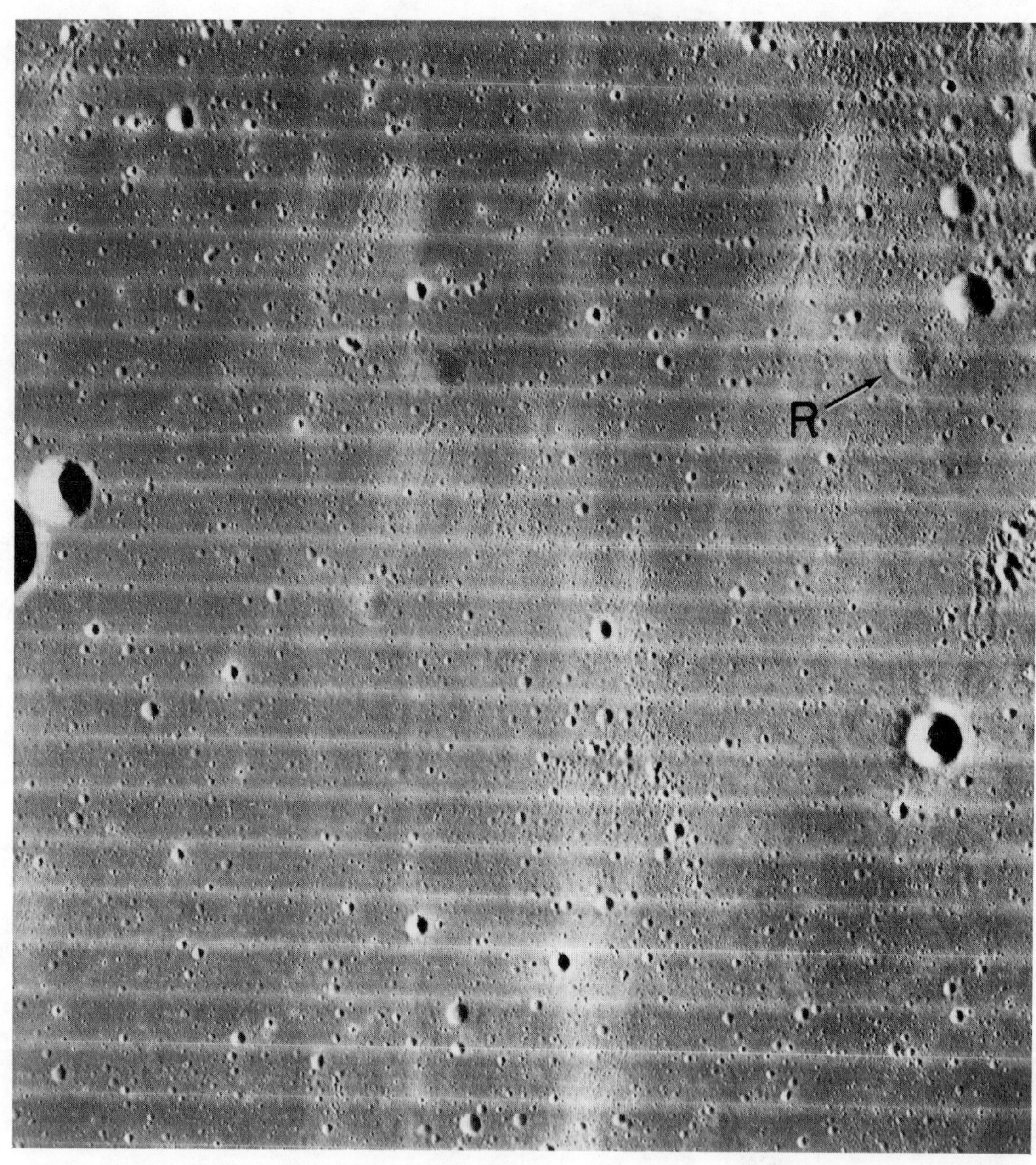

Figure 9-16 A portion of the *Orbiter 2* medium resolution photo used to prepare the large-scale map of the Maestlin G region of southern Oceanus Procellarum shown in Figure 9-17. Note the ring structure marked R.

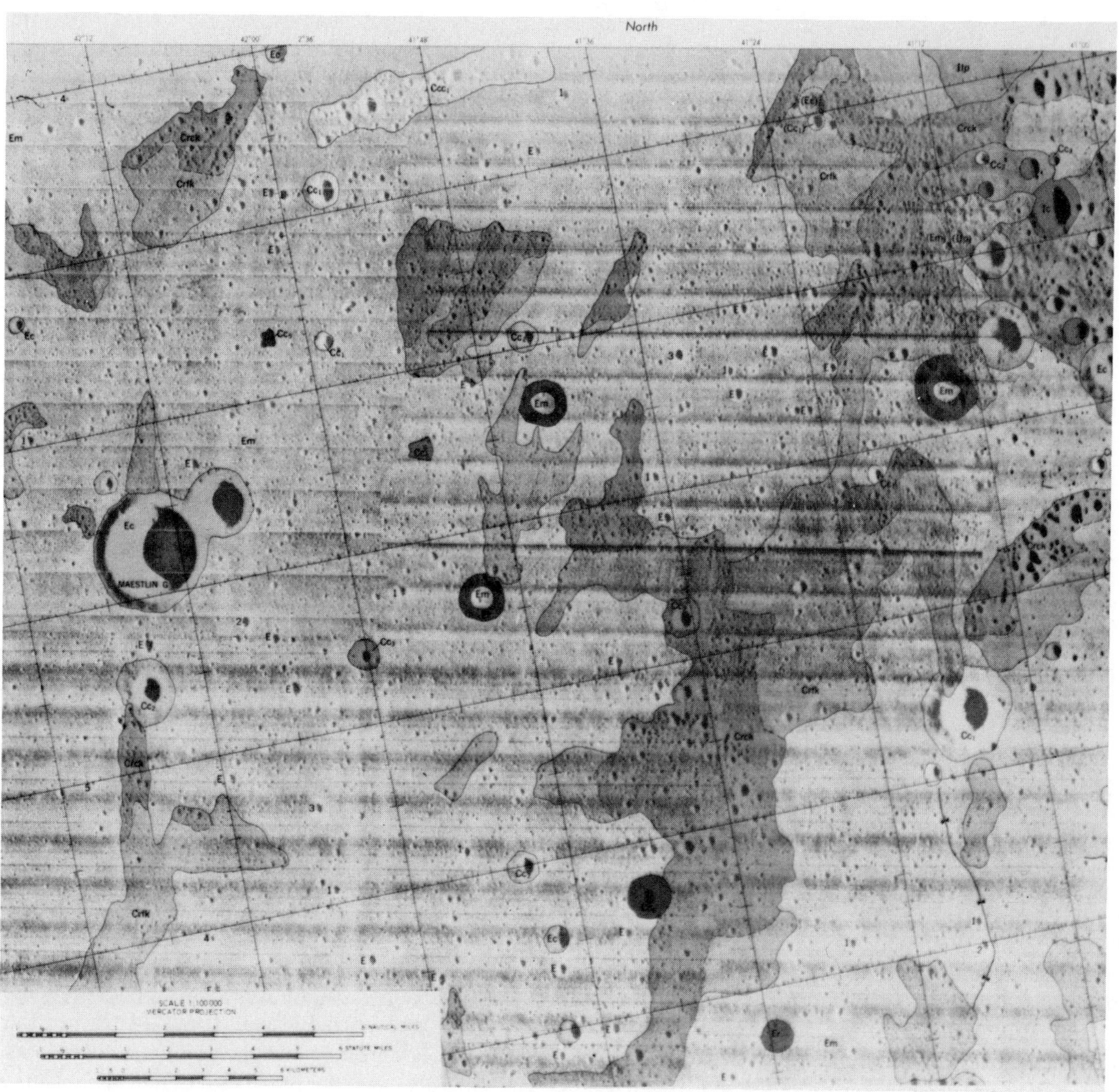

Figure 9-17 Part of the 1:1,000,000 map of the Maestlin G region, drawn on the *Lunar Orbiter 2* medium resolution photo (see Figure 9-16) as a base. The principal map units are (oldest first): Itp, Imbrium terra plains; Ic, Imbrium crater material; Em, Eratosthenian mare material; Er, Eratosthenian ring material (volcanic extrusion or ejecta); Ec, Erastosthenian crater material; Crck and Crfk, Copernican ray material (course and fine) from the crater Kepler (includes clusters of secondary craters); Cc_1 through Cc_5, Copernican age craters (primary and secondary) with 1 denoting the oldest and 5 the youngest in the sequence (using the Trask classification to determine relative ages). Consult U.S. Geological Survey Map I-622 for more details on these units as described in the Explanation section of that map.

CHAPTER 10

Man on the Moon

In mid-afternoon (EDT) of July 20, 1969, nearly 103 hours after leaving Cape Kennedy (Figure 10-1), the historic words that "the Eagle has landed" in the southwest sector of Mare Tranquillitatis confirmed to waiting millions on Earth that men could be placed safely on other planetary bodies. When astronaut Neil Armstrong emerged 6½ hours later from the *Apollo 11* lunar module (LM) to take his famed "one small step for a man" onto the lunar surface, mankind itself began a new "giant leap" in the exploration of space. For more than 2 hours, Armstrong and Edwin ("Buzz") Aldrin, Jr., collected rocks at "Tranquillity Base," cored into the regolith, deployed scientific experiment packages, took hundreds of photographs, and in rare moments delighted a worldwide TV audience by cavorting about the Moon in reduced-gravity "leaps and bounds." The two astronauts then joined Michael Collins in the orbiting command and service module (CSM) and returned to Earth in a splashdown in the Pacific Ocean on July 24, carrying with them 22 kilograms (nearly 49 pounds) of rocks in tightly sealed evacuated containers.

The success of this first manned mission to the Moon was duplicated on November 18, 1969, at a site in the Ocean of Storms some 320 kilometers south of Copernicus. *Apollo 12* astronauts Charles Conrad, Jr., and Alan Bean, separated from Richard Gordon in the CSM, engaged in two separate extravehicular activity (EVA) periods collecting samples, setting up instruments, and photographing the area around the site. On February 5, 1971, some nine months after *Apollo 13* came back to Earth following failure to land on the Moon, *Apollo 14* reached the primary landing target of that earlier ill-fated flight in a highlands like area near the Fra Mauro Crater. Astronauts Alan Shepard, Jr., and Edgar Mitchell (leaving Stuart Roosa above in the CSM) culminated their activities outside the LM with a climb up a steep-sloped crater. On July 30, 1971, astronauts David Scott and James Irwin (with Alfred Worden remaining overhead in the CSM) landed the *Apollo 15* LM just south of Palus Putredinus in the northern Apennines next to Hadley Rille. The pair conducted extensive geological and geophysical tasks during three excursions in which a lunar roving vehicle (LRV), nicknamed the "Moon buggy," was used for the first time. The *Apollo 16* mission placed astronauts John Young and Charles M. Duke, Jr., on the surface of the lunar highlands near the crater Descartes on April 20, 1972; during their stay Thomas ("Ken") Mattingly carried out orbital experiments

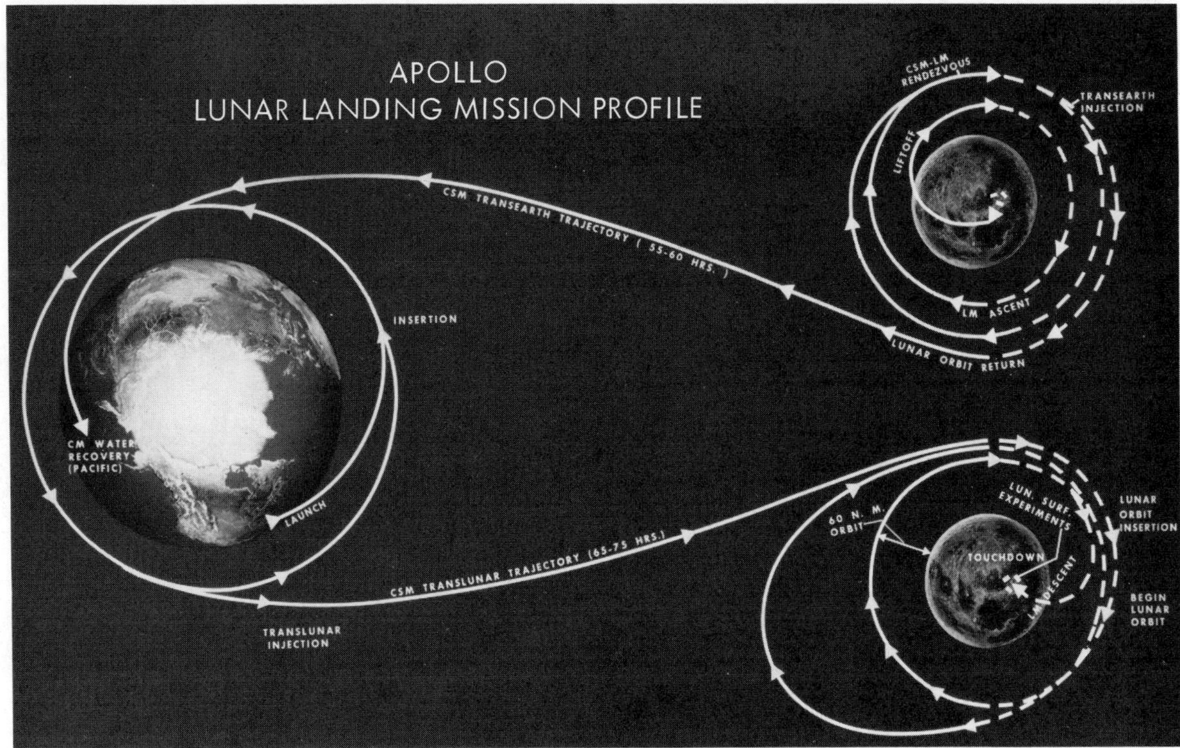

Figure 10-1 A schematic representation of the mission profile leading to the landing and return of *Apollo 11* on its historic flight to the Moon. Subsequent *Apollo* flights follow similar configurations, but the Command and Service Module (CSM) orbits at lower altitudes while the Landing Module (LM) is in operation. (NASA photo.)

in the CSM. *Apollo 17*, the last of the series, resulted in the most comprehensive surface exploration and sample collection in the *Apollo* program. Astronauts Eugene Cernan and Harrison ("Jack") Schmitt (the first professional *geologist* to set foot on the Moon!) landed on December 11, 1972, in a narrow, mountain-enclosed valley of the Taurus-Littrow region in highlands-type terrain east of Mare Serenitatis. Astronaut Ron Evans, circling in the CSM, completed extensive observations that included the use of several new and sophisticated instruments.

Rationale for Selection of Apollo Sites

The various sites visited during the *Apollo* program are located in Figure 1-3 in relation to the sites reached by U.S. and Soviet unmanned probes. A high-altitude view of each site is depicted in Figure 10-2. A summary of the characteristics of each site and the reasons for selecting it are given in Table 10-1.

The *Apollo 11* and *12* sites were chosen after evaluation of high-resolution *Lunar Orbiter* photographs because they were located on smooth surfaces with few craters and hence offered safer landing places. Preliminary studies had indicated a regolith of about 3 to 6 meters thickness at the *Apollo 11* site and perhaps 1.5 to 3 meters at *Apollo 12.* The surface at the *Apollo 12* site—and by assumption its underlying bedrock—was believed to be slightly younger than at *Apollo 11* because of this thinner regolith cover, the slightly lower total crater population, and the relative sharpness of craters in the <1-km size range. Bedrock at each site has been mapped as Imbrian in age, representing lavas of the Procellarum Group. Scientific interest in the *Apollo 11* locality was heightened by its proximity to the highlands and the presence of a series of rays and secondary craters from the Theophilus Crater to the southeast and possibly other craters. The *Apollo 12* site lay on a broad ray traced to Copernicus as its source. Thus, it was hoped that rock fragments excavated from that large crater, including some derived from considerable depths within the mare lavas and perhaps from older basin fill or the ancient lunar crust, might be found at the site.

The spot selected for *Apollo 14* presented a greater challenge to navigation at the moment of touchdown because of the rolling nature of the terrain and a higher crater density. This site was chosen primarily to sample supposed ejecta (Fra Mauro Formation) from the Imbrium Basin. The *Apollo 15* site provided an opportunity to sample ancient pre-Imbrian terrain exposed in one ring and to examine at close hand a large sinuous rille in mare lavas. *Apollo 16* offered a chance to investigate two depositional units of supposed volcanic origin, together with what seemed to be domes, cones, depressions, and cracks

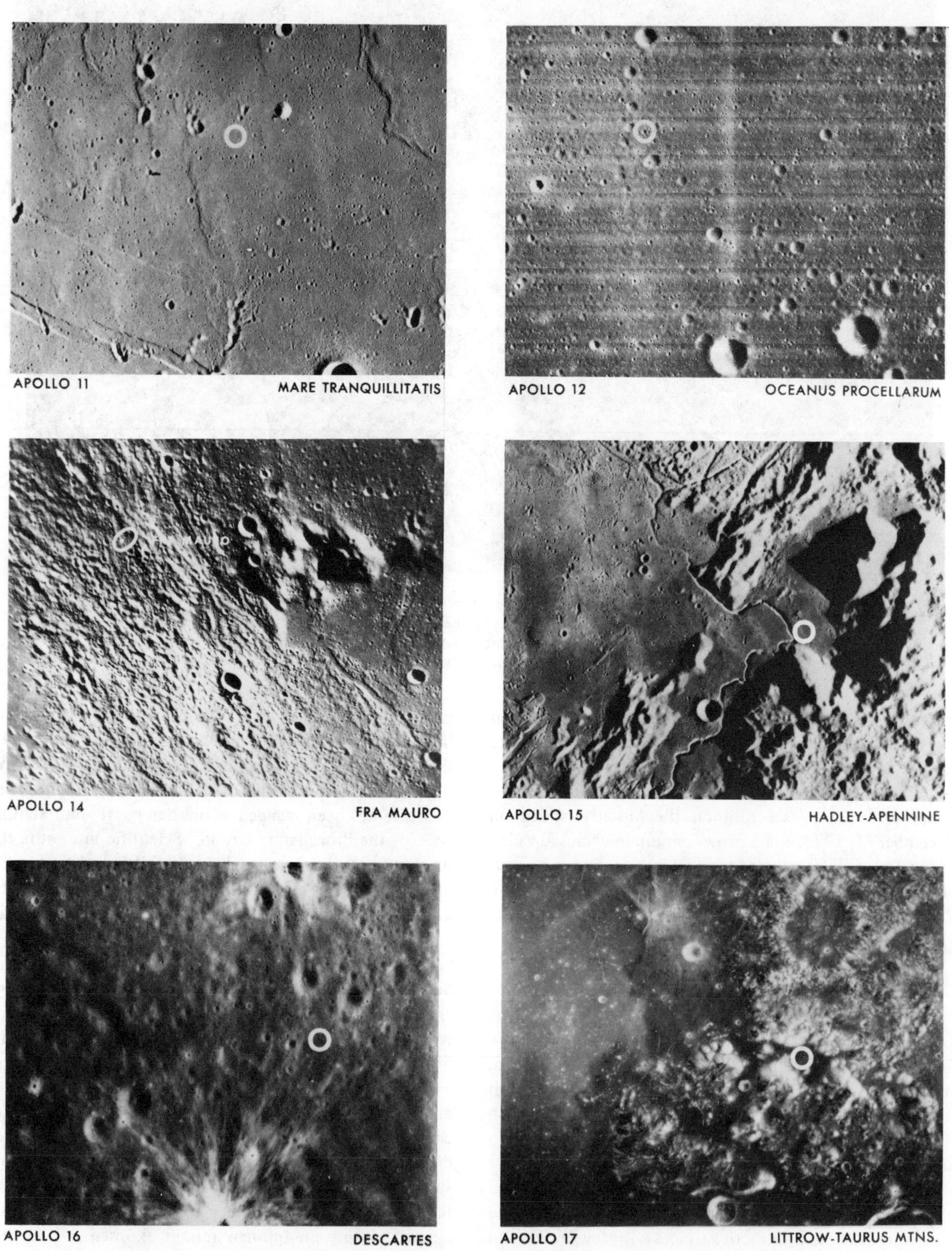

Figure 10-2 High altitude views, taken from *Lunar Orbiter* and *Apollo* spacecraft, of the region around each Apollo landing site (indicated by white circle). (NASA photos.)

TABLE 10-1

Site Science Rationale

	Apollo 11	*Apollo 12*	*Apollo 14*	*Apollo 15*	*Apollo 16*	*Apollo 17*
Type	Mare	Mare	Hilly upland	Mountain front/ rille/mare	Highland hills and plains	Highland massifs and dark mantle
Process	Basin filling	Basin filling	Ejecta blanket formation	• Mountain scarp • Basin filling • Rille formation	• Volcanic construction • Highland basin filling	• Massif uplift • Lowland filling • Volcanic mantle
Material	Basaltic lava	Basaltic lava	Deep-seated crustal material	• Deeper-seated crustal material • Basaltic lava	Volcanic highland materials	• Crustal material • Volcanic deposits
Age	Older mare filling	Younger mare filling	• Early history of moon • Premare material • Imbrium Basin Formation	• Composition and age of Apennine Front material • Rille origin and age • Age of Imbrium Mare fill	• Composition and age of highland construction and modification • Composition and age of Cayley Formation	• Composition and age of highland massifs and possibly of lowland filling • Composition and age of dark mantle • Nature of a rock landslide

all of endogenetic character, set within hilly, cratered terrain carved out of the ancient highlands. *Apollo 17,* the final mission in the *Apollo* series, explored still another region of the highlands in the rugged Taurus Mountains near the crater Littrow along the southeast edge of Mare Serenitatis. It was surmised that rock slides along steep mountain fronts would provide samples of some of the oldest rocks exposed anywhere on the Moon. In the valleys conspicuous dark deposits associated with relatively low crater densities were thought to represent lavas mantled by younger, explosively ejected volcanic cinder and ash units, possibly traceable to apparent cinder cones some 10-15 km from the site.

Geologic Exploration of the Apollo Sites

Each successive mission on the lunar surface resulted in increases in total payload of scientific instruments, dura-

Figure 10-3 Bar diagrams comparing the successful *Apollo* landings in terms of equipment delivered, time spent in surface activities, distance covered during excursions, and rock, soil, and core materials collected. (NASA diagram.)

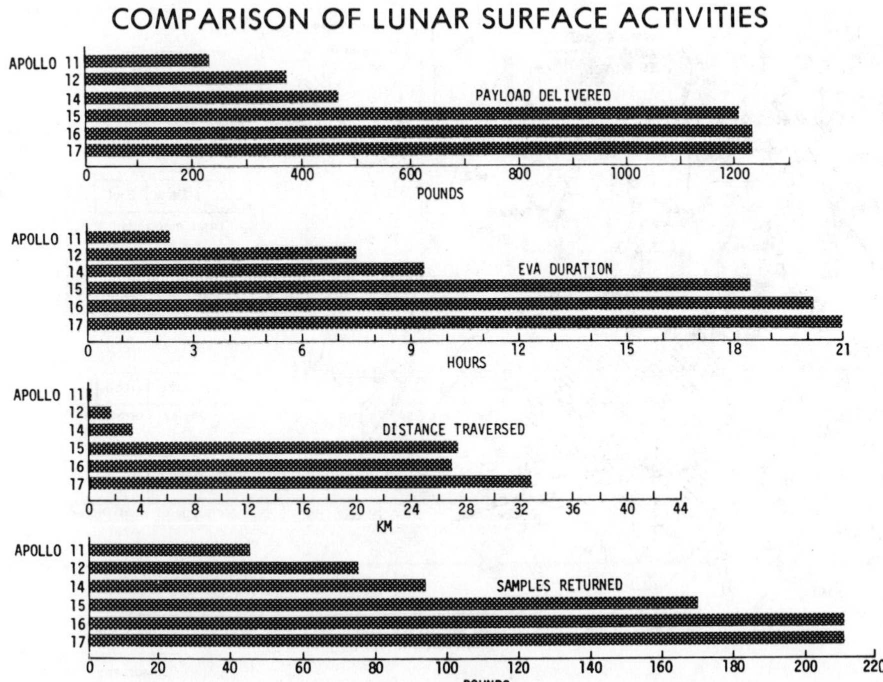

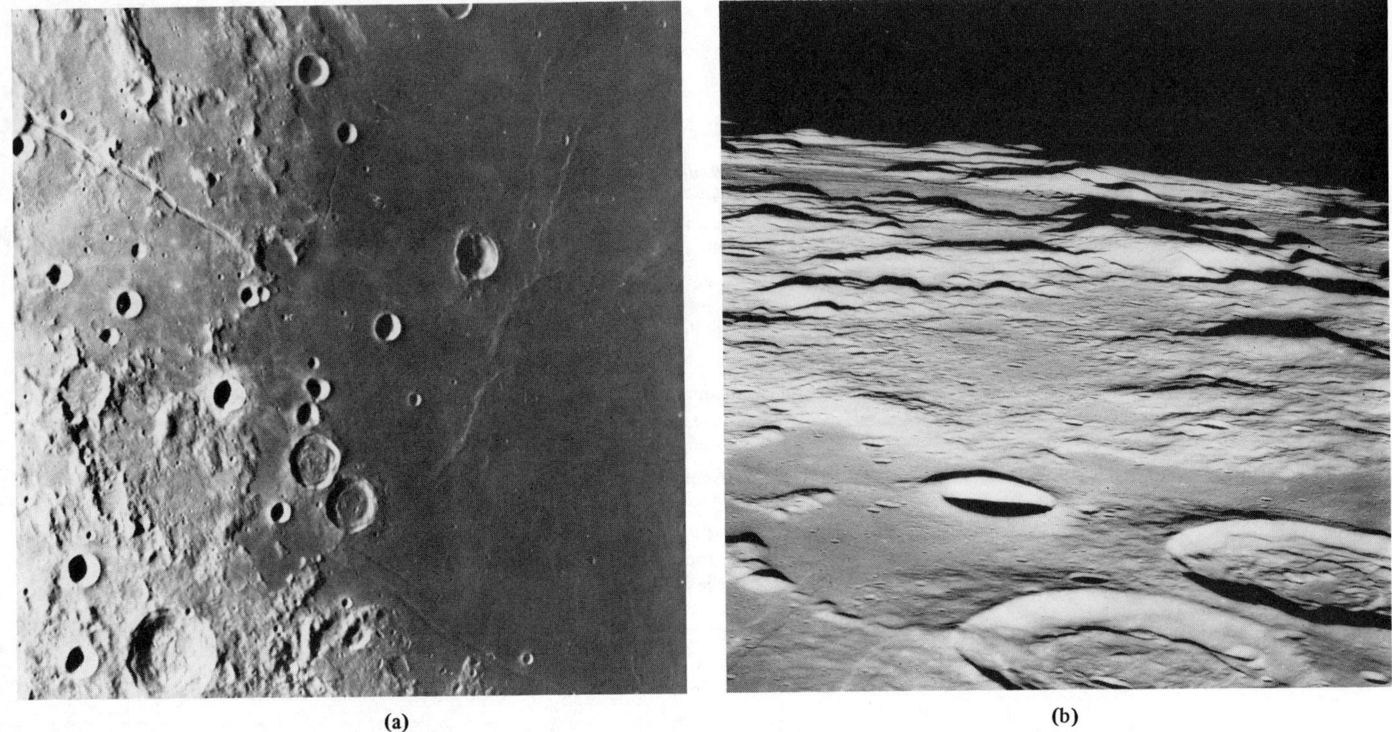

Figure 10-4 (a) A telescope photo (taken by the Lick Observatory 120-inch reflector) showing the southwestern corner of Mare Tranquillitatis, part of the central highlands, the large craters Julius Caesar (upper left) and Dionysius (lower left), and the Ariadaeus Rille (center left). The twin craters Sabine and Ritter appear in the lower center. *The Apollo 11* site is located in the lower right corner area of the picture north of the small crater Moltke above the rille. (b) a view of the uplands beyond the southwest edge of Mare Tranquillitatis as illuminated by a low-angle rising sun. The two large craters in the foreground are Sabine and Ritter. (NASA photo AS11-41-6123.)

Figure 10-5 A generalized geologic map of the region of southern Mare Tranquillitatis that includes the *Apollo 11, Surveyor V,* and *Ranger 8* sites. An explanatory key appears on the right. The ray deposits probably come from the crater Theophilus (located about 200 km south of the site). (Courtesy of the U.S. Geological Survey.)

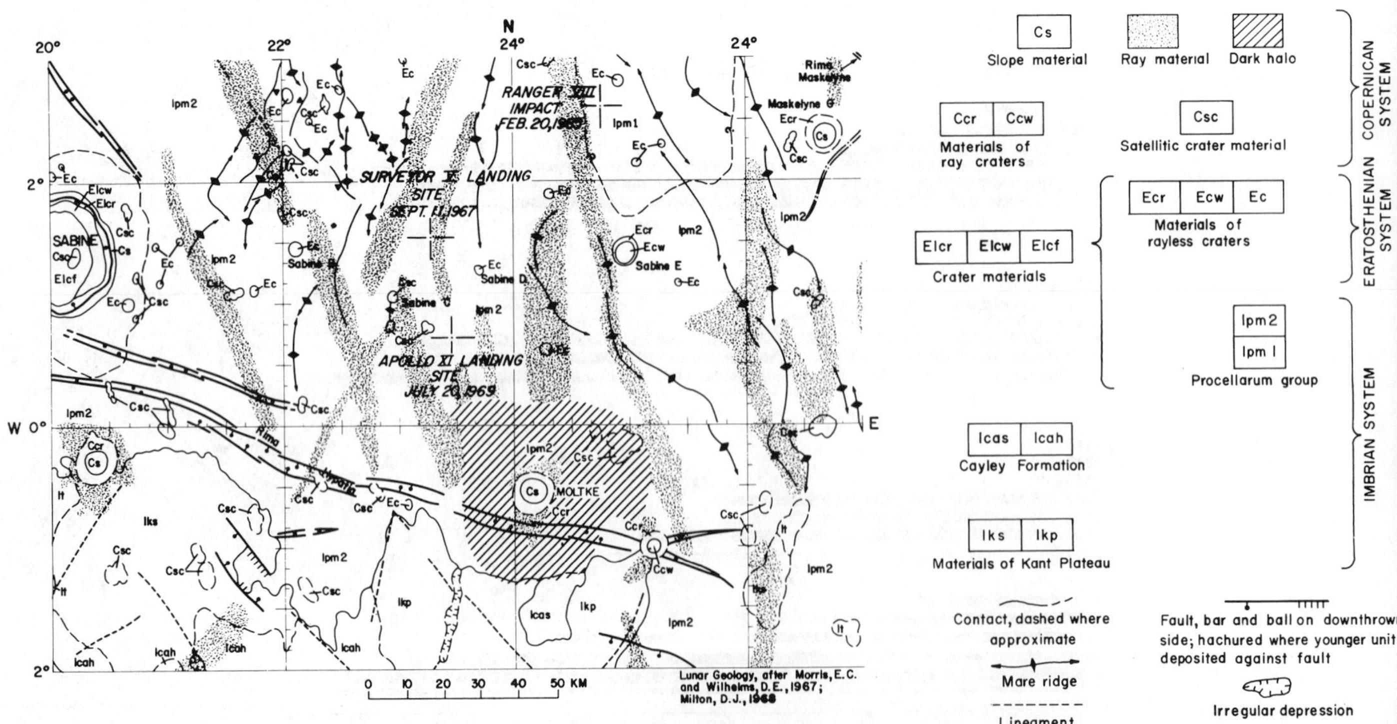

tion on the surface, distance covered during EVAs, and weight of samples brought back to Earth (Figure 10-3). A synopsis of the regional and local geology around each site, astronaut activities, and some of the principal scientific accomplishments from crew observations on the surface are presented in this section.

Apollo 11

The *Apollo 11* site (lunar coordinates 0°40′N and 23°30′E) lies close to the edge of Mare Tranquillitatis near rugged highlands (Figure 10-4a and b) constructed of older pre-Imbrian and Imbrian terra units representing a mixed assemblage of impact and volcanic deposits. Some lineations and ridges around the crater Julius Caesar are remnants of large scour marks and depositional features produced by rapidly moving ejecta expelled during formation of the Imbrium Basin. Other lineations and possible fault troughs may relate to fracture systems radial to this basin. The Ariadaeus Rille, in two offset segments, is one of the largest straight-walled features on the lunar frontside. The Tranquillitatis lavas fill a somewhat elliptical lowlands which may be the remnant of a degenerate impact basin, but this area is not now underlain by a mascon. Concentric rilles and numerous wrinkle ridges occur throughout Mare Tranquillitatis; those around crater Lamont appear to reflect the influence of an older buried crater rim. Rays, recognizable as lighter streaks and patches on the dark lava surfaces, have been traced to various fresher-looking craters.

A geologic map of the region around the *Apollo 11* landing site (Figure 10-5) points up many of the features of interest and potential obstacles that concerned the astronauts during their descent. A "feel" for the level-to-gently-rolling terrain over which the LM had to be piloted enroute to touchdown can be gained from inspection of Figure 10-6a. The high-resolution (about 5-meters) *Orbiter* photo of the landing site area (Figure 10-6b) illustrates the problem facing the crew in avoiding clusters of those very small craters not visible from the higher altitudes. Similar areas in this part of Mare Tranquillitatis had been investigated earlier by *Ranger 8* and *Surveyor 5*.

The composite thickness of lava flows in this region of the basin probably does not exceed a few hundred meters. Irregularities in the premare surface, including ejecta from Imbrium and other basins and furrows and ridges extending from crater Lamont, influence the distribution of the mare ridges on the lavas as well as other linear features that result directly from faulting within the cooling lavas or foundering of surface-cooled flow units. Based on differences in crater densities, two Imbrian flow subunits can be mapped throughout the area. The *Apollo 11* LM set down on a narrow band of the older, slightly more heavily cratered subunit (I_{pm} 1 in Figure 10-5, but not depicted in the immediate site area). Both subunits have a greater number of larger (>500 m) subdued, pan-shaped craters

Figure 10-6 (a) View of the approach to the *Apollo 11* landing site (lying just in front of the dawn terminator) as seen by the astronauts from the CSM window. The site, several craters, and other landmark features (identified in colloquial terms) are labeled. (AS II-37-5437). (b) Part of an *Orbiter V* high resolution photo that includes the *Apollo 11* LM landing site (arrow). The width of each strip is approximately 200 meters. (NASA photo.)

(a)

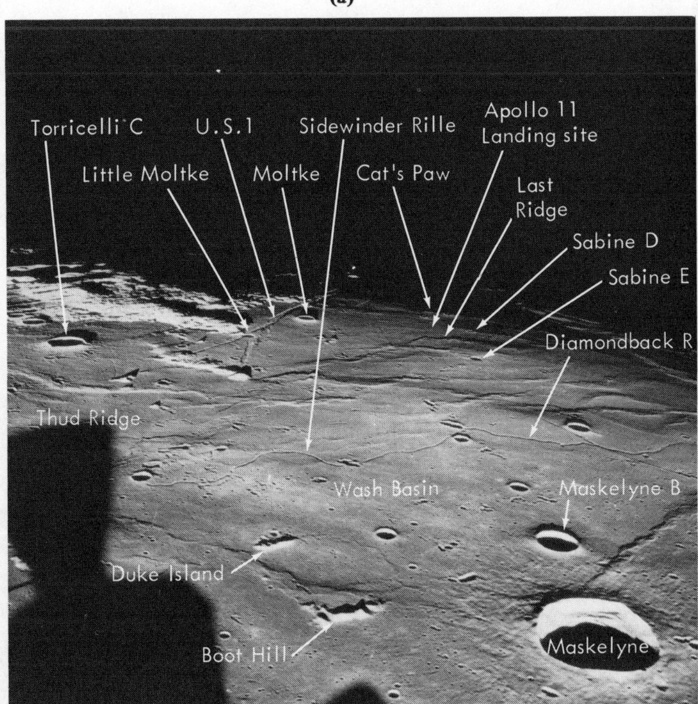

(b)

(a)

(b)

Figure 10-7 (a) A view of the rock-strewn and cratered regolith to the immediate south of the *Apollo 11* landing site. Several of the rock fragments visible here were later collected by Astronaut Armstrong. (b) Part of the interior of a 33-meter-wide crater located about 65 meters east of the *Apollo 11* LM. A special close-up camera (adjusted to photograph small objects on the lunar surface) stands in the lower left corner. Both of these pictures were taken with a hand-held Hasselblad camera having a 70-mm focal length. (NASA photos.)

than some other potential *Apollo* sites considered for the first landing; this excess distribution suggests a relatively older age for the Tranquillitatis lavas. There is, however, a deficiency of craters in the 50-to-125-m class owing in part to destruction of many smaller craters over time and in part to a "smoothing out" by burial under postmare ejecta deposits. Many secondary craters from Theophilus (Copernican age) occur throughout the region; this association is particularly evident in the high-resolution *Orbiter* photos.

The views (Figure 10-7a and b) awaiting Armstrong amd Aldrin as they scouted the landing site were much like those scanned by the *Surveyor* TV cameras in other mare settings. Many small craters are shallow, rimless pits or depressions whereas others have low raised rims. Some larger, generally younger craters—including the unvisited West Crater (180 m wide and 30 m deep) about 400 m east of the site—display subdued to sharp rims and, in some instances, blocky debris within their interiors. Careful mapping from returned photographs reveal irregular to elongate depressions. Some of these represent chance groupings of craters, but others appear to be troughs that extend preferentially into northwest and northeast directions and were probably formed by drainage or collapse into joints controlled by the lunar grid system. Fragments and blocks up to 5 m in width are scattered about Tranquillity Base. However, clusters of larger blocks are found in the ejecta apron around West Crater and in ray deposits that pass through the site.

The *Apollo 11* astronauts performed many preplanned scientific tasks during their EVA period (as indicated on the traverse map of Figure 10-8). These included photo-documentation of surface features, collecting samples of lunar materials, and activating instruments in the ALSEP (*Apollo* lunar surface experiments package). The astronauts on *Apollo 11* and subsequent missions were equipped with a variety of tools (Figure 10-9) for collecting and storing rock and soil materials under the difficult conditions of lower gravity and of body movements constrained by space suits. Shortly after stepping onto the Moon, Aldrin made two scoops into the soil to provide the *contingency* or "grab" sample (~1.4 kg) that guaranteed at least some returned materials in the event the excursion had to be aborted early. Most (~16 kg) rocks were collected later as *bulk* samples; descriptions of these were radioed back to the Manned Spacecraft Center (now known as the Lyndon B. Johnson Spacecraft Center) in Houston, Texas (locations of a few samples were later reconstructed from returned panoramic photography). Plans to obtain *documented* samples (each photographed in place) were abandoned toward the end of the EVA except where the two core-tube samples were removed. In later missions, documented samples made up a much larger fraction of all returned lunar materials. Besides hand-collected rocks, other samples are obtained with a scoop or by raking through the soil, by digging shallow trenches, and by driving core tubes into the regolith with a hammer or a motor-powered drill.

Some rocks lie more or less completely exposed on the *Apollo 11* surface, but many are partially buried. *Fillets* of loose, powdery soil are deposited on one side of some rocks as though surging ejecta had piled up against obstruc-

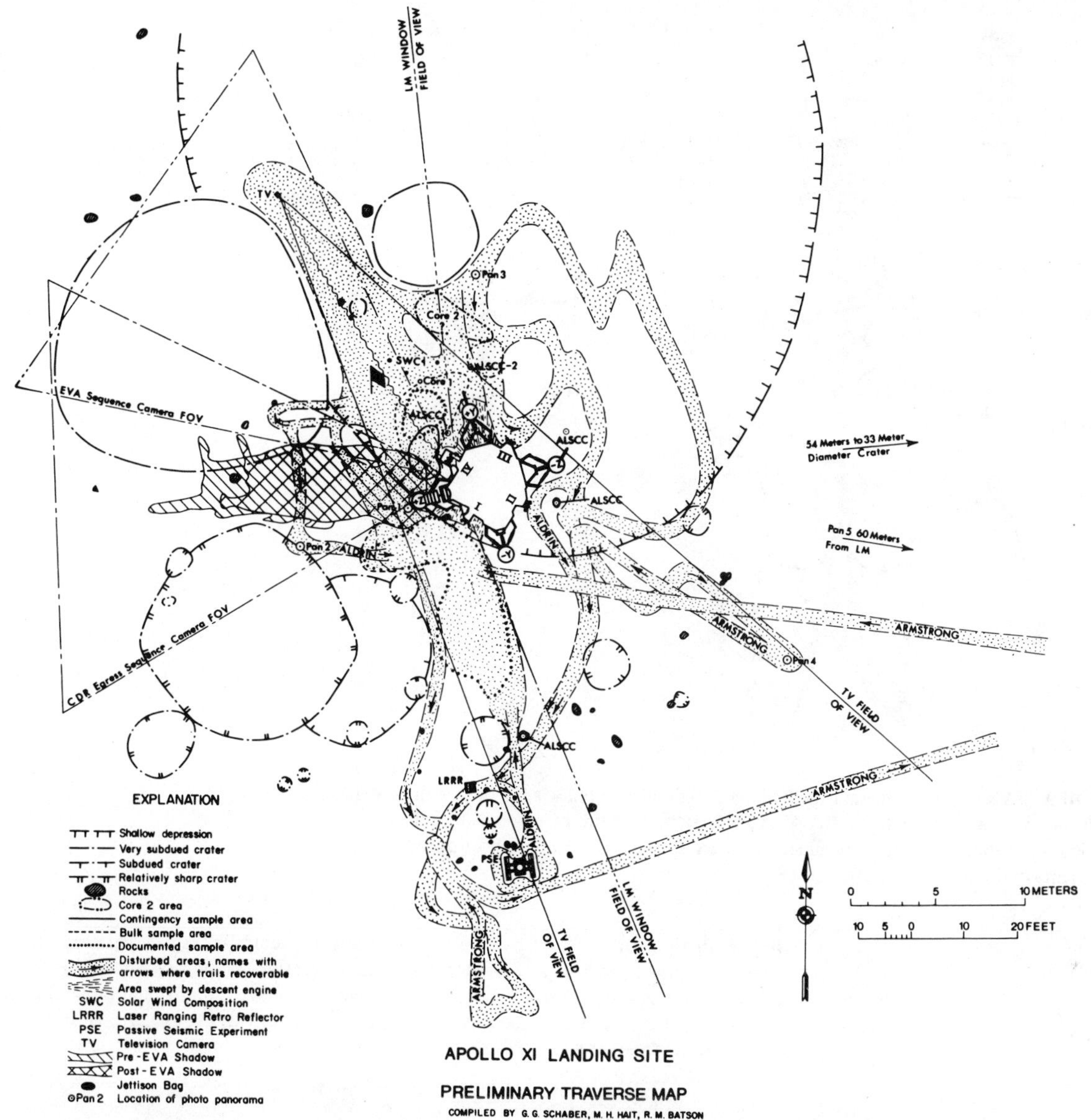

Figure 10-8 A traverse map which records the principal surface activities (EVA) of Astronauts Armstrong and Aldrin at the *Apollo 11* site as reconstructed from photographs, from radio communications received during the mission, and from debriefings after their return to Earth. (Illustration courtesy U.S. Geological Survey.)

tions. While the soil is weakly cohesive, as proved by the sharp footprint impressions (Figure 1-13), in places thicker and softer masses have accumulated in depressions and outside rims or rises. Fine-grained upper layers are readily disturbed and transported by rocket exhaust and astronaut activities, as evidenced by darker streaks and patches superimposed on a lighter gray surface. Drawing upon television observations by the *Surveyors*, some scientists had predicted that the near surface would be darkened by such processes as radiation damage and ion sputtering. But, at all *Apollo* sites, the topmost layer (~1 mm thick) is commonly up to 40% lighter (higher albedo) than layers immediately below. Close-up stereo views of an undisturbed soil surface (Figure 10-10) reveal an irregular, granular microtexture resulting from numerous small clumps of aggregated particles. Some lumps are conspicuously coated or glazed with dark glass; elsewhere pits and pockets in both soil and rock contain similar glass deposits. Several individual rocks (Figure 10-11), whose initial positions were fixed from photographs, show, when examined later on Earth, a rounded surface where exposed above ground and an angular surface where buried.

These features, also found at the other sites, show the regolith to be in a state of extremely slow but continuous turnover that leads to a fairly uniform layer over most of the Moon. Upper regolith densities of 1.6 to 1.8 (repre-

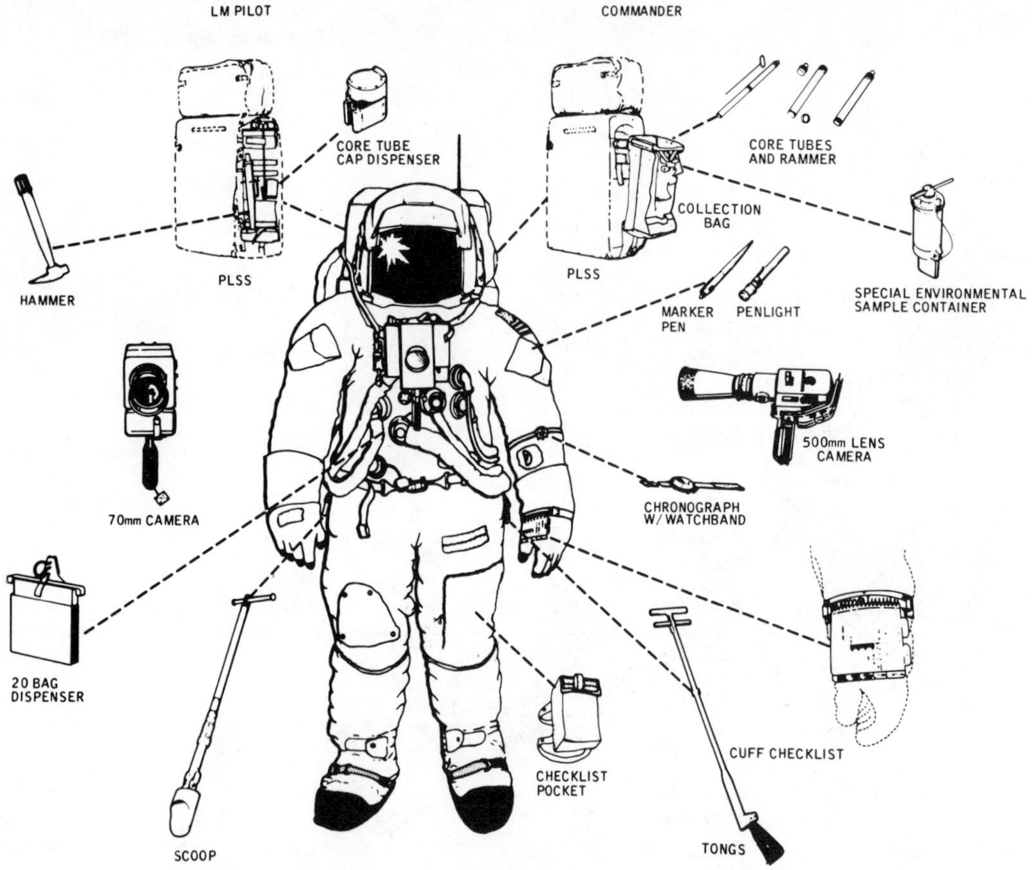

LM PILOT

COMMANDER

CORE TUBE CAP DISPENSER

CORE TUBES AND RAMMER

COLLECTION BAG

HAMMER

PLSS

PLSS

SPECIAL ENVIRONMENTAL SAMPLE CONTAINER

MARKER PEN

PENLIGHT

500mm LENS CAMERA

70mm CAMERA

CHRONOGRAPH W/WATCHBAND

20 BAG DISPENSER

CUFF CHECKLIST

CHECKLIST POCKET

SCOOP

TONGS

Figure 10-9 A schematic diagram depicting an astronaut suited up for on-site lunar exploration during extravehicular activity (EVA). Various items of equipment carried by the astronaut for sample collection and photo-documentation are shown. The backpack or portable life support system (PLSS) is also indicated. (NASA illustration.)

Figure 10-10 One of a pair of stereoviews showing clumps of aggregated particles and rock fragments, the largest clearly coated with glassy spatter, as photographed by the close-up camera at the *Apollo 11* site. Some of the smaller particles are individual mineral grains. This surface may have been disturbed by astronaut traverses or the LM exhaust. Horizontal field of view is approximately 8 cm. (NASA photo.)

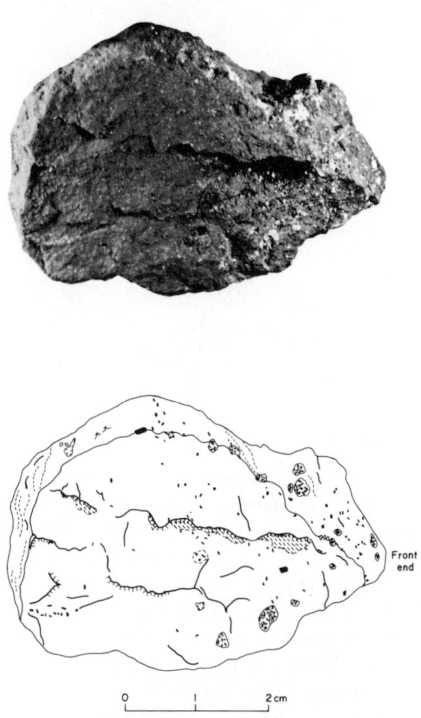

Figure 10-11 A photograph of Apollo 11 sample 10023 taken at the Lunar Receiving Laboratory at NASA's Johnson Spacecraft Center in Houston, along with a sketch "map" showing in detail some of its surface features (including pits and fractures). When in place on the Moon, the upper part of the specimen was exposed above the "soil" surface and was thus more rounded by micrometeorite bombardment (and possibly solar wind erosion). The lower buried portion remains more angular and rough and has flatter faces, as expected when protected from these erosion processes. (NASA, LRL photograph S-69-45393.)

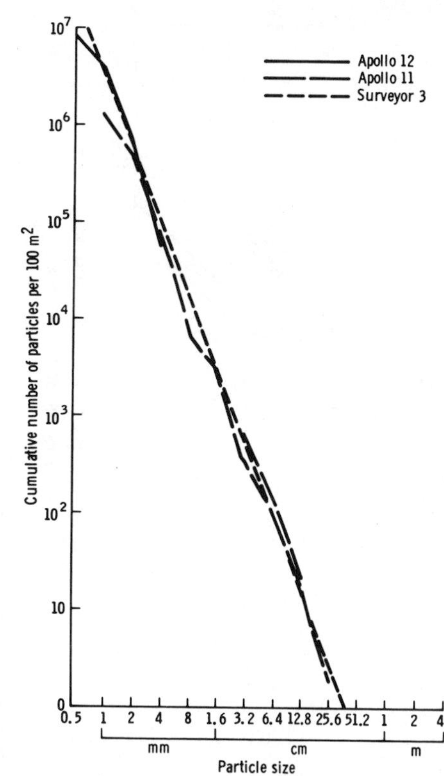

Figure 10-12 Cumulative size-frequency distribution of smaller particles and fragments (as measured and counted in Hasselblad and closeup camera photographs) on the surfaces of the *Apollo 11* and *Apollo 12* sites. A previous distribution curve made from *Surveyor 3* pictures of areas near the *Apollo 12* site indicates that (a) it is possible to make accurate measurements with a remote TV system and (b) the size distributions are nearly identical at all three sites, which implies that particle reduction processes over time tend to produce about the same degree of comminution in different mare regions. (Diagram courtesy U.S. Geological Survey.)

senting porosities of 40-50%) at the various *Apollo* sites are consistent with earlier estimates of densities made from radar studies over large areas of the lunar surface. The cumulative frequency distributions of small craters (Figure 7-23b) and visible surface fragments (Figure 10-12) at the *Apollo 11* and *12* sites and at selected *Surveyor* sites are similar. This implies that impacts act everywhere on the Moon to reduce particles within the upper layers of the regolith to comparable size spreads. The flux of small meteorites and micrometeorites should now be almost equal over the entire lunar surface, although it may have varied over geologic time.

Repetitive impact cratering causes both *comminution* and *progressive turnover* or "gardening" of a regolith that gradually increases in thickness since the first exposure of its source beds. Thus, older lavas will be overlain by thicker soil covers. The evidence for this gardening is inferred from the increase in mean grain size downward with depth and the greater mixing and better sorting of fragments of diverse compositions upward toward the surface. Rocks initially brought by natural processes from hard, underlying layers to the surface environment break and erode mainly from the impacts of small meteorites and micrometeorites. Grains smaller than about 22 micrometers are further reduced in size primarily by irradiation with solar wind particles (such as ionized gases and electrons). The history of any given fragment depends on its initial size, on when it was first introduced into the regolith, on the number of cratering events affecting it, and on its depth in the regolith at various times.

Shoemaker (1970) has derived a crater production function determined from observed distributions at the *Apollo 11* site by which he calculates the turnover time versus initial size (Figure 10-13) for a regolith that began to accumulate 3.6 billion years ago as underlying lava rocks of that age were bombarded. He concludes from his statistical model that (1) fragments initially near the base of the regolith are likely to have been reexposed only once, on average, at the up-building surface (resulting, in effect, in one complete turnover with additional mixing); (2) the upper layers within the thickening regolith experience several partial turnovers; (3) the effect of three major cratering events can be recognized in the regolith at the site; (4) a 10-cm fragment, after first exposure on the surface at

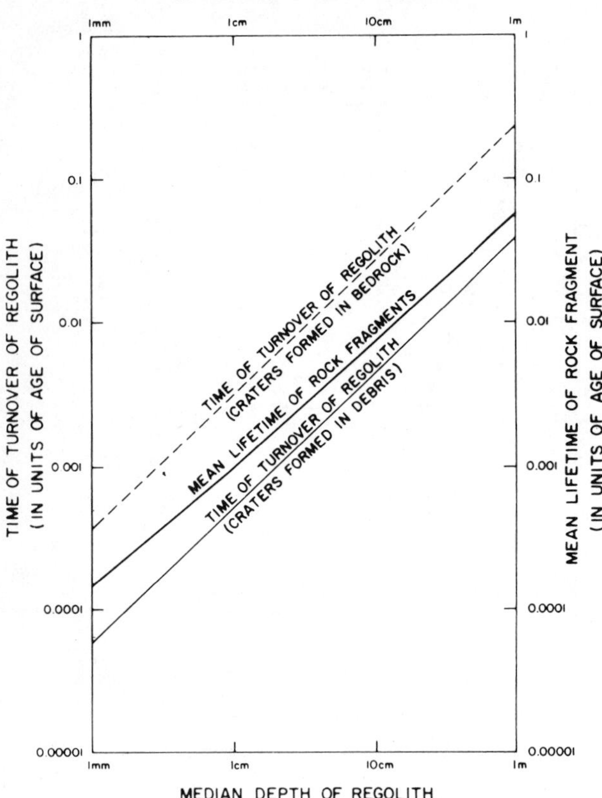

Figure 10-13 Plots (in log–log scale) used to determine turnover times for the regolith and the mean lifetimes of rock fragments, calculated by Shoemaker et al. (1970) from observations applicable to Tranquillity Base. The time units are given as fractions of the age (3.6 billion years) of the lava rocks assumed to underlie the *Apollo 11* site. Because craters formed in bedrock are smaller (for a given energy release) than those produced in the regolith, turnover times will be longer if a development cycle begins in bedrock. (Reprinted with permission from E. M. Shoemaker et al., *Proceedings of the Apollo 11 Lunar Science Conference* [A. A. Levinson, ed.], 1970, Pergamon Press.)

Tranquillity Base, will be worn down by small-particle bombardment in about 40 million years; and (5) as much as 50% of the rock material around the site was derived from an area within a 5-km radius, 5% came from distances beyond 100 km, and 0.5% from distances greater than 1000 km.

Apollo 12

Both the *Apollo 12* and *14* sites lie in the lunar equatorial belt south of Copernicus. The *Apollo 12* site (coordinates 3°12′S and 23°24′W) is located within Mare Cognitum approximately 120 km southeast of Lansberg Crater (Figure 10-2). The lava surface is an extension of flows of Imbrian age from the southeast end of Oceanus Procellarum. The surface was judged to be younger than the Tranquillitatis units because of lower crater frequencies, a greater number of smaller craters, and a somewhat thinner regolith (1 to 3 m). A cluster of craters–some of which are secondaries from Copernicus–between 50 and 400 m wide occurs in the vicinity of the site. The *Apollo 14* site (coordinates 3°40′S and 17°29′W) is about 160 km due east of the *Apollo 12* site at a point some 80 km north of the center of Fra Mauro Crater. The landing region is in hummocky uplands terrain (see Figure 10-2) comprising a north-south belt of materials (Fra Mauro Formation) ejected from the Imbrium Basin, now almost completely surrounded by younger mare lavas. Maps recording the traverses followed by the astronauts at both sites appear in Figure 10-14.

Apollo 12 came down almost exactly on target near the large crater containing the *Surveyor 3* spacecraft which had landed almost two-and-a-half years earlier (Figure 10-15a). Parts of this *Surveyor* were retrieved for later study on Earth to determine the short-term effects of the lunar environment on an object of known characteristics. Presence of nearby large craters account for the comparative ease with which rock samples (ejecta) were collected. Many of these samples were picked up on or near crater rims. Most were crystalline rocks presumably derived from lava units beneath the thin regolith. Of unusual interest were two small peaked mounds (Figure 10-15b), interpreted to be large clumps of indurated regolith (possibly tossed out as projectiles from craters elsewhere). The regolith itself was penetrated by coring to a depth of some 70 cm. The resulting core-tube sample (Figure 10-16), in which a total of 41 cm of lunar soil was recovered (in contrast to the 15 cm from *Apollo 11*) preserved an undisturbed sequence of ten thin layers distinguishable by subtle color and morphological changes. The gradual increase in grain size with depth accords with the expectation of greater comminuation in the more reworked top layers, but the stratification argues against any major turnover and mixing at least within the sample interval. These layers may represent successive base surge and throwout deposits from nearby craters. However, the time at which these deposits were emplaced cannot be precisely specified; they may result from relatively young events and have not yet been disturbed by extensive gardening. Another *Apollo 12* sample, collected from a trench, contains coarse glass

(a)

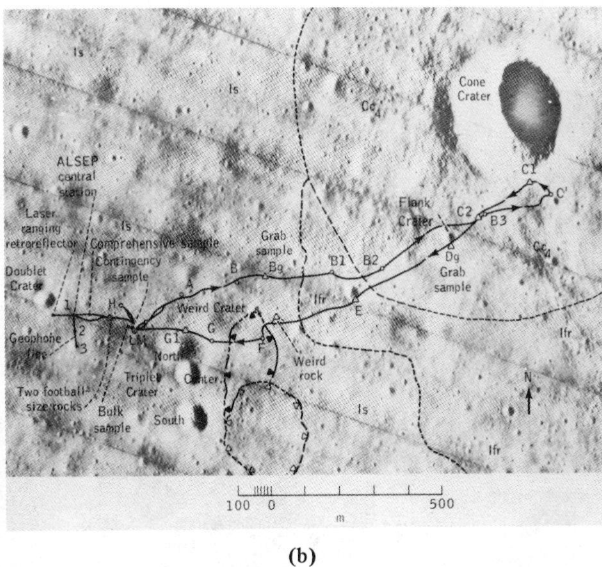

(b)

Figure 10-14 (a) Map of the *Apollo 12* site (superimposed on a high resolution *Orbiter 4* photograph) showing the EVA traverses, major sampling points, and location of the ALSEP instruments (map prepared by R.L. Sutton, U.S.G.S.). (b) A mock-up model of the *Apollo 14* site on which the two principal lines of traverse during EVA (unbroken line is first excursion and broken line is second) have been plotted. (NASA photo.)

Figure 10-15 (a) The *Surveyor 3* spacecraft within the large crater at which it landed in 1967, as photographed by the *Apollo 12* astronauts. The LM is visible on the horizon at a point 183 meters away. (b) One of the peaked mounds on the lunar surface southwest of the *Apollo 12* LM. (NASA photos.)

(a)

(b)

(a)

Figure 10-16 (a) Photograph of the double-tube core samples from the *Apollo 12* site exposed in the Lunar Receiving Laboratory after their sealed container was opened. (b) A schematic "stratigraphic column" showing the nature and thickness of various units defined from the study of the layered material shown in (a). (NASA photos.)

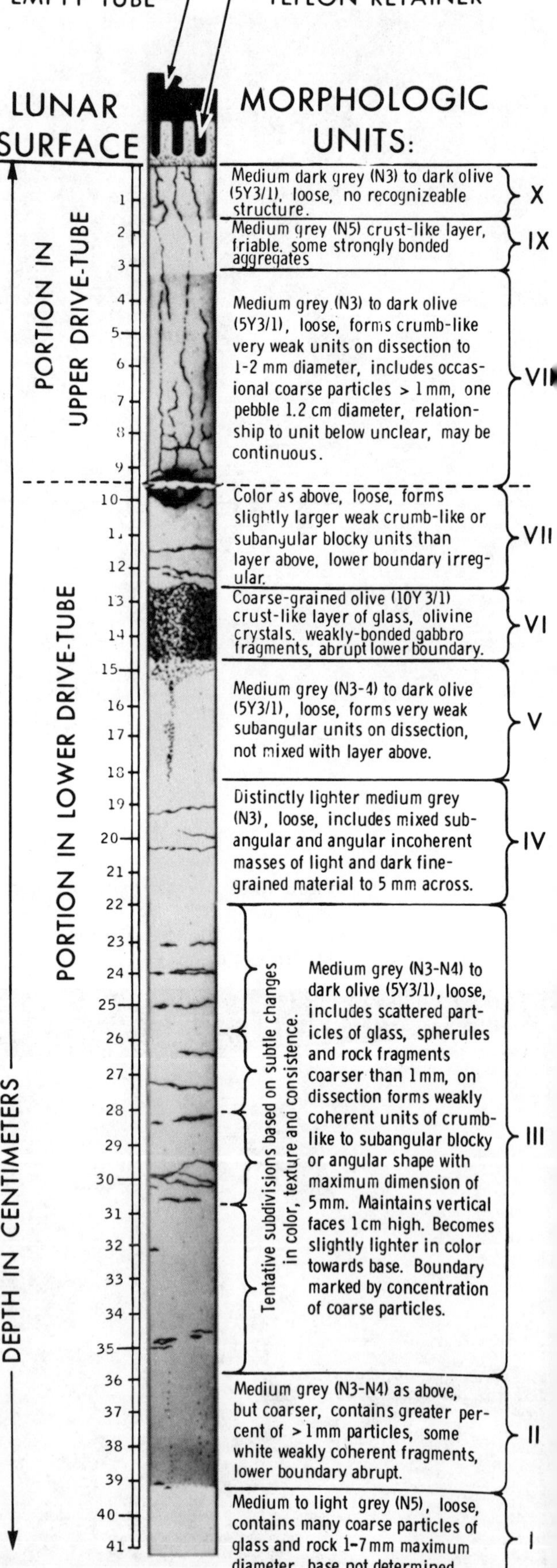

(b)

fragments with pumiceous texture, flow bands, and stretched vesicles mixed with finer irregular glassy shards—features commonly found in volcanic ash but also known in impact ejecta.

Apollo 14

Apollo 14 landed close enough to its prime target at Cone Crater, a steep-sloped impact structure nearly 340 m wide at the rim, for most of the two planned EVAs to be conducted on schedule. The local topography around the site consists of relatively subdued hillocks and crater walls and a rolling surface dotted with ejecta blocks as large as 3 to 5 meters on a side (Figure 10-17). This hummocky terrain developed initially from ejecta deposits derived from the Imbrium Basin and has been subsequently modified by local cratering. It is reasonable to assume that many of the largest rock blocks, some as much as 15 m in diameter, were tossed from Cone Crater and other nearby excavations. An 8-to-10-m thick regolith underlies the site.

Nearly all the rocks brought back by *Apollo 14* are breccias composed of well-consolidated to friable fragmental debris held together by glass and recrystallized material. Some of these specimens are made up of at least two distinct phases—one characterized by a darker gray color and the other by a lighter grayish white that the astronauts described during EVA as "white rock." Three, and possibly four, distinct stratigraphic units of fragmental materials, each distinguished by a characteristic population of particle types, have been recognized at the Fra Mauro site. The upper unit(s) contain higher proportions of lighter (whitish) fragments or *clasts* than those underneath that display more of the darker materials. Blocks of the light-and-dark breccias occur commonly around Cone Crater and are believed to have been excavated from the deeper layers penetrated by that structure. Both clasts and matrix in these breccia units are considered to be pre-Imbrian crustal materials carried to the *Apollo 14* site during the Imbrium Basin event by base surge and trajectory paths. The units together now make up the Fra

Figure 10-17 (a) A composite photomosaic that covers a 360-degree panoramic view of the *Apollo 14* landing site, taken during EVA-2 from the boulder field near Cone Crater. (b) Composite photo, extending through 90°, that shows a cluster of boulder-sized rocks ejected from Cone Crater (whose rim is about 15 m in the background) at the *Apollo 14* site. The rocks are breccias with large clasts and crude layering. Planar fracture sets are also visible. (NASA photos.)

(a)

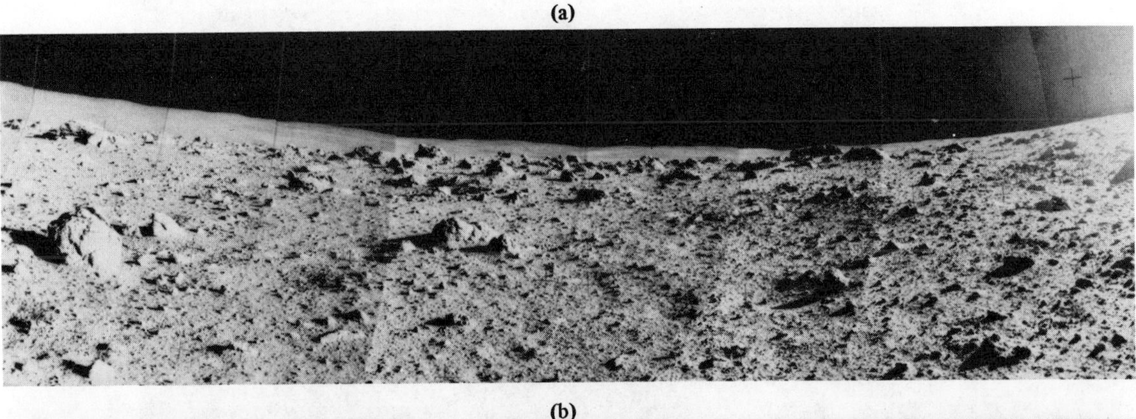

(b)

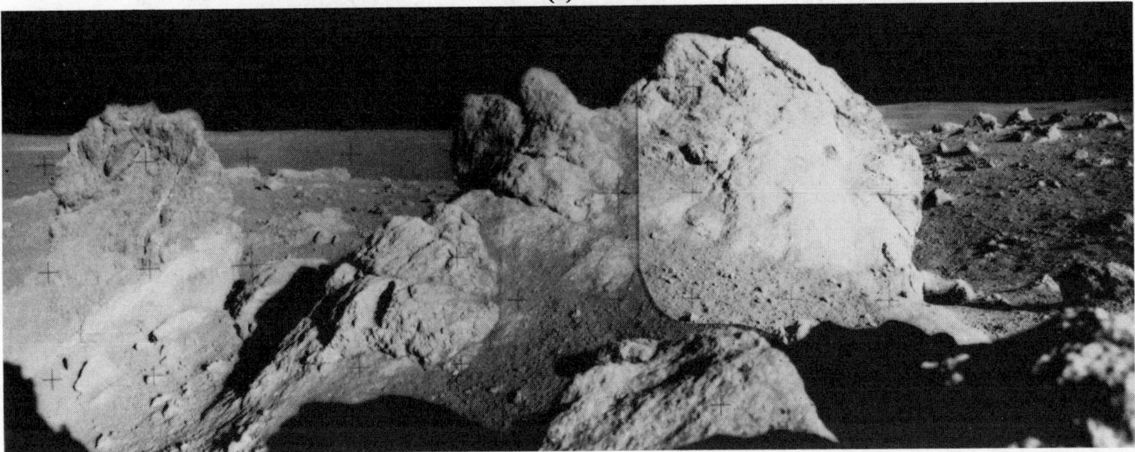

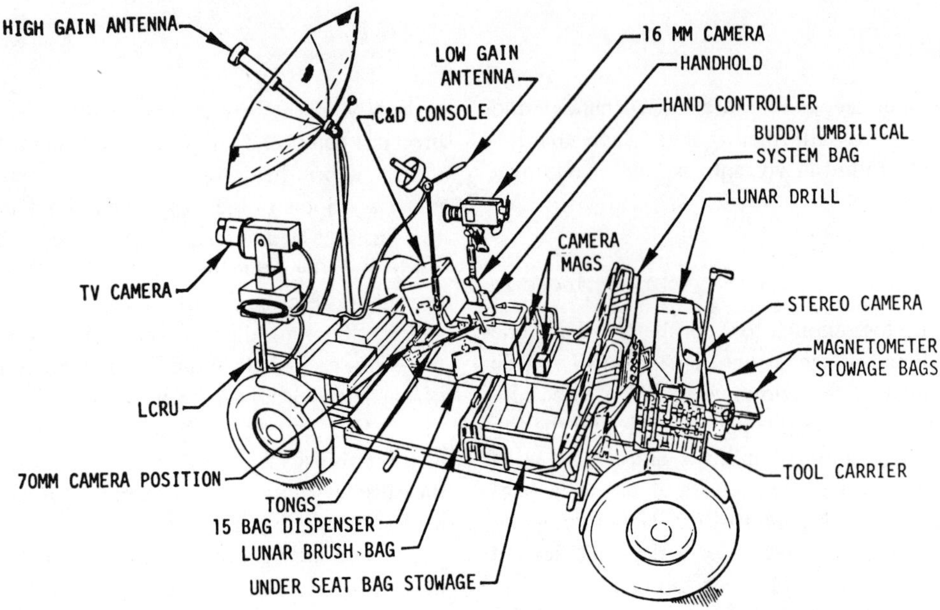

LRV STOWED PAYLOAD INSTALLATION

Figure 10-18 The lunar roving vehicle (LRV), or "Rover", used to transport the astronauts to distant sampling stations during EVA. Powered by batteries, the LRV can reach speeds up to 17 km/hr downslope. Equipment for sample collecting and storage, for photodocumentation and TV transmission to Earth, and for measuring the magnetic field at each station is carried on this vehicle. (NASA illustration.)

Figure 10-19 A traverse map on a photobase showing the EVA routes and stations at the *Apollo 15* site. (Courtesy G. Swann, U.S. Geological Survey.)

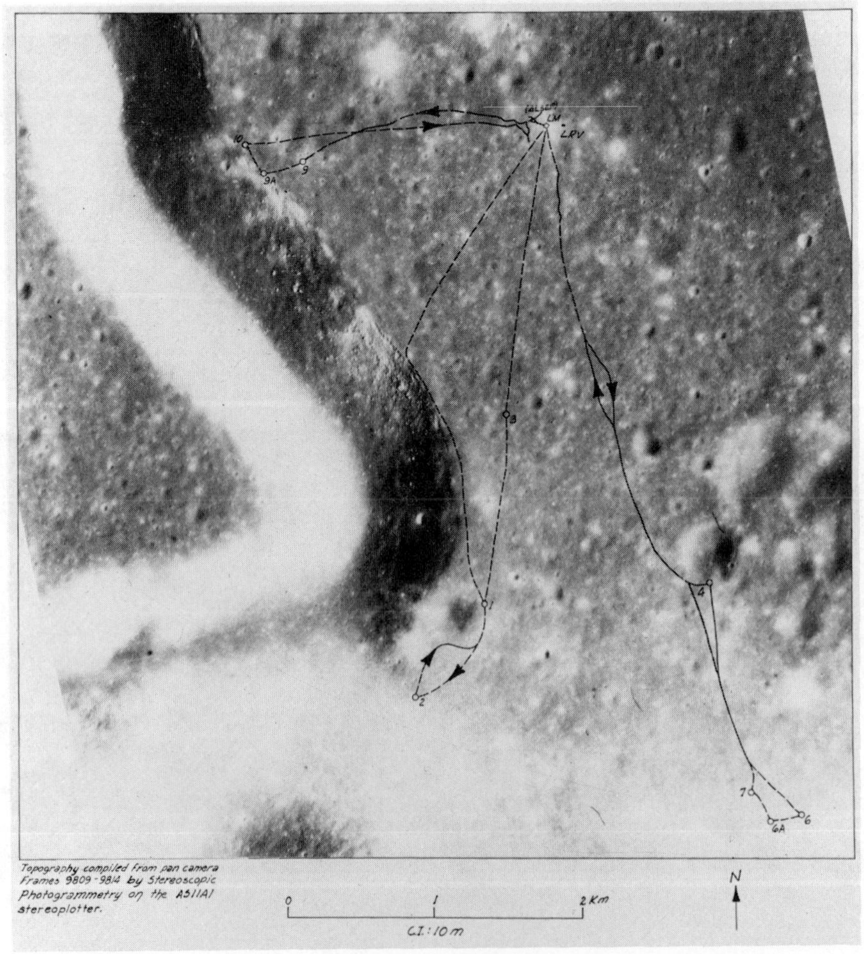

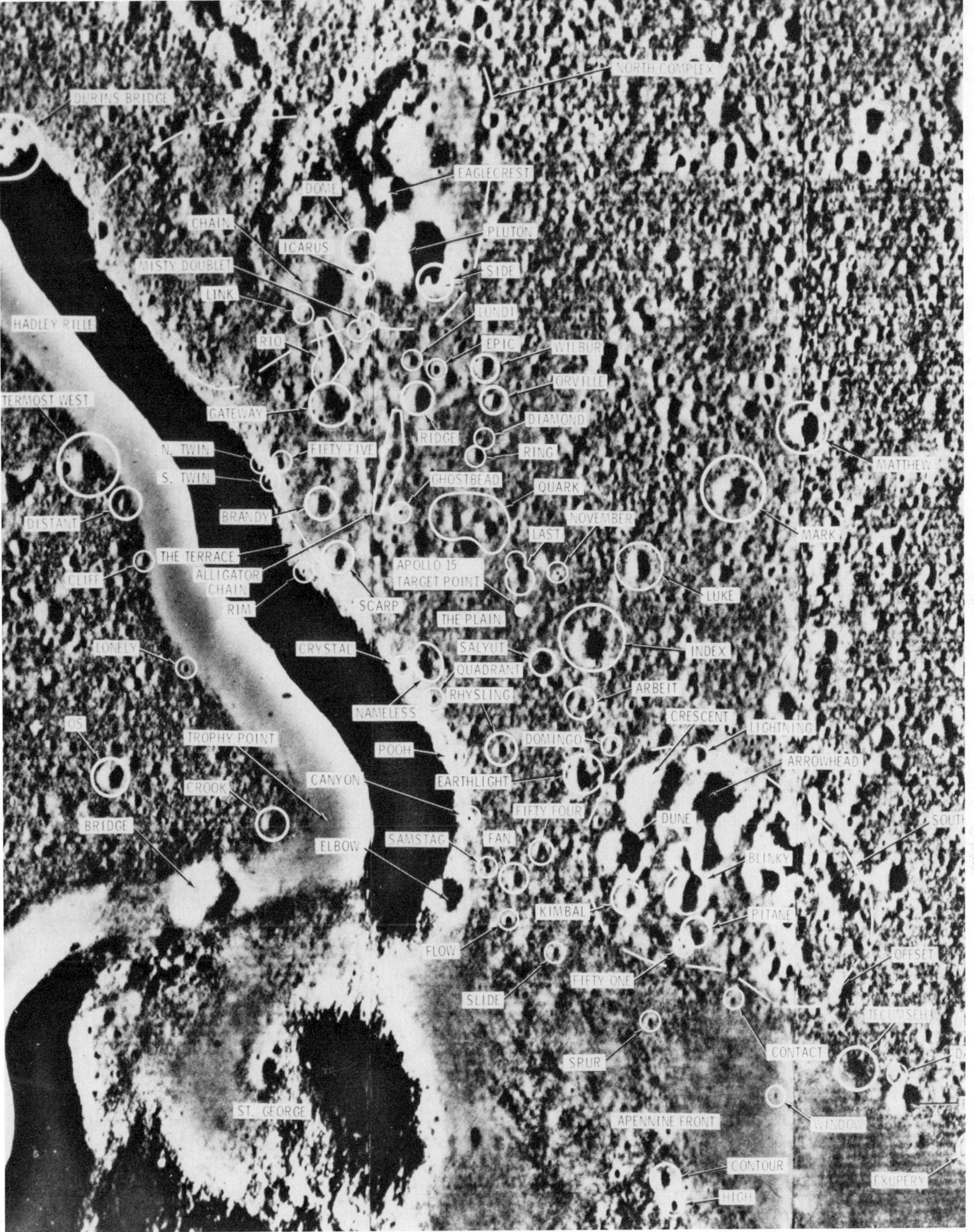

Figure 10-20 A working photomap (prepared from *Lunar Orbiter* photos) used in planning for operations around the *Apollo 15* landing point. Prior to the mission, the astronauts and others give simple descriptive "pet" names to distinctive craters and other interesting features at the site so that they can communicate their observations while on the Moon to Earthbound teams at Mission Control in Houston by means of these reference points. (NASA photo.)

Figure 10-21 A two-photo composite looking to the east at the *Apollo 15* site. The Apennine Mountains are visible below the horizon. Mount Hadley lies to the left and the Hadley Delta to the right (both out of this view). (NASA photo.)

Mauro Formation. A few (nine of the ninety-seven larger than 1 cm) of the *Apollo 14* samples are crystalline rocks representing either mare-type lavas or remelted soils.

Apollo 15

The *Apollo 15* mission was a landmark in manned geologic exploration of the Moon. Use of the lunar roving vehicle (LRV), shown on the front dust jacket and in Figure 10-18, allowed the crew to carry out three EVAs lasting over eighteen hours. With this electrically-powered "auto" they covered a total of 28 km in their traverses as they photographed and sampled along the Apennine Front and Hadley Rille (Figure 10-19). Figure 10-20 illustrates the detail to which individual craters in a complex landing site such as *Apollo 15* are given descriptive or colloquial names for reference purposes. This is done to aid in identifying, to the astronauts and to Earthbound mission control and geologist support teams, the approximate locations of observation or sampling sites during traverses.

Among many exciting discoveries was the recognition of what seem to be prominent lineations exposed in the mountains (Figure 10-21) of Hadley Delta and Mount Hadley to the north (Figure 10-22a-d) rising 3 to 4 km above the plains. The Apennine Front is generally held to be part of the uptilted middle ring raised during formation of the premare Imbrium Basin, and thus it might contain some of the oldest and deepest rocks, perhaps including crystalline layers of ancient crust, brought up to the surface. Some workers had considered these massifs to consist of, or at least be covered by, ejecta from the Imbrium Basin to the west and the Serenitatis Basin to the east, overlying even older ejecta complexes from still other basins. Close inspection along the Apennine Front indeed confirmed that the lower slopes are covered by a thick debris layer of fragmental rocks carried down as "talus" eroded from higher exposures.

In general, the lineations look like a repeating sequence of layers with relatively uniform thicknesses between 20 and 50 m (Figure 10-22b). In some photographs the linear features in the mountains resemble light and dark materials. Where exposed, these lineations or striations show apparent dips of 20 to 30°. In Mount Hadley and elsewhere, a second set of near-horizontal lineations is evident. From a distance many of those cross lineations resemble patterned ground (see p. 110) and may prove to be related. Similar lineations are evident on hill slopes at the *Apollo 16* and *17* sites. Comparable lineations have been noted in much of the highlands, in both plains and more rugged terrain, by the astronauts during orbital observations.

These linear features have elicited various explanations. To some astrogeologists, they are comparable to "strati-

(a)

(b)

Figure 10-22 (a) Mount Hadley, north of the *Apollo 15* site, rising approximately 4.5 km above the mare plains and showing two sets of prominent lineations (inclined to the left and subhorizontal) along its slopes. (b) Close-up, taken with a 500-mm lens, of the lineations exposed along the face of Mount Hadley.

(c)

Figure 10-22 (continued) (c) The Hadley Delta looking southeast toward Silver Spur along the Apennine Front. (d) Close-up, through the 500-mm lens, of Silver Spur showing terraces and lineations that may relate to layering.

(d)

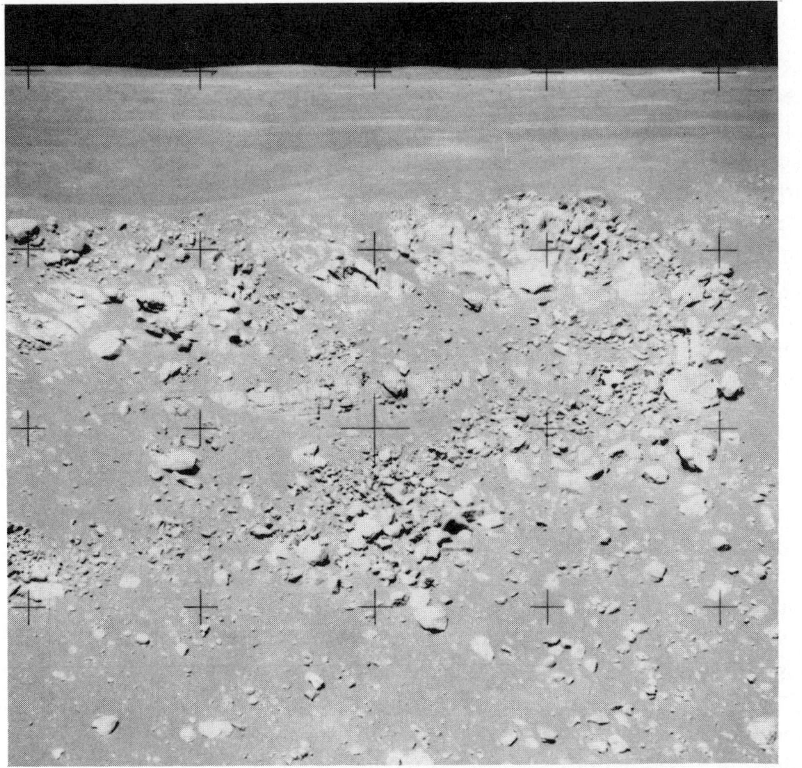

(e)

Figure 10-22 (continued) (e) The west wall of Hadley Rille, viewed through the 500-mm lens, showing well-defined bedrock layers and erosional talus. (f) A fresh, shallow rock-strewn crater near Hadley Rille, with the Apennine Front and Hadley Delta in the background. (NASA photos.)

(f)

fied" igneous sequences such as occur on Earth in layered intrusions of the Skaergaard in Greenland (Figure 10-23) or of the Bushveld complex in South Africa. This type of layering could have developed in the upper lunar crust but well enough below the existing surface at the time to be shielded from the effects of contemporaneous impacts. Another view considers the lineations to be a thick sequence of successive, regularly spaced lava flow layers extruded on a primitive lunar crust prior to formation of the Imbrium Basin. Or if the Apennines consist mostly of ejecta, the layers could be deposits piled up from numerous base surges, although it is difficult to explain the uniformity of thickness and regularity of spacing by this process. Still others have suggested that both the inclined and horizontal lineations are regular sets of joints whose north and northeast local trends align with the lunar grid system (see p. 110). However, some convincing observations and model experiments have forced a cautious reassessment of the reality of these lineaments as geologic structures. Instead, Howard and Larsen (1972) propose that most of these sets of close-spaced linear features are instead illusions brought about by the oblique illumination when the Sun is at low angles (the usual case during lunar exploration). Evidence cited by Wolfe and Bailey (1972) in support of this conclusion includes (1) shifts in trends and inclinations of the lineations with changing Sun angle, (2) symmetry of trends around the advancing Sun's rays, (3) persistence of linear trends even as slope orientations vary, (4) absence of any control of topography by these linear features (Silver Spur at *Apollo 15* seems to be an exception, where benches apparently are coincident with some linear expressions), and (5) persistence of slopes which are covered by thick mantling rubble (talus or regolith). They have successfully reproduced conjugate sets of linear features in a mound of cement powder subjected to lighting conditions similar to those during the *Apollo 15* mission. They deduce that fine-scale irregularities in the powder itself cause the apparent lineations; on the Apennine Front or Mount Hadley, a surface covered with debris and pockmarked by craters could have the same effect.

Multiple, nearly horizontal layers crop out along the wall slopes of Hadley Rille (Figures 10-22e-f) and Figure 10-24). These layers are presumed to be exposures of mare lava flows that underlie the plains around the landing site (Figure 10-25). At least nine distinct flow units, some up to 10 m thick, are visible in the uppermost 100 m of the

Figure 10-23 Layers (light and dark bands convex upward) and open joints (inclined down to the left) in the Gabbrofjaeld, part of the layered igneous complex in the Skaergaard intrusion, southwest Greenland. The face of the mountain exposed here is over 300 meters high. (From L. E. Wager, and G. M. Brown, *Layered Igneous Rocks*, W. H. Freeman & Co., San Francisco, 1967, 587 p.)

Figure 10-24 A two-photo composite looking north into Hadley Rille and the hills beyond. (NASA photo.)

Figure 10-25 Reconstructed cross-section, based on photointerpretation, through the west wall of Hadley Rille near the *Apollo 15* landing site. Rille is approximately 300 m deep. (From Howard, Head, and Swann, 1972.)

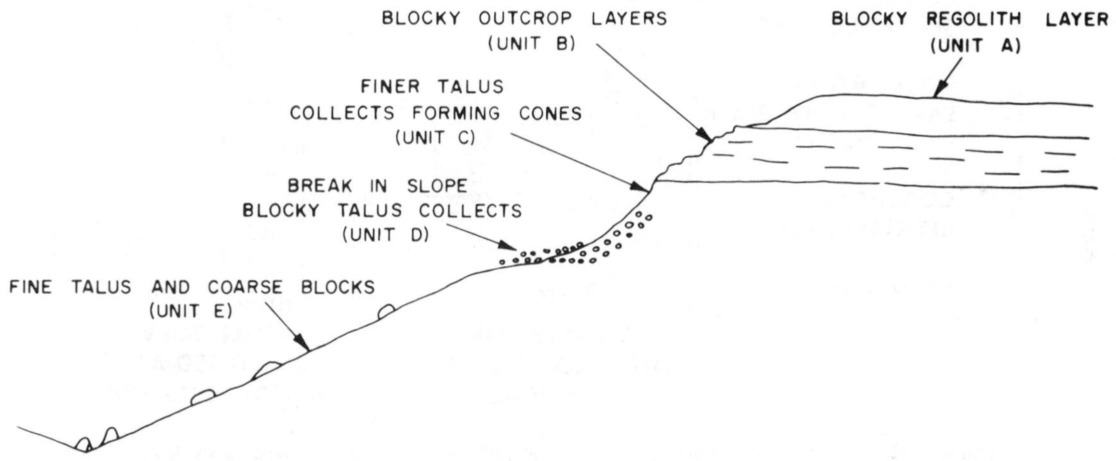

west slope. Most units are remarkably free of extensive jointing, in contrast to most terrestrial basalt flows, although sparse crude joints can be picked out in close-up photos. Thus, blocks that break loose from flow faces and roll inward toward the base of the rille some 300 m below the rim can be as large as 10 to 15 m on a side.

The origin of this rille, even after direct examination, is still in doubt. Because its deepest parts are also the widest, it is unlike an abandoned river channel. The fact that the east rim is 30 to 40 m higher than the west one, coupled with a tendency for segments of the rille to parallel the mountains-mare junctures, suggests to some that the rille is developed and controlled by faulting. But a prevailing view treats the rille as a large-scale trench formed by collapse of lava tubes, possibly related to drainback and shrinkage of the mare lavas, even though such collapse is unusual where multiple overlying flow units are present. The idea is further supported by observation of a "high lava mark" or bench above the plains at the base of Mount Hadley. Similar lava marks elsewhere on the Moon, previously thought to be terraces related to talus slopes, have been recognized by the CSM astronauts during orbital observations.

More than 350 individual samples, weighing collectively about 78 kg, were returned from ten sampling areas at the *Apollo 15* site. Most rock samples were breccias,

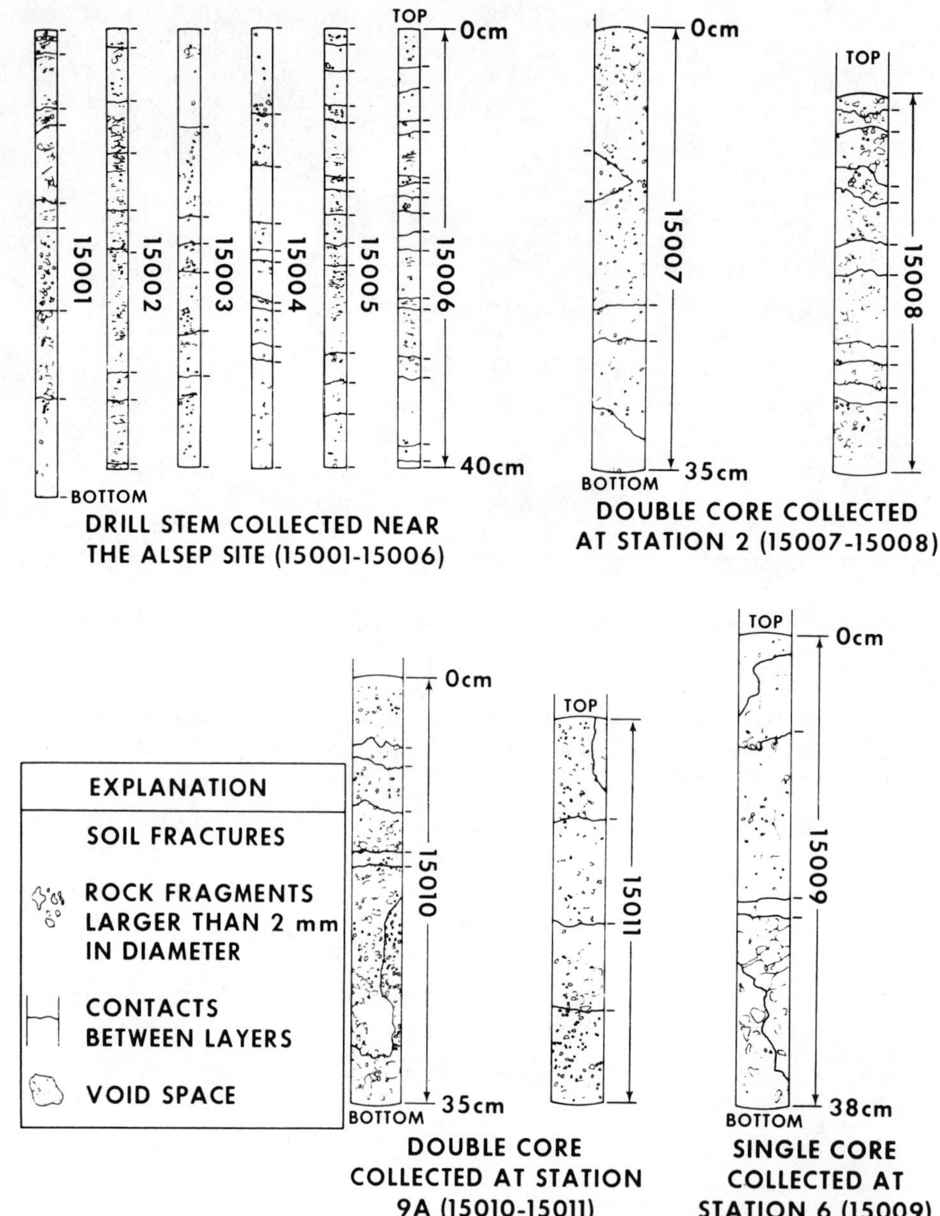

Figure 10-26 Sketches of essentially undisturbed soil material within the core tubes driven into the *Apollo 15* regolith by a rotary-percussion drill (samples 15001-15006 in six sections) and hammer drive (samples 15007-15011). Rock fragments, fractures, contacts, and voids were recognized in photographs obtained through the metal casing by X-ray radiography. (NASA illustration.)

plus a variety of crystalline basalts. Many of these were picked out of the soils as fragments by raking. Core samples were extracted from the regolith by use of a power-driven drill capable of cutting into bedrock. The general appearance of materials, brought out essentially undisturbed within these cores, is illustrated in Figure 10-26. The longest core, extending 242 cm into a regolith more than 5 m thick, contained forty-two recognizable layers defined by variations in grain size, texture, and sorting. These have built up layer by layer from successive base surges. The density of the cored regolith there increases from 1.36 g/cm^3 in the top half to 2.15 g/cm^3 in the bottommost section.

Apollo 16

The *Apollo 16* LM landed on April 20, 1972 at a site (9°00′S, 15°31′E) about 60 km north of the 50-km-wide crater Descartes. This is the only *Apollo* site located well within the lunar highlands. Lying west of the high Kant

Plateau, the landing area falls just within the edge of the undulating Cayley Plains where that unit meets the more rugged Descartes Mountains. The site was selected to allow exploration of hilly, furrowed terra units and cratered terra plains. Proximity to the Descartes Mountains seemed to offer an excellent chance for sampling parts of the ancient uplands crust built up by volcanic outpourings after the stage of maximum impact bombardments but prior to emplacement of the maria in the lowlands. Materials making up each topographic unit had been interpreted by photogeologic techniques to be volcanic in nature. The Cayley Formation of Imbrian age was thought to consist of very old (premare, ~3.8-4.0 billion-year-old) fluid lava flows intermingled with ash flow or fall deposits. The Descartes materials unit appeared to be generally contemporaneous and gradational with the Cayley unit. The hilly terrain within the Descartes Mountains was interpreted before the mission to be constructed of a more viscous equivalent of the Cayley materials that probably also included some pyroclastic (explosive) breccias. The hills would thus be made up of domes, volcanic cones, and shields that likely piled up late in the development of the highlands. Ridges and furrows, perhaps related to faulting and to extrusion of lavas from elongate vents, cut across the hills. Terraces with steep faces that extend on to the smooth plains units around the *Apollo* 16 site (Figure 10-27a) were interpreted to be flow fronts or landslide slumps. Linear bands visible in *Orbiter* and *Apollo* photos were attributed to stratified outcroppings of flow layers broken by fractures controlled by the lunar grid system.

Using the lunar roving vehicle (LRV), astronauts John Young and Charles M. Duke, Jr., covered a traverse distance of 20.3 km and thus visited all major stops scheduled dur-

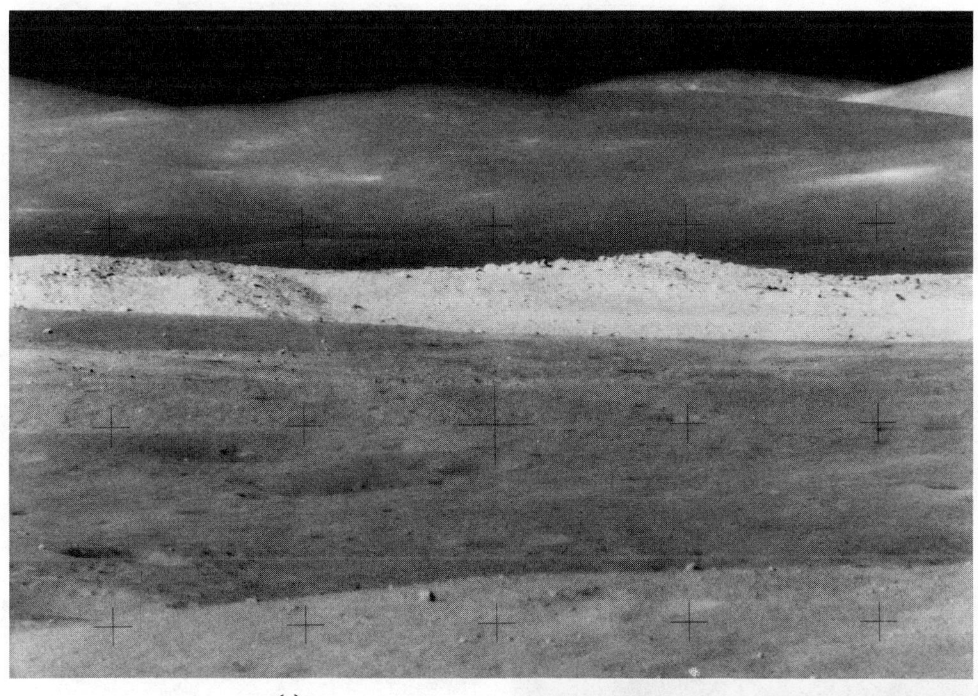

(a)

(b)

Figure 10-27 Scenes around the *Apollo 16* site. (a) Terrace deposits (bright ridge) between Stone Mountain (background) and the heavily cratered smooth plains (foreground). (b) A view typical of that seen by the automatic photo camera mounted on the LRV during an EVA.

(c)

(d)

Figure 10-27 (continued) (c) View into a boulder-strewn crater near South Ray. (d) Part of a large, boulder-free crater with a subdued rim which, from crater-age morphology, should be older than the crater shown in (c). (e) Astronaut preparing to sample "House Rock" within the North Ray crater. (f) A small sample being removed by the tongs from a dark breccia containing numerous white clasts. (NASA photos.)

(e)

(f)

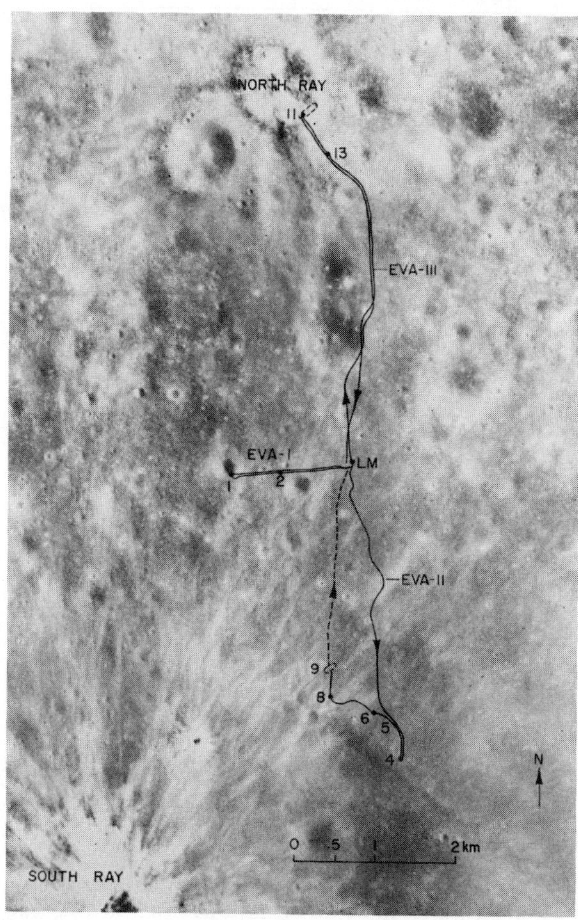

Figure 10-28 The traverse routes and sampling stations occupied during the three EVAs of the *Apollo 16* mission. (Courtesy G. Swann, U.S. Geological Survey.)

ing the three EVAs. They remained a total of more than twenty hours outside the LM. Most of the traverse stops (Figure 10-28) were designed to sample the Cayley flow units, expected to be examples of the highlands basalts (see p. 197). The only accessible Descartes materials were sought along the front of Stone Mountain. Between stops, the astronauts took many photographs of the scenes ahead with an automatic camera mounted on the LRV (Figure 10-27b). One prime stop was North Ray Crater, a 1-km-wide depression between 150 and 200 m deep, which showed up to seven discernible layers, an apparent central mound, and a rocky ejecta blanket when viewed in pre-mission photos. South Ray Crater displayed even fresher features (Figure 10-27c); as measured from calibrated photographs, light-toned ejecta rays from this crater locally produce albedo readings as high as 0.55. Cosmic-ray exposure ages (see p. 227) of rocks ejected from the North and South Ray Craters date their excavation at 50 and 2 million years ago respectively. Both North and South Ray Craters cut into deeper parts of the Cayley Formation. Other large craters around the site are presumably much older (Figure 10-27d).

Young and Duke collected approximately 95 kg (209 lb) of rocks and soil. Several samples were nearly football-sized. The largest bounder examined and sampled was "House Rock" (Figure 10-27e)–10 m high by more than 20 m long–lying near the bottom of North Ray Crater. A fractured surface in this boulder, described by the astronauts as a shatter cone, later proved to be a "percussion" cone caused by a low-velocity impact. Undisturbed soil (shielded from the solar wind) was collected from under several large rocks as a source of material whose isotopic composition of gases would more nearly reflect the natural state of the lunar lithologies.

The great surprise during this mission was the absence of volcanic-like rocks that fitted the preconceived expectations for Cayley and Descartes materials units. Crystalline rocks, mainly feldspathic gabbros and anorthosites, were sparse; some contain evidence of later thermal metamorphism. A few igneous-looking rocks have subsequently been classed as recrystallized shock-melted soil. Almost all samples brought back to Houston were heterogeneous breccias composed of light and dark components (Figure 10-29). Fragmental rocks from the Cayley Formation come from five crudely stratified zones. As reconstructed from photos and sample distributions, these form this sequence from the surface downward:

(1) Regolith and ejecta debris around craters; 20 to 30 m thick.
(2) Blocky breccias with a light-colored matrix; 110 to 140 m thick.
(3) Friable, weakly consolidated unit; 50 to 100 m thick.
(4) Coherent, dark-matrix breccias; 30 m thick.
(5) Breccias with whitish anorthositic gabbro.

Breccias in the vicinity of Stone Mountain, representing Descartes units, are mostly light-matrix, fine-grained materials. Heiken et al. (1973) have found that the ratio of light to dark clasts is 4.2:1 in Descartes units and from 1.1:1 to 2.2:1 in Cayley units.

Many of these breccias are now thought to be consolidated aggregates of surface rubble and ejecta deposits. Glass and other signs of shock damage indicate impact fragmentation. Many fragments seemingly were derived from local and regional sources, but much of even that material may have been introduced earlier as ejecta from Imbrium, Serenitatis, and other basin-forming events. Photogeologic reinterpretation of the Descartes site by Chao et al. (1973) after the mission indicates the Cayley to be somewhat younger than the Descartes Formation based on crater counts. The Cayley is held to be a facies of the contemporaneous Hevelius Formation around the south-western edge of Oceanus Procellarum. In this view, both units were derived simultaneously during the Orientale basin-forming event. The Hevelius Formation represents base surge ejecta which come to rest as hummocky deposits

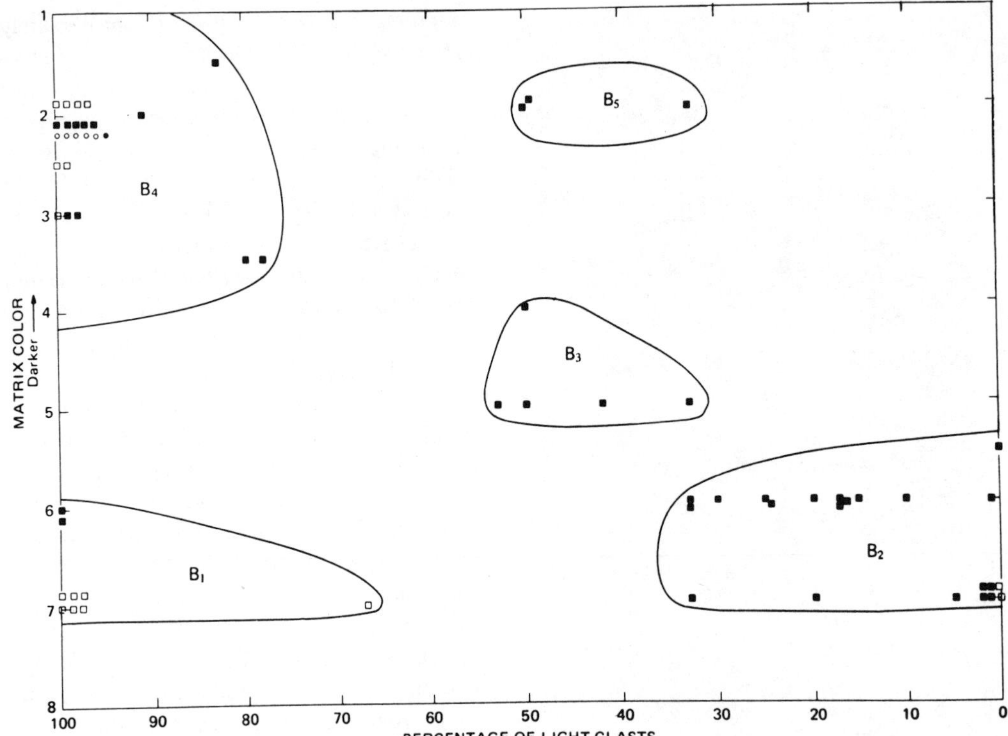

Figure 10-29 Classification of breccias from the *Apollo 16* highlands site based on percentage of light clasts and matrix color. (From H. G. Wilshire et al., 1973, courtesy Lunar Science Institute.)

within the inner ejecta blanket. The Cayley is the corresponding suborbital ejecta tossed much farther from Orientale which is then "ponded" into depressions that form the light-colored plains units of the eastern highlands.*

Most investigators have since rejected the prelanding model that predicted volcanic flows and ash deposits still existing intact in the Descartes region. Rimless craters, suggested as vents for internal activity, are now interpreted as secondary craters from large impacts. Conceivably, some breccias could represent pyroclastic deposits that were later shock-metamorphosed by subsequent impacts. If so, then parts of this highlands region had indeed been covered by volcanic deposits emplaced after explosive ejection from vents. More likely, the highlands—even if first built up from lava extrusions—have been so thoroughly cratered that any older volcanic units are now effectively comminuted into a thick (1- to 2-km) debris pile that mantles the outer crust (Short and Forman, 1972). This is supported by failure to observe bedrock in the rugged terrain of the Descartes Mountains. In this new concept, only deeper zones within the impacted premare crust would have survived the breakups brought about by the high rates of impact inferred for the early history of the Moon. Rarely, these layered units were thrust up to the surface, as witnessed at the *Apollo 15* site.

*V. R. Oberbeck et al., at the Fifth Lunar Science Conference in March, 1974, present arguments for a largely local origin of the smooth plains deposits in the highlands. In their view only a small proportion (10–20%) of the Cayley Formation at *Apollo 16* is derived from distant sources such as the Orientale Basin. The bulk of material in this and other plains units is emplaced as base surge and ejecta from large, nearby primary craters and, more especially, as great masses thrown out and transported from numerous secondary craters developed during primary events.

Apollo 17

The landing point for *Apollo 17* (20°10′N, 30°45′E) was a dark-floored valley between two massive hills within the Taurus Mountains that form the highlands bordering the eastern edge of the mare-filled Serenitatis Basin. The site lies some 35 km south-southwest of the crater Littrow. The nearest encroachment of Serenitatis lavas is about 50-60 km to the west. The regional setting is displayed in the U.S. Geological Survey's *Moon Atlas Map* (I-800, sheet 1) of the Taurus-Littrow-*Apollo 17* region. The landing area (depicted geologically in map I-800, sheet 2) was photographed by the mapping camera on *Apollo 16* (Figure 10-30) and by the *Apollo 17* astronauts before descent.

A geological sketch map of the *Apollo 17* area (Figure 10-31), made prior to the mission, describes the major materials units distributed in the vicinity of the landing site. Rugged, blocky mountains, called *massifs*, rise 2 to

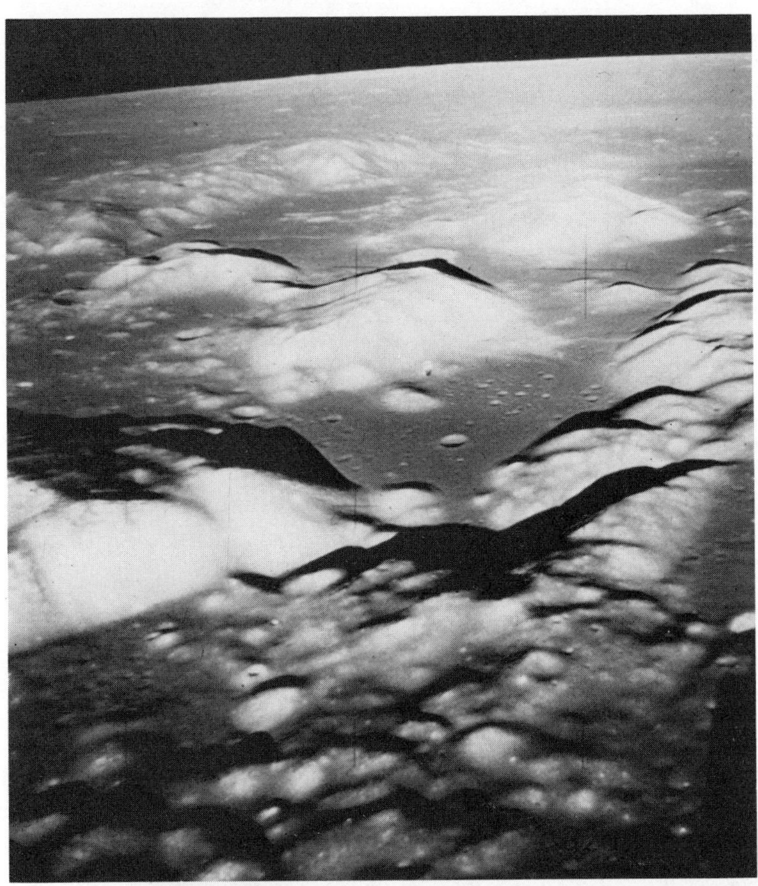

(a)

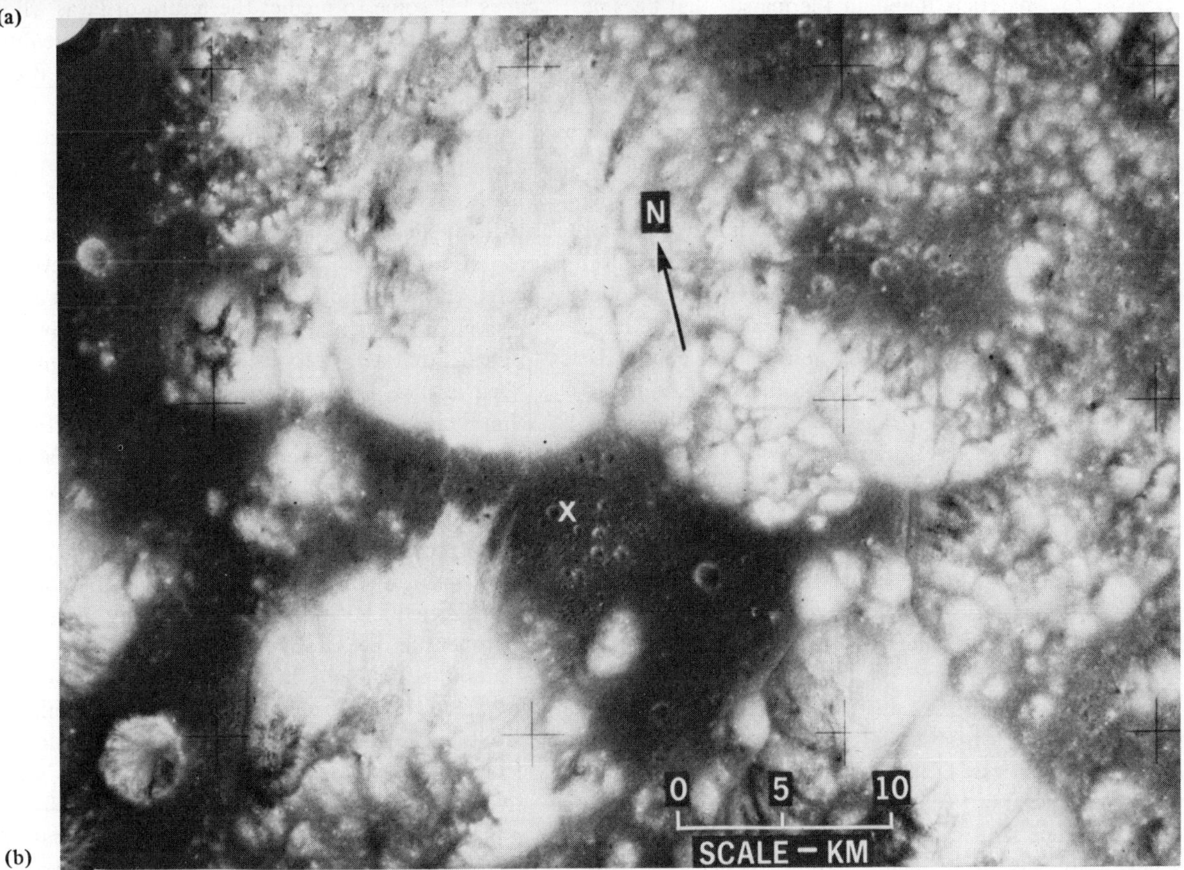

(b)

Figure 10-30 (a) Oblique view looking west at *Apollo 17* landing area, photographed from the LM one revolution prior to touchdown. (b) Enlargement of mapping camera view of *Apollo 17* landing area. (NASA photos.)

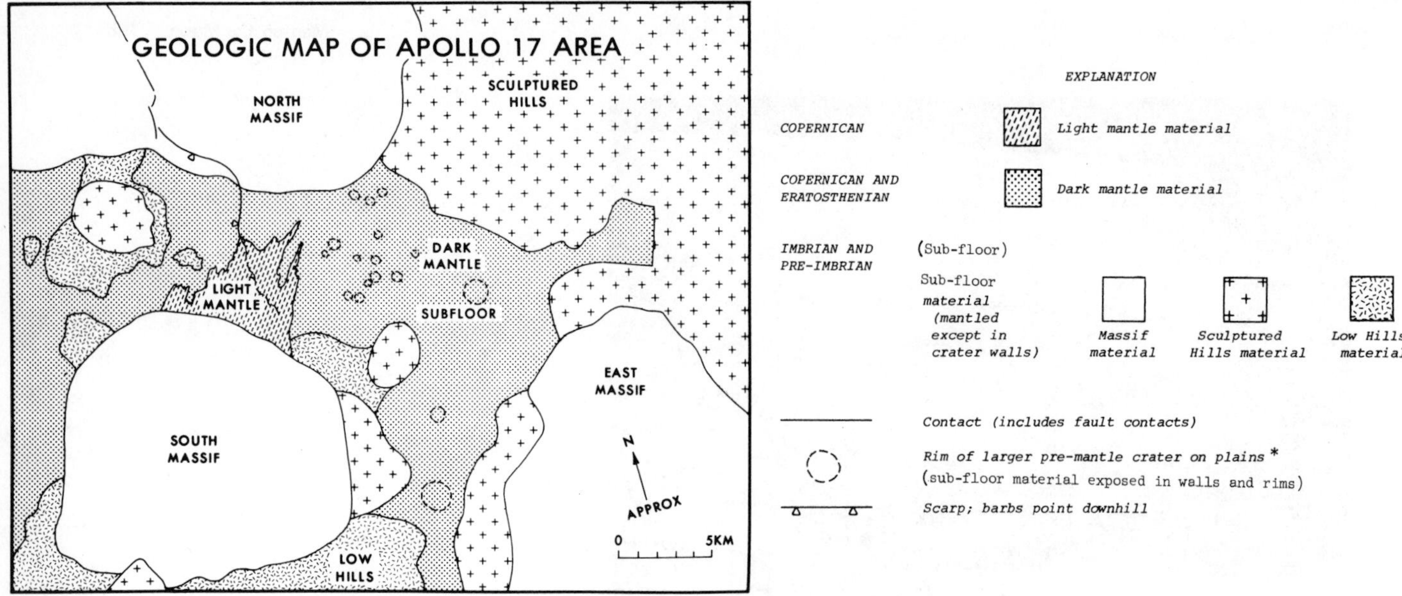

Figure 10-31 Geological sketch map and legend for *Apollo 17* site. (NASA illustration.)

3 km above the dark plains. These units, characterized by higher albedos, were predicted to be similar to the breccia units making up the highlands around *Apollo 16*. The massifs graded into more subdued topographic features, referred to as the *Sculptured Hills* to the east and the *Low Hills* protruding from the dark units, that presumably are also composed of brecciated rocks. These hills may be ejecta from the Serenitatis event or may be just lower levels of the materials found in the massifs that became partly inundated by later lavas. The relative brightness of both massifs and hills suggested that light-colored (whitish) fragments are a principal component of the breccias. The dark units, by analogy to the low-albedo mare surfaces, were expected to be lavas of basaltic character. A series of flow units made up the *subfloor* in the valleys, but the immediate surface was believed to be covered with a *dark mantle* of loose, fragmental material. Although a typical mare regolith could match most of these properties, the extreme darkness of the mantle, the scarcity of large boulders, and the proximity to known cinder cones (see Figure 10-55a) led to the assumption that these materials are cinders (perhaps comparable to scoria) and ash ejected from volcanic vents and fissures.

Two other unusual features were identified at the site. An extension of brighter deposits appears to spread from South Massif on to the valley west of the touchdown point. Because this light mantle passes over small craters in the subfloor whose outlines are still visible, it must be a layer as thin as a few meters or less. Its relation to the steep-sloped massif is precisely that expected of an avalanche of loose debris or talus carried partway across the valley. Both this mantle and the adjacent subfloor are affected by an east-facing escarpment whose relief in places reaches 80 m. The scarp evidently is a surface trace of a high-angle fault cutting deep into the valley fill. The valley itself may be bounded by other faults paralleling the massifs, such that it acts as a graben later invaded by lavas. The visible cross-fault would be younger than these subfloor lavas but probably older than the dark mantle material that, in turn, has been covered by the avalanche. These relationships are shown schematically in Figure 10-32. Although no actual ages could be assigned to the different units at Taurus-Littrow, the paucity of craters on the valley surface and the seeming freshness of nearby cones led some to predict the return of lavas whose ages would be less than *Apollo 12* rocks and, in one optimistic view, conceivably billions of years younger than those from any sites previously visited.

In most respects, the interpretative predictions made before the flight of *Apollo 17* were proved to be accurate. This is a tribute to the level of sophistication developed by the teams of geologists from the experience of five previous missions. There were, however, several unexpected surprises—both as exploration progressed and after rocks and instrumental data were analyzed following return to Earth—that helped to justify the site selection.

Astronauts Eugene Cernan and Harrison ("Jack") Schmitt accomplished all major objectives of the mission essentially on schedule in this most ambitious of *Apollo* trips. All stations preselected for investigation were visited (Figure 10-33). Additional samples were picked up en-route between stations without dismounting from the LRV by using a long-handled scoop sampler at unscheduled stops (labeled LRV in the figure). Every instrument and experiment in the ALSEP was properly activated.

One important objective of this last mission was to test the "value" of utilizing a *scientist* as an astronaut. Jack Schmitt, a geologist trained at the California Institute of Technology and later employed by the U.S. Geological Survey, performed to the highest standards. His field and sample descriptions were precise and informative. His close interactions with the geology team at the Manned

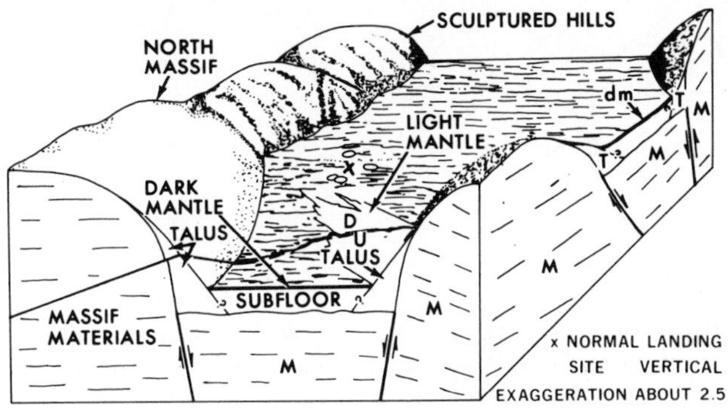

Figure 10-32 Schematic cutaway diagram of the *Apollo 17* Taurus-Littrow region. (NASA illustration.)

Spacecraft Center optimized acquisition of geological data by allowing near-realtime decisions to be made from the rapid exchange of communications between himself and colleagues having a common background. Perhaps the only "problem" arising from placing a geologist on the Moon was the difficulty in controlling Schmitt's eagerness, so as to hold him to the timelines set for the EVAs. This enthusiasm, shared by both of the other astronauts, paid off in the larger-than-expected return of lunar samples—some 86 kg (39 lb) more than the anticipated 460 kg. It should be said, too, that just as Schmitt, the geologist, functioned effectively as an astronaut, so also did Cernan, the pilot-engineer, excel as a geoscientist in his own right. Only by close scrutiny of the reports recorded from the astronauts at work could the professionalism of Schmitt be distinguished from the evident competence of Cernan. *Apollo 17*, as several previous missions, emphasized the effectiveness of the many months of training in the geosciences given to all astronauts by the permanent geological staff of NASA and by distinguished geoscientists from the U.S. Geological Survey, the universities, and other sources.

Figure 10-34 presents some of the views witnessed by the astronauts around the *Apollo 17* areas. Some general conclusions concerning the site include the following:

(1) Larger craters in the valley tapped subfloor materials consisting of coarse- to fine-crystalline dark basalts similar to those collected at the *Apollo 11* site.

(2) Most such craters, however, failed to expose bedrock units; these may be covered.

(3) The regolith in the valley is, in places, up to 20 m thick.

(4) Cinders and other signs of volcanic (pyroclastic) materials apparently are absent; the nature of the dark mantle remains uncertain, but may just be regolith containing a high proportion of very dark basalt fragments, or may be a thin deposit of black glass (beads and/or coatings).

(5) A "dustlike" deposit appears to drape over craters and blocks everywhere in the valley.

Figure 10-33 Close-up photograph of the *Apollo 17* landing site area showing the principal craters and traverse stations. (NASA illustration.)

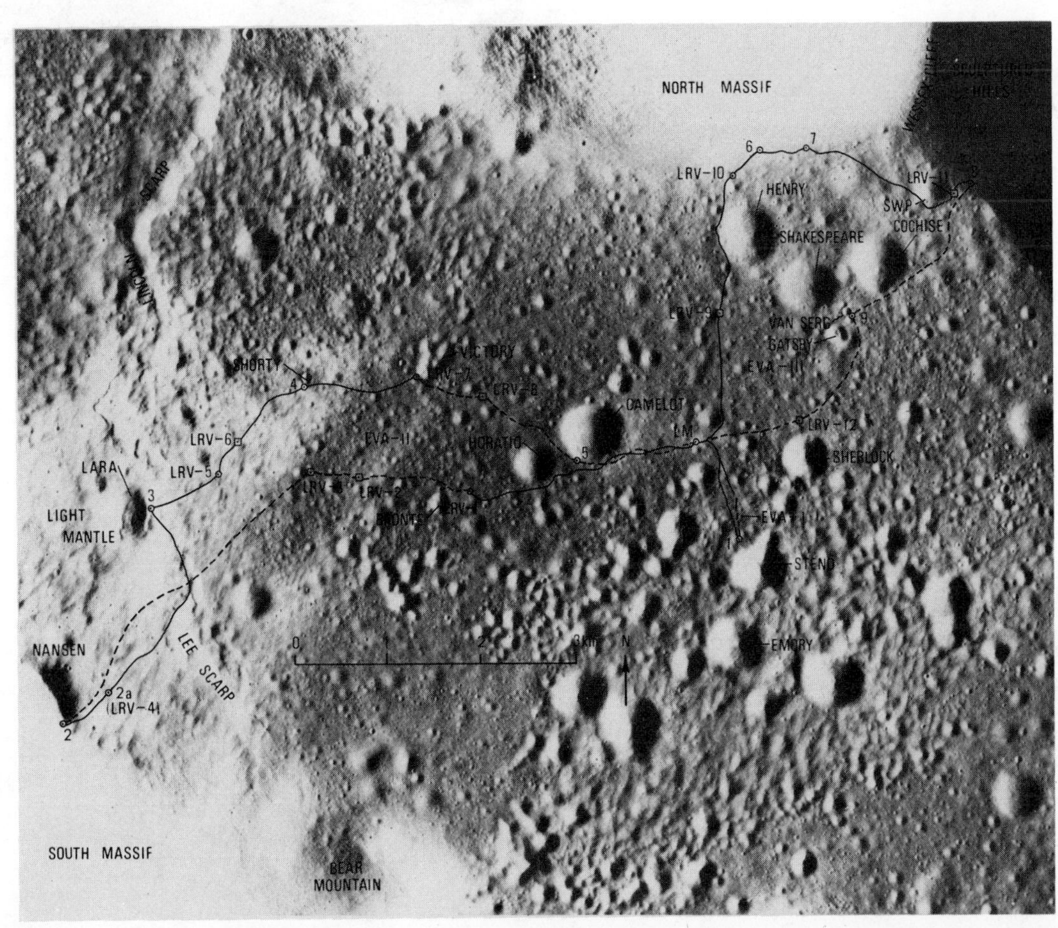

(a)

Figure 10-34 (a) The LM site at *Apollo 17,* looking southwestward toward South Massif. In the foreground is one of the explosives launching devices, with its signal antenna. (b) North Massif in the background and a large boulder field near a shallow crater in foreground. (c) Split Rock, a huge boulder at *Apollo 17* Station 6. East Massif is in upper left and Bear Mountain is at the right. Astronaut Jack Schmitt is preparing to sample this breccia boulder that has rolled down from North Massif.

(b)

(c)

(d)

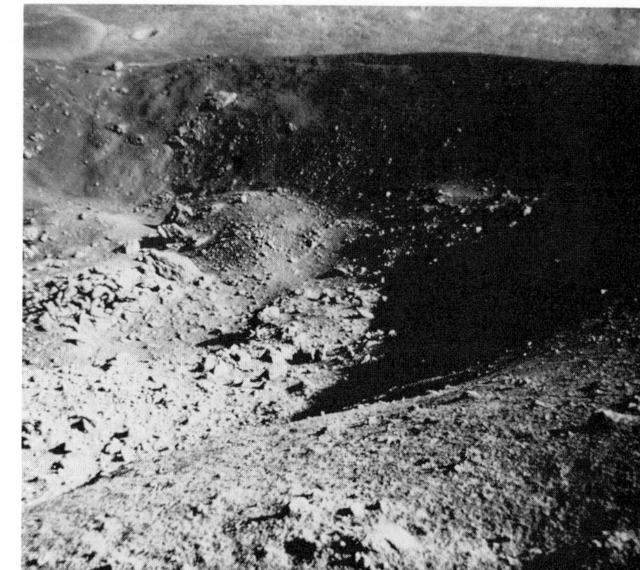

Figure 10-34 (continued) (d) Shorty crater, at Station 4 of *Apollo 17.* (e) The rim of Shorty crater. A patch of medium gray soil is exposed just below the gnomon. To its right is the outcrop of orange-colored soil that caused considerable excitement when first discovered by Jack Schmitt. A large boulder appears behind the gnomon. (f) A vesicular boulder composed of subfloor basalt from Station 1 at *Apollo 17.* (NASA photos.)

(e)

(f)

(6) Craters larger than 2 m are uncommon in the valley units.
(7) More blocks and boulders than expected occur in the valley.
(8) Soil clods in the valley are local impact-derived breccias.
(9) Both North and South Massifs are composed almost entirely of breccia units whose thickness must be 2 km or more; these breccias are as complex as those at *Apollo 16* and are similar in many ways.
(10) Broad lithologic "layering" is evident in the massifs; these may be different ejecta deposits.
(11) Several sets of lineations, some dipping up to 30°, appear in both massifs.
(12) The nature of the Sculptured Hills is still unsettled; it may be a mass of ejecta blocks from the Serenitatis event.
(13) The massifs were pushed up by the Serenitatis event.
(14) The light mantle unit is indeed probably caused by an avalanche.
(15) Some light mantle material may be ray material from the Tycho event.
(16) The fault across the valley appears as a series of talus-draped lobes, representing an echelon fault.

Other specific comments are best fitted to observations at each station:

At the LM site (station 1) all samples are either subfloor basalts or soil breccias. Blocks up to 4 m are present. Two core-tube samples (one from a 3.2-m-deep hole) were obtained. Station 2, next to South Massif, was dominated by complex crystalline breccias. The outermost layers of soil here, and at station 8, consist of a thin blue-gray cover overlying a 5-10-cm lighter gray zone that changes into marbled material below. Station 3 was close to the fault scarp; only breccias and loose rubble were collected there. At station 5, Camelot Crater (600 m) excavated well into the subfloor basalts (titanium-rich or ilmenite-bearing). Stations 6 and 7 lie along the edge of the North Massif. Boulders there had rolled down the slopes leaving distinct tracks or trails. Split Rock (Figure 10-34c) appears to be a gabbroic rock but may be a recrystallized breccia or impact melt. Light gray breccias and darker blue-gray breccias, often with a tannish matrix, are the prevalent rock types coming from this massif. Several boulders contain two breccia types with a well-defined boundary between the layers. These may represent the contact between the visually different units well up on the side of the massif. "Intrusive" dikelets of melt and fragments penetrate some rocks. Some breccia boulders contain a vesicular crystalline phase that may be wrapped around large clasts and blocks as though injected as a liquid (impact melt) during emplacement of the fragmental complex. Samples taken at station 8, in the Sculptured Hills, were mainly subfloor basalts and gabbros instead of the highlands breccias as had been suspected. Almost no crystalline basalts were found around Van Serg Crater (90 m) at station 9, well within the valley. The breccias found there originated, however, locally by compression of a thick regolith. Nearby, the larger Cochise Crater had penetrated to the lava units and possible bedrock layers were exposed.

The most exciting find of the mission was made at station 4. There, Shorty Crater (110 m) had penetrated through the light mantle well into subfloor material. Shorty has a raised rim, a central mound but with local steepening of slopes into depressions, a hummocky floor, and benches along the inner slope. This dark-haloed crater had been typed as a possible volcanic vent and, seen on site, does resemble certain maar craters. While sampling near the rim crest, Jack Schmitt suddenly spotted a patch of material that clearly was a bright orange to orange-brown. This zone, about a meter wide and 2 meters in exposed length (Figure 10-34e), was covered by less than a centimeter of gray soil and bottomed in a shiny blackish layer composed of ilmenite-bearing brown glass spheres. Yellowish and brownish streaks could be observed within the orange zone. Individual clumps extracted from the zone showed some color banding when broken open. The zone grades laterally into gray soil consisting largely of light gray ropy glass. Search elsewhere around Shorty turned up several other patches, suggesting that this may have been a continuous layer in the vicinity of the crater.

Speculation almost immediately occupied the attention of the Houston scientists and of the general public even as the mission progressed. Such coloration is a common alteration effect in the aureoles around volcanic vents. The color usually indicates ferric iron resulting from oxidation and hydration of iron-bearing rocks brought about by escaping steam and other vapors. Sulphides and arsenides can also have similar tints. The appearance of Shorty Crater, the preservation of distinct layers, and the occurrence of the orange soil together led to hopeful postulates (or perhaps "wishful thinking") of young volcanism at *Apollo 17*. The observations promoted exciting implications for hot-Moon advocates and for supporters of volcanism as a major cause of craters.

Less than 3 months after the mission, evidence from analysis of the orange soil had dashed this hope. The returned sample was found to consist of glass spherules and shards whose color derives from a high TiO_2 content (see p. 220). The smaller (<100 μm) sizes have transparent orange tones while larger particles are darker, more opaque, and variably crystallized. Similar orange and black glasses made up a small proportion of the regolith at *Apollo 11* and other sites. Proponents of the volcanic view contend that the pronounced homogeneity of this glass is characteristic of volcanic "beads" formed by "fire-fountaining" (spurts of lava at the surface driven by gases). But, homogeneity of impact-melted tektite glass and of glass at Lonar Lake, India—an impact crater in basalt—overrule this objection. Further, several other characteristics point to the glass as impact in origin. A fallout of blackish glass was followed by the orange-brown glass and finally by the thin gray deposits of ropy glass and other particles. Age dating of the glass (see p. 235) places this event at 3.71 billion years ago. However, exposure ages of rocks collected at Shorty Crater are much younger than the glass suggesting that the crater has been generated long after the orange glass was deposited.

Shorty Crater is a microcosm of the history of dashed expectations, changing opinions, and continuing controversies that have been a hallmark of the *Apollo* program. Even before *Apollo 11*, most planetary geologists had accepted the impact origin of the majority of lunar craters, but some continued to seek support for a volcanic origin. Solid evidence for volcanism had been anticipated in the *Apollo 16* region, but no such signs were found there. *Apollo 17* again was selected to prove that crater- and vent-forming volcanism was an important surface-modifying process. Failure to find any specific examples of cinders from not-too-distant cones weakened the prospects for such proof, but the discoveries at Shorty Crater momentarily breathed new life into the arguments of the volcanic proponents. Ultimately, science eventually proved superior to speculation, and Shorty itself has been relegated to the list of millions of other lunar impact craters.

The ideas concerning regolith generation—particularly those involving mixing through repeated turnover—that were first set forth after *Apollo 11* (see p. 147) were also being reevaluated in light of data from subsequent missions. The great age of Shorty Crater, with its unbroken layers just below the surface, seemed to contradict the hypothesis of continual disturbance and reworking through constant bombardments by meteorites and micrometeorites. Layering was observed elsewhere at *Apollo 17* and also in core-tube samples from other sites. These observations, supplemented by detailed studies of the trench and core samples (see Chapter 11), have forced a new appraisal of the earlier ideas of Shoemaker and others. The answer now seems simple: the regolith is not continuously churned up and tossed about in a random manner on a large scale; instead, it builds up by successive deposition of thin to thick (millimeters to centimeters) layers of ejecta mostly from the myriads of craterlets and small craters that populate the immediate neighborhood of a sampling site. Micrometeorites will disturb each layer to some extent and may rework the particles on top. On the whole, then, a regolith profile is constructional—that is, it remains intact as it

Figure 10-35 The "ellipse of fire," defined by localities on the lunar nearside that show signs of volcanism. The numbered localities are: (1) Littrow area; (2) Sulpicius Gallus Formation, southwestern edge of Mare Serenitatis; (3) Haemus-Apennines foothills; (4) Rima Bode area; (5) Copernicus CD area; (6) Fra Mauro crater; (7) Alphonsus floor; (8) Theophilus rim; (9) Langrenus C; (10) Picard and Pierce; (11) western edge of Crisium; and (12) Macrobius area. (Courtesy F. El-Baz, 1973).

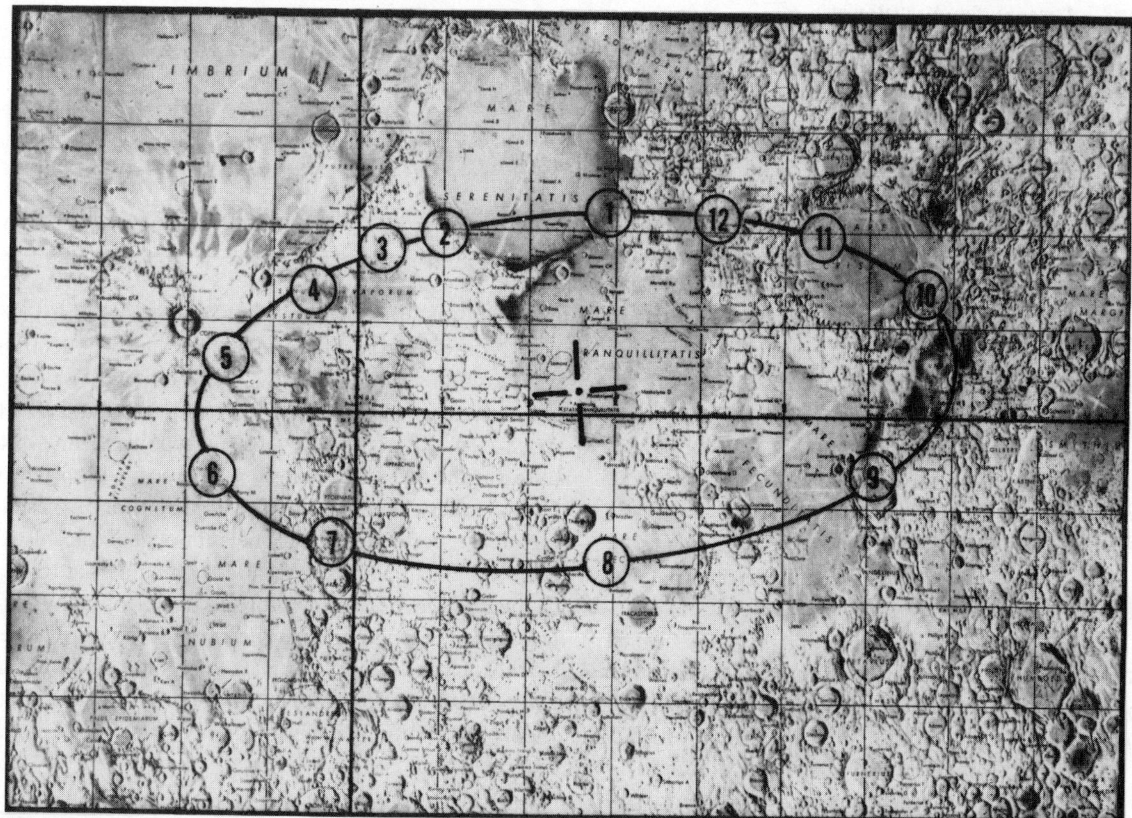

TABLE 10-2

Apollo Mission Experiments

		Mission and Landing Site						
	Experiment	*A-11 Sea of Tranquillity*	*A-12 Ocean of Storms*	*A-13 Mission Aborted*	*A-14 Fra Mauro*	*A-15 Hadley-Apennine*	*A-16 Descartes*	*A-17 Taurus-Littrow*
Number	*Orbital Experiments*							
S-158	Multispectral Photography (Goetz)		X					
S-176	CM Window Meteoroid (Cour-Palais)				X	X	X	X
S-177	UV Photography–Earth and Moon (Owen)					X	X	
S-178	Gegenschein from Lunar Orbit (Dunkelman)			X	X	X		
S-160	Gamma-ray Spectrometer (Arnold)					X	X	
S-161	X-ray Fluorescence (Adler)					X	X	
S-162	Alpha-particle Spectrometer (Gorenstein)					X	X	
S-164	S-band Transponder (CSM/LM) (Sjogren)			X	X	X	X	X
S-164	S-band Transponder (Subsatellite) (Sjogren)					X	X	
S-165	Mass Spectrometer (Hoffman)					X	X	
S-169	Far UV Spectrometer (Fastie)							X
S-170	Bistatic Radar (Howard)			X	X	X	X	
S-171	IR Scanning Radiometer (Low)							X
S-173	Particle Shadows/Boundary Layer (Subsatellite) (K. Anderson)					X	X	
S-174	Magnetometer (Subsatellite) (Coleman)					X	X	
S-209	Lunar Sounder (Ward)							X
	Surface Experiments							
S-031	Passive Seismic (Latham)	X	X	X	X	X	X	
S-033	Active Seismic (Kovach)				X		X	
S-034	Lunar Surface Magnetometer (Dyal)		X			X	X	
S-035	Solar Wind Spectrometer (Snyder)		X			X		
S-036	Suprathermal-ion Detector (Freeman)		X		X	X		
S-037	Heat Flow (Langseth)			X		X	X	X
S-038	Charged-particle Lunar Environment (O'Brien)			X	X			

builds up except in the closest proximity to an impact, where it may become disrupted by lateral distortion before subsequent burial by infalling ejecta from later events.

Apollo 17 also proved the value of orbital observations. Astronaut Ron Evans, alerted by the discovery of the orange soil during the second EVA, claims to have been able to spot the coloration at Shorty and other parts of the site when the CSM passed directly overhead during an orbit. He also noted orangish- to reddish-brown patches beyond the southwestern edge of Mare Serenitatis (associated with Sulpicius Gallus Formation). After Cernan and Schmitt rejoined Evans, all three astronauts began to note light orange to brown tones around craters in other parts of Serenitatis and at the southwest edge of Mare Crisium. A check with several astronauts who had previously orbited the Moon elicited other recollections of similar color tints. *Apollo 14* photographs also show rust-colored tones near Langrenus and Theophilus east of the *Apollo 16* site. It is not yet clear whether some of these observations were psychosuggestive or will prove valid after further examination. The explanation of these sightings must be revised from one supporting volcanism to a more general hypothesis that holds this material to be melt droplets falling out after impacts into titanium-rich lava rocks. One intriguing hypothesis proposed by E. Roedder of the U.S. Geological Survey, based in part on finding euhedral olivine crystals in the orange glass, is that the widely distributed patches of these colored deposits repre-

TABLE 10-2 (continued)

	Experiment	*A-11* *Sea of Tranquillity*	*A-12* *Ocean of Storms*	*A-13* *Mission Aborted*	*A-14* *Fra Mauro*	*A-15* *Hadley-Apennine*	*A-16* *Descartes*	*A-17* *Taurus-Littrow*
		Mission and Landing Site						
S-058	Cold-Cathode Ion Gauge (Johnson)		X		X	X		
S-059	Lunar Field Geology*	X	X	X	X	X	X	X
S-078	Laser Ranging Retroreflector (Faller)	X			X	X		
S-080	Solar Wind Composition (Geiss)	X	X	X	X	X	X	
S-151	Cosmic-ray Detection (Helmets)	X						
S-152	Cosmic ray Detector (Sheets) (Fleischer)						X	
S-184	Lunar Surface Close-up Photography (Gold)		X	X				
S-198	Portable Magnetometer (Dyal)				X		X	
S-199	Lunar Gravity Traverse (Talwani)							X
S-200	Soil Mechanics (Mitchell)				X	X	X	X
S-201	Far UV Camera/Spectroscope (Carruthers)						X	
S-202	Lunar Ejecta and Meteorites							X
S-203	Lunar Seismic Profiling (Kovach)							X
S-204	Surface Electrical Properties (Simmons)							X
S-205	Lunar Atmospheric Composition (Hoffman)							X
S-207	Lunar Surface Gravimeter (Weber)							X
M-515	Lunar Dust Detector (Bates)		X	X	X	X		
S-229	Lunar Neutron Probe (Burnett)							X

Names in parentheses are of the principal investigators responsible for planning, gathering data, and interpreting the experiment.
**Apollo 11* and *12*: Shoemaker.
Apollo 14 and *15*: Swann.
Apollo 16 and *17*: Muehlberger.
Source of table: Simmons, G. (1972), *On the Moon with Apollo 17*, NASA EP-101.

sent dispersed lava melt or "splash" tossed out by impacts into the maria even as these were being filled up with still liquid basalt.

Still, the arguments for some form of volcanism that can produce ash deposits may not be totally dead. El-Baz (1973) has found that dark mantle deposits in valleys, around low-rimmed craters, and associated with cones and domes tend to line up on the lunar nearside in an "ellipse of fire" (Figure 10-35), in analogy to the terrestrial "circle of fire" defined by the volcanic chains surrounding much of the Pacific Ocean Basin. The center of this ellipse (near 4°N, 23°E) approximately coincides with the nearside surface projection of the displaced center of mass of the Moon (see p. 186). Nevertheless, the only sure signs of purely endogenetic volcanism on the Moon remain those clearly associated with the maria–rilles, domes, flow fronts and wrinkle ridges, dark-haloed craters (in some instances), and certain types of ring structures.

Surface Experiments

Even after the astronauts leave the Moon on each mission, their efforts continue to pay off as data are returned by radio from instruments left behind. The *Apollo* lunar science experiment package (ALSEP) consists mainly of geophysical experiments that are proving invaluable as clues to the nature and history of the lunar interior. Table 10-2 gives a list of those experiments conducted on

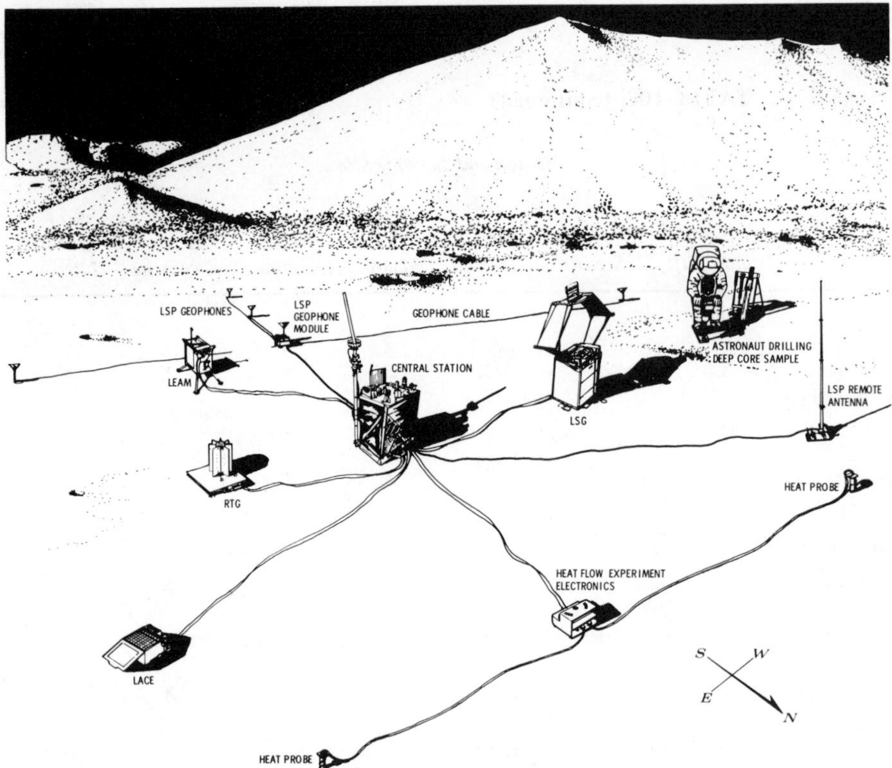

Figure 10-36 Cartoon showing deployment of instruments in the ALSEP at the *Apollo 17* site. Clockwise: LACE (Lunar Atmosphere Composition Experiment); RTG (Radioisotope Thermoelectric Generator, which powers the instruments); LEAM (Lunar Ejecta and Meteorites Experiment); LSP (Lunar Seismic Profiling), with geophone assembly); LSG (Lunar Surface Gravimeter); and HFE (Heat Flow Experiment, attached to probes). Other experiments at *Apollo 17* not tied to the central station were the CRD (Cosmic Ray Detector), LNP (Lunar Neutron Probe), LTG (Lunar Traverse Gravimeter), SME (Soil Mechanics Experiment), and LNP (Lunar Neutron Probe). Instruments used in earlier missions include: LSM (Lunar Surface Magnetometer); PSE and ASE (Passive and Active Seismic Experiments); SWS (Solar Wind Spectrometer); SIDE (Suprathermal Ion Detector Experiment); LRRR (Lunar Ranging Retro-Reflector). (NASA diagram.)

Figure 10-37 Part of the ALSEP array of scientific instruments in position around the *Apollo 12* LM. The magnetometer is set up in front of the large crater (center). Behind it (left to right) are the solar-wind spectrometer, the suprathermal ion detector, thermoelectric generator, central control and transmission station (with astronaut), and passive seismometer. (NASA photos.)

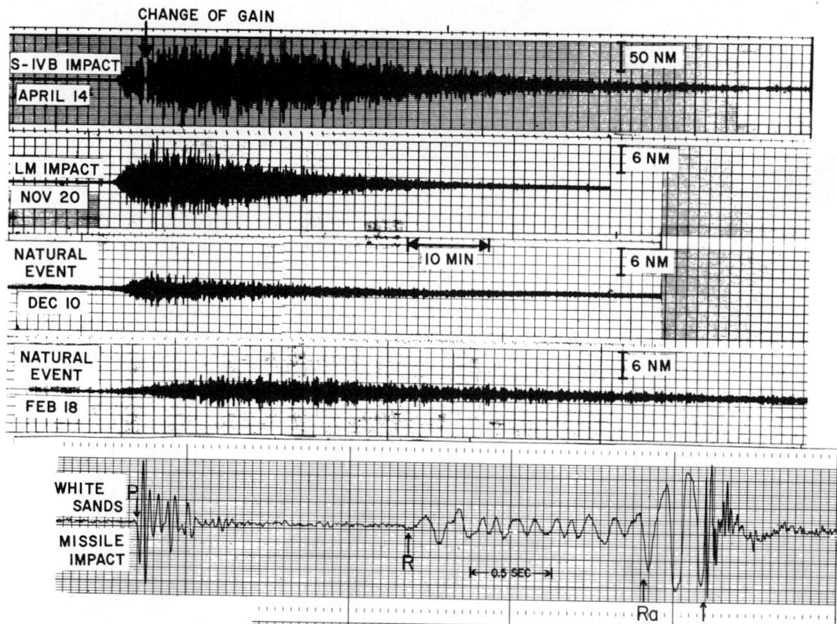

Figure 10-38 The seismic signals recorded by the long-period vertical component seismometer in the Passive Seismic Experiment sensor set up at the *Apollo 12* site. Seismograms from two natural events, the impact of the *Apollo 12* LM and the *Apollo 13* expended Saturn-IVB rocket casing, are shown. These signals continued to be recorded for 60 to 80+ minutes. Their peak amplitudes (measured in nanometers [nm] or milli-micrometers) were small, indicating low-energy events. A terrestrial seismic signal from a rocket missile impact (note different time scale) is included for comparison. (From G.V. Latham et al., *Science*, 1970, copyright 1970 by the American Association for the Advancement of Science.)

each mission. A schematic view of the deployment of the ALSEP at *Apollo 17* is shown in Figure 10-36, and a view of some of the instruments on the lunar surface at *Apollo 12* appears in Figure 10-37.

Seismological Experiments

Among the most informative instruments are the seismometers set up to record moonquakes, impacts, and other natural or artificial vibrations. These are disclosing evidence on the internal structure and thermal state of the Moon, its tectonic stability, and the frequency of meteoroid collisions or other surface-disturbing events.

Most of the seismic records telemetered back to Earth have proved strikingly different from almost any seismograms obtained from terrestrial earthquakes or explosions. Tracings of signals generated by two large natural events on the Moon, as recorded by the *Apollo 12* vertical component seismometer, are shown in Figure 10-38. The lunar signals are characterized by a gradual buildup to a maximum intensity followed by an extremely long, steadily diminishing decay in amplitude over a period of one hour or longer for some events. In contrast, a typical terrestrial seismogram usually shows a pronounced first arrival of a P or longitudinal wave, then a train of additional P waves, followed by erratically vibrating S or shear waves, and later a series of L or surface waves; the recorded durations of most earthquakes as monitored on these records last only seconds to a few tens of minutes at most. This succession of body and surface waves is not immediately evident on examining the seismograms from the Moon. When the first portions of the lunar seismic signals are analyzed in detail by filtering out various wave frequencies and expanding the time scale, some of these P- and S-wave oscillations can be identified. The implications of these strikingly different seismograms are discussed below.

At first, the unusual lunar seismograms received from the *Apollo 11* site were the subject of considerable speculation. Later, invaluable clues to the nature of some of the recorded events resulted when the abandoned LM from *Apollo 12* and the spent Saturn 4B third-stage booster rocket from the aborted *Apollo 13* flight were deliberately allowed to crash onto the Moon. These controlled impacts (whose time and location were accurately tracked) were equivalent to detonating explosions of 0.88 and 12.4 tons of TNT, respectively, on the lunar surface. The seismic signals (Figure 10-38) occurring shortly after each crash were almost identical to those from the natural events except for their greater amplitudes and durations (over four hours of detectable vibrations after the S-4B impact). Thus, most of this class of similar lunar signals are inferred to be meteoroid impacts. One large impact event occurred twenty three days after the *Apollo 16* touchdown. The impact point was only a few tens of kilometers from the *Apollo 14* seismic array. From energy calculations, the resulting crater must be as much as 100 m in diameter. On the whole, however, the number of impact events has been one to three orders of magnitude lower than had been predicted to happen over the same time interval based on best estimates from meteorite flux curves (see Figure 3-8).

Over the first thirteen months of operation, the *Apollo 12* seismograph picked out 272 natural events of both internal and impact origin. Later records from that station indicate an annual rate of ~600 internal events. Even greater frequencies of occurrence—up to several hundred events per month—have since been recorded by the net of

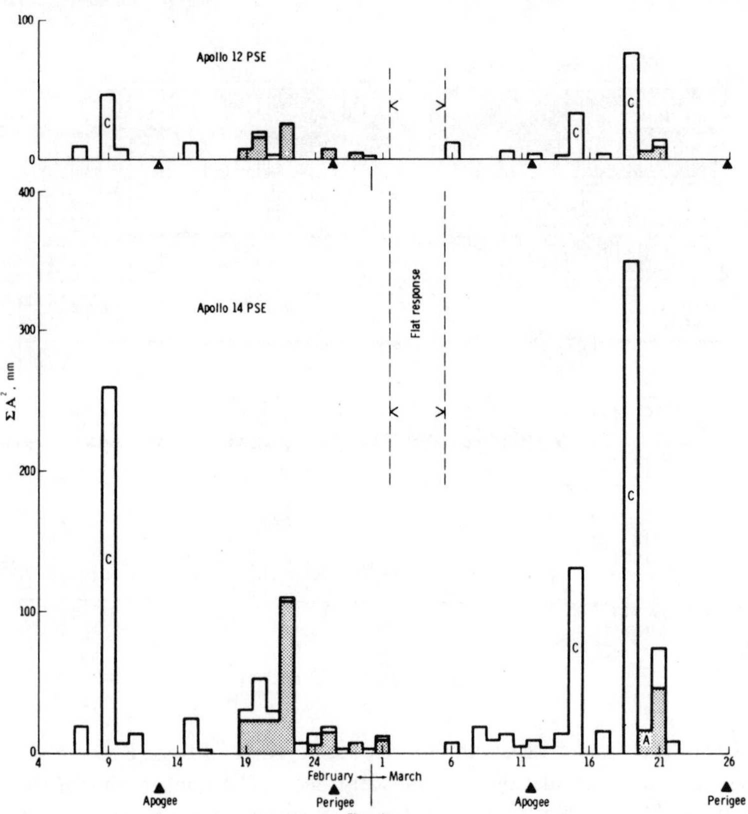

Figure 10-39 Plots of relative amplitudes (signal power) of the LP components recorded by *Apollo 12* and *14* seismometers for lunar seismic events during a two (lunar) month period in 1971. The white bars represent events having impact-like characteristics; the shaded bars relate to events having moonquake-like characteristics. Note the tendency for the strongest seismic signals to occur at specific times relative to apogee and perigee. (NASA diagram, from Latham et al., in NASA SP-272.)

seismic stations emplaced during the next three landings. The Moon averages up to 3200 internal quakes per year (as recorded at the *Apollo 16* station; fewer quakes are recorded at the other stations). These quakes amount to about 90% of all events recorded annually. Nearly all of these appear to be very-low-energy quakes (between 10^9 and 10^{12} ergs, corresponding to 1 to 3 on the Richter scale). These moonquakes (A events) are distinguished from those seisms caused by impact (C events) by the presence of strong waves that have sharper first-signal rise times, horizontal shear motions (H-phase), and lower frequencies in the wave train.

Although certain quakes take place at random times (as expected) from impacts, others occur with much greater frequency near times of perigee and, to a lesser extent, apogee when the Moon is respectively closest to and farthest from the Earth (Figure 10-39). In fact, after accumulation of many seismic records from the Moon over the last three years, it is now possible to predict the recurrence of moonquakes from a known source to an accuracy of just a few hours or even minutes (Figure 10-40). Tectonic strains within the lunar interior (confined almost exclusively to the outer half of the Moon) should approach a maximum at perigee owing to increased tidal forces and are then most likely to be released as moonquakes.

Much of the internal seismic activity monitored thus

Figure 10-40 Plots of the large amplitude category A_1 moonquakes recorded through 26 months by the *Apollo 12* seismograph. The recurring pattern is that of the strongest quakes at perigee each lunar month. A long-term cycle, spread over 7 calendar months, is suggested by the highest maxima occurring about every 206 days. (Courtesy G. Latham, The University of Texas Marine Biomedical Institute, Galveston, Texas.)

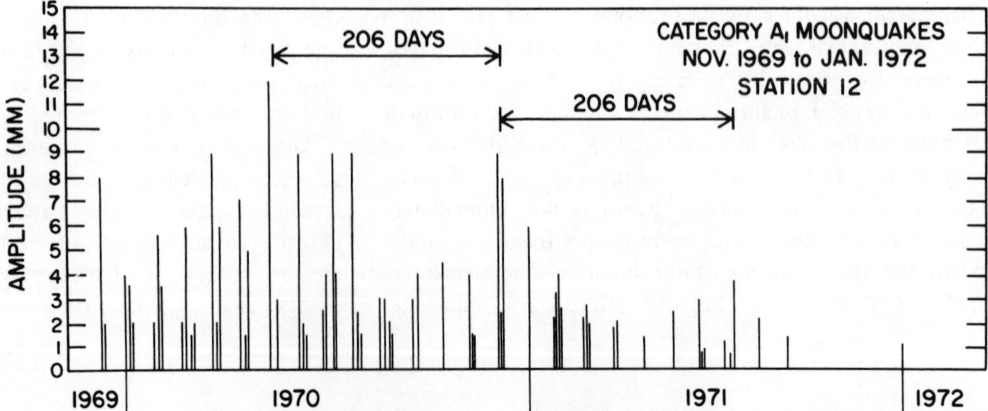

far produces signals of the type expected from movements of magma at depth into cracks or pockets that open up as tidal stresses flex the lunar crust each month. The pattern of occurrence of many moonquakes, starting with a build-up of weak foreshocks followed by larger tremors, is similar to that observed prior to and during some terrestrial volcanic events. The moonquakes, though, appear to originate in much deeper interior zones where heat may be accumulating because of higher concentrations of radioactivity.

Seismic data derived simultaneously from the triangulation networks of operating seismographs at the *Apollo 12, 14, 15,* and *16* sites indicate that a large number of these periodic moonquakes originate below a relatively few epicenters (small areas on the surface over the foci). The foci beneath these epicenters have been calculated to be deeper than 600 km, with a few as deep as 1000 km (Figure 10-41). The width of an average focus appears to be only about 10 km. One prominent epicenter lies in the Mare Humorum-Mare Nubium region about 600 km southwest of the *Apollo 14* site, while another is east of the Apennine Front around Mare Serenitatis. These two principal sources account for about 20% of the total seismic energy detected by the *Apollo* seismographs. In all, forty one deep-focus zones had been pinpointed by early 1973 (Figure 10-42). Their epicenters define two distinct belts each up to 2000 km in length. One runs generally north-south along a line some 20° west of the prime meridian passing through Sinus Medii. A second line extends north-east-southwest toward the eastern limb of the frontside. The cause of this localization of epicenters into belts—somewhat analogous to the circum-Pacific and trans-Mediterranean earthquake belts—is not yet known. Speculations center on rifting at depth or divergence of convection cells.

Analysis of these repetitive moonquakes over a long time span has disclosed a broad, 7-month cycle of maximum frequencies that ties in with long-term gravitational effects from the Sun (see Figure 10-40). Still another type of moonquake has been correlated with the monthly cycle of solar heating. Weak signals representing a traveling disturbance have been recorded at *Apollo* stations by the short-period component of the seismographs. These signals begin approximately 48.25 hours after local sunrise. The signals appear to result from thermal stresses induced by the diurnal temperature changes that affect the upper regolith. Finally, some seismic signals of noncyclic nature probably are produced by surface slumping and other mass-wasting movements that occur randomly.

The peculiar form of the lunar signals is more the result of the internal characteristics of the transmitting media than of the types of events that release the seismic energy. The long reverberations that typify these signals imply a very low attenuation of the seismic waves down to depths of 20 km or greater. Seismic waves arriving from distant sources after passing through the interior become trapped ("bounce around") in this surface region and only gradually release their energy to the seismometers. This behavior can be quantified by the Q or *quality factor*, which measures energy dissipation; high values imply little dissipation and hence efficient transfer of energy. Calculated Q values for near-surface materials around the *Apollo* sites are as high as 5000 compared with 10 to 300 for most terrestrial rocks in the continents. Such lunar values suggest either dispersion or scattering (or both) of waves passing through fluid-free, fractured and/or fragmented materials marked by extreme physical heterogeneity at a scale of a kilometer or greater. The presence of numerous craters of this size and larger, perhaps together with widely spaced cracks

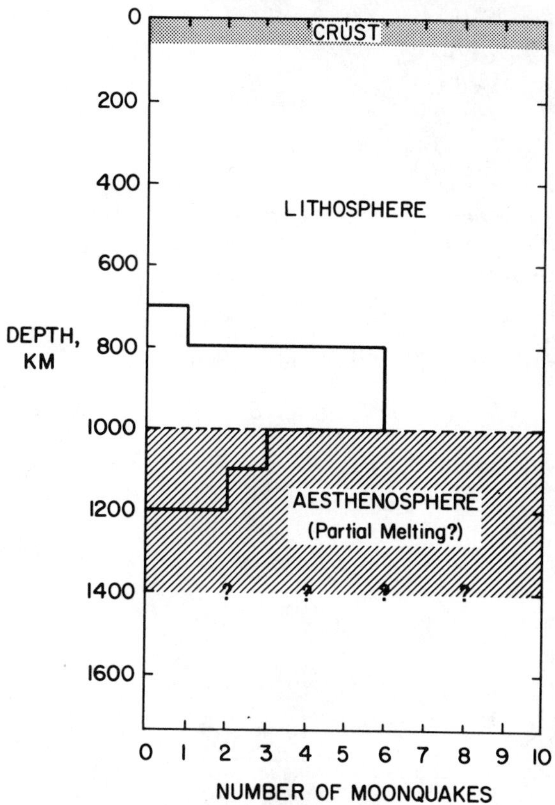

Figure 10-41 Frequency diagram indicating the number of moonquakes as a function of depth. On the diagram most foci plot between 800–1000 km below the surface; a later interpretation by G. Latham (personal communication, April, 1974) indicates these foci most probably lie in a somewhat shallower zone between 600–800 km. A zone of lower shear wave velocities, indicative of possible partial melting, occurs between 1000–1400 km. This zone is analogous to the asthenosphere on Earth that extends from about 100 to 200 km below the geoidal surface. Another view considers this asthenosphere to extend to the center of the lunar "core," forming a continuous region of reduced seismic velocities because of some degree of melting. (Courtesy G. Latham, The University of Texas Marine Biomedical Institute, Galveston, Texas.)

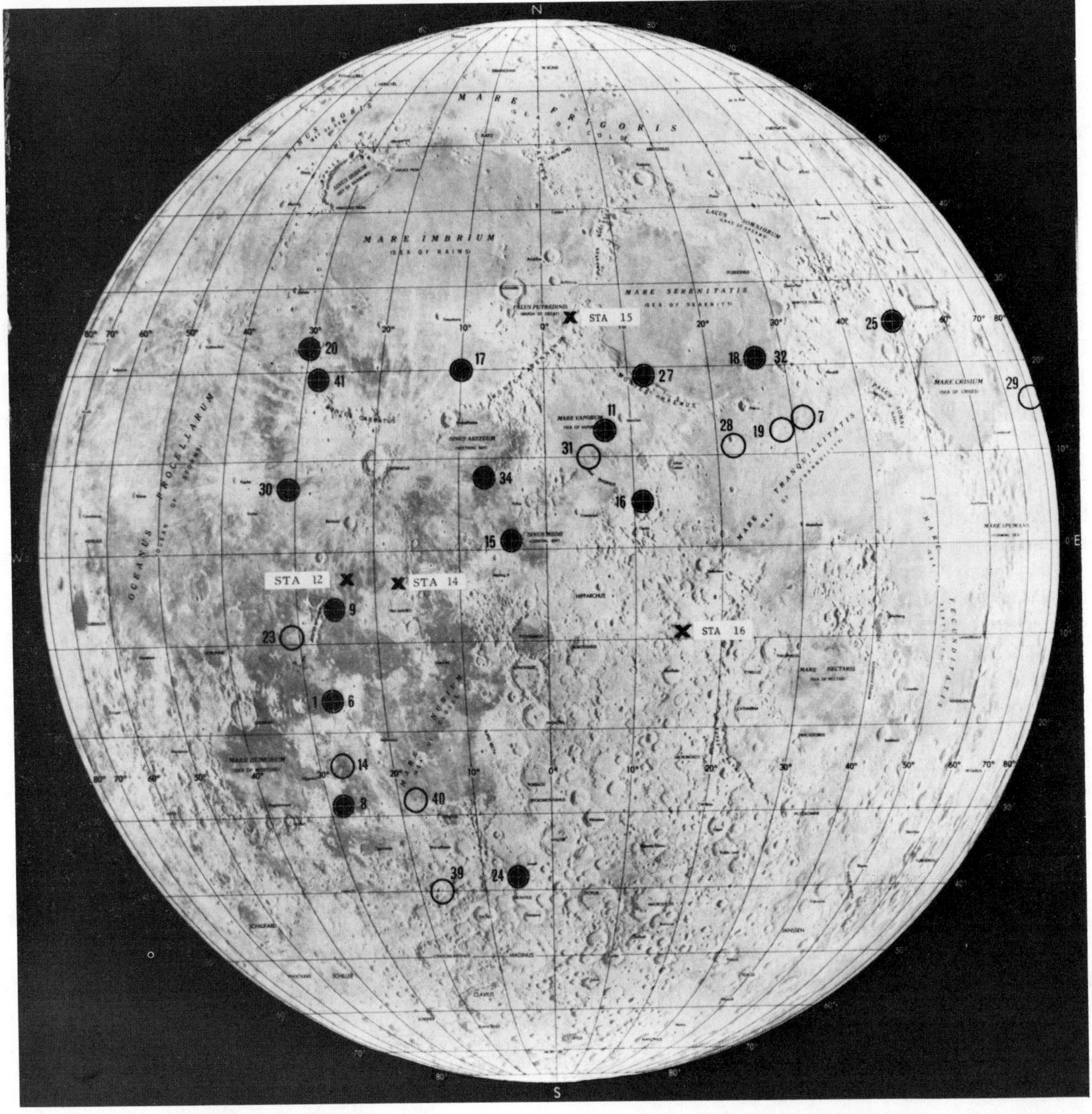

Figure 10-42 The hypocenters or foci of the principal moonquakes located at depths of 600–1000 km in the mantle below the nearside surface of the Moon. A total of 41 such hypocenters has been detected over the entire Moon. Those shown in this plot appear to be arranged in two distinct belts. Foci indicated by open circles are less definitively located than those marked by dark field crosses. Not shown are 11 epicenters fixed at widely scattered locations which are correlated with high-frequency seismic signals (HF moonquakes) that may result from tectonic events occurring at depths of less than 200–300 km from the surface. (Courtesy G. Latham, The University of Texas Marine Biomedical Institute, Galveston, Texas.)

and joints in the mare lavas, can account in part for the signals received at the *Apollo 12* and *14* stations in the southern Oceanus Procellarum region.

By *Apollo 15*, seismologists had established a satisfactory seismic model of the internal structure for the upper parts of the Moon already examined (Figure 10-43; see also p. 314). The outer shell of the Moon consists of a scattering zone of heterogeneous material (most likely, a mixture of ground-up, compacted rubble grading downward into more competent rock) in the top 20 km. This zone has velocities characteristic of *dry* basalts containing microcracks which are progressively closed with increasing load pressure. Below this there is a more homogeneous crustal zone that continues to a crust-mantle discontinuity as deep as 60-70 km. Compressional (P) wave velocities increase rapidly with depth from a low of 100-200 m/sec near the surface to about 5 km/sec at a 10-km depth and thereafter rise more slowly to 6 km/sec at 20 km. Beneath that lower depth there is a sharp increase in velocity to 6.7 km/sec; velocities below this discontinuity seem to remain nearly constant until undergoing another abrupt increase to at least 7.7 km/sec at the boundary with the top of the mantle.

Since installation of the *Apollo 16* seismic station, additional data resulting from moonquakes and the impacts of the LM and the spent S IVB rocket stage from *Apollo 17* led to the postulate of a thin, even higher velocity (~9.0 km/sec) zone that lies directly below the crust-mantle boundary. Some uncertainty still remains as to whether velocities just below the boundary are closer to 7.7 or to 9.0 km/sec. In places, both zones may exist at different depths. In fact, the higher velocity zone could be an apparent effect resulting from a regional dip in the interface of the crust-mantle boundary or, if real, it could either be a localized, non-Moonwide feature or a continuous layer less than 40 km thick. If the higher velocity zone is confirmed, this layer should be a high density rock unit, such as eclogite, containing garnet or spinel or perhaps some other dense mineral phase. In the upper mantle at a depth between about 65 and 150 km, P wave velocities show a negative gradient regardless of the actual magnitude of the average velocity. A decrease in velocity of ~0.2 km/sec in this interval below the base of the mantle may represent a compositional and/or density change or, conceivably, a zone of "softening" or viscosity reduction analogous to the inferred state of rock materials in the terrestrial asthenosphere.

Analysis of signals received at the *Apollo 12* and *14* stations after the S-4B impact indicates the crust in that vicinity may be only 25-30 km thick—much less than the 70 km first reported. Present opinion leans toward a Moonwide crust of variable rather than uniform thickness. It is reasonable to postulate also that, by analogy to terrestrial continents and ocean basins, the thickness of the farside highlands crust is greater than that of mare regions on the frontside. As indicated from laser altimetry (see p. 186), the mean height of the surface above the lunar spheroid reaches a maximum in the farside highlands. To maintain isostatic equilibrium, the highlands crust there would require deeper "roots"—hence greater thickness—much as the extended roots at the base of a continental mass on Earth. This hypothesis has not been verified by seismic or gravity measurements on the farside. The frontside basins may owe much of their diminished altitude to removal of crust by impact followed by infilling with denser lavas that are not fully compensated isostatically.

Energy released from two post-*Apollo 16* meteorite impacts and several larger moonquakes generated seismic waves that penetrated to and reflected from the deep interior of the Moon at detectable amplitudes. Preliminary interpretation of the data indicates a major discontinuity below 1000-1100 km that could be the boundary of a core or, more likely, a region whose seismic properties are

Figure 10-43 Compressional velocity profile as a function of depth within the Moon based on seismic data from the *Apollo* sites. Measured velocities in lunar rocks returned from the Moon and in several types of terrestrial rocks are also plotted. Region 2 (stippled) is defined by velocities from *Apollo 16* aluminous basalts loaded to different pressures. Velocity discontinuities at 25 km (within the crust) and at 60–70 km (crust–mantle boundary) are indicated. The insert at lower left records the velocities of regolith, basalt fill, and subjacent breccia units at the *Apollo 17* site (Kovach et al., 1973). The rapid increase in velocity from the lunar surface to ~ 25 km depth is attributed to the effects of load pressure on fractured basalt flows (this portion of the profile is typical of a mare fill such as covers Oceanus Procellarum). Between 25 and 60 km the observed velocities are similar to those determined for KREEP and anorthositic rocks. The region below 60 km is characterized by one or two velocities depending on the location of the recording station relative to the seismic source. The lower velocity is appropriate to a pyroxene or pyroxene–olivine upper mantle. The higher velocity can be accounted for by the presence of a layer (~ 40 km thick) of a garnet-rich high-pressure equivalent of anorthosite or a spinel–olivine-pyroxene assemblage. (From Toksöz et al., 1973.)

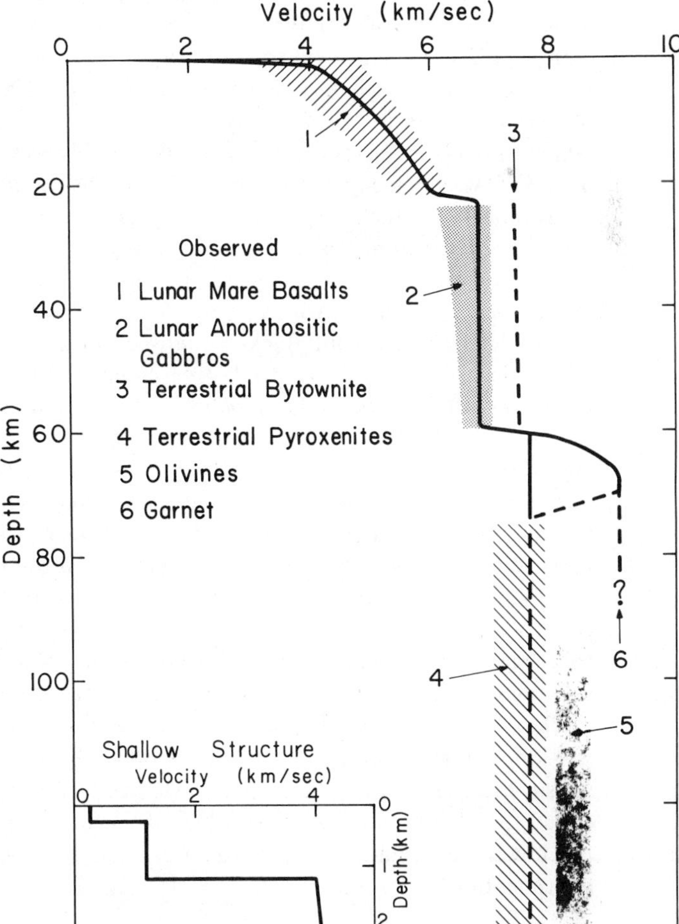

similar to the Earth's asthenosphere. Failure to detect S waves from the larger events raises the possibility of a fluid or partly molten region, although other evidence suggests that this deep zone should be solid. The Q factor of ~500 for this zone is consistent with rock materials that have been "softened". These seismic data, therefore, generally favor a hot lunar interior but have not fully defined those zones that are not now still solid.

Total seismic energy released annually from all lunar events is estimated to be about 10^{15} ergs, a value considerably lower than that produced by an occasional single large event—a major earthquake or explosive volcanic eruption—recorded on Earth and well below the 5 X 10^{24} ergs generated by all terrestrial earthquakes each year. The rate of this energy release in the Moon is about 10^{-6} that of Earth. This low level of seismic activity, together with the ability of the lunar crust to support mascons and the absence in circular structure of offsets caused by strike-slip (wrench) faulting, all point to a rigid, stable outer shell and a solid interior that prevent convective motions on a large scale. Thus, tectonism now plays little or no active role in a static Moon which is "cold" by comparison with the Earth.

Seismic techniques can also aid in determining the thickness of the regolith and other near-surface layers at the different sites. The *Apollo 14* astronauts pounded the ground with a portable "thumper" (a device which strikes with a predetermined force) to send near-surface seismic waves to an array of buried geophones (detectors). This revealed a regolith of ~9 m thickness underlain by 20 to 70 m of Fra Mauro ejecta debris deposited in rubbly layers having a P-wave velocity of ~100 m/sec. After the *Apollo 16* astronauts left the Moon, a series of explosive grenades were launched automatically to distances some hundreds of meters away from the seismograph station. The energy imparted on explosion at contact generated seismic waves used to determine a regolith thickness between 50 and 100 m—the deepest found at any *Apollo* site. A lunar seismic profiler (LSP) was used in the *Apollo 17* valley at Taurus-Littrow. Explosive charges up to 6 pounds, implanted during the geologic traverses at points from 150 to 2500 meters from an array of four geophones, were detonated remotely after the astronauts had departed. A very-low-velocity layer (100-250 m/sec) corresponds to a regolith of 15-20 m thick underlain by a zone of broken rock about 300 m thick. Below this is a 1-2-km-thick layer with an average velocity of 1200 m/sec (but showing directional variations between 560 and 1510 m/sec) that relates to the subfloor basalt flows. Below this is a higher-velocity (~4-km/sec) layer which probably is a continuation of the massif breccias beneath the valley. This zone can be traced to a depth of 2 km, below which the signals from the low-energy explosions grow too weak for effective interpretation.

Magnetometer Experiments

Prior to the *Apollo* missions, unmanned satellites, including *Explorer 35*, had indicated virtually no lunar magnetic field (i.e., less than 0.00001 that of the Earth).* This would imply that the Moon has no counterpart to the terrestrial magnetic dipoles resulting from electric currents produced by dynamo motion in a liquid outer core. Lines of magnetic force related to the geomagnetic field (see Figure 2-5), the solar wind, and other interplanetary fields appeared to pass almost unimpeded through the Moon as though its interior consisted of nonmagnetizable components unsuited for generation of induced fields.

However, for Apollo 12 through 16, magnetometers placed at each landing site (see Figure 10-37) have now measured weak but steady fields (Figure 10-44) representing a *remanent* (permanent) magnetism associated with the area around the site. There is, in addition, a small time-varying magnetic response superimposed on the remanent magnetism that is related to fluctuations in interplanetary magnetic fields (Figure 10-44) and hence indicative of a slight induced magnetic field within the Moon. The fields vary from one site to another. Values reported by Dyal and Curtis (1972) are 38 ± 3 γ at *Apollo 12*, 103 ± 5 γ and 43 ± 6 γ at two *Apollo 14* stations, and 6 ± 4 γ at *Apollo 15*—suggesting that these are local or regional rather than Moonwide anomalies. Stronger magnetic fields were measured by a portable magnetometer at the *Apollo 16* site. Readings there reached 240 γ in the LM area; 120 γ, 180 γ, and 210 γ at three other stations; and 313 γ—the highest value recorded from any site—at North Ray Crater.

Various investigators have reported detection of weak remanent magnetism in lunar crystalline rocks, residing mainly in grains of native iron metal, iron oxide, and iron sulphide. This permanent or "hard" magnetization is stable and does not result from present exposure to the Earth's or the Sun's magnetic fields. A "soft" (much weaker) component of the magnetization can be caused by stray fields from the LM. Its existence was verified during *Apollo 16* by returning a lunar sample from a previous mission and remeasuring any changes in its known magnetic strength after it lay near the LM for two days; however, the small observed changes do not account for the total field strength inherent to the rock.

Magnetic field strengths somewhat larger than those of the lava rocks are associated with breccias from the *Apollo* sites. These breccias contain 5 to 10 times as much metallic iron. The total iron content of soils and derivative breccias is influenced by their thermal history, in which loss of volatiles and other processes lead to enrichment of magnetic

*The unit of magnetic force used in this section is γ (gamma). One $\gamma = 10^{-5}$ gauss (see a standard Physics or Geophysics text for definition). The Earth's magnetic field varies from 0.3-0.7 gauss.

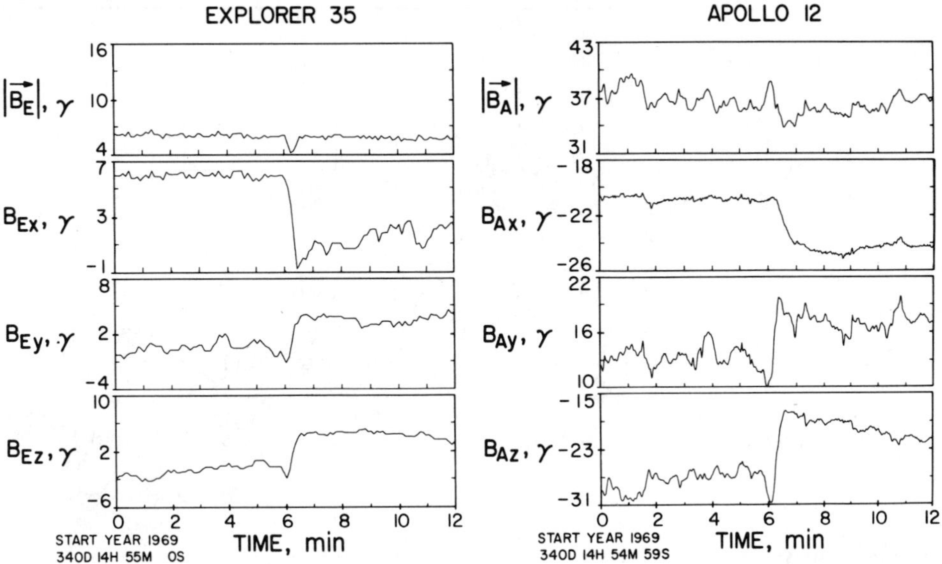

Figure 10-44 Magnetic field measurements made simultaneously by the *Apollo 12* Surface Magnetometer and the magnetometer on board the lunar orbiting *Explorer 35* satellite. B refers to the magnetic field vector. The field is measured by three magnetic fluxgate sensors mounted at the tips of three orthogonal booms designated arbitrarily as the X-, Y-, and Z-axis components. Measurements of the external magnetic field mainly from the Sun by *Explorer 35* (3-component system) show broad correspondence with variations noted at the *Apollo 12* site. Difference ($B_{Apollo\ X,Y,Z} - B_{Ext}$) and step-function transient plots show this interrelationship to better advantage. (From Dyal and Parkin, 1971.)

components. Thermoremanent magnetism (TRM) at temperatures as low as 300°K is possible in the fragmental materials at the time they were subjected to shock from impacts or even now during diurnal temperature cycles.

Just before the *Apollo 15* CSM blasted off toward Earth, a small subsatellite equipped with a magnetometer was inserted into orbit 110 km above the surface. Another such subsatellite was ejected during the *Apollo 16* mission, but crashed about forty days later after returning some data. The magnetometer measures interactions of the solar wind and interplanetary magnetic fields with the Moon and can also detect effects of remanent magnetism associated with near-surface sources. Variations in field strengths of more than 1 γ at orbital altitude have been correlated with some of the larger lunar craters. From a calculated model it is known that such a small disturbance would result from magnetized bodies of unspecified nature assumed to be spherical. Bodies of this kind between 20 and 200 km in diameter would give rise to corresponding field strengths from 10 to 1000 γ if measured on the ground. The distribution of variations in magnetic field strengths within a swath beneath the subsatellite has now been mapped for the nearside and farside of the Moon along *Apollo 15* orbital paths (Figure 10-45). The regional fields on the lunar frontside are generally weaker and less structured than those on the farside. The most prominent magnetization center on the Moon lies between the craters Aitken and Van de Graff (20°S, 172°E).

The present magnetic state of the Moon as a whole is consistent with a concentration of up to 4% metallic iron in the outer 400 km of the mantle-crust. However, it is likely that the metallic iron content of rocks in this region is very much lower than that upper limit. Although a very weak magnetic dipole (~2 γ) originating in the Moon can account for part of the surface magnetic field strength, most of the remanent magnetism is "fossil" or ancient in origin. A minimum magnetizing field of 500 to 1000 γ is required to impress the observed remanent magnetism on lunar rocks at the *Apollo* sites. Since no such field exists today, it is suggested that the magnetization was impressed some 3 to 4.5 billion years ago. Possibly the source of the magnetizing force was the interplanetary (solar) field or a strong (~100 gauss) terrestrial field acting on a Moon much closer to the Earth before its recession. But the most likely explanation is that the Moon had a modest dynamo field capable of inducing the stable paleomagnetism now observed. Crustal rocks and mare lavas solidifying in that interval would gain remanent magnetism as susceptible minerals passed through the *Curie point* (temperature below which a substance can enter into a permanently magnetized state; for iron, the critical temperature is 780°C) following crystallization and cooling. A small metal core (whose diameter is estimated by Runcorn to be about 700

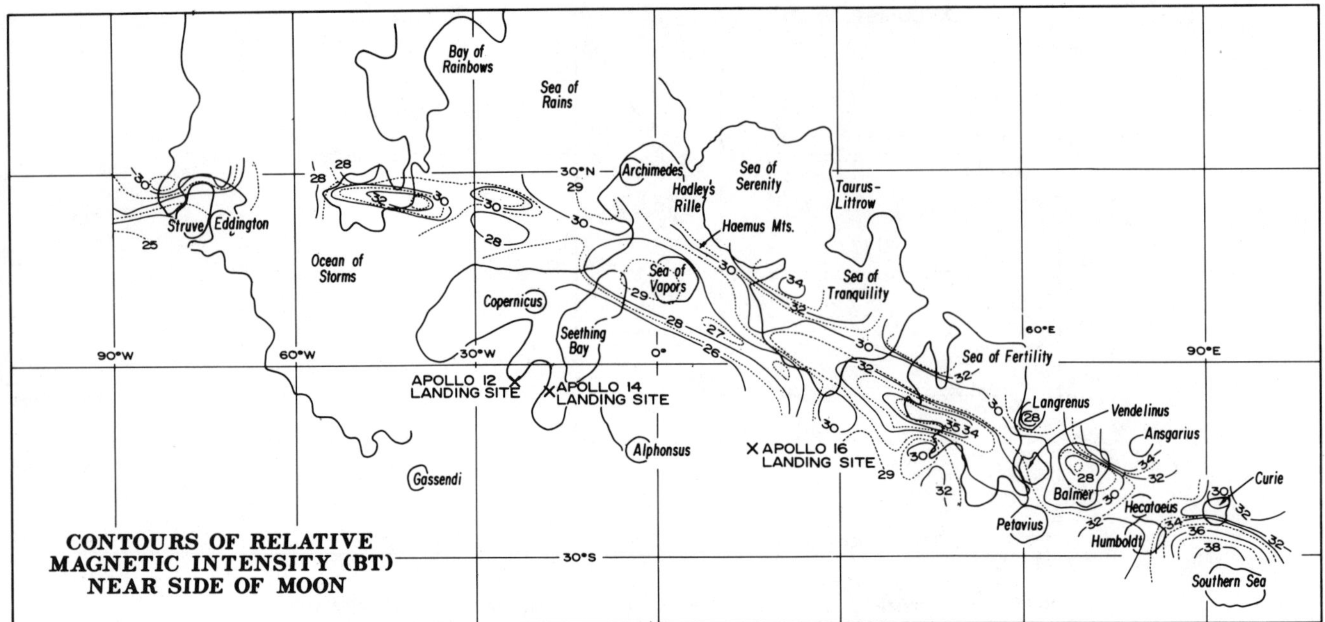

Figure 10-45 A map of contoured values of the relative magnetic field strength B_T (measured transverse to the spin axis of the spacecraft) determined by the magnetometer on the *Apollo 15* subsatellite along a track on the nearside of the Moon. (From Coleman et al., 1972.)

km or 20% of the Moon's average diameter) may have existed in a fluid state during that period but is now solid or too viscous for circulating motions to produce a continuing field. These ideas gain some support from the higher local field strength readings found at the *Apollo 16* highlands site and the higher regional magnetization on the lunar farside. At the *Apollo 16* site, the Cayley units are magnetized but the Descartes units are not, a difference explained by assuming the Cayley had been emplaced at a temperature of ~800°C. The older average age for crustal rocks in the highlands implies their formation when the internal field of the Moon was stronger prior to slowdown of the dynamo. However, the higher metallic iron content of *Apollo 16* soils may also explain these larger magnetic fields.

When the existence of weak, time-variant fields at the several *Apollo* sites was considered together with the subsatellite data, Dyal and Parkin (1971) and others concluded that interactions with the Sun's magnetic field produces a weak magnetic field in the Moon by electromagnetic induction. Variation in the solar field strength causes the lunar magnetic field to become dipolar and also generates electric (eddy) currents within the Moon's interior. The resulting fields tend to be localized. As the Moon's orbit carries it to a position between the Sun and the Earth (and well beyond the front of the terrestrial magnetosphere), the weak lunar field can interact with the solar wind. The stream of solar particles tends to deflect around the lunar sphere on the daylight side forming a conical cavity (devoid of solar wind) behind on the nightside; a boundary zone of turbulent wind expands outward because of compression at the lunar limbs. Any lunar magnetic field on the Sunward side is itself compressed by the solar wind, but a field on the darkside remains essentially unaffected. On orbiting through the terrestrial magnetosheath and magnetosphere, the lunar magnetic field is extensively altered and thoroughly distorted.

Differences in magnetic permeabilities of materials within the Moon influence the passage of solar and geomagnetic fields through the lunar body as it moves through space. The solar wind-related magnetic field can be separated from the other magnetic fields by comparing daytime and nighttime measurements made by the surface and subsatellite magnetometers. The ability of the solar wind to pass through the interior depends on the electrical conductivity of components within the Moon. A profile showing variations in conductivity with depth can be constructed (see Figure 13-11) from analysis of magnetic data. This conductivity also varies with composition and temperature, so that it is possible to determine temperature gradients—whose values are influenced by the compositional models chosen—inside the Moon. Interpretation of data from *Apollo 12, 14,* and *15* magnetometers has led Dyal and Parkin (1972) to propose a generalized three-layer structure for the Moon's interior. In this model, the near-surface shell has temperatures below 440°K, the intermediate shell has an average temperature around 800°K, and the inner core reaches approximately 1240°K (see p. 310 for a discussion of assumed compositions). Sonett and others (1971) have concluded that the present magnetic state of the Moon requires it to be essentially solid throughout (local pockets of melt are permissable), with thermal gradients holding maximum temperatures in the core to a range well below the melting points of constituent minerals at the higher interior pressures.

Surface Gravity Measurements

A lunar traverse gravimeter (LTG) was used in the valley at Taurus-Littrow (*Apollo 17*) to measure minute variations in the value of lunar gravity at the surface caused by differences in densities and distribution of the several major units at the site. The depth to the valley base below the basalt flows was a particular objective. The instrument, mounted on the back of the lunar rover, consists of two masses suspended between two springs and coupled to a central spring. Variations in gravity alter the frequency of vibration as the mass accelerates freely after release. In operation, the LTG measured gravity values (after applications of free air and Bouguer corrections) along the edges of North and South Massifs that were 20 mgals* lower than those at the LM station. A thickness of basalt flows of 1 km is calculated for a density contrast of 0.8 g/cm^3 between breccias and basalt. This instrument was also capable of establishing an Earth-Moon gravity tie of 162,694 mgals (preliminary value).

*A milligal (sea level mgal) is 0.001 of a gal—a unit named for Galileo—which is the sea level value of g (gravity), or an acceleration of 978.05 cm/sec^2 at the equator.

Heat Flow Studies

Both seismic and magnetic evidence suggest that the amount of heat reaching the surface would be low—consistent with a now-cold (unmelted) Moon. Actual values of heat flow determined on the Moon at the *Apollo 15* and *17* sites were two to three times higher than had been predicted.

The *Apollo 15* heat flow experiment was conducted by emplacing thermal sensors as probes in two holes drilled into the regolith to depths of 1 and 1.6 m, respectively, below the surface (Figure 10-46). In each, the temperature (252°K and 250°K) at a depth of 80 cm is about 32°K higher than that at the surface average over a single lunation cycle (a maximum surface temperature of 348°K was reached at lunar noon; temperatures dropped by almost 255°K by the end of the lunar night). At depths greater than about 50 cm, the temperature remains almost constant through a lunar day. Thus, below 0.7 m in undisturbed regolith a steady thermal gradient of ~1.75 ± 1°K/m has been calculated. Small heaters within the probes that give off known quantities of heat were used to determine the thermal conductivity as a function of depth. The measured conductivities in one hole, at depths

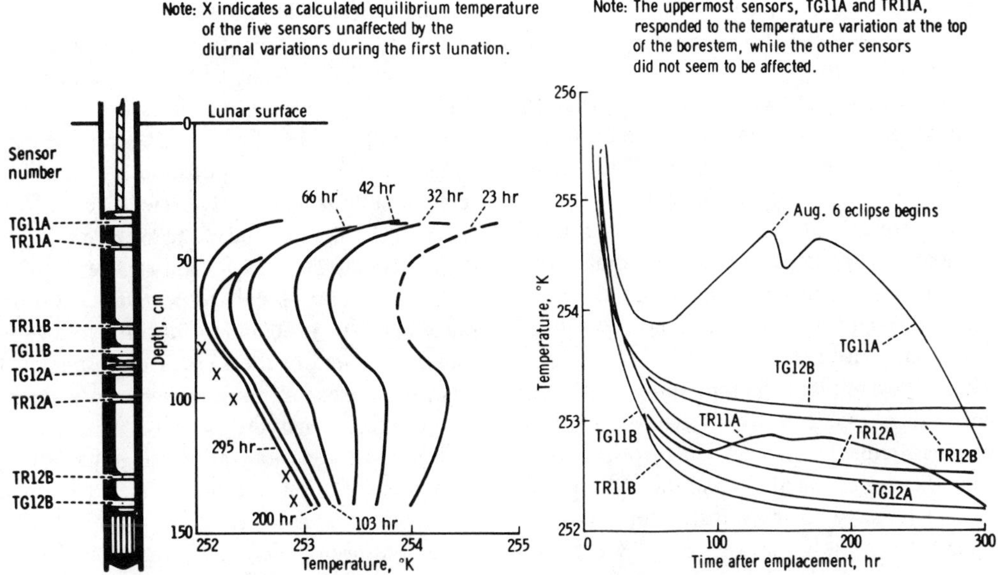

Figure 10-46 Temperature histories in the hole occupied by heat probe 1, measured over the first 300 hours after emplacement at the *Apollo 15* site. The location of the different sensors (thermocouples and differential thermometers) in the probe appears on the far left. The left diagram is a plot of temperatures as a function of hole depth, measured at different times; the right diagram shows temperatures plotted against time after emplacement, measured by the different sensors. (Reprinted from M.G. Langseth et al., "The *Apollo 15* Lunar Heat-Flow Measurement", *The Moon,* Vol. 4, No. 3/4, pp. 390-410. Copyright 1972 by D. Reidel Publishing Company.)

of 49, 91, and 138 cm, were 1.4×10^{-4}, 1.6×10^{-4}, and 2.5×10^{-4} W/cm-°K (watts per centimeter–Kelvin degree), respectively. Combining conductivity and thermal gradient data, an average heat flow value of 3.0×10^{-6} W/cm^2 (also given as 0.7×10^{-6} cal/cm^2-sec or 30 ergs/cm^2-sec) applies to the *Apollo 15* site.

The heat flow experiment was repeated at *Apollo 17.* The two probes, placed 11 m apart, had their sensors at depths of 1.29 and 2.33 m. One probe measured nearly constant heat flows of $\sim 2.8 \times 10^{-6}$ W/cm^2 between 66 and 233 cm, with an average conductivity of 1.7×10^{-4} W/cm-°K for that interval. The temperature at depths of 66 and 233 cm were 253.6°K and 256.5°K, respectively. The second probe obtained a range of heat flow values from 4.6×10^{-6} W/cm^2 in the 66-130-cm depth interval to 1.9×10^{-6} W/cm^2 between 186 and 233 cm. These variations may arise from the influence of a buried boulder. However, the average of 2.8×10^{-6} W/cm^2 (28 ergs/cm^2-sec) is close to the value found at *Apollo 15*, and after corrections it is expected to become nearly identical. The mean surface temperature just before sunrise is 103°K or ~10°K higher than at *Apollo 15*.

The same experiment was attempted at the *Apollo 16* site, but was abandoned because of an instrument accident. This failure was unfortunate because it is impossible now to determine whether the similar values obtained at two different sites are characteristic of the entire Moon, or typical only of mare basalts, or perhaps just coincidental. The nature of the highlands terrain and rock types there afforded a notably different set of conditions by which these alternatives could have been checked.

The lunar heat flow results represent more than one-half the typical values ($\sim 6.2 \times 10^{-6}$ W/cm^2) found in terrestrial land measurements, a somewhat surprising outcome considering the size of the Moon and the assumption that most of its heat energy had been lost. The values are also notably greater than the values of $1.0\text{-}2.5 \times 10^{-6}$ W/cm^2 calculated for a small planetary body of chondritic composition. These unexpectedly high lunar values seem to indicate the Moon's interior is much hotter than most thermal models had anticipated. If the temperature gradient in the lower regolith is extrapolated to great depths, the lunar interior would appear to be at least partly molten—a condition contradicted by other evidence. Langseth et al. (1972) have computed the present heat production Q required to account for measured heat flows in the range $0.57\text{-}1.0 \times 10^{-13}$ W/cm^3. Potassium- and uranium-poor chondrites generate heat in amounts in the range $0.17\text{-}0.22 \times 10^{-13}$ W/cm^3, but the more radioactive lunar basalts have much higher Q values around 3.5×10^{-13} W/cm^3. Langseth, the principal investigator in the heat flow experiment, thus subscribes to the higher concentrations of long-lived radioactive elements in the outer shell(s) of the Moon (up to 10X greater than chondrites) as the prime source of the "excess" heat energy still emanating from the interior. The similar heat flow values measured at both *Apollo 15* and *17* are an average of heat produced from the surficial basalts and the underlying, less radioactive layers of presumed feldspar-rich nature.

Particle and Atmosphere Detectors

Among instruments deployed in the ALSEP array in most *Apollo* missions were several designed to detect and analyze any "atmosphere" of uncharged and ionized gases introduced naturally both from the solar wind and the lunar surface or interior or released artificially from the spacecraft.

The solar wind spectrometer measures the variations in directions of arrival, velocity, density, and energy of incoming protons and electrons from the Sun and the interaction of this solar plasma with the lunar environment. The plasma recorded at the lunar surface is superficially indistinguishable from the solar plasma detected by satellites some distance from the Moon, and no measurable plasma is found in the Moon's shadow. The solar wind composition experiment consists of a mounted panel of aluminum foil (similar to a windowshade) set up on the surface to trap ionized particles in the plasma. Following return of the soil to Earth, the entrapped ions are released for analysis by mass spectrometry. Absolute fluxes and relative abundances of He^3, He^4, Ne^{20}, Ne^{22}, and Ar^{36} can then be determined. The average solar wind He^4 flux at the *Apollo 11, 12, 14,* and *15* sites is 6.2, 8.1, 4.2, and 17.7×10^6 cm^{-2}-sec^{-1}, respectively. Ratios of He^4/He^3 range between 1860 and 2450 compared with 7×10^5 for values measured in the Earth's atmosphere. Variations in this ratio appear to correlate with observed changes in the level of solar activity.

A cold-cathode ionization gauge has measured a nighttime concentration of ionized gases that amounts to $\sim 2 \times 10^5$ ions/cm^3. These gases consist mostly of helium, with some hydrogen, neon, and argon, introduced from the solar wind. The gases produce a nighttime surface pressure of less than 10^{-12} torr (a pressure unit, equal to 1 mm of mercury at 0°C in Earth gravity), but in the lunar daytime this pressure rises to $\sim 10^{-10}$ torr. This suggests that some of the helium and other inert gases are adsorbed on lunar soil particles at night and later released during daytime heating. In the suprathermal-ion-detector experiment (SIDE), a mass analyzer is able to sense bursts of ions and classify them according to mass per unit charge. Aside from LM discharges and solar imputs, the SIDE picked up one major event that has been interpreted as a cloud of volatiles apparently released from the lunar interior. This event, lasting fourteen hours on March 7, 1971, was detected by SIDE arrays at both *Apollo 12* and *14* sites 180 km apart, suggesting a distant source from which a large cloud spread to both localities. This cloud seems to have been largely water vapor, as deduced from identification of H_2O^+, OH^+, and O^+ ions. Unless an alternate explana-

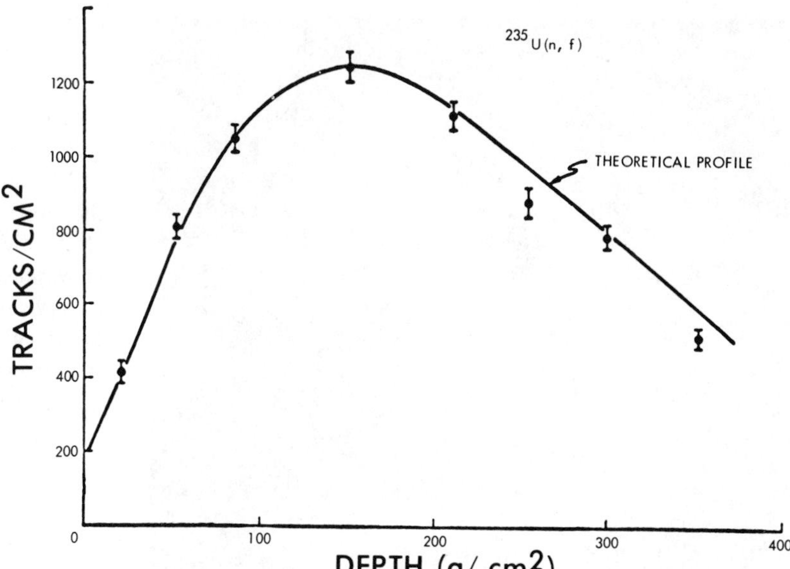

Figure 10-47 Theoretical neutron flux depth profile (solid line) and observed values (points with error bars) of relative fission track densities as measured by mica detectors in the Lunar Neutron Probe at various depths in the *Apollo 17* deep drill-hole emplaced during EVA 1. (From D. S. Woolum et al., 1973, courtesy Lunar Science Institute.)

tion is later accepted, this event constitutes the first, and so far only, detection of minute quantities of water, possibly released as volcanic steam, indigenous to the Moon.

The lunar atmosphere composition experiment (LACE) in *Apollo 17* uses a sensitive magnetic sector field spectrometer to search for very small quantities of gases. Mass numbers rather than species are detected. These densities were much less than had been predicted on the basis of cold-cathode gauge measurements. The decrease in density over time represents dissipation of gas contaminants from the LM. The dominant species present were H^1, He^4, and a gas with a mass number of 28 (CO_2?). The LACE detected essentially no water and much less neon than anticipated. Species of mass numbers appropriate to F and HF were also found. Argon, produced on the Moon by decay of potassium-40, tended to disappear afternightfall, but then increases during the day to a maximum of 10^4 atoms/cm^3; this and certain other gases appear thus to freeze at night and outgas in the daytime. In contrast, the daytime He^4 concentration of 3 X 10^3 atoms/cm^3 increases during the lunar night to about 6 X 10^4 atoms/cm^3.

Surface Electrical Properties

The surface electrical properties (SEP) experiment on *Apollo 17* utilizes radio waves sent out at six frequencies to determine some of the subsurface properties of shallow layers in the Taurus-Littrow valley. A signal generated by a transmitter is dispersed through a dipole antenna. A receiver mounted on the lunar rover permits measurements to be made for different parts of the valley as various stations are occupied. The transmitted signals pass to the receiver (1) directly along a line of sight above the surface, (2) directly through near-surface materials, and (3) indirectly through deeper layers by reflection. The arrival of signals along any path depends on the velocity and attenuation properties of the mediums encountered. Differences in these properties and in their geometric distribution lead to delays in arrival that produce signal interference. The specific interference patterns are analyzed to unravel the nature and location of subsurface structures. First results from this experiment indicated the dielectric constant of the surficial materials increased from 2.5-3.0 at the top to 4.0-4.5 at a depth of 50 m. These values were somewhat lower than expected. No water above the detection threshold (0.01%) was found. The contact between regolith and subfloor rocks was inclined.

Lunar Neutron Probe

The rates at which fragmental materials build up and become intermixed in the regolith and the ages of distinguishable layers within this unit can be assessed by the lunar neutron probe (LNP). Detectors extending the length of the probe measure nuclear reactions involving neutrons released by fission of natural uranium and by interactions between incoming cosmic rays and certain elements. One detector consists of a target of boron which, by neutron capture, ejects alpha particles that leave distinctive tracks on passing through a plastic counter. A second detector uses mica to record the tracks produced by fission of U^{235}. A cadmium absorber in the probe obtains information on the neutron flux spectrum. The probe was lowered into a deep drill hole during the first EVA at *Apollo 17* and was removed at the end of EVA 3 after forty-nine hours of exposure. Track densities were determined in the detectors after their return to Earth. Distribution of fission track densities in the mica as a function of depth is shown in Figure 10-47. Such data establish a neutron flux-depth profile which aids in reconstructing the irradiation history of the regolith. Preliminary interpretation considers the regolith to be reasonably well mixed but still stratified.

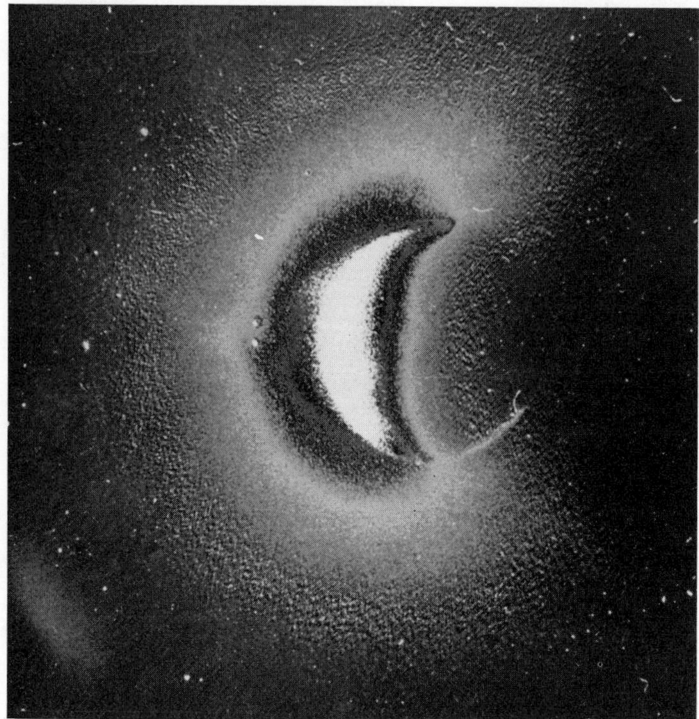

Figure 10-48 Black and white copy of a color-enhanced photo showing the Earth's geocorona, a halo of low-density gases excited in the ultraviolet portion of the spectrum by solar radiation. A UV camera/spectrograph responds to induced radiation emitted at 1216 Angstroms. The broad, diffuse gray belts represent a region extending out to 60,000 km in which hydrogen is the principal excited gas. The bright crescent on the Earth's daylight side is caused primarily by excited oxygen and nitrogen. The arcuate streak on the lower right (toward the night side) is the result of auroral activity over the south magnetic pole. This image was obtained from the lunar surface at the *Apollo 16* site by Astronaut John Young. (NASA photo.)

This experiment should be compared with the results of determination of variations in gadolinium-158 with depth made on core materials from previous missions (see p. 229).

Earth-Moon Experiments

A laser retroreflector array was pointed toward the center of the Earth's libration at the *Apollo 11, 14,* and *15* sites to provide a triangulation network of targets. Each target contains numerous small fused silica corner cubes. A ruby laser light beam directed from the Earth will, upon finding a target, be reflected along a path parallel to the line of incidence if alignments are correct. Remarkably, all three of these small targets (rectangles less than a meter in longest dimension) have been intercepted by the beam. Together, the targets form a reference base for high-precision measurements of Earth-Moon distances accurate to better than ±4 m. This experiment also provides data useful in determining the motion of the Moon in orbit, details of lunar librations, and such terrestrial movements as the drifting of continental masses.

The *Apollo 16* astronauts used a far ultraviolet camera/spectrograph to photograph the Earth's outer atmosphere (geocorona) and some selected stellar objects. Thousands of kilometers above the Earth's surface, bombardment of the thin terrestrial atmosphere by solar particles induces electronic excitations that give off light energy in the UV at wavelengths below 1300 angstroms (the aurora borealis is a visible counterpart influenced by the geomagnetic field). Figure 10-48 shows a first from space–a view of the diffusely glowing hydrogen in the outer geocorona, the day air glow of atomic oxygen and molecular nitrogen in the inner geocorona, and part of the aurora borealis on the Earth's nightside.

Orbital Experiments

During *Apollo 11* through *14* the astronauts performed a series of specialized orbital experiments, such as taking selective photographs of the surface, from the CSM. In the last three missions, the astronaut remaining in the CSM while surface exploration was underway conducted many complex experiments (see Table 10-2) that added significant data to the understanding of the Moon's outer regions. A recessed bay in the CSMs of *Apollo 15, 16,* and *17* carried several scientific instrument modules (SIMs) for conducting remote sensing measurements during lunar orbits (Figures 10-49a and b).

Selenodetic Measurements

Photographs obtained with both the panoramic and metric cameras in the *Apollo 15, 16* and *17* SIM bays were used to set up a unified selenodetic coordinate system for the Moon. These photographs, together with data from the laser retroreflector triangulation experiments, provide a network of control points from which an accurate cartographic base has been prepared. This aids in developing topographic maps of improved accuracy on which geologic and geophysical information can be plotted.

Additional topographic data was obtained with a laser beam altimeter mounted in the SIM bays. Altitude profiles along orbital paths (Figure 10-50) indicate that the mare-covered lowlands on the frontside of the Moon are generally 2 to 5 km lower than the lunar spheroid (a hypothetical surface approximating an ellipsoid of revolution), while parts of the highlands on that side are about 3 km higher. On the farside, however, most of the highlands are as much as 5 km above this spheroid. This means, then, that the average height of the farside is several kilometers farther from the Moon's center of mass than are terra regions of the frontside. This center of mass is, furthermore, about 2 km closer to the Earth than is the center of the Moon's figure (spheroid). As projected on the front surface, this

(a)

Figure 10-49 (a) The open SIM Bay in the *Apollo 15* CSM showing the various scientific instrument packages, as photographed from the approaching LM. (b) Location of the SIM (Scientific Instruments Module) Bay used for orbital experiments from the CSM during the *Apollo 15* and *16* missions. When in operation, the gamma ray and mass spectrometer sensors are extended about 8 meters outward on a boom. The subsatellite is released into its own orbit prior to CSM blastoff to Earth in order to measure magnetic and particle fields around the Moon in subsequent months. Film casettes must be retrieved enroute to Earth from the Service Module before separation from the conical Command Module carrying the astronauts to their splashdown landing. (NASA illustrations.)

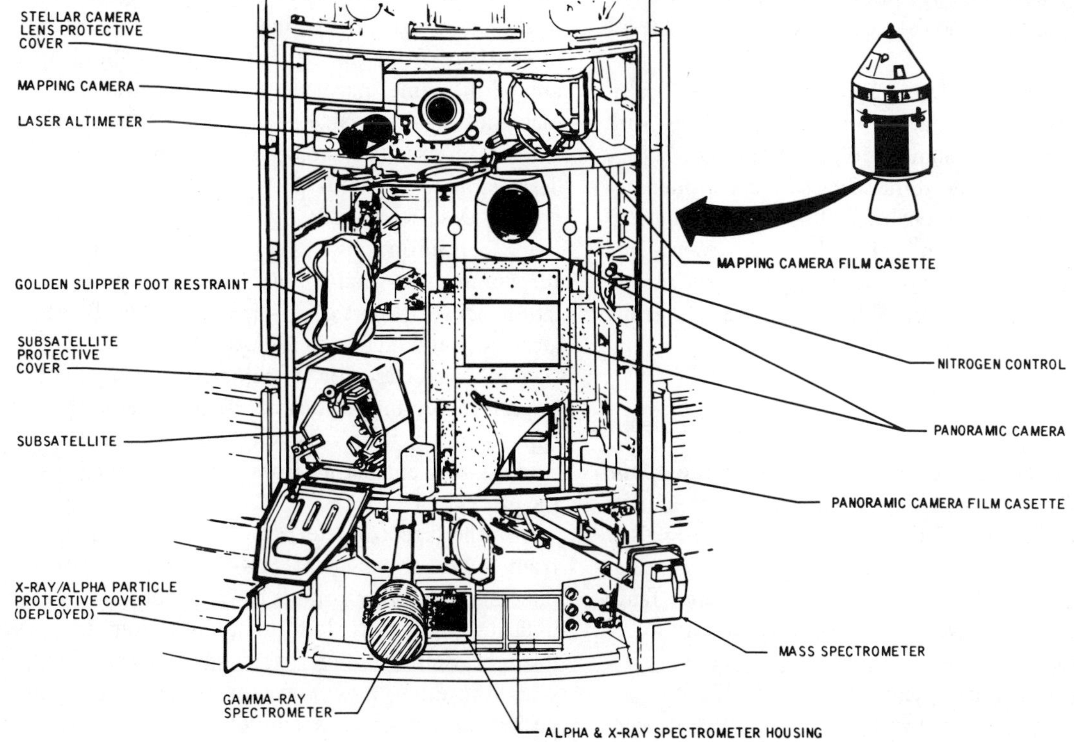

(b)

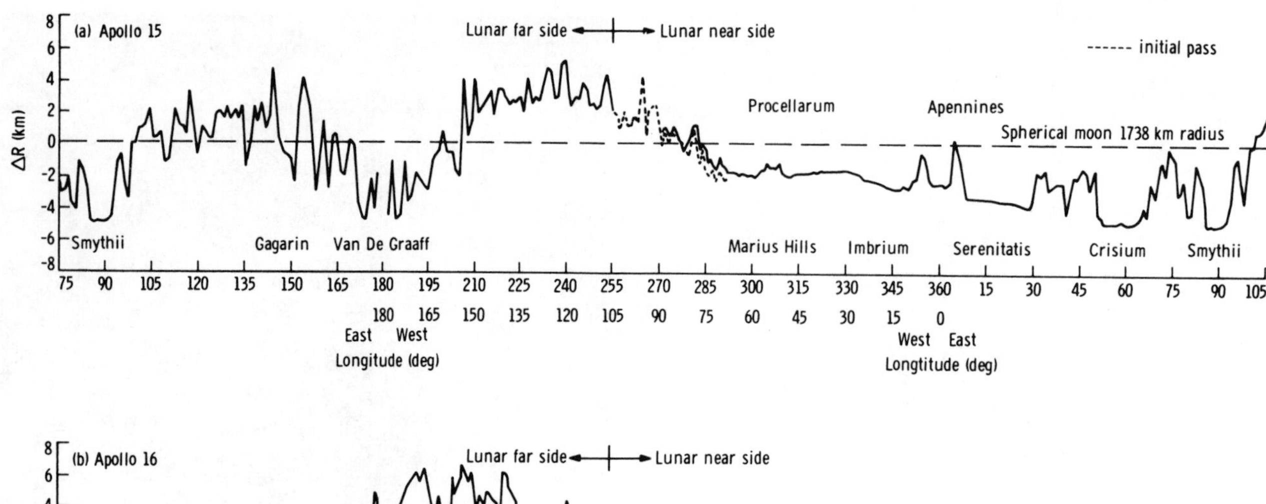

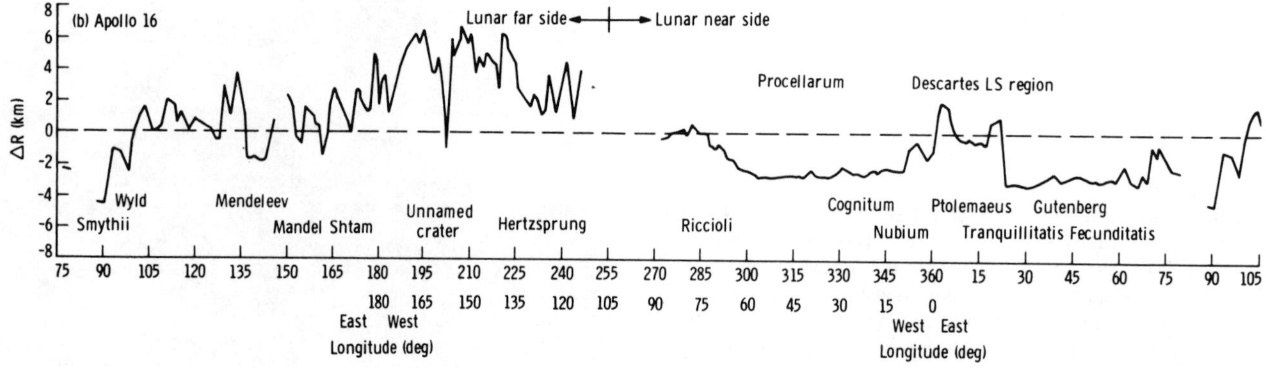

Figure 10-50 Altitude profiles (expressed as height differences Δ R from the mean elevation of the lunar spheroid) determined by the laser beam altimeter in the SIM bays of *Apollo 15* and *16*. (Courtesy W.L. Sjogren, Jet Propulsion Laboratory.)

center of mass emerges to the east of the 0° latitude and longitude intersections. The altimeter data also allowed recalculation of the Moon's mean radius to a value of 1738.1 km.

These topographic determinations with the laser altimeter show the mare surfaces to be generally level, with relief changes averaging about 150 m over areas of 200 to 600 km on a side. The terra surfaces are considerably rougher, with relief variations of 600 m to as much as 1.5 km over areas of 1600 km^2. The altimeter experiments also showed that the depths of ringed basins (e.g., Crisium, Smythii) vary inversely with width; one crater (Gagarin) on the farside is more than 6 km deep.

Radiation Spectrometers

An X-ray fluorescence spectrometer mounted in the *Apollo 15* and *16* SIM bays picked up secondary X-rays emitted by Si, Al, and Mg from the top millimeter of materials at the lunar surface as these elements are excited by solar X-rays. Ratios of Al/Si obtained by this method range from about 1.1 to 1.5 over typical highlands terrain to 0.6 to 0.9 over the maria. Ratios of Mg/Si vary between 0.7 and 1.0 in these highlands and rise to 1.1 to 1.2 over corresponding maria (Figure 10-51). These results are consistent with feldspar-rich (high alumina) rocks in the terrae and pyroxene- and olivine-rich (high-magnesium) basalts in the mare regions. The high values of 1.4 to 1.5 for Al/Si obtained from the Descartes region around *Apollo 16* are of the same magnitudes observed over much of the highlands on the farside. This supports the choice of that site as well suited to exploration of typical terra materials. The estimate of about 26% Al_2O_3 in surface materials around Descartes, made by converting the Al/Si data to oxide concentrations, agrees with the reported analyses for alumina in several *Apollo 16* breccia samples. A correlation between increasing A1/Si ratios and higher optical albedos in the highlands indicates that lighter-toned rocks rich in plagioclase are a major component in the more brightly reflecting regions of the terra. During the return trip to Earth, this X-ray spectrometer was also used to study X-rays emanating from several distant stars outside the galaxy.

Another spectrometer, designed to pick up gamma rays from radioactive-element (U^{238}, Th^{235}, K^{40}) decay in minerals within the top 10 to 20 cm of the surface cover, indicated extremely low uranium-thorium and generally low potassium contents over most of the Moon. Considered together, these low contents rule out uranium- and potassium-rich granites and other silicic rocks as major lithologic types within the orbital belts surveyed—and, by inference, in the rest of the Moon. However, notable increases in detected gamma rays (corresponding to thorium concentrations as high as 10 ppm) in the northwest quadrant of the lunar frontside—mainly in southern Mare Imbrium and northern Oceanus Procellarum, with particularly high values around the Aristarchus plateau (Figure 10-52)—and a similar increase on the farside in an area antipodal to the frontside high are indicative of local to regional near-surface enrichments of radioactive elements in some parts of the upper lunar crust. The gamma-ray spectroscopy results also reveal that some maria have 1.15 to 1.20 times the concentrations of U, Th, and K found in the surrounding terrae. However, in contrast to the Imbrium and Oceanus Procellarum regions, lava surfaces in the Fecunditatis, Tranquillitatis, and Smythii lowlands show very low K, U, and Th values. This may indicate that the lavas in the lowlands of the western frontside are distinctly more radioactive than those on the eastern side. The gamma-ray spectrometer also measured higher radioactivity readings in the vicinity of the *Apollo 16* site than noted on the farside or in most other parts of the frontside highlands. There is thus an inverse correlation between gamma-ray readings and rise in topography. Intermixing of K-, U-, and Th-rich ejecta from the Imbrium Basin with Al-rich highlands materials could plausibly account for this. Many of the same radioactivity anomalies just described were picked up as well by an alpha-particle spectrometer, which senses small concentrations of radon gas (containing polonium-210 and other daughter isotopes from the uranium series decay) escaping from the surface.

Taken together the X-ray, gamma-ray, and alpha-particle spectrometer results have given the first broad, regional picture of the geochemistry of the lunar surface and, inferentially, its outer crustal layers. The lack of an absorbing atmosphere made it possible to survey the Moon by these instruments, which normally cannot function from aircraft on the air-shielded Earth. In general, these geochemical devices have now demonstrated the tendency of the Al contents to vary inversely with the K and U contents of rocks exposed on the lunar surface. The highlands, on this basis, are evidently composed of plagioclase-rich rocks (anorthositic) that are decidedly unlike most granite or metamorphic rocks in Precambrian shield areas of the Earth's crust. These element distributions also seem to eliminate primitive (undifferentiated) chondritic material as a crustal component.

Gravity Measurements

While orbiting the Moon, the *Apollo* CSMs and LMs (as well as the small subsatellites released from *Apollo 15* and *16*) provided a means of charting lunar gravity variations in much the same way as the *Lunar Orbiters* were used in the analysis that discovered mascons. Changes as small as 1 mm/sec in spacecraft speeds are measured by a Doppler system that employs radio tracking (S-band transponder) to determine precise orbital positions. Computer reduction of the data leads to calculations of varying accelerations associated with orbit perturbations caused by local gravitational effects. Results can be plotted as gravity profiles along given orbits (Figure 10-53). Typical mascon-related anomalies can be produced by assuming some type of material distributed near the surface that can produce a mass-per-unit-area value of 800 kg/cm^2. For Mare Serenitatis, as an example, this material could be grouped into a discoid body of 245-km radius and about 20-km thickness whose density is 0.4 g/cm^3 higher than surrounding mate-

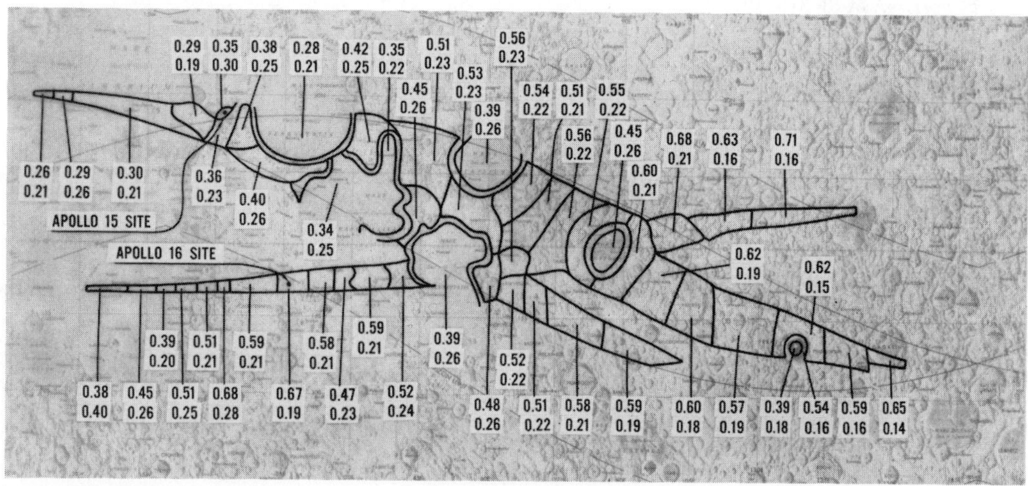

Figure 10-51 Map of the lunar frontside on which ratios of AlSi (top number) and Mg/Si (bottom) have been determined by X-ray fluorescence spectrometry for various sampling regions (bounded areas indicated by arrows) along the orbital paths of the *Apollo 15* and *16* CSMs. The spectrometer is mounted in the SIM bay. Generally, the lunar highlands have higher Al/Si and lower Mg/Si values than mare-filled basins and lowlands. These values actually are "averages" that apply to large areas (20,000 km^2—owing to the low resolution of the instrument. (From I. Adler et al., 1973.)

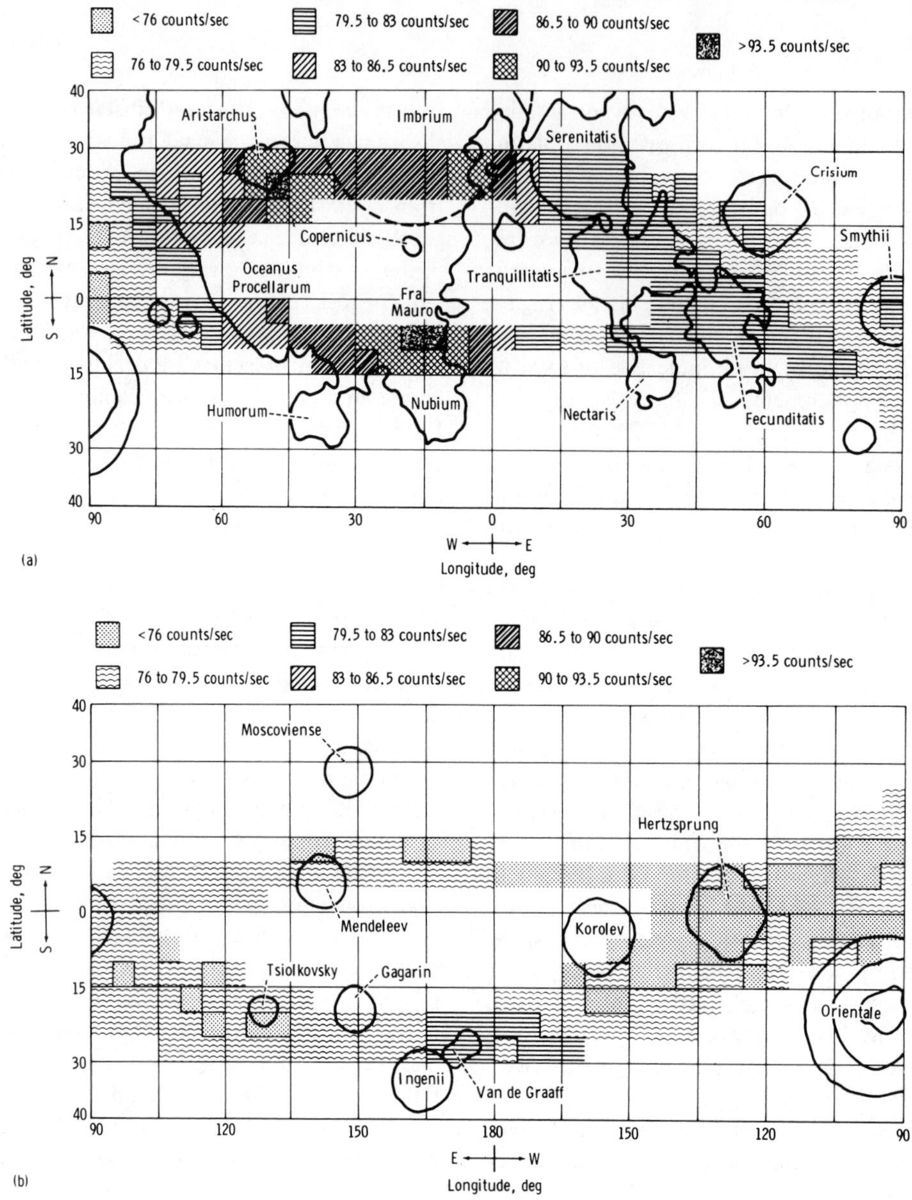

Figure 10-52 Plot of gamma-ray spectrometer readings, in counts per second (cps), obtained from one of the SIM bay experiments during *Apollos 15* and *16*. Values from several orbits have been averaged over sectors of the lunar frontside and farside bounded by latitude and longitude coordinates as shown. Radioactivity highs (>91 cps) occur just west of the *Apollo 15* site near the edge of Mare Imbrium and in a region of Oceanus Procellarum east of the crater Aristarchus. (From J.R. Arnold, A.E. Metzger, L.E. Peterson, R.C. Reedy, and J.I. Tromka, "Gamma Ray Spectrometer Experiment," in Chapter 18, *Apollo 16 Preliminary Science Report*, NASA SP-315, 1972.)

rials. To support such a body, a rigid crust having a minimum thickness of 100 km is required.

S-band transponder data from the last three missions have brought to light the presence of smaller positive mass concentrations than those established for many large basins by *Orbiter* data. These generally are associated with very dark (low–albedo) surfaces in mare-covered regions. Examples of these irregular mascons include dark areas southeast of Copernicus and in Sinus Aestuum and areas near Traesnecker Rille and the crater Lamont. One interpretation holds these mini-mascons to result from near-surface accumulations of solidified igneous rocks having higher densities than the overlying mare lavas. It is not yet clear whether these are intrusive masses or undifferentiated material from deeper levels of the Moon that filled feeder channels without reaching the surface.

Apollo 17 Experiments

In addition to repeating some of the orbital experiments flown on previous missions, *Apollo 17* executed several new ones.

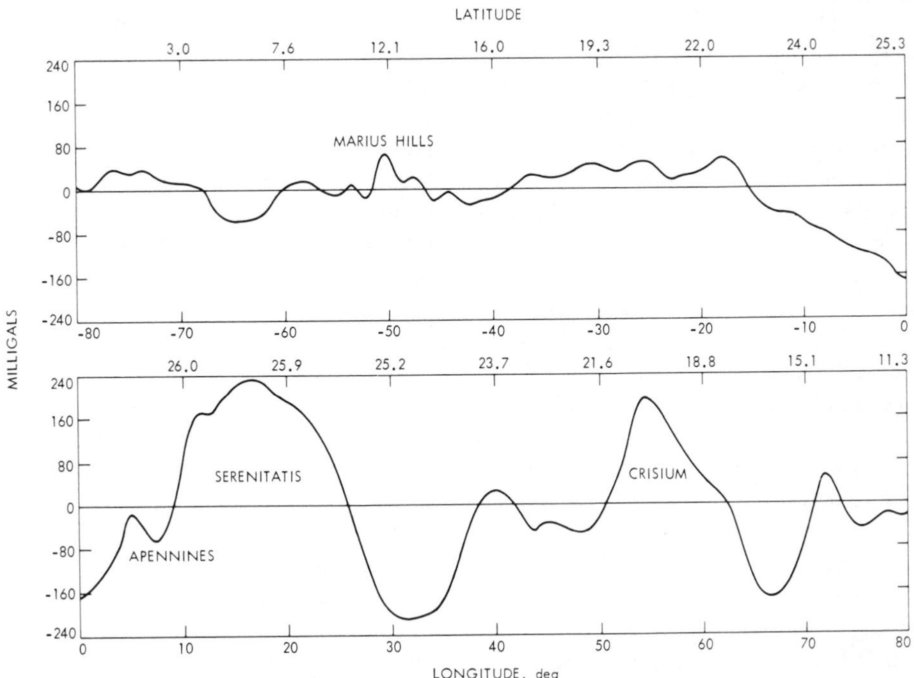

Figure 10-53 Profile of gravity variations measured in the S-Band Transponder experiment (using radio tracking to determine changes in spacecraft speeds) during Orbit 4 of the *Apollo 15* CSM. Large positive gravity anomalies are indicated over Serenitatis and Crisium. (Courtesy W.L. Sjogren, Jet Propulsion Laboratory.)

Figure 10-54 A radar (film) image of a part of Mare Tranquillitatis near the Jansen craters, produced by the Lunar Sounder on the *Apollo 17* CSM. The top of the image is the surface profile determined with pulses reflected from the traveling nadir points along the CSM orbital track. Most of the remaining image represents time-dependent backscatter received from surfaces at increasing distances out from the nadir points (similar to the swaths viewed by Side-Looking Aircraft Radar or SLAR). A prominent mare ridge appears in the top center of the image and a second ridge shows at the extreme left. A bright line below the central ridge is interpreted as a subsurface specular reflection caused by an interface (base of flow layer?) at a depth of 24 meters. (Courtesy R.J. Philips, Jet Propulsion Laboratory.)

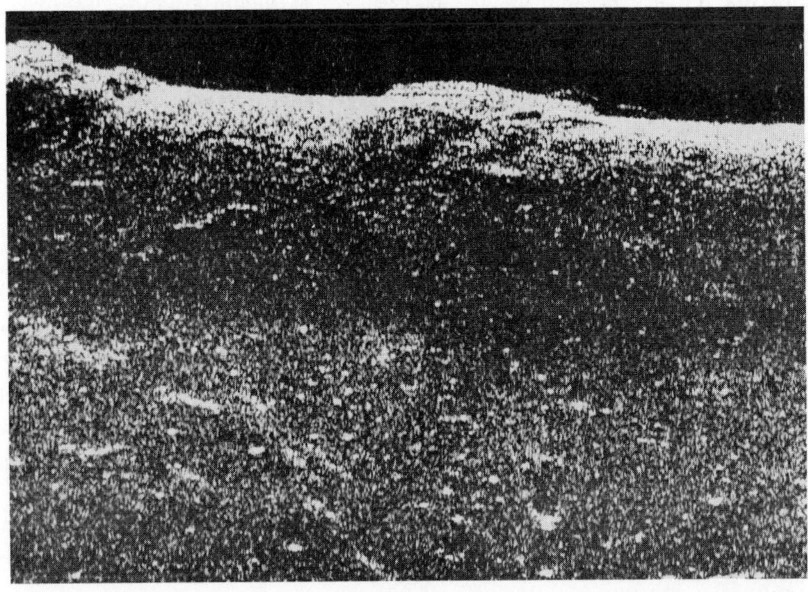

A lunar sounder applied a technique similar to radar by which surface profiles and subsurface structure in the top 1.0 km could be discerned along a path beneath the orbiting CSM. Radio signals at three frequencies (5, 15, and 150 MHz) were directed downward in short pulses and then reflected from both surface and shallow layers. Because the signals picked up by the receiver are influenced by the dielectric properties of these layers (further calibrated by the SEP experiment—see p. 185), something of their nature, composition, distribution, and thickness can be derived from later analysis of recorded (on film) data (Figure 10-54). Results indicate that no subsurface water was detected; this greatly weakens arguments for a possible permafrost zone near the lunar surface—a condition sought as a possible supply of water for an eventual manned lunar base designed for long occupancy. No recognized concentrations of metals were detected—a fact that vitiates hopes for ore deposits of economic value as a future incentive to return to the Moon.

The *Apollo 17* SIM bay also contained an infrared scanning radiometer (ISR) capable of making accurate temperature measurements at a 2-2.5-km spatial resolution (width of the ground track), almost an order of magnitude better than possible with Earth-based instruments. The instrument can measure over a temperature range between 80 and 400°K at a ±2°K sensitivity. During its nighttime operational periods, the ISR detected many small positive thermal anomalies or "hotspots" whose magnitudes were quite similar to those determined from Earth (see p. 64). The number and extent of the anomaly fields within the orbital swaths were about the same as those expected from comparisons with telescope observations. Most of the Moon appears to be thermally featureless, but a few large positive anomalies were located. There are fewer anomalies on the farside than on the front; those on the nearside are associated with the maria, especially at the contacts of basins with the terrae. Boulder fields, in particular, produce some fine structure in the thermal profile. One unexpected result was detection of "coldspots" 5 to 10°K below surface averages that may be caused by greater rates of heat loss from areas of regolith that have higher porosities.

Some of the characteristics of the lunar atmosphere and the surface were investigated by the far ultraviolet spectrometer (FUS), which detects excitations of gas species near the terminator that are energized by sunlight and which can also sense fluorescence of these gases on the darkside. The UV range covered falls between 1180 and 1680 angstroms, within which atomic and molecular excitations of H, N, C, O, H_2O, CO_2, Kr, and Xe can be detected even where concentrations of some are as low as 10-300 atoms/cm^3. With the FUS in operation, a gas concentration of ~3000 atoms/cm^3, mostly as H_2, was determined at the surface. This density was less than predicted on the basis of measurements at several LM sites, where local contamination probably affected the results. The 50 atoms/cm^3 of atomic hydrogen was only about 1% of that expected if solar wind protons were exchanging at the surface and reentering the atmosphere by thermal processes. Most other species were present in concentrations of 10^3-10^4 atoms/cm^3 or less. No fluorescent excitations were noted in dark parts of the Moon; passes near Aristarchus failed to find any evidence of escaping gases that might tie in with transient events described by Kozyrev and others (p. 64).

Astronaut Observations from Orbit

Although instrumental analyses from orbit—analogous to airborne geophysical and remote sensing operations on Earth—provided much exciting and invaluable data that greatly expanded our understanding of the Moon, the role of the man during lunar exploration was effectively resubstantiated by the specialized tasks performed uniquely by the CSM astronauts. In particular, the visual observations and photography with Hasselblad, panoramic, and metric cameras conducted over both predetermined sites and targets of opportunity have led to many new facts that otherwise would have remained unknown. While the astronauts on the surface provided descriptive details of a small region, the orbiting astronauts were able to integrate their sightings into a synoptic conception of large areas where generalizations could be validated. This dual approach resembles the geologic methodology commonly used in terrestrial field studies: a regional reconnaissance, often from an airplane, coupled with close examinations at selected outcrops where careful descriptions and sampling provide the necessary control.

The list of important observations from orbit is long and impressive. Some typically informative scenes are shown in Figure 10-55. Other examples are the steeply dipping layers in the central peak of Tsiolkovsky (Figure 9-13) and the cluster of overlapping domes (Figure 8-9b) of probable volcanic origin noted by *Apollo 15* astronaut Worden. Still other accomplishments include the recognition of multiple, overlapping flows in the Imbrium and Serenitatis basins, localization of source vents for some sinuous rilles, pinpointing of dark or orange-colored deposits, and description of acceleration and deceleration dune deposits associated with ejecta blankets from some large craters (e.g., King Crater). Of particular value were the detailed observations made by astronauts on earlier missions of some of the candidate landing sites for the later missions.

The landing of men on the Moon has resulted in a spectacular increase in knowledge of our planetary satellite. The on-site studies made by the astronauts and their care-

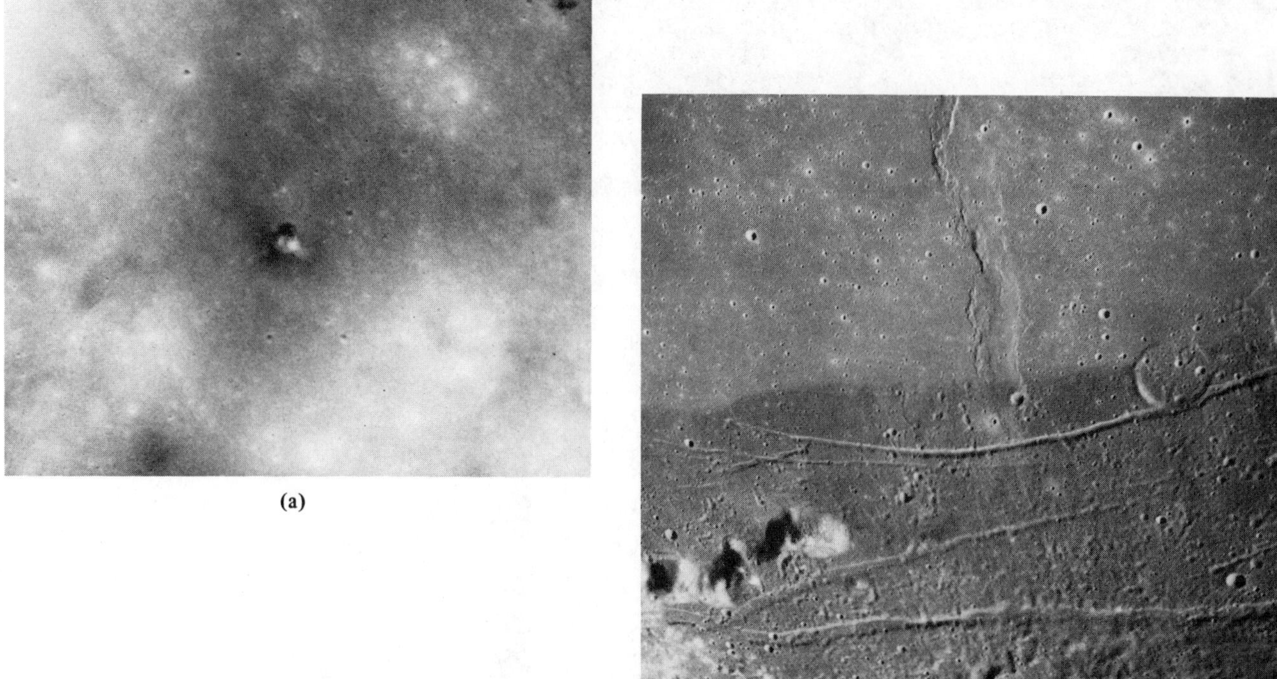

(a)

(b)

(c)

Figure 10-55 (a) A small cinder cone located very near the proposed landing site for *Apollo 17* in the Littrow-Taurus Mountains area. This portion of the pan camera photo taken from the *Apollo 15* CSM (photo: pan AS 15-9559) has been considerably enlarged to show the conical crater surrounded by a dark ash deposit. (b) The contact between two lava flows (light and dark units) in Mare Serenitatis photographed by the Hasselblad camera during *Apollo 17* (AS 17-23069). (c) King Crater, on the lunar farside, a much studied crater showing probable volcanic activity (Hasselblad AS 16-122-19580).

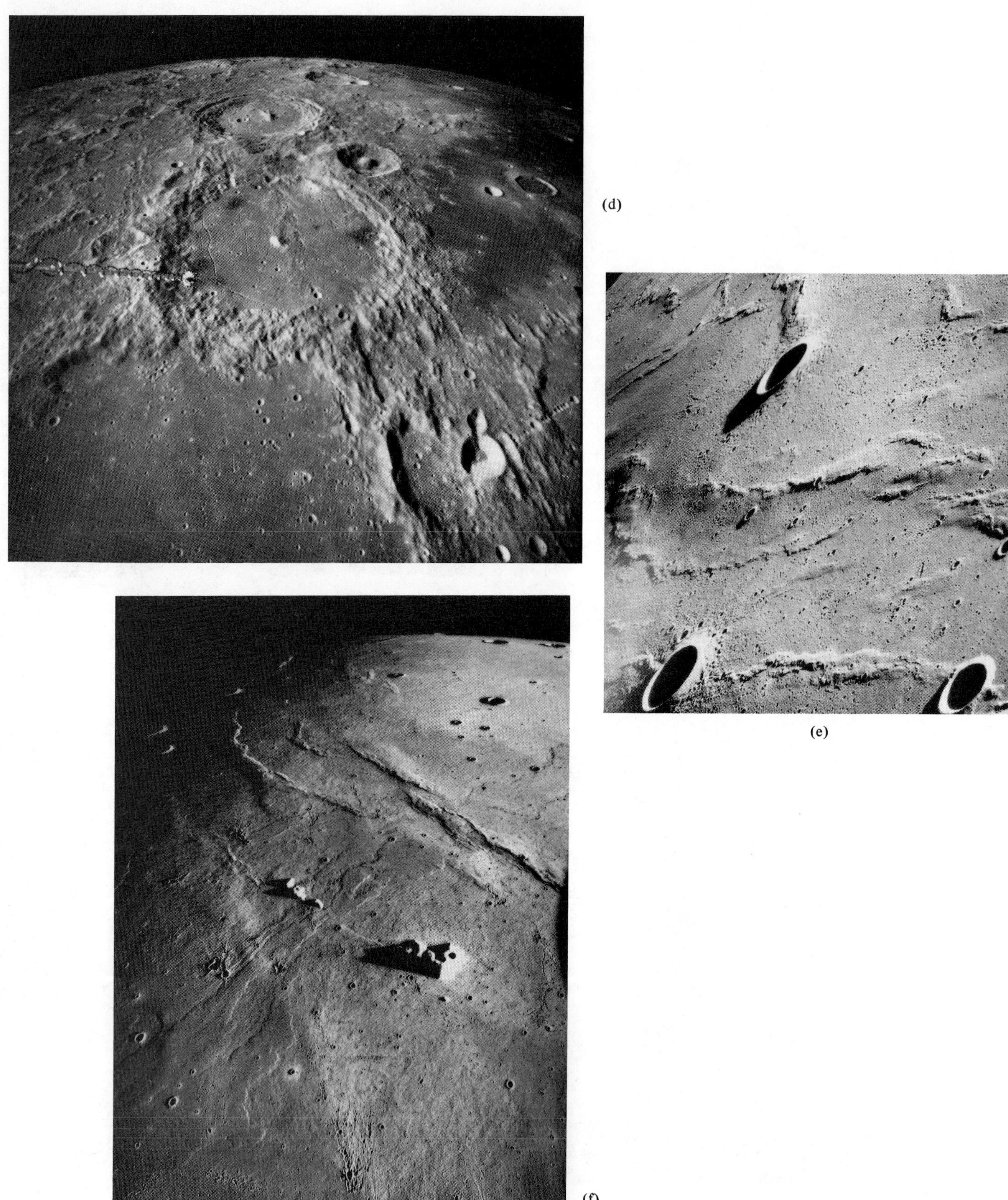

Figure 10-55 (continued) (d) Oblique view toward the south of craters Ptolemaeus (bottom), Alphonsus, and Arzachel in the lunar highlands (metric AS 16-2478). (e) The crater Selecus, photographed by the Hasselblad from *Apollo 15* in low angle illumination near the terminator (AS 15-9813355). (f) Lava marks in a farside volcanic flow, photographed by the metric camera on *Apollo 15* (AS 15-1556). (NASA photos.)

ful emplacement of sensitive instruments are unique contributions that simply could not be duplicated with such high success and quality by automated landers or orbiting space probes. But more than any other achievement, the ability of trained, perceptive humans to selectively pick out a variety of rock and soil samples for later analysis must stand out as the prime justification for sending astronauts to this new, unexplored "continent"—the Moon.

REVIEW QUESTIONS

1. Identify: LM, EVA, CSM, ALSEP, LRV, SIM bay.
2. Name the scientific instruments that are set up in the ALSEPs and the phenomena they are designed to measure.
3. What are the principal results of the X-ray fluorescence and gamma-ray spectrometer experiments in *Apollo 15* and *16*?
4. Discuss the significance of the principal rock type(s) returned from the *Apollo 14* and *16* sites.
5. Review the explanations proposed to account for the lineations observed in the mountains around the *Apollo 15* site.
6. What new ideas resulted from direct examination of the Hadley Rille at *Apollo 15?*
7. How has the Moon itself been used as an observing laboratory in solar astronomy?
8. In what ways do the records of lunar seismic events differ from those of terrestrial earthquakes?
9. What types of lunar seismic events have been identified? When and where do these tend to occur?
10. Describe the model for the lunar interior that has developed from the seismic investigations.
11. List several ways in which a magnetic field can be impressed on lunar rocks.
12. What are the implications of the magnetic field measurements made at four *Apollo* sites in relation to the Moon's past thermal history?
13. How are electrical conductivity data converted to information about composition and temperatures within the Moon?
14. What is the significance of the heat flow data obtained from the *Apollo 15* site?
15. What is the nature of the lunar "atmosphere"? What might have caused "water" to be detected at the *Apollo 14* site?

CHAPTER 11

The Scientific Payoff from the Lunar Samples

Valuable as the on-site explorations and instrument emplacements have been, the more far-reaching contributions from the *Apollo* landings and from the unmanned *Luna 16* and *20* sampling probes continue to come from the intensive analyses of rock samples returned to the Lunar Receiving Laboratory (LRL) in Houston, Texas, and to the Soviet facilities. Teams of experts in laboratories all over the world have been enlistedto study these precious lunar materials. Never before in the history of geoscience have so many diverse, exacting, and careful measurements of nearly every conceivable physical and chemical property been made by specialists using almost every approach to rock analysis devised.

Characteristics and Classifications of Lunar Rocks

During the preliminary examination of *Apollo 11* rocks, four categories of specimen types were arbitrarily established:

Type A:	Fine-crystalline (grain size < 1 mm) vesicular volcanics	16 samples (Types A and B)
Type B:	Medium-crystalline (grain size > 1 mm) vuggy volcanics	
Type C:	Breccias—rock and mineral fragments (clasts) and glassy spatter held in a matrix of fine material and glass; the bulk of these rocks appear to be *microbreccias*, a term which indicates that most visible fragments have maximum dimensions between 0.1 and 1.0 cm.	20 samples
Type D:	Loose "soil" (subdivided into a "coarse fraction" > 1 mm and "fines" < 1 mm; the term coarse fines between 1 and 4 mm), applies to particulates including small rock fragments and glass within the regolith.	2 samples (12.5 kg)

Type A rocks are further distinguished from type B by somewhat higher K_2O and Na_2O and lower FeO and TiO_2 contents, suggesting that these two types derive from different source units and are not simply textural variants of one another.

This working classification also applies to rocks from the other sites, but a wider range of crystal sizes and more varied textures occur in some of the volcanic rocks. Microbreccias appear to be scarce (two samples) at the

Apollo 12 site, but predominate at the *Apollo 14* site (only two crystalline rock samples there). At the *Apollo 15* site breccias are the prevalent rock type near the Apennine Front, but crystalline rocks are characteristic of the mare plains around the LM. Brecciated rocks, again, are the most common types found around the *Apollo 16* and *17* sites. Examples of typical lunar rock samples are reproduced in Figure 11-1.

At first glance, most of the crystalline rocks can be described as basalts in appearance and composition. More detailed examinations have resulted in recognition of several varieties of basic igneous rocks defined by their mineral contents and textures. Thus, in addition to the general term *basalt*, various workers have applied the following rock names, from familiar or less common terrestrial representatives of the gabbro clan, to specific hand specimens and to clasts in the breccias and soil from all landing sites:

Ferrobasalt (iron-rich) Titaniferous basalt (titanium-rich) Feldspathic basalt (plagioclase-rich)	Chemical variants
Olivine basalt (ilmenite) Cristobalite basalt Quartz basalt (normative* quartz only)	Characteristic mineral variants
Gabbro and microgabbro (coarse-medium) crystalline rocks of basaltic composition Norite (hypersthene-plagioclase gabbro) Troctolite (olivine-plagioclase) Peridotite (coarser-crystalline, olivine-rich, pyroxene-bearing, and feldspar-free) Dunite (mostly olivine)	Petrologic and textural variants

Anorthosite, a plagioclase-rich igneous rock with subordinate amounts of pyroxenes, olivine, and other minerals, is an occasional (2 to 57%), but genetically important, type of fragment found mainly in the breccias and soils at each site and as one large rock fragment at the *Apollo 15* site. In contrast to terrestrial anorthosites, those from the Moon are usually much finer-crystalline (although many are also crushed) and have feldspars with very high Ca/Na ratios. However, as on Earth, the anorthosites grade into gabbroic anorthosites, then into anorthositic gabbros (with about 70% plagioclase), and finally into gabbros as the ratio of pyroxene to plagioclase progressively increases to 1.0 or higher. Rock samples from the highlands and/or from depths within the crust that are characterized by Anorthosites (and gabbros), Norites, and Troctolites are collectively referred to as the *ANT* group.

As crystalline rock samples accumulated from the six *Apollo* sites, it gradually became apparent that there are three significantly different representatives of the basaltic clan. These types could be separated on the bases of chemical and mineral composition. Typical *mare basalts*, or *ferrobasalts*, such as those underlying the regolith at Tranquillitatis (*Apollo 11*), Procellarum (*Apollo 12*), Fecunditatis (*Luna 16*), and Palus Putredinus (*Apollo 15*), are characterized chemically by high FeO and low Al_2O_3 contents and mineralogically by abundant clinopyroxenes, some olivine, and relatively low amounts of plagioclase. The *nonmare, highlands, feldspathic* or *high-alumina basalts* have been found so far almost entirely as fragments in *Apollo 11, 12, 14, 15, 16,* and *17* samples, but also as a few igneous rock samples at *Apollo 15, 16,* and *17.* The rock type has high Al_2O_3 and CaO and lower FeO and MgO contents and is characterized by abundant calcium feldspar, clinopyroxenes, and some orthopyroxenes. Gabbros, anorthositic gabbros, and troctolites are coarser-grained examples of this second type. A third basaltic type, which includes some of the norites, was first clearly identified in *Apollo 12* materials and is especially common as fragments in *Apollo 14* breccias. This type contains one or more mineral phases that are notably enriched in potassium (K), the rare earth elements (REE), and phosphorus (P)—to which the acronym KREEP has been applied. These *KREEP-rich basalts*—also called *Fra Mauro basalts*—are characterized by roughly equal amounts of plagioclase, and orthopyroxenes, with minor amounts of potassium feldspar. Taylor et al. (1973) have subdivided all nonmare basaltic rocks (including norites, anorthositic gabbros, and troctolites) into *KREEP-rich* and *KREEP-poor.* They contend that *KREEP* is a residual component that is mechanically mixed into breccias and regolith in differing proportions to produce one or the other of the two general types. According to H. Wilshire (personal communication), *KREEP*-rich norites commonly have a characteristic hornfels texture.

Nearly all the major rock types listed or defined above have now been identified at each *Apollo* site. Reexamination of *Apollo 11* soil fragments confirms the presence of just about every crystalline rock type found at the later sites, including a specimen of KREEP-rich basalt originally termed LUNY Rock 1 prior to its correct appraisal. In fact, the best estimates of the varieties and abundances of the principal crustal rock types on the Moon are obtained from the careful analysis and identification of rock and glass fragments from the regolith at each landing site. This widespread occurrence of diverse fragments in the

*A *normative* mineralogy refers to the hypothetical mineral assemblage calculated to form in some specific proportion during crystallization of a melt with the same composition as the rock if cooling proceeds under equilibrium conditions. The actual or observed (modal) mineralogy may be missing one or more given mineral phases that do not form even though the chemistry of the melt would favor their production. The term normative quartz implies that the silica content of the melt should require some of this mineral to crystallize; other factors may have prevented its development.

Figure 11-1 (a) Close-up view of coarse-grained, cristobalite basalt (A 11: 10044). (b) Fine-grained, vuggy olivine basalt (A 11: 10072). (c) Microbreccia showing pits, glass-lined pits, glass splash, and white anorthosite fragments (A 11:10019). (d) Heterogeneous breccia consisting of a dark mafic phase (plagioclase and pyroxene) and a lighter, silicic phase (alkali feldspar, plagioclase, quartz) (A 12: 12013; see Figure 11-20).

(f)

(g)

(h)

(i)

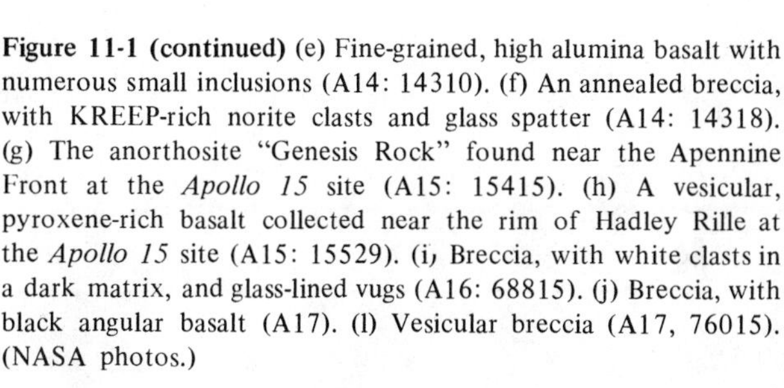

Figure 11-1 (continued) (e) Fine-grained, high alumina basalt with numerous small inclusions (A14: 14310). (f) An annealed breccia, with KREEP-rich norite clasts and glass spatter (A14: 14318). (g) The anorthosite "Genesis Rock" found near the Apennine Front at the *Apollo 15* site (A15: 15415). (h) A vesicular, pyroxene-rich basalt collected near the rim of Hadley Rille at the *Apollo 15* site (A15: 15529). (i) Breccia, with white clasts in a dark matrix, and glass-lined vugs (A16: 68815). (j) Breccia, with black angular basalt (A17). (l) Vesicular breccia (A17, 76015). (NASA photos.)

(j)

soil materials, in contrast to the fewer rock types recognized in the larger hand specimens, indicates extensive mixing of rocks from both local and distant (foreign) sources.

Because crystal or grain sizes and fabric patterns are distinctive variables by which the lunar crystalline rocks can be further subdivided, a number of textural terms (Table 11-1) frequently used in petrography (the description and classification of rocks) have proved convenient

TABLE 11-1

Textures Applicable to Lunar Rocks

Porphyritic:	Phenocrysts of pyroxene and/or olivine in a fine-crystalline matrix of plagioclase, pyroxene, and opaque minerals
Ophitic:	Larger subhedral plagioclase laths (elongate crystals) included in or surrounded by pyroxenes
Poikolitic:	Smaller plagioclase crystals (and other minerals) within pyroxene and/or olivine crystals
Diabasic:	Plagioclase laths surrounding pyroxene minerals within interstitial areas between the plagioclase crystal network
Intersertal:	Euhedral plagioclase laths and pyroxenes in a network surrounding an interstitial mesostasis containing fine-grained ferromagnesian minerals, cristobalite, and glass
Gabbroic:	Coarser-crystalline, usually granular
Equant:	Equidimensional (non-elongate) crystals
Variolitic:	Small radial or sheaflike aggregates of crystals (commonly plagioclase) grouped around common centers, in a fine-crystalline to dense groundmass
Vitrophyric:	Crystalline phenocrysts in a conspicuous groundmass of glass
Terms referring to voids	
Vesicular:	Smaller openings, more or less spherical, usually numerous
Vuggy:	Larger openings, commonly irregular in shape, fewer in number
Pitted:	Cavities and "craterlets" found only at the specimen surface

Figure 11-2 Examples of the range of textures observed in some *Apollo 11* and *12* crystalline rocks. (NASA photo.)

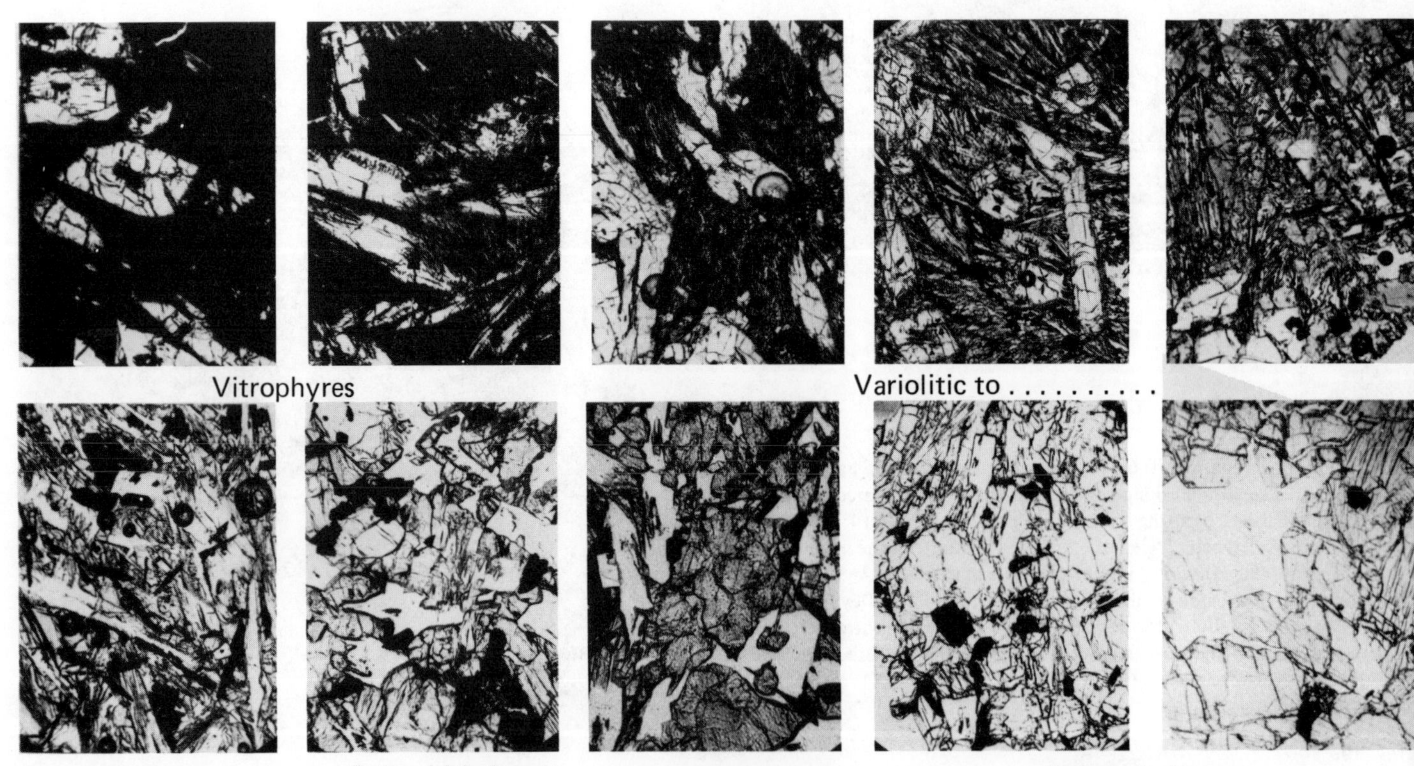

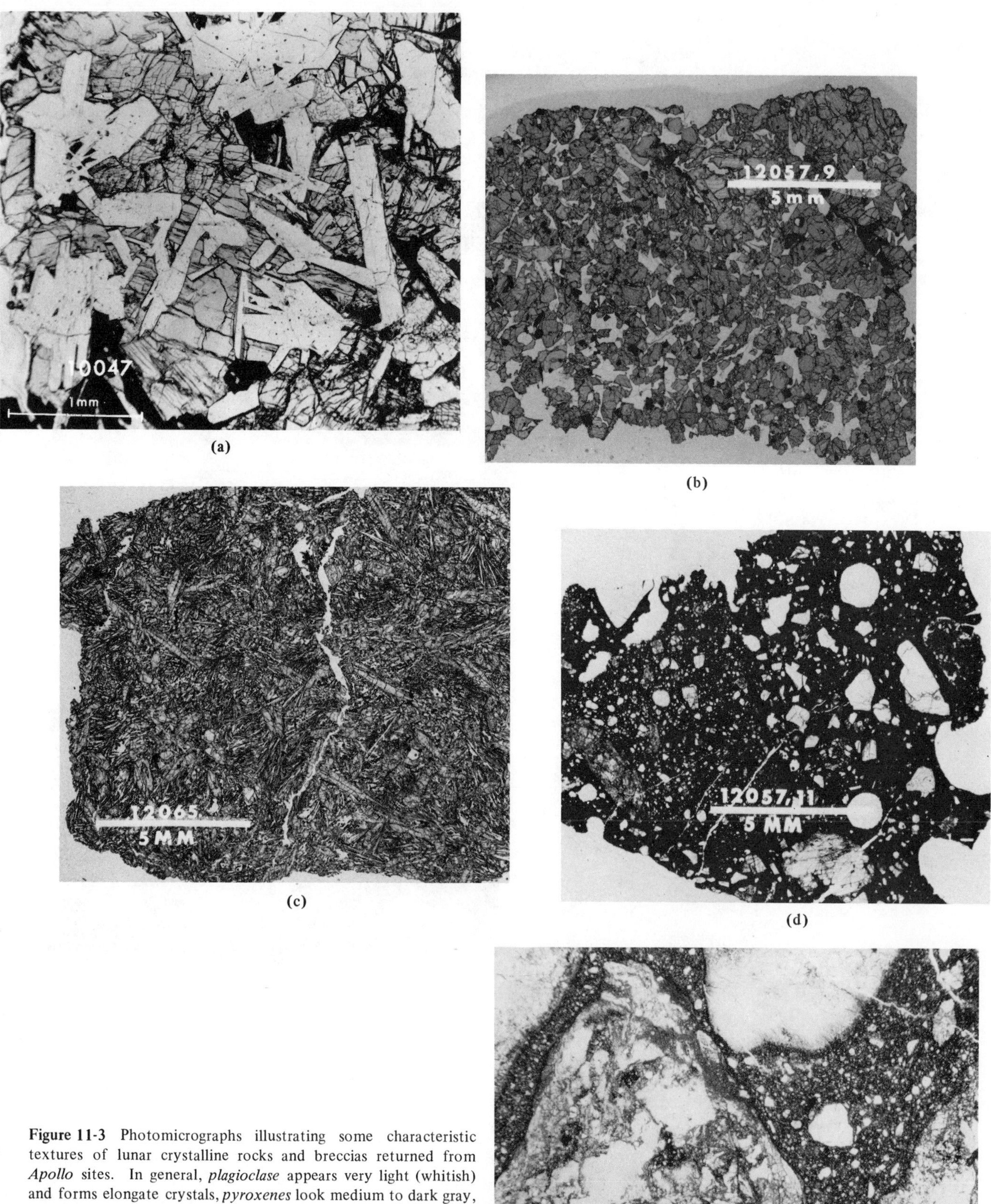

Figure 11-3 Photomicrographs illustrating some characteristic textures of lunar crystalline rocks and breccias returned from *Apollo* sites. In general, *plagioclase* appears very light (whitish) and forms elongate crystals, *pyroxenes* look medium to dark gray, and *ilmenite* is opaque (black). (a) Ophitic basalt (A11: 10047). (b) Equant texture in basalt (medium gray crystals are olivine) (A 12: fragment from soil sample 12057). (c) Variolitic basalt (A 12: 12065). (d) Typical fragmental texture of a microbreccia. Rock fragments, mineral grains, and glass spheres are all held in a finely ground matrix with some glass binder (A12: fragment from soil sample 12057). (e) Breccia clasts with a fine-grained matrix (breccia-within-breccia) (A15: 15435).

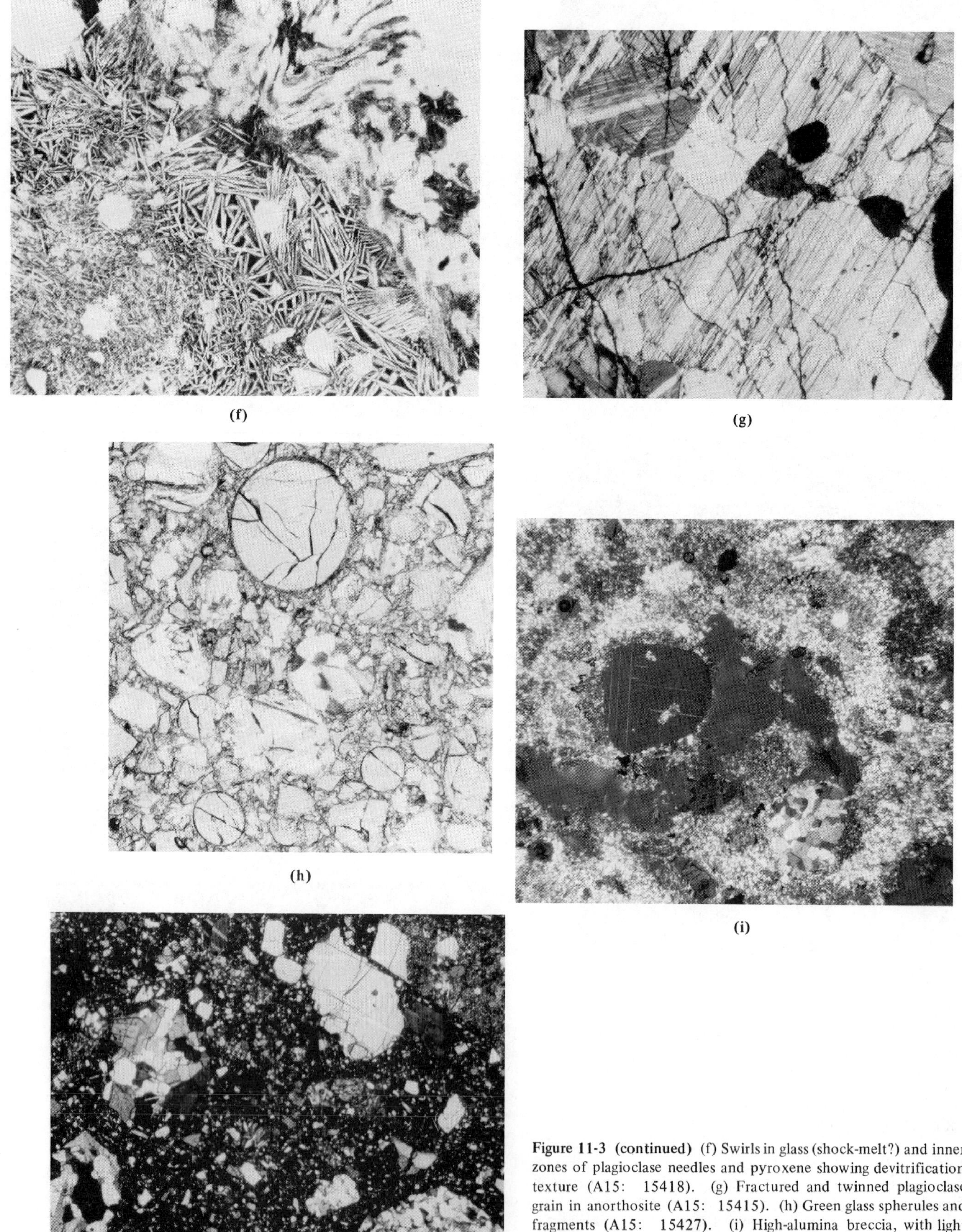

(f) (g) (h) (i) (j)

Figure 11-3 (continued) (f) Swirls in glass (shock-melt?) and inner zones of plagioclase needles and pyroxene showing devitrification texture (A15: 15418). (g) Fractured and twinned plagioclase grain in anorthosite (A15: 15415). (h) Green glass spherules and fragments (A15: 15427). (i) High-alumina breccia, with light clasts (plagioclase) and dark matrix; poikolitic texture (A16: 61156). (j) Breccia, with medium-gray matrix (A16: 61295). (NASA photos.)

descriptors for establishing *differentiation* sequences and genetic histories. Some of these textures are illustrated in Figures 11-2 and 11-3. Within a suite of coassociated rocks, changing grain sizes and the geometric distributions of different minerals provide information on the rates of cooling and shifts in composition within a parent magma.

Mineralogy of Lunar Rocks

To date, more than thirty minerals have been reported from the lunar rocks (Table 11-2). Most of these are uncommon or rare but still may be genetically significant. Six groups make up the bulk of all minerals present in the Moon rocks:

(1) *Feldspars:* Calcic plagioclases, whose anorthite (An) contents range from about 60 to 99% (Figure 11-4), occur in nearly all lunar rocks. Most lunar plagioclases normally are strongly zoned (enriched in K and Na in their outer parts), but usually are not extensively twinned. In hand specimens, the feldspars tend to be translucent gray to light brown; under the microscope the feldspars are seen to be lathlike (Figure 11-6d) to equigranular and show little internal alteration. Potassium-sodium-rich feldspars are rare in the crystalline mare lavas, but have been observed as tiny grains in the *mesostasis* (crystal and/or glass-filled interstices between larger, earlier formed crystals) that represents the last, silica-rich (up to 75-85% SiO_2 by weight) material to crystallize in the cooling lavas. Alkali (K-Na) feldspars are more common in the KREEP basalt fragments.

Figure 11-4 A trapezoid diagram (representing part of a ternary diagram) showing the composition of plagioclase species in terms of the relative proportions of Anorthite (Ca-feldspar), Albite (Na-feldspar), and Orthoclase (K-feldspar). Any feldspar composition can be expressed by the percentage of each end member (for example, $AN_{60}AB_{30}OR_{10}$). Feldspar compositions in *Apollo 11* are plotted as dots, *Apollo 12* as stars, and those representative of terrestrial basalts as squares. This diagram indicates that feldspars in terrestrial basalts typically have about equal proportions of sodium and calcium feldspars whereas those in the lunar rocks are much more calcic. (From W. G. Melson and B. Mason, 1971.)

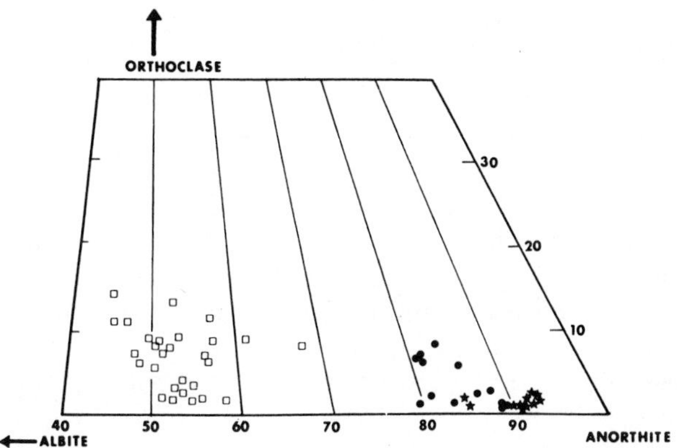

TABLE 11-2

Lunar Mineralogy

Abundant	
Pyroxenes	$(Mg,Fe,Ca)_2(Si_2O_6)$
Plagioclase	$(Ca,Na)\ (Al,Si)_4O_8$
Olivines	$(Mg,Fe)_2(SiO_4)$
Accessory	
Ilmenite	$FeTiO_3$
Chromite	$FeCr_2O_4$
Ulvospinel	Fe_2TiO_4
Spinel	$MgAl_2O_4$
Cr-pleonaste	$(Fe,Mg)\ (Al,Cr)_2O_4$
Perovskite	$CaTiO_3$
Dysanalyte	Ca,REE,TiO_3
Rutile	TiO_2
Nb-REE-rutile	$(Nb,Ta)\ (Cr,V,Ce,La)TiO_2$
Baddeleyite	ZrO_2
Zircon	$ZrSiO_4$ + REE,U,Th,Pb
Quartz	SiO_2
Tridymite	SiO_2
Cristobalite	SiO_2
Potash Feldspar	$KAlSi_3O_8$ + Ba
Apatite	$Ca_5(PO_4)_3(F,Cl)$ + REE,U,Th,Pb
Whitlockite	$Ca_3(PO_4)_2$ + REE,U,Th
Zirkelite	$CaZrTiO_5$ + Y,REE,U,Th,Pb
Amphibole	$(Na,Ca,K)\ (Mg,Fe,Mn,Ti,Al)_5Si_8O_{22}(F)$
Iron	Fe
Nickel-iron	(Fe,Ni,Co)
Copper	Cu
Troilite	FeS
Cohenite	Fe_3C
Schreibersite	$(Fe, Ni)_3P$
Corundum	Al_2O_3
Goethite	$HFeO_2$
New minerals	
Armalcolite	$(Fe,Mg)Ti_2O_5$
Tranquillityite	$(Fe,Y,Ca,Mn)\ (Ti,Si,Zr,Al,Cr)O_3$
Pyroxferroite	$CaFe_6(SiO_3)_7$

From U. B. Marvin, 1973, *The Moon After Apollo,* Technology Review, no. 8, vol. 75, pp. 12-23. Copyright © 1973 by the Alumni Association of the Massachusetts Institute of Technology.

(2) *Pyroxenes:* These calcium-magnesium-iron silicates are usually the most abundant mineral constituents. Common species among the *clinopyroxenes* (those with crystal structures in the monoclinic system) include augite, ferroaugite, and pigeonite, distinguished by the proportions of Ca, Mg, and Fe cations in each (Figure 11-5). Species low in calcium, referred to as *subcalcic clinopyroxenes*, are particularly common in mafic lavas. Hypersthene, the iron-rich *orthopyroxene* (crystallizing in the orthorhombic crystal system), occurs in norite-KREEP basalt fragments. Orthopyroxenes crystallize at higher temperatures than clinopyroxenes and thus tend to occur in rocks that solidify at greater depths. A pyroxenoid not found on Earth, now named *pyroxferroite*, constitutes a new mineral

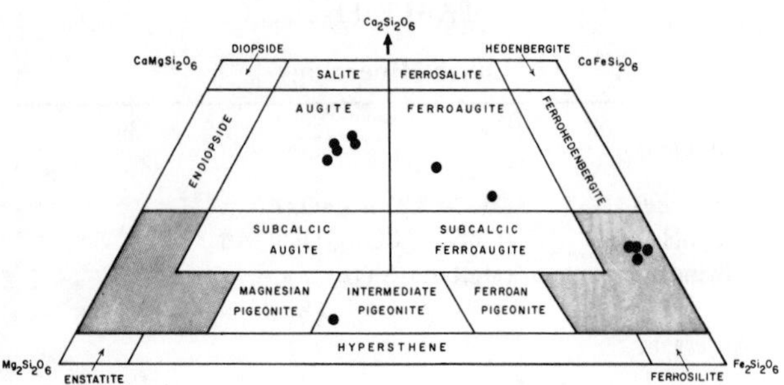

Figure 11-5 A trapezoid diagram defining the different mineral species in the pyroxene according to the proportions of four end member chemical "molecules" (at corners of trapezoid). The composition of each pyroxene species can vary within the limits of its field (indicated by bounding lines). The shaded areas represent compositional ranges for which no terrestrial pyroxenes are known. The dots mark typical compositions of some of the pyroxenes identified in *Apollo 11* samples; those in the shaded field on the right refer to pyroxferroite, the newly discovered pyroxenoid from the Moon. [Reprinted with permission from B. Mason et al., *Proceedings of the Apollo 11 Lunar Science Conference*, (A. A. Levinson, ed.), 1970, Pergamon Press.]

species discovered in the lunar samples. Most of the pyroxenes are cinnamon-brown to yellow-brown, transparent and clear (little internal alteration), and without well-developed cleavage. Under the microscope, pyroxene crystals often have characteristic "hourglass" microstructure (Figure 11-6a) or show regular intergrowths of two distinct pyroxene species between exsolution lamellae (Figure 11-6b). Point-by-point analysis with an electron microprobe (see footnote, p. 35) reveals considerable variations in element (and hence mineral) composition within individual pyroxene crystals (Figure 11-7). Where these variations follow regular distribution patterns, the crystals are said to be *zoned*—a phenomenon usually associated with rapid crystallization that indicates continuing reactions between growing crystals and the chemically changing liquid magma during extensive differentiation. The zoning commonly consists of iron-rich outer rims and a magnesium-rich inner core. Thus, a typical single pyroxene crystal may be zoned mineralogically from rim to core in this way: pyroxferroite→hedenbergite→subcalcic augite→pigeonite→hypersthene. Differentiation trends within a related rock suite can be defined by plotting the compositional variations within individual pyroxene crystals from a single or several specimens on a ternary diagram (see Figure 11-5).

(3) *Olivine:* This magnesium-iron silicate is a subordinate or minor mineral in *Apollo 11* rocks. The ratio of Mg to Fe by weight ranges from 1:1 to about 4:1. Olivine becomes an important major constituent in most *Apollo 12* rocks (Figure 11-6c) and can make up as much as half of the volume of some samples. It is found in some rocks from all other sites as well.

(4) *Silica phases: Cristobalite* is found in the mesostasis of many lunar volcanic rock samples. *Tridymite*, a lower-temperature form of silica, is sometimes present as needle-shaped crystals (especially in more silicic rocks). *Quartz*, normally crystallizing at even lower temperatures, is a very rare constituent in lunar rocks.

(5) *Opaque oxides:* One of the major findings in the *Apollo 11* rocks was the high abundances of titanium (Ti), especially in some basaltic lava rocks. This element resides chiefly in *ilmenite*, an iron-titanium oxide that is fairly common in some terrestrial basic igneous rocks. Typically, *Apollo 11* crystalline rocks contain 20 to 50% (by volume) of this mineral, whose crystal habits include platy, tabular, and equant forms (see Figure 11-3). *Apollo 12* rocks, with lower TiO_2 contents, usually have only 1 to 20% ilmenite. Volumewise, this mineral is even less common in both *Apollo 14* lavas and breccias, is generally present in only low percentages in *Apollo 15* basalts, and is very low in *Apollo 16* brecciated rocks. However, ilmenite once again becomes abundant in the basaltic rocks from the valley-filling flows at the *Apollo 17* site. In the lunar rocks, ilmenite commonly replaces (or is replaced by) other opaque minerals to form distinct intergrowths (Figure 11-6e). These minerals, making up the *spinel series*, include *pleonase,* (chromium-bearing, commonly pinkish), *chromite* (Al- or Ti-rich), *ulvospinel* (another iron-titanium oxide, commonly with some chromium), and a new mineral named *armalcolite* (after *Arm*strong-*Al*drin-*Col*lins), a titanate containing magnesium in place of some of the iron.

(6) *Other minerals:* Among accessory phases in the lunar rocks are *zircon* and *apatite*, both of which occur as small grains mainly in the mesostasis. Both *garnet* and an amphibole have also been reported. *Whitlockite*, a calcium phosphate, is a principal host for the phosphorus and rare earth elements in the KREEP basalts. *Tranquillityite*, another new mineral known only from the Moon, is an Fe-Ti-Zr silicate that also contains some of the rare earth elements found in those basalts. Another zirconium-rich mineral, similar to zirkelite but rich in rare earth elements, also seems to be a new species. The iron sulphide *troilite* and native iron metal with a low nickel content are primary, although minor, constituents in the volcanic rocks. Native iron with variable amounts of kamacite and taenite is found as dispersed grains and fragments in the regolith. Taenite, in particular, indicates "contamination" by meteoritic materials added during the cratering process.

All of the above minerals are devoid of water. None contains hydrogen as an essential element; some hydrogen has been introduced into the samples, however, from the solar wind. The mineral *goethite* (a hydrous iron oxide which is the chief constituent of limonite) has now been

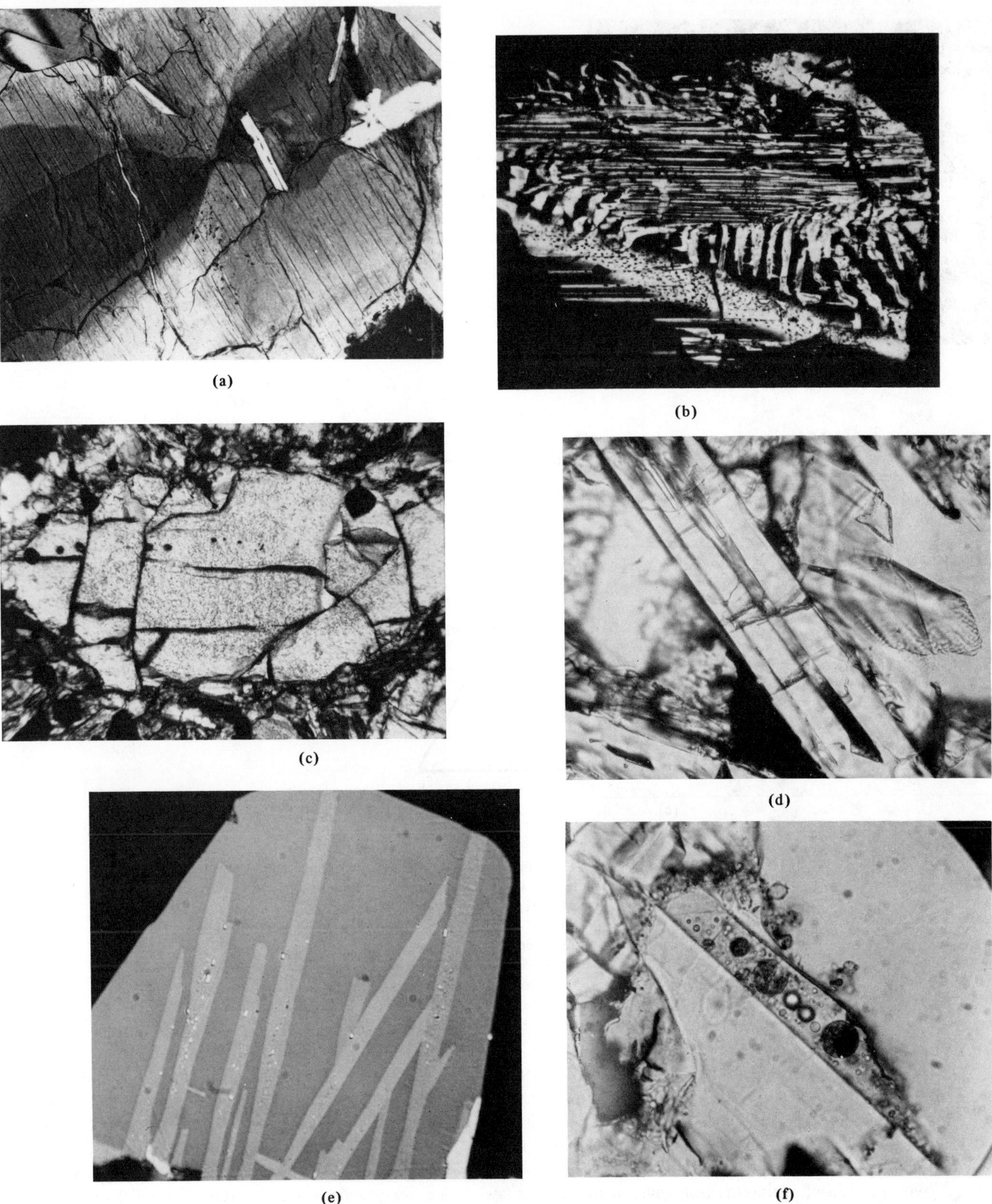

(a) (b) (c) (d) (e) (f)

Figure 11-6 (a) Hourglass zoned pyroxene in sample 10057 (width of view in photo equivalent to 2 mm) (courtesy E. Roedder). (b) Lamellar intergrowths of two pyroxene phases exsolved in a crystal from sample group 12057 (courtesy B. Mason). (c) An olivine crystal with variable iron content (sample 12004) (courtesy L. Walter). (d) Part of a single crystal lath of calcic feldspar (anorthite). The darker band in its center is a "core" composed of very finely crystalline minerals (sample 12021) (courtesy L. Walter). (e) Exsolution lamellae of ilmenite (lighter gray) in a chromite crystal (darker). Metallic iron appears in the upper left (white) (sample 12040) (courtesy L. Walter). (f) Blebs of high silica glass (rounded about 5-10 m in diameter, in plagioclase resulting from the immiscibility of two incompatible liquids). Tiny spheres of iron-rich glass occur within the blebs (sample 10057) (courtesy E. Roedder).

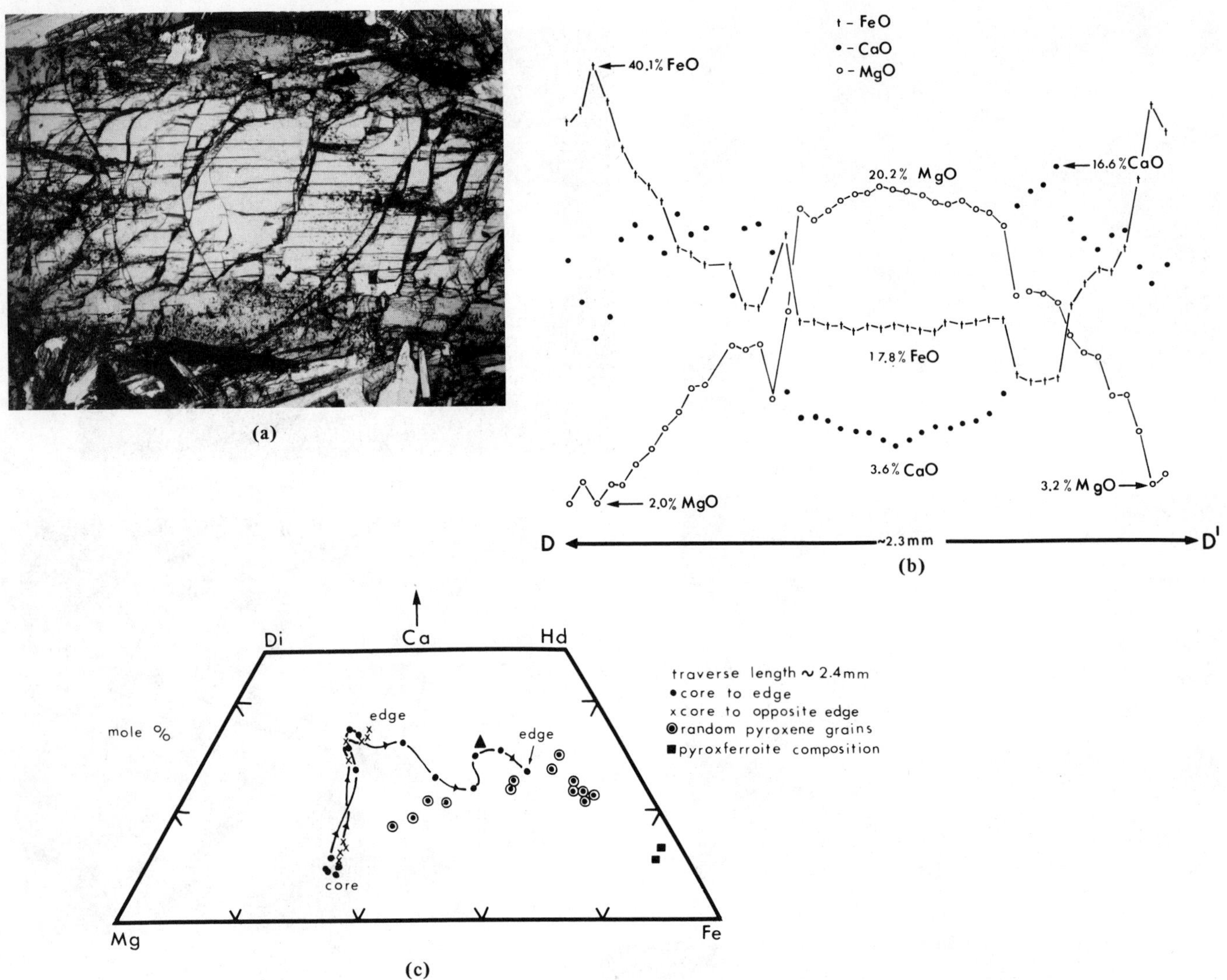

Figure 11-7 (a) Photomicrograph of a zoned clinopyroxene grain (about 2.3 mm wide, top to bottom) in sample 12021, 23. The light brown central area (core) has a pigeonite composition; the darker brown edge region is close to ferrohedenbergite in composition. (b) Variations in the concentrations of Fe, Mg, and Ca (recalculated as oxides) across the width (edge-core-edge) of another clinopyroxene grain in 12021. These compositional changes are determined by electron microprobe analysis. (c) Plot of chemical data from the analysis of many clinopyroxene grains in sample 12021, 137 on the trapezoid diagram of Figure 11-4 as a base after recalculation to appropriate mineral compositions. Compositional zoning trends are clearly indicated, especially in the single grain referred to in (b) above (shown by connected lines). (From Klein et al., 1972.)

reported from an *Apollo 15* rock and several *Apollo 16* samples (the so-called "Rusty" rocks). However, some investigators have questioned whether this phase is truly indigenous to the Moon or was instead formed after return of samples to the Earth by reaction of atmospheric water with introduced meteoritic iron constituents such as lawrencite ($FeCl_2$). No evidence has been claimed as yet for OH^- in the very few grains of a phase with mica structure and rare crystals of amphibole found in the lunar rocks; terrestrial counterparts normally have OH^- as a component in their compositional formulas, but other anions (possibly Cl^- or PO_4^{-3}) presumably substitute for OH^- in this lunar mineral.

No liquid inclusions of any kind were noted in any lunar crystals. Olivine grains, however, show small, rounded inclusions of silicate glass whose compositions are probably very close to the original homogeneous melt that became entrapped in skeletal crystals of this early-crystallizing mineral. Larger (>50-μm) inclusions in the olivine often are recrystallized into mattes of needlelike pyroxenes, plagioclase, and ilmenite. Other lunar minerals, such as pyroxene, plagioclase, and ilmenite, contain clusters of inclusions that are either silica-rich glass (Figure 11-6f)—normative in quartz plus potash feldspar—or iron-rich glass, commonly now recrystallized to pyroxene. These inclusions are interpreted as positive evidence of *liquid immiscibility*, that is, separation or unmixing from a melt of two chemically incompatible liquids that could not

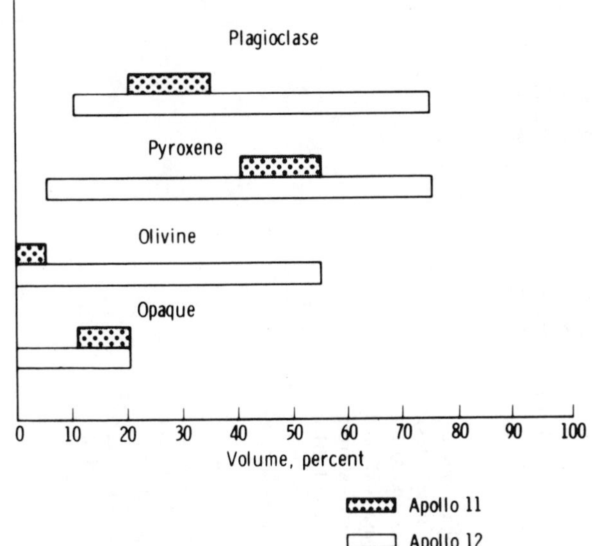

Figure 11-8 Variations in the proportions (in volume percent) of the different major minerals (opaques are mostly ilmenite) in the *Apollo 11* (clear) and *Apollo 12* (stippled) crystalline (basaltic) rocks. (Illustration from NASA Document SP-235.)

produce a single series of crystallizing minerals. This phenomenon, considered theoretically as a possible mechanism by which granites could be derived from primary basic rocks, had not been previously recognized in basalts on Earth. Now, guided by recognition of immiscibility in the lunar basalts, similar examples have at last been found in terrestrial basalts—a discovery which may have important implications for the petrogenesis of magmas and ore deposits.

Although the average chemical compositions of the principal mineral species in the *Apollo 11* and *12* lava rocks are nearly the same, the proportions of minerals from one sample to another vary much more in *Apollo 12* than in *Apollo 11* (Figure 11-8). This wide variability is one line of evidence for a more complete differentiation in the parent magmas of lava rocks emplaced around the *Apollo 12* site. Olivine is particularly common in *Apollo 12* crystalline rocks. Crystalline rocks and rock fragments from the *Apollo 14* site differ from both *Apollo 11* and *12* rocks by having more feldspar and less ferromagnesian and opaque minerals. Most *Apollo 15* lava rocks contain olivine, tridymite, or cristobalite in higher proportions than similar rocks from *Apollo 11* and *14*.

Lunar Rock Chemistry

The chemistry of the lunar rocks offers other clues to their origin. Variations in major element contents (as oxides) among typical crystalline rocks, soils, breccias, and glass from all *Apollo* sites and the *Luna 16* and *20* sites are plotted in Figure 11-9. Compared with similar terrestrial rocks, the lunar crystalline rocks (as well as soils and breccias inferred to derive from them) as a whole are characterized by lower silica contents, higher ferrous iron oxide (FeO) and nearly complete absence of ferric iron oxide (Fe_2O_3), depletions in potassium and particularly sodium but higher K_2O/Na_2O ratios, and total absence of water. For all sites, the mare lavas are distinguished from nonmare types by higher Fe and Cr contents and higher Ca/Al ratios and lower Ni. Chemical distinctions among the three principal classes of terra or highlands crystalline rocks (the ANT suite) are indicated in Table 11-3.

There are however, some notable differences in major element contents among the various rock samples collected at the *Apollo* sites. Thus, *Apollo 11* and *17* crystalline rocks contain much more TiO_2 than basalts from the

TABLE 11-3

Chemical Characteristics of the Three Classes of Terra Rocks*

Elemental Unit	*(A) Anorthositic Rocks*	*(N) KREEP-poor Norites and Troctolites*	*(T) KREEP-rich Norites*
Al_2O_3 (%)	25	20-25	15-20
FeO (%)	0-5	4-9	8-10
MgO (%)	2-8	8-16	7-13
P_2O_5 (%)	0-0.06	0.1-0.3	0.3-2
K_2O (%)	0.01-0.2	0.05-0.1	0.2-2.0
U (ppm)	0.4	0.4-1.0	2-6
La (ppm)	0.1-4.5	10-30	40-80
Eu (ppm)	0.6-1.2	1-2	2-3
Eu anomaly	Positive	Negative	Negative
Hf (ppm)	0.01-1.5	4-10	10-30

*Condensed from Table 1 (p. 616) of LSAPT Report on the Fourth Lunar Science Conference, in *Science,* **181,** 17 Aug. 1973, pp. 615-621.

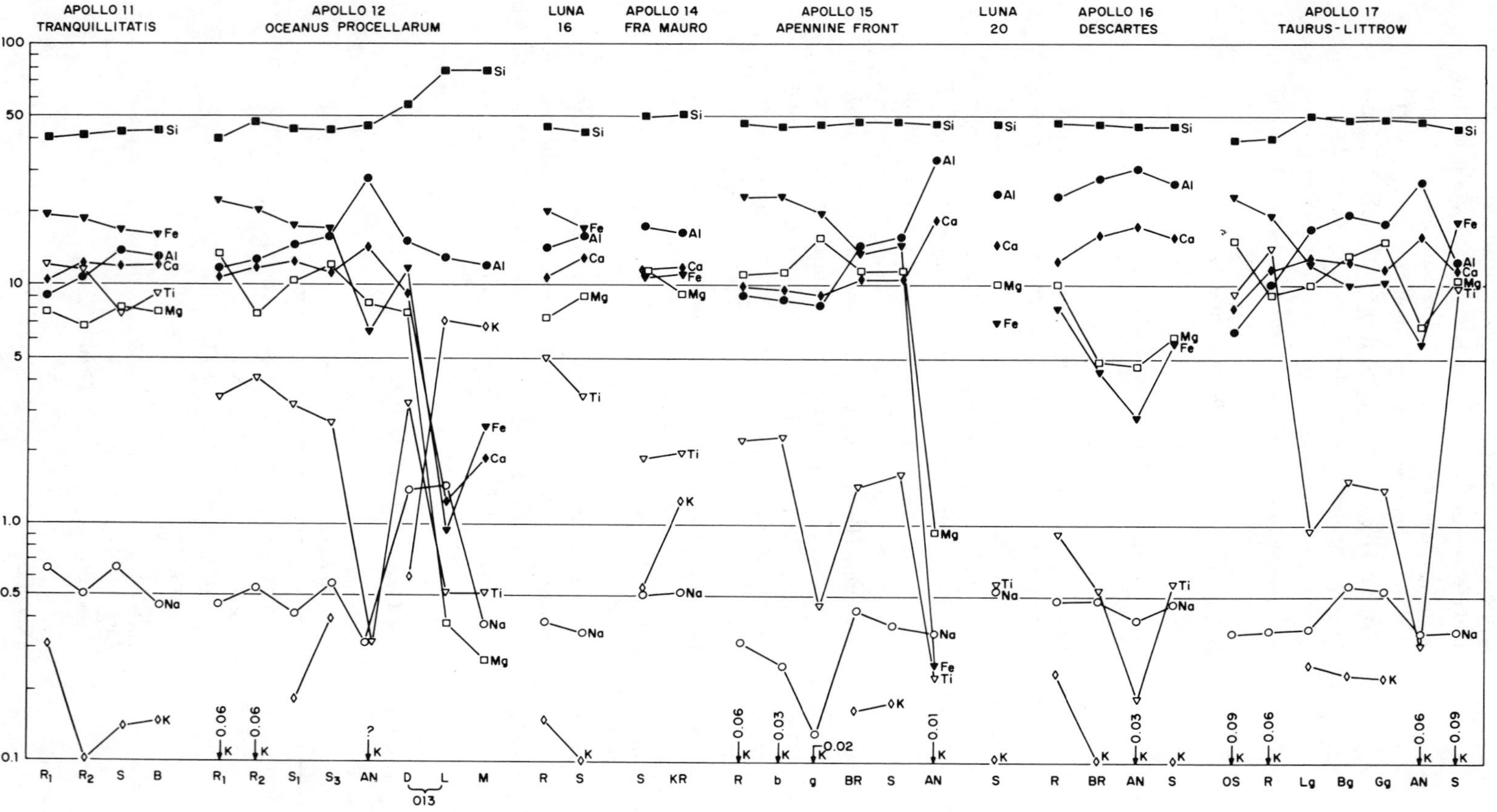

Figure 11-9 Major element contents (as oxides) as determined for selected lunar materials from *Apollo* and *Luna* sites plotted on a log scale. General symbols (applicable to all sites): R = crystalline rocks, usually basalts; Br = breccias; S = soil. Specific identifiers for *Apollo 11*: R_1 & R_2 = high and low-K basalts; *Apollo 12*: R_1 and R_2 = Type I and Type II basalts; S_1 and S_2 = normal and light-colored (KREEP-rich) soils; An = anorthositic fragments extracted from soils; D and L dark and light materials in 12013; M = mesostasis or "granitic" phase in 12013; *Apollo 14*: Kr = KREEP glass. (The diagram that includes the preceding plots has been modified from Hinners, 1971; sources of data are listed in his Figure 3. The diagram consisting of *Apollo 15, 16, 17*, and *Luna 20* values has been prepared by N. M. Short; sources listed.) *Apollo 15*: R = basalts and gabbros (from Christian et al., 1972); g = green glass sample 15301 (from Cavarretta, et al., 1972); b = basalt sample 1555; An = anorthosite "Genesis Rock" sample 15415 (b, Br, S, and An from *Apollo 15* Preliminary Science Report NASA SP-289); *Luna 20*: S = soil analysis from Vinogradov, 1972; *Apollo 16*: R = metaigneous rocks; Br = breccia types I, IV; An = crushed anorthositic breccia type II; S = soils from stations 1, 4, 5, 6 (from *Apollo 16* Preliminary Science Report NASA SP-315); *Apollo 17*: Os = orange soil; B = basalts, several stations; Lg = Light-gray breccias; Bg = blue-gray breccias; Gg = green-gray breccias; An = Anorthositic rocks; S = Soils (from Summary of *Apollo 17* Preliminary Examination Team Results, W. C. Phinney; unpublished handout at Fourth Lunar Science Conference).

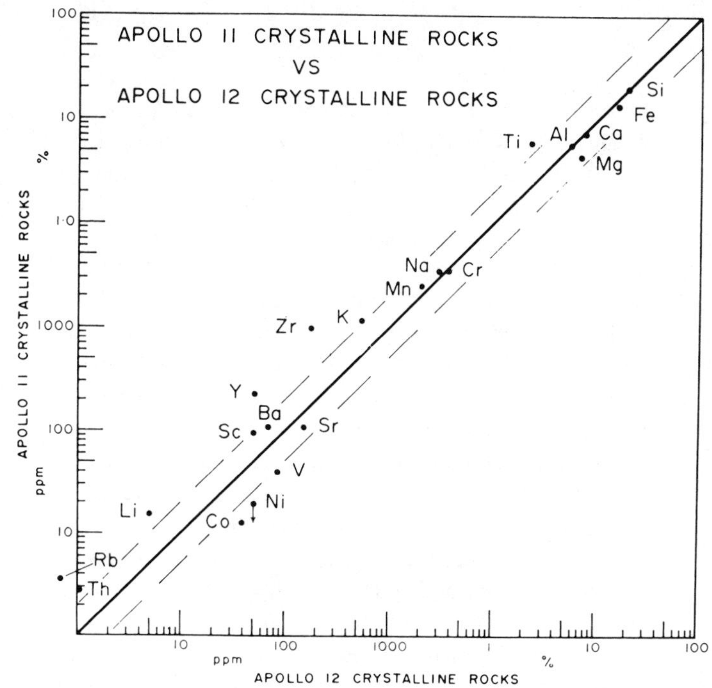

Figure 11-10 Log-log plot which compares the average concentrations of major and trace elements in selected samples of crystalline rocks (basalts) from both *Apollo 11* and *12* sites. Points lying along the heavy diagonal line indicate equal concentrations. Points plotting to the right of this line represent enrichments (and to the left, depletions) of concentrations in *Apollo 12* with respect to *Apollo 11* samples. Thus, *Apollo 11* basalts contain more K, Ti, Ba, Li, Zr, Y, Sc and Th and less Mg, Ni, Co, and V than *Apollo 12* rocks. [Reprinted with permission from S. R. Taylor et al., *Proceedings of the Apollo 11 Lunar Science Conference* (A. A. Levinson, ed.) 1970, Pergamon Press.]

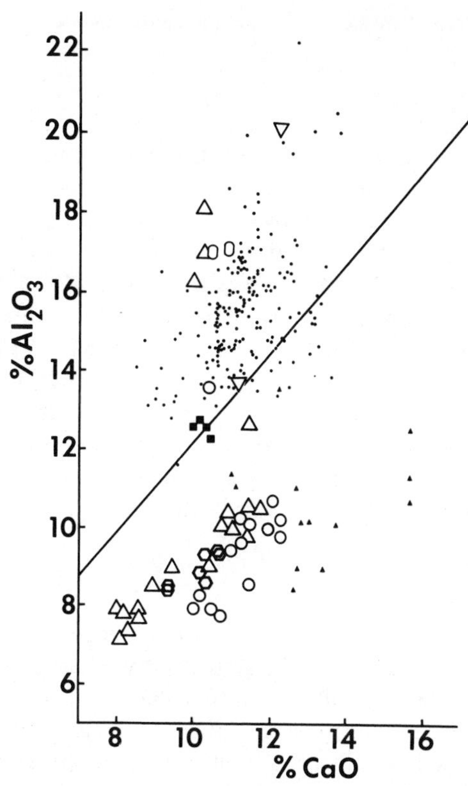

Figure 11-11 Variations of Al_2O_3 with respect to CaO in lunar and terrestrial basalts and in stony meteorites. Symbols as follows: open circles, triangles, and hexagons = *Apollo 11, 12,* and *15* mare basalts respectively; solid circle = average of *Luna 16* soil fragments; solid triangles, apex down = *Apollo 14* crystalline rocks; solid triangles, apex up = averages of noritic or KREEP basalt fragments from *Apollo 12* soils; solid oblong circles = average of several hundred rock particles from *Apollo 14* soil; solid star = *Apollo 15* sample 15418; small dots = terrestrial oceanic ridge basalts and Icelandic basalts; small triangles = silica-poor Hawaiian basalts; solid squares = typical eucritic meteorites; straight line = best fit average of Al_2O_3 and CaO contents in chondrites, eucrites, and howardites. (From "*Apollo 15* Preliminary Examination Team," *Science*, 1972, copyright 1972 by the American Association for the Advancement of Science.)

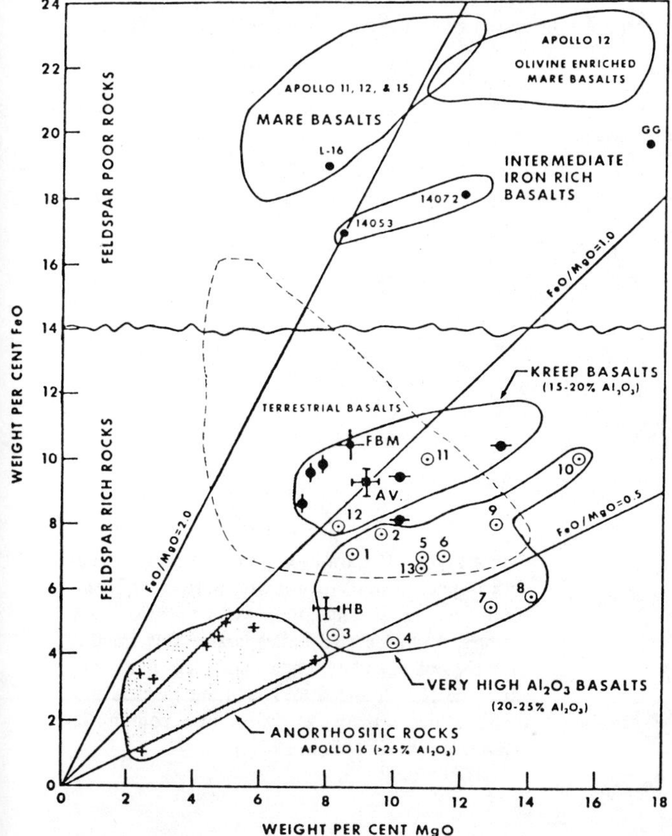

Figure 11-12 Compositional groups of lunar crystalline rocks as defined by plotting FeO versus MgO. For some rocks a range of Al_2O_3 content is also given. These symbols are used: + , rock with more than 25% Al_2O_3; ⊙, rocks with 20-25% Al_2O_3; ●, *Apollo 15* KREEP basalts; ●, *Apollo 16* KREEP basalts; FBM and HB, Fra Mauro and Highlands basalts; GG, green glass from *Apollo 15;* L16, *Luna 16* basalt. Numbered points refer to individual sample numbers as identified in Figure 1 of the article by Gast (1973).

other sites. Only rarely are such high titanium contents found in any basic igneous rocks on Earth. Most *Apollo 12* lava rocks are notably enriched in iron and somewhat depleted in titanium relative to *Apollo 11* rocks. Overall, *Apollo 14* rocks have lower Fe and Ti and higher Si, Al, and K (as oxides) than either *Apollo 11* or *12* rocks. Rocks from the *Apollo 15* site have a high FeO/MgO ratio and lower average TiO_2 than rocks from the earlier sites. Breccias from *16* and *17* are particularly high in CaO and Al_2O_3 and very low in TiO_2. The *Apollo 16* rocks are readily divided into four distinct groups by their Al_2O_3 contents: (1) 30-34% (cataclastic anorthosites); (2) 26-29% (breccias); (3) 23-25% (metamorphosed KREEP-rich basalts); and (4) 17-19% (meta-igneous).

Geochemical Trends

The compositional differences within the crystalline rocks noted above are related to systematic geochemical changes or trends that result from differentiation of one or more parent magmas into several distinct rock types. Differences among soils and breccias from one site to another depend both on the compositions of local, underlying crystalline rocks and on ejecta of variable bedrock lithologies contributed from more distant sources. One way to compare compositions of similar rocks at any two separate sites is exemplified in Figure 11-10. Geochemical trends are usually revealed by making graphical plots of the changing amounts of one element with respect to one or more of the other elements analyzed for in samples from related rock suites (such as the gabbro-anorthosite series). Examples of variations of CaO versus Al_2O_3 and of FeO versus MgO in various rock materials from *Apollo 11, 12, 14,* and *15* and *Luna 16* are shown as two-component plots in Figures 11-11 and 11-12. Trends in variations of FeO relative to Al_2O_3 within crystalline rocks, microbreccias, and soils, and glass from a single site, *Apollo 12,* are depicted in Figure 11-13. Plots of FeO versus Al_2O_3 for the full range of rock types found on the Moon (Figure 11-14) show clearly the inverse relationship of these two oxides. The contents of Fe and Al are therefore sensitive indicators of the compositional differences between the suites of rocks that characterize the highlands and those common to the mare plains. A typical multicomponent (ternary) diagram representing changes in composition among the principal types of rocks and rock fragments at the Apollo 12 site is illustrated in Figure 11-15.

These diagrams, and others for different element combinations, disclose two general facts about the behavior of elements in the lunar (as well as terrestrial) rocks that originate by differentiation. First, the concentrations of some compared element combinations for a given rock type tend to plot in clusters, indicating that the spread or range of values for these concentrations is relatively small; this means that a rock type can frequently be identified by the amounts of certain elements present. Second, changes of some elements in related rocks tend to covary directly—that is, the systematic increase (or decrease) of one element in rocks from different members of a series is commonly accompanied by a corresponding increase (or decrease) of one or more of certain other elements. For other compared elements, the changes covary inversely—that is, an increase (decrease) by one is matched by a decrease (increase) of another. However, lack of correla-

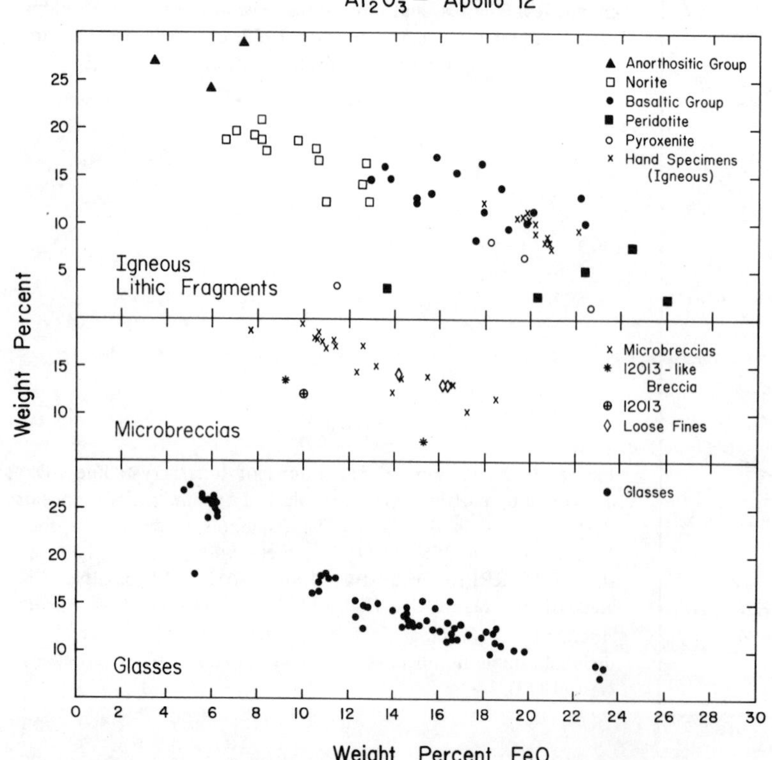

Figure 11-13 Variations in percent FeO (abscissa) with respect to Al_2O_3 (ordinate) in rock fragments (lithologic types indicated) extracted from soil samples, microbreccias and separated fine fraction from the soil, and glass spheres and blebs in samples from the *Apollo 12* site. The trend in each plot shows a decrease in Al_2O_3 as iron content increases. (From Keil et al., 1971.)

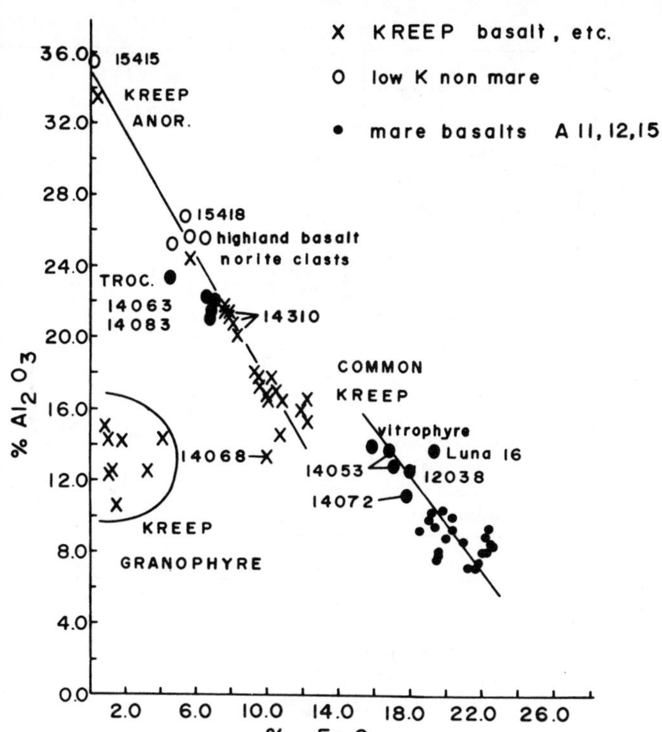

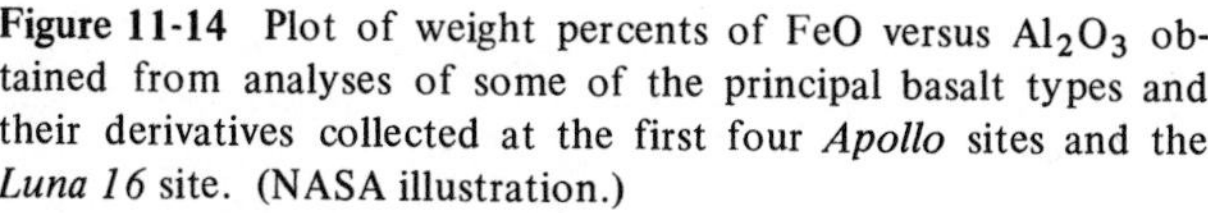

Figure 11-14 Plot of weight percents of FeO versus Al_2O_3 obtained from analyses of some of the principal basalt types and their derivatives collected at the first four *Apollo* sites and the *Luna 16* site. (NASA illustration.)

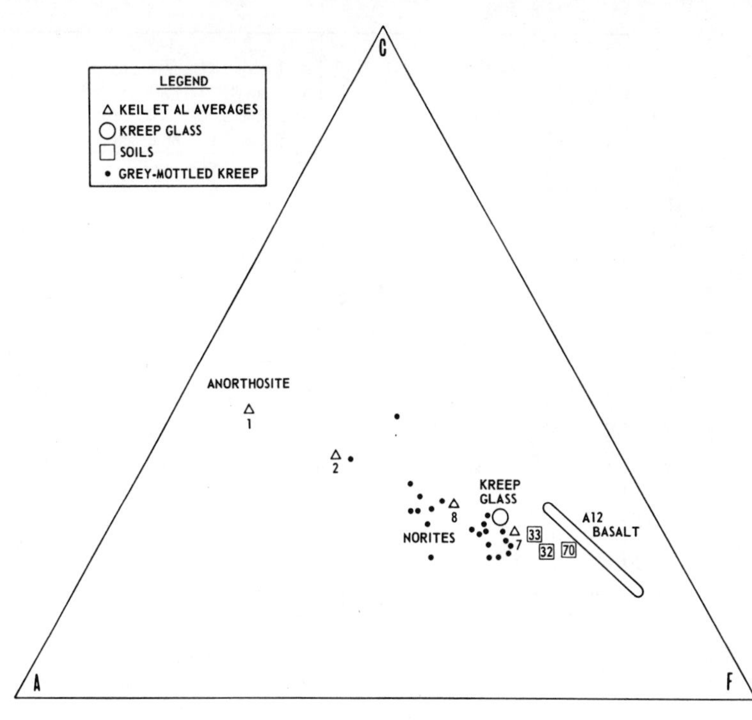

Figure 11-15 Ternary diagram showing compositional variations among several rock types found in the soil and as large samples at the *Apollo 12* site, A = Al_2O_3 + alkalis; C = CaO; F = FeO + MgO. (From C. Meyer et al., 1971.)

tion or covariance among compared elements could mean either that the elements are geochemically unrelated or that the rocks themselves have no genetic ties.

Minor Elements in Lunar Rocks

Concentrations of many elements are lower in the lunar rocks than in either their terrestrial equivalents or in carbonaceous chondrites (see Figure 13-1). These include more volatile elements such as the alkalis sodium, potassium, and rubidium as well as the minor elements lead, bismuth, arsenic, mercury, copper, cadmium, barium, strontium, zinc, chlorine, and bromine. The proportions of most volatile elements are 10 to 100 times less, and of many nonvolatile, siderophile elements 1000 to 10,000 times less, in lunar surface rocks than in the carbonaceous chondrites. Such metallic elements as gold, platinum, iridium, osmium, rhenium, nickel, cobalt, and germanium are depleted in lunar basalts to a similar or greater degree than in terrestrial basalts. Some elements, particularly the low-volatile, or refractory, ones, are enriched in lunar lava rocks with respect to their terrestrial equivalents. In the lunar basalts these elements exceed solar (chondritic) abundances by factors of about 3 to 10. Zirconium, hafnium, and yttrium have the highest values and appear to concentrate within the late-crystallizing mesostasis where the silica content can range between 60 and 80% and potash rises above 1%.

Concentrations of some trace elements in crystalline rocks of different kinds (e.g., anorthosites and basalts) vary significantly at any one *Apollo* site and may show differences within similar rock types (e.g., basalts) from several sites (Figures 11-16a and 16b). Variations are evidenced as well in soils from different *Apollo* sites (Figure 11-17). Trace element contents in lava rocks, breccias, and soils from the same site will also differ. Enrichments of silver, zinc, cadmium, thallium, and other trace elements in the soil relative to the volcanic rocks underlying a site have been attributed to admixture with 1 to 2% meteoritic materials. Some trace element concentrations are dissimilar in types A and B lava rocks from the *Apollo 11* site—an indication that these types come from separate flows. Despite notably different mineral proportions, many trace elements show relatively little variation in *Apollo 12* crystalline rocks; this suggests derivation from a single flow that experienced local fractional crystallization. In contrast, variations in both refractory and volatile element contents are indicative of different sources or degrees of melting—a condition observed in samples from other *Apollo* sites.

Plots of the concentrations of a group of the rare earth elements (REE) and certain other diagnostic elements in a selected group of rocks and soils from *Apollo 12, 14*, and *15* are shown in Figure 11-18. These materials show a moderate to strong enrichment of nearly all rare earth elements, along with Li, K, Rb, Sr, Ba and Zr, relative to chondritic meteorites presumed by many geochemists to be the parent or source materials for both the Earth and

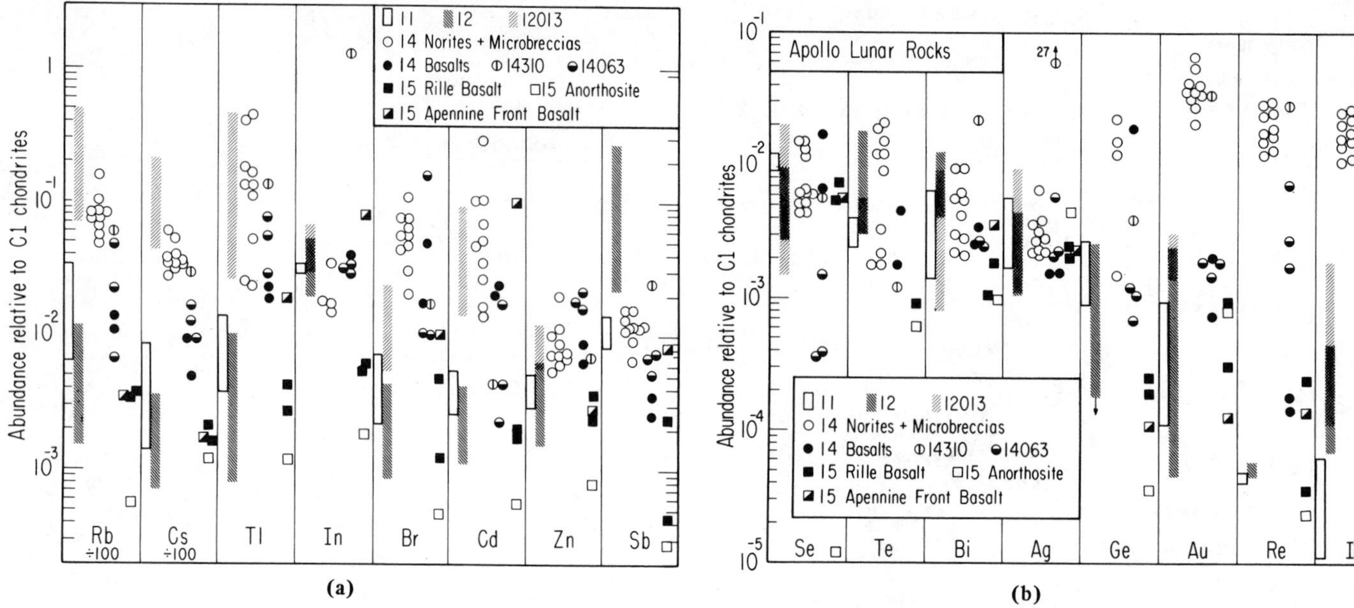

Figure 11-16 a and b: Trace element contents in crystalline rocks from *Apollo* sites *11* through *15*. All concentrations are expressed as a ratio of an element in lunar rocks to the average value of the same element in Type I Carbonaceous Chondrites. In general, the *Apollo 14* norites and *Apollo 15* anorthosites show the highest and lowest contents of these trace elements, respectively. Symbols next to 11 and 12 refer to mare basalts from the *Apollo 11* and *12* sites. (From J. W. Morgan et al., 1972.)

Figure 11-17 Enrichments of nine trace elements in the fine fraction of soils from the *Apollo 11, 12,* and *14* sites relative to mare lavas from the first two sites. KREEP-basalts and norites are the principal factors for the higher trace element contents in *Apollo 14* fines. (From G. Brown et al., 1972.)

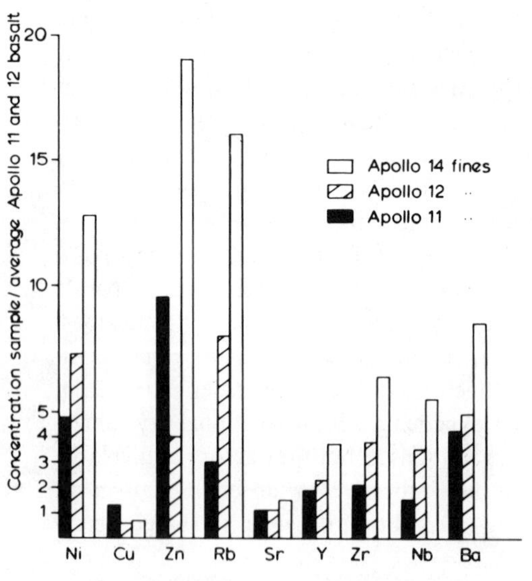

Figure 11-18 The relative abundances of rare earth elements (REE), Li, K, Rb, Sr, Ba, an in rocks and soils from *Apollo 12, 14,* and *15* sites. The ordinate units represent the rati the concentration of each element in the lunar samples normalized to the concentratio that element averaged from the analyses of 29 chondritic meteorites as a reference stan (for example, Li in 15555 is approximately 50 times more abundant in this lunar sample in the chondrites). Note the sharp increase in Sr and Eu and the lower relative abundance other REE in the plagioclase from sample 15555 compared with basaltic rocks and soils. A thosites also show this behavior. (From C. C. Schnetzler et al., *Science*, 1972, copyright 1 by the American Association for the Advancement of Science.)

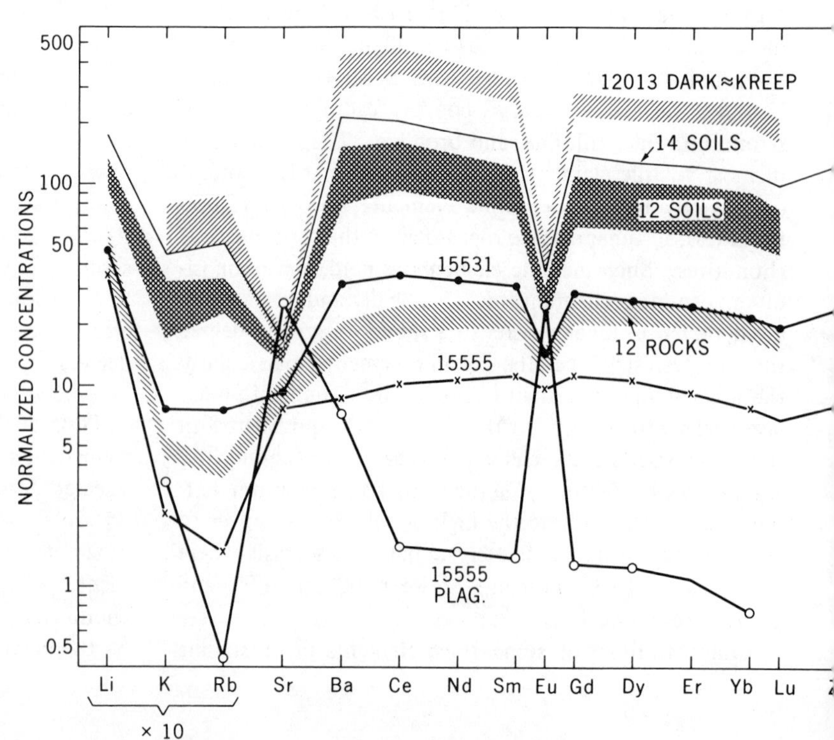

the Moon. Most rare earth elements (principally in the clinopyroxenes) and Sr and K are also somewhat enriched in the lunar basalts with respect to terrestrial basalts, andesites, and average continental crust. However, europium is conspicuously less enriched than the other rare earths in the lunar lavas and soil. On Earth, europium exists in basaltic rocks as both Eu^{+2} and Eu^{+3} in a fixed ratio at equilibrium; on the Moon, the strongly reducing chemical environment shifts this equilibrium toward a predominance of Eu^{+2}. In lunar basalts, the ratio of Eu^{+2}/Eu^{+3}, obtained from whole-rock (unseparated-mineral) analysis, has values (~5) intermediate between lower terrestrial basalts (~0.2) and higher achondritic meteorite values. Divalent europium should concentrate mainly in the feldspars of the lunar rocks; enrichment in the lunar anorthosites and in separated plagioclase from basaltic rocks, where Eu^{+2} can substitute for Ca^{+2}, confirms the relation (Figure 11-18).

There are other signs that the state of oxidation in the lunar lavas was much lower than in comparable terrestrial lavas. Expressed as partial pressures of oxygen, the P_{O_2} calculated for lunar basalts is ~10^{-13} atm compared with 10^{-5} atm for terrestrial basalts. Thus, the purity of ilmenite (possibly with minor amounts of its titanium reduced to Ti^{+3}), the dominance of Fe^{+2} (ferrous or reduced state), an almost total absence of Fe^{+3} (ferric or oxidized state), and the occurrence of troilite and iron metal all point to a highly reduced state.

Presence of small amounts of Ni-poor metallic iron in lava rocks from the *Apollo* sites confirms that this phase forms directly in rocks indigenous to the Moon. However, the soils and some breccias at all sites contain from 4 to 6 times more metallic iron by weight than do the crystalline rocks. Most of these metal particles are typically high in nickel (3 to 15%) and comprise from 0.3 to 0.6% of all the small grains present. As much as 1% of the *Apollo 16* core sample grains consists of metallic nickel-iron. Changing proportions of nickel-rich iron in the fragmental materials result from mixing of dispersed meteoritic debris with smaller amounts of the iron native to the crystalline rocks during impact production of the regolith. Thus, a rising nickel abundance correlates with soil maturity, as defined by increased comminution and greater thickness brought about by continued bombardment. Most *Apollo 14* breccias are especially enriched in such siderophile elements as nickel, gold, iridium, and rhenium. This has been interpreted as evidence for intermixing of fragments of a large iron meteorite responsible for excavating the Imbrium Basin with the ejecta debris that now makes up the Fra Mauro Formation.

The amounts of radioactive uranium and thorium in the lunar basalts range from about 0.03 to 1.2 ppm (0.000003 to 0.0002%) for U and from 0.13 to 4.45 ppm (0.000013 to 0.00044%) for Th. Most of this radioactivity is localized in the mesostasis phases in these lunar lava rocks. Much higher values for U (up to 10 ppm) and Th (35 ppm) are reported from the KREEP-rich basalts and from the lighter-colored (more granite like) rock materials found at the *Apollo 12* and *14* sites. The anorthosite sample 15415 contains 0.0015 ppm uranium and 0.0035

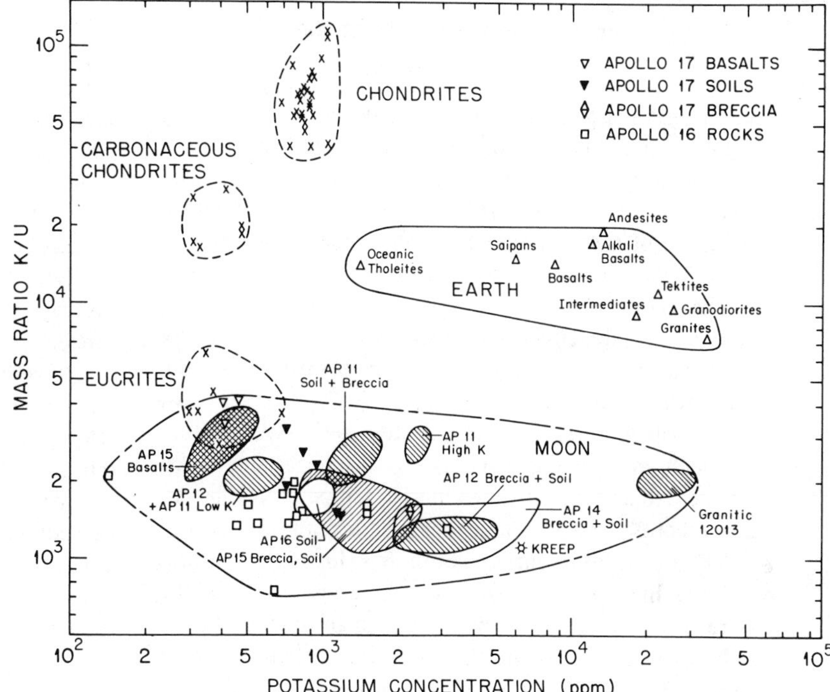

Figure 11-19 Variation in the mass ratio of K/U relative to potassium concentration (in parts per million or ppm) in lunar rocks compared with stony meteorites and terrestrial igneous rocks. Almost all rocks from these three planetary sources can be distinguished chemically on this basis. (From G. D. O'Kelley et al., *Science*, 1972, copyright 1972 by the American Association for the Advancement of Science; updated 1973.)

ppm thorium—one of the lowest values yet measured. The average (unredistributed) uranium content for the entire Moon is estimated to be 0.06 ppm. The K/U ratios in lunar rocks of all kinds are lower and more variable than in terrestrial igneous rocks and much lower than ratios obtained from the chondrites (Figure 11-19), presumably owing to the conspicuous loss of volatile potassium. Mainly because of variations in K_2O content, the K/U ratios found in the highlands and Fra Mauro breccias and the KREEP-rich basalts fall between 1500 and 2000, while those in mare basalts typically cluster around 3000. The Rb/Sr ratio is also low. Unless potassium concentrations increase sharply with depth, uranium and thorium remain the chief sources of heat energy from radioactive decay.

Petrogenetic Aspects of the Crystalline Rocks*

So far, all samples of lunar materials (crystalline rocks, breccias, soils) consist of rock types that have passed through one or more sequences of *magmatic differentiation*. No primary (and primitive) undifferentiated rocks, indicative of the initial states of lunar materials when the Moon formed, have been recognized.

The crystalline rocks solidified, then, from hot, dry, fluid, and highly reducing magmas in which water probably never was present and volatiles were abnormally low compared with most terrestrial magmas. This absence of water, which serves as a flux (lowering the temperature at which melting commences) in most terrestrial rocks, requires that generally high initial temperatures be attained before the lava source rocks can become molten and move up toward the surface. Both these high temperatures and the high iron content reduce the viscosities of the mare lavas to values lower by more than an order of magnitude (factors of 10 to 30) than for terrestrial lava flows of comparable basic composition. With lower viscosities, individual lava flows filling the mare basins and lowlands were able to spread over much larger areas to distances as much as 1000 km beyond their eruptive centers, even though slopes commonly were as gentle as 1°. The finer-sized textures of most of the crystalline rocks indicate rapid cooling and solidification, perhaps owing to accelerated heat losses to the surface vacuum brought about by continual convection within the highly fluid lavas.

The temperature range through which crystallization of mare lavas proceeded following extrusion has been estimated by various workers through determination of the melting relationships for the mineral assemblages in the lunar basalts. Smith et al. (1970) found experimentally that ilmenite begins to crystallize from a ferrobasalt magma at ~1150°C, followed by clinopyroxene at 1130°C and plagioclase at 1120°C; iron-chromium spinels begin to solidify at temperatures between 1150° and 1200°C. This narrow interval of about 30°C between appearance of the first- and the last-formed crystals of the major silicate minerals is notably less than the 150-200°C determined experimentally for terrestrial basalts, which undergo complete solidification between 1000° and 1100°C. Other experimenters have set similar limits on the crystallization temperatures as low as 1085°C for plagioclase crystals and 950°C for the alkali-silica-rich mesostasis. Green (1971) has determined a somewhat different crystallization sequence (olivine-spinel-clinopyroxene-plagioclase-ilmenite, in order of successive formation) for certain *Apollo 12* lavas in which the first crystals begin to appear as cooling proceeds from 1380° to 1210°C. Melts with lower MgO contents start to crystallize at lower temperatures. As MgO increases, the proportion of olivine to other minerals also rises; thus, olivine-rich rocks represent those which crystallized at higher temperatures. Green and Ringwood (1972) consider the lavas at the *Apollo 11, 12,* and *15* mare sites to originate from a common parent material in the interior at depths greater than 300 km where temperatures exceeded 1350°C. Mineral and chemical differences in these mare basalts resulted from variability in the degree of partial melting of the source rocks over a range of depths (changing pressures and temperatures).

Probably the most accurate values of crystallization temperatures are those determined by Roedder and Weiblen (1971) using the technique of heating inclusion-bearing olivine grains in vacuum and observing the temperatures at which other mineral inclusions melt and homogenize (to be examined later as quenched glass). Melting temperatures obtained in this manner are equivalent to crystallization temperatures. Ilmenite inclusions disappear at 1210 (±5)°C, pyroxene melts at 1130 (±5)°C, and plagioclase is lost at 1103 (±3)°C.

In the opinion of many investigators, the observed variations in chemical composition, mineral proportions, and textures for mare lavas from the *Apollo 11, 12, 15,* and *17* sites are consistent with moderate to extensive magmatic differentiation. Most of the samples lying on the surface at each site probably were excavated locally from craters that penetrate the topmost flow unit (a distinguishable layer resulting from a single extrusion event) of unknown thickness. One or more additional units representing different stages or end products of fractional crystallization have been postulated to underlie the upper unit in order to account for compositional variations in samples from a given site. A smaller fraction of crystalline fragments, including those in microbreccias and regolith, are "exotic," that is, have been introduced to the site from distant sources.

According to some lunar samples investigators, the mare lavas are outpourings from multiple eruptions into large basins. In this respect they behave much like lava

*See Chapter 13 for a fuller treatment of this subject.

lakes that fill pit craters and calderas of the Hawaiian type. Cooling proceeds from the surface downward to form an early crust, but this commonly founders or is engulfed by later extrusions, so that new fluid materials reach the surface by means of turbulent or convective flow. Chemical variations can be pronounced because of mixing or solid-liquid reactions until the system stabilizes as a permanent crust develops. Continued cooling is accompanied by solidification of olivine, ilmenite, and pyroxene crystals that, being denser, sink to lower depths within the basins. Later, more pyroxenes begin to crystallize along with some of the feldspar. Owing to the rapidly changing chemistry within the lavas, individual crystals that remain in contact with the liquid experience zoning and partial replacement. The final liquids, now higher in potash and silica, occupy any interstitial space and harden into mesostasis minerals. In the course of cooling, solidifying crystals and liquid now separate further by crystal settling and flotation of heavier and lighter fractions respectively to form layers and pockets of compositionally distinctive segregates (*cumulates*) enriched in certain minerals. This "density separation" could lead to local stratification of various differentiated lithologic types within the mare lavas and, on a larger scale, might account for anorthosites, norites, and the presence of layering in the original lunar crust.

A variant of this considers the mare lavas to have completed most of their chemical changes deeper within the Moon. At some time before emplacement, melting began by partial fusion and differentiation of a parent "mantle" rock, not yet positively identified as a component within returned lunar samples. Extrusion into basins and lowlands could impose further chemical variations as described in the preceding paragraph. Most *Apollo 12* rocks, in particular, are interrelated to a single differentiation event in which the products are contrasted mainly in the proportion of olivine to other major minerals.

Thus, the main distinctions between these two differentiation schemes—near-surface versus depth—rests on the stages and conditions under which certain chemical changes proceed. These, in turn, relate to the time(s) when initial melting began or remelting took place and the interval required thereafter for differentiation and emplacement to be completed. The initial magma at depth may have been richer in alkali metals and other volatile elements or it could have been derived from parent materials originally depleted in these elements at the time the Moon formed. In the first instance, most of the volatiles mobilized during melting would likely be lost eventually by outgasing from lavas as these approached the surface. Higher percentages of sodium in the finer-crystalline basalts seem to support this postulate; rapid cooling that leads to smaller crystal sizes would entrap most of this element before it could escape. The end products, in any case, differ from terrestrial basalts in their much lower Na_2O and higher FeO contents. However, when the effect of the abnormally high TiO_2 content in *Apollo 11* samples is subtracted out of the total composition, the amount of SiO_2 increases relatively to reach values typical of many terrestrial basalts and also approaches that of *Apollo 14* crystalline rocks (SiO_2 ~46-51%, TiO_2 ~1.5%).

The terra and subcrustal (nonmare) rocks are also products of differentiation that began earlier (a few hundred million years after the Moon's accretion) and ended one to two hundred million years before the inception of mare lava emplacement. However, the volume of materials undergoing differentiation that eventually led to several moonwide outer layers of crystalline rocks must have been considerably greater than was involved in production of the mare lavas. The enormous scale and extent of the melting during this early stage no doubt affected much of the outer part of the Moon. Extreme differentiation proceeded long enough to bring about an efficient separation of plagioclase from other crystallizing, denser minerals. Flotation carried this plagioclase to the surficial region where it accumulated as anorthositic rocks. Somewhat later, further melting and additional differentiation fostered separation of the KREEP-rich norites and troctolites from KREEP-poor norites. These two groups were effectively rearranged into distinct subcrustal layers. From phase equilibria considerations, it appears that the KREEP-rich rocks represent the lower-melting fraction of a more primitive subcrustal material. Partial fusion of an anorthosite-capped outer crust could have mobilized this fraction upward to then solidify as shallow layers or as surface extrusions.

Much of the resulting layered structure consisting of these ANT rocks no doubt remains intact at depth (as suggested by geophysical evidence). However, intense (cataclysmic) meteoroid bombardments following crustal development have destroyed the continuity of most of the outermost units. These now are found mainly as breccia fragments redistributed and thoroughly mixed within ejecta blanket and base surge deposits now covering the surface to varying depths of several kilometers or more. These widespread deposits seemingly preclude an opportunity to sample older, undisturbed crystalline layers anywhere in the lunar highlands by direct access.

Some Lunar Rocks of Special Interest

One crystalline rock from the *Apollo 12* group showed so many unusual differences from all other larger samples that it has merited extra attention from many investigators. Sample 12013 (Figures 11-1d and 11-20) is a heterogeneous agglomeration of light and dark phases distributed in patches and veins and mixed with rock fragments. The two darker phases, distinguished in hand specimen by shades of gray, contain orthopyroxenes and plagioclase (An_{80}-An_{96}), some ilmenite, minor alkali feldspar, and

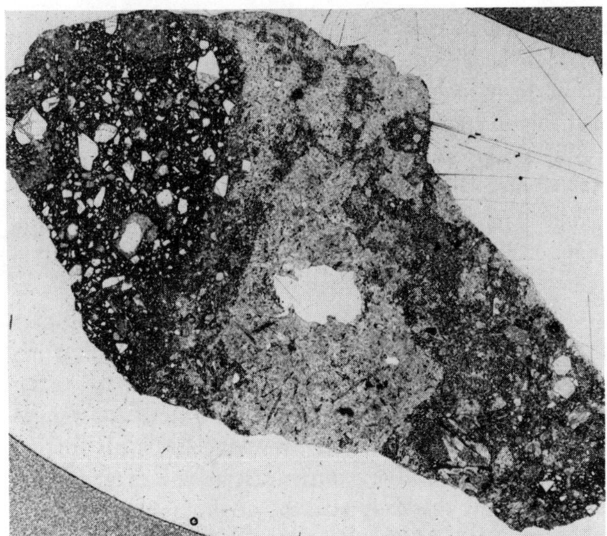

Figure 11-20 Surface view of a slab face cut through sample 12013 revealing a mixed breccia texture in which rock fragments within two darker phases intersperse with a lighter (felsitic) phase (mainly in the center of the specimen). Slab is 16 cm in maximum dimension (see Figure 11-1d). (NASA photo.)

small amounts of olivine and other minerals. These phases are thus similar to the norites. The lighter phase is characterized by alkali feldspar (typically, about 75% K, 17% Na, 5% Ba, and 2% Ca mole fractions), some plagioclase (An_{50}-An_{55}), and needlelike crystals of quartz as the dominant minerals, along with some pyroxene, ilmenite, apatite, and zircon as subordinate constituents. Both dark phases are made up of fragmental aggregates now in a fine-crystalline matrix. The light phase, termed a felsite (fine-crystalline equivalent of some silica and potash-rich igneous rocks) by James (1971), has a variolitic texture. Its spatial relation to the dark phases suggests that it was intruded into an earlier-existing breccia or possibly fragmental rubble. The silica content for the entire rock averages 61% by weight, but can run as high as 75% in the interstitial fillings in the darker phases; the potash (K_2O) content approaches 2%. Such elements as Ba, Li, Zr, Rb, Pb, and especially U are 10 to 40 times more abundant than in typical *Apollo 11* and *12* samples.

The origin of this peculiar rock has been hotly contested, but the consensus now holds that it arrived at the *Apollo 12* site as a thermally metamorphosed breccia fragment ejected from a crater that cut deep into the lunar crust. Copernicus is a prime candidate for the source event. This sample resembles the "suevite" from the Ries impact structure in Bavaria, in which fragmental material is intimately mixed with shock-melted glassy to partially recrystallized "igneous-looking" phases. The light phase in sample 12013 is, so far, the most silicic or "granitelike" large rock material brought back from the Moon. Its composition is broadly similar to the mesostasis of the crystalline lavas. This lithologic type may, of course, be a locally anomalous variant of this mesostasis, but, alternatively, it may comprise a sample of a more general surface or subcrustal layer—possibly present in mixed ejecta debris from the Imbrium event that was reejected from Copernicus or elsewhere.

Several of the crystalline rocks returned from the *Apollo 14* and *15* missions have also been given special consideration because of certain singular characteristics. Sample 14310 (Figure 11-1e) is a fine-crystalline, high-alumina basalt consisting of calcic plagioclase and pyroxenes (augite, pigeonite, orthopyroxene) in a ratio of 2 to 1. This rock also contains numerous small, finer-grained inclusions. One view treats these inclusions as foreign rock bodies or *xenoliths* that represent broken-up surface crust engulfed during emplacement of a lava flow. Another considers the inclusions to be ejecta fragments that became mixed with a now-crystallized shock melt. Sample 15415 (Figure 11-1g) was called the "Genesis Rock" by the astronauts when first found at the base of the Apennine Front because it appeared to be an anorthosite, hopefully from the very old original lunar crust. The rock was exposed as a large clast perched on a block of breccia. After return to Earth, analysis showed it to indeed be composed almost entirely of a calcium-rich plagioclase (An_{96}) that makes up almost 97% of the rock by volume. Although very old (see p. 237), this anorthosite is still a half-billion years younger than the accepted time of formation of the Moon. Sample 15426 (Figure 11-3h) was nicknamed the "Green Rock" as it was being collected. When later examined on Earth, the rock was, in fact, a yellowish-green (similar in color to bottle glass) owing to a predominance of numerous glass spheres, many of which have devitirified. This rock has a normative mineralogy (see footnote, p. 197) of ~45% pyroxene, ~30% olivine, and ~20% plagioclase. The rock and similar fragments from the soil may represent ejecta from the large craters Aristillus and/or Autolycus; if so, it is probably a quenched phase from a deeper layer excavated by these cratering events. Rock 15555 is one of the first lunar lava samples whose location, in a flow unit exposed on the east edge of Hadley Rille, has been pinpointed and stratigraphically fixed. This rock is a prophyritic olivine basalt much like the mare lavas brought back from the *Apollo 12* site.

Further discussion of possible types of parent rock for the lunar lava rocks, their relation to the anorthositic gabbros, norites, KREEP basalts, and felsites, and suggested models for producing layered crystalline rock differentiates will be deferred until Chapter 13.

Composition of the Regolith

When typical lunar soil material is passed through a train of sieves of different mesh sizes, the amounts (in weight percentage) collected on these sieves can be expressed as cumulative frequency plots of particle size distributions

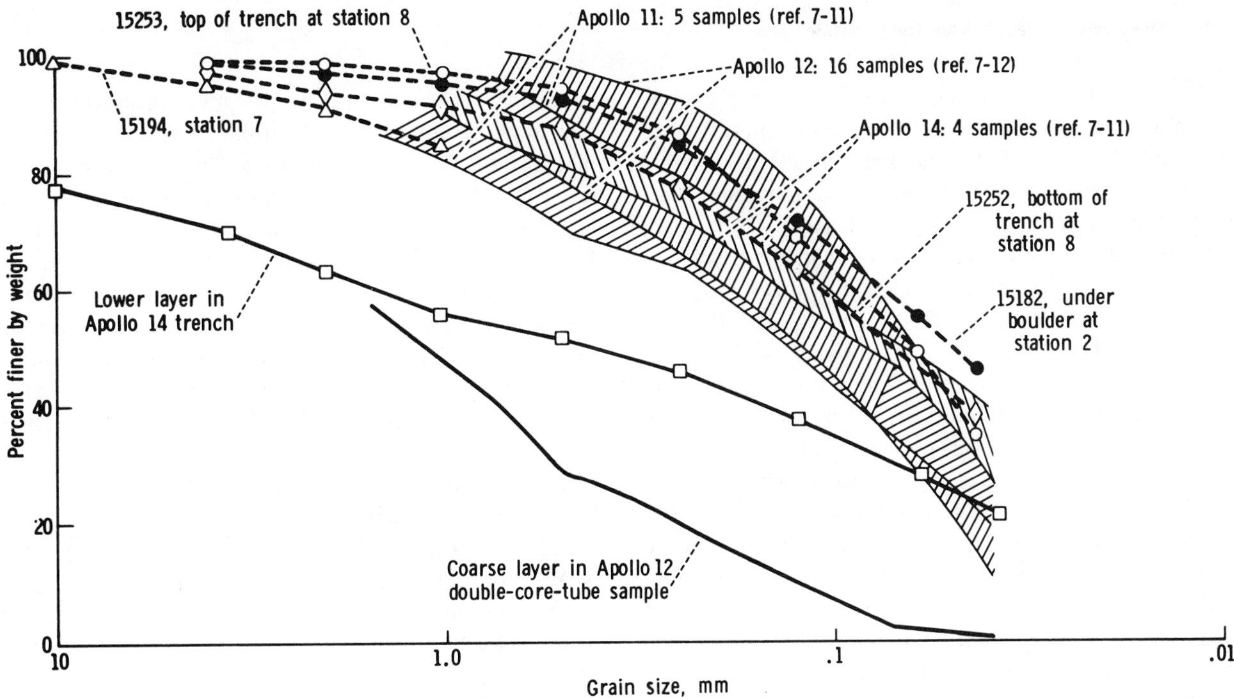

Figure 11-21 A series of curves showing the distribution of grain (particle) sizes in soils from the *Apollo 11, 12, 14,* and *15* sites. The variations found in multiple samples from a given site are indicated by boundary envelopes with different cross-hatching patterns for *Apollos 11, 12,* and *14*. *Apollo 15* size distributions, obtained from single samples, appear as dashed lines. (From *Apollo 15* Preliminary Science Report; NASA SP-289, 1972, p. 7-4.)

Figure 11-22 A grouping of coarser particles separated from *Apollo 11* soil by sieving. The large particle on the far right is a coherent soil breccia fragment. The light particle to its left is shocked anorthosite. Coarse-grained basalt lies just above it. Below and left of this basalt is a glass-spattered rock fragment. Glass spherules are visible near the bottom of the photo (the one on the farthest left is 1 mm in diameter). Angular chips of dark glass are evident near the edge of the picture. (Courtesy J. R. Wood, Smithsonian Astrophysical Observatory.)

(Figure 11-21). These curves define the degree of comminution and mixing characteristic of the regolith history at each site. Most of the "coarse fines" fraction separated by sieving consist of rock fragments, agglutinates (fragments and grains bound by a glass "cement"), and other glassy particles (Figure 11-22). The fragments are closely related in lithologies and textures to the larger rock samples. The fragments in *Apollo 11* and *12* regolithic material are identified as mostly mare basalts and breccias coming from nearby sources, but about 5% are anorthositic gabbros and anorthosites carried to these sites from distant sources (Figure 11-23). The regolith at *Apollo 12* contains varying amounts of the following rock fragments: KREEP-rich basalts 28-68%; mare basalts 26-67%; anorthosites 4-9%; a siliceous component 1-3%; meteorite material 0-1.5%. Marvin et al. (1971) estimate that half of these fragments were derived from a circular area within 3 km of the landing point, that the next quarter came from the area between the inner circle and a circle of 19 km radius, and that the remaining one-quarter was brought in from distant sources in the area between circles of 19 and 500 km radii. Coarser fragments in the *Apollo 14* soils are mainly KREEP-rich norites and anorthositic rocks similar to rock fragments in the breccias of the Fra Mauro Formation (Imbrium ejecta) collected at that site.

The fine sieve fractions (<1 mm) are composed of tiny rock fragments and individual mineral grains that obviously are comminution products from both the lunar crystalline rocks and breccias. The mean (average) particle size in the combined fine fractions of different soil samples from the *Apollo 12* site ranges from ~52 micrometers (μm) to 108 μm; the range limits in the *Apollo 11* fines are slightly smaller. Mean grain sizes for *Apollo 15* soil fines from 18 sampling locations fall between 42 and 98 μm. The coarsest regolith is found at the *Apollo 14* site where mean sizes ranging from 75 μm to 612 μm (averaging 157 μm) in 12 samples have been measured. Thus, the bulk of the regolithic fines corresponds to coarse silt to medium-fine sand in a standard soil particle size classification.

For a single vertical profile, plots of particle size distribution within distinct layers in the stratified double core sample from the *Apollo 12* site (Figure 11-24) show similar means and degrees of sorting in most layers. This is to be expected in parts of a regolith subject to extensive reworking by "gardening." Layer VI, however, has a very different size distribution, with a preponderance of coarse, lithologically similar fragments. This layer may be an interspersed ejecta or base surge unit, probably from a nearby cratering event, that subsequently escaped turnover.

Lunar Glasses

The most unusual feature in the regolith is the presence of large amounts of glass. This glass occurs as spatter or glazings on fragments, agglutinates, discrete vesicular

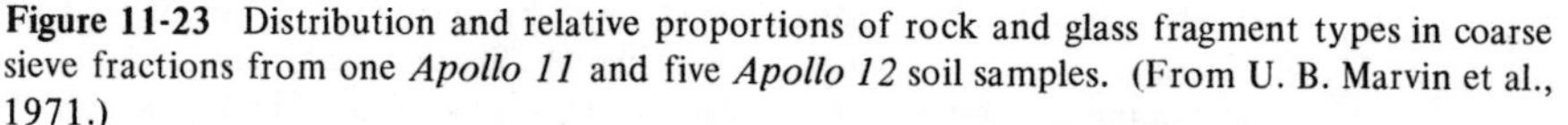

Figure 11-23 Distribution and relative proportions of rock and glass fragment types in coarse sieve fractions from one *Apollo 11* and five *Apollo 12* soil samples. (From U. B. Marvin et al., 1971.)

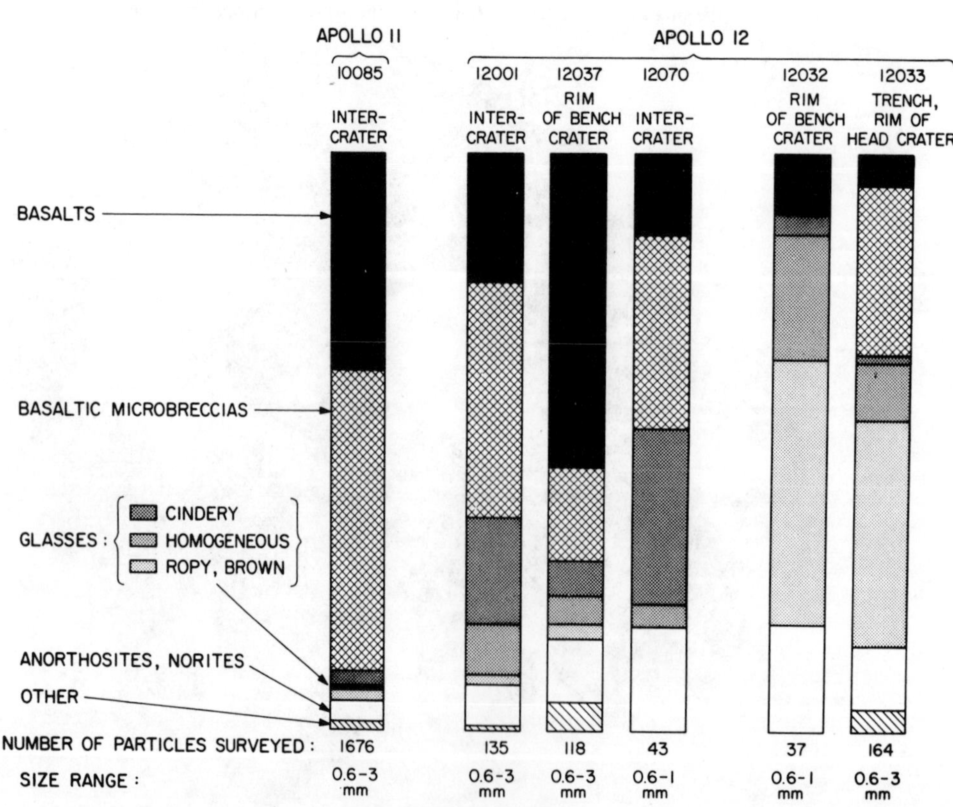

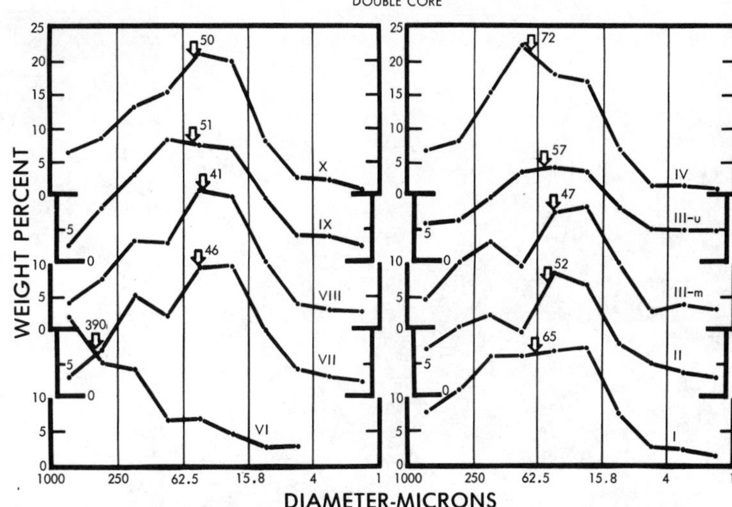

Figure 11-24 Distribution of particle sizes in terms of weight percent in each grain size interval (determined by weighing sieve fractions) for each of the 10 stratigraphic units recognized in the double core tube samples from the *Apollo 12* site (see Figure 10-16). The arrows point to the median grain size in each unit. (From Sellers, et al., 1971.)

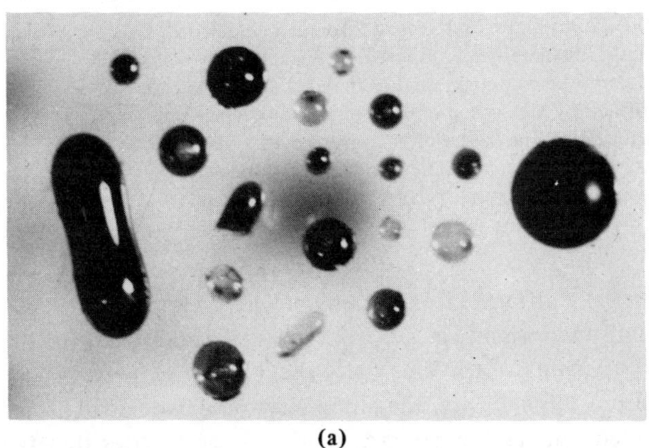

(a)

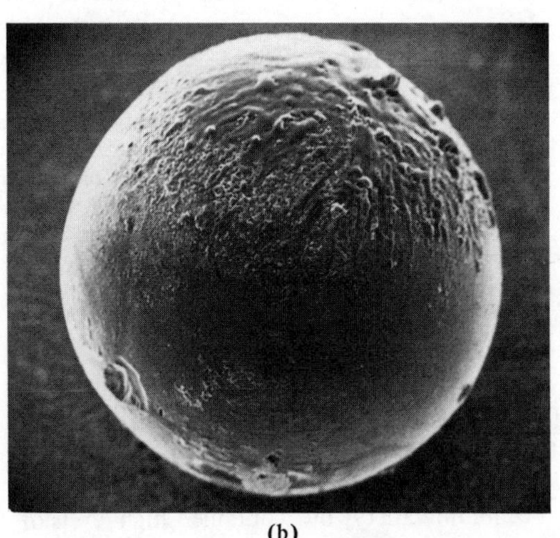

(b)

Figure 11-25 (a) Glass beads separated from lunar soil, showing a variety of shapes and colors (courtesy C. Klein, Jr. and C. Frondel). (b) A glass spheroid (yellow-brown; about 0.5 mm in diameter) from the *Apollo 11* soil, marked externally with spatter ridges and micropits (courtesy B. Glass, University of Delaware).

blebs of irregular shape, as stringers and shards, and as regular-shaped bodies. This last category (Figure 11-25a) consists of spheres, ellipsoids, tear-drops, and dumbbells (shape analogs) of clear to opaque glass whose color varieties include colorless, green, gray, yellow, orange-brown, red-brown, and wine-red. Most of the rounded forms are less than 1 mm in diameter (Figure 11-25b). Chemically, these glasses show a wide range in composition in which FeO and MgO contents vary at least as much as the extremes shown by the crystalline rocks. The compositions are thus similar to those expected from direct melting (with some loss of volatiles?) of different lava or crustal rocks or, uncommonly, of the individual mineral phases of pyroxene and plagioclase (usually contaminated with small amounts of iron and magnesium). The color varieties are closely tied to element percentages (e.g., wine-red indicates TiO_2 contents of 30% or more; yellow-brown, high FeO, with TiO_2 content down to 15-20%). The green glass spheres found in samples 15426 and 15923 and in most *Apollo 15* soil samples, different chemically from most other glasses returned from the Moon, have especially

Figure 11-26 A ternary plot of alkali and MgO contents of four groups of tektites, terrestrial volcanic glasses, and lunar glasses extracted from the soils. With the exception of a small fraction of the Australasian tektites, the compositional fields of tektites do not coincide with the field defined by the lunar glasses. (Courtesy B. Glass, University of Delaware.)

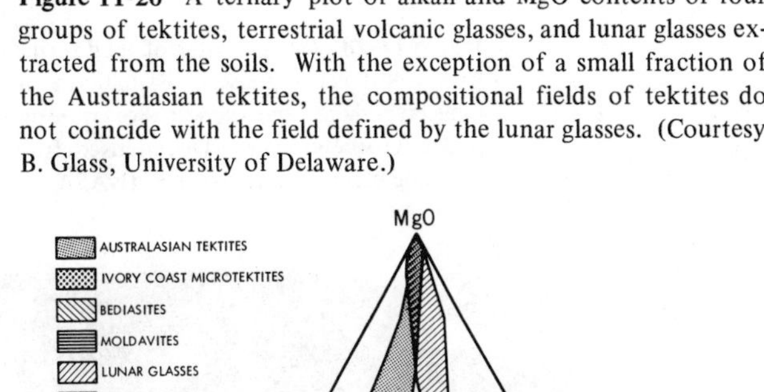

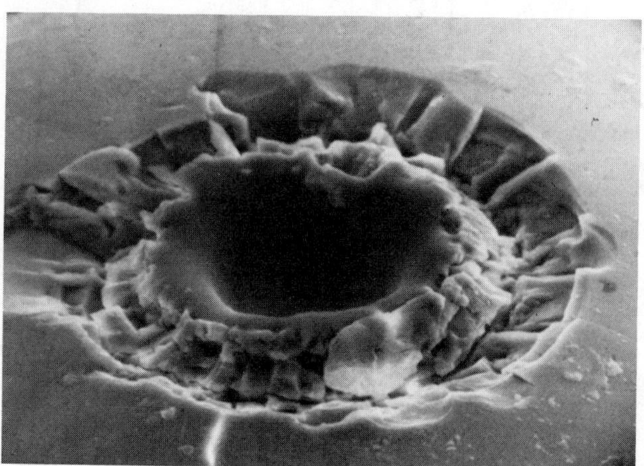
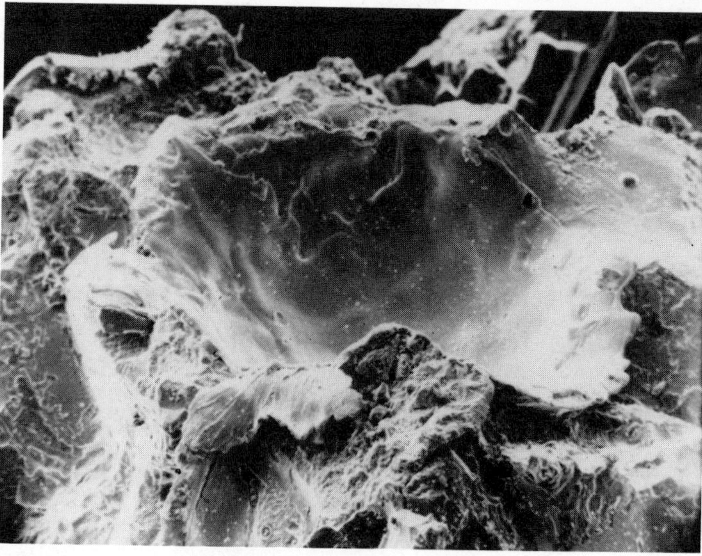

(a) (b)

Figure 11-27 (a) A high magnification (X 1000) photomicrograph (made through an electron microscope) showing a "zap" crater some 200 μm in diameter superimposed on the surface of a glass sphere. Note the inner spatter rim with fluted septa running radial to it and the irregular, chipped outer rim. Zap craters or micropits are formed by high-speed impacts of micrometeorites on rock surfaces. The region immediately surrounding the craterlet is converted to a glassy state. Nickel-iron spheres and blebs (remnants of metallic micrometeorites) are sometimes found lining the cavity and mixed with glass (courtesy J. Hartung, NASA, Johnson Spacecraft Center). (b) Another zap crater formed directly on a crystalline rock surface (courtesy C. Klein, Jr., Harvard University.)

high MgO contents and contain tiny crystals of orthopyroxenes. When the orange soil collected at Shorty Crater during the *Apollo 17* excursion was sieved and examined at the Lunar Receiving Lab at Houston, it was found to consist largely of spheres and irregular fragments of orange-brown to burnt umber glass. This glass was a typical high-titanium variety, but contained high levels of such volatile elements as zinc (up to 100 ppm—a significant enrichment compared with the average Zn contents of other lunar glasses or rocks) even though depleted in sodium and potassium.

The chemical compositions of lunar glasses are characteristically different from terrestrial igneous (volcanic) glasses and, more significantly, from tektites associated with the four main strewnfields (p. 44) on Earth (Figure 11-26). From this evidence, the Moon must be downgraded, if not ruled out, as the primary source of the tektites that fall on Earth.

Examination of lunar glass "beads" at high magnifications reveals them to have pits, grooves, spatter ridges, and Ni-Fe metal coatings. Some pits are elaborately sculptured, with central bowl-shaped craters, rims, and flanges and radiating septa (Figure 11-27a). Similar pits and cavities are noted on the surfaces of lunar rocks that have been ex-

Figure 11-28 (a) A cluster of well-formed (euhedral) crystals that appear to have precipitated from a hot vapor phase in a small vug or cavity within a breccia from the *Apollo 14* site. The long crystals are pyroxenes; shorter, stubby ones are plagioclase. (b) Iron crystals (the largest is 3 micrometers across) condensed from vapors. These photographs were taken through a scanning electron microscope. (NASA photos.)

(a) (b)

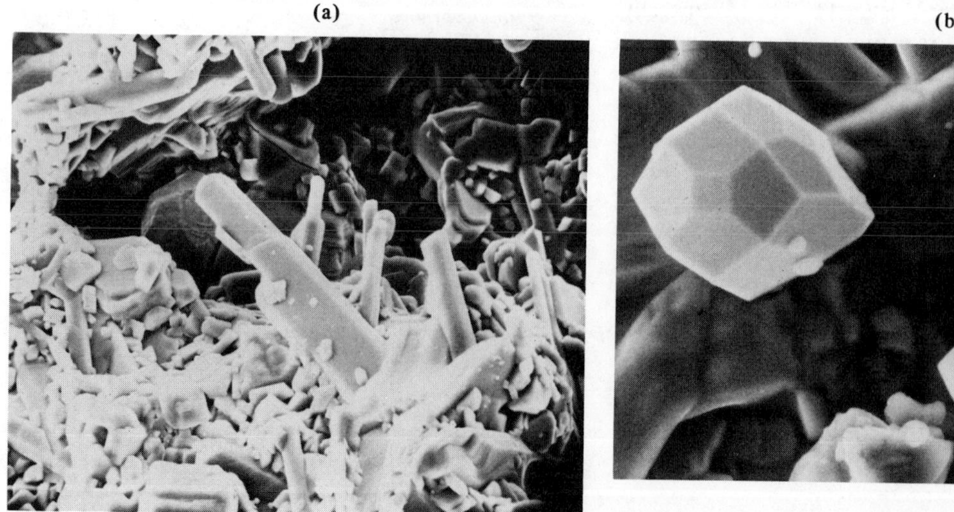

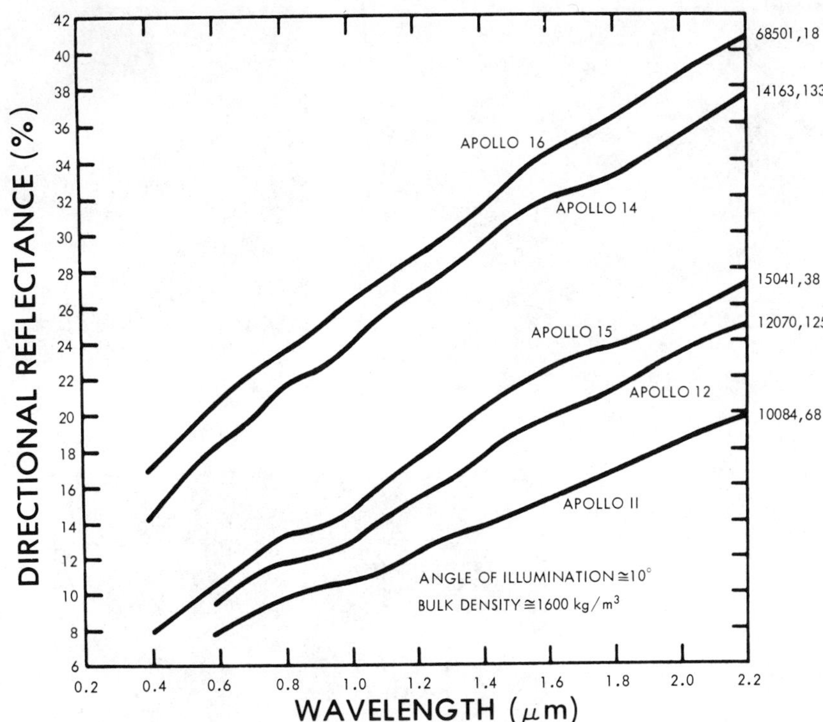

Figure 11-29 The spectral reflectances of the fine fraction, normalized to a nearly constant density, of soils from five *Apollo* sites. Measurements were made in the laboratory at a single angle of illumination. (From R. Birkebak and J. Dawson, 1973, courtesy Lunar Science Institute.)

posed at the surface (Figure 11-27b). Such markings, associated as they are with droplets and splashes of meteoritic iron, are clearly established as miniature impact craters (many being only a few micrometers in diameter) caused by micrometeorites striking the glass surfaces after they had cooled and accumulated as particles in the regolith. Those micrometeorites that hit glass or rock surfaces at low velocities leave shallow craters without rims, whereas those at high velocities produce deeper, rimmed craters and much glass splash.

The lunar glasses are inferred to result from shock-melting and vaporization (at pressures above ~500 kbars) of rocks and soil by impacts at all scales. Thus, as large (meters to kilometers) craters are formed, up to ~2% of the total crater volume is converted to silicate vapors and melt splash that are distributed in the ejecta and base surge deposits. Part of the vapor and melt forms globules in flight. The remainder collides with rock fragments and sticks as crusts and coatings or cements these particles together. Some of the vaporized rock will actually condense into crystalline silicate and metal phases (Figure 11-28a and b). Still other glass develops in place by micrometeorite impacts that shock-melt the material immediately around the pits and craterlets on rocks and rock fragments. A higher proportion of the surroundings is melted and vaporized in these smaller craters.

Glass coatings deposited from shock-induced vapor phases and micrometeorite impacts lead to reduction of the intrinsic ability of particles in the lunar regolith to reflect light; this results in a general lowering of albedo. Variations both in overall composition and in glass content are responsible, in part, for differences in specific reflectance from one *Apollo* site to another (Figure 11-29). Glasses in the regolith tend to darken because of loss of volatile elements during shock-heating, with a corresponding rise in residual proportions of the more refractory Fe and Ti compounds (often occurring as microcrystallites) that absorb and scatter light. The solar wind can also convert crystalline material at particle surfaces to a glassy state by disturbing orderly atomic arrangements and disrupting bonds. Adams and McCord, and Hapke point to development of thin films of dark glass, colored by dispersed Fe and Ti, on particles and agglutinates on the lunar surface as a possible cause of the lower albedos observed in the uppermost regolith layers.

The amount of glass present in the regolith is an index of its maturity (total exposure time and extent of reworking). The "youngest" soil, around Cone Crater at *Apollo 14*, contains less than 10% glass. In order of increasing maturity: *Apollo 12* soils contain ~20% glass; *Apollo 11* soils have glass contents around 20 to 30%; and *Apollo 14* soils near the LM show glass contents between 40 and 75%. Results from *Apollo 15* soils indicate glass contents to range between 35 and 87%, largely depending on location of the samples relative to the Apennine Front. *Apollo 16* soils contain markedly variable amounts of glass particles and agglutinates, ranging between 25 and 85%. Loss of glassy coatings by abrasive "scrubbing" from dust impacts and/or radiation bleaching causes the lightening of albedo observed in the topmost "skin" layer of the regolith.

Shock Metamorphism in Lunar Materials

Other shock effects are conspicuous in individual fragments and grains in the soil, the breccias, and some crystalline rock samples. These effects include planar features in feldspar and silica phases (Figure 11-30a), thetomorphs (Figure 11-30b), incipient melting (Figure 11-30c), quench

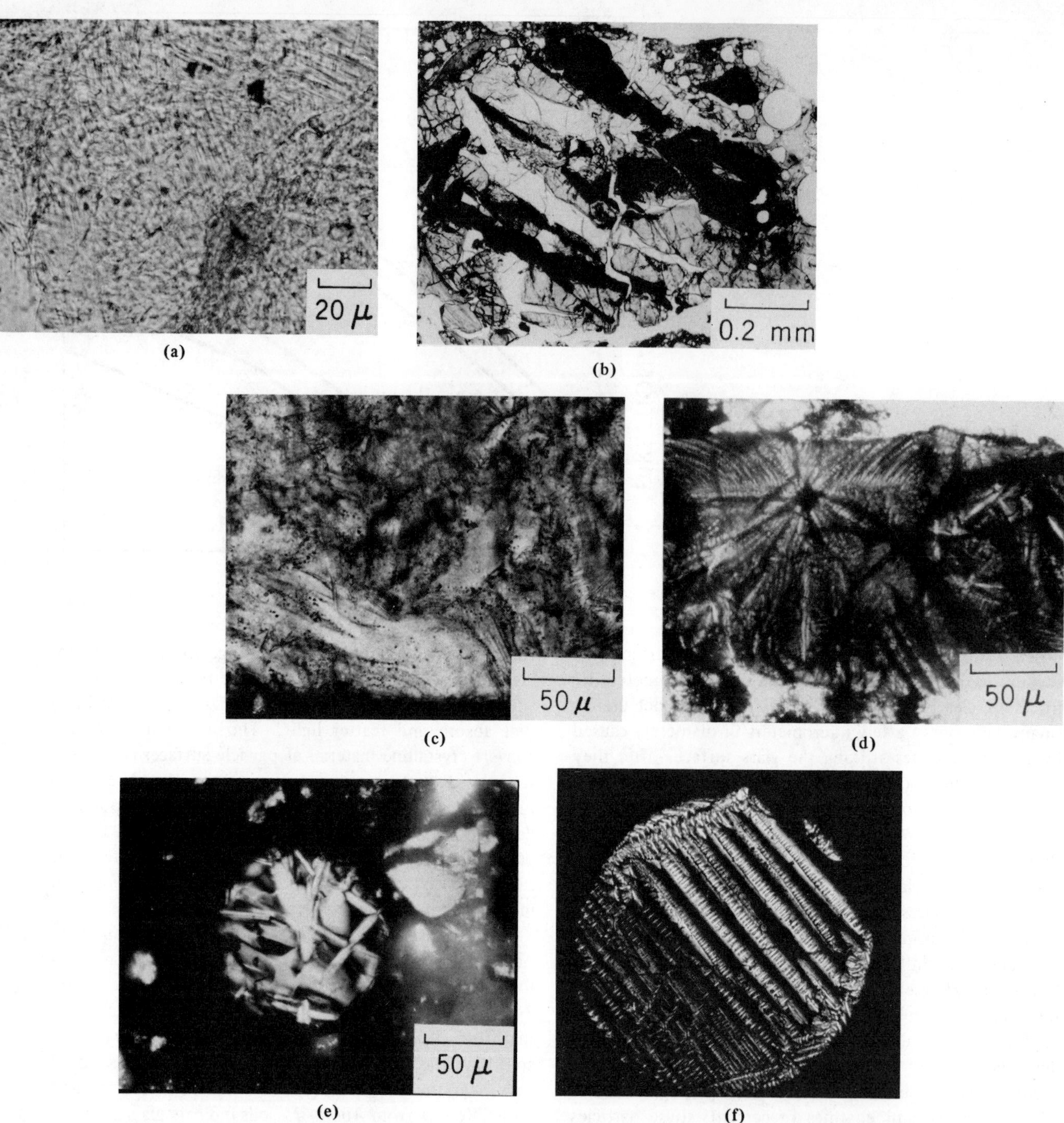

Figure 11-30 (a) Planar features in a silica (tridymite?) grain from microbreccia sample 10060. (b) A crystalline rock fragment in microbreccia sample 10065 showing typical ophitic texture; however, the feldspar (light gray) is now shock-isotropized (thetomorphs) to maskelynite and the pyroxene (darker gray) contains numerous tiny fractures. (c) A partly fused crystalline rock fragment in microbreccia 10060. The plagioclase has begun to melt (lower left), but the pyroxene is still crystalline, although highly shocked. (d) A radiating cluster of quench crystals of pyroxene that grew during rapid cooling of the surrounding glass (microbreccia sample 10023). (e) A matte of intergrown feldspar laths formed by recrystallization of a clear glass spherule of plagioclase composition (microbreccia sample 10065). (f) Photomicrograph of a barred olivine "chondrule," one of many found in soil sample 14259. The sphere, seen here in thin section under crossed nicol prisms, is 0.2 mm in diameter. The lunar "chondrules," which bear a strong resemblance to those characteristic of chondritic meteorites, may have developed by rapid quench crystallization of glass spheres. (Figures a through e from Short, 1971; f from Nelen et al., 1972.)

ESTIMATED SHOCK PRESSURE
PHASE
0
200
400
600
(kb)
Cristobalite (/Quartz)
Plagioclase
Pyroxene
Olivine
Ilmenite
NORMAL
Fractured
planar features ➧ glass
pl. f's ➧ maskelynite
planar f's
planar f's
lamellae
MIXED GLASSES
HOMOGENEOUS GLASSES

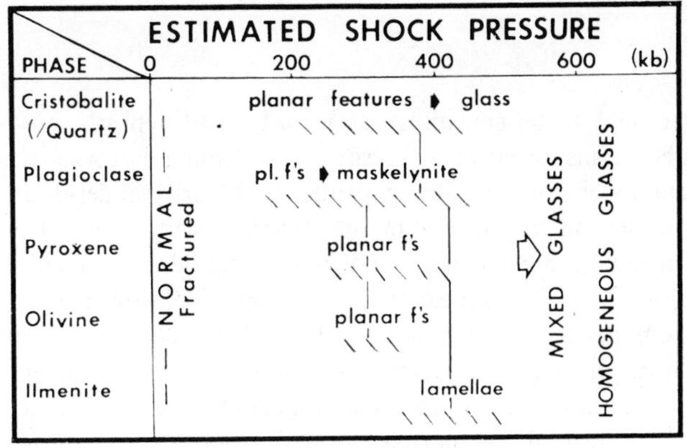

Figure 11-31 A scale of shock metamorphism (with estimated peak pressures) applied to the principal minerals in *Apollo 11* rocks, together with various characteristic shock features formed over the pressure intervals indicated. (Reprinted with permission from M. R. Dence, et al., *Proceedings of the Apollo 11 Lunar Science Conference* [A. A. Levinson, ed.], 1970, Pergamon Press.)

crystallization (Figure 11-30d), and later recrystallization of rapidly cooled glasses (Figure 11-30e) (Short, 1971). Drawing upon studies of shocked terrestrial rocks, Dence et al. (1970) have devised a general scale of shock damage for the lunar rocks (Figure 11-31).

Certain soil and breccia samples contain spheres that have recrystallized into Mg-rich olivine whose texture is remarkably like that of the barred olivine chondrules (Figure 3-2b) found in many ordinary chondrites. The spheres were first noted in *Apollo 14* materials, but are now recognized in samples brought back from other sites. These unusual spherules (Figure 11-30f) are thought to have formed as liquid droplets carried along in base surge clouds, such as the one accompanying the Imbrium Basin event now preserved in *Apollo 14* breccias. This observation, if proved, suggests that at least some of the chondrules in meteorites may have been produced by impacts within condensing dust clouds or on asteroidal bodies.

In principle, every rock fragment exposed near the surface at a sampling site could have been involved in one or more cratering events. But most lava bedrock excavated from a large crater will not be subjected to pressures high enough to impose permanent changes. Therefore, only a few rocks should show definite evidence of shock-induced features. However, all the rubble which builds up within a dispersed ejecta blanket or the regolith on the cratered lunar surface is likely to be repeatedly bombarded by small meteorites until the debris is comminuted to a fine sand and powder mixed with local and exotic rock fragments. The differing proportions of glass in the regoliths from the several *Apollo* sites represent different stages of progressive buildup from recurring impacts, each one of which produces only a small quantity of glass for local distributions. French (1971) has traced through the complex history of rocks and minerals from the *Luna 16* regolith

Figure 11-32 A schematic diagram outlining the progressive development of shock metamorphic features in different grains and rock fragments from the *Luna 16* soil sample. (Courtesy B. M. French, NASA Goddard Space Flight Center.)

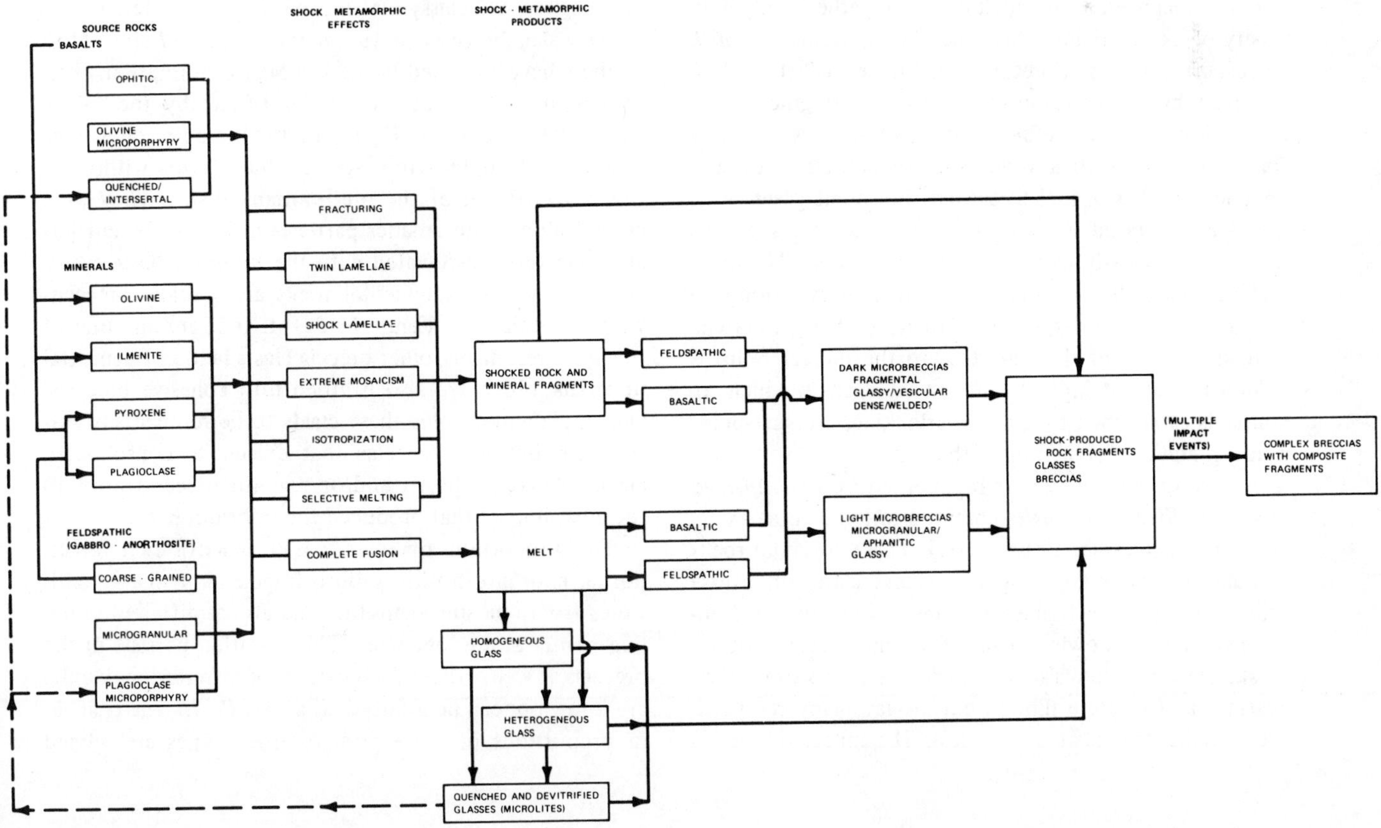

SOURCE ROCKS
BASALTS
OPHITIC
OLIVINE MICROPORPHYRY
QUENCHED/ INTERSERTAL
MINERALS
OLIVINE
ILMENITE
PYROXENE
PLAGIOCLASE
FELDSPATHIC (GABBRO - ANORTHOSITE)
COARSE - GRAINED
MICROGRANULAR
PLAGIOCLASE MICROPORPHYRY
SHOCK METAMORPHIC EFFECTS
FRACTURING
TWIN LAMELLAE
SHOCK LAMELLAE
EXTREME MOSAICISM
ISOTROPIZATION
SELECTIVE MELTING
COMPLETE FUSION
SHOCK - METAMORPHIC PRODUCTS
SHOCKED ROCK AND MINERAL FRAGMENTS
MELT
FELDSPATHIC
BASALTIC
BASALTIC
FELDSPATHIC
HOMOGENEOUS GLASS
HETEROGENEOUS GLASS
QUENCHED AND DEVITRIFIED GLASSES (MICROLITES)
DARK MICROBRECCIAS FRAGMENTAL GLASSY/VESICULAR DENSE/WELDED?
LIGHT MICROBRECCIAS MICROGRANULAR/ APHANITIC GLASSY
SHOCK-PRODUCED ROCK FRAGMENTS GLASSES BRECCIAS
(MULTIPLE IMPACT EVENTS)
COMPLEX BRECCIAS WITH COMPOSITE FRAGMENTS

as these are subjected to continuing shock metamorphism in multiple impact events (Figure 11-32).

Origin of the Regolith

The various lines of evidence described in the preceding sections taken together argue convincingly for the hypothesis that the regolith is generated by impacts between extralunar bodies and materials at the lunar surface. In sum, this evidence includes:

(1) Incorporation of fragments of underlying basalt (lava rock) units
(2) Small mean particle sizes ($<60\ \mu m$) and correlation of sorting with extent of turnover
(3) Wide range of expected shock features
(4) Shock-lithified breccia fragments (see below)
(5) Glass spatter from shock melting
(6) Occurrence of nickel-iron meteorite debris
(7) Presence of microcraters on rock fragments and glass sphere surfaces

Lunar Breccias

The microbreccias, in particular, are also genetically tied to the effects of shock-generating impacts. The degree of induration is controlled in part by the amount of glass binder (ranging from 10 to 50% by volume, as seen under the microscope) in the fine-grained matrix (see Figure 11-3). Variations in glass content, degree of cohesion, nature of incorporated clasts, textural fabric, and distribution and magnitudes of shock effects together imply a diversity of likely origins. One general type, termed *regolith breccias*—or, also, soil breccias (see Figure 11-3d)—is characterized by a high proportion of glassy fragments and cinders, lack of a well-defined matrix, weak cohesion (more friable and porous), a wide range of particle sizes, and notable amounts of nickel, carbon, and noble gases. This group is generated by impacts into preexisting regoliths that contain mostly locally derived fragments. The loose regolith materials are converted by shock-lithification into "instant" rocks. Thus, during formation of larger craters, a small fraction of the soil close to the impact point is compressed and compacted into an aggregate in which cohesion is established by grain interlocking, pressure-point welding, and partial melting of the matrix.

A second general type is represented by the *annealed* or *recrystallized breccias* (sometimes referred to as *ejecta blanket breccias*) (see Figure 11-1f). These polymict rocks contain numerous larger light and dark clasts of diverse lithologies, are less friable, and are marked by an even-grained matrix showing signs of varying degrees of recrystallization. They also have little or no glass other than coatings and fracture fillings, but display many recrystallized fragments that were once glass. The annealed breccias are held to be consolidated ejecta blanket deposits, possibly transported as base surges containing hot vapors. The extent of annealing depends on the original depth of the breccia sample within the deposit. With progressive annealing, both matrix and fragmental glass gradually diminish and disappear while the matrix texture changes from detrital to crystalline with well-formed plagioclase laths. In many respects, this type of breccia resembles, both in mode of emplacement and manner of lithification, the terrestrial ignimbrites formed from volcanic *nuées ardentes*.

Regolith breccias are the more common type at the *Apollo 11, 12,* and *15* sites. Not all components in these are from local sources, as indicated by inclusions of other breccias whose textural fabrics are not due to shock induration, by absence of glass spatter in some breccias, and by occurrence of exotic rock types. These foreign components are added as ejecta and base surge materials from impacts beyond the immediate *Apollo* sites, as well as from volcanic deposits of pyroclastic nature.

Petrologically and chemically both the regolith and regolith breccias are not equivalent simply to the weighted averages of various proportions of minerals and elements in the local crystalline lava rocks. In general, these fragmental rocks are impoverished in Al and contain excess Mg with respect to the lava rocks. Elements like Zn, Ag, Cd, and Bi are enriched relative to averages in the crystalline rocks. These and other chemical differences can be achieved by adding only 1 to 2% chondritic and Ni-Fe meteorites, along with small amounts of anorthosite, norite, and KREEP components as rock fragments, yellow-brown glass, and glassy matrix, to local soils materials.

Annealed breccias prevail at the *Apollo 14* site. Most of these have been tied to the Fra Mauro Formation. This formation, typified at the *Apollo 14* site by the "black and white" rocks (see Figure 10-17b), incorporates large fragments disrupted from several thick layers within the top 5 to 10 km of the pre-Imbrian crust. They were carried along with smaller particles in hot, turbulent jets and base surges associated with the Imbrium Basin event. Some *Apollo 14* fragmental rocks are actually breccias-within-breccias (see Figure 11-3e)—that is, are mixtures of smaller, presumably older breccia clasts held in fragmental materials which are also lithified into cohesive breccias. One explanation holds these clasts to be remnants of surficial pre-Imbrium breccias that existed as consolidated ejecta blanket deposits and/or regolith breccias prior to the great impact that produced the Imbrium Basin.

The proportion of mare basalt, KREEP-rich basalt, norite, and anorthositic gabbro fragments in the regolith varies from one site to another and also at different sampling points at any one site. This variation persists in the breccias as well. Where the locations of different rock units are known or can be inferred, as at *Apollo 15*, the changes in proportions of these principal rock types are related

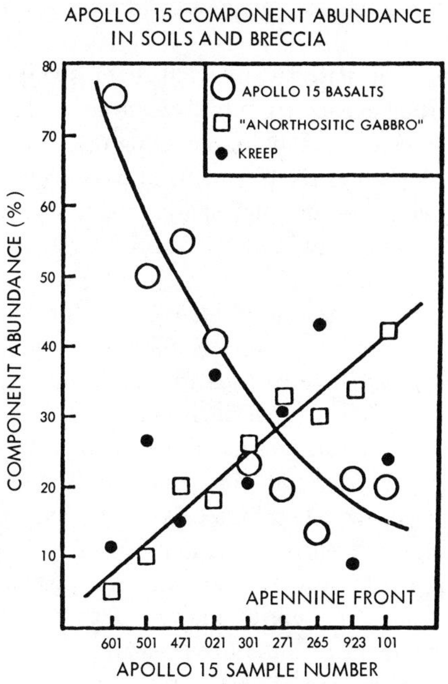

Figure 11-33 Plot of the relative abundances of three principal rock types present as fragments in soil samples (indicated by last three numbers of the 15000 series) collected at various distances from the Apennine Front (closest sample point is 101) during an *Apollo 15* EVA. Both gabbros and KREEP basalt fragments increase as the Front is approached, suggesting that these rock types outcrop along the Front and slide downward into the nearby regolith. At increasing distances from the Front, mare basalt flows from the vicinity of Hadley Rille contribute progressively greater amounts of that rock type to the soils. (From Schonfeld and Meyer, 1972.)

systematically to proximity of a regolith breccia to nearby source areas (Figure 11-33). As the Apennine Front is approached and mare lava rocks exposed near Hadley Rille become more distant, the proportion of anorthositic rocks increases at the expense of basaltic rocks. KREEP-rich rocks are more broadly distributed but appear to follow the trends pursued by anorthosites derived from the Apennine Front. Breccias collected near the Apennine Front are dominantly KREEP rock fragments held in a brown glass matrix; these rocks are derived by lithification of regolithic debris carried downslope.

Breccias having similarities to annealed types predominate at the *Apollo 16* site, but some are derived by lithification of a heterogeneous regolith. In hand specimen, four combinations of clasts and matrix can be discerned: (1) light-colored (whitish) clasts in a white to gray crushed-rock matrix, (2) light clasts in a dark matrix, (3) dark clasts in a light matrix, and (4) light and dark clasts in a dark matrix rich in plagioclase. Most light clasts are of anorthosite and anorthositic gabbros. The dark clasts include rock fragments with higher proportions of olivine, pyroxene, and glass spatter. The matrix is mainly ground-up or remelted anorthositic material with variable amounts of glass. Petrographic examination has led to recognition of three major rock groups.*

(1) *Polymict breccias (type I in the* Apollo 16 *Preliminary Investigation Team classification):* Contain rock clasts (up to a few centimeters in width) of anorthositic, troctolitic, or noritic composition (ANT suite) and glassy fragments set in a matrix of crushed materials. One group of polymict breccias contain poikiloblasts–large (up to 2 mm) crystals of pyroxene which enclose small plagioclase and olivine crystals–that grew during subsequent thermal metamorphism of the breccias.

(2) *Cataclastic anorthosites (type II)*: Consist predominantly of clasts of crushed and shattered anorthosites and troctolites interspersed with annealed zones whose rock character is nearly identical to the clasts except for stringers of thetomorphic and devitrified glass; plagioclase comprises as much as 70% of the minerals present.

(3) *Partially molten breccias (type IV)*: Composed of white to gray clasts and matrix *mixed in varying proportions*; some clasts show degrees of partial melting or recrystallization; the matrix contains finer fragments held by glass that commonly has recrystallized into a matte of interlocking plagioclase needles; some specimens show large tubular vesicles.

Type II and IV rocks are essentially monomict breccias (clasts of uniform lithology) that form an intergrading continuous suite ranging from melt-free to incompletely melted specimens of a common parent material. They assumed their present character from–and are transported by–impact-related mechanisms. The rocks may come from original source areas in the adjacent highlands that consist principally of a coarse-grained igneous complex of anorthosites, troctolites, and some feldspar-rich gabbros. If these are representative of the regional lithologies in the frontside terra plains and mountains, then the older outer crust is dominated by calcic feldspars with variable amounts of mafic minerals and spinels. The type I rocks, by contrast, are polymict breccias (clasts of different lithologies) derived by induration of regolithic materials that include 10-20% KREEP basalt components introduced during basin-forming events. Some type III rocks may be recrystallized impact melts that formed in pools on crater floors or walls.

The *Apollo 17* breccias are broadly similar to those of *Apollo 16*. The kinship among several major varieties of rock samples, including basalts, from these two sites is

*The type III specimens from the *Apollo 16* site are crystalline igneous rocks containing up to 70% plagioclase. A diabasic subgroup (plagioclase, orthopyroxene, olivine, spinel) is chemically similar to KREEP basalts. A poikolitic subgroup (pyroxene, plagioclase, ilmenite) shows signs of thermal metamorphism.

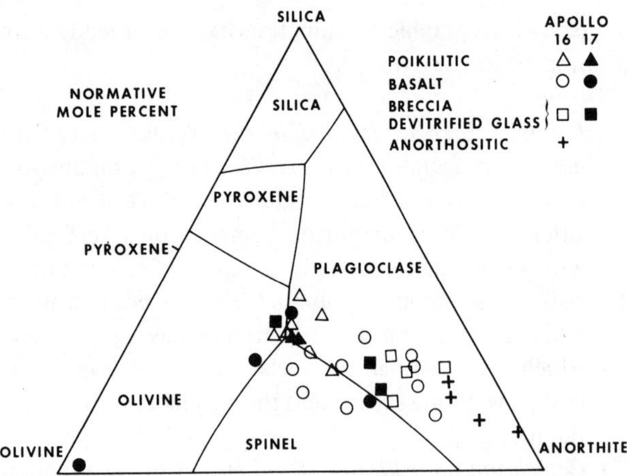

Figure 11-34 Pseudo-ternary phase diagram showing major mineral fields for lunar rock types characteristic of the highlands crust. Most rocks fall within the plagioclase field (containing varying normative mole percents of calcic plagioclase, pyroxene, olivine, and spinel). Other assemblages include plagioclase, spinel, and olivine (within spinel field), and pyroxene, olivine, and spinel (olivine field). The similarity of *Apollo 16* and *17* rocks is evident. (From Warner, Simonds, and Phinney, 1973.)

evident from a plot of their normative mineralogies on a ternary diagram (Figure 11-34). Many breccias are darker than the light-dark types found at *Apollo 16*. Six groups of *Apollo 17* breccias have been defined: (1) dark matrix breccias, most containing mainly basalt clasts; (2) large masses of glass-bonded agglutinates that bind fragments of dark matrix breccias; (3) vesicular, greenish-gray breccias (identified on-site by the astronauts as tan-gray breccias or anorthositic breccias) dominated by a matrix rich in poikolitic orthopyroxene; (4) foliated, layered light-gray breccias, with variable coherence, in which many clasts are composed of feldspathic cores mantled by a dark glass and mineral fragment rind; (5) blue-gray breccias, a complex assemblage of anorthositic clasts, mineral grains, and a fine-grained blue-gray matrix, all sometimes enclosed within coarser tan matrix material; and (6) crushed and brecciated gabbroic to anorthositic rocks, including norites and troctolites. Partially remelted breccias are much in evidence. Some breccia blocks are notably vesicular, with large, irregularly elongate vugs.

The Soviet Luna Landers

In September 1970, the Soviet automated probe *Luna 16* landed in Mare Fecunditatis on the eastern edge of the Moon (see Figure 1-3) and returned with a 103-g sample of lunar soil obtained by drilling 35 cm into the lunar regolith. Fecunditatis is one of the oldest ringed basins on the Moon, but the mare material that fills it is probably underlain by layers of ejecta from somewhat younger basins. In addition, the landing site is close to rays from large craters such as Langrenus, Taruntius, Theophilus, and Tycho.

The regolith material collected there is very similar to the soil brought back from the *Apollo* sites. The *Luna 16* material consists largely of basaltic rock fragments and impact-produced glassy fragments of basaltic composition. A small amount of more feldspathic rocks, including anorthosite, is present and may be debris carried in from the adjacent highlands. The basalts, presumably from mare bedrock, are similar but not identical to *Apollo 11* and *12* basalts. They are composed dominantly of clinopyroxene, plagioclase, olivine, ilmenite, and mesostasis minerals (Figure 11-35). Silicate-melt inclusions within crystals are different in composition from those in other *Apollo* rocks. Chemically, the rocks are much like *Apollo 12* material in that TiO_2 is low. The basalt fragments have the lowest Rb/Sr ratios and europium anomalies of any samples returned from the Moon. Trace element analyses indicate 1-2% content of meteoritic matter added to the *Luna 16* soils, comparable to amounts found at the *Apollo* sites. Microbreccia fragments are even more common than crystalline ones. Shock-deformation effects are conspicuous in most fragments. These particulates are identical to impact-produced fragments at the *Apollo* sites. This further demonstrates that regolith formation by meteorite impact is a general, Moonwide process.

The *Luna 20* lander, which set down in more highlands-like terrain to the north of *Luna 16*, brought back about 50 g of soil. Petrographic studies indicate that KREEP basalts and anorthositic gabbros make up more than two-thirds of the particles in the sample. Glassy materials comprise only about 10% of this soil. The concentrations of Al, K, Ba, Sr, Zr, and REE are similar to, but somewhat lower than, characteristic highlands materials from the *Apollo 15* site. Taken with the *Apollo 16* and *17* results, this extends the growing belief that old crustal rocks in the

Figure 11-35 A small fragment of crystalline basalt extracted from *Luna 16* soil material obtained from Soviet scientists for study by U.S. investigators. (Courtesy B. M. French, NASA Goddard Space Flight Center.)

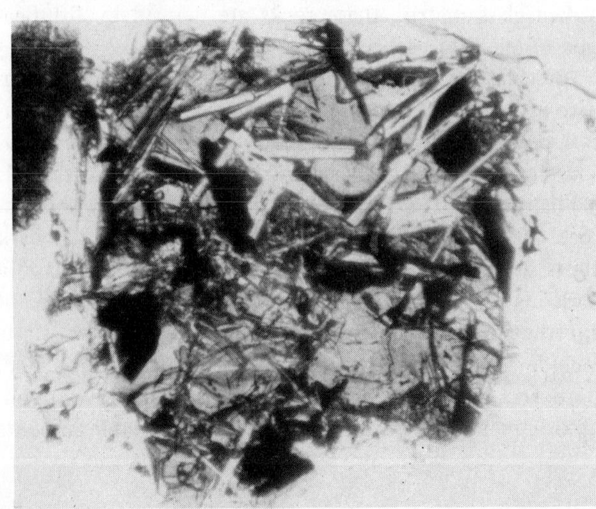

highlands consist largely of anorthositic gabbros and related rock types.

Exposure Ages of Lunar Materials

The *particle track* method has been used to estimate the length of time or *exposure age* during which rock fragments remain at or just below the regolith surface and the average rates at which the surfaces of these fragments are worn away and removed by micrometeorite particles and by flaking off of radiation-damaged layers. While near the surface, lunar materials are continually irradiated by energetic particles from the Sun *and other* regions of the galaxy. Low-energy solar radiation—mainly the heavy-element Fe-group nuclei in the solar wind—can penetrate rock surfaces to depths up to a centimeter or more. This inflicts crystal structure damage (atomic disorganization) within mineral grains along the line of penetration. The radiation damage can be made visible under the microscope as dark, elongate streaks or tracks (Figure 11-36) when a polished rock surface is etched by a strong solvent. The number of these tracks decreases exponentially with depth in a grain or rock layer (Figure 11-37). Thus, track density (tracks per cm^2) is proportional to the duration of exposure to the solar wind which, despite fluctuations associated with solar flares and other disturbances, is assumed to have a flux rate that averages out to a single value over long time intervals. Although solar wind flux rates had been measured previously by satellites (e.g., the *Pioneer* series), an improved production curve (rate) was determined from track density counts in the glass filter on the lens of the *Surveyor 3* TV camera brought back by the astronauts. The filter, subjected to seven recorded solar flare events during its nearly thirty-one months of residence on the Moon, contained track densities in excess of 10^6 tracks/cm^2 in its outermost atomic layers. Using the improved rate estimate, the exposure ages of surfaces of rock fragments lying on top of the regolith can be deduced by counting track densities (numbers per unit area) at different depths. Such counts on rocks returned from the several sites can often reveal whether a rock has been moved or turned over several times, with differing durations of exposure for different parts of the rock. The distribution of tracks within some larger *Apollo* specimens indicates that these typically occupy the same position on the regolith surface from less than 1 million up to 30 million years.

Other tracks, with longer path lengths, are produced by more energetic cosmic rays from galactic sources (outside the solar system) that can penetrate much deeper than solar wind particles into a rock or soil material. Iron-group cosmic rays that induce tracks to depths of several centimeters give rise to exposure ages of 10 million to 30 million years for rocks remaining close to the surface. These values are consistent with those obtained by Shoe-

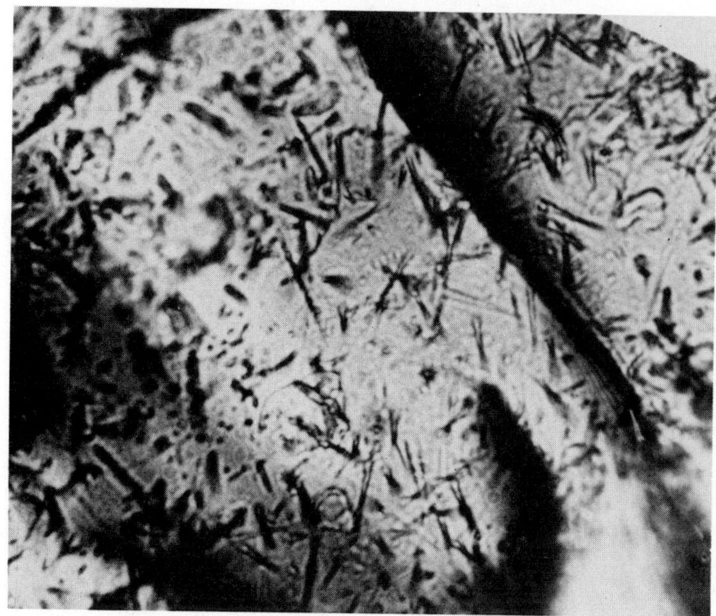

Figure 11-36 Nuclear particle tracks (which can be produced by solar wind ions of heavy mass, high energy cosmic rays, spallation nuclei, and fission of radionuclides) "developed" in a crystal of lunar feldspar by etching with NaOH solution, as they appear in the optical microscope. (Courtesy R. Walker, Washington University, St. Louis.)

Figure 11-37 Variations in track density relative to depth below exposed surfaces of rocks from three *Apollo* sites. These densities are attributed primarily to solar wind irradiation, with most increases during solar flares. (Crosaz, et al., 1972.)

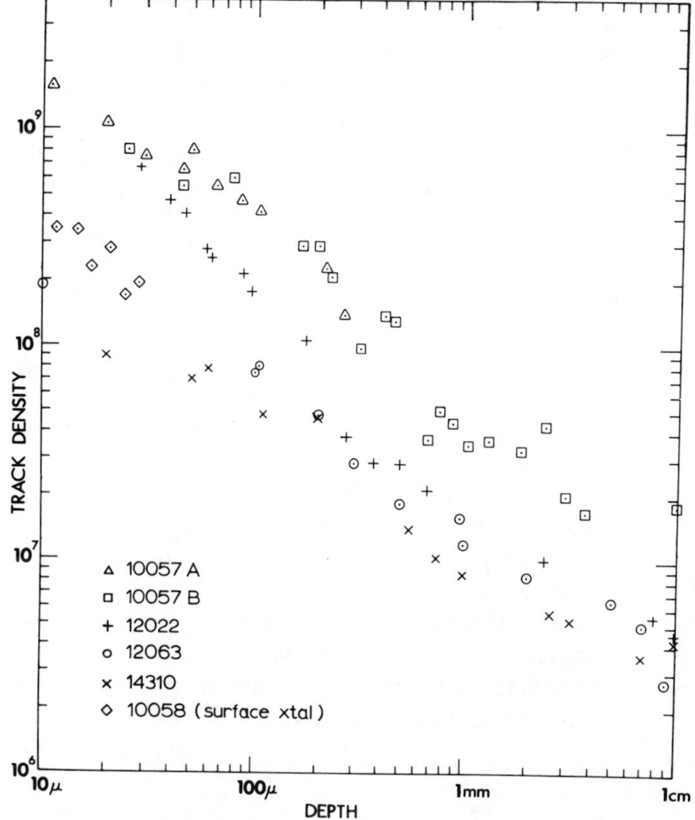

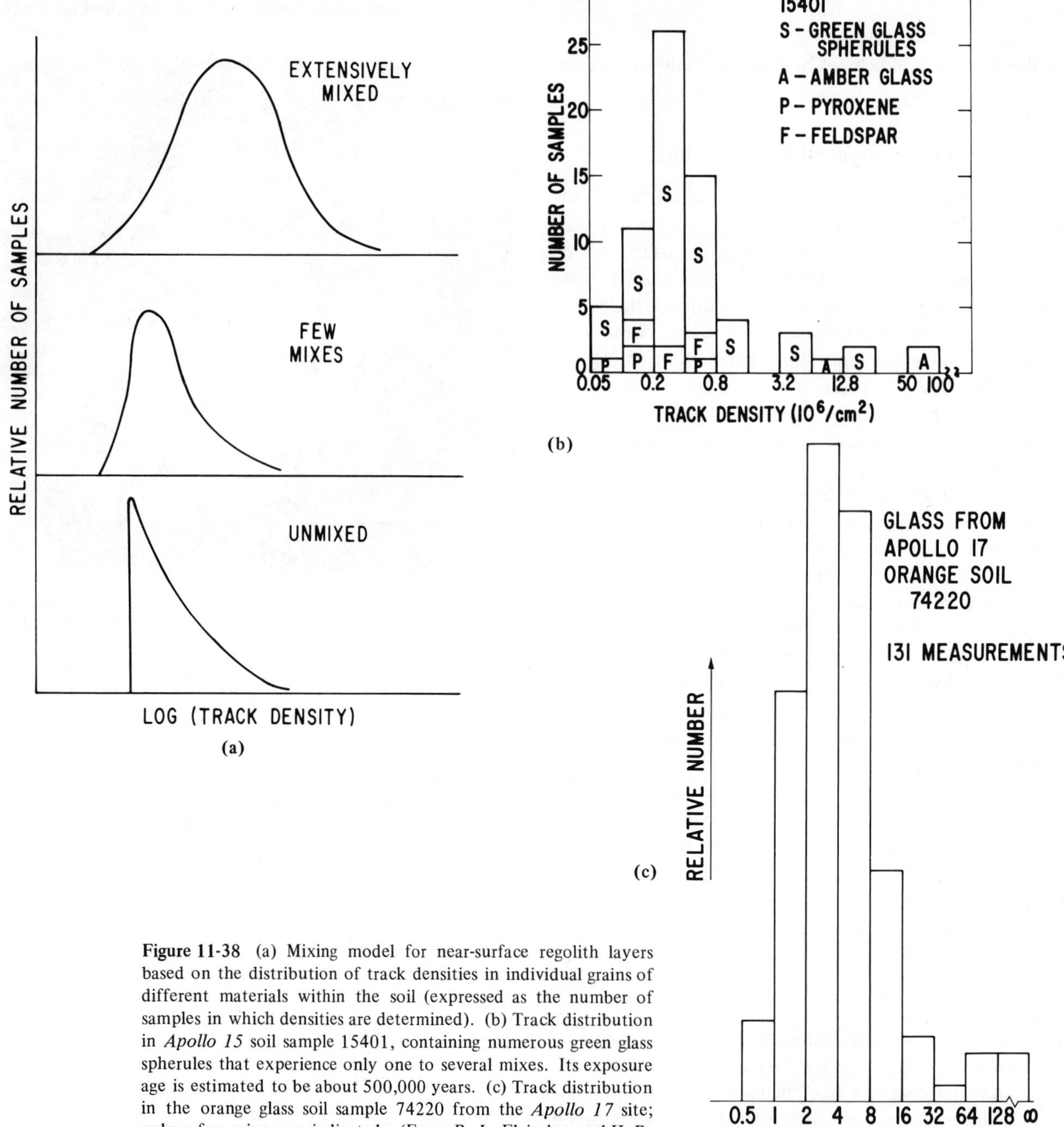

Figure 11-38 (a) Mixing model for near-surface regolith layers based on the distribution of track densities in individual grains of different materials within the soil (expressed as the number of samples in which densities are determined). (b) Track distribution in *Apollo 15* soil sample 15401, containing numerous green glass spherules that experience only one to several mixes. Its exposure age is estimated to be about 500,000 years. (c) Track distribution in the orange glass soil sample 74220 from the *Apollo 17* site; only a few mixes are indicated. (From R. L. Fleischer and H. R. Hart, 1973, courtesy Lunar Science Institute.)

maker et al. (1970) based on rates of turnover brought about by meteorite bombardment. Very-high-energy cosmic rays composed of alpha particles and protons can affect buried rock materials in the upper meter of the regolith. The residence time beneath the surface for any given rock depends on its present depth, its previous positions, and the number of turnovers or stirring events to which it was subjected. Analysis of rare gases and radionuclides induced by these cosmic rays shows that rock fragments have resided in the topmost meter of the *Apollo 11* soil for times between 30 million and 500 million years. In contrast, rock material collected below the surface along the slope of Cone Crater at the *Apollo 14* site has occupied near-surface positions for no longer than 14 million years. Cosmic-ray exposure ages can also fix the time when a relatively young (fresh-looking) impact crater was generated, provided the dated samples can be tied directly to rock units ejected from the same crater.

Fleischer and Hart (1973) have used track densities to determine the extent to which layers in the upper regolith have been mixed. The tracks are used as a measure of surface exposure or residence time of individual grains. Plots of the number of samples containing given track densities are made and then compared with calculated distribution models for different degrees of mixing (Figure 11-38a) as the number of turnover-redeposition events increase. Application of this method to the green glass from *Apollo 15* (Figure 11-38b) and the orange glass from *Apollo 17* (Figure 11-38c) shows that only a few mixes of the layer(s) sampled have occurred. This result implies a short and relatively recent exposure history.

Studies of particle tracks also enable some aspects of solar- and cosmic-ray flux histories to be reconstructed back millions of years before the present—an extremely difficult determination on the atmosphere-shielded Earth. If the time(s) when a rock fragment first appeared at the top of the regolith can be assessed independently from other evidence, crystals within the rock may incorporate a record of the flux that acted during the exposure period. More deeply buried (and hence shielded) grains and fragments in the soil below the top 60 to 100 cm that experience repeated stirring can thus provide information about the flux rates and composition influencing the lunar surface during the earlier stages of development of the regolith some 2 to 4 billion years ago.

Heavy concentrations of particle tracks in the other parts of grains and the outer layers of rock fragments, together with corresponding increases in induced radioactivities, have been associated with solar flare activity. Coupled with exposure age data, measurements of these solar flare effects lead to the conclusion that flares have occurred persistently in the last 500 million years and that solar activity has not changed appreciably in the last few million years. Using both solar- and cosmic-ray tracks, calculated rates at which lunar rocks are eroded by micrometeorites and by flaking off of grains damaged by solar wind and cosmic radiation fall between 10^{-7} and 10^{-8} cm/year (from less than 1 to about 10 angstroms per year, or from a single to a few atomic layers). Thus, the time required to remove 1 centimeter of surface material is, on average, not shorter than 10 million years. If we assume that most of the eroded material remains in the immediate locality where it is sampled, then the thickness of regolith developed by destruction of a hardrock layer through radiation and micrometeorite processes alone would build up by approximately 1 meter every billion years. The fact that regolith thicknesses of 3-4 m are present at *Apollo 11* and *12* sites, both underlain by mare lava rocks whose ages are near 3½ billion years, is evidence that these processes play a larger role in the thickening of the regolith than do contributions of ejecta from major impacts elsewhere.

Exposure and turnover ages can also be determined from examination of chemical effects brought about by the interaction of incoming radiation with the regolith. For example, as galactic rays impinge upon these materials, they produce neutrons that cause changes in the isotopic composition of such rare elements as gadolinium (Gd) and samarium (Sm). The isotope Gd^{157} is converted by neutron capture into Gd^{158}. Variations in the ratio Gd^{158}/Gd^{157} depend on the rate of the neutron flux over time and on exposure conditions within the bombarded materials. Figure 11-39 shows the measured Gd^{158}/Gd^{157} ratios in soil samples collected close to the surface at several *Apollo* and the *Luna 16* sites. These ratios correspond to the calculated neutron fluence (a measure of dosage based on 10^{16} neutrons/cm^2 at the specified low-energy value) shown on the abscissa. Ratios determined from the *Apollo 15* regolith plot in a distinctive pattern with peak values in the core interval centering at 100 cm (depth ~210 g/cm^2) below the surface. Several theoretical models, each leading to different gradient patterns, can be devised for continuously mixing, eroding, accreting, or undisturbed soils, or some combination of these cases. The best fit between observed and theoretical ratio variations is that in which the cored regolith unit (considered as though a slab of soil) is either instantaneously deposited or accretes in about 400 million years and then remains undisturbed thereafter except for addition of a small amount of material at the top. The lower

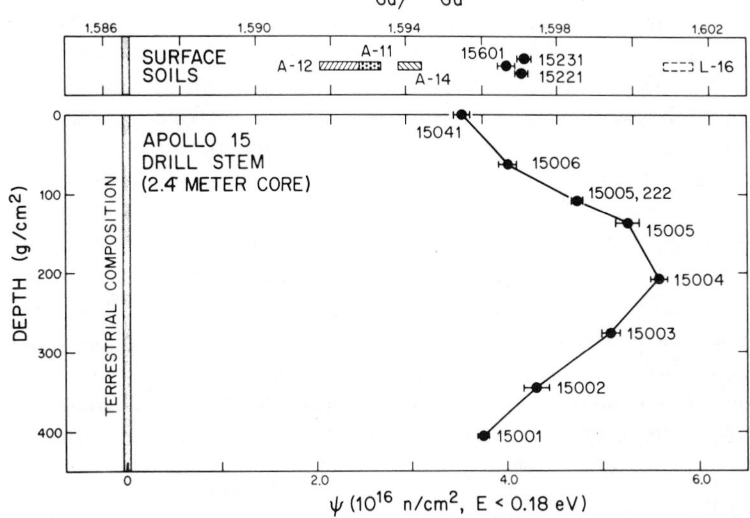

Figure 11-39 Values for measured Gd^{158}/Gd^{157} ratios (also recalculated as neutron fluences) in a number of surface soil samples from the *Apollo 11, 12, 14,* and *15* sites and the *Luna 16* site (upper graph). The variations in neutron fluences measured in sample intervals of the *Apollo 15* double core (see Figure 10-26) are plotted as a function of adjusted depths (after normalizing to a constant density), also obtained by gadolinium isotope determination (lower graph). (From Russ, et al., 1972.)

part of the *Apollo 15* core material, according to this interpretation, has stayed in an undisturbed state for the last half-billion years. Measurements of certain isotopic ratios (e.g., Kr^{81}/Kr^{83}, Ne^{20}/Ne^{21}, Ar^{37}/Ar^{39}) permit calculation of the exposure ages during which their host materials were accessible to the solar wind and cosmic rays. These calculations are based on known production rates of these isotopes by the several processes involved. Depending on isotopes used, values ranging from about 20 million to 500 million years have been reported for individual rocks collected at the surface. Because the solar wind gas contents become nearly constant with depth in core samples from the *Apollo 11* and *12* sites, complete mixing through turnover takes place on average about once every 100 million years. A complex history for both fragments and soils is suggested by these results.

Other Applications of Isotopic Distributions in the Lunar Rocks

Evidence bearing on the composition of the solar wind, of cosmic-ray particles, and of gases released from the lunar interior results particularly from analyses of soils and breccias. Some aspects of the Sun's recent history can be reconstructed from isotope distributions in this material, while information about the Sun's activity millions of years ago may be stored in fragments held in the breccias. Hydrogen (and small amounts of deuterium and tritium), carried in almost exclusively by the solar wind, extensively impregnates the outer surfaces of fragments in the regolith. In addition, rare or noble gas isotopes such as He^3, Ne^{20}, Ne^{22}, Ar^{36}, Ar^{38} and some isotopes of Kr and Xe are, in large part, implanted by absorption or occlusion in lunar rock and soil materials by solar wind irradiation and/or cosmic-ray spallation (reactions involving fission of heavy-nuclei elements brought on by collisions with energetic particles). Radiogenic He^4 and Ar^{40} are products of decay of U, Th, and K, although some He^4 can be cosmogenic or solar. Selected isotopic ratios (e.g., He^4/He^3, He^4/Ne^{20}, He^4/Ar^{40}) are indicative of the relative contributions of gases from solar, cosmogenic, and radiogenic sources.

The present abundances and isotopic composition of the noble gases in lunar materials (as well as meteorites) are often complex and require careful work to pinpoint the origin and sources of the different isotopic fractions. Plots of concentration gradients of noble gas isotopes as a function of depth within a grain (Figure 11-40) assist in reconstructing the history of lunar materials subjected to bombardment by extralunar particles of many kinds. Measurements of noble gases in extraterrestrial solids offer invaluable clues for determining element productions and mass fractionation processes in the evolution of the early solar system, the present state of the Sun, and the exposure ages of particles and fragments lying on the lunar surface.

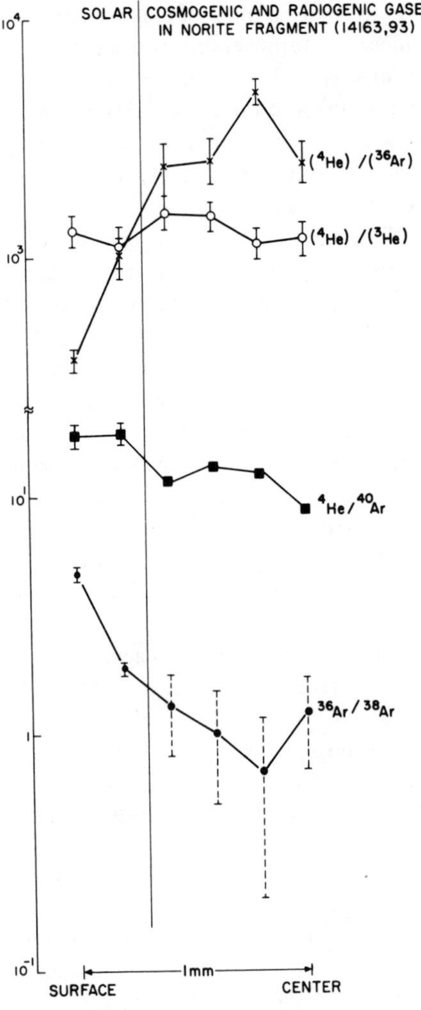

Figure 11-40 Concentration gradients, in terms of isotopic ratios, of several noble gases (from solar, cosmogenic, and radiogenic sources) measured from the surface to the center of a norite fragment extracted from *Apollo 14* soil sample 14163. (From G. H. Megrue and F. Steinbrunn, 1972.)

Some radiogenic argon-40 derived by decay of potassium-40 appears to have been released from the Moon's interior by degasing. Thus, ratios of Ar^{40}/Ar^{36} are higher than those expected from solar or spallation processes alone because of this excess argon-40 added as a radiogenic component. Instead of escaping from the lunar environment, this argon becomes trapped in the soil by re-implanting through interaction with the solar wind. Information received by the suprathermal-ion detector operating on the Moon indicates that there is an active lunar "atmosphere" in extremely low concentrations comprised principally of neon released after implanting by the solar wind plus some argon and other radiogenic gases (e.g., xenon, radon).

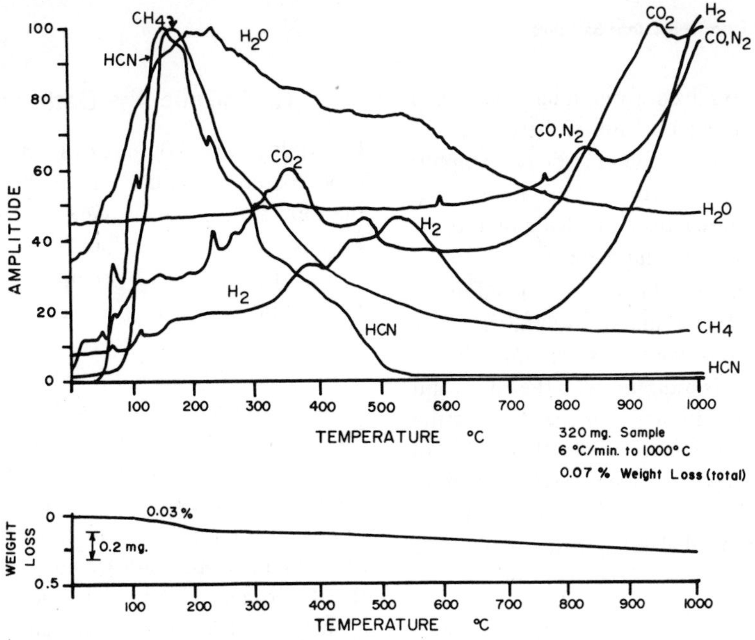

Figure 11-41 Patterns of gas evolution and weight loss in subsurface soil sample 61221 from the *Apollo 16* site. Thermal analysis-gas release measurements were on the indicated gaseous compounds as the sample was heated at 6°C per minute to 1000°C. The actual amounts of gas lost were small; total weight loss was only 0.07%. The relative losses as a function of temperature have been normalized for each gas to 100% amplitude over the temperature interval for which loss was greatest. The amounts of volatile constituents in this sample are notably greater than in any samples from earlier *Apollo* missions. Gibson and Moore (1973) interpret these measurements as an indication that cometary debris was admixed with the soil during the impact event that produced North Ray crater. (From E. K. Gibson and G. W. Moore, *Science*, 1973, copyright 1973 by the American Association for the Advancement of Science.)

Carbon, Organic Matter, and Inorganic Gases

Another announced major objective of the lunar landings was to search for evidence of organic matter and carbon compounds as possible "missing links" in the reconstruction of the origin and evolution of life in the solar system. To date, no positive signs of preexistent (fossil) or viable life forms have been found in any lunar sample. The lunar environment apparently has been hostile to the inception or survival of living organisms since the Moon approached its present state. Small amounts of certain carbon and nitrogen compounds of nonbiogenic nature (methane, simple hydrocarbon compounds, amino acid precursors, carbides) are reported in samples analyzed by some investigators. Presently, however, some of these are of terrestrial nature and most likely were introduced as contaminants sometime after the spacecraft returned to Earth. Other molecules may have reached the lunar surface as components in carbonaceous chondrites; still others are produced by the solar wind. Thermal analysis of volatile constituents in lunar soils released by heating in a vacuum shows a large fraction to consist of inorganic gases (Figure 11-41) derived from these sources: (1) atmospheric contaminants, (2) solar wind implants, (3) chemical reaction

Figure 11-42 Total carbon contents in rocks, soils, and breccias from *Apollo 11, 12, 15,* and *16* sites. (From "*Apollo 16* Preliminary Examination Team," *Science*, Vol. 179, pp. 23-34, copyright 1973 by the American Association for the Advancement of Science.)

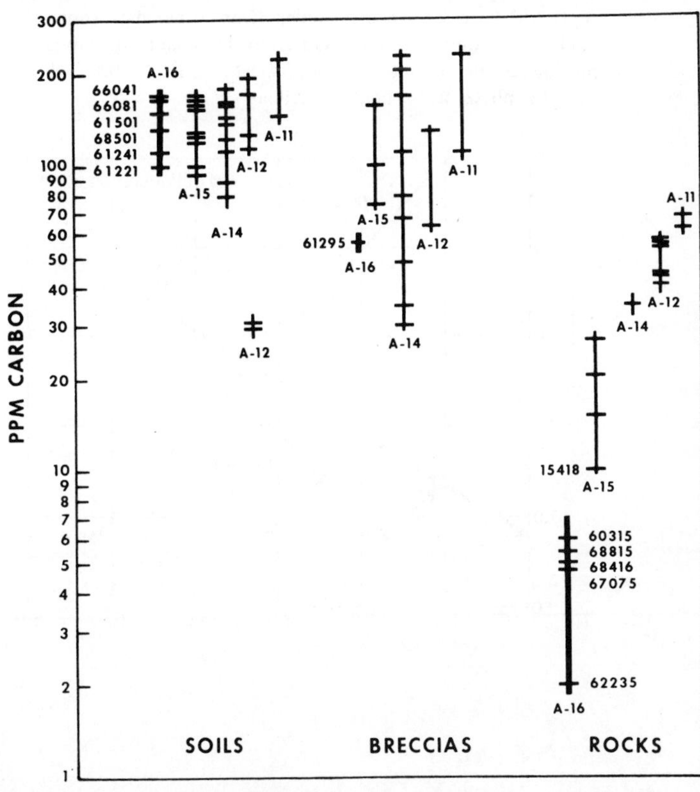

products, and (4) gases exsolved from lunar melts and trapped in vesicles. The actual amounts of these gases are quite small. A fraction of the carbon dioxide and sulphur compounds released during laboratory-controlled heating most probably derives from reactions involving meteoritic carbon and troilite in the regolith and breccias.

Most of the elemental carbon (Figure 11-42) and nitrogen, along with methane (CH_4), found in the lunar samples represents direct contributions from the solar wind. Some of the carbon, nevertheless, is indigenous to the rocks and must have been incorporated in materials that formed the Moon. Mare lavas contain much more carbon than is found in highlands crystalline rock (those from *Apollo 16* commonly have from 2 to 10 ppm C). Degasing of molten materials making up the primordial outer crust is a plausible explanation.

Figure 11-43 Development of an isochron line from analysis data for radiogenic rubidium-87 and its daughter strontium-87 ratioed to nonradiogenic strontium-86 which remains constant. The horizontal dashed line refers to some previous time at which each of four just-crystallized minerals (gray dots) in a rock had acquired equal ratios of Sr^{87}/Sr^{86} following homogenization (e.g., by melting); however, the initial distribution of Rb^{87} differed in each mineral depending on its ability to incorporate this nuclide in its crystal structure. Over some period of time (from crystallization to the present) a definite proportion of radiogenic rubidium (depending on its decay constant λ) in each mineral decayed into Sr^{87}, thereby lowering Rb^{87}/Sr^{86} and raising Sr^{87}/Sr^{86}. When these new ratios for each mineral are plotted, the points fall in a straight line whose slope a can then be inserted in the equation $a = \tan^{-1} (e^{\lambda t} - 1)$ to solve the remaining unknown t. This slope a increases (becomes steeper) with increasing time. If only the whole rock (without mineral separation) had been analyzed, its rubidium and strontium isotopic ratios would fall somewhere on this isochron. If the initial Sr^{87}/Sr^{86} ratio (on the ordinate at the dashed line intercept) had been known or assumed (from other data sources), a model isochron age would result from the a value appropriate to the line connecting the two points. (From Faul, 1966; adapted from Lanphere, Wasserberg, and Albee, 1966.)

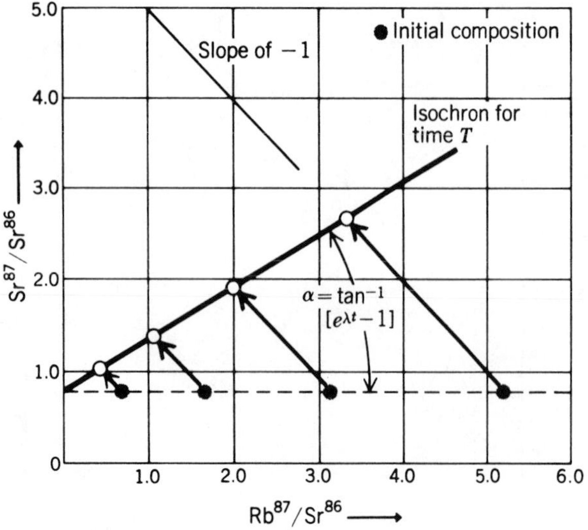

Radiogenic Age-Dating of Lunar Rocks

Determination of the ages of both rocks and soils by methods based on natural radioactive decay ranks as one of the outstanding achievements in the lunar samples investigations. Never before had such a large array of very old rocks, of which many experienced rather simple histories even while undergoing modifications unusual in relation to terrestrial rocks, been so painstakingly subjected to almost every analytical method developed for age-dating. These methods include determination of uranium-thorium-lead (U-Th-Pb), rubidium-strontium (Rb-Sr), potassium-argon (K-Ar), and argon-argon (Ar^{40}-Ar^{39}) isotopic compositions by mass spectrometery. Some of the general principles and techniques used to establish rock and mineral ages from isotopic data are best illustrated by reference to the Rb-Sr method as an example below.

When individual minerals carefully extracted from one or a group of related rock samples are separately analyzed for rubidium and strontium isotopes, the plotted analysis points (consisting of ratios of the isotopic concentrations of the radiogenic parent Rb^{87} and daughter Sr^{87} each to the nonradiogenic isotope Sr^{86}) can be connected by best fit to form an *isochron line* (Figure 11-43). The slope of the line affords a direct measure of the time from the present at which a particular event (e.g., crystallization or later metamorphism) reaches a stage allowing retention of newly produced daughter isotopes. This age value can be accurate to within the experimental errors of the determination; these may be as low as ±1 million to ±5 million years. The intersection of the isochron line with the ordinate gives the initial value I of the appropriate isotopic ratio (Sr^{87}/Sr^{86}) at the time the rock was crystallized.

Whole-rock analyses, made on fragmental material in which individual minerals are too fine to be separated or on ground-up crystalline rocks, result in single plotted points. A second point, lying on the ordinate, represents the best value (assumed or determined from other studies) for the initial isotopic ratio I_0 of Sr^{87}/Sr^{86} at the time strontium and other elements fractionated in the nebula to form the solar system and/or the parent planetary body. The current value of I_0 is 0.69898 ± 0.00003 for this ratio, termed BABI (*b*asaltic *a*chondrite *b*est *i*nitial), has been derived from numerous analyses of achondritic meteorites assumed to be samples of primordial solids which differentiated shortly after planetary accretion. The isochron line connecting the single analysis and the initial ratio points permits calculation of a *model age*. Determination of a model age generally fixes the time of origin of the initial materials making up a rock without regard to any subsequent history.

If a rock were formed at the same time as the solar system and remained unchanged in a closed system (no loss of parent or daughter), then its model age would be the age of the solar system. Likewise, if a rock had crystallized

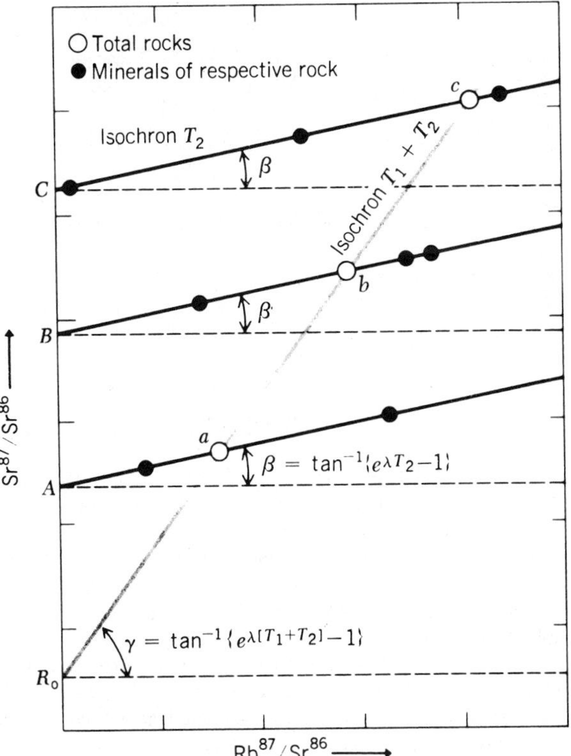

Figure 11-44 A rubidium-strontium diagram showing the effects of a homogenizing event (at T_2) on three rocks (a, b, c) that had formed at the same time T_1. Although proximate, the three rocks are never in direct contact (so do not mix) and do not receive or lose rubidium and/or strontium from other sources (i.e., are not open systems). Because of simultaneous formation prior to homogenization, these rocks share the same initial Sr^{87}/Sr^{86} ratio (R_0). Whole rock analysis of each would produce three points from which the isochron $T_1 + T_2$ can be drawn. Note, however, that each rock has its own characteristic Sr^{87}/Sr^{86} ratio at the time of homogenization (shown by dashed lines with intercepts at A, B, and C). Thereafter, each rock develops new rubidium and strontium isotopic ratios in its constituent minerals (black dots) as soon as conditions causing homogenization cease to operate (e.g., rocks cool below some critical thermal level). When the individual minerals in each rock are analyzed, the resulting T_2 isochrons will date the homogenization event. When whole rock analyses are carried out on each, the resulting $T_1 + T_2$ isochron can indicate the original time of formation of these rocks, provided each remained a closed system during the subsequent event. (From Faul, 1966; adapted from Lanphere, Wasserberg, and Albee, 1964.)

later from melted primordial materials without undergoing fractionation, the same result would be obtained. But if the rock solidified at a later time after experiencing an increase of parent element relative to the daughter element, then the model age would be less than the solar system but somewhat greater than the actual age of crystallization. Conversely, if the rock were formed at a later time and were enriched in the daughter element, then the model age would be greater than both the actual age of crystallization and the age of the solar system. The model age is thus a measure of the effective fractionation processes which took place in generating the magmas from which the rocks formed.

It is important to note that the term *age* can have several other meanings. One "age" may refer to the time in which the primordial materials in a planet were first organized; if different isotopes later become selectively "differentiated" (e.g., by contamination or volatilization), evidence for this age may be lost. A second "age" can describe a time of crystallization after melting; if this occurs after initial planet formation, radioactive decay since then will have altered the original parent and/or daughter ratios. A third "age" can result when some of the parent and/or daughter isotopes are lost or added to by some modifying process such as heating, which can drive off variable amounts of gaseous species (e.g., Ar^{40}) or cause selective diffusion of isotopes from one mineral phase to another. Both crystallization and thermal metamorphism ages are examples of equilibration or "resetting the clock"; homogenization by melting or diffusion leads to the same isotopic ratios of daughter elements in all phases, but the subsequent decay of the parent isotope, redistributed in different concentrations in these phases, thereafter will alter these once-constant daughter product ratios to different degrees (Figure 11-44).

For the lunar rocks, age determination by the Rb^{87}/Sr^{87} method has proved highly reliable. A typical isochron plot developed from Rb-Sr analyses of minerals in a single rock appears in Figure 11-45. The isochron age for this

Figure 11-45 A Rb-Sr isochron developed from analysis of individual minerals in lunar lava sample 10044 from the *Apollo 11* site. The slope of the isochron line leads to an age (of crystallization) of 3.7 billion years and an Sr^{87}/Sr^{86} intercept of 0.69909. A line characteristic of 4.6 billion years is also shown. Some *Apollo 11* lava samples, when analyzed by the whole rock method, plot along this 4.6 b.y. isochron despite crystallization ages of about 3.6 b.y. (From Papanastassiou and Wasserberg, 1970.)

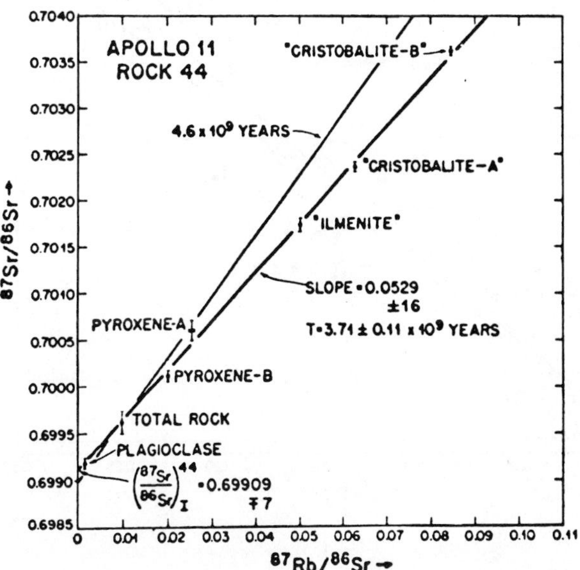

Apollo 11 sample is 3.7 AE (1 AE = 1 aeon, or 10^9 years, 1 billion years, also abbreviated 1 b. y.). Mare lavas from other sites (*Apollo 12, Apollo 15, Apollo 17, Luna 16*) date between 3.8 and 3.1 AE. However, when *Apollo* crystalline rocks from each site are ground up for whole-rock analyses, the resultant model ages cluster around values between 4.3 and 4.6 AE. A most remarkable result is found for series of whole-soil analyses representative of soils from five sites (Figure 11-46). These soils, which are mixtures of a wide variety of rock fragments, breccias, and glasses of different ages, *act* as if they are samples that have remained unfractionated for ~4.6 AE.

The rubidium-strontium ages reported by several investigators commonly show a small spread in values owing in part to differences in analytical techniques. Thus, values for sample 14310 range from 3.88 to 4.00 AE, for 15415 from 4.05 to 4.15 AE, and for 15555 from 3.30 to 3.53 AE. The reliability of these values is indicated by analytical precision or error statements—as, for example, 14503 = 3.94 ± 0.04 AE, or 15555 = 3.53 ± 0.18 AE. To some extent, the variations in ages reported by different investigators also reflect sampling problems; the amounts of material taken from a given sample for analysis by individual laboratories is small, not thoroughly homogenized, and therefore not necessarily representative. Different clasts from individual breccia samples usually show varying ages which are real values rather than experimental or sampling errors. This indicates that rock units of several ages have been incorporated within the sample; the youngest clast age sets a lower limit to the time at which the regolith or ejecta blanket deposit was lithified into a breccia.

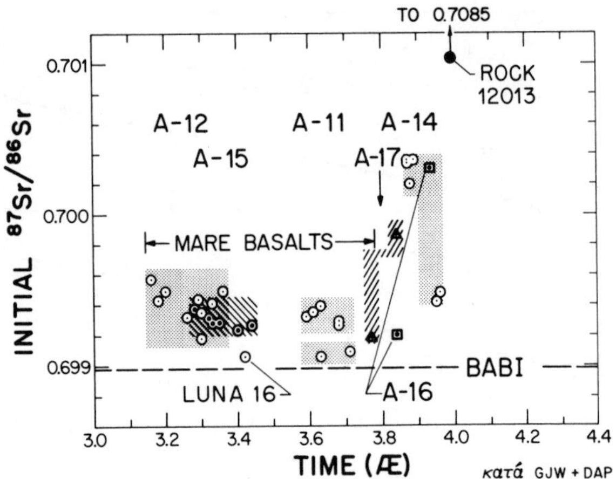

Figure 11-47 A T-I plot (time of crystallization versus initial ratio of Sr^{87}/Sr^{86} as measured by extrapolation of the isochron line of a sample to the ordinate) for lunar rocks from six *Apollo* missions and the *Luna 16* unmanned probe. Symbols: ⊙ rocks from the *Apollo 11, 12,* and *14* and *Luna 16* sites; ◉ *Apollo 15* rocks; ▣ *Apollo 16* rocks; △ *Apollo 17* rocks; ● *Apollo 12* granite sample 12013. The mare basalts and basaltic clasts have low I values near that of BABI, incompatible with a source material of chondritic composition. Some breccia rocks from *Apollo 14, 16,* and *17*, mainly older than mare lavas, have higher I values probably because of contamination with older crustal materials having high Rb contents. (Courtesy G. J. Wasserburg and D. A. Papanastassiou, California Institute of Technology.)

A plot of initial ratios (*I*) for Sr^{87}/Sr^{86} versus time (Figure 11-47) for samples from all *Apollo* sites provides a convenient summary of isochron ages and also discloses some interesting genetic information. All mare lavas have *I* values between 0.6990 and 0.6996, with a rock fragment from the *Luna 16* soil being the least radiogenic. These isotopic ratios are nearly primitive—that is, are very close to the value accepted for BABI. Such low ratios indicate that the magma source(s) for the basalts consisted of primitive rocks that had formed almost simultaneously with the accretional organization of the terrestrial planets.* These rocks also had low rubidium contents. In contrast, those lunar crystalline rocks (nonmare lavas) older than 3.88 AE have *I* values greater than 0.7000.

Potassium-argon ages as first reported were generally lower than those for the same samples dated by the Rb-Sr method. K-Ar ages for *Apollo 12* lava rocks fall between 1.4 and 2.8 AE as compared with the 3.3 AE ages obtained by Rb-Sr dating. This wide range of younger ages reflects subsequent thermal events—including reheating from impacts—that outgased some of the argon. Turner (1970)

Figure 11-46 Evolution diagram for Rb-Sr constructed from data obtained by whole rock analysis of soils from *Apollo 11, 12, 14, 15* and *Luna 16* sites. The ratio points define an isochron line which extrapolates almost exactly to the BABI value. (From Wasserburg, et al., 1972.)

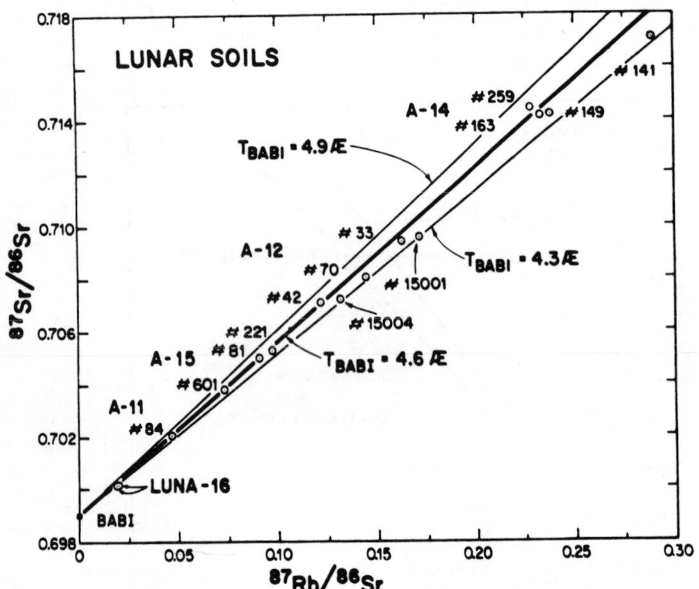

*This conclusion is further supported by detection in some lunar rocks of traces of decay products from short-lived radionuclides such as Pu^{244}, which could be incorporated only if accretion had occurred in the few tens of millions of years (several half-lives) before these nuclides became extinct.

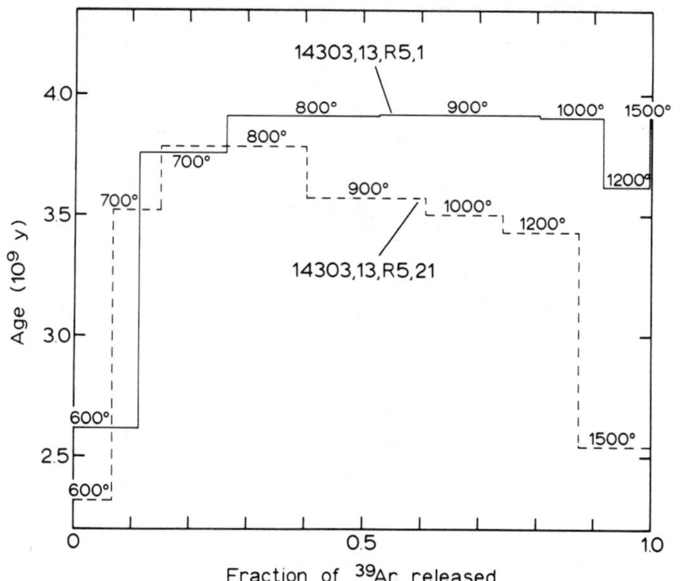

Figure 11-48 Release from two *Apollo 14* fragments of argon-39, produced earlier by irradiation with neutrons from a Co^{58} source, by stepwise heating over 100°C intervals. An Ar^{40}-Ar^{39} age is then obtained for each fraction released. The solid line plot for a basalt fragment from breccia sample 14303 reaches an age "plateau" in which the age (3.90 AE) remains almost constant as heating to higher temperatures drives off remaining Ar^{39}. The dashed line plot for a microbreccia fragment in the same sample marks a steady decrease in age with continued temperature rise, owing perhaps to losses from different minerals or rock particles in the fragment. The age reported for this fragment (3.78 AE) is that of the highest plateau level reached (at 800°C). (From Kirsten, et al., 1972.)

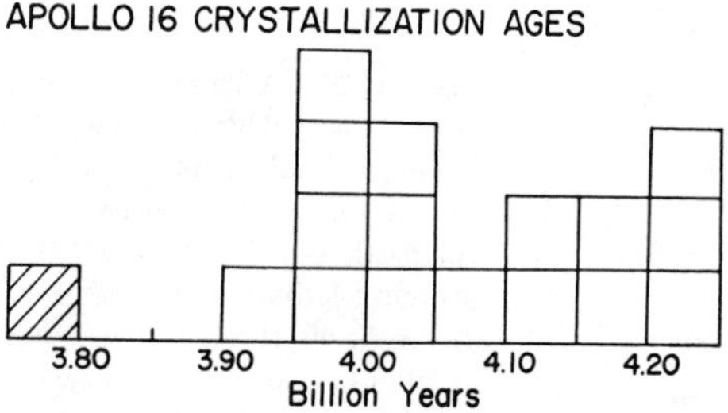

Figure 11-49 Histogram showing the crystallization ages of 17 samples from the *Apollo 16* site as determined by the Ar^{40}-Ar^{39} stepwise heating technique. Most samples are rock clasts and fragments recovered from soil "course-fines." The spread of ages between 3.9 and 4.25 AE suggests that crustal rocks of somewhat different crystallization ages have become intermixed in the Cayley and Descartes units sampled in the group. (From Husain and Schaeffer, 1973.)

and others have used the Ar^{40}-Ar^{39} technique to adjust K-Ar ages for any argon losses and for contamination by cosmic-ray-produced Ar^{40}. The technique involves conversion (by neutron activation) of K^{39} to Ar^{39}, step-increased heating to drive off loosely held Ar^{40} from crystal sites until a constant Ar^{40}-Ar^{39} ratio is obtained, and calculation of the amounts of K^{40} and Ar^{40} that thus remain at these stable (tightly held) sites. Both whole-rocks and mineral separates are suited to analysis by this technique (Figure 11-48). Application of the Ar^{40}-Ar^{39} technique usually results in corrected K-Ar ages that generally agree with those obtained by the Rb-Sr method and sometimes reveals subtle age variations not found by Rb-Sr dating. Thus, after certain Ar^{40}-Ar^{39} corrections are applied, new average ages for *Apollo 11* rocks include 3.8 AE for type A (high-K) rocks and 3.5 AE for type B (low-K) rocks. Ar^{40}-Ar^{39} ages for *Apollo 15* mare lavas cluster around 3.3 ± 0.1 AE. Lavas from *Apollo 17* give Ar^{40}-AR^{39} ages of 3.74 to 3.80 AE.

The Ar^{40}-Ar^{39} technique has also worked successfully in reconstructing initial ages of fragments in breccias and soils despite often large losses of argon from shock heating. However, other argon-dating techniques, such as measuring the ratio of Ar^{40} to Ar^{36}, aid in determining the amount of Ar^{40} driven off by some singular heat-producing event. Soils and regolith from both the *Apollo 12* and *14* sites show lower K-Ar ages that appear to derive from mixtures of two components of differing age. Basaltic fragments in *Apollo 14* breccias date at ~2.8 AE before Ar^{40}-Ar^{39} adjustments. Some KREEP-rich melt fragments separated from the soil date at 0.9 to 2.8 AE, the range depending on differing proportions of the intermixed components. Breccias collected at the *Apollo 15* site give Ar^{40}-Ar^{39} ages varying from 3.2 to almost 4.0 AE. Some fragments and green glass from this site date at ~3.8 AE, a value consistent with a lower limit to the age of the Imbrium event. Breccias and soil fragments from *Apollo 16* yield Ar^{40}-Ar^{39} age dates between 3.85 and 4.20 AE (Figure 11-49). Recrystallized norite breccias at the *Apollo 17* site give ages that cluster around 3.95 AE but with the youngest thermal event at 3.84 AE. The orange soil collected around Shorty Crater during *Apollo 17* was, at first, expected to be one of the youngest materials returned from the Moon (estimates ranged from a few million years to 2 billion years at most). Age-dating by the Ar^{40}-Ar^{39} technique produced a value of 3.71 AE—essentially the same age as the mare lavas in which the crater was emplaced (see p. 170 for proposed explanations).

Uranium-thorium-lead isochron ages are determined from analyses of different isotope pairs (Pb^{206}/U^{238}, Pb^{207}/U^{235}, Pb^{208}/Th^{232}) as well as lead-lead ages

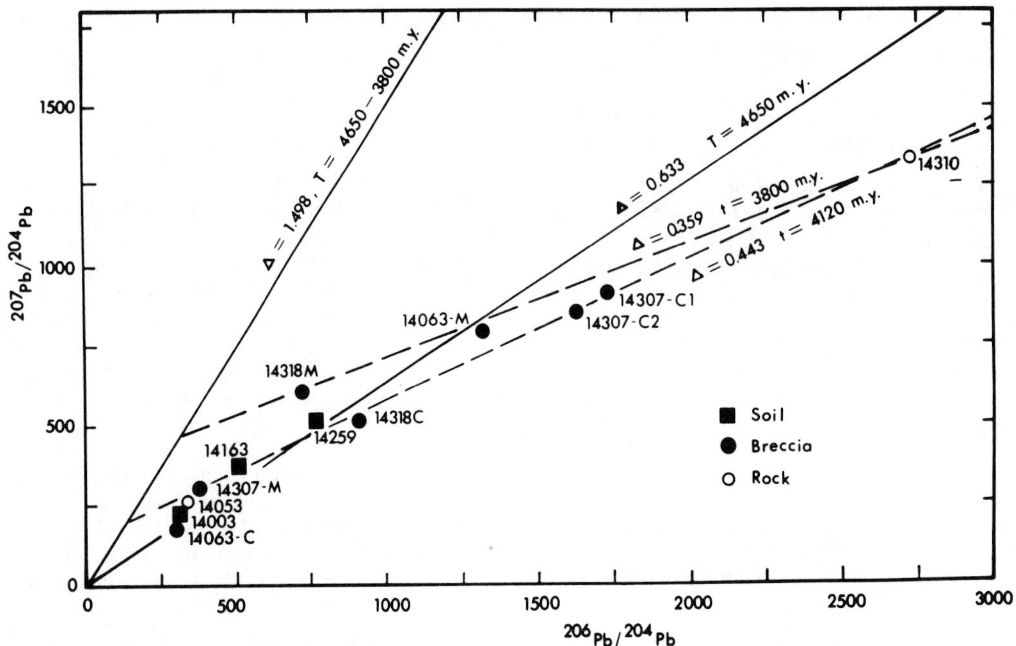

Figure 11-50 A Pb-Pb isochron plot for a series of samples from *Apollo 14* using the isotope ratios shown. The solid line labelled with Δ = 0.633 (slope value) is the primary isochron for all samples that began producing radiogenic lead isotopes when the Moon was assumed to have formed 4650 million years ago. Samples such as 14003, 14053, 14063M and C (for *m*atrix and *c*last), 14163, and 14318M have ages between 4500 and 4650 million years (depending on position relative to curve Δ = 1.498). Samples 14163, 14307, 14310, and 14318 define secondary isochrons (broken lines) of different ages in which disturbances of the U-Th-Pb system (additions or losses of radiogenic lead and/or uranium) occurred to "reset the clock." Events at 4120 and 3800 million years are indicated. (From Tatsumoto, et al., 1972.)

(Pb^{206}/Pb^{207}; and Pb^{206}, Pb^{207}, and Pb^{208} each ratioed to common lead, Pb^{204}). Isochron ages for some *Apollo 11, 12, 15,* and *17* mare lavas obtained from two or more isotope-pair analyses of the same sample commonly are *concordant* (mutually in close agreement).

For other rocks, age values determined from different isotope pairs may be *discordant* (not agreeing) (Figure 11-50). Thus, for the crystalline rock 12063, Cliff et al. (1971) obtain values of Pb^{206}/U^{238} = 3.68 AE, Pb^{207}/U^{235} = 4.14 AE, Pb^{208}/Th^{232} = 3.38 AE, and Pb^{207}/Pb^{206} = 4.40 AE. The older values represent the approximate time of introduction of the original radiogenic uranium and thorium. The younger values are interpreted to reflect selective fractionation of U and Pb at the time of crystallization of the lava about 3.4 billion years ago.

Evidence for two or more events (stages) in the history of the material (whether crystalline or fragmental) is frequently revealed in a *concordia diagram* (see Figure 11-51 for explanation). Thus, using U-Th-Pb systematics, basaltic samples from the *Apollo 11, 12, 14, 15,* and *16* sites usually show an older (primordial or source) age near 4.6 AE and a younger (crystallization) age which tends to correspond to that determined by other methods (Figure 11-52). Soils sometimes have a third, even younger age as well, indicating a record of "events" after their parent lava rocks had solidified.

When model (whole-rock) ages are determined for *Apollo* basalts, the U-Th-Pb results do not fall near the 4.6 AE values that characterize the Rb-Sr results. For example, basalt sample 14310 has a generally concordant age of ~4.3 AE (Pb^{206}/U^{238} = 4.24 AE, Pb^{207}/U^{235} = 4.27 AE, Pb^{208}/Th^{232} = 4.10, Pb^{207}/Pb^{206} = 4.28 AE; from Wasserburg et al., 1972). Thus, somewhat younger ages are found, as would be expected in a chemical system open to enrichment in uranium and thorium accompanied possibly by removal of some early-formed lead isotopes.

U-Th-Pb model (whole-rock) ages between 4.3 and 4.9 AE have been assigned to various breccia samples from the *Apollo* sites. Most U-Th-Pb model ages for the soils at the mare sites are somewhat less than 4.6 AE. Many mare soil ages are interpreted as representing mixtures of rock materials that have not lost uranium or lead (and thus retain the primordial ratios established at the time of the Moon's formation 4.6 billion years ago) with rocks subjected to varying losses (such as lead-depleted KREEP-rocks from the Copernicus event). However, soils from the *Apollo 16* and *Luna 20* highlands sites give anomalously older ages ranging from 4.9 to 6.1 AE. These highlands soils are

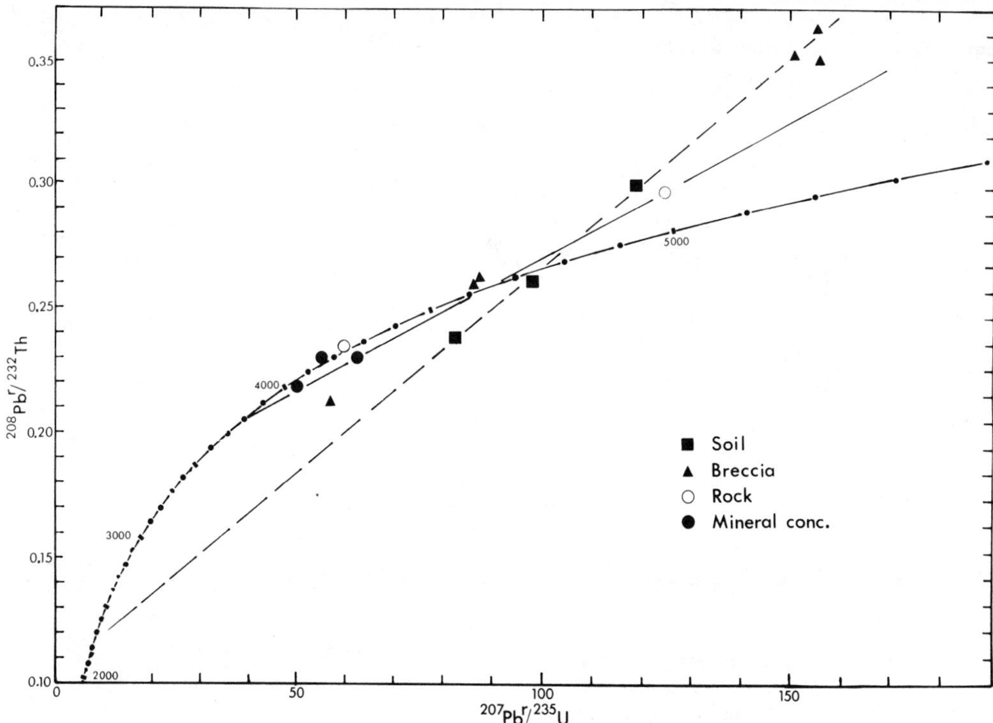

Figure 11-51 U-Th-Pb concordia diagram for samples from *Apollo 14*. The principal curve (line with dots) is generated from values of Pb^{208}/Th^{232} and Pb^{207}/U^{235} ratios as these change with geologic time (numbers next to curve, in millions of years). This concordia curve represents the locus of all concordant U-Th-Pb ages. When some radiogenic lead is lost (or added) because of a later modifying event (e.g., remelting and crystallization in an open system or incorporation of labile lead), the plotted ratios will tend to move off the curve. This may happen to different degrees in a series of related materials. These new plotted points commonly can be connected by a straight line. Extrapolation of this line produces two intercepts on the concordia curve. The older intercept value relates to the time in which the system was first activated while the younger value indicates the time of the modifying event. In the plots constructed here, both discordia lines have one intercept between 4500 and 4650 million years—a time related to incorporation of U- and Th-bearing materials at the Moon's origin. Rock and mineral samples produce a discordia line that intercepts at 3800 million years—an age approximating the time of crystallization of noritic and basaltic layers later involved in the Imbrium event. Breccias and soils define a discordia line which crosses between 2000 and 3000 million years. The nature of this young "third event" has not been identified. (From Tatsumoto, et al., 1972.)

dominated by crustal materials that must have been enriched in Pb early in lunar history.

It is evident, then, that U-Th-Pb dating methods are producing a wide range of age values from the major lunar materials. Ages obtained from samples collected at most of the sites are summarized in the concordia diagram of Figure 11-47. Some of these ages are valid indicators of the event(s) recorded, but others are influenced by certain discrepancies.

As first proposed by Silver (1971), some of these apparent discrepancies in the uranium-lead ages are explained by differential fractionation among lead isotopes transferred as volatiles at high temperatures. Thermal events that involve outgasing of lavas or heating during impact-related shock metamorphism will selectively volatilize the low-vapor-pressure elements (e.g., Pb, Bi, Rb), making them *labile* (easily transferred as gases) in the lunar environment. Escape of some labile lead from the Moon may account, in part, for the generally low concentration of total lead in the lunar rocks compared with terrestrial basic rocks. These *parentless* leads (separated from their uranium-bearing host minerals) can accumulate as "contaminants" in abnormal concentrations.

Ages of mare materials obtained by U-Th-Pb methods are generally older than Rb-Sr or K-Ar ages for the crystalline rocks* and younger for the fragmental rocks (Figure 11-53). This is another example of discordant ages, in which different values are found for the same rock when different isotope pairs are used. This is sometimes due to

*One exception is the "Genesis Rock" 15415 found near the Apennine Front at the Apollo 15 site. This sample has a Pb^{207}/Pb^{206} age between 3.8 and 4.1 AE (Tatsumoto, et al., 1972; Tera and Wasserburg, 1972), an Ar^{40}-Ar^{39} age of 4.05 AE (Turner, 1972), and a Rb^{87}-Sr^{87} age of 4.15 AE (Wasserburg, 1972).

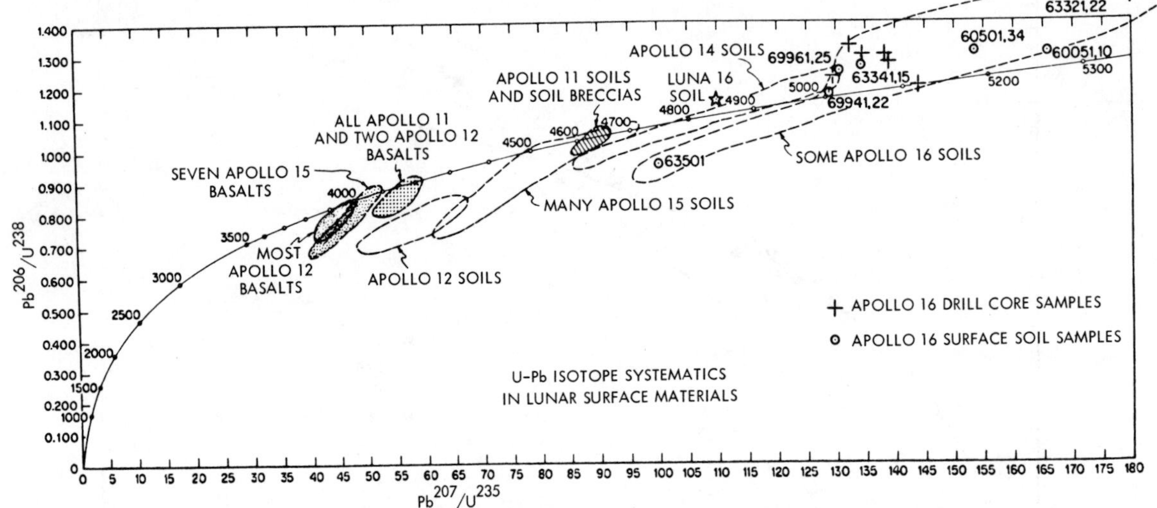

Figure 11-52 Summary concordia diagram showing values of Pb^{206}/U^{238} versus Pb^{207}/U^{235} for lunar samples from various *Apollo* sites. Some sample groups fall close to the concordia line at the ages indicated, but others plot off this line. (From L. T. Silver, 1973, courtesy Lunar Science Institute.)

Figure 11-53 A plot comparing whole rock Pb-Pb isochron ages and Rb-Sr mineral isochron ages determined in the same samples of lunar basalts. Boxes outline the range values for groups of rocks from each individual *Apollo* site. The solid line extending from the origin would pass through concordant ages if both dating methods had yielded the same values. A systematic discordancy is indicated by the dashed line such that most samples have Pb-Pb ages older than ~ 600 million years relative to Rb-Sr ages. (From Tatsumoto, et al., 1972; see p. 1541 of that paper for sources of analytical data.)

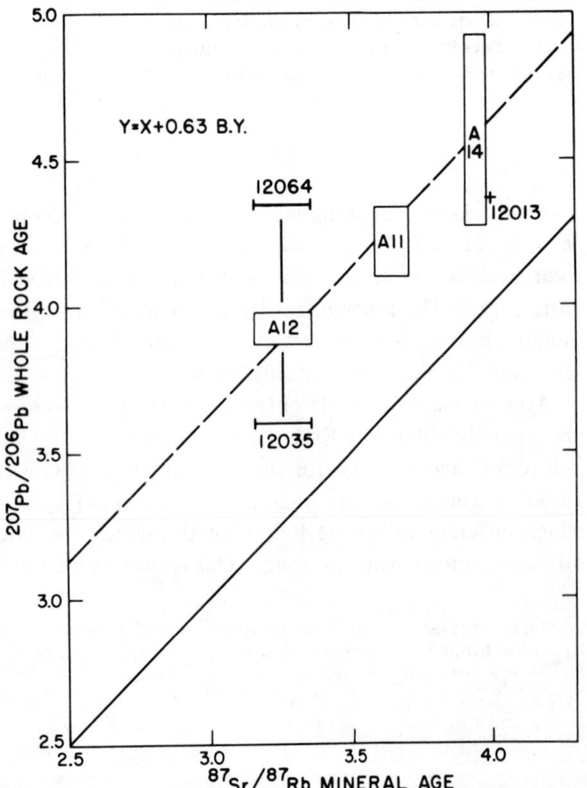

analytical errors, but more frequently there are real differences that result from separate events affecting isotopic distributions to different extents. However, appropriate corrections can be applied which may bring the U-Th-Pb values more or less into accord with Rb-Sr and K-Ar (also corrected) values. For example, by assuming a loss of about two-thirds of the primordial lead (and/or gain of some uranium) from *Apollo 12* lava rocks during melting, the U-Ph age of 4.1 AE is adjusted to correspond to the 3.3 AE age found by other methods.

The data for the crystalline rocks are interpreted as follows. The mineral isochron ages clearly define the time when each rock unit actually *crystallized*, during which Rb and Sr ions were partitioned into specific minerals whose crystal structures would accept them. The oldest Rb-Sr (and U-Th-Pb) values, given as model ages, are a consequence of the initial isotopic composition of the primordial materials characteristic of the Moon at the time of its formation after chemical fractionation of parent and daughter elements. These materials soon accreted into the lunar body and were subjected to melting in at least the outer shell. Apparently, the lunar materials at that time were "in communication" by mixing of the melts, so that isotope ratios were initially homogeneous. These initial ratios remained as a "baseline" even after the primordial source rocks were remelted and emplaced as mare lavas. At least two evolutionary paths for changing Sr^{87}/Sr^{86} ratios can be reconstructed.

One trend is evidenced by the type B (low-K) lavas from the *Apollo 11* site and some of the *Apollo 12* lavas. Their source underwent remelting without significant *additions* or *losses* of any radiogenic isotopes. This implies that the melt remained approximately a *closed* system and did not differentiate extensively, so that the Rb/Sr ratio *in the melt* stayed almost the same as in the initial 4.6-AE source rock. The lower isochron ages found for these basaltic lavas merely date the time they crystallized after rehomogenizing (but not when they began to melt unless melting and crystallization happened over a short time span). As the lavas solidified, the radiogenic parent isotope redistributed (partitioned) among the several mineral phases that formed. Each phase thereafter accumulated additional amounts of Sr^{87} from Rb^{87} decay.

This "resetting of the clock" generally parallels the scheme for a two-stage Sr evolution as diagramed in Figure 11-43. Thus, for the crystalline lavas, whole-rock analyses reveal a primordial (model) age whereas mineral-separates analysis establishes a younger (isochron) age for a later redistribution into individual minerals. The same approach does not lead to primordial ages of terrestrial basalts by whole-rock analysis because these rocks have experienced subsequent differentiation that altered the initial isotopic composition; when model ages using a Sr^{87}/Sr^{86} value of 0.699 are calculated, the values can be much less than 4.5 AE or, in some instances, considerably older than 5 billion years.

A second trend is evidenced by model ages around 3.9 to 4.0 AE obtained from type A (high-K) rocks at the *Apollo 11* site. This implies that their parent materials experienced some modification of the initial isotopic composition prior to final crystallization. The present Rb/Sr ratio is higher than before melting. Differentiation with selective changes in Rb and/or Sr could account for this. More likely, the melt behaved as an *open* system. Thus, as the melt upwelled toward the surface, the assimilation (incorporation) of rocks with higher abundances of Rb, K, Ba, and other trace elements may have raised the Rb/Sr and K/Ar ratios of the lavas that reached the surface.

A variant of these ideas has also been postulated. This holds that differentiation into two magma sources, one with a high Rb/Sr and the other with a lower ratio, occurred early in the Moon's history. These sources or reservoirs occupied different positions within the lunar mantle. As each is later tapped, it supplies lavas that give rise to crystalline rocks of both type A and type B found at the *Apollo 11* site.

The concept of a two-stage process by which radiogenic elements undergo concentration changes is illustrated by the U-Th-Pb systematics. During the first stage—involving an early melting of the Moon—uranium and thorium and some primitive leads are distributed in the crystalline rocks of the mantle and crust in fixed initial ratios. Then, in the second stage, this lead, together with radiogenic lead formed in the subsequent 200-300 million years, is removed during later melting from the magma reservoirs and becomes concentrated in rocks of the upper crustal layers. The ages of these rocks are thus apparently older than the crystallization ages determined by Rb-Sr. Rocks in underlying layers, enriched relatively in U and Th, would date at somewhat younger ages. This, in part, can account for the discordant ages obtained by the U-Th-Pb and Rb-Sr methods from some lunar samples.

Rb-Sr isochron ages of ~4.0-4.1 AE for KREEP-rich norites and sample 12013 indicate that these rocks come from subsurface sources (deeper in the outer shell?) which had already crystallized before the emplacement and crystallization of the basaltic lavas. This may indicate an earlier, unrelated melting event. Radiogenic ages of the *Apollo 15* anorthosite fragment also suggests that it had crystallized prior to the period of general outpourings of the basalts.

These results for the crystalline rocks bear on the seeming paradox of older Rb-Sr and U-Th-Pb ages for regolith and microbreccia materials which appear to consist mainly of rocks derived from underlying, younger (3.2-3.7 AE) mare lava units. This at first looked to be a major mystery, but is now understood in principle if not in detail. Two factors have been proposed to account for this apparent age reversal (i.e., the "deposit" seems older than its source materials).

The first relies on the model-versus-isochron-age argument just developed. In effect, the natural grinding and intermixing of the lava rocks during buildup of the regolith duplicate the laboratory operation involved in preparing these crystalline samples for whole-rock analysis. Since it has been demonstrated that whole-rock analysis of the lavas gives model ages older than the isochron ages derived from individual mineral analyses, it follows that whole-rock analysis of the naturally mixed regolith would also give comparable model ages. However, isotopic compositions determined for some soil samples yield Sr^{87}/Sr^{86} values near 0.703-0.704 and 0.708-0.710 from the *Apollo 11* and *Apollo 12* sites, respectively—notably higher than the 0.699-0.700 actually measured in the crystalline rocks. The soils also have higher Rb/Sr ratios than the basalts.

To account for this, a second factor is proposed. Values for Rb^{87}/Sr^{87} and Sr^{87}/Sr^{86} from the lunar soils plot in a linear array on the 4.6-AE isochron. This scatter of points shows the soil to be a somewhat open system in the geochemical sense—that is, materials from outside the immediate area are introduced during formation of the regolith. Thus, these plotted differences result from intermixing of variable amounts of one or more *exotic* (foreign) components high in radiogenic Rb^{87} and Sr^{87} with debris from local sources. That is, while the lava fragments volumewise make up the bulk of the regolith, a subordinate amount of some "cryptic" or "magic" (terms used by lunar sample investigators!) components are postulated as also present. These components are either very old (~4.6 AE) or have been selectively enriched in Sr^{87} and/or depleted in Rb^{87}. If this enrichment is quite large, the cryptic component will, in effect, "swamp" the influence of the more abundant lava fragments and hence lead to apparently older ages. These components have not yet been isolated and identified, owing largely to the finely comminuted particle sizes in the soil fines or breccia matrix. However, KREEP glass and older KREEP-norite fragments are prime candidates for being magic components because of their high Rb/Sr and U-Th-Pb contents. The KREEP materials represent lunar rocks rich in radioactive isotopes that have not undergone fractionation since about 4.4 billion to 4.6 billion years ago. Labile parentless leads, and possibly strontium, may be concetrated in the glassy fractions of the fine materials. The existence of such components is supported by chemical data for the fragmental materials; these show compositional differences (especially in certain trace elements) with respect to the crystalline rocks which require that exotic components be present.

Thus, regolith-breccia ages all pertain to primordial formation or crystallization events. There is, however, no direct way by which the usual radiometric dating methods will determine the time interval during which the regolith itself developed as a sedimentlike deposit, although by measuring the nuclides produced by cosmic-ray bombardment it is possible to determine the time of deposition of sedimentary layers in the regolith. This was illustrated earlier in this chapter by the use of gadolinium isotopes to reconstruct the ages of layers in the *Apollo 15* deep core. Such techniques may also be used to determine the age of formation of deep craters. Material buried deeper than about 10 m is shielded from cosmic rays, but when excavated by impact and deposited on the surface the released rock fragments begin to acquire cosmic-ray exposure ages.

Looking back on *Apollo*, the sample-analysis program stands as the most intensive and best-coordinated effort of its kind in the history of science. The tremendous variety, accuracy, and ingenuity of techniques applied to the returned samples provided a striking demonstration of the power and range of modern analytical methods. Within months after the first samples were collected by Armstrong and Aldrin, scientists had developed a remarkably good picture of the composition and age of a planet on which man had never before set foot. The lunar-samples program serves as a lasting tribute to dedication and excellence in scientific endeavor–an achievement to be accepted with pride by the entire Earth sciences community.

REVIEW QUESTIONS

1. Define these terms: microbreccia, clast, ophitic, mesostasis, cumulates, BABI, AE.

2. List the principal crystalline (igneous) rock types found among the *Apollo* samples, and give some identifying characteristic of each.

3. Give the names of the *new* minerals discovered in the lunar rocks.

4. Discuss the implications of changing chemical (and mineralogical) compositions of the lunar pyroxenes.

5. Review in broad terms the major element geochemistry of the lunar lavas.

6. Cite the evidence for chemically reducing conditions in the Moon.

7. Compare the uranium contents of lunar rocks to terrestrial ultrabasic rocks. What is the reason for the low K/U ratio in lunar rocks?

8. Outline the major arguments favoring magmatic differentiation as a cause or factor in developing lunar lavas of varied compositions.

9. What is the significance of these samples: 12013? 14310? 15415?

10. How are most lunar glasses produced? What evidence tends to discredit the Moon as the prime source of tektites?

11. Describe the various effects of shock metamorphism observed in lunar samples.

12. Review the line of reasoning for impact origin of the lunar regolith.

13. Distinguish between lunar regolithic and annealed breccias; discuss the origin of each.

14. What have the *Luna* probes contributed to our knowledge of the Moon?

15. How are particle tracks formed? What do they tell us?

16. What elements make up the lunar "atmosphere"?

17. Review the results of organic matter studies of the lunar materials.

18. What is the difference between an isochron age and a model age? How is each determined?

19. List, in sequence of increasing age, the isochron ages obtained for crystalline (lava) rocks at each U.S. and Soviet landing site on the Moon.

20. State the "paradox" of the lunar soil-crystalline rock ages and summarize the explanations put forth to resolve it.

CHAPTER 12

Mars, Venus, and the Planets Beyond

In the previous seven chapters we have seen the extent to which unmanned and manned space probes are adding fundamental data drawn from observations, measurements, and sample collection to our knowledge and understanding of the Moon. This exploration program has clarified and advanced many of the basic concepts regarding our nearest neighbor and has generated important new ideas and models which further our insight into the nature and characteristics of the terrestrial planets. The same methods of investigation—measuring particles and fields with sensitive detectors, scanning surfaces with sensors that pick up visible, infrared, and ultraviolet radiation, and analyzing materials in situ with instrumental systems mounted on landing devices—can be adapted with suitable modifications for examination of the other planets. This approach succeeds in transferring the base of observation from resolution-limited Earthbound telescopes to mobile, close-in sensor platforms that fly-by, orbit, or land on the planetary targets. Since 1962, the United States has launched two probes to Venus, four others to Mars, and one to Mercury in its *Mariner* program, and two toward Jupiter in the *Pioneer* program. The Soviet Union has sent five spacecraft to Venus and six to Mars.

Pre-Mariner Knowledge of Mars

In this chapter we shall concentrate on the results of the *Mariner* missions, with emphasis on how they are leading to improvements in the conceptual models previously devised for the two planets whose orbits are nearest the Earth's. Let us begin by reviewing some of the major known or surmised facts about Mars that had been established prior to the *Mariner* flights.

Mars shows variable albedo in which the brightest areas (polar) are whitish, larger bright regions are reddish-orange to yellow-brown, and darker areas have a reddish-brown hue. A range of other colors have been reported for different regions of Mars. Several parts of the planet appear bluish-green through the telescope to some observers; because these same areas show up as shades of gray when examined photometrically on color film, it is considered unlikely that they are related to living vegetation in any way. The boundaries between colored regions are usually diffuse. Seen in total perspective, the regions appear in the telescope as patches, broad bands, streaks, and spotty mottling. As the seasons change, the dark areas become smaller or larger (some areas disappear and redevelop), vary in brightness, and spread out in a wave or zone between poles and equator. Using both telescopic observations and pre-*Mariner 9* photography, cartographers have produced a general map of Mars in which major patterned areas and distinct surface features are identified by Latin names (Figure 12-1). As late as 1965 Martian maps included darker connecting strips that depicted the "canals," which appeared to some viewers as narrow, more or less straight bands extending in a network joined at dark spots ("oases"). Exceptional viewing conditions through the telescope resolve some of these bands and streaks into series of close-spaced dots.

The lighter areas, called "deserts," are considered by some workers to consist of fine silt or dust of uncertain composition. Spectral analysis from Earth shows this material to match limonite (hydrated iron oxide) in such properties as photometric function, brightness, and color index. Hematite (ferric oxide), limonite-coated silicate grains, and weathered igneous rocks (andesite, basalt) have also been proposed as possible principal constituents in the desert regions. The darker areas (sometimes referred to as maria) are even less well understood than the lighter areas. A once-common belief held them to be manifestations of some form of primitive life such as algae or lichens. Another idea is that they are less-oxidized rock surfaces (basaltic?) periodically covered or swept clean of dust coatings by circulating wind currents. No distinct topographic relief is visible from Earth. Some dark areas may occupy depressions. Many of the brighter areas also occur in apparently lower areas; others extend onto possibly higher ground or plateaus. Whitish areas beyond the polar regions may be "frosted" uplands.

The polar caps are accepted as evidence of frozen substances that are partially thawed or sublimated as the seasons change. The larger southern cap initiates its annual growth in the winter by condensation of atmospheric substances, but the resulting ground deposits are obscured by mists that remain until the beginning of local Martian spring. The exposed ground "snow" begins to disappear in late spring. The frost breaks up into white patches interlaced with dark areas—including a dark band around its periphery. By summer the cap usually shrinks to a small near-polar covering or disappears entirely. The northern cap follows a similar cycle of change during its seasons. Considerable disagreement over the composition and thickness of the caps persisted up to the Martian flybys. Frozen water and frozen carbon dioxide (dry ice) were both suggested; prior to *Mariners 6* and *7*, opinions based on spectral studies favored the water-ice view, although some theoretical work supported carbon dioxide ice. Most interpretations leaned toward a very thin covering (millimeters to centimeters) comparable to hoar frost.

The pre-*Mariner* consensus held the Martian atmosphere to consist primarily of a mixture of nitrogen and carbon dioxide, together with very small quantities of argon, water vapor, oxygen, and other constituents. Most estimates placed surface pressures between 20 and 125 mbars, with the preferred average being about 8% of the terrestrial sea level value of ~1000 mbars. Nitrogen was considered the most likely major constituent; carbon dioxide would exceed nitrogen only if the total pressure were to prove less than 12 mbars.

Little is known about the interior of Mars, owing in part to uncertainty about its surface composition. Using a mean density in the range of estimates between 3.94 and 4.02 g/cm^3 and a value of 0.389 for the moment-of-inertia constant (compared with 0.400, the constant in the moment-of-inertia equation of a solid sphere), one can fit the mass distribution within Mars into two plausible models: (1) a two-layered interior consisting of an outer mantle of silicates (estimated to be from 200 to 2000 km thick) whose assumed densities lie between 3.7 and 4.0 and an inner core composed of substances whose densities have values between 8.1 and 10.3 (if nickel-iron) and between 4.2 and 4.4 (if metallic silicates); or (2) a three-layered interior in which a lighter outer crust (density around 3.3) of 200-600-km thickness occurs in addition to mantle and core layers. The proportion of total mass present in the core ranges from 0.09 to 0.19, depending on the model used. Estimates of maximum temperatures in the region of the core boundary fall between 2100 and 2700°C.

The Mariner Mars Program

Mariner 4, the first probe to Mars, was launched on November 28, 1964, and swung past the planet on July 14, 1965. This spacecraft took twenty two TV pictures during a 26-minute interval starting at a distance of 16,800 km and continuing until it closed to within 9850 km of the Martian surface. At that time, Mars was about 210 million kilometers from Earth. *Mariner 6* was launched from Cape Kennedy on February 24, 1969, and commenced to transmit pictorial data on July 29, 1969, at a distance of 1,240,800 km from the planet (*far-encounter series*). Data were collected through its closest approach (*termination or near-encounter series*) at 3430 km on July 31, 1969. A companion spacecraft, *Mariner 7*, took off on March 27, 1969, some thirty one days after *Mariner 6*. It arrived in the vicinity of Mars on August 2, 1969, only five days after *Mariner 6* because the planet had moved some 74 million kilometers closer to Earth in the month between the two launches. *Mariner 7* started its far-encounter sequence when the spacecraft was about 1,716,000 km from Mars and completed its near-encounter sequence at a distance of 3430 km. Both spacecraft performed their tasks when Mars was about 100 million kilometers from Earth, or less than half the separation at the time of the *Mariner 4* flyby.

The improvements in detail brought about by the close proximity of high-resolution cameras on these probes is

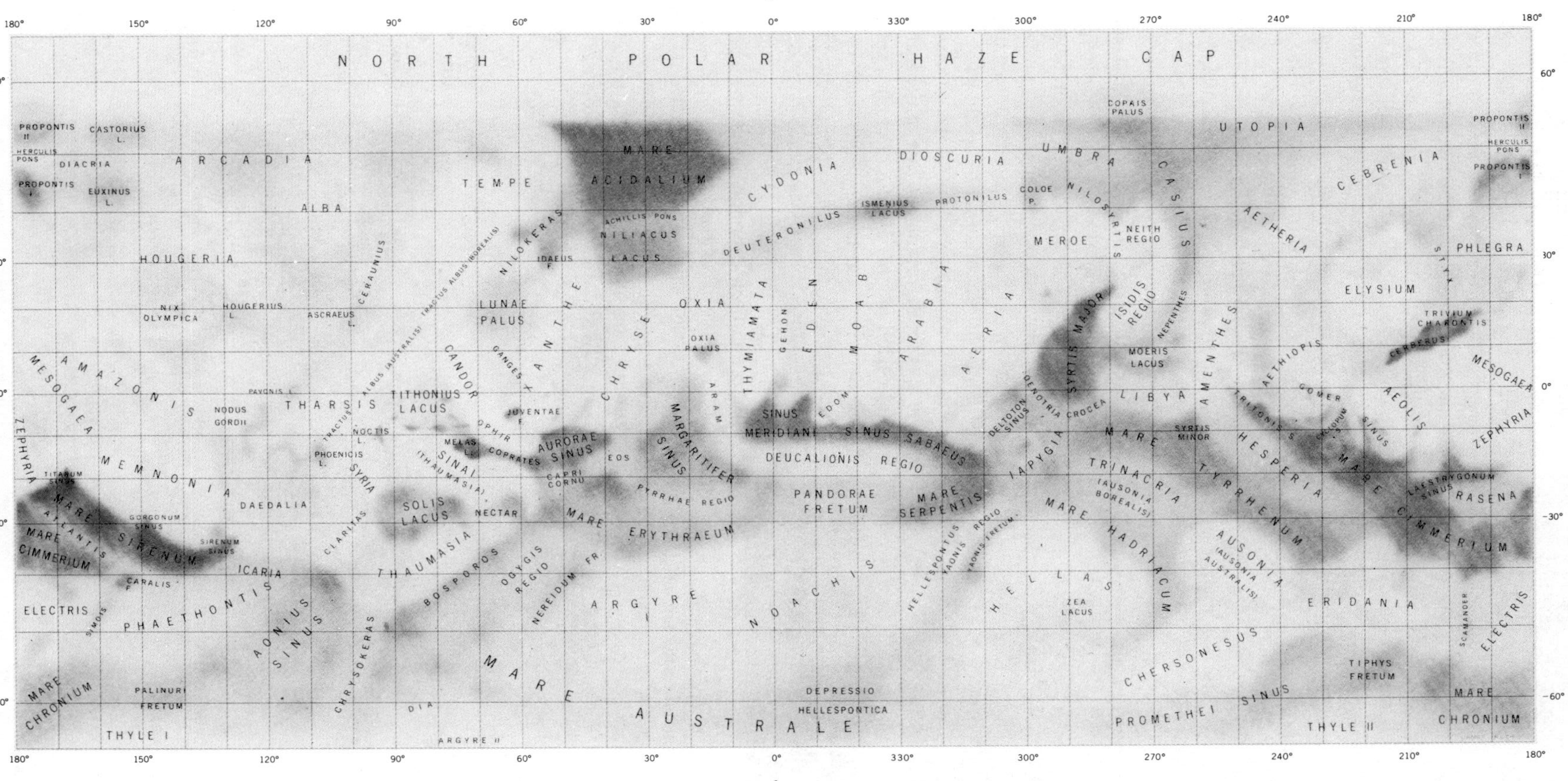

Figure 12-1 A general map of Mars, prepared by G. de Vaucouleurs and others, used in planning the *Mariner* 1971 missions. On this Mercator projection, the scale at the Martian equator (0°) is 1:25,000,000 (a 10° latitude interval there represents 570 kilometers). (Courtesy NASA, Jet Propulsion Laboratory.)

Figure 12-2 A far-encounter view of Mars taken when *Mariner 7* was 514,810 km from the nearest point on the surface. The small, dark patches near the apparent center of the surface observed here lie between Thaumasia and Tharsis (see Figure 12-1). Several narrow, dark linear markings (suggestive of the so-called canals) can be seen in the vicinity of Ogygis Regio. (Photo by Jet Propulsion Laboratory [JPL]).

readily apparent by comparing even a far-encounter image from *Mariner 7* (Figure 12-2) with one of the best telescope photos ever taken from Earth (Figure 2-8a). Many of the larger patterns representing the first-order surface features depicted on the Mars map (Figure 12-1) are clearly defined by the strong photographic contrasts. Certain areas have notably lower albedos suggestive of lava-filled maria.

In May 1971, NASA launched two more spacecraft toward Mars on missions designed to map 70% of the red planet from orbit with a variety of instruments. *Mariner 8* was lost due to a failure after launch, but on November 13, 1971, *Mariner 9* became the first probe to be successfully inserted into orbit around another major planet. Because of the loss of its companion, the original exploration pattern was changed to allow the single spacecraft to perform most of the tasks which were designed for the spacecraft pair. The orbit was highly elliptical and was initially inclined at 64°20′ to the Martian equator. The periapsis altitude (point of closest approach) initially had been set at 1380 km, but later the orbit was trimmed to a 1650-km periapsis altitude in order to obtain wider viewing to complete the mapping coverage in the remaining mission time. The spacecraft completed two orbits each Earth day (orbital period: 11 hr. 59 min. 28 sec). Typically, thirty one TV pictures were taken during each orbital revolution. Thus, in a single day, this probe produced almost as many high-resolution pictures as either *Mariners 6* or *7* had obtained in their flybys. In March 1972, after some 225 revolutions had been completed, the spacecraft was temporarily "shut down" as part of its orbit passed into the Mars shadow zone. The *Mariner 9* resumed its data gathering in May 1972 on a limited acquisition schedule until final shutdown on October 27, 1972, after 698 orbits around Mars. By then, more than 7300 large- and small-scale pictures of the Martian surface had been received—a

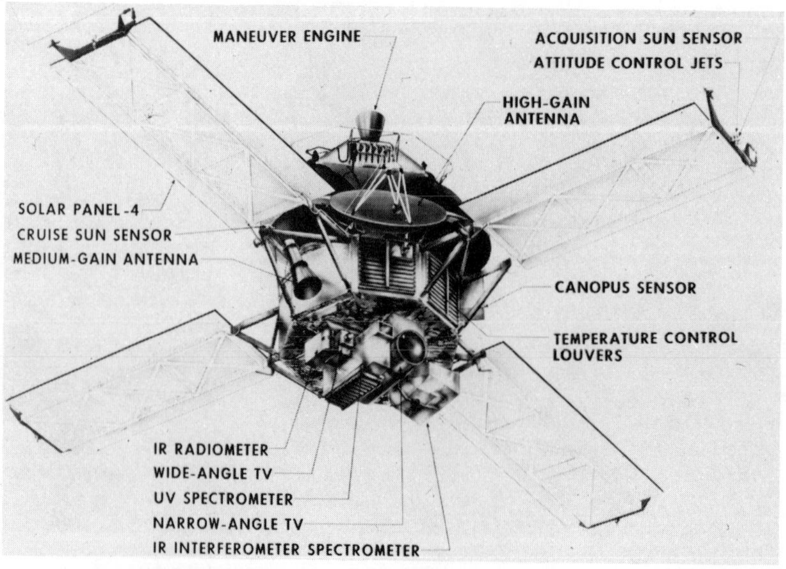

Figure 12-3 A diagram of the underside of the *Mariner* Mars 1971 spacecraft showing the various sensors used to obtain UV, IR, and visual data during detailed mapping of the Martian surface. This probe is a more advanced model of the earlier Mars *Mariners.* (NASA photo.)

bonanza that rivals the achievements of the *Lunar Orbiter* program.

The four *Mariner* Mars probes are similar in design (Figure 12-3). Powered by solar radiation (collected at the panels) and auxiliary batteries, various sensors on each were installed to perform specific tasks. *Mariner 4* carried a magnetometer for measuring the Martian and interplanetary magnetic fields, an ion chamber and a cosmic-ray telescope for cosmic-ray studies, and a single narrow-angle-lens TV camera system (focal length of 305 mm; alternating red and green filters; resolution at closest approach of ~5 km). *Mariners 6, 7,* and *9* were equipped with an ultraviolet spectrometer capable of detecting such gas species as H, O, N, N_2, N_2^+, CO, CO^+, CO_2^+, CN, Kr, and Xe in the Martian atmosphere and an infrared radiometer (one channel in the 8.5-12.4 μm band and another in the 18-25-μm band) used to detect thermal emission for surface and polar cap temperature determinations. *Mariners 6* and *7* also had an infrared spectrometer (IRS) operating between 1.9 and 14.3 μm in two channels. It was designed to look for compositional data from the atmosphere (particularly in the 1.9-6.0-μm range) and from the surface (near 10 μm) and to determine surface temperature and topography along the ground track. An infrared interferometer spectrometer (IRIS), sensing between 5 and 50 μm, replaced the IRS on *Mariner 9*. Its prime function was to obtain vertical temperature profiles in the atmosphere and measure water vapor, but it could also provide information on composition of atmospheric materials and on surface elevations.

There were two TV cameras mounted on *Mariners 6* and *7*. One had a wide-angle lens (focal length of 52 mm) and a red, green, and blue multispectral filter wheel by which scene contrasts (related to albedo) in the far-encounter series are emphasized and with which color photos could be reconstituted; its optimum resolution approached ~3 km. The second camera used a narrow-angle lens (focal length of 508 mm); optimum close-in resolution was near 300 m.* The 50 mm wide-angle lens on the *Mariner 9* camera was equipped with a filter wheel containing five

*The wide-angle lens on all *Mariners* has an effective field of view of 11° by 14°; the narrow-angle lens on *Mariners 6, 7,* and *9* has a field of view of 1.1° by 1.4°.

(a)

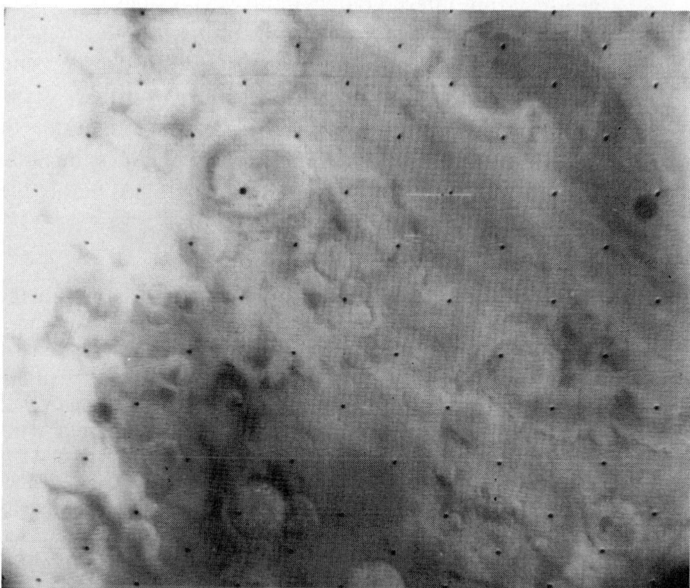

(b)

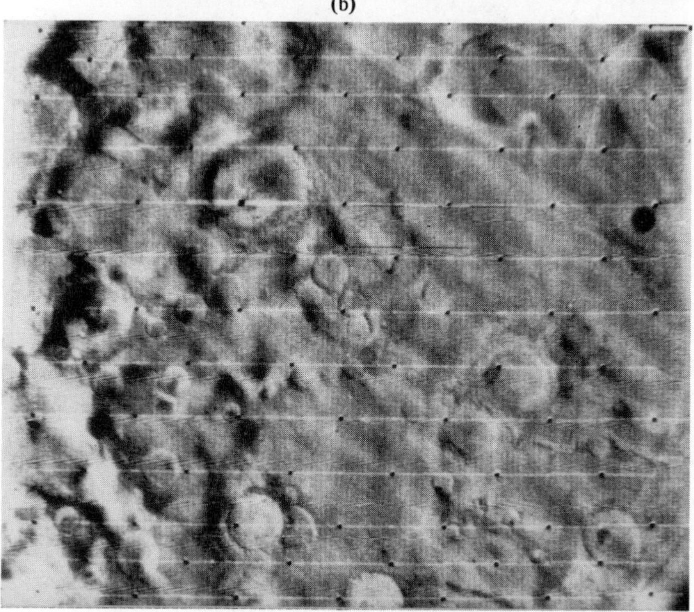

(c)

Figure 12-4 A region in the south polar latitudes, observed by *Mariner 9* in December, 1971, during the waning phase of a great dust storm. This photo sequence illustrates the improvement in surface details that can result from controlled reprocessing of the original digital data to remove certain geometric distortions, filter out undesired noise from the received signals, and correct for non-uniform photometric response. These enhancements lead to better photo contrasts and apparent improvement in resolution. (a) "Raw" uncorrected image. (b) Photo with shading corrected. (c) Final reprocessing through a high-pass filter. (NASA, Jet Propulsion Laboratory.)

color and three polarization filters. At periapsis in the trimmed orbit this camera could image an area about 450 km by 570 km (A-frame) at a resolution of about 1 km. The narrow-angle lens (500-mm focal length; equipped with a single yellow filter) provides close-ups (B-frame) of about 1-2% of the area of the broader, synoptic views at resolutions ranging between 100-300 m. After being received on Earth, the TV pictures are reprocessed to enhance low-contrast surface details (Figure 12-4). Although albedo differences rarely exceed 15%, the photographic products are made to show a much greater dynamic range in contrast to emphasize subtle differences.

All *Mariner* spacecraft were directed by small on-board electronic computers programmed to carry out the operational sequences. The *Mariner 6, 7,* and *9* computers could be reprogrammed on command by radio to vary the sensor tasks in response to new target selections based on near-real-time interpretations of images previously received. The radio transmission itself proved useful as a scientific experiment for determining pressure, gas density, and electron density within the Martian atmosphere. Variations in these properties can be measured in terms of changes in both frequency and signal strength as the radio waves pass through the Martian atmosphere just before the *Mariner* flight path curves behind Mars (giving rise, in effect, to an *occultation*, in which direct line of sight to the spacecraft is lost). Because the atmosphere acts as a refracting lens in which signal attentuation and delay at each altitude are related to density, pressure, etc., information about the structure of the atmosphere can be directly obtained.

Figure 12-5 Cumulative curves showing crater abundances in the Deucalionis Regio area, measured from *Mariner 4* near-encounter frames, compared with crater frequency distributions in two lunar maria (from *Ranger* photos) and two highlands regions (south pole, from *Orbiter 4* and Tsiolkovsky on the farside, from *Orbiter 3*). The crater distribution on Mars appears to include more larger craters and fewer smaller craters than typical lunar maria and less craters than lunar highlands unaffected by mare lavas. Other studies indicate that the cratered Martian surface approaches saturation for larger sizes but becomes progressively less saturated as sizes decrease. (B. C. Murray et al., *Journal of Geophysical Research*, Vol. 76, pp. 313-368, 1971.)

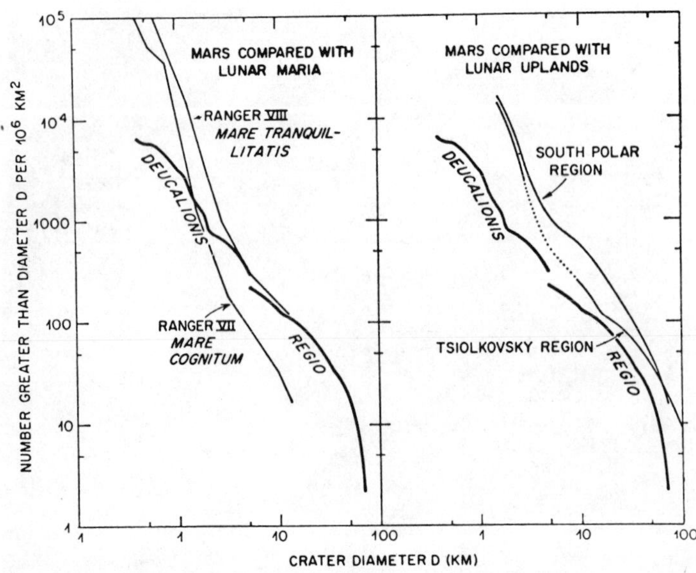

Achievements of Mariners 4, 6 and 7

The principal results from the first three probes in the *Mariner* Mars program are summarized as follows:

Evidence of Cratering

Prior to *Mariner 4* few workers had seriously considered the predictions by Öpik and by Tombaugh some fifteen years earlier that Mars would be heavily cratered and thus would more closely resemble the Moon than Earth. Thus, most scientists were genuinely surprised when the first TV pictures from *Mariner 4* revealed a relatively large number of shallow circular structures from a few kilometers up to 120 km in diameter (see Figure 1-10). Cumulative frequency plots (Figure 12-5) of all visible craters in the twenty two analyzed frames from *Mariner 4* (supplemented now by data from *Mariners 6* and *7*) indicate diameter distributions that fall between those for average lunar highlands and maria. This observation soon generated a lively debate on the age of the cratered surfaces. Reasoning tends to follow the line adopted for the estimates of lunar ages—namely, various flux rates were postulated and equilibrium and saturation conditions were assumed where applicable. A complication enters when the comparative shallowness (low depth-to-diameter ratio) of the craters relative to those on the Moon is considered. Isostatic adjustments, infilling by lavas, dust, or desert sand, and initially low excavation depths owing to softer surface materials were put forth as possible factors. Two schools of opinion arose. One view favored a young surface whose formation age ranged from less than 300 million years to perhaps 800 million years to account for an apparent deficiency of craters on Mars compared with the lunar highlands; however, a numerical error in the preliminary announcement of crater densities on Mars led to large underestimates of these ages. The other view proposed an ancient surface, whose age falls between 2 billion and 5 billion years, that has subsequently been modified by erosion processes resulting from atmospheric and/or water "weathering" and by deposition of windblown particles. The "old surface" interpretation now prevails, since the age dates of 3.2 billion to 3.6 billion years have been assigned from *Apollo* sample studies to the somewhat less cratered lunar maria. This, of course, assumes similar infall rates for meteorites and comets within the region through which both Moon and Mars passed during the

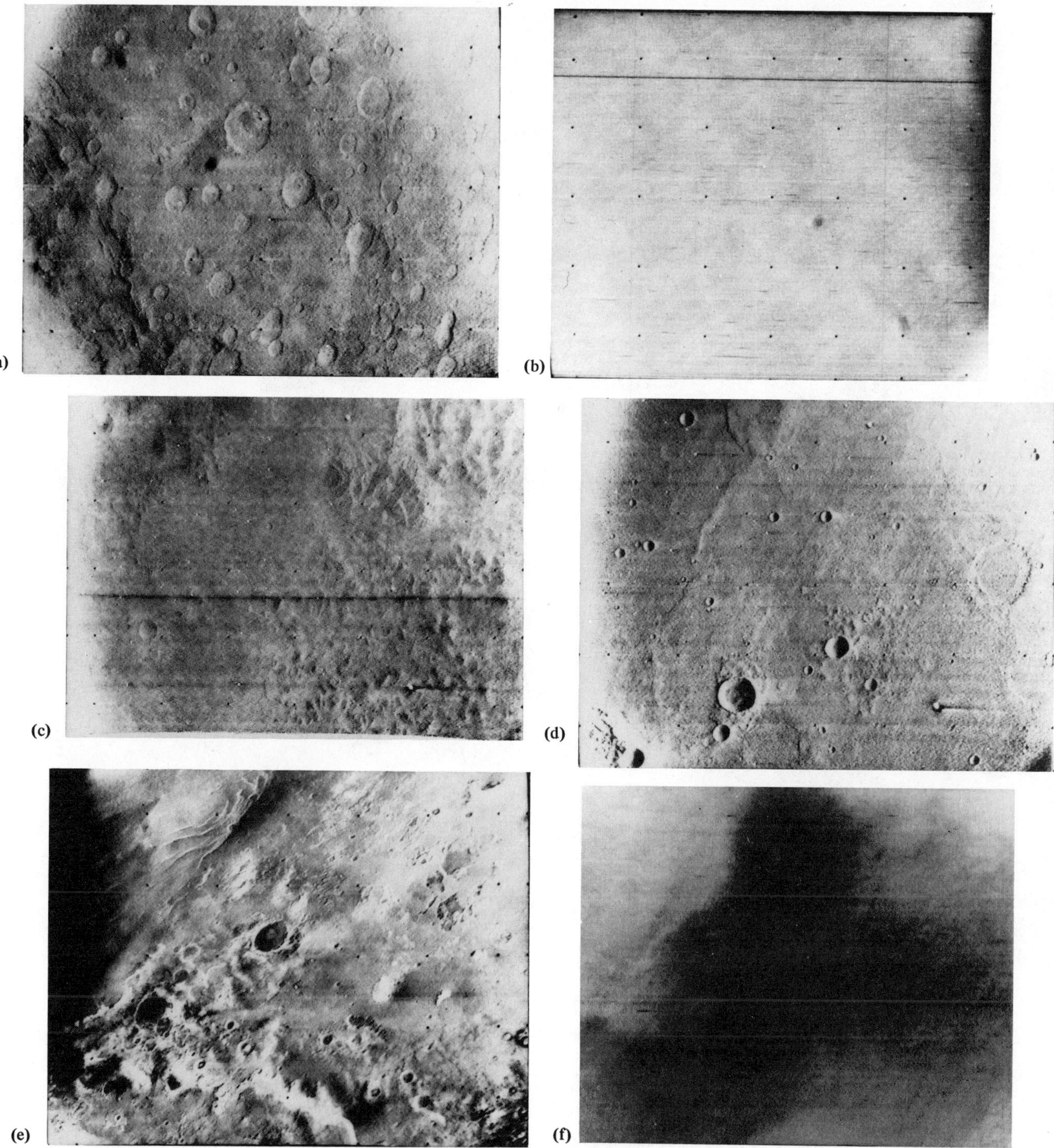

Figure 12-6 (a) The heavily cratered Hellespontus region from frame 7N25, covering an area of 1228 km width. Note the linear ridges in the lower left. (b) "Featureless" terrain in the Hellas desert to the east of the Hellespontus region, from frame 7N29. (c) Chaotic terrain within Pyrrhae Regio, consisting of knobs, stubby ridges, and troughs, frame 6N6, covering a lateral width of 128 km. (d) Terrain typical of the Deucalionis Regio area, showing a large steep-walled, flat-bottomed crater (right), several bowl-shaped craters, a smoother inter-crater region, and a sinuous ridge. Frame 6N20, width of photo equivalent to 89 km. (e) Part of the south polar region (pole near lower right) showing a CO_2 frost-covered surface over much of the area. Dark-floored craters (some over 100 km wide) and broad, dark "pits" and "furrows" extend through this "snow." Near the bottom right is a cluster of "quasi-linear markings" which may be depressions, ridges, scarps, or dunes. Frame 7N17, covering a width of 1546 km. (f) A portion of the Martian surface, including part of Syrtis Major, Deltoton Sinus, and Iapygia, seen as a far-encounter view (frame 7F88) from *Mariner 7* at a distance of 155.270 km. This photo illustrates the sharp contrast between the dark, mare-like surface and the lighter "desert" region. (Photos by NASA, Jet Propulsion Laboratory.)

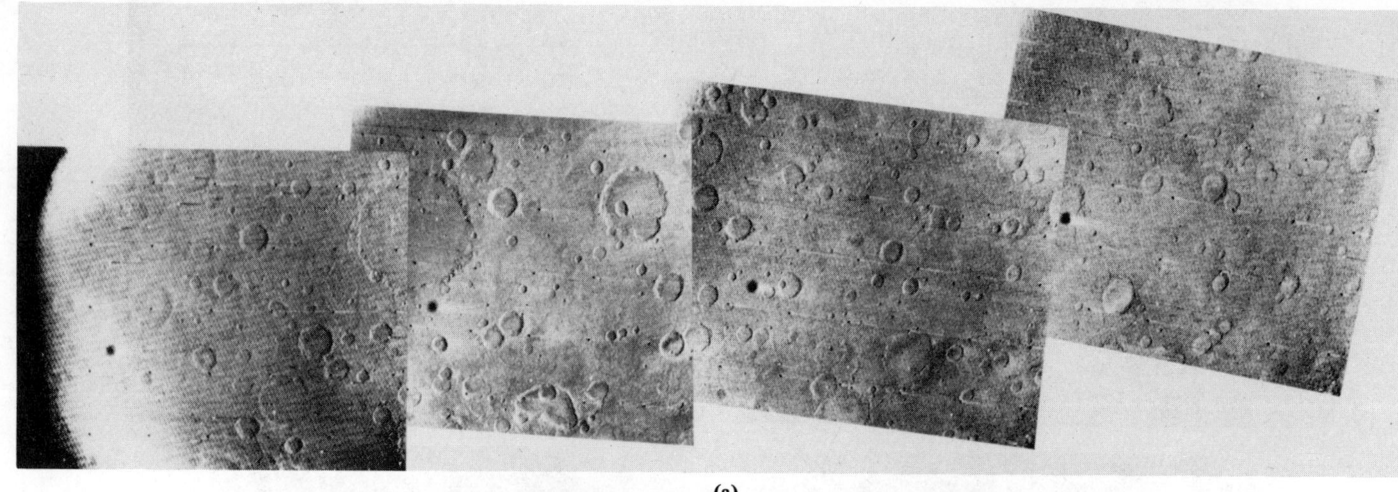

(a)

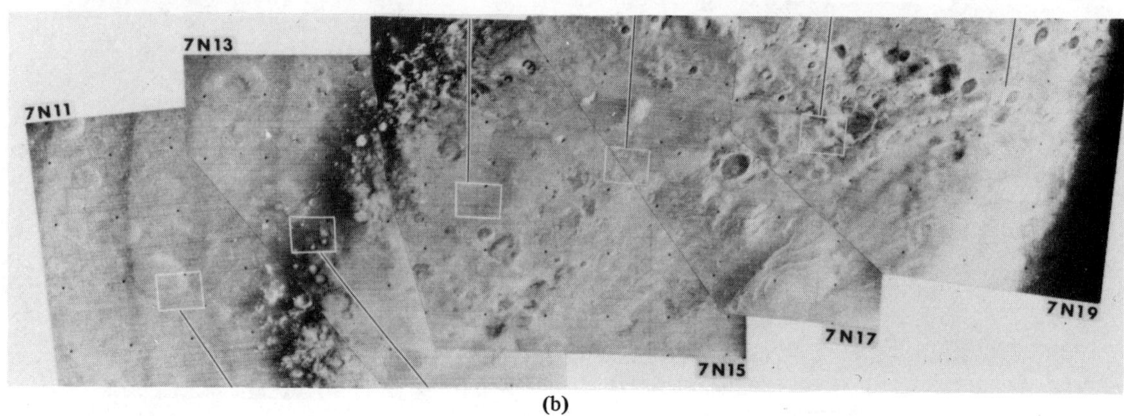

(b)

Figure 12-7 (a) A mosaic strip constructed by overlapping four near-encounter *Mariner 6* frames (6N17-19-21-23), each equivalent to about 100 km in the long direction, to show the character and distribution of the larger craters in Deucalionis Regio. (b) A panoramic strip made from the indicated *Mariner 7* frames to depict the terrain around the Martian south pole. An automatic gain control (AGC) in the onboard TV system enhanced brightness contrasts to allow maximum discrimination of surface features, but the inability of AGC to respond to abrupt changes in brightness level produced a false dark band near the frost-cap edge. This photo version indicates that frost tends to concentrate in low areas such as crater bottoms and along slopes out of direct sunlight. Without the effect of AGC, reprocessed images show a uniformly bright polar cap, devoid of interior features, spread mainly between 60° and 80° south (much diminished at the pole itself). (Photos by NASA, Jet Propulsion Laboratory.)

early history of the solar system. Furthermore, the conclusion favoring an ancient surface for Mars implies the absence of a subsequent ocean, of intense subaerial erosion, and of major mountain-building processes. Both views have received support from the *Mariner 9* photography which shows heavily cratered regions interspersed with smooth, sparsely cratered plains.

Mariners 6 and *7* expanded the area of coverage from the 1% seen by *Mariner 4* to about 10%. A greater diversity of terrain types (Figure 12-6) was revealed by the thirty wide-angle and twenty-eight narrow-angle near-encounter frames taken by the spacecraft. Panoramic strips (Figures 12-7a and b) prepared as montages of the individual frames aid in defining the interrelationships among these terrains.

Thus, as in *Mariner 4* views, heavily cratered surfaces extend over much of the desert areas and in some of the darker regions. Craters from 50 to 75 km in width are common, and one approximately 500-km-wide structure named Nix Olympica, an oval ringed blotch clearly visible in telescope photos, in the far-encounter photos superficially resembles the basin underlying Mare Orientale on the Moon. Most large craters are flat-bottomed and shallow, but almost no volcanic fillings, such as occur in Tycho and Aristarchus, are discerned within them. Central peaks are sometimes developed, but well-defined rays and blankets are covered or absent. Less conspicuous are the (generally smaller) circular depressions whose comparative sharpness and bowl-shaped interiors mark them as younger craters not destroyed by erosion or infill. In contrast to these impacted regions, a previously unobserved terrain type, flat and smooth, nearly devoid of craters, and hundreds of kilometers in extent was noted in the Hellas desert. If cratering were once Marswide, as generally presumed, this low, broad, "featureless" area may be a large impact basin now extensively filled with erosional debris. In some areas scanned by the *Mariners* an assemblage of irregular, intersecting knobs, ridges, and valleys make up a third terrain type–called "chaotic"–in which some linear features appear to follow preferred orientations while others are distributed in a random, jumbled pattern. This landform, found over much of the equatorial region of Mars, is topographically lower than the terrain around it. Its origin is still unclear, but frost-heaved ground, volcanic collapse, or flow structures are plausible inferences. Still other features, which resemble some sinuous rillelike depressions, domes, fault scarps, and low-albedo units (lavas?) seen on the Moon, have been recognized on the Martian surface.

None of the *Mariners* observed any regular, continuous markings that would correspond to the canals. Instead, areas once identified with these features are resolved in patches of unconnected but close-spaced dark surfaces or rows of large craters. However, other light and dark patterns visible in the far-encounter photos lack any obvious correlations with the topography indicated in the close view of the same areas.

Surface Temperatures

Various measurements of the temperature of the surface material indicate an average of 290°K at Martian noon near the equator, dropping to about 200°K on the nightside of the planet. During maximum illumination, temperatures varied with latitude to a low of ~230°K near the polar caps. There is also a seasonal variation (owing to the inclination of the Martian axis of rotation) such that temperatures decrease from summer to winter by about 20 to 40°K in the mid-latitudes and about 30°K in the polar regions. If the Martian surface were to behave as an ideal blackbody (perfect radiator), its surface temperature (if determined solely by solar heating) would be ~225°K.

Atmosphere Characteristics

The results from *Mariner 4, 6,* and *7* experiments concerning the Martian atmosphere are internally consistent. They indicate the atmosphere to be composed almost entirely of CO_2, with small amounts of H_2O and some CO and H. Surface pressures typically are ~6 to 7 mbars but locally are higher over some depressions. Neither CH_4 nor NH_3 were detected, and nitrogen is low. The atmospheric CO_2 does not condense into visible clouds, but there is particulate matter in the "air" which may account for the haze noted in some *Mariner 7* views of the planetary limb. This aerosol makes up a layer 8 to 15 km thick that begins from 10 to 50 km above the surface. Several diffuse bright patches near the edge of the southern polar cap that appeared to change as the spacecraft approached Mars may be ice fog formed as water experienced diurnal volatization. The *Mariners* also found evidence of an ionosphere (a region of ionized gases and electrons built up as solar radiation ionizes some atoms within the very thin upper atmosphere) whose maximum charge density lies near the 125-km level.

Polar Cap

Mariner 7 closely inspected the southern polar cap as it appeared late in the Martian southern winter (Figure 12-7b). Thermal and composition measurements confirmed the presence of either solid CO_2 or a fog of CO_2 particles suspended above the surface. A temperature of 150°K

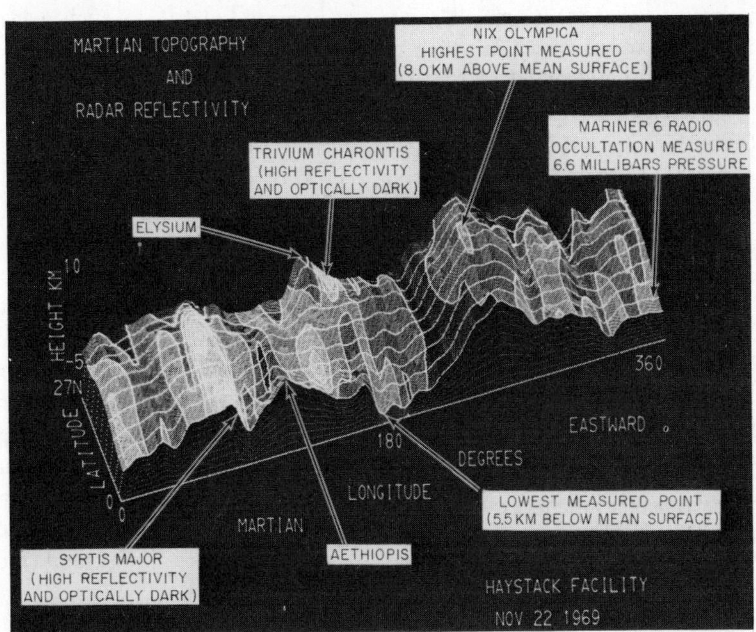

Figure 12-8 A schematic diagram of Martian topography for a circumferential belt (from the equator to 27°N) in which relative heights (relief) have been determined from radar reflectivity data obtained at the M.I.T. Haystack Observatory on November 22, 1969.

suggests that the cap is solid CO_2. The ground "snow" only thinly veneers the underlying cratered surface, so that thicknesses less than 100 cm (but piled locally into drifts exceeding 10 m) have been estimated. Elsewhere, the central depressions of some craters appear dark, as though the solid CO_2 had been removed, whereas their rims are conspicuously whitened owing to retention or reprecipitation of the snow at higher, colder elevations.

Surface Relief

Interpretation of radar data indicates maximum relief ranging up to 17 km (Figure 12-8) over the entire planet. Variations in atmospheric density at the surface, determined from *Mariner* data on CO_2 distribution, can be produced by differences in elevation from 4 to 5 km within the limited areas sampled.

Magnetic Fields

Mariner 4 detected no magnetic fields, although a moderately strong field (~10 to 15% that of Earth) had been predicted for Mars based on comparison to Earth and assuming some metal core. This at least places an upper limit on the strength of the magnetic field as less than the sensitivity of the magnetometer, which was able to respond to a field about 0.03% that of the average for the Earth. The trapped radiation and solar-plasma detectors showed values consistent with a Martian magnetic field strength between 0.1 and 0.2% that of the Earth's field. With such a weak field, no radiation belts would develop in the spatial environment around Mars.

Martian Interior

The spacecraft offered no direct new evidence on the interior of Mars. Absence of a detectable magnetic field indicates either (*a*) no metallic core, or (*b*) a solid nickel-iron core, or (*c*) a core in which the rate of movement of viscous or liquid metal is insufficient to generate strong electromagnetic fields. Binder (1969) calculates a hot (800-1500°C) but solid core of diameter between 1600 and 1900 km, comprising no more than 5% of the total planetary mass.

The Mariner Orbiting Mission*

The early Mars *Mariners* presented a picture of a planetary body which seemed to be as static and inactive (i.e., as "dead") as the Moon. In this view, the Martian surface was heavily cratered at the onset of its history, but its Moonlike terrain is gradually being obliterated by impact erosion and modified by wind scouring and dust covering that result from movements of a thin, low-pressure atmosphere of carbon dioxide. It now appears coincidental that these probes chanced to photograph parts of Mars dominated by ancient impact topography. *Mariner 9*—with its far more extensive survey of Mars, forced a reconsideration of the assumption of an extinct planet. A wide variety of unexpected surface features was revealed. Many seem quite young relative to the old cratered regions, and some show signs of continuing activity. These signs include volcanic shields, calderas, and flows; faulting of the outer crust; great canyons apparently enlarged by flowing fluids; areas with characteristics reminescent of widespread glaciation; vast regions affected by such aeolian action as scouring, grooving, and dune deposition; and rapidly changing transient deposits of windblown origin. Although many terrestrial phenomena associated with marine sedimentation, geosynclinal deformation, and water erosion seem to be absent on Mars, the planet still must be rated as a dynamically changing body whose surface has been experiencing a transitional growth between that of the passive Moon and a very active Earth. As a guide to the evolutionary sequence followed by Earth, Mars may well be in a stage analogous to that which marked our planet about 3 billion years ago as it first accumulated a primitive atmosphere. That Mars will continue to develop along lines now established on the Earth is precluded by its inability to retain an organized hydrosphere.

The Great 1971 Dust Storm

As *Mariner 9* sped toward Mars, astronomers observing the planet by telescope noted in late September of 1971 the buildup of a yellowish cloud in the mid-southern latitudes. By early October this dust storm had spread over all of Mars, totally obscuring the surface except for bright areas around the polar caps (Figure 12-9a). Since the last such planetwide storm had occurred sixteen years earlier, Martian experts had not anticipated this undesirable interference in planning the mission, even though these storms are more likely near perihelion. The severity of the storm—the longest and most extensive ever documented—threatened most major objectives in this flight. Fortunately, the storm was strongly abating by late December, and good viewing conditions thereafter continued to improve, allowing almost all tasks to be carried out in the following two months. In fact, this storm ranks as a serendipitous event in that considerable insight into atmospheric dynamics and other meteorological phenomena on Mars emerged along with a better understanding of the interrelations between aeolian processes and the variable surface markings that had puzzled astronomers in the past.

The magnitude of this storm is hard to imagine by Earth standards. Winds of CO_2 gas blew continuously for several months at velocities estimated to exceed 200

*A comprehensive review of scientific results from this mission is provided by the 34 papers appearing in the July 10, 1973 issue of the *Journal of Geophysical Research*, v. 78, No. 20.

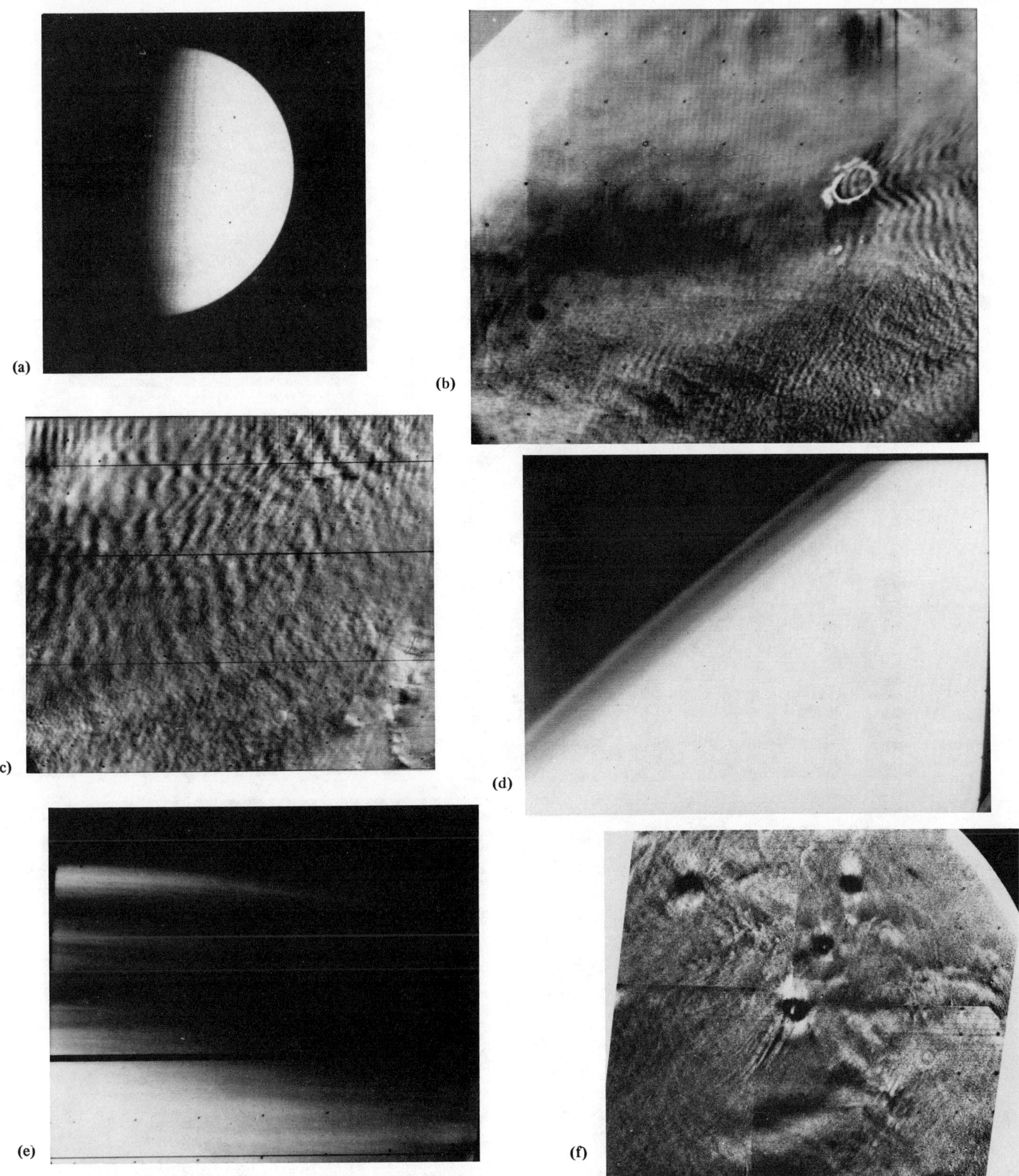

Figure 12-9 (a) Obscuration of the Martian surface by a planetwide dust storm, as viewed by *Mariner 9* in a far-encounter photograph taken at a distance of 710,000 km. (b) Wave clouds of CO_2 in the atmosphere of Mars' northern hemisphere, in the Mare Acidalium region. The waves appear "diffracted" around the frost-covered rim of a circular structure (upper right) (MTVS number 4208-90). (c) Close-up details of wave clouds. These typically have crest-to-crest wavelengths of about 40 to 50 km (4203-93). (d) Cloud layers at various altitudes observed in a limb-view across the Martian surface, most of which was still covered by dust-laden clouds (4056-1). (e) Detailed view of different cloud decks in the atmosphere (4158-23). (f) Enhanced photos (mosaic of 4) of the Martian surface above Tharsis showing four prominent spots breaking through the atmospheric dust storm. The single spot in the upper left coincides with Nix Olympica. The three aligned spots (North, Middle, and South Spot—the latter coincident with Nodus Gordii) were later identified as volcanic structures. (NASA, Jet Propulsion Laboratory.)

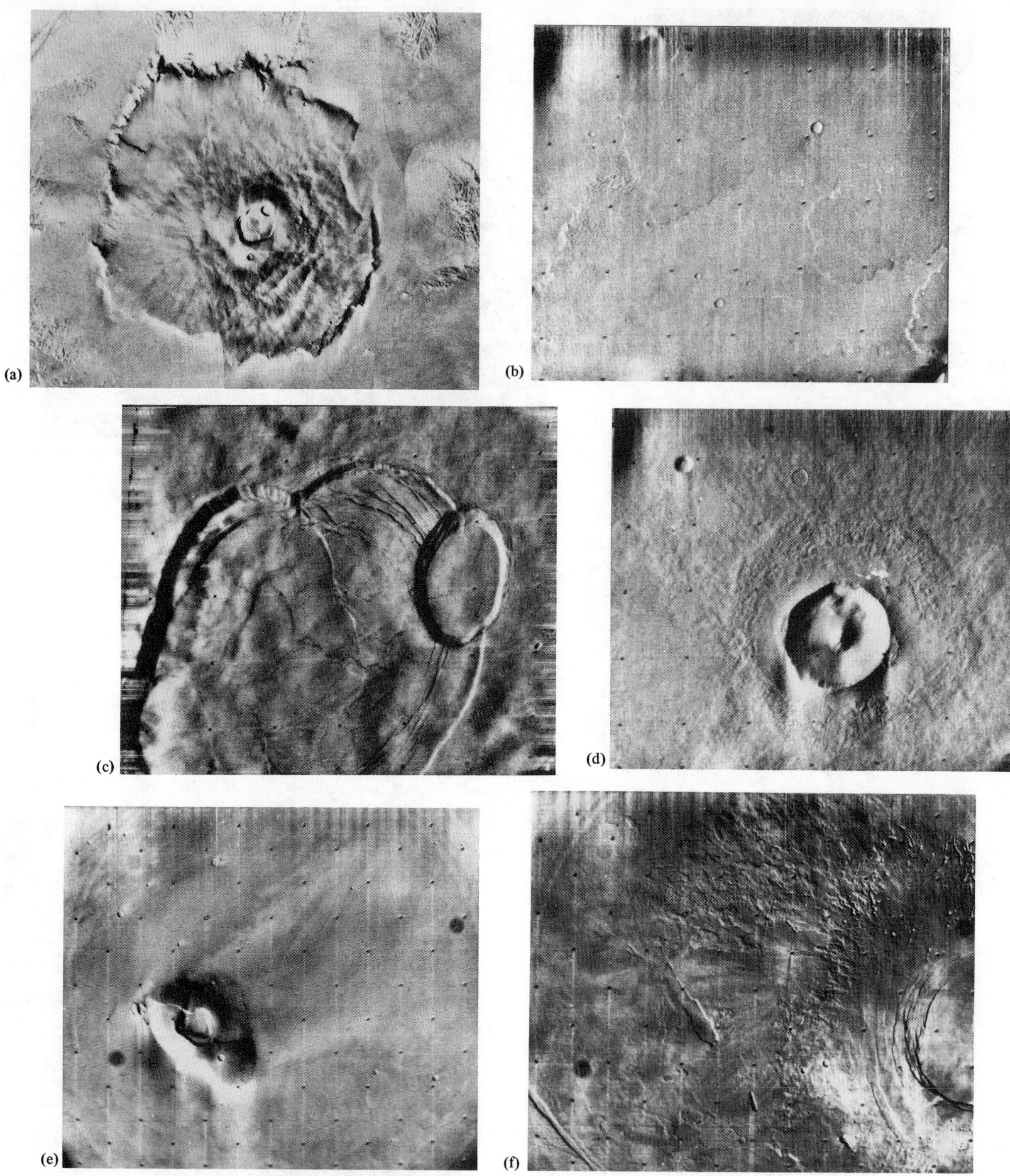

Figure 12-10 Four-photo *Mariner 9* mosaic of Nix Olympica, a huge shield volcano south of Hougeria, with a summit caldera rising almost 25 km above the mean surface of Mars, flank channels and flows, and a prominent 3-km-high scarp (wave-cut?) at the base (Orbit 146). (b) Lobate lava fronts, similar to those in basaltic fields on Earth and the Moon, on the western slopes of Nix Olympica (4179-30). (c) Details of the central caldera in Nix Olympica (4265-52). (d) Shield volcano with a broad stratocone in its central caldera, Casius region (4230-66). (e) A domed volcano constructed from several coalescing bulbous cones capped with a summit crater showing several levels of collapsed floors. One or more large faults offset the rim (4189-72). (f) South Spot, or Nodus Gordii, another shield volcano with a large central caldera. Concentric horsts and grabens surround the rim; and flow ridges, channels, and tubes appear along the outer flanks (4182-45). (NASA, Jet Propulsion Laboratory.)

km/hr—comparable to strong hurricane winds. These lifted particles as large as 10 μm into suspension many kilometers above the surface. The opacity of the dust in the atmosphere, as well as its substantial scattering properties for both incident and reflected light, resulted in an almost total obscuration of surface details. Albedos of 0.3 to 0.4 were measured on the dust envelope. Even as the Martianwide storm subsided, individual dust clouds as large as 500 km in diameter persisted for some time thereafter. Still other cloud types (diffuse, streaky, and streaming) were evident in close-up pictures (Figures 12-9b and 9c). Some of these clouds formed distinct wave trains resulting from a simple harmonic-motion effect as the moving atmosphere encountered obstructions such as craters and prominences, rose and cooled, and then moved downward as a heavier "air" parcel into the surrounding gases. After the dust had settled, whitish clouds could still be seen, especially in the winter climate of the northern hemisphere. Discrete cloud layers were particularly evident in limb views (Figures 12-9d and 9e). The lower cloud layers were probably composed of water-ice, but higher haze layers could have been condensed CO_2.

During the peak of the storm, four large dark spots in the general vicinity of the Amazonis-Tharsis region stood out against the background of the obscuring dust layers (Figure 12-9f). These spots appeared to be topographic features protruding through the dust clouds and, indeed, were later shown to rise at least 6 to 8 km above the general level of the Martian surface.* The single isolated spot (upper left in the figure) is identified with Nix Olympica, and the three spots in a row (now referred to as North, Middle, and South Spots) coincide with darker areas on the Tharsis plateau noted from previous Martian observations.

Volcanism on Mars

As the dust cleared, these spots were gradually resolved into broad, shieldlike (convex-upward) prominences that contained one or more irregular, intersecting craters at their summits. Occasional small, transient water-ice clouds, appearing as diffuse bright patches, have been noted near the tops of Nix Olympica and South Spot. When fully seen, Nix Olympica (Figure 12-10a) appears as a huge volcanic pile (500 km wide at its base) whose form resembles the island of Hawaii (itself a basaltic landmass some 225 km wide at its suboceanic floor). Close-up views of Nix Olympica's gentle slopes and the flatlands beyond reveal lobate flows of basaltic character (Figure 12-10b) and lava channels (see Figure 12-13e). Similar flows are observed beyond the volcanoes in the Amazonis basin. The steep-walled central crater, some 65 km wide, closely matches in morphology the large calderas in the Hawaiian and Galapagos islands (Figure 8-2b) of the Pacific that form by collapse as lava is extruded from or is withdrawn into its source chambers. This, and several other craters on the Martian surface with irregular walls and multiple depressions, bear even more striking resemblances to some well-known terrestrial calderas (Figure 12-10c). A few craters of probable volcanic origin contain one or more broad central peaks with shapes typical of stratocones (Figure 12-10d). However, the floors of the central or summit depressions at Nix Olympica and the three Spot shield volcanoes are flat and devoid of central peaks, as is normal for most calderas that eventually are filled with lava lakes. The inner walls of Middle Spot (Pavonis Lacus) show large grooves and flutes (see Figure 12-12d) caused by landsliding during subsidence. The rim beyond each caldera wall inclines gently away without rising abruptly at the edge into an inverted-V-profile. The lack of a sharp rim and absence of hummocky terrain along the caldera flanks, together with an overall positive profile, distinguish these shield structures from large, young impact craters. An exceptional example of a smaller, domed volcano with a central pit crater that develops several benchlike levels is shown in Figure 12-10e. Other signs of volcanism on the Martian surface include shield volcanoes in the Elysium region, domes with summit craters (similar to those in the Marius Hills on the Moon), aligned calderas, ridges produced by squeezeups (analogous to lunar wrinkle ridges), and polygonal fracturing of lava fillings in craters.

*Changes in the names of some prominent features on Mars have been proposed to the International Astronomical Union (I.A.U.). Subject to I.A.U. approval, Nix Olympica will be called Olympus Mons and Coprates Canyon will be renamed Valles Marineris.

The comparatively youthful appearance of these volcanic edifices, together with the sharpness of many of the fracture systems described in the next section, had led to speculation by Martian investigators over the present thermal state and evolutionary stage of Mars. There is, as yet, no positive evidence of recent or contemporaneous volcanic activity. But, the signs of volcanism are consistent with the hypothesis that Mars may be, just now, "turning on," that is, heating up from radioactivity decay with accompanying regional melting and differentiation. One view considers the large volcanoes to have erupted in the last billion years. In places the crust has been stretched and has foundered on a subjacent melt zone. In this concept, Mars may be entering an early stage of tectonic activity that will result in moving plates driven by upwelling lavas and formation of deep basins similar to those underlying Earth's oceans. The thin Martian atmosphere would be the product of volcanic emanations. In time, water oceans might be expected.

Martian Faults and Canyons

South Spot (Nodus Gordii) develops another type of structural deformation associated with volcanism. Concentric grabens resulting from internal subsidence occur

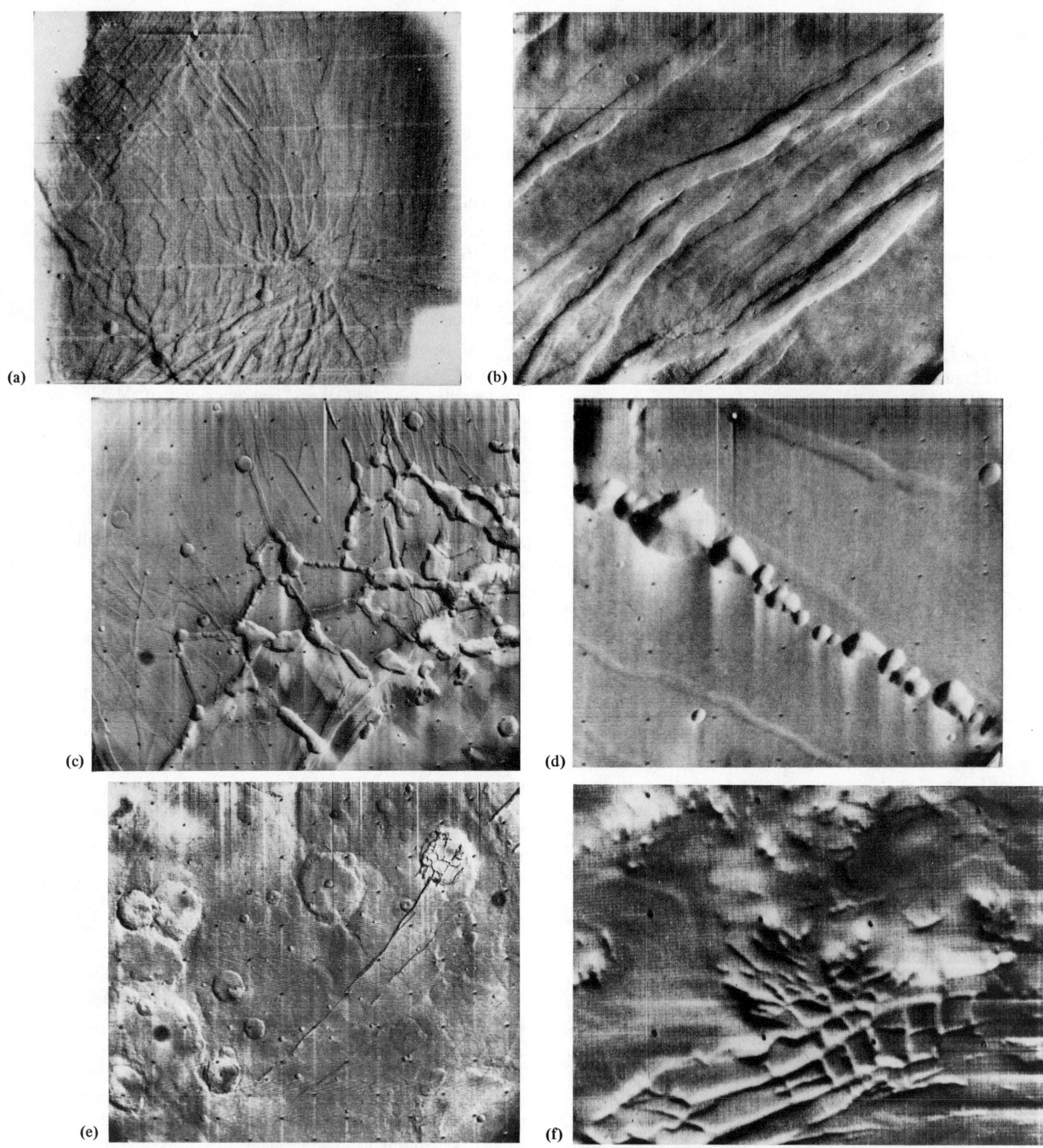

Figure 12-11 (a) A high plateau in the Phoenicus Lacus region, now broken by a series of radiating and criss-crossing fault troughs typically about 2 to 3 km wide (4083-42). (b) Close-up view of narrow, subparallel grabens separated by broader, fault-bounded horsts on a Martian plateau at the northeast edge of Aonius Sinus (4224-60). (c) Numerous interconnected troughs and expanded "valleys" and several sets of chain craters near the westernmost end of the Coprates Canyon system (4187-45). (d) Detailed view of a row of irregular-shaped craters, some connected to neighbors, parallel to nearby linear rille-like valleys. This crater chain may be volcanic in origin (maars?) (4144-90). (e) A fault zone, consisting of several sets of parallel step faults along which vertical movement is suggested. One major fault passes through a crater apparently filled with congealed lava that developed irregular fractures ("turtle-back" structure) on cooling (4174-27). (f) Transecting ridges, some 2 to 5 km apart, whose origin is uncertain. They may be fractures filled with dike material or ice. The surrounding material, weaker or less cohesive, could have been removed by wind deflation. The long dimension of the photo covers 48 km on the ground (4212-15). (Photos: NASA, Jet Propulsion Laboratory.)

outside of its inner rim (Figure 12-10f). Great fractures and faults up to several thousand kilometers long are found over much of the Martian surface, particularly in regions such as Mare Sirenum, Tharsis, and Arcadia that show volcanic characteristics. At Phoenicus Lacus (Figure 12-11a), for example, updoming of the crust induces circumferential tension that is relieved by formation of numerous intersecting fault valleys (grabens) separated by higher, flat uplands (possibly horsts). A detailed view of this type of terrain (Figure 12-11b) indicates the boundary walls of the grabens—typically depressions at least several kilometers wide—to be steeper than 45°, as is characteristic of a normal fault. A maze of crisscrossing canyons (Figure 12-11c) east of South Spot lies within an elevated plateau that includes the Tharsis ridge. These canyons appear to be grabens that may have been enlarged by later erosion processes. Many resemble the straight-walled lunar rilles; others grade into disconnected, linear crater chains (Figure 12-11d), similar to some aligned craters on the Moon, that may represent vents or maars. Parallel step-fault zones (Figure 12-11e) extending hundreds of kilometers across the Martian surface produce steep cliffs above downdropped blocks. Other fracture systems may become filled with resistant materials that stand above their surroundings (Figure 12-11f). These may be lava intrusions (dikes) or possibly windblown deposits carried into crevasses of now-vanished ice masses.

A major system of faults and fractures, enlarged by subsequent erosion, has developed into a vast east-west-trending canyonland beginning southeast of Tithonius Lacus near the Martian equator and extending eastward to Aurorae Sinus (Figure 12-12a). This remarkable example of troughed terrain—named the Coprates Canyon—runs for more than 2700 km, reaches a width of four to five hundred kilometers, and in places is nearly 6 km deep. The presence of downdropped fault blocks, some causing recognizable topographic offsets, has prompted comparison with the great East African Rift Zone which attains a similar length on Earth. Coprates Canyon indeed may have begun by tensional rifting resulting from magma withdrawal or incipient spreading of crustal plates, but its present dimensions were affected by erosional processes. Landslide deposits lie near the base of the walls of the main canyon and move out onto its floor (Figure 12-12b). To the south, the rim is cut by deep gashes that are likened to eroded box canyons. To the north, subsidiary canyons extend as alcoves well into the uplands and branch into numerous deep tributaries that seemingly have experienced headward erosion. The western end of Coprates Canyon grades into an array of narrow zigzagging, disjointed depressions (see Figure 12-11c) that remind one of great cracks formed in a cooling lava (Figure 11-12f). Proximity of these incipient canyons to the volcano field around Tharsis suggests that they began as fractures in the crust during volcanic upwelling and later were enlarged by some other process.

Evidence of Water Erosion on Mars

These views of the Martian surface define a crust that is thoroughly fractured and segmented into mosaics of blocks and slabs. This implies a mobile subcrust over which a more rigid crust has fragmented and foundered. Lava extrusions are associated with some of these great fractures and in places fill the fractures with sinuous ridges similar to those observed on the lunar maria. The structural patterns on Mars bear some resemblance to major patterns in the Earth's continents and ocean basins developed from active plate tectonics; however, no zones of subduction and crumpling of accompanying sediments are evident in the *Mariner* photos.

The eroded surfaces on Mars, where examined in detail (Figure 12-12c), have much in common with terrestrial "badlands" topography developed in weak rocks by water and landslide actions. Although wind erosion could conceivably have sculpted out these dendritic chasms, water erosion is preferred by many Martian experts as the explanation for this type of topography. This consensus, although largely intuitive, is based on comparison with water-formed canyons and channels in terrestrial plateaus and badlands. There is, of course, no evidence for active river systems on Mars, and acceptance of the hypothesis thus is predicated on existence of flowing water some time in the Martian past. The interpretation is further complicated by the occurrence of other, smaller canyons elsewhere on Mars that have no stream inlets or outlets in the usual sense (Figure 12-12e).

Several drainage forms were disclosed in TV pictures from *Mariner 9*. Those with distinct fluviatile characteristics are especially common in the Chryse. Mare Erythraeum, Rasena, Memnonia, and Sinus Sabaeus regions. A streamlike channel over 410 km long and 5 to 6 km wide is shown in Figure 12-13a. This feature is somewhat like the sinuous rilles on the Moon, many of which are assumed to have a volcanic origin. A variant of this type of irregular channel, with jagged, anastomosing branches (Figure 12-13b), is similar to large braided stream systems on Earth. An even broader sinuous valley, with well-defined tributary sections, is noted in the Rasena region of Mars (Figure 12-13c). This depression continues for over 700 km and at places is several tens of kilometers wide. This type of channel displays riverlike features such as headward tributaries, "downstream" widening, and terminal distributaries (Figure 12-13d). A number of the larger sinuous channels are traceable into regions of chaotic terrain, thought now to be collapse or landslide topography related to permafrost melting. However, many smaller gullies clearly are lava channels commonly found

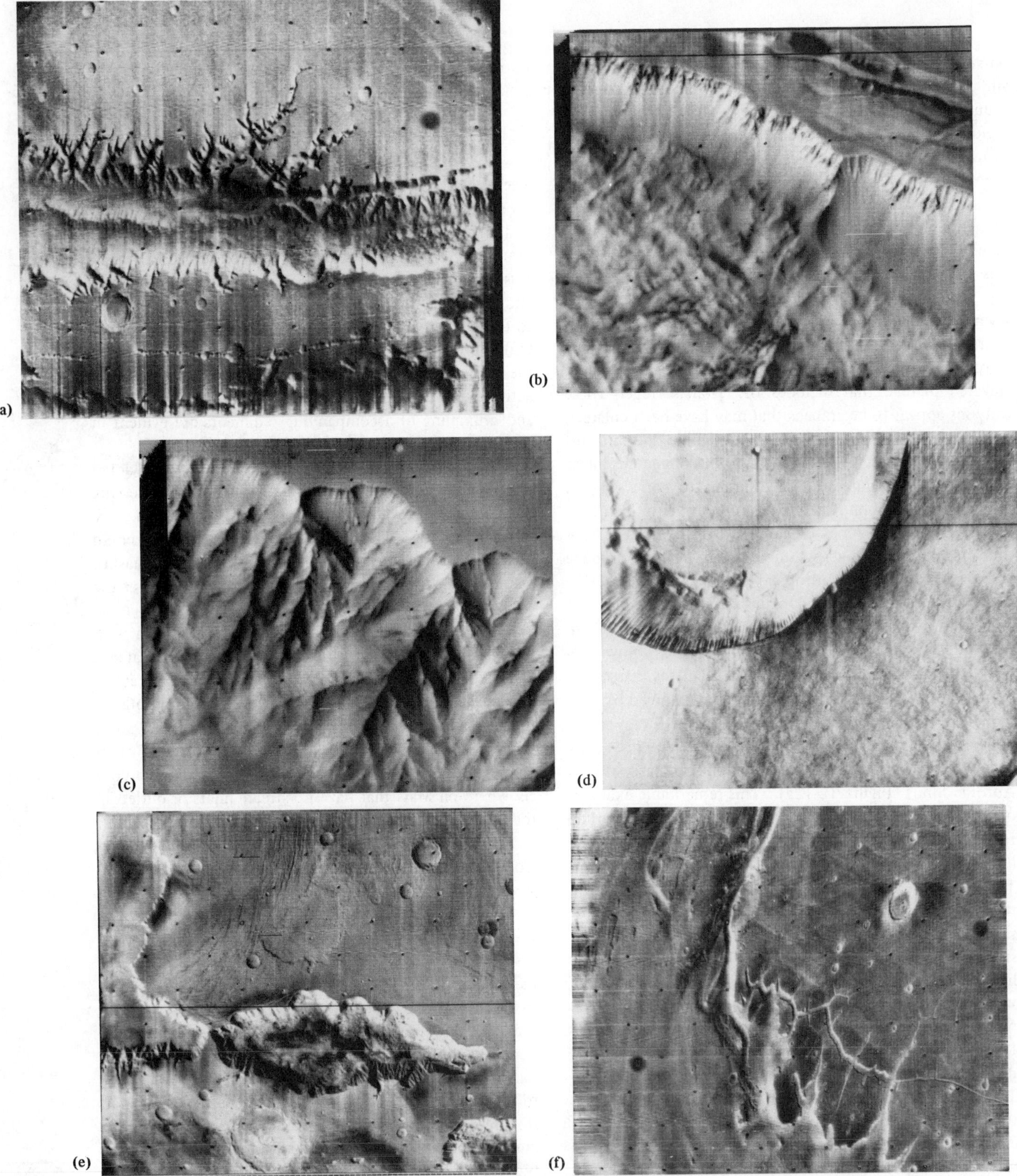

Figure 12-12 (a) Part of the east-west Coprates canyonlands running parallel to the equatorial belt south of Tithonius Lacus (4144-86). (b) Part of the wall of the main Coprates Canyon showing scalloped and grooved slopes and landslide debris at the base (4193-42). (c) "Badlands" topography formed in a segment of the Coprates Canyon system (4197-36). (d) Grooves in the scarred inner slope of the Middle Spot volcano caused by repeated sliding of rocks and loose material (4142-93). (e) An elongate depression several hundred kilometers in length, showing canyonlike erosional characteristics but without signs of stream drainage. A subsidiary feature just north of the main Coprates Canyon system (4193-54). (f) A narrow, canyonlike crack, developed on a plateau of possible volcanic flows, in the Lunae Palus region east of East Spot (4237-64). (NASA, Jet Propulsion Laboratory.)

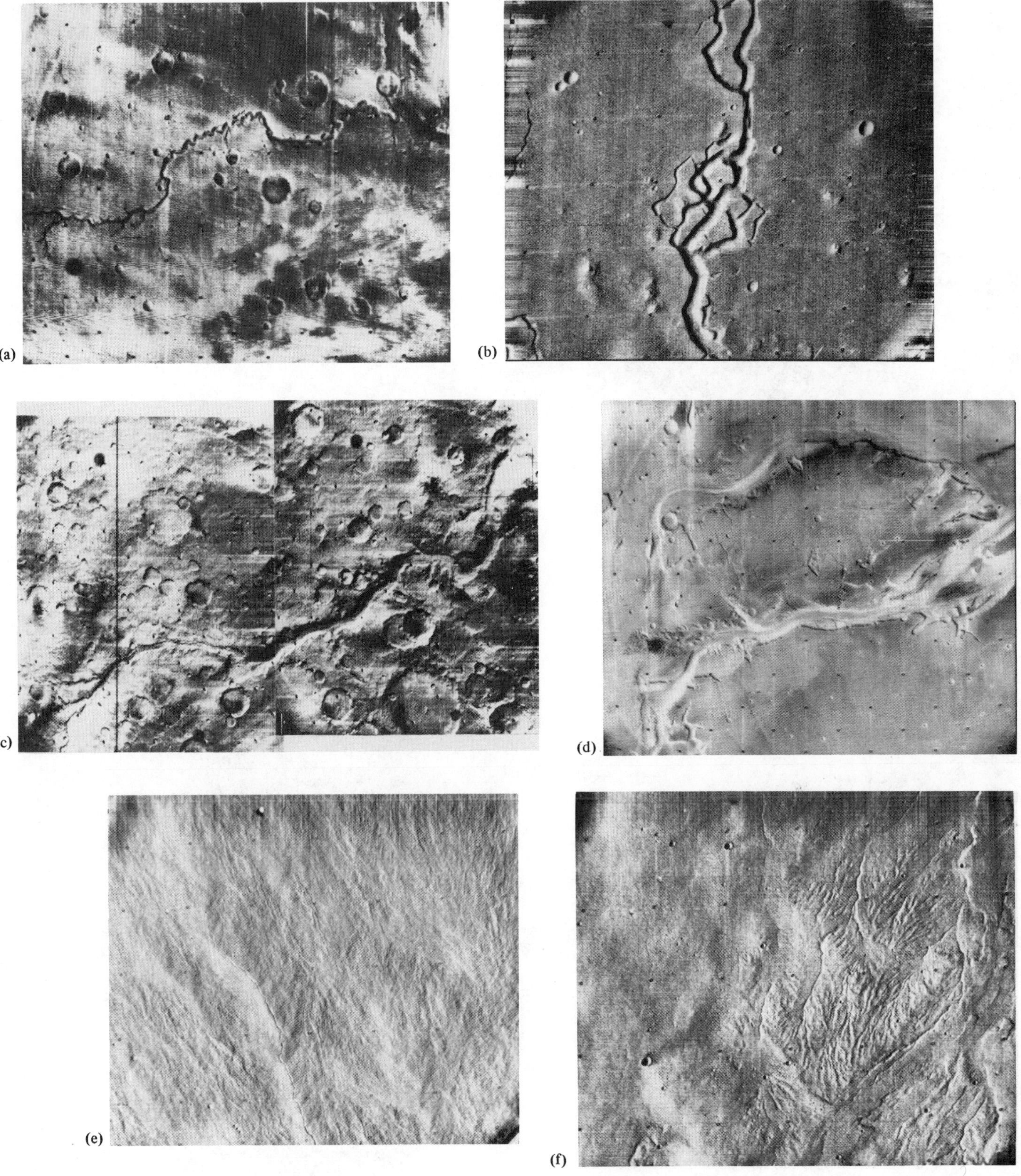

Figure 12-13 (a) A giant gully over 400 km long and up to 6 km wide in Mare Erythraeum. This feature, of unknown origin, resembles many lunar sinuous rilles (Orbit 133). (b) An irregular, branching valley of possible volcanic origin, north of Aethiopis (4276-24). (c) A broad, irregular sinuous valley in the cratered Rasena region of Mars. This feature, over 700 km long, may be a lava channel, a collapse tube, or, possibly, was produced by fluid (water?) erosion (Composite photo; Orbit 140). (d) Part of a channel-like valley in the Lunae Palus area (4193-90). (e) Lava channels and other flow markings, and a raised ridge with a central fracture, on the northwest flank of Nix Olympica (4133-96). (f) Incipient dendritic drainage on a recently exhumed irregular terrain. These features may be either volcanic or fluvial in origin (4182-96). (NASA, Jet Propulsion Laboratory.)

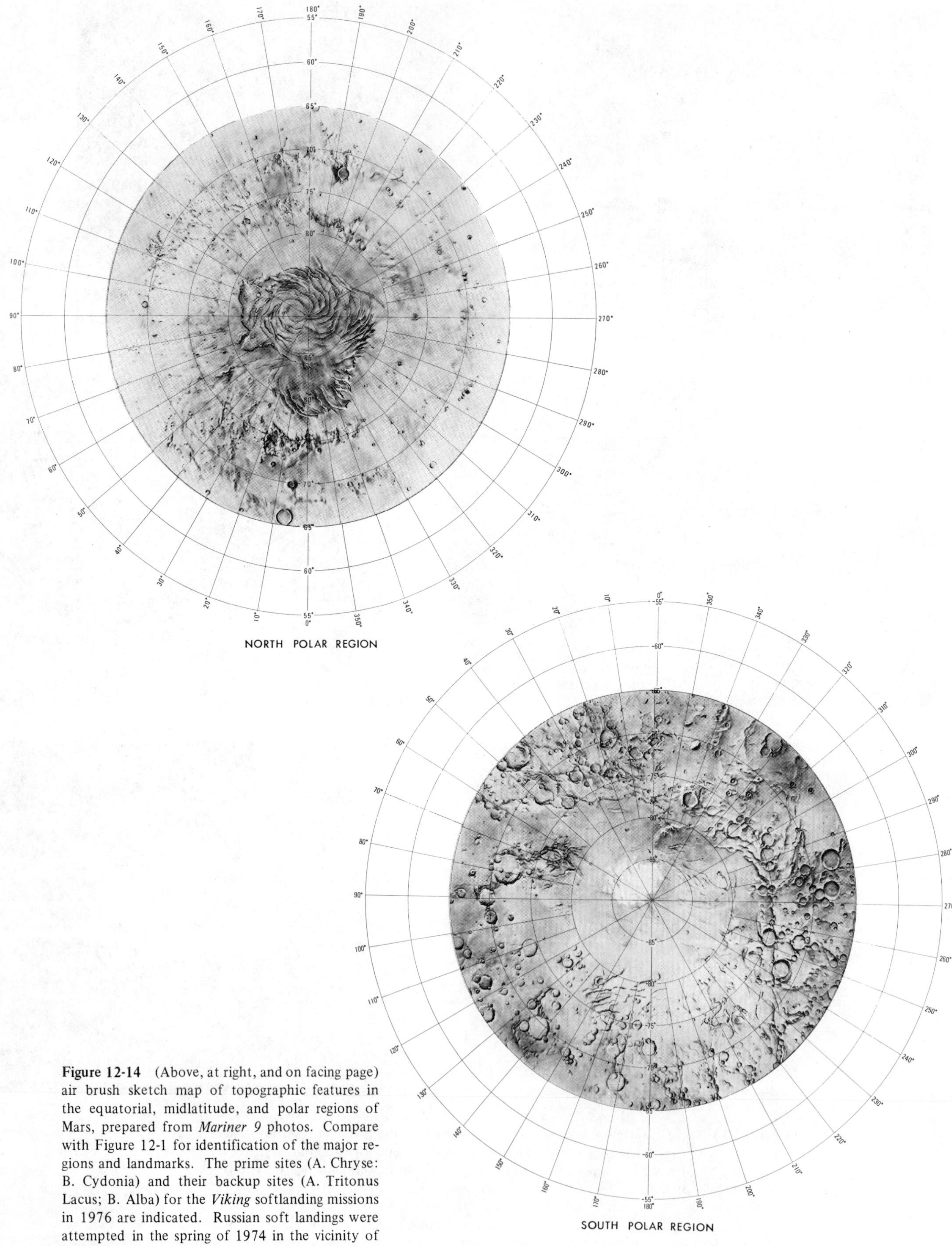

Figure 12-14 (Above, at right, and on facing page) air brush sketch map of topographic features in the equatorial, midlatitude, and polar regions of Mars, prepared from *Mariner 9* photos. Compare with Figure 12-1 for identification of the major regions and landmarks. The prime sites (A. Chryse: B. Cydonia) and their backup sites (A. Tritonus Lacus; B. Alba) for the *Viking* softlanding missions in 1976 are indicated. Russian soft landings were attempted in the spring of 1974 in the vicinity of the Hellas Basin. (NASA photo.)

along the flanks of some Martian shield volcanoes (Figure 12-13e). Features described as "incipient dendritic drainage" (Figure 12-13f) may possibly be a form of lava channeling, but the currently preferred interpretation holds them to be fluviatile in origin.

There is now a growing belief among scientists that many, if not the majority, of Martian channels not directly tied in with volcanic structures have originated by fluvial processes. This view hinges on the channels' close physical resemblance to drainage networks on Earth, especially in weaker materials, on the tendency for these streamlike channels to follow downslope surfaces, and on their widespread occurrence over most of the ancient terrains of Mars. There are no clearcut signs that oceans ever covered the Martian surface, although the 3-km-high scarp circumscribing Nix Olympica is most easily explained as a wave-cut cliff produced by a surrounding sea. Presently the amount of water in the atmosphere of Mars is insufficient (see p. 276) to produce any active rainfall capable of developing any persistent runoff. However, the observations from *Mariner 9* strongly imply that at times in the past rain must have been extensive enough to have produced ubiquitous stream flow. These periods of precipitation may have been episodic and recurring. The water, probably originally released to the Martian surface from volcanoes, would likely become stored as permafrost throughout the planet but with maximum concentrations in the polar regions. The polar icecaps are still being affected by a 50,000-year precessional cycle in which first one cap become enlarged over that interval at the expense of the other and then the cycle reverses repetitively. This sequence involves transfer of sublimed CO_2 and water between the poles across the equatorial belt. During the transfer, erosion by both aeolian and fluvial processes is active up to higher latitudes. Conditions during past cycles may have led to condensation of water at some stage. Quantities of water in excess of the miniscule amounts now in the atmosphere could result from thawing of permafrost during a warming trend or from periods of increased volcanic activity. Regardless of the mechanisms involved, the verification of concentrations of water in actively flowing systems—in the present or the past—would have tremendous influence on hypotheses supporting the occurrence of living or fossil organisms in the Martian environment.

Panoramic Views and Maps from Mariner 9

Mariner 9 imaged a great variety of morphological and structural surface features. The improvement in knowledge of the Martian surface that resulted from this mission is effectively demonstrated by comparing a sketch map of the topography of the equatorial-mid-latitude belt of Mars (Figure 12-14) with the same region shown in the premission 1971 planning chart (Figure 12-1). A composite photomosaic covering part of this map area (Figure 12-15) illustrates the actual appearance of much of the

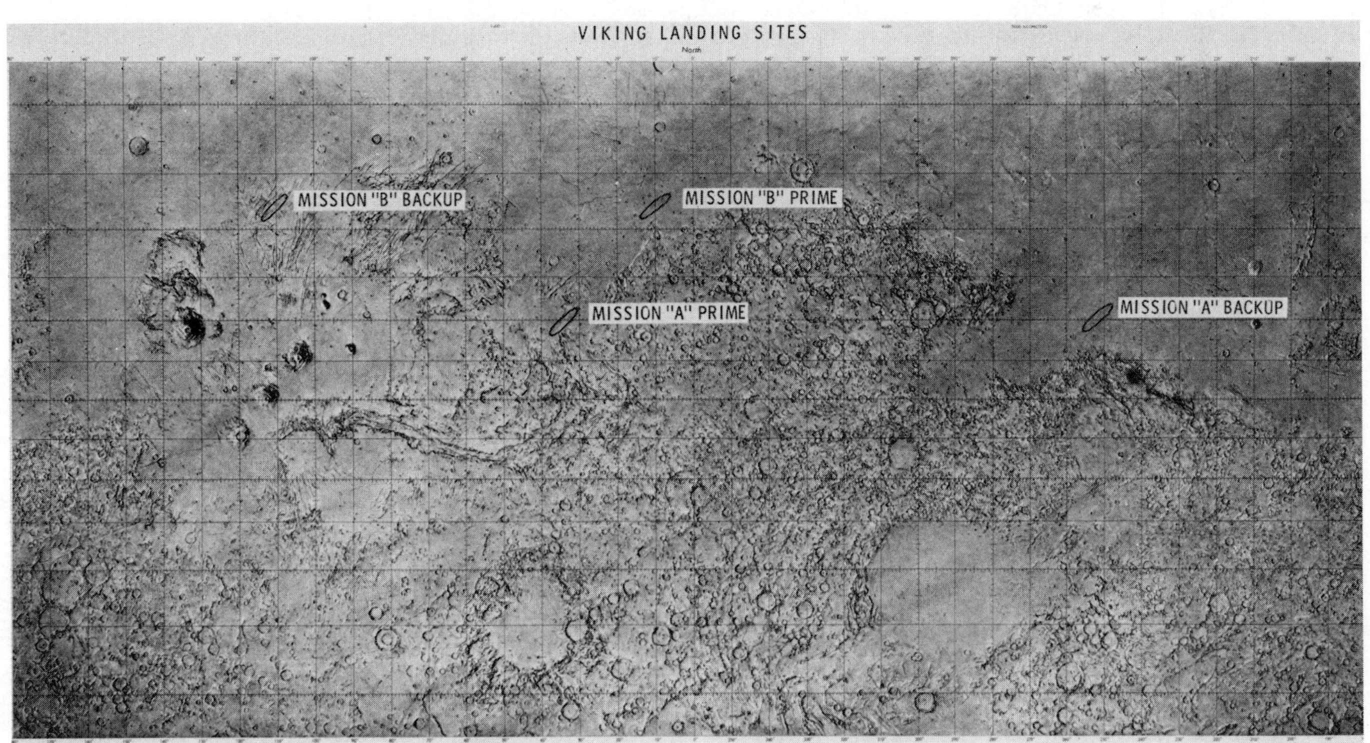

Figure 12-15 Panoramic view of the equatorial region of Mars from 30°N to 30°S and 10°W to 140°W. This composite photomosaic is produced from over 250 individual TV frames (A-camera) and images about 20% of the entire planet. Prominent features shown here include Nix Olympica, the three Spot shield volcanoes, and the Coprates Canyon region. The entire continental United States would fit in this mosaic with the west coast at the western end of the great canyon system and the east coast at the eastern end of this system. (NASA photo.)

Martian surface. An impressive view of most of the northern hemisphere of Mars appears in Figure 12-16. The U.S. Geological Survey has now mapped the entire planet using the pictures returned from the post-dust storm coverage. Individual photos are assembled in uncontrolled mosaics mounted in Lambert conformal projection. An example of one of the Mars charts that subdivide all of the planet between 65°N and 65°S into twenty nine map quadrangles is reproduced in Figure 12-17.

The U.S. Geological Survey has also devised a classification of terrain units, grouped into broader catagories of *cratered, regional,* and *structural* units, to facilitate the preparation of the quadrangle series of geological maps of Mars. A preliminary *Geologic-Terrain Map of Mars,* using these terrain units, has now been completed at a 1: 25,000,000 scale (Figure 12-18a). A variant of this map, using a stereographic projection and tone patterns for the terrain units, appears as Figure 12-18b. The terms applied to many of the units are simply descriptive (e.g., "chaotic terrain," "grooved terrain") and do not connote specific materials identifications or causative processes. Certain terms, however, have some genetic significance (e.g., "densely cratered terrain," "volcanic shields"). Pictorial examples of some of these terrain units are given in Figure 12-19; more examples appear in several other figures in this chapter. Sharp (1972) has described the characteristics of some of these terrain types and has suggested possible origins (Table 12-1). He cites as potential genetic processes (1) magma withdrawal, (2) tensional forces in the crust, (3) wind action, aeolian (4) melting of ground ice. Another type of terrain, generally smooth but marked with sinuous channels and aeolian features, is a stripped surface that may have developed from planation by running water (Milton, 1973).

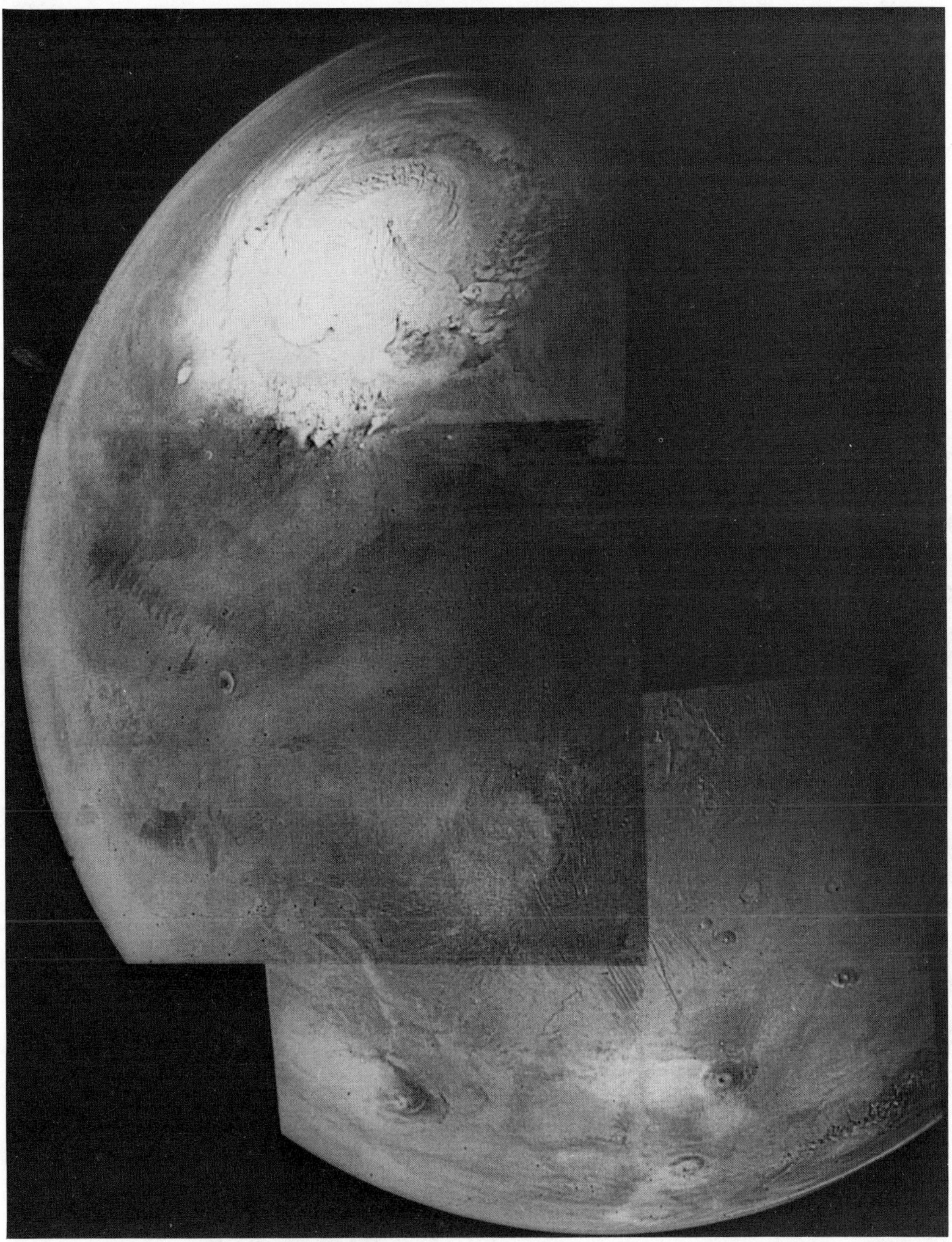

Figure 12-16 A composite photomosaic of much of the northern hemisphere of Mars. The northern polar cap, shrinking during the late Martian spring, is shown at the top while such landmarks as Nix Olympica and the North-Middle-South Spots are near the bottom. The three computer-enhanced images used in this mosaic were taken by *Mariner 9* on August 7, 1972, at an average distance of 13,700 km from the surface. (NASA photo; composited at the Jet Propulsion Laboratory.)

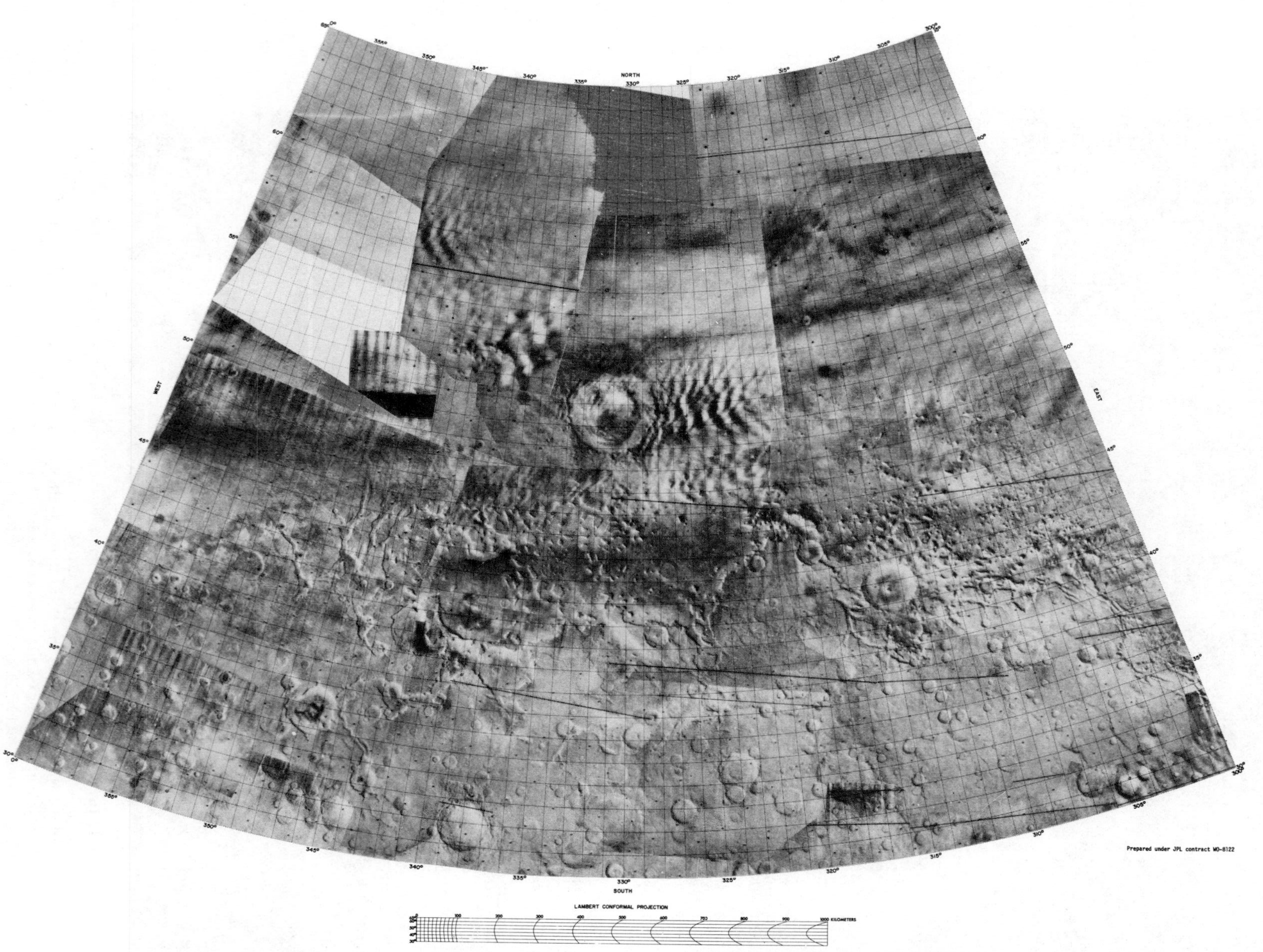

Figure 12-17 An uncontrolled photomosaic (using parts of 33 pictures) of Mars Chart MC-5 covering the regions of Cydonia, Dioscuria, Ismenius Lacus, and Protonilus. The terrain includes both impact and volcanic structures, knobby terrain, channels and canyons, and linear terrain. (Prepared by U.S. Geological Survey.)

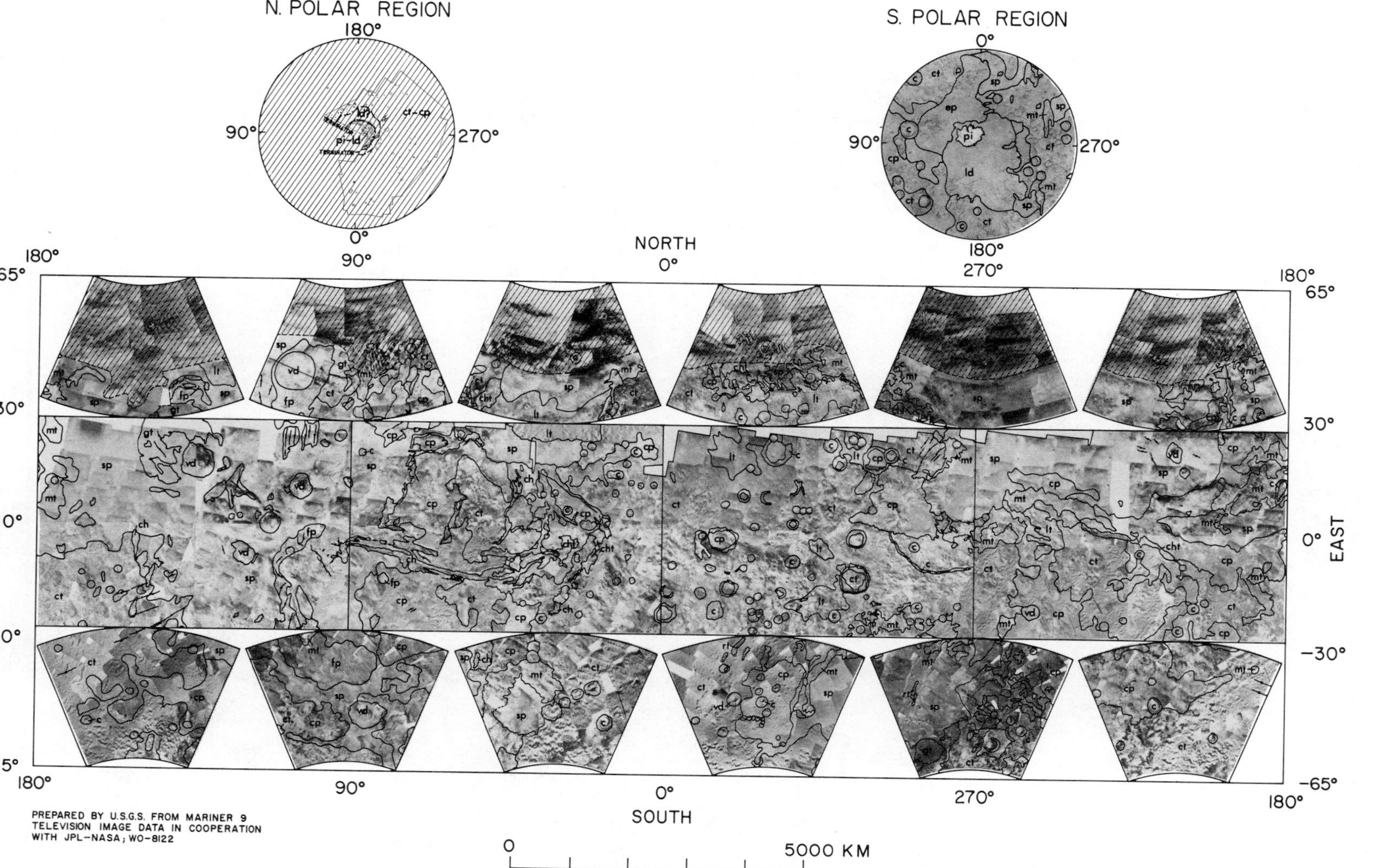

Figure 12-18 (a) Preliminary version of the Geologic-Terrain Map of Mars on which prominent surface features visible in the photobase have been classified in terms of major terrain units. Each of the four large rectangles is subdivided into four quadrangles (Mars Chart series) in the more detailed map version. The map section covering the north polar region in detail is not completed in this illustration. A color version of this map appears on p. 4032 of the *Journal of Geophysical Research*, vol. 78, no. 20. The units marked here by letters are identified as follows: c, craters; sp, smooth plains; cp, cratered plains; ch, channel or canyon deposits; ct, cratered terrain; lt, lineated terrain; mt, mountainous terrain; vd, volcanic deposits; gt, grooved terrain; fp, fractured plains; cht, chaotic terrain; rt, ridged terrain; pi, polar ice; ld, laminated deposits; ep, etch-pitted plains. (Courtesy U.S. Geological Survey.)

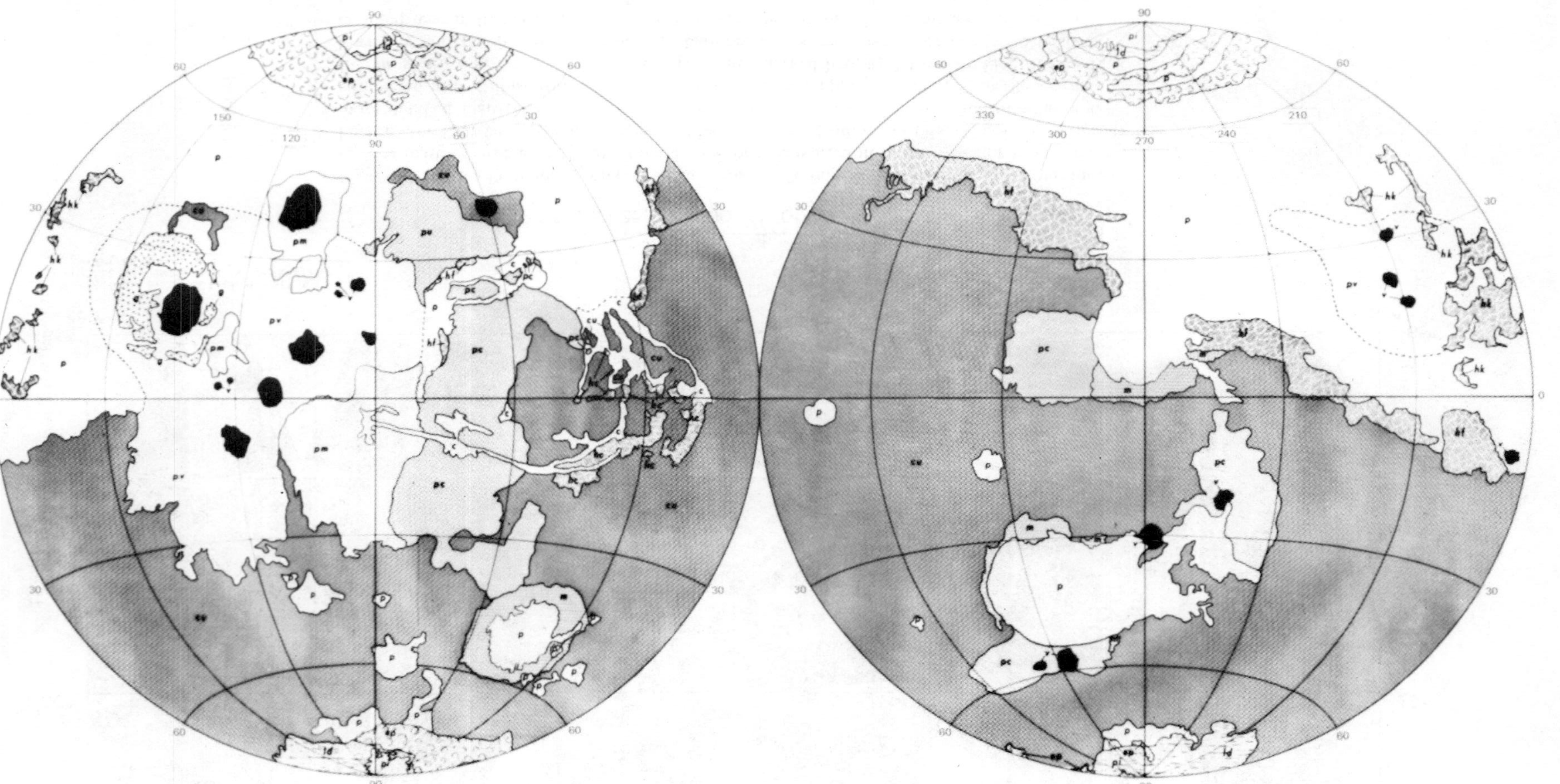

Figure 12-18 (continued) (b) A generalized physiographic province map of Mars, shown in a stereographic projection prepared by T. A. Mutch and R. S. Saunders of Brown University as a simplified version of the map shown in Figure 12-18(a). The units identified by letters are: *Ancient units*: cu, cratered terrain, undivided (densely to moderately cratered); m, mountainous terrain (including basin ejecta); *Modified Units*: p, plains, undivided (sparsely to moderately cratered); c, channel deposits (smooth floors composed of wind-blown and mass-wasted materials); hk, hummocky terrain, knobby; hf, hummocky terrain, fretted; hc, hummocky terrain, chaotic; *Volcanic Units*: pc, cratered plains (analogous to lunar maria); pm, moderately cratered plains (without volcanic surface structures); pv, volcanic plains (few craters); v (black), volcanic constructs (shields, domes, or cones); *Polar Units*: ep, etched plains (irregular, often coalescing pits); ld, laminated deposits (benches and layers less than 100 m thick); pi, permanent ice (CO_2; water-ice).

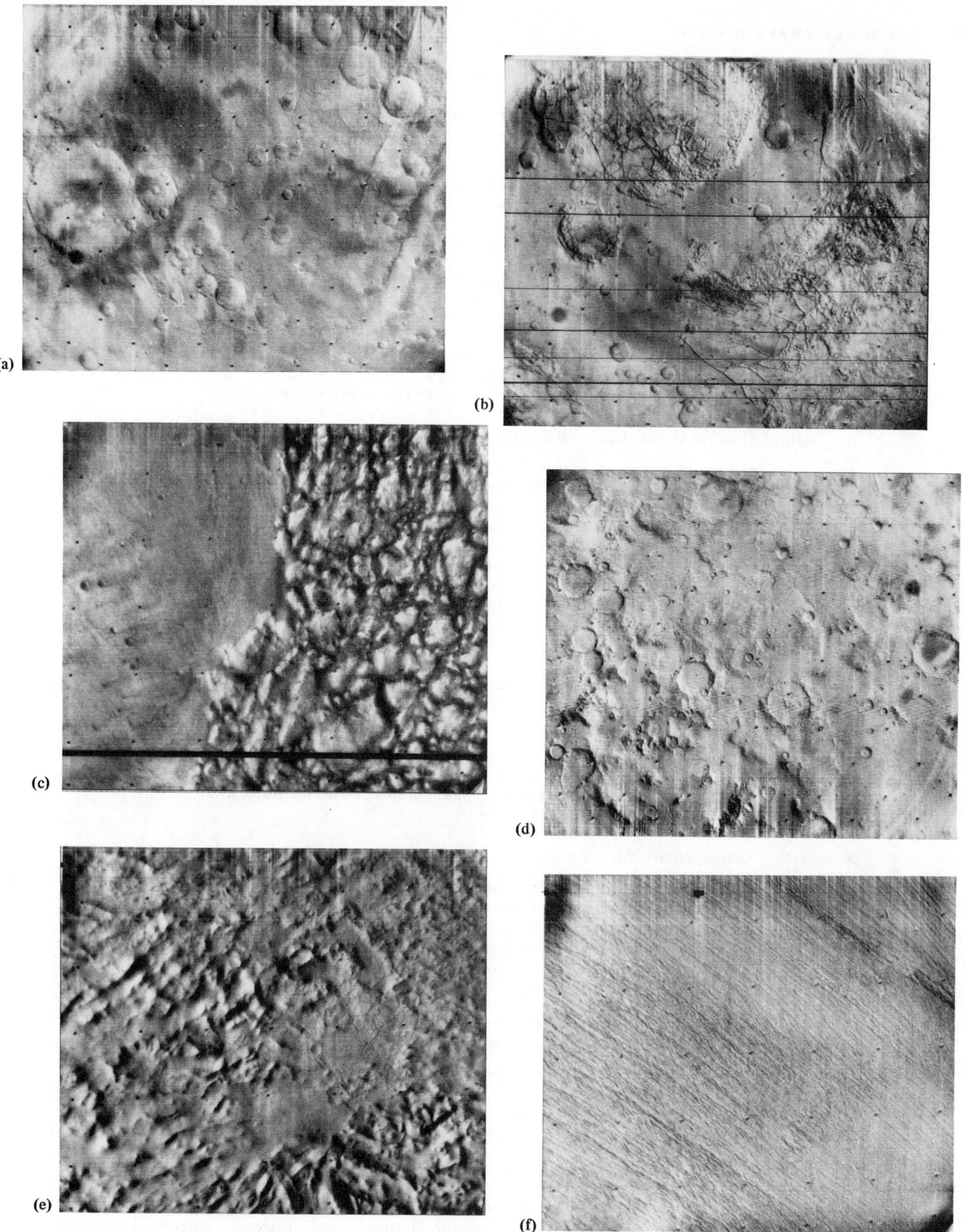

Figure 12-19 (a) Densely cratered terrain (lower left; upper right) and moderately cratered terrain (central area) north of Syrtis Major (4183-90). (b) Chaotic terrain, consisting of a complex mosaic of broken slabs and blocks, near the boundary between Chryse and Margaritifer Sinus (4206-57). (c) Region of chaotic terrain (right) and smooth plains (left), east end of Mare Cimmerium (4218-24). (d) Mountainous terrain (upper half) associated with uplift and fracturing of the Libya basin and cratered plains terrain (lower half) (4188-54). (e) Knobby terrain, consisting of hills and ridges with sharp divides (4218-60). (f) Grooved terrain in the vicinity of Nix Olympica. Area shows close-spaced low ridges and parallel linear troughs (4174-63). (NASA, Jet Propulsion Laboratory.)

TABLE 12-1

Some Major Terrain Types on Mars

I. Pitted and Etched Terrains (South Polar Area)

- Mostly within blankets of airborne particulate and volatile material
- Associated dunes and wind-scoured surfaces
- Floors of stripped basement rocks
- Produced by aeolian deflation and ablation; not subsidence

II. Fretted Terrain (Mid-Northern Latitudes)

- Smooth floors, highly irregular bounding escarpments, abundant outliers, tributary chasms
- Base level imposed by depth of frozen ground
- Escarpment recession by ground-ice sapping
- Removal of finely weathered debris by deflation or fluid transport?

III. Troughed Terrain (Equatorial Region)

- Primarily within a zone from 0° to 16°S latitude and 52° to 90° longitude
- Individual units up to 100-200 km wide and several hundred kilometers long
- Largest example 2700 km cumulative length and widths up to 500 km
- Some, possibly all, are closed depressions
- Trough margins irregular in detail; scalloped, embayed, sharply indented
- Strongly developed dendritic pattern of tributaries
- Initiation of troughs strongly controlled by structures (fractures)
- Paucity of floor craters suggests relative youth and/or active floor deposition
- Commonly associated with, or grading into, pit craters
- Caused by separation of fracture walls (spreading), withdrawal of magma, or deterioration of ground ice
- Trough development may be continuing at present, but the process may have extended over hundreds of millions of years

IV. Hollowed Terrain (Equatorial Region)

- Occurs mostly in region to west of troughed terrain
- May be a less advanced stage of development of troughed terrain, guided by a more complex system of fractures

V. Chaotic Terrain (Equatorial Region)

- Mainly to east of troughed terrain (Coprates Canyon)
- Floored by chaotic jumble of large angular blocks
- Arcuate slump blocks on walls; arcuate fractures in uplands
- May involve only deterioration of ground ice without deep subsidence caused by spreading or magma withdrawal
- Alternatively, may be related to withdrawal or spreading

Source: Adapted from series of tables developed by R. L. Sharp, California Institute of Technology. (See p. 4074 in special Mars issue, J. Geophys. Res., v. 78, July 10, 1973.)

Craters and Basins on Mars

Judging from surface characteristics alone, both the Moon and Mars appear to have many morphological features in common. Thus, several large circular basins, ringed by uplands, have been recognized on Mars. These basins include Argyre (Figure 12-20), Libya, and Hellas (whose 2000 km diameter exceeds the Moon's Imbrium Basin by 50%), and two smaller depressions, named Iapygia and Edom, about the size of Mare Crisium. By analogy to lunar basins, the Martian basins probably are impact in origin and presumably formed by infall of planetesimals early in solar system history. Like some depressions on the Moon, the Martian basins contain intermittent darker materials, forming one type of smooth plains unit, that are thought to be lava fillings, presumably of a basaltic nature. Unlike the Moon, this unit is sporadically distributed over the entire Martian sphere in irregular patches. Even more frequent are light-colored smooth-plains units. These may be volcanic ash deposits, but more likely are materials laid down by aeolian processes. Ejecta blanket deposits outside the basin are absent or poorly preserved.

Mariners 6 and *7* confirmed that much of the 10% of the total Martian surface which they photographed is

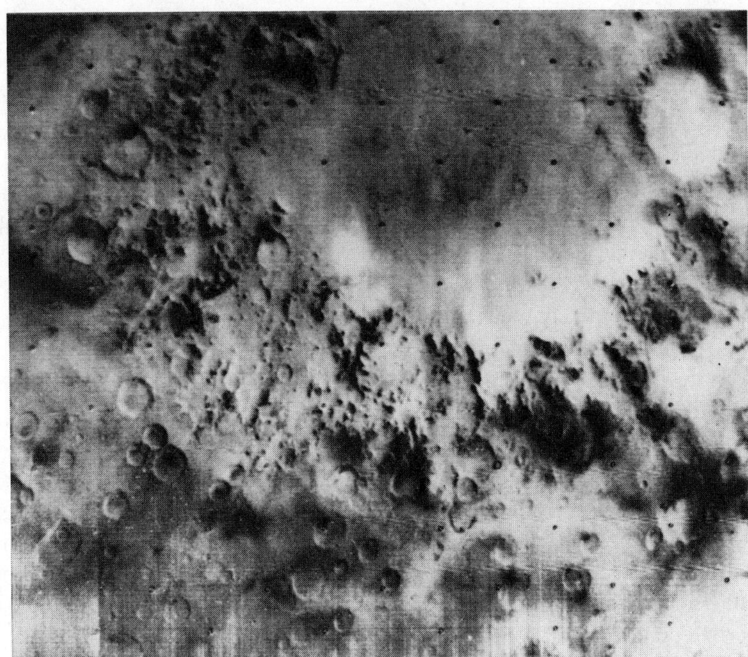

Figure 12-20 Argyre, a Martian basin more than 600 km wide near the western end of Mare Australe. The ringed basin shows some of the characteristics of lunar basins, such as Crisium and Serenitatis. Note the rim and the hummocky debris-strewn surface of ejecta deposits beyond. (NASA photo.)

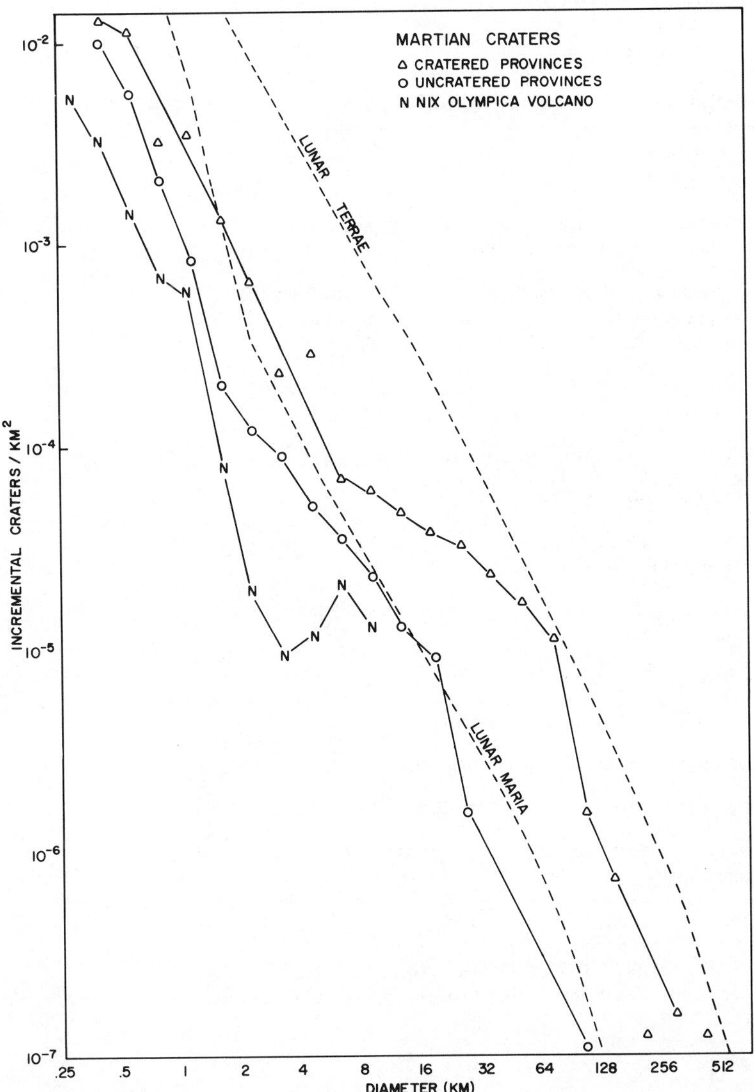

Figure 12-21 Cumulative frequency curves for crater counts, using *Mariner 9* photos in densely to moderately cratered provinces (Sinus Sabaeus. Deucalionis Regio, and Mare Sirenum), uncratered provinces (Solis Lacus, Elysium, and regions near Nix Olympica), and the flanks of the shield volcano, Nix Olympica, representing older, intermediate, and younger terrains, respectively. (From J. McCauley et al., 1972.)

heavily cratered. After *Mariner 9* sent back views of almost 100% of the planet, it became obvious that the earlier coverage had biased scientists into believing that the Martian surface resembles the lunar highlands. The new coverage emphasizes a diversity of landforms on Mars not duplicated on the Moon. Even so, cratered terrain makes up more than 40% of the equatorial region and dominates the southern hemisphere. More than 50,000 craters larger than one km have been recognized in this region. Some parts are extensively crated, with surfaces holding the larger craters approaching saturation; others contain relatively few craters, with frequencies comparable to or less than the lunar maria. Many craters appear quite old, with faint rims, shallow interiors, and no evident ejecta blankets. Other craters have younger, less degraded features with scalloped rims and exterior ejecta deposits. For even these, however, terracing is inconspicuous, and none show the rays and crowded secondary crater fields found around lunar craters of Copernican age. Less than 5% of the Martian craters contain central peaks. Most Martian craters look as though they have been modified by wind erosion and dust deposition (see, for comparison, the terrestrial craters formed in deserts as shown in Figure 7-1c and d). Some craters retain signs of having once been buried but are now undergoing exhumation as loose surface materials are removed by deflation.

It is highly probable that most circular craters on Mars were impact-produced. Other Martian craters with more irregular outlines are better explained as volcanic collapse pits and calderas; some of these commonly lie near the summits of shields and uparched terrain and are associated with such volcanic landforms as domes and channels. Mars differs from the Moon in possessing extensive terrains that seem to have developed long after major cratering had ceased and some smooth-plains units were emplaced. Thus, much of the volcanic terrain may be geologically recent, as indicated by the relative sparsity of impact craters and the freshness of flow features. Fractured plains, including chaotic terrain, and canyonlands also seem to have been developed later in Martian history.

The relative ages of formation of major terrain units can be assessed by the same methods used in lunar surface mapping. Crater frequency counts and overlapping relationships are particularly useful. Frequency counts of crater distributions in three Martian provinces are plotted in Figure 12-21 from *Mariner 9* data. A general sequence

TABLE 12-2

Geologic History of Mars (Tentative Sequence of Events)*

- Buildup of windborne sediments in polar basins and further cyclic effects related to permafrost and "interglacial" periods.
- Increasing wind action by Martian atmosphere leading to aeolian erosion and deposition effects (etch-pitted and laminated terrains).
- Formation of chaotic terrain (associated with frost action?) and the various fluviatile sinuous channels (time of origin of Martian atmosphere not fixed but gas density may have reached maximum during Tharsis volcanism).
- Growth of Nix Olympica and the Spot shield volcanoes and development of nearby lava plains; estimates of the age of this volcanism are as young as 0.3-0.5 AE ago.
- Epeirogenic upwarping of the Tharsis region; rifting in Coprates region.
- Buildup of plains units in Phoenicus Lacus and Elysium areas.
- Widespread volcanism around Nix Olympica area, spreading to Arcadia and Alba regions; plains development from basalt flows, later faulted.
- Further modifications of ancient cratered terrain and beginning of older plains units in Lunae Palus, Hesperia, and elsewhere.
- Volcanism in the Amazonis region.
- Decrease in cratering rates; inception of volcanism near Hellas.
- Formation of Argyre and Libya basins; continued impact cratering.
- Formation of the Hellas basin and cratering of the general surface.
- Generation of outer crustal materials, presumably by melting, differentiation, and solidification into one or more layers of igneous rocks.
- Formation of Mars by accretion (?) approximately 4.6 AE ago.

*Adapted from synopsis in paper by H. Masursky, 1972, "An Overview of Geological Results from Mariner 9", *J. Geophys. Res., 78,* 4009-4030. Events listed from top in order of increasing ages.

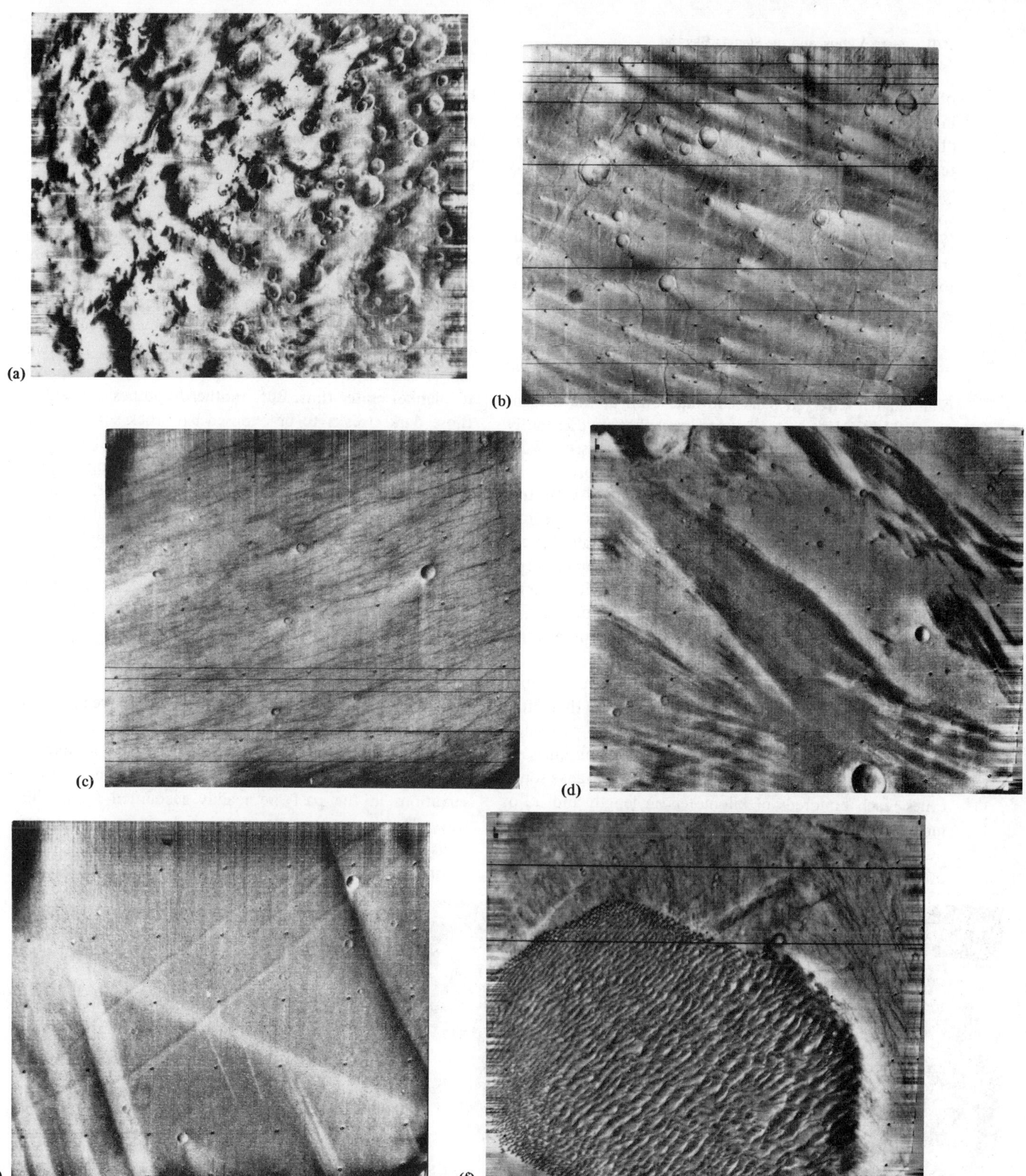

Figure 12-22 (a) A cratered region showing dark blotches and other markings both within craters and on the surrounding uplands. Most, if not all, of these dark areas and some of the lighter areas are related to deflation and deposition brought about by the Martian dust storm of 1971 (4131-99). (b) Light plumes and streaks generally originating at and extending beyond craters and other surface barriers. This wind-produced effect is in the Hesperia-Mare Tyrrhenum region (4198-21). (c) Numerous thin, elongate dark streaks developed by acolian action just south of Elysium (4202-69). (d) Broad, dark surface markings that cross and orient in several directions in Syrtis Major (4268-36). (e) Unusual light markings that are straight, remarkably regular and long, and oriented in two distinct directions, in the Tharsis region just north of Phoenicus Lacus (4271-31). (f) Dune-like ridges and thin, dark markings, believed to result from aeolian processes, in the Hellespontus west of the Hellas basin (4264-16). (NASA, Jet Propulsion Laboratory.)

of development, from oldest to youngest, of terrain units in the equatorial regions has been proposed by McCauley (1972): (1) densely cratered terrain; (2) moderately cratered terrain; (3) cratered plains; (4) volcanic terrains (shields, domes, pit craters), more or less contemporaneous with (5) fractured and grooved terrains; (6) chaotic terrain; (7) channeled and canyonland terrains, and (8) certain smooth plains (wind deposits). The times of formation of these terrain units certainly overlapped. A summary of the geologic history of Mars is given in Table 12-2.

Aeolian Features on Mars

Significant events on Mars continue even at the present time. Most relate to aeolian processes that are constantly removing and redepositing loose surficial deposits. Wind action is now the dominant surface-modifying process and may have been a major influence well into the Martian past. Substantial changes in surface markings appear to result from the violent dust storms that periodically besiege the planet. Earth-based telescopes have observed large-scale changes in albedo that were attributed either to redeposition of surface materials or to seasonal spreading of vegetative growth. The latter explanation is now ruled out by the *Mariner* observations.

Most of these variable or transient features are described as "blotches," "streaks," "plumes," or "tails" that often are directly correlated with topographic features such as craters (Figure 12-22a). Both light and dark markings are exemplified in Figure 12-22b and c. These streaks sometimes reach hundreds of kilometers in length and 50 or more kilometers in width and are surprisingly regular in outline (Figure 12-22d). In a given local region, they are often strikingly parallel and presumably orient in the direction of the winds that prevailed at the time each group was formed. In some areas, overlapping and crossing streaks suggest wind shifts and possibly particle-size fractionations (Figure 12-22e). Dark streaks frequently emanate from the interiors of craters, while white streaks more commonly extend tangentially around crater rims. These white tails seem to be more stable than darker ones. At present, it is not clear whether the tails are mainly depositional or erosional in nature. Opinion favors formation of the white tails by deposition of finer-sized dust, in which the smaller particles characteristically are also lighter in color, but it is then difficult to account for dark tails and streaks in the same way. Some dark markings appear to blanket crater rims. But, another hypothesis considers those dark streaks to be exposed bare rock surfaces (basaltic?) swept clean of lighter, weathered dust that deposits elsewhere as white tails. However, strong support for deposition of wind-transported materials as the principal active process responsible for most markings comes from coassociation of streaks with swarms of dunelike hills (made up, most likely, of coarser materials) (Figure 12-22f).

When the same areas of Mars covered by both *Mariners* 7 and 9 are compared, tails and other markings are observed in the post-dust storm period in 1972 that were not present or were different in shape in 1969 (Figure 12-23). Some of these markings experienced notable boundary changes over periods of less than 3 weeks when viewed during successive *Mariner 9* orbital passes. Variations in albedo patterns are often complex and widespread; the shifts around Syrtis Major noted from telescope observations in the past are readily accounted for in this way. That some deposits on crater floors or along crater walls also shift in time has been confirmed by comparing *Mariner* 7 and 9 photographs of the same cratered areas.

Figure 12-23 A part of the Noachis region of Mars viewed by *Mariner 7* in 1969 and again by *Mariner 9* in early 1972. Splotches and other markings are present in the 1972 picture which, despite lower resolution and poorer image quality, appear to be absent or subdued in the 1969 views. (NASA photo.)

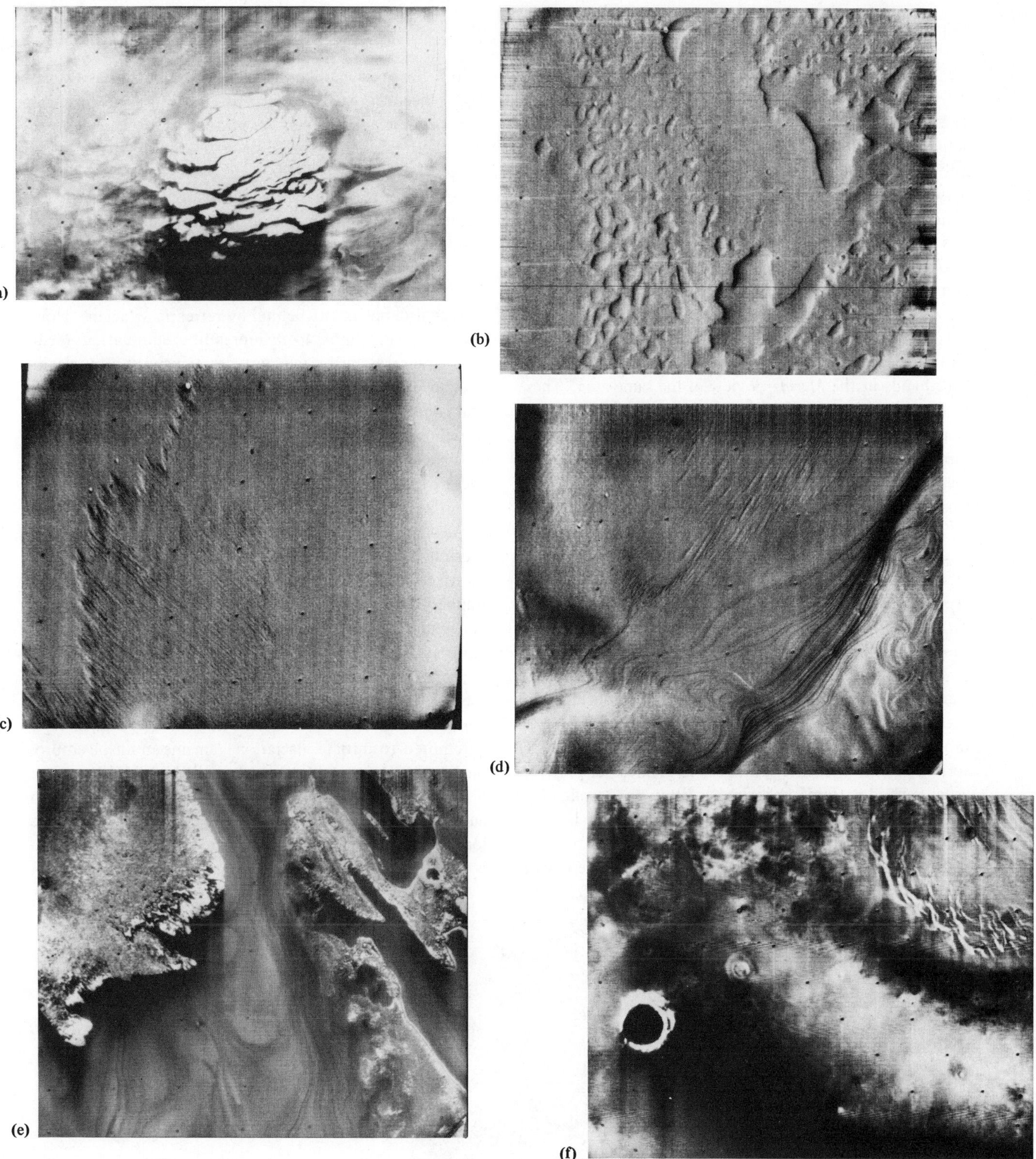

Figure 12-24 (a) The residual ice cap around the Martian south pole, seen by *Mariner 9* in late 1971 (4160-33). (b) Etch-pit terrain, about 800 km north of the south pole, marked by numerous depressions and hollows that result from wind deflation (4132-21). (c) A fluted and grooved plateau-like surface in a serrated escarpment located in the south polar latitudes. These markings most probably represent wind-carved grooves (4174-56). (d) Laminated terrain near the south polar ice cap. Note the elongate ridges (4213-21). (e) Co-association of laminar deposits and spurs of the residual ice cap at the Martian south pole (4230-12). (f) The north polar region of Mars showing a depressed basin centered about the pole (upper right corner) that contains inward tilted rock units. The bright band beyond is part of the shrinking CO_2 ice cap. The frost-rimmed crater (left) may be covered by water ice. The older terrain underlying the polar deposits is less heavily cratered than the corresponding region around the south pole. (NASA, Jet Propulsion Laboratory.)

Characteristics of the Polar Regions

The Martian polar regions are also centers of contemporary activity. By late 1971, the northern hemisphere was well into its winter period, while the southern polar region had passed into late summer. Somewhat earlier in a corresponding season, a broad CO_2 frost cap with ragged edges was seen around the south pole by *Mariner 7* (see Figure 12-7b); *Mariner 9* saw only a small residual cap whose boundaries were straight and sharp (Figure 12-24a). Curvilinear features that stand out in the *Mariner 7* images (at the bottom of photo 7N17 in Figure 12-7b) because of "decorating" by the thin frost cover are recognized only as faint, dark bands in the *Mariner 9* view of the same area. These features are interpreted as circumpolar ridges whose topographic forms were emphasized by frosting. The residual cap itself seems to be a thin layer of water-ice left behind (because of a higher sublimation point) when the CO_2 frost sublimed.

The southern polar region surrounding this cap contains three distinctive types of terrain. Old *cratered surfaces* predominate at lower latitudes. Closer to the poles, *smooth depositional terrain* is interspersed with *etch-pitted areas.* Although some of the pits (Figure 12-24b) resemble collapse kettles characteristic of glacial outwash, they are generally believed to be caused by wind deflation, perhaps aided by frost ablation (evaporation). Other surfaces are extensively grooved by huge striations (Figure 12-24c), as though gouged or scoured by wind (or less likely, glacial) action. At even higher latitudes, large areas display unusual banded patterns, described as *laminated* terrain (Figure 12-24d). High-resolution photos reveal these laminae to be alternating light and dark layers from 10 to 30 m thick exposed along steplike cuesta ridges that face inward into a broad saucerlike depression. As many as twenty layers occur in each of the 6 to 12 cuestas recognized in the basin. The dark layers are possibly pyroclastic deposits but their periodicity suggests a relationship to a cyclic process such as the 50,000-year precession or wobble of the Martian rotational axis, the seasonal changes in the polar icecaps, or the occasional dust storms. Around the residual cap, the layers are intimately interleaved with whitish water-ice (?) masses (Figure 12-24e). Where the ice appears to be receding, the bands superficially resemble morainal deposits left behind by retreating glaciers. However, the bands most likely represent accumulations (sediment-traps) of dust-sized particles transported by glacio-aeolian processes from source areas closer to mid-latitude and equatorial regions. The near-absence of impact craters in deposits close to the pole implies that the regional surface there is very young by Martian standards. However, these deposits cover older, variably cratered bedrock units at both poles.

Most polar terrain features result from strong wind erosion, assisted by surface frost and subsurface permafrost action. The youthful appearance of the polar regions implies that the surface is being modified even now by continuing cyclic processes which both scour off older features and cover irregularities by deposition. Whether the present thin polar caps, even when enlarged during the Martian winters, are capable of producing some of the features seen, or a much thicker mass of CO_2 and/or H_2O is required to initiate glaciation, remains an unsolved problem. The depressed basin topography—visible also in the northern polar region (Figure 12-24f)—is consistent with an isostatic sinking of the Martian crust under a heavy ice load. At present, the remnant cap region at the south pole is only slightly lower than the surrounding terrain. But

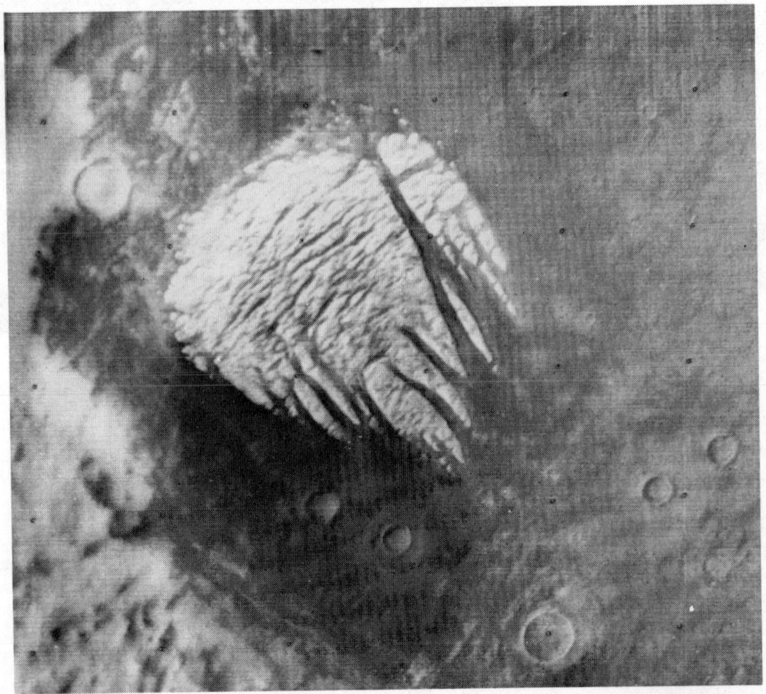

Figure 12-25 Cratered, splotchy terrain in the Martian mid-latitudes. The apparently elevated region in the left center, known as White Rock, is a high albedo surface broken by grooves and cracks. This area may be covered by a frozen gas condensate or some chemical sublimate (4216-54). (NASA, Jet Propulsion Laboratory.)

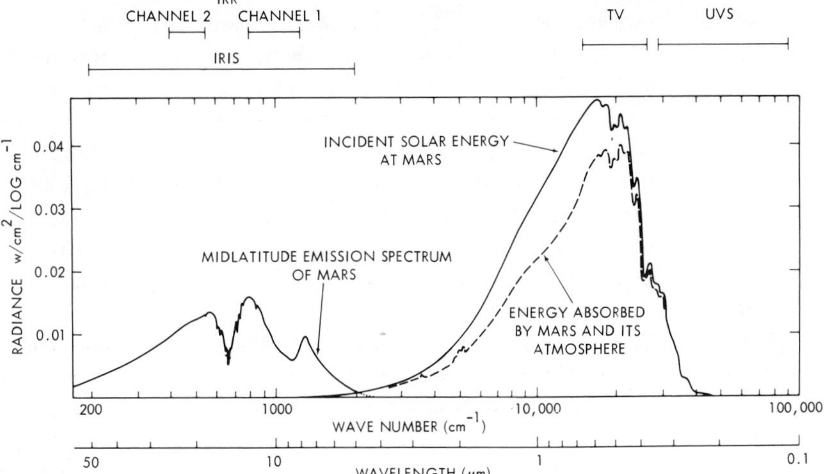

Figure 12-26 Distribution and behavior of solar energy in the vicinity of Mars. A fraction of this energy is absorbed by the thin Martian atmosphere while much of the remainder between 0.1 and 3 μm is reflected. At higher wavelengths, incoming radiation is reradiated in the infrared giving rise to an emission spectrum. Distinctive emission and absorption peaks and troughs in the energy curves, resulting from interactions with specific materials, are labelled. The operational wavebands of several *Mariner 9* sensor systems are indicated. (Courtesy Laboratory for Planetary Atmospheres, NASA Goddard Space Flight Center.)

the cumulative thickness of the laminated materials is estimated to be 4 to 6 km. These layers must then occur as stacks of bowed plates that comprise a series of "stair-steps" which has now been distorted upward by some rebounding of the depressed crust.

A large, bright topographic feature much nearer the equator, nicknamed the White Rock (Figure 12-25), superficially resembles the residual cap. Infrared radiometer measurements prove that its surface temperature is too high for either CO_2 or H_2O ice.

Mariner 9 Instrumental Measurements from Orbit

In the geologic realm, the direct viewing in the *Mariner-9* television experiment garnered the greatest scientific re-

Figure 12-27 (a) Thermal emission spectra from a nonpolar region of Mars (18°S, 13°W; obtained during orbit 8 at 12:00 local time). Emission curves for three different blackbody temperatures are plotted for comparison. The IRIS curve nearly fits the calculated curve for a temperature of 230°K. (b) Spectra obtained over the south polar cap during revolutions 29 and 30. The blackbody temperature over the bright residual cap is calculated to be 140° ± 10°K.

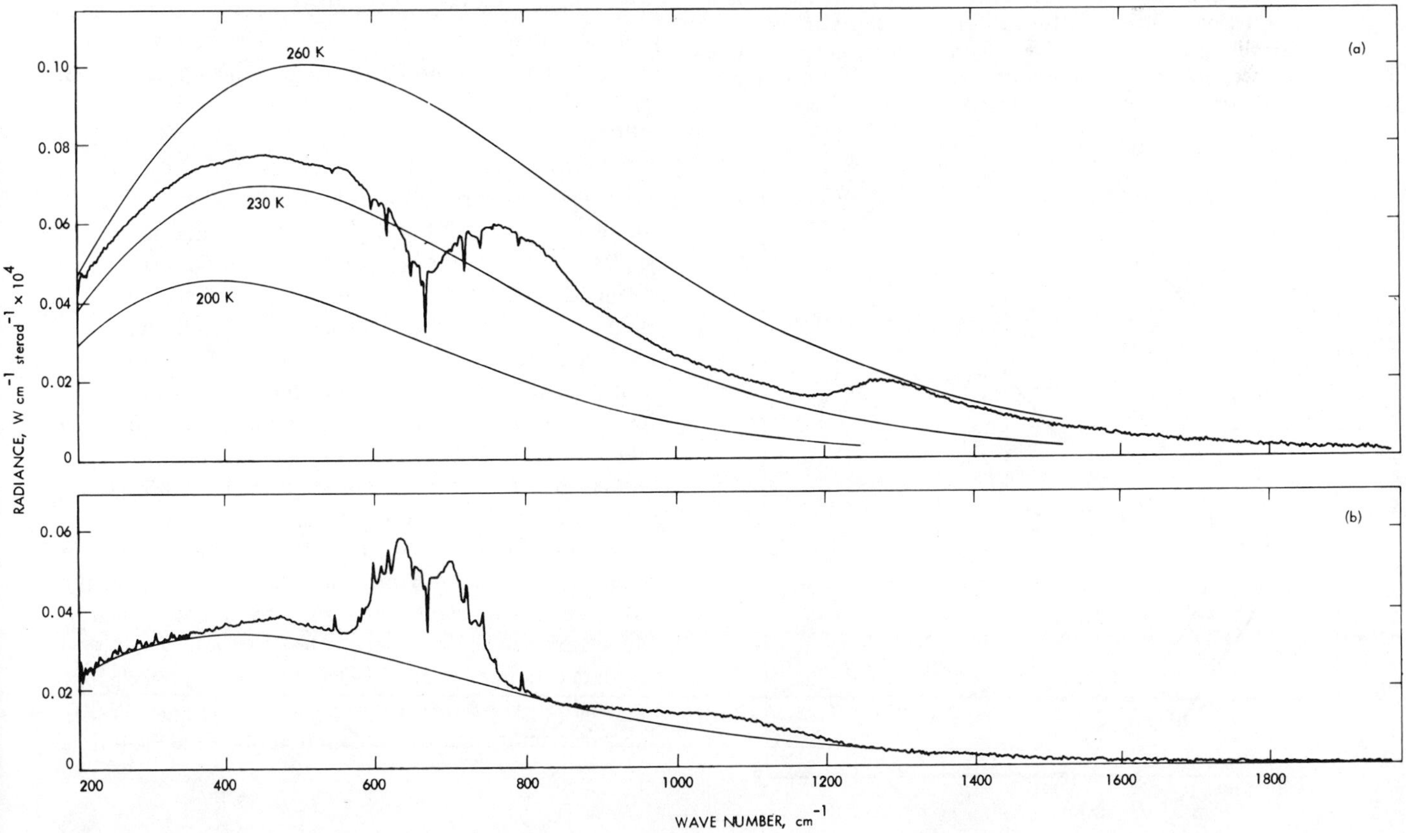

wards. However, the other instrument packages onboard also contributed data vital to a more comprehensive understanding of Mars as a planet. These sensors provided information on the composition of the atmosphere and dust; on temperature and pressure distribution within the Martian atmosphere as a function of longitude, latitude, and local time; and, through the interrelation of pressure variation and surface heights, on the topography of the surface over large areas as well as along individual traverse profiles. The operating spectral regions of these sensors and the general characteristics of the incident solar radiation at Mars, the absorbed portion of that spectrum, and the Martian thermal emission spectrum are summarized in Figure 12-26.

Infrared Spectrometer

The infrared interferometer spectrometer (IRIS)—essentially a Michaelson interferometer—recorded the thermal emission spectrum of the atmosphere and the underlying surface through the spectral interval from 5 to 50 μm. Typical spectra obtained early in the mission at low (equatorial) and high (polar) latitudes appear in Figure 12-27. The pronounced downward trough in the equatorial spectrum between 540 and 800 cm^{-1} (from approximately 19 to 13 μm) is characteristic of molecular-band excitation of gaseous CO_2 in an atmosphere which is cooler at high altitudes than near the ground. In the same wavelength region, CO_2 radiation measured near the south pole produces an upward peak that represents an emission effect resulting from principal excitation in a warmer gas layer above a cooler surface region. The polar spectra also show faint but definite emission lines of H_2O vapor in the interval from 200 to 350 cm^{-1}; during spring over the north polar cap as much as 20-30 precipitable micrometers of water is present in the atmosphere.

At values near 1200 and 1300 cm^{-1} the broad trough (absorption) and peak (emission) responses in the spectral curves for nonpolar and polar regions, respectively (in Figure 12-27), are attributable to excitation of SiO_2 bands in dust-sized materials (diameters between 1-10 μm) within the atmosphere. Silicates derived from surface weathering of rocks as diverse as basalts and granites could give rise to this observed spectral response. A mixture of silicates with varying compositions has been inferred from these data, but an average value of ~60% SiO_2 is suggested by the maxima and minima in the emission feature. If this rather high SiO_2 content, compared to values of 40 to 50% for terrestrial and lunar ultrabasic and basaltic rocks, holds up in more detailed analysis, the Martian crust may well have been geochemically differentiated into rocks at least of andesitic composition. A more precise determination of dust composition must await comparisons with terrestrial silicate rock spectra of various compositions obtained for different particle-size ranges.

The temperatures within the Martian atmosphere depend on location, altitude, local time, seasonal effects, and the amounts of suspended dust at the time of measurement. Figure 12-28 shows a planar cross section of temperature variation with pressure (and hence with height) made from a single orbital pass across the southern polar region. When atmospheric temperatures at different elevations above much of the planetary surface are averaged over many spacecraft revolutions, isobaric temperature maps can be constructed to record changes in temperature during diurnal solar heating. Revolution-dependent examples that define the temperature at various given times of day along latitude lines are presented in Figure 12-29 for the case in which the pressure level is 2 mbars (corresponding to an altitude of 12 km above an arbitrary 6.1-mbar pressure surface). These examples also demonstrate the changes in atmospheric temperatures that occurred as the dust progressively settled between December 1971 and February 1972. The maps indicate that, during the storm, temperature maxima were attained near Martian sunset (~1800 hours) because of thermal lag in the atmosphere caused by the dust. With abatement of the storm, the maximum values decreased, shifted toward the equator, and occurred earlier in the day. The temperature and pressure data acquired by IRIS have also been used to calculate a generalized wind field diagram for Mars (Figure 12-30).

Surface temperatures as a function of latitude and time

Figure 12-28 Temperature distribution (contoured isotherms) in the Martian atmosphere as a function of pressure and corresponding altitudes, obtained by IRIS during a single pass over the south polar region (indicated by latitude; the heavy line corresponds to the south polar cap). Note the effects of a temperature inversion near the surface in which temperatures rise with height to a maximum near 15 km and then drop thereafter with increasing altitude.

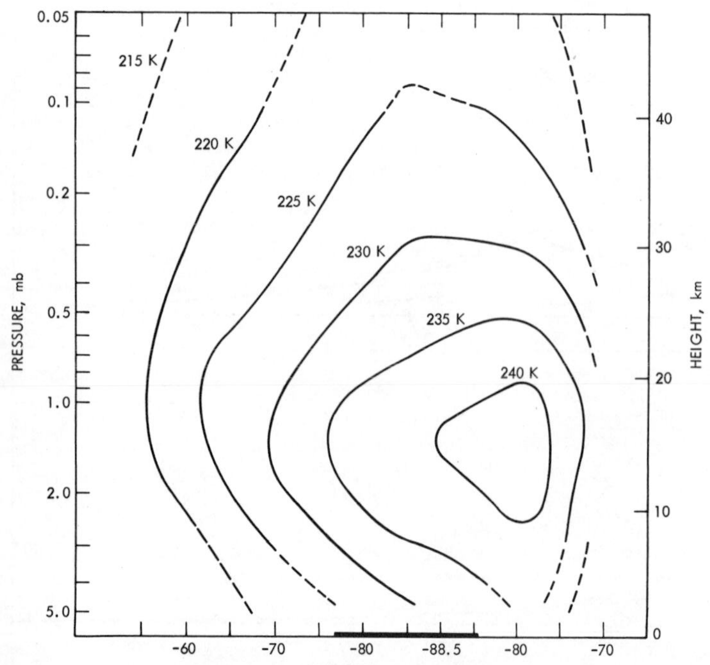

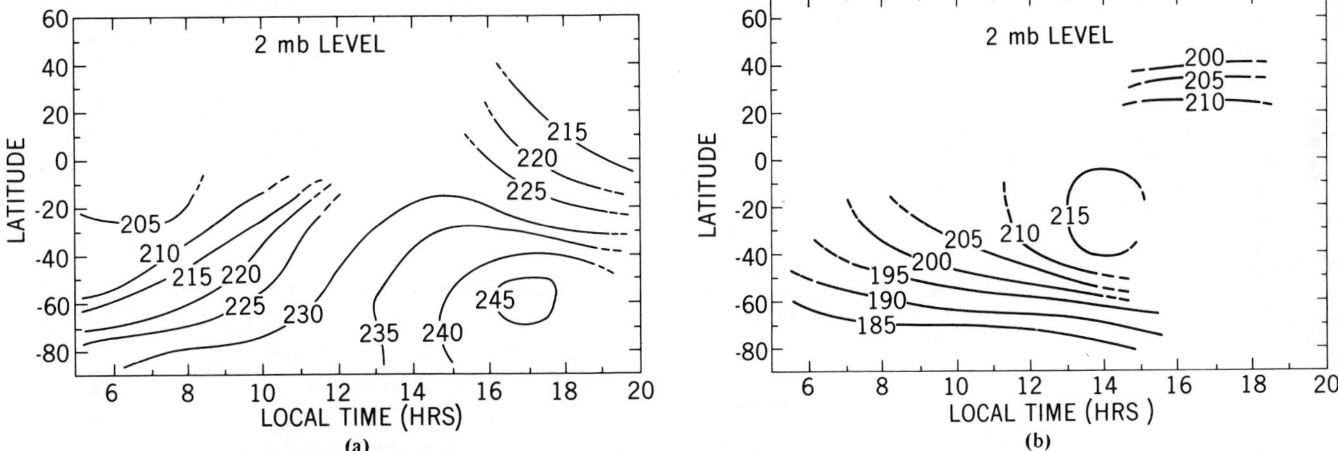

Figure 12-29 Atmospheric temperatures above Mars at the 2 mb level as functions of latitude, local time during the Martian day, and observing periods (in terms of orbital revolutions). Temperatures (a) during the early revolutions, taken during the dust storm, are notably higher than those (b) characteristic of a clear atmosphere after most of the dust had settled. (Courtesy Laboratory for Planetary Atmospheres, NASA Goddard Space Flight Center.)

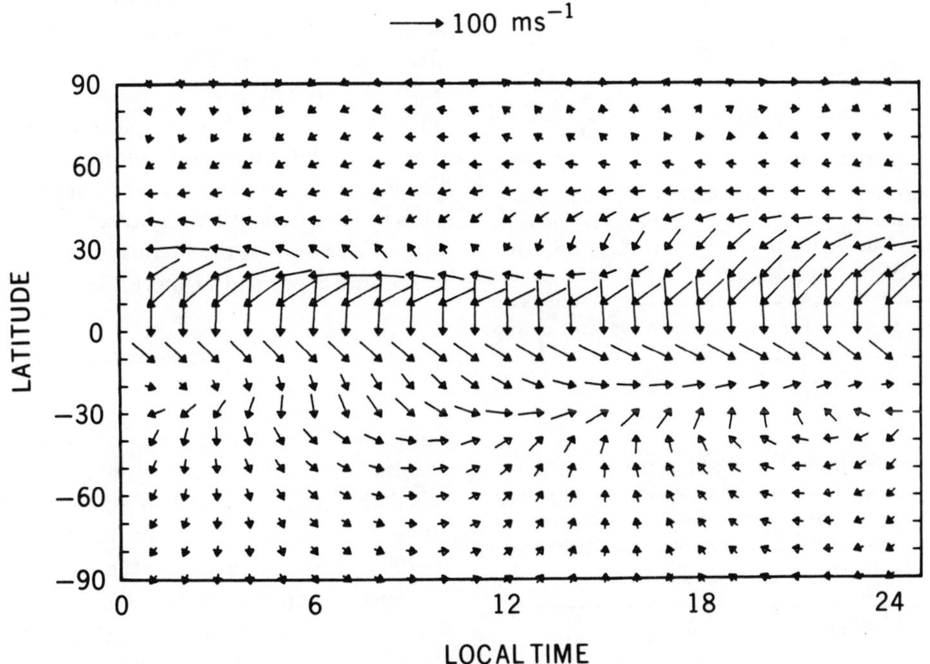

Figure 12-30 A general wind field pattern derived for the Martian surface using temperature and pressure data taken during the first 85 orbits by the Infrared Interferometer Spectrometer (IRIS) on *Mariner 9*. The diagram shows an average velocity and direction of winds at one scale height (10 km above mean surface level) as a function of latitude and local time (thus, horizontal translation along a latitude line will show the vectorial shift in a wind during a full Martian diurnal cycle). The arrowhead on each wind vector symbol indicates the direction towards which the moving gases are flowing. The length of each vector is proportional to the length of the 100 meter/sec arrow shown at the top. Large wind field values measured near the equator are abnormally high relative to actual surface values owing to a numerical instability in the calculations. This diagram, although not a circulation map over the entire planet, does successfully predict some of the observed patterns of many of the Martian surface transients (tails, plumes, etc.) produced during the great dust storm of 1971. (Hanel et al., 1972.)

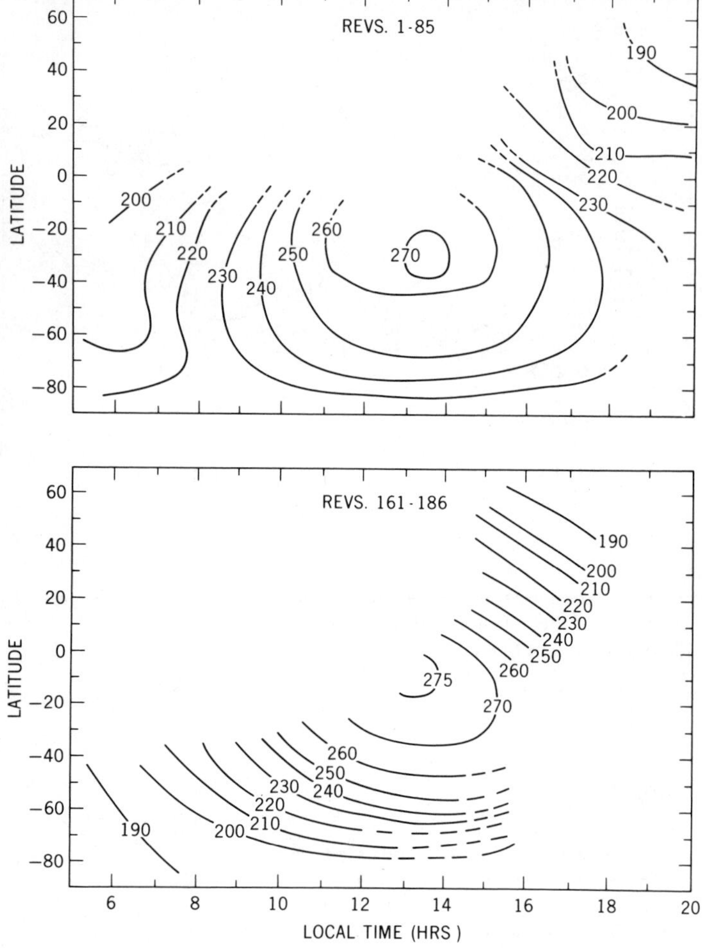

Figure 12-31 Surface temperatures on Mars as functions of latitude, local time during the Martian day, and observing periods. Measurements were made during the dust storm (Revs 1-85) and afterwards (Revs 161-18). The main effects of abatement of the dust storm were to slightly increase the surface temperature maximum, shift it closer to the equator, and steepen the temperature gradient between equator and the polar regions. (Courtesy Laboratory for Planetary Atmospheres, NASA Goddard Space Flight Center.)

Figure 12-32 Theoretical (dashed line) and observed (solid line) curves of atmospheric temperature as a function of time of day obtained during the dust storm (orbits 60-70) by the infrared radiometer onboard *Mariner 9* as it passed over -25° latitude regions of Mars. (NASA illustration.)

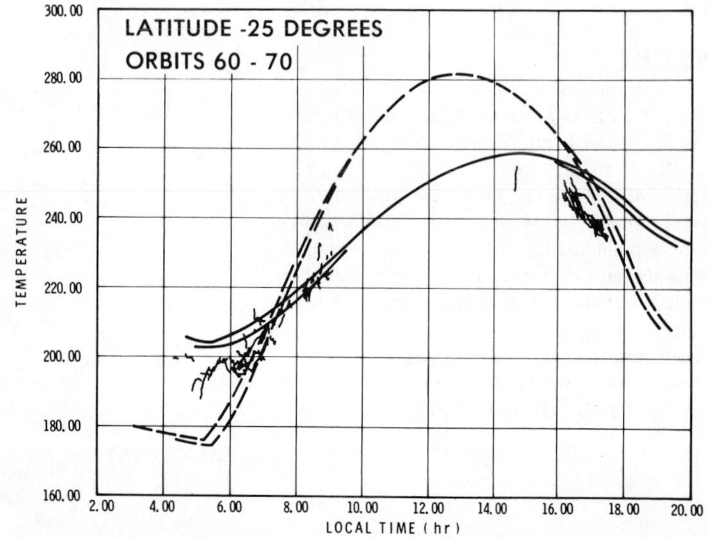

of day have also been obtained by IRIS for periods during and after the dust storm (Figure 12-31). These temperatures exhibit substantially less variation over the three months than did the atmospheric temperatures. A maximum shortly after Martian noon (~1300 hours) is located in the southern equatorial latitudes near the subsolar point.

Infrared Radiometer

Surface temperature and atmospheric data were also provided by the other sensor systems. Figure 12-32 shows a theoretical surface temperature-time plot at 25° latitude predicted for a clear Martian atmosphere. The actual temperature distributions measured by the infrared radiometer are also plotted in this figure. The observed 25° to 30°K decrease relative to expected values at noontime results in part from greater reflectance of thermal radiation from the dust-laden atmosphere. However, the dust blanket also absorbs thermal energy, so that at night the observed temperatures are higher than those of a clear atmosphere subject to radiation cooling after solar illumination ceases. Other factors responsible for significant departures in actual temperature distributions from those calculated in theoretical models include (1) variations in surface albedo (a darker surface absorbs more heat and hence is more effective in warming the atmosphere above it), (2) differences in slope (topography), and (3) changes in surface materials (composition and/or particulate state). The IR radiometer has also produced temperature profiles over parts of the surface. The radiometer scans across a narrow field of view so that the thermal data are localized. A typical profile appears in Figure 12-33.

Ultraviolet Spectrometer and S-Band Occultation Experiments

Both the UV spectrometer and the S-band (radio-wave) occultation technique can be used to determine compositional changes in the Martian atmosphere up to very high altitudes. Electron densities in the Martian ionosphere up to 300 km were measured in the S-band experiment. The UV spectrometer confirmed the *Mariner 7* discovery of small amounts of ozone or O_3 (~0.1 ppm) concentrated above the polar and high-latitude regions. The UV spectrometer, looking edgewise through the atmosphere, also detected both atomic hydrogen and oxygen out to distances of 20,000 km from the surface. These species come from photodissociation of CO_2 and H_2O. The amount of hydrogen observed at that altitude can be translated into the quantity of water escaping from Mars each day. Approximately 100,000 gallons (enough to fill a large swimming pool) leave the planet every day. If all this water were precipitated at once from the atmosphere, a continuous layer from 10 to 20 μm thick would envelop the entire surface of the planet. Presumably, this much water is being

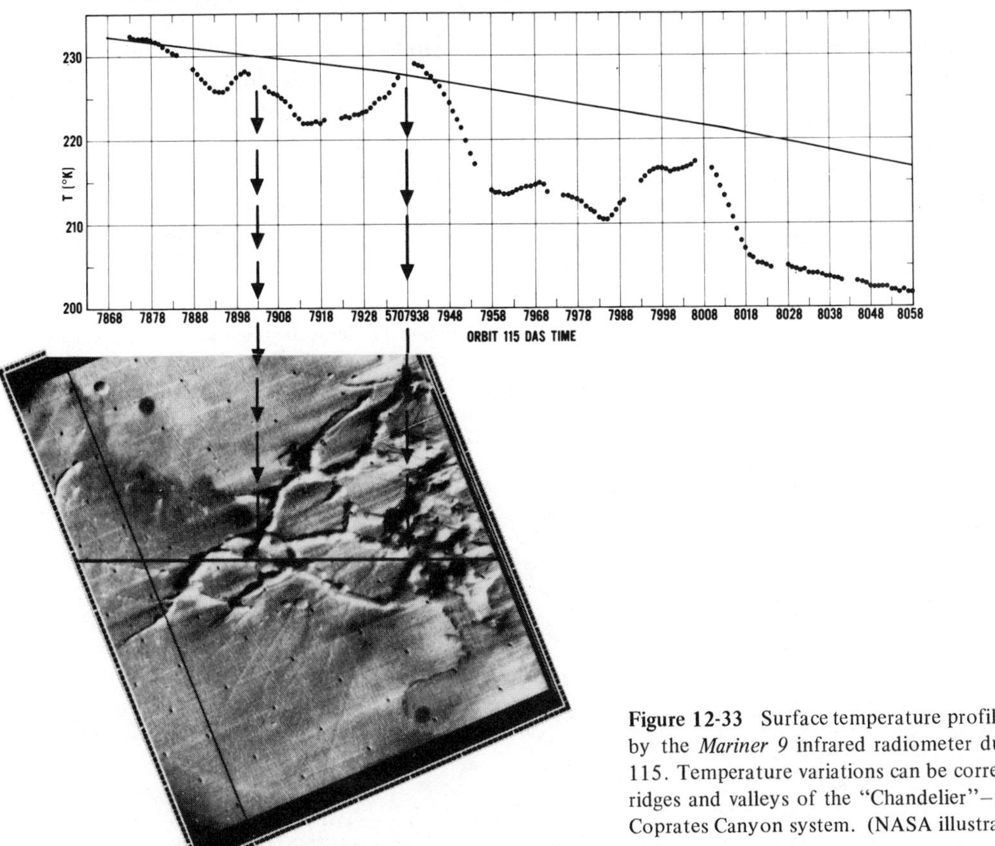

Figure 12-33 Surface temperature profile obtained by the *Mariner 9* infrared radiometer during orbit 115. Temperature variations can be correlated with ridges and valleys of the "Chandelier"—part of the Coprates Canyon system. (NASA illustration.)

Figure 12-34 Differences in elevation across part of the Coprates Canyon in the Tithonius Lacus region determined indirectly from pressure measurements above the surface made by the ultraviolet spectrometer on *Mariner 9*. (NASA illustration.)

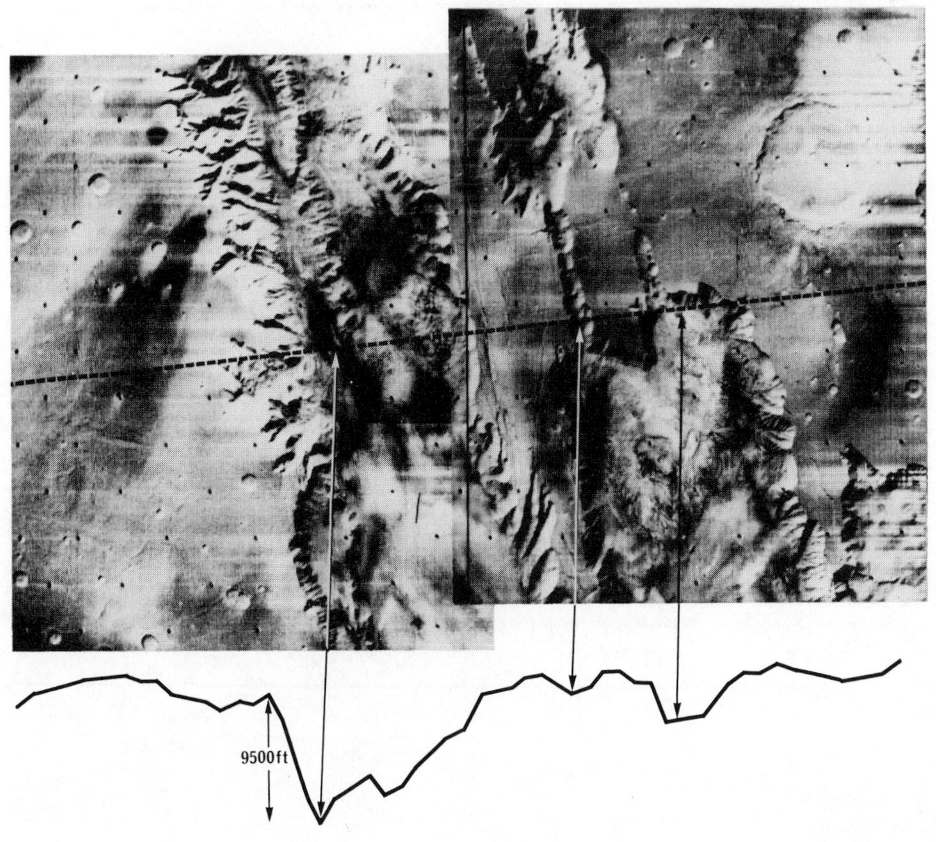

replenished at the surface to maintain a steady loss rate in the Martian exosphere.

Elevation differences in Martian topography were determined indirectly from data obtained by the UV spectrometer, the infrared interferometer spectrometer, and the S-band occultation experiment. The UV spectrometer measures variations in intensity or brightness of the ultraviolet airglow at 3050 A (angstroms) along traverse lines under the spacecraft. This intensity varies in proportion to the total amount of atmosphere within a vertical column below the sensor's line of sight. Surface pressures rise directly with the increase in the mass of this excited atmosphere. As on Earth, higher pressures are recorded in low regions and lower pressures are coincident with rises in

Figure 12-35 (a) Generalized (smoothed) differences in elevation (km) in the equatorial region of Mars as determined from ultraviolet spectrometer (UVS) data obtained by *Mariner 9* (courtesy C. Barth, University of Colorado). (b) Martian topography (km) as calculated from Infrared Interferometer Spectrometer (IRIS) data acquired by *Mariner 9* (Courtesy J. Pearl, Laboratory for Planetary Atmosphere, NASA Goddard Space Flight Center.)

(a)

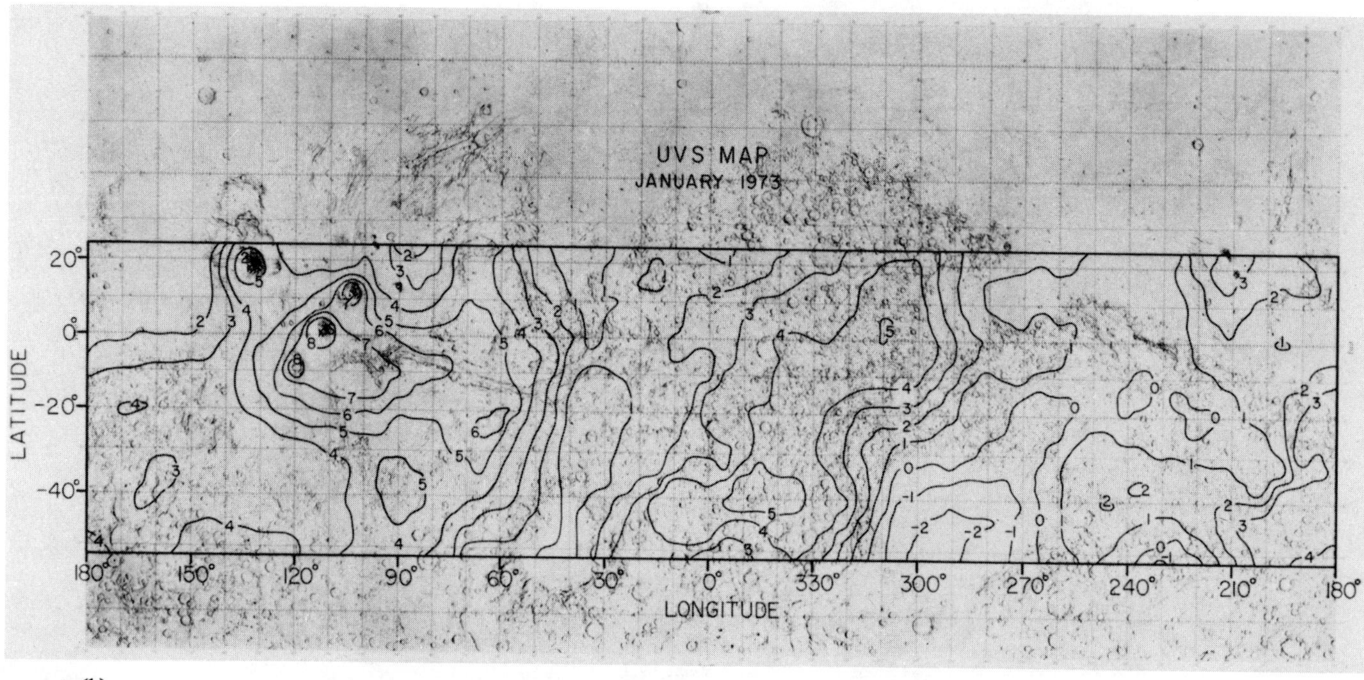

(b)

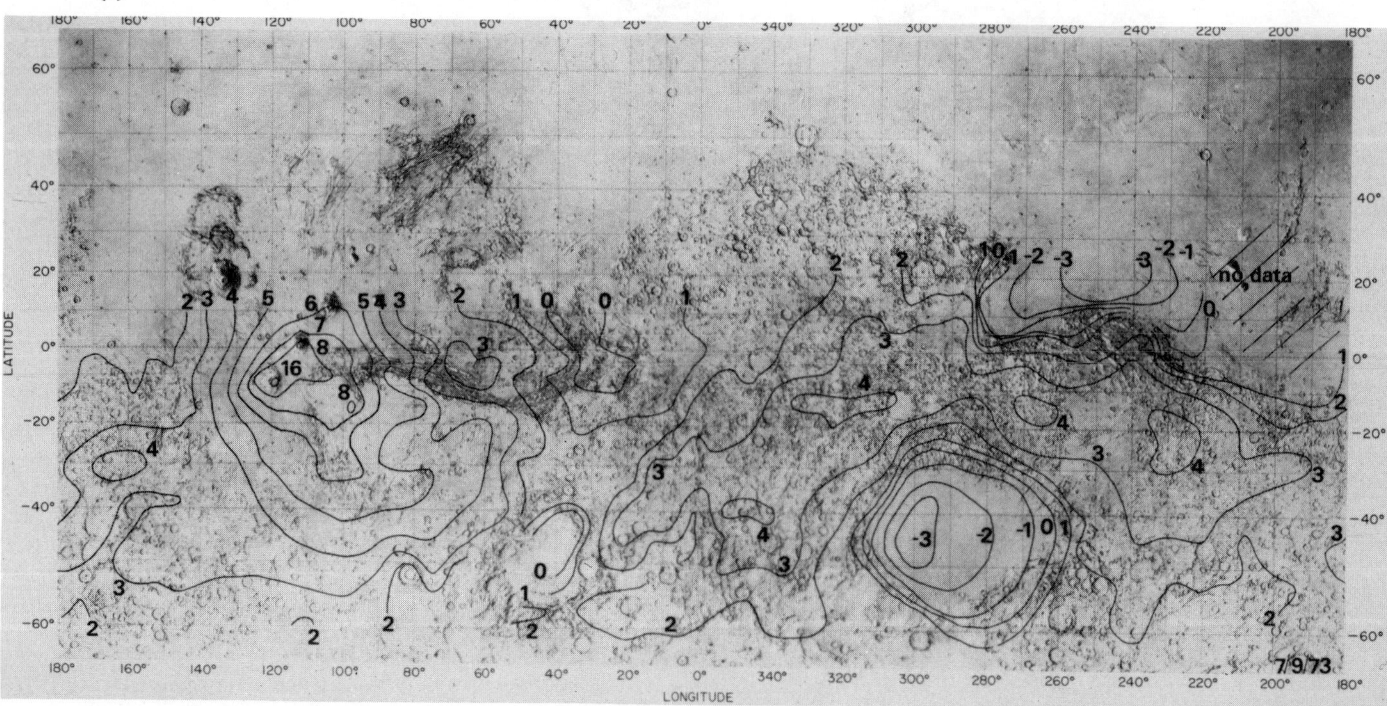

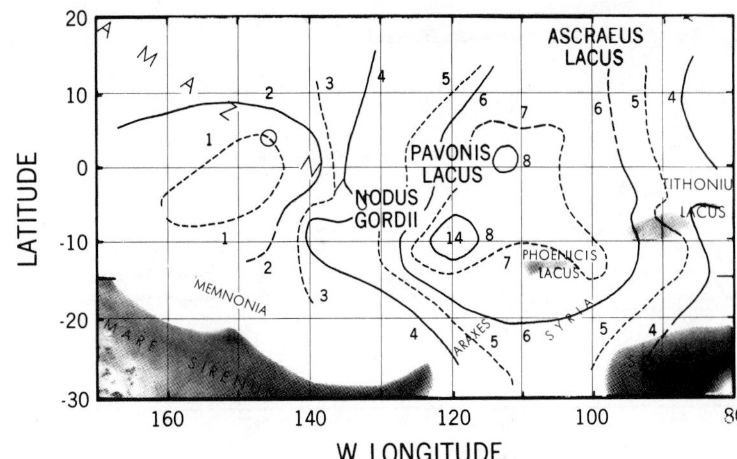

Figure 12-36 Topography (contoured in 1 km intervals) of the Martian surface in the Amazonis-Tharsis region as determined from pressure data obtained by IRIS on the *Mariner 9* spacecraft. (Courtesy Laboratory for Planetary Atmospheres, NASA Goddard Space Flight Center.)

surface elevation. An example of a topographic profile made along a narrow track (~15 km wide) across the great Coprates Canyon in the Tithonius Lacus region is given in Figure 12-34; depths of 5 to 6 km below the surrounding plains have been recorded. A generalized topographic map of much of the equatorial region of Mars based on UV data appears in Figure 12-35. Note that there is almost no correspondence between contour highs and the prominent dark markings in the *Mariner* 1971 map of Mars (see Figure 12-1). The zero contour line marks an approximate altitude below which liquid water can potentially exist on or beneath the surface at a triple-point pressure appropriate to Martian temperatures and atmospheric pressures.

Values of atmospheric opacity (expressed as the degree of transmittance of CO_2 as the radiation passes through the atmosphere from surface to spacecraft) are estimated by the infrared interferometer spectrometer using measured intensities within the 15-μm absorption band (near 670 cm^{-1}). Surface pressures are then derived as a function of these opacity values. Variations in pressure over large areas are converted into equivalent altitudes above or below a reference surface set at 6.1 mbars. In this way, contoured maps of elevation (in kilometers) can be produced. Results are shown in Figure 12-36 for the Tharsis region, which includes Middle and South spots. The summit of South Spot lies 17 km above the floor of the Amazonis basin and 6 km above its base; the average slope along the flanks of this shield volcano is thus about 3°.

The S-band occultation measurements provide vertical pressure profiles that can be converted to surface pressure values. Changes in elevation are expressed as variations in the Martian radius. The variations are inversely proportional to surface pressure—i.e., the pressure decreases as the radius increases outward. This is illustrated in Figure 12-37 for the Hellas region, along with pressure-mapping

Figure 12-37 Variations in surface pressure (dashed line) and surface topography in terms of the radius of Mars (solid line) in the Hellas region as determined by the S-Band occulation experiment on *Mariner 9*. The mean radius in that part of Mars (measured across longitude lines) appears as a short-long dashed line. Measurements of pressure (circles) and altitude (squares), made by the infrared spectrometer (solid black) and ultraviolet spectrometer (open) on the *Mariner 9* spacecraft, are shown for comparison.

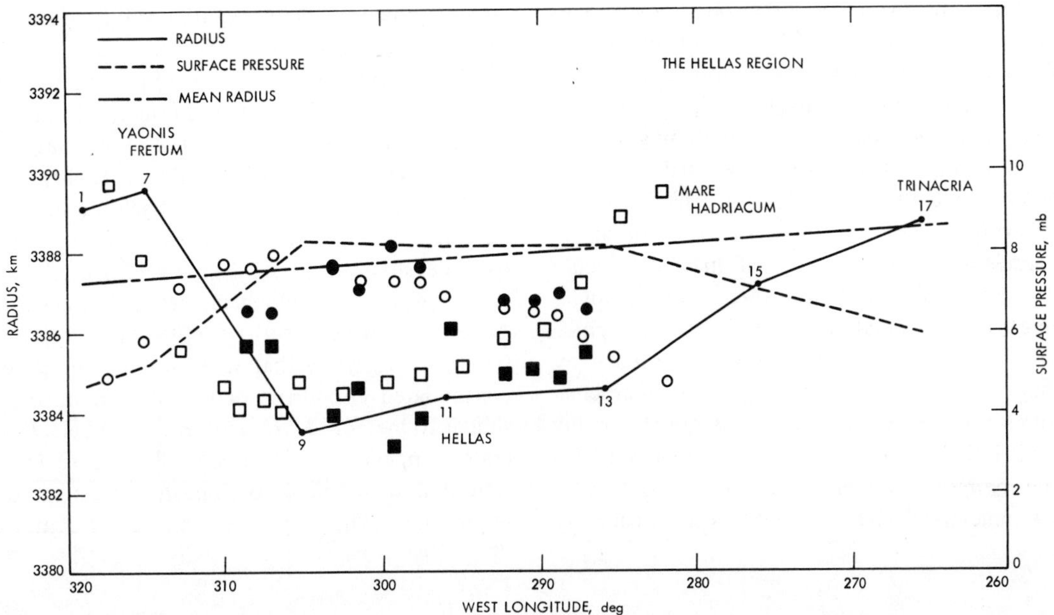

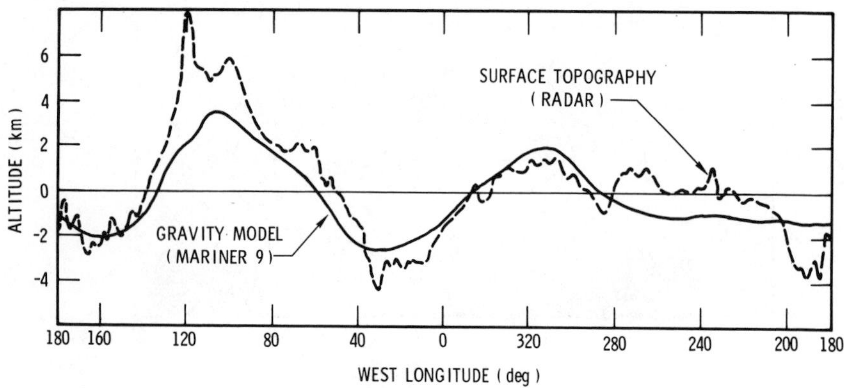

Figure 12-38 Elevation profiles along the 15°S latitude across Mars. The solid line represents differences in elevation required to produce the observed variations in the gravity field measured by a celestial mechanics experiment using the *Mariner 9* spacecraft. The dashed line describes the surface-height variations determined by radar from Earth in the summer of 1971. Note the agreement between the two profiles. (NASA illustration.)

results obtained in 1969 by the UV spectrometer and IRS systems on *Mariner 7.* Hellas is shown to have a relief of nearly 7 km, with its deepest part about 4 km below the mean radius in that part of Mars. Its western rim near Hellespontus lies about 2 km above the mean radius. The steeper western slope grading into a more gradual eastern rise to Trinacria was also noted by IRS mapping. In contrast, the region between Solis Lacus and Mare Sirenum remains 4 to 8 km above the Martian mean radius. The south polar region (and much of the southern hemisphere) is ~3.0-3.5 km higher than the north polar region (and northern hemisphere). Pressure variations ranging from less than 3 to nearly 10.3 mbars (relative to an average of 5.5 mbars for the entire planet) correspond to planetwide elevation differences of more than 18 km. High pressures (7 to 10 mbars) in the northern latitudes indicate a polar radius of 3372.7 km compared with equatorial radii varying between 3397.8 and 3400.4 km. The figure of Mars may be approximated by a triaxial ellipsoid with a polar flattening ratio of 0.0075—a somewhat larger value than the 0.0051 obtained from gravitational effects on the *Mariner 9* orbit. Thus, the shape of the solid body of Mars is more oblate, by a maximum of 8 km at the poles, than the hypothetical shape assumed by an equipotential surface that would result from the Martian gravity field.

The gravity field over Mars was measured by a celestial mechanics experiment similar to those carried out on *Lunar Orbiter* spacecraft that resulted in detection of the mascons. Perturbations of the orbital path of *Mariner 9* owing to accelerations by gravitational differences could be detected by analysis of radio-tracking data for Doppler changes (measured as variations in the round-trip phase delay of the radio signals). The data are reduced by spherical harmonic analysis to yield appropriate mathematical solutions for the models chosen. A gravity profile along a 15°S latitude traverse is shown in Figure 12-38. This profile represents calculated deviations of surface heights (in kilometers) from the surface of a spherical body of assumed uniform density. These deviations relate in part to internal inhomogeneities in mass distribution. However, a comparison with the surface topography constructed from radar data for the same latitude demonstrates that the observed gravitational changes correlate well with the physical surface of Mars. Thus, when the gravitational values along a series of orbital traverses are contoured, the resulting map of the gravity field over much of the southern hemisphere (Figure 12-39) in effect describes the surface height variations needed to produce this field. For its size, Mars shows substantially greater gravitational—and corresponding topographic—variability ("roughness") than either the larger Earth or the smaller Moon. This has led to some intriguing speculations about a considerable strength for the Martian crust, lithologic differences resulting from geochemical differentiation and density stratification, and possible isostatic disequilibrium within the Martian crust and mantle.

Martian Satellites

Aided by the remarkable pointing ability of the instrument platform on *Mariner 9* in orbit, the spacecraft was able to acquire an extraordinary series of photographs of the two small moons of Mars which are in synchronous (locked) rotation as they orbit the planet. The outlines and surfaces of both Deimos and Phobos (Figure 12-40) were imaged by TV with surprising clarity. Each moon proved to be irregular in shape. Deimos is rounded but not spherical; its dimensions are 13.5 ± 2 km by 12.0 ± 0.5 km. Phobos is somewhat more elongate, being 25 ± 5 km long by 21 ± 1 km wide. Both satellites have very low albedos (~0.05), suggesting that they may be composed of well consolidated chondritic or basaltic achondritic materials similar to meteorites from the asteroidal belt. Conspicuous craters appear on each moon. Phobos' surface is nearly saturated with round to elongate craters surrounded by outer rims. The very small size of these moons should

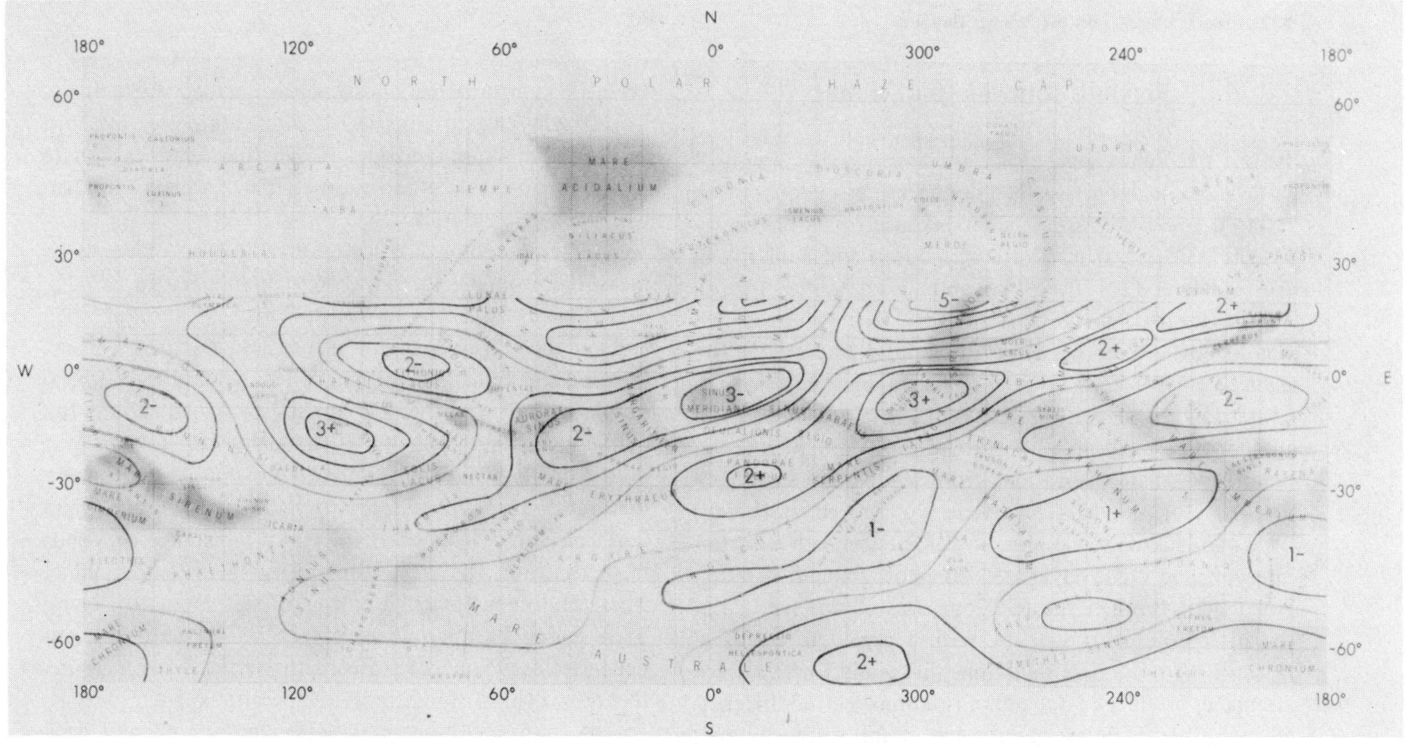

Figure 12-39 Contours (in kilometers) of equivalent surface heights across most of Mars determined from a spherical harmonics analysis of Martian gravity data obtained by *Mariner 9*. The contours express the deviations from sphericity of a uniformly dense body. (NASA illustration.)

preclude volcanism as the cause of these depressions, leaving impact as the probable explanation for their origin. The crater densities indicate the moons to be very old—perhaps contemporaneous with the first cratered surfaces of primitive Mars if they existed then as Martian satellites. The possibility that the moons are captured asteroids cannot presently be excluded.

In sum, Mars can rightly be described as a fantastic and challenging planet with far more variety than geoscientists dared hope for. It shares a kinship with the Moon, but is marked by significant differences as well. Mars reveals signs of wind and water erosion and of tectonic deformation such as one might envision for some earlier day on Earth. Yet, above all, Mars possesses its own characteristic features that set it apart as a planetary body unique to itself.

Missions to Venus

The story of the exploration of Venus parallels that of the Martian probes in many respects. Most of the new information concerns the Venusian atmosphere and the outer planetary environment above its surface. So far, little direct knowledge about the surface or interior of Venus has been obtained by the two American *Mariner* space vehicles that passed close to the planet or by the five Soviet *Venera* space probes that penetrated through its gaseous envelope (the first three apparently crashed or burned up instead of achieving hoped-for soft-landings).

Figure 12-40 Phobos, the larger of the two Martian moons, as imaged by the TV camera on *Mariner 9* during the 34th orbit, when the spacecraft passed to within 5540 kilometers of this cratered, irregularly shaped satellite. (NASA photo.)

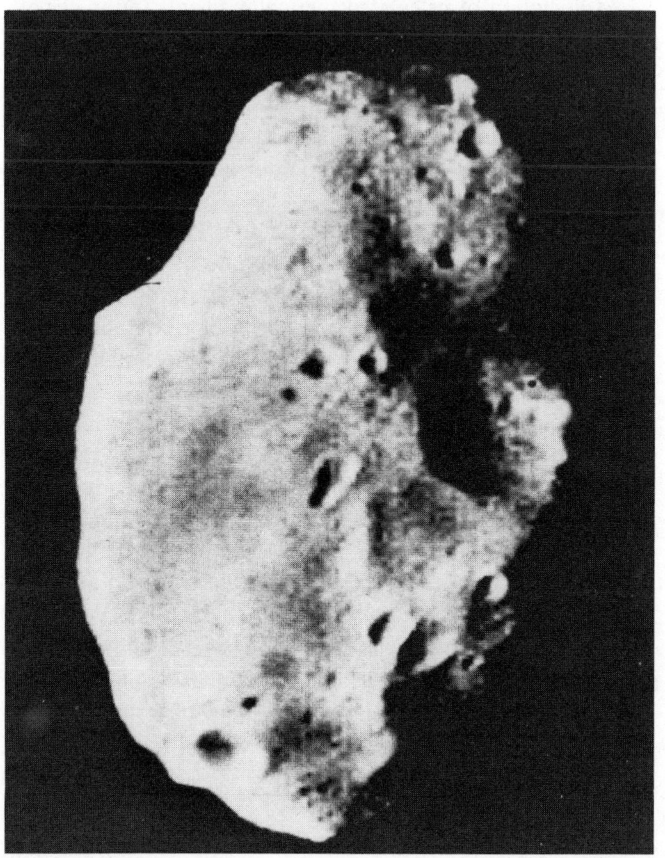

Previous Knowledge of Venus

Before the Venus space probe missions, a fair amount of knowledge had been gathered concerning Venus—often referred to as our *sister* planet because of its proximity (44,000,000 km at inferior conjunction) and similarity in size (diameter 12, 170 ± 20 km and mass 4.8678 × 10^{27}g). Venus' mean density of 5.158 ± 0.010 g/cm^3 is just slightly lower than that of Earth (about 5.52 g/cm^3). The planet's motions are unusual in the solar system because of its slow *retrograde* rotation (clockwise as viewed from the north celestial pole). Estimates of its rate of rotation based on visual observations from Earth were extremely unsatisfactory (some values were close to an Earth day) owing to lack of visual reference points beneath a continuous opaque cloud layer. When photographed in ultraviolet light through the telescope, this cloud layer is observed to rotate very rapidly with a period of ~4 days. But in 1962 radar determinations that could fix on surface anomalies produced estimates of rotational periods bracketed by 180 to 570 days as extremes. Improved techniques led to values around 250 ± 5 days by 1964. The best present value is given as 242.98 ± 0.04 days.

The Venusian interior is, by analogy to Earth, assumed to consist of a thin crust (~1% by mass), a mantle (68-78%), and a core (21-31%). The core in this model is presumed to be Ni-Fe in composition. Internal temperatures may reach 2000-2700°C at a depth near 1700 km. This should lead to considerable melting and production of near-surface magmas with accompanying volcanism. Extrapolation of calculated thermal gradients to the planet's center points to conditions favoring a very fluid core. If these assumptions about the thermal state in the interior are correct, then motions in this core ought to generate electric currents that cause a strong, readily detectable magnetic field which would trap solar particles in well-organized radiation belts.

A distinct Venusian atmosphere is evidenced by occultations of bright stars passing beyond its limbs. Prior to the U.S. and Soviet space probes, this atmosphere was assumed to reach out to 100 km or farther. Carbon dioxide was the only compound recognized with certainty from spectral analysis, but nitrogen was presumed to be the principal constituent (perhaps 80% by weight) along with minor amounts of argon, a little water, and very small quantities of oxygen, HF, and HCl. The total pressure (at the surface) was thought to be from 10 to 50 times higher than terrestrial values (~1 atm), although some interpretations placed it as low as a few atmospheres. The planet appears to be perpetually shrouded by a cloud layer whose height is variously placed at 50 to 90 km above the surface. This layer is the main reason for an albedo of ~0.71, indicating strongly reflective substances, assigned to the whole planet. The spectral data available were inadequate for determining the cloud composition: speculative interpretations on principal constituents included water drops, ice, frozen carbon dioxide, carbon suboxide, mercury halides, ammonium nitride, silicate particles, carbonate particles, formaldehyde, hydrocarbon droplets, and anhydrous ferric chloride.

The temperature data for observable parts of Venus, as measured from Earth, show considerable ranges in value that depend both on the instruments and methods used and on the portions of the planet actually being sensed. Infrared spectral analysis in the 8-12-μm band gives a measured value of ~234°K that applies specifically to the top of the cloud layer region. The planet seems to emit heat both from its surface and from levels within its atmosphere that raises the temperatures to values much higher than that calculated for a blackbody at the distance of Venus from the Sun if it were warmed by solar heating alone. These equivalent blackbody brightness temperatures have been sensed by microwave detectors; the resulting values depend closely on the wavelengths selected for measurement. Typical brightness temperatures are ~380°K, 550°K, and 680°K at millimeter, centimeter, and decimeter wavelengths, respectively. The different values may arise within the atmosphere at different altitudes.

Without valid composition and temperature data, explanations of the high observed temperatures relied in part on speculation. The model of Sagan and Pollack depends on the "greenhouse effect": (1) the atmosphere is transparent to visible solar radiation; (2) the surface is strongly heated by insolation; (3) considerable thermal energy is reradiated in the infrared; (4) the atmosphere is largely opaque to infrared radiation that interacts with CO_2 and/or H_2O by absorption; (5) temperatures therefore build up within the atmosphere; and (6) further surface heating releases still more CO_2 and water from the lithosphere, leading to a "runaway" condition in the greenhouse effect. The model proposed by Goody and Robinson assumes that, because of uneven day/night heating of the atmosphere, dynamical motions are set up which bring about a circulation and mixing of CO_2 and other gases. This establishes a gradient of decreasing temperatures with altitude (adiabatic lapse rate, at constant energy). Using a lapse rate of ~80°K/km, extrapolation of upper atmosphere temperatures to the surface produces values generally in accord with observations. Other models, such as the ionosphere model of Jones (heating within a solar wind-ionized upper atmosphere) or the aeolosphere model of Öpik (heat added from friction between particle-laden strong winds and the Venusian surface), have encountered theoretical difficulties that limit their applicability.

The nature of the Venusian surface has been inferred from radar and passive microwave (radiotelescope) observations and from chemical calculations assuming mantle-crust materials analogous to those of Earth. Radar waves have been beamed over the entire planet at wavelengths ranging from millimeters to centimeters. Information

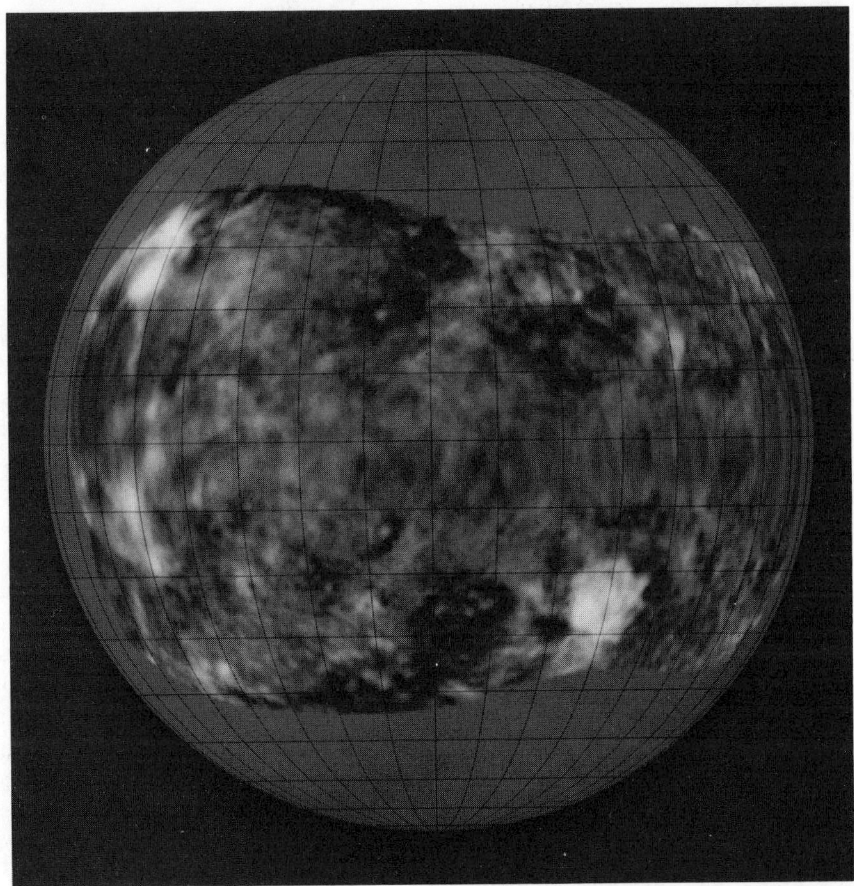

Figure 12-41 A map showing variations in intensity of a backscattered radar beam transmitted to Venus at a wavelength of 12.5 cm. The brighter areas represent marked departures from the mean surface reflectivity. These may be "rough spots" (rises and depressions) on an otherwise generally smooth surface. The radar signals were sent from the Jet Propulsion Laboratory Goldstone Tracking Station and the returns were received on two separated antennae which allowed more precise measurements by an interferometry technique. (Courtesy R. M. Goldstein, Jet Propulsion Laboratory.)

about the surface comes from interpretation of Doppler shifts in the frequency of incoming waves and from calculation of the dielectric constants (electric-charge capacity) of materials that can give the observed results. The beam is backscattered from the surface because of differences in the dielectric constant, surface roughness, slope changes, and, in some instances, elevations. One recent result from radar studies is depicted in Figure 12-41. Reflectivity variations emanating from the surface region appear as intensity differences (related to beam power reflected back to Earth), but these variations do not necessarily coincide with actual ground features inasmuch as any of the preceding factors may be the prime contributor to the intensities. In general, Venus seems to be a relatively smooth planet, with little definite topographic structure but with notable small-scale roughness. However, radar examination of a small sector of the Venusian surface (an area about the size of Alaska) at a resolution approaching 10 km has revealed a considerable number of circular depressions ranging up to 100 km in diameter. Apparently Venus, like the other inner planets, has been peppered by planetesimals and meteorites throughout its history and its gaseous envelope is relatively inert as a major weathering agent and base leveler. Mountains of high relief (as might be caused by upwelling convection currents within the interior) are unlikely to remain stable on the Venusian crust because rock materials at the high surface temperatures would tend to flow plastically over time and thus could not be supported at large differences in elevations. Calculation of dielectric constants for the Venusian surface gives values consistent with those of silicate "soils" but, curiously, are also matched by constants applicable to certain fluid hydrocarbons (leading to a proposal for a "sea of organic compounds" covering the lithosphere). The constants, however, appear incompatible with liquid water at the surface.

If CO_2 and water were released from the interior of Venus before atmospheric and surface temperatures rose to present levels, then reactions with silicates would have yielded extensive deposits of calcium and magnesium carbonates. With rising temperatures, however, these carbonates would be unstable in the presence of free silica, bringing about such reactions as:

$$\underset{\text{(Calcite)}}{CaCO_3} + \underset{\text{(Quartz)}}{SiO_2} \rightleftharpoons \underset{\text{(Wollastonite)}}{CaSiO_3} + CO_2$$

$$\underset{\text{(Magnesite)}}{MgCO_3} + \underset{\text{(Quartz)}}{SiO_2} \rightleftharpoons \underset{\text{(Enstatite)}}{MgSiO_3} + CO_2$$

$$\underset{\text{(Dolomite)}}{(CaMg)CO_3} + \underset{\text{(Quartz)}}{SiO_2} \rightleftharpoons \underset{\text{(Diopside)}}{(CaMg)SiO_3} + CO_2$$

These reactions, if continuing, would even now be removing CO_2 from storage in rocks to replenish the atmosphere and maintain the high temperatures. But, if excess silica is absent, mixtures of carbonates and silicates probably do not interreact in the outer Venusian crust. Also, the reaction

$$CaCO_3 + \text{Heat} \rightleftharpoons CaO + CO_2$$

does not take place at temperatures estimated for the surface. Other possible minerals suggested as surface constituents include corundum (Al_2O_3), hematite (Fe_2O_3), and limonite ($Fe_2O_3 \cdot 3H_2O$), plus those in certain igneous rocks.

American and Soviet Exploration of Venus

The *Mariner* and *Venera* spacecraft have answered several key questions about Venus. The microwave and infrared radiometers on *Mariner 2* (closest approach at 41,000 km) and *Mariner 5* (closest approach at 10,150 km) confirmed the general brightness-temperature values previously deduced from Earth. Almost simultaneous with the *Mariner 5* flyby in October 1967, the Soviet probe *Venera 4* entered the dense atmosphere, descended by parachute, and attempted to reach the surface. Later studies indicated it to have "disappeared" at an altitude some 25 km above the surface. Two more *Venera* probes experienced similar fates. As they descended, their instruments detected an atmosphere of 90 to 95% CO_2, very little water, and less than 3% nitrogen (below the detection limit of the sensor). At 25 km this gas mixture had a temperature of $323(\pm 10)^oK$ and a pressure of almost 2 atm. From the gradients determined during descent, surface temperatures as high as 600^oK (too hot for liquid water) and pressures near 100 atm are estimated by extrapolation. Such a dense atmosphere has the property of ***super-refractivity:*** light within the bottom 30- to 35-km layers is so sharply "bent" toward the surface that it must remain within these layers; in principle, an observer would see continuously across the entire surface curvature of the planet and, from certain positions, would view himself as part of a series of expanding rings. The first three *Venera* probes failed to reach the Venusian surface probably because the harsh atmospheric conditions caused malfunctions within the instrument capsule, although possible adverse effects on radio transmission could account for loss of signals.

Soviet space scientists now claim that in late 1970 *Venera 7* actually set down on a solid Venusian surface and transmitted data to Earth for 23 minutes. This probe measured a total surface pressure closer to 90 atm than to the 100 atm estimated earlier and a surface temperature near 750^oK. These data decrease the likelihood of the greenhouse effect being the best explanation for the high surface temperatures. The convective mixing that should result from the steep temperature gradient in the atmosphere would prevent solar radiation from being trapped near the surface. Many researchers now consider the clouds of Venus to be ice and water caught in a high-altitude cold trap where some condensation is possible.

The *Venera 7* feat was exceeded by *Venera 8* on July 27, 1972, when a 495-kg descent module from the spacecraft operated successfully on the Venusian surface for 50 minutes after landing. Unlike its predecessors, this module set down on the "day" or solar-illuminated side of Venus. As it descended, wind velocities in the Venusian atmosphere up to 50 m/sec were measured at a 45-km altitude; velocities diminished to 2 m/sec some 10 km above the surface. A surface temperature of $\sim 745^oK$ was sensed at the touchdown site. This nearly identical value relative to the earlier nighttime temperature measured by *Venera 7* indicates a minimum horizontal temperature gradient, a condition explained by assuming extensive mixing and cross-circulation of a Venusian gas envelope having a very high heat capacity. A more exact compositional analysis of the atmosphere, showing $\sim 97\%$ CO_2, less than 2% N_2, less than 0.1% O_2, NH_3 between 0.01 and 0.1%, and less than 1% H_2O vapor in the high clouds, resulted from this mission. Despite its density and the increased opacity owing to the clouds, the atmosphere was transparent enough to sunlight to permit detection of some visible solar radiation at the surface. An analysis of the surface itself, made by a gamma-ray spectrometer, disclosed up to 4% potassium, ~200 ppm uranium, and ~650 ppm thorium. On Earth, concentrations of these elements at that level are characteristic of granitic rocks and some shales. It is inferred from this that the outer crust of Venus may well consist of a highly differentiated, silica-rich igneous material. The layer immediately beneath *Venera 8* had a bulk density of ~ 1.5 g/cm^3. By analogy with the lunar surface, this low value suggests the presence of loose, poorly consolidated particles, but not necessarily of a regolithic nature.

The *Mariner 10* space probe, enroute to Mercury as its prime target, passed within 5800 km of the Venusian surface on February 5, 1974. Its two television cameras are equipped with both visible light and special ultraviolet filters. During the Venus fly-by, more than 3400 images of the planet (Figure 12-42a and b) were acquired in an 8-

day interval. When operating in the UV range around 3550 A to detect excitation of carbon monoxide, the cameras discerned distinct light and dark markings (up to 30% difference in brightness) in the upper troposphere-lower stratosphere region that define mass motions of CO in a circulating atmosphere. These motions produce slowly changing "cloud" patterns as much as 1000 km wide that, in general, are symmetrically distributed with respect to the rotational axis and equatorial plane. Spiral streaks appear to diverge from the equatorial region and open upward toward the poles in the direction of rotation. Over one region of the planet, the circulation patterns outline a dark Y-shaped feature which had been poorly observed in UV light from Earth-based telescopes (having, at best, only about ½ the resolution of the Mariner images). Closer to the poles, the streaks merge into broader rings. A large open area, nicknamed the "Eye," lies astride the equator near the subsolar region (but displaced about 3500 km westward). As seen in high resolution images, this region reveals a traveling disturbance in the atmosphere consisting of cellular features shaped as whorls and polygons. These features apparently are large-scale convection cells (100 to 500 km wide) produced by upward circulation resulting from solar heating. The interference of the turbulent "Eye" with the smoother zonal flow (at a velocity of ~100 m/sec along the equator) of Venus' gaseous envelope gives rise to widely spaced, dark bowlike waves concave in the direction of planetary rotation. The observations just described suggest a global circulation system in which a pressure excess around the subsolar disturbance drives the zonal winds in uniform flow toward the poles. The spiral streaks represent associated jet streams, deflected by angular momentum in opposite directions in each hemisphere. Following convergence at the polar energy sinks, the gases return to the equatorial belt.

No spacecraft has as yet found signs of the internal magnetic field predicted by some scientists for Venus because of the presumably hot interior; however, weak fields attributed to solar influences were detected. This may mean that (1) the very slow rotation of the planet fails to produce the motions needed to generate the fields; or (2) a moon (satellite) is necessary as an energy source to drive

Figure 12-42 (a) Composite view of Venus taken from 720,000 km by the *Mariner 10* television camera with UV filter in place. Individual images acquired on February 6, 1974, have been computer-enhanced to increase contrast between the swirling "clouds" of carbon monoxide (excited in the near ultraviolet) and a dark background (normally very bright in visible light owing to reflection from a water vapor layer). (b) Details of UV-excited gas "clouds," showing whorls, mottled patterns, and spiral streaks that result from convection and divergent "wind" flow in the Venusian atmosphere. *Mariner* spacecraft was about 740,000 km from the surface when this view was obtained. (NASA photos.)

(a)

(b)

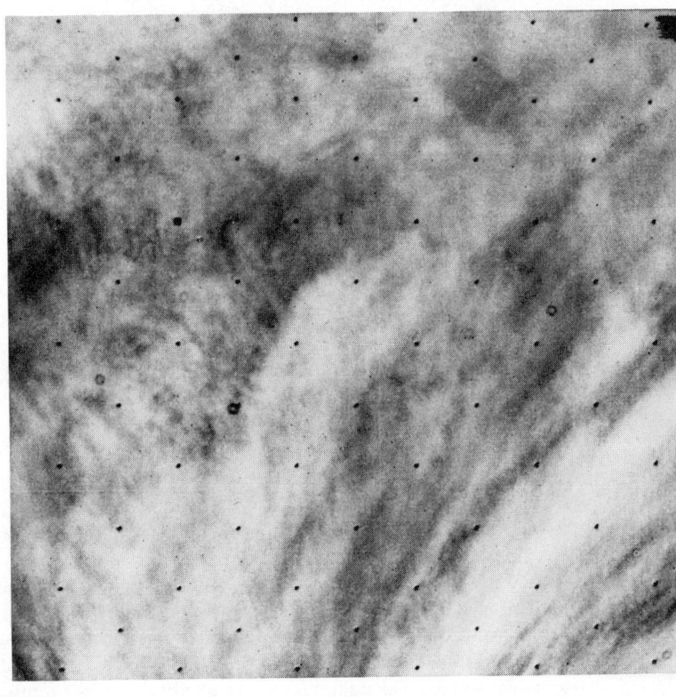

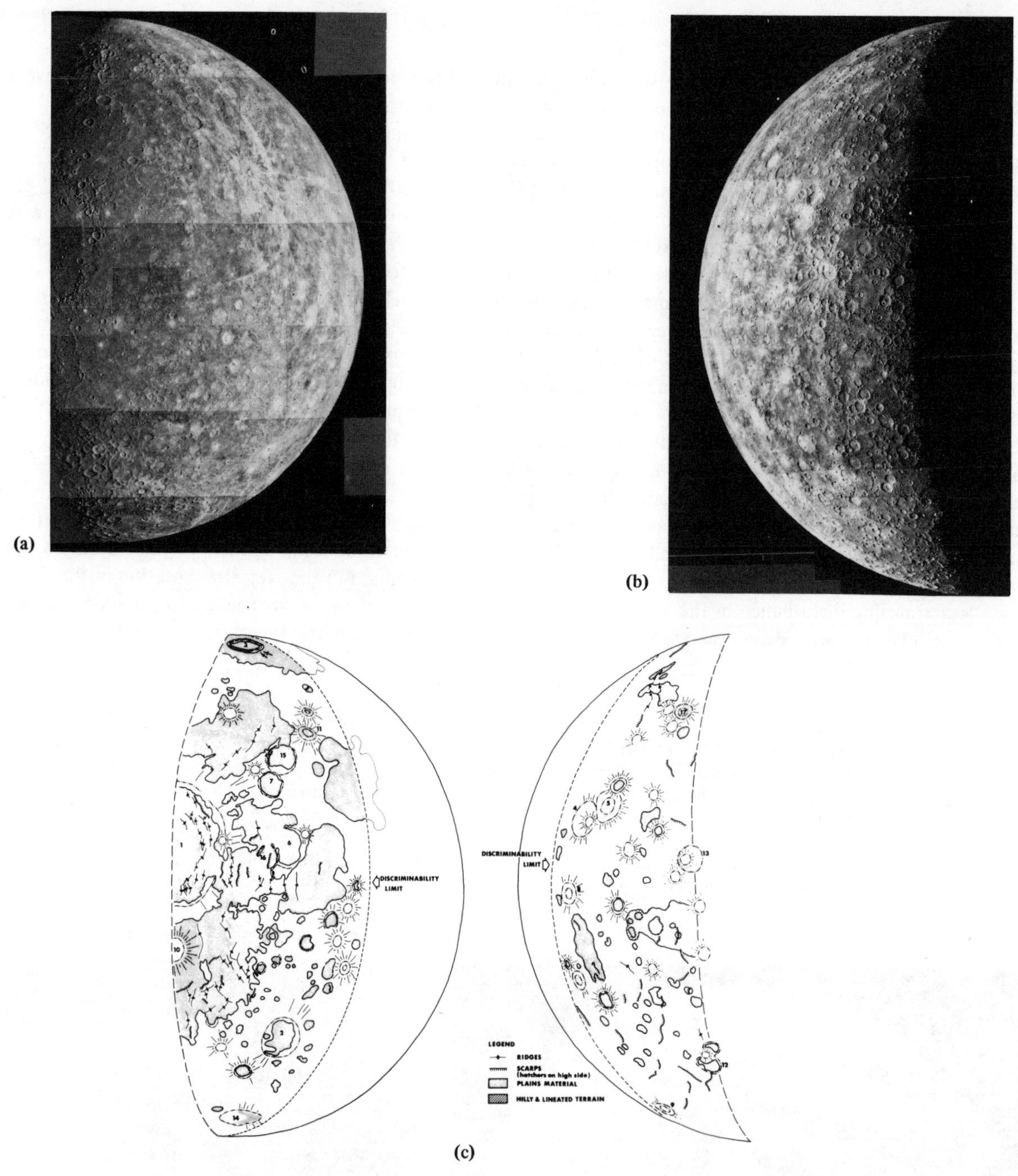

Figure 12-43 (a) Mosaic of 18 pictures of Mercury obtained by *Mariner 10* six hours after closest approach when the spacecraft was 210,000 km beyond the farside of the planet. Near the day-night terminator (left edge of view), the low sun angle shadowing brings out a series of one or more rings (just above center) that form a 1300-km diameter depression provisionally named the Caloris Basin. Smooth areas within the basin and out to more than 1500 km beyond the ring appear to be mare-like in appearance and contain numerous ridges and zig-zagging fractures (see Figure 12-44d). Note the radial grooves associated with the plains material in and around the basin. Other, more obscure basins have been identified in this part of Mercury. Bright rayed craters, resembling Copernicus (upper right) and Tycho (near bottom, with dark rim) show up more readily as the subsolar region is neared. (b) Mosaic made from 18 computer-enhanced, high-resolution TV pictures taken at 42-second intervals from a distance of about 200,000 km from Mercury showing the sunlit face of Mercury (evening terminator on right, north at top) made during approach to the planet. Most of the scene is in the southern hemisphere. An intensely cratered surface, comparable to the lunar highlands, contains individual craters as large as 200 km. Only four poorly defined basins, unfilled by mare lavas, are evident. Note double-ringed craters, deeply grooved terrain, and extended scarps. The conspicuous rayed crater just above center has been named Crater Kuiper (Figure 12-44a). (c) Sketch maps of the illuminated areas of Mercury appearing in the mosaics above showing the distribution of plains materials, hilly and lineated (furrowed) terrain (see Figure 12-44c), wrinkle ridges, and principal scarps. The numbers refer to identified basins; number 1 is the Caloris Basin. (NASA photos; pictures received and mosaicked at Jet Propulsion Laboratory; maps courtesy Newell Trask, U.S. Geological Survey.)

the core currents through precessional torque; or (3) the core of Venus has solidified (unlikely, in view of the high internal temperatures). The probes have also discovered that the ionosphere is considerably weaker than expected. *Mariner 10* demonstrated peak ionosphere densities of ~300,000 electrons per cubic centimeter on the sunlit side of Venus that drop to about 10,000 electrons/cc on the night side. One explanation considers the solar wind to "blow off" any ionized gases almost as quickly as they form. The solar wind also introduces considerable amounts of hydrogen to the Venusian atmosphere; this serves to stabilize the CO_2, thus preventing extensive dissociation into CO and atomic oxygen (both present in minor amounts along with some neon and argon).

Mariner to Mercury*

Following launch on November 3, 1973, *Mariner 10* continued beyond Venus and then sped past Mercury's surface some 705 km away on March 29, 1974, when that planet was almost 149 million km from Earth and about 70 million km from the Sun near the aphelion point. The spacecraft has swung around the Sun and returned to Mercury in September, 1974, to gather more data. The array of sensors used in this flyby mission include a wide- and narrow-angle lens telescopic TV camera, a magnetometer, an ultraviolet spectrometer, and an infrared radiometer. Despite periodic problems with instruments and power, the probe has performed admirably in providing the first detailed examination of the innermost planet in the solar system. Two panoramic views of Mercury taken during the inbound and outbound passes appear in Figure 12-43a and b, together with sketch maps of each in Figure 12-43c. Approximately 55% of the planet has been imaged by 2300 pictures at resolutions ranging from 20 kilometers at 4,500,000 km away to 0.15 km at 10,000 km.

At first glance Mercury looks remarkably like the Moon, but this planet (radius 2439 km, one-and-a-half times larger than its lunar counterpart) shows some important differences. When seen in TV images from about 2,000,000 km away, the Mercurian surface reveals numerous light and dark spots having broadly circular outlines. Mercury's general albedo is similar to the Moon's nearside highlands; it shows less overall color differences than the Moon. Albedo contrasts of ~20% are observed; however, *dark* mare plains or crater fillings are almost entirely absent. Higher albedo patchy areas and streaky markings resemble Tycho-like crater complexes similar to those of the Moon. When a mosaic is prepared from high resolution images of sections of both the inbound and outbound sunlit areas (Figure 12-43a and b), the resulting scenes remind one of the heavily cratered lunar highlands. The bright-spot areas are commonly identified with fresher-appearing bowl-shaped craters surrounded by ejecta aprons and light-toned rays. The majority of craters are less than 200 km in diameter. Many between 100 and 200 km tend to be double-ringed. Most have flat floors and are notably shallower than those on the Moon (Figure 12-43b). Many of the intermediate-sized craters have central peaks, but these peaks appear to be less common in the largest circular features.

Ever-closer views of Mercury (Figure 12-44) from *Mariner 10* confirm a crater frequency distribution similar to the lunar terra. Most craters appear as even-floored unterraced depressions with worn rims and obscure ejecta blankets (Figure 12-44a). Occasional craters have hummocky or benched floors and irregular, poorly developed terraces (Figure 12-44c). Craters as small as 200 meters are visible in the closest approach images. Variations in morphology and superposition effects attest to a relative cratering sequence with a full range of form modifications.

Some surfaces are so densely cratered that most circular features are strongly distorted (Figure 12-44b and c). These surfaces contain elongate depressions, ridges, and knobs, especially prominent in the rough, hilly intercrater areas. The interiors of larger craters cut into such terrain are usually much less cratered. Linear troughs and grooves cross through craters or emanate from tangential contacts with rims (Figure 12-44b). Many of these markings pass indiscriminately through individual craters as though regional in extent; others distribute in crudely radial patterns around certain craters. Long, often high scarps are more frequent than noted on the Moon. Some prominent scarps may be 2-3 km high and commonly extend for many hundreds of kilometers. Most such scarps show lobate fronts with scalloped faces and are notably non-linear over long distances. These scarps are interpreted to indicate compression-induced faults resulting from structural adjustments during shrinkage of the crust as Mercury's core was formed.

On the inbound side (Figure 12-43b), only four indistinct basins between 230 and 385 km wide can be discerned. Thirteen basins occur on the outbound side of Mercury; all but one are less than 440 km in diameter. Most basins have ill-defined outlines that are most easily seen near the terminator. These basins are best recognized from the two or more sets of topographic arcs that are parts of uplifted rings. The one very large depression—named the Caloris Basin—has an outer ring more than 1300 km in diameter.

These basins and many larger craters are covered with smooth, relatively uncratered plains which bear strong resemblance to the Moon's maria (Figure 12-44d). Lunar-like wrinkle ridges and long, polygonal fractures (Figure 12-44e) are common, especially where illuminated near the terminator (Figure 12-44f). The Mercurian plains

*This section on Mercury was written from data presented at a special symposium of the American Geophysical Union meetings held just 13 days after Mercury encounter and from the reports presented in the July 12, 1974 issue of *Science*, volume 185.

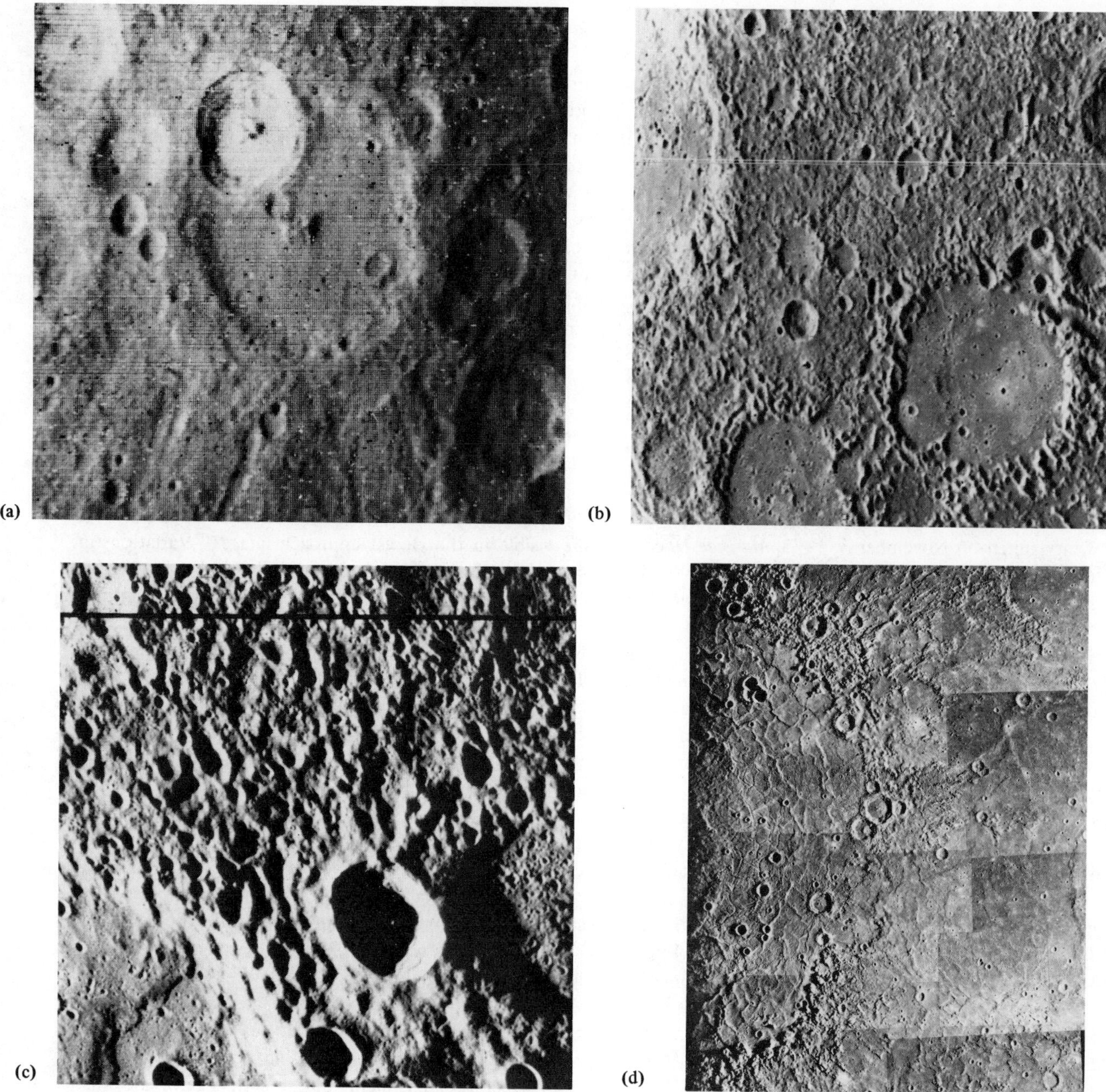

Figure 12-44 (a) View of a small part of the inbound side of Mercury taken by the *Mariner 10* TV camera when the spacecraft was about 88,500 km from the surface. A large (80 km wide) crater with a subdued rim and central peak has been cut by a smaller (41 km), fresher appearing crater having a well-defined central peak and terraces. This crater, provisionally named Crater Kuiper after the astronomer Gerard P. Kuiper, a Mercury investigator who died three months before *Mariner* reached its target, is the same one with the long rays evident in Figure 12-43a. (b) Another view of the inbound side from a distance of 35,000 km covering an area of 290 by 220 kilometers. The surface is heavily cratered and also dotted with numerous low, irregular ridges and furrows that make the hill and lineated terrain type. An elongate valley (7 km wide and more than 100 km long) is one of many linear grooves emanating from the 80-km wide flat-floored crater in the bottom center. (c) A part of the densely cratered approach side of Mercury as viewed by *Mariner 10* when it was 18,200 km from the planet during approach. A 25-km wide fresh crater with raised rim and deep interior and a 61-km wide older crater with an apparent flow front in its central floor are conspicuous among the many smaller (some 1 km across) craters. The black line at top is a TV "dropout" line. (d) A mosaic showing wrinkle ridges, polygonal fractures, radial furrows, and plains materials around the great Caloris Basin (left) on the outbound side of Mercury.

materials have albedos of 0.08-0.12, suggesting a more silicic composition than lunar mare lavas, and are usually lighter (more reflective) than the surrounding cratered terrain; no dark floor fillings have been seen. The plains materials are generally believed to be melted rock emplaced as flows, but some may be base surge deposits similar to the Cayley Formation materials on the Moon.

No volcanic domes, cones, or lava flow fronts have been identified in the images returned from Mercury. Straight rilles—deep and up to 6 km wide—have been noted, along with chains of close-spaced craters (Figure 12-44e). Sinuous rilles are very rare on the plains-poor side of Mercury, but some twisting channels have been identified in the plains terrain.

Preliminary estimates indicate about as many craters in the Mercurian rough terrain as on the lunar highlands. The crater frequency in plains material within and around the younger Caloris Basin is about 3.5 times greater than in the lunar maria, but craters larger than 10 km are scarce. These observations, coupled with the extensive cratering noted on the inbound side, imply that the Mercurian surface is at least as old as much of the nearside of the Moon.

Mariner 10's geophysical instruments have disclosed some vital facts about Mercury. The planet's mean density of 5.44 g/cc is consistent with an interior composed mainly of iron. A metallic core extending outward to as much as 75-80% of the planet's radius is implied. This is supported by recognition of a moderate magnetic field whose strength is about 1% that of Earth. The presence of a magnetic field has led to a magnetosphere, as indicated by *Mariner 10*'s encounter with a strong bow shock wave, that has trapped solar particles in the planetary environment. As the planet was approached, the field rose from the interplanetary level of 1 to 4 γ to about $18 \pm 2\gamma$ and then from 40γ in the bow shock region to a maximum of 98γ at a distance of 704 km; extrapolation to the surface gives values between 100 and 200γ. The origin of the magnetic field is still speculative: the slow rotational period of Mercury (57 days) may be insufficient to generate dynamo electric currents in the iron core even if it is hot enough to be molten.

A very thin Mercurian atmosphere ($\sim 2 \times 10^{-9}$ millibars, as determined by the UV spectrometer) consists mainly of helium, argon, and neon. The first two gases could be derived from radioactive decay of uranium and potassium within Mercury's crust, but the planet's low gravity seems to have held these substances for a long period. Accumulation of neon and especially helium by trapping from the solar wind is plausible, although the sparsity of detected hydrogen argues against this origin. While continuing degasing could re-supply the atmosphere, the high daytime surface temperatures (700°K at the subsolar point) measured by the IR radiometer would further prompt their escape. The mid-afternoon temperature drops to about 460°K and reaches a low of about 90°K just before Mercurian dawn. Thus, the range of temperature variations in the equatorial region is about 610°K (about 1000°F), far greater than any other planet. This great

(e)

(f)

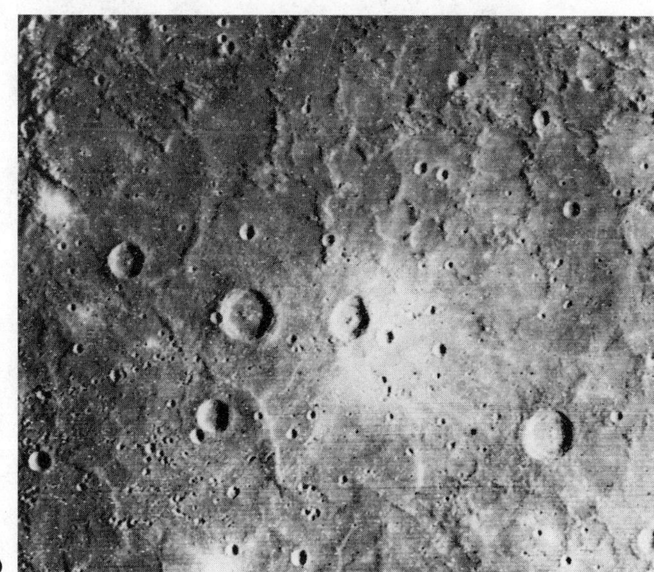

Figure 12-44 (continued) (e) A close-up of the Mercurian surface obtained shortly after closest encounter, showing a moderately cratered mare-like region cut by scarps and wrinkle ridges and peppered with craters as small as 500 meters. Some craters are aligned in apparent "chains." (f) A darker, smooth, relatively uncratered area on the outbound side of Mercury as imaged from 86,000 km away. "Ghost" craters, wrinkle ridges, and rough, hilly terrain are evident in this view. The bright rayed crater in the lower center is 10 km across. Some of the smaller craters (especially at lower left) resemble secondary crater swarms. (NASA photo, Jet Propulsion Laboratory.)

decrease in temperature during cooling in the night phase implies a rapid heat loss from a porous, insulating surface layer which is probably much like the lunar regolith.

Mercury, then, may have the most primitive surface of the inner planets. By analogy to the Moon, it melted shortly after formation and an iron-rich silicate crust soon formed. Differentiation at that time brought about extreme separation of heavier constituents leading to a thick iron core which still may not be solidified. Massive bombardments by planetesimals persisted thereafter to produce the cratered crust that now envelops much of the planet. This continual impacting has added considerable material by accretion to the outer crust. The rough intercrater regions may be among the oldest surviving planetary surfaces in the solar system. The heavily cratered nature of the Mercurian surface suggests that large numbers of sizeable planetary particles existed near the Sun in early solar history. This may indicate a more isotropic distribution of meteoroid material throughout the inner solar system than previously believed, but the high crater densities can also be explained as simply the inevitable effects of the late stages of the accretionary process by which a terrestrial type planet is born. Like the Moon, the only processes available to affect Mercury after these stages would be the degradational ones associated with further bombardments by objects ranging from small meteoritic bodies through micrometeorites and solar wind and cosmic particles.

The successful second pass by Mercury on September 21, 1974, after *Mariner 10's* swing around the Sun, produced more than 2000 new images. Surface covered was extended by 15%, including the first views of the South Polar region. Two additional Mercury passes by this probe now seem feasible.

The Pioneer Missions

The *Grand Tour* missions to the outer planets, originally designed to take advantage of the infrequent near-conjunction (alignment) between Jupiter, Saturn, and Pluto in 1977 or between Jupiter, Uranus, and Neptune in 1979, have been replaced by flyby probes aimed primarily at Jupiter and Saturn. *Pioneer 10* began its flight in March of 1972, followed by the launch of *Pioneer 11* in April 1973. Using a "gravity kick" from Jupiter as it moves by (Figure 12-45), *Pioneer 10* will then head out of the solar system towards the constellation Taurus to become man's first galactic probe (without, however, any communications link by then to retrieve data). *Pioneer 11* will travel another route, nearing Jupiter on December 3,

Figure 12-45 Schematic summary of the *Pioneer* missions to Jupiter. *Pioneer 10* was launched in March of 1972. (NASA illustrations.)

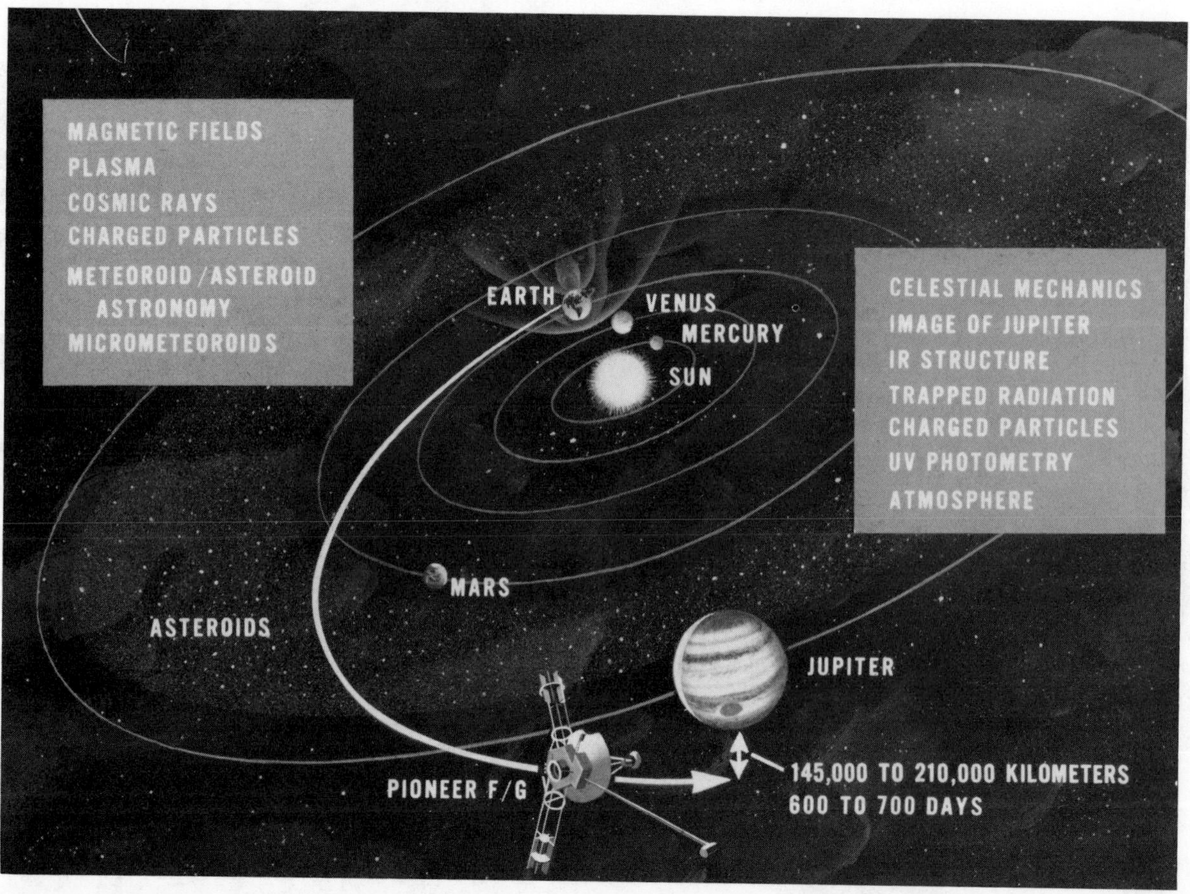

1974, and then Saturn in September of 1979. Two spacecraft, with more sophisticated instrumental packages, will be launched in 1977 towards Saturn in the *Mariner* Jupiter-Saturn mission.

The pair of *Pioneer* spacecraft (each weighing 270 kilograms) gather data on Jupiter's surface, interior, atmosphere, magnetosphere, and radiation belts using an array of instruments that include an image-producing photopolarimeter, a UV photometer, an infrared radiometer, a magnetometer, and several radiation detectors. The cosmic-ray telescopes and other high-energy particle counters, along with a meteoroid detector, plasma counter, and magnetometer, sensed solar and interplanetary particle and radiation fields and dust fragments enroute to the planet.

After departing Cape Kennedy, *Pioneer 10* journeyed more than 800,000,000 km (greater than four times the distance to the Sun) before it skimmed past Jupiter on December 3, 1973, at speeds exceeding 50,000 km/hr when only 130,000 km away. The most significant results of the first mission are summarized below.

(1) During the successful pass through the asteroid belt in late 1972, *Pioneer 10*'s meteoroid detector determined the number of small particles in this region to be considerably less than feared. This implies that most of the dust in the belt has been swept into the larger asteroidal bodies or perhaps never existed. However, the region near Jupiter registered dust-meteoroid levels one to two orders of magnitude greater than anticipated, evidenced by approximately 300 times as many micrometeoroid impacts on the spacecraft counter as occurred in interplanetary space. This concentration of fine materials may represent a localized dust belt around Jupiter itself, attracted by its large mass, or may be a solarwide residual belt in a heliocentric orbit.

(2) *Pioneer 10* first encountered the bow wave of Jupiter's outer magnetic field when the spacecraft was about 7.6 million kilometers from the planet. A stronger inner magnetic field, somewhat compressed by the solar wind, commences about 2,900,000 km from Jupiter. After its pass, the spacecraft reached the outer limit of the magnetosheath at about 240 Jupiter radii (16.6 million kilometers) following its outbound deflection "sidewards" toward the constellation Taurus. The magnetic field strength at the Jovian equator was found to fluctuate between 2 and 12 gauss. This magnetic field is reversed relative to the Earth's; that is, a terrestrial compass would point toward Jupiter's south pole as the source of existing field lines. Jupiter's magnetic moment is approximately 2000 times greater than Earth's.

Intense radiation belts surround Jupiter, with the maximum region of trapped protons lying some 3 to 4 planetary radii from its center. The radiation belts, populated largely by electrons and protons, exceed the intensities of Earth's two belts by 10,000 to one million times. Surprisingly, in places the Jovian belts contain 100 times as many high energy electrons (5×10^8 electrons/cm^3 sec having energies of > 3 Mev) as high energy protons. The energetic electrons apparently increase near the Jovian surface; a closer approach by the spacecraft almost certainly would have wiped out most on-board instruments.

The apparent shape of the radiation belts is that of a flat disc, likened to a "pancake"; the magnetic field forms a thicker envelope, likened to a "squashed doughnut," with the planet occupying the "hole." The radiation belt and magnetic disc are inclined in relation to Jupiter's equator, causing the magnetic pole to tilt about 15° from the pole of planetary rotation. This angular departure produces an up-and-down wobbling of the belt and disc over about 30° during each 10-hour rotation of Jupiter, as evidenced by periodic variations in instrument readings during *Pioneer 10's* passage through the Jovian environment.

(3) Interaction between the solar radiation and Jupiter's outer gaseous envelope has produced an ionosphere of charged particles consisting of several layers having a total thickness of ~ 600 km. Within the outermost atmosphere of Jupiter, the UV photometer detected a hydrogen-alpha glow of about 1000 Rayleighs (1 Rayleigh measures an emitted radiation of 1×10^6 photons per square centimeter per second) and a helium glow of ~ 10–20 Rayleighs. The previously assumed ratio of H/He of ~ 0.5 has been revised to 0.84, a significant finding bearing on Jupiter's origin and its present state of energy production. Results from a mass distribution experiment show the outer atmospheric zone, from the tops of its cloud deck downward through a thickness of 2300 km to contain only about 1% of the Jovian mass; densities in this interval increase uniformly to a value of 0.26 (relative to 1.0 for water) at the 2300-km depth.

(4) Jupiter is re-radiating almost 2½ times as much energy as it now absorbs from the Sun. Temperature measurements show the darker brownish-orange belts in Jupiter's gaseous envelope to be warmer [~ 126°K (-233°F)] than the brighter, more bluish belts [~ 119°K (-245°F)] in the atmosphere; temperatures in the dark zones rise to ~ 160°K at a depth of 50 km. The bright zones overlie the darker belts, giving rise to the characteristic banded visible "surface" of the thick atmosphere so beautifully photographed by *Pioneer 10* (Figure 12-46). The Great Red Spot, rising some 8 km above its surroundings, appears relatively colder at top and hotter below, a possible convective effect. The bright and dark belts are apparently quasi-

Figure 12-46 Two views of Jupiter taken by the imaging photopolarimeter on *Pioneer 10* when the spacecraft was about 2,500,000 km from the giant planet on December 1, 1973. The picture on the left was made through a blue filter mounted in the photopolarimeter; the picture on the right was made through a red filter. The Great Red Spot, radiating light strongly in the visible red, appears as a dark ellipse (approximately 45,000 by 12,000 km in dimension) in the left picture but is relatively brighter (and hence shows less contrast) in the right picture. The satellite Io is visible in both pictures. Note also some of the differences in structure of the bands of light and dark gases that surround the planet. (NASA photo; processed and rectified at the Lunar and Planetary Laboratory of the University of Arizona.)

permanent temperature cells in the nonturbulent atmosphere that are stretched out into girdling bands by Jupiter's rotation. The spacecraft could detect no differences in the average temperature (irrespective of the belt differences and a somewhat warmer equatorial region) on the day and night sides of Jupiter; this implies a high heat capacity for its atmosphere as well as an indication of some internal heat-producing mechanism(s).

(5) The flattening of Jupiter's polar diameter, owing to its rotation, was determined to be some 300 km larger than the 4600-km polar shortening previously measured through optical telescopes. The overall mass of the Jovian planet–satellites system exceeds earlier measurements by almost twice the mass of the terrestrial Moon.

(6) Most of the mass increase resides in the moons of Jupiter. The new density values (in g/cc) of the four planet-sized satellites are: Callisto, 1.65; Ganymede, 1.93; Europa, 3.07; and Io, 3.50. For Io the density increase is about 20% higher than the previously determined value. To account for such a density, this satellite—between the Moon and Mercury in size—almost certainly must consist of iron-rich silicates. Its surface may be covered with frozen ammonia, which could account for its very high reflectance. *Pioneer 10* demonstrated Io to have a tenuous atmosphere (20,000 times thinner than Earth's) extending out at least 110 km; an ionosphere has formed near the top of this atmosphere, as shown by the ultraviolet glow observed from the spacecraft.

Other Programs for the Planets

All of the non-lunar probes excepting *Mariner 9* have performed only "quick look" operations. A balanced long-range program for planetary exploration should involve a continuation of flyby missions where appropriate, together with probes that are placed in orbit for extended periods and/or that soft-land with instrument packages. A synopsis of past and future exploration plans within the U.S. space program is presented schematically in Figure 12-47. A summary of a balanced American program for planetary exploration in the decades ahead, as defined in mid-1972, is reviewed in chart form in Figure 12-48. Target dates for advanced lunar exploration programs including permanent bases, large Earth-orbiting space stations even more sophisticated than Skylab, and manned landings on Mars—all conceptually feasible with present space technology—are no longer specific and will depend

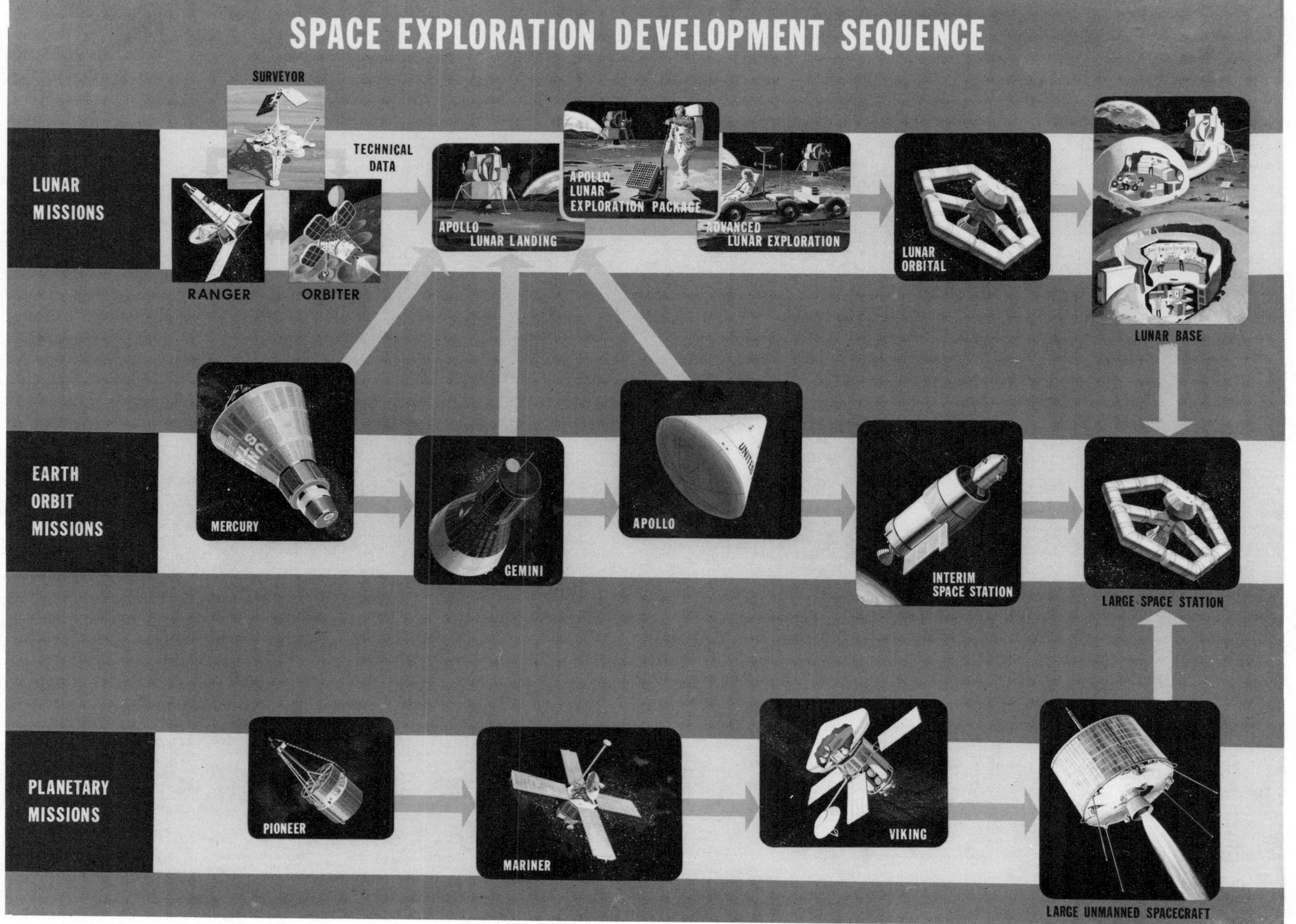

Figure 12-47 A pictorial diagram in which panels depict the general sequence of missions around the Earth, to the Moon, and into outer space over a period beginning in the early 1960s and extending (according to proposed long-range plans) into the 1980s. (NASA illustration.)

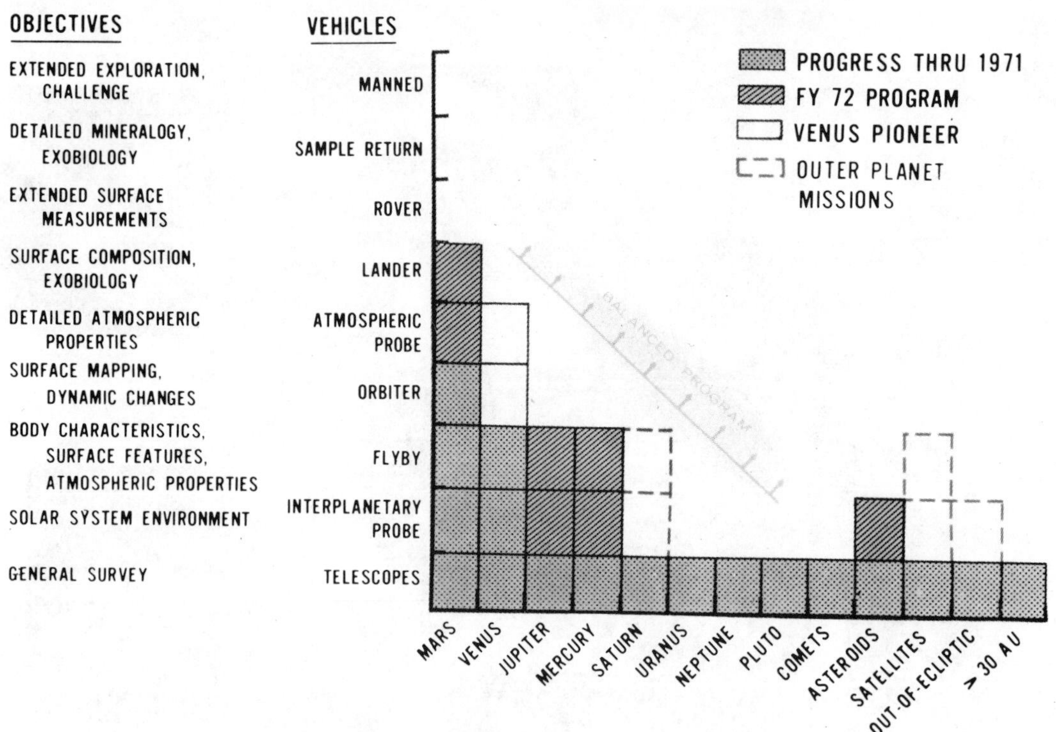

Figure 12-48 Status report and planning schedule, as of July 1972, for the exploration of the solar system by manned and unmanned probes. Future programs represent provisional plans that inevitably will be modified as exploration proceeds and budgetary priorities are reassessed. (NASA diagram.)

Figure 12-49 An artist's representation of the exploration sequence being planned for the *Viking* 1976 mission to Mars. (NASA illustration.)

Figure 12-50 Typical view of the Earth's surface from *ERTS-1* (the first Earth Resources Technology Satellite) orbiting 920 km above the geoid. The scene is imaged by the red band (0.5-0.6 micrometers) sensor of a Multispectral Scanner; optimum resolution is about 70 meters. The area shown is north-central Utah, with the Great Salt Lake at lower left, Bear Lake near top center, Salt Lake City at center bottom margin, and the northern extension of the Wasatch Mountains running vertically through center of this view. Eventually, most of the major land areas of the Earth will be imaged by ERTS-1. (NASA photo.)

on trends, rates, and levels of development of the NASA programs as authorized and funded in subsequent years. Some of these goals and missions will likely also be accomplished by the Soviet space efforts.

Beginning in 1975, the United States will launch its first soft-landers to Mars as part of the *Viking* program (Figure 12-49). Candidate sites for the attempted landings are located in Figure 12-14. These surface probes will make geochemical measurements and will search for signs of living organisms or evidence of past life. Four Soviet spacecraft reached Mars in February and March of 1974. One of these successfully orbited the planet to take photos and measure physical properties. A second probe attempted a soft-landing but failed to attain a stable touch-down, perhaps because of high near-surface winds. Before radio transmission ceased, the lander reported an atmospheric water content several times higher than indicated by *Mariner 9*. The remaining two spacecraft were unable to properly complete their tasks.

Although this quest for scientific knowledge extends to the limits of the solar system and beyond, the Earth itself is not being neglected as an object for practical studies and applications. On July 23, 1972, NASA launched the first in a series of Earth Resources Technology Satellites (ERTS) designed to monitor and survey geological, geographic, hydrologic, agricultural, forest, land use, and pollution aspects of the terrestrial surface (Figure 12-50). Also, in 1973 the first manned space station (*Skylab*) began to perform similar tasks as well as astronomical and geophysical experiments. Thus, the exploration of space once again is turning full cycle: planetary geology in the broadest sense begins and ends with the Earth.

REVIEW QUESTIONS

1. Would Mars or Venus be a better location from which to study the Earth through telescopes, radar, etc.? Why?

2. How have the Martian color markings and their shifting positions been explained?

3. In what ways do Martian craters differ from those on the Moon?

4. Review the evidence for volcanic activity on Mars; compare Nix Olympica to the Hawaiian shield volcanoes.

5. Suggest several explanations for the formation of canyonlands on Mars.

6. What types of landforms on Earth can be compared with surface features observed on Mars?

7. Discuss the Martian polar caps in terms of their composition, seasonal changes, and effects on their surroundings.

8. What are the Martian transient or variable markings? Speculate on how they are produced.

9. Discuss the general characteristics of the Martian atmosphere and its effects on surface features.

10. Describe several experiments that have provided information about Martian topography.

11. Describe the two Martian moons as seen by *Mariner 9* and discuss the significance of these observations.

12. If you were an observer at the surface of Venus, in its northern hemisphere, from which direction (east or west) would the Sun appear to be rising (through the clouds)? Why?

13. Examine the thermal state and atmosphere of Venus and Jupiter.

14. How is the runaway greenhouse effect considered to have acted to develop the present state of the Venusian atmosphere?

15. Compare and contrast the surface of Mercury with that of the Moon.

CHAPTER

13

Epilogue: What Does It All Mean?

We are now ready to collect many of the ideas introduced in previous chapters into a synthesis of concepts bearing on the nature and evolution of the planets. We shall also reexamine some of the general observations about the Moon in the light of recent interpretations of *Apollo* data. The emphasis, of necessity, is placed on the Moon because it is presently the only planetary body other than Earth that geoscientists have now explored comprehensively by direct sampling and on-site instrumentation.

One broad conclusion emerges when we consider the updated views on the genesis and history of the planets that stem from the first decade of space exploration. Take the Moon as an example. Three seriously held hypotheses for its origin had been proposed prior to the *Surveyor-Lunar Orbiter-Apollo* sequence of direct investigations on or near the lunar surface. Each hypothesis was based on certain commonly known facts, but these were interpreted differently with respect to the ultimate questions of origin. No firm, convincing set of arguments that compel acceptance of one hypothesis over the others is evident after evaluating the three and their variants. This is due largely to lack of definitive evidence indicating which ones are at odds with the significant facts and therefore should be eliminated as implausible or unlikely. Many scientists had hoped that the wealth of data gathered from the Moon during the 1960s would spotlight one hypothesis as most likely to be correct.

New papers on the Moon's origin and history in the light of *Apollo* results have appeared periodically in the early 1970s. Yet despite several years of intensive examination of the lunar rocks, there have been no fundamental shifts in position by the advocates of each major hypothesis. Almost without exception, every investigator found something in the data that *appears* to support his favored view. The problem of how the Moon came into being remains unsolved, although various aspects of its subsequent history clearly are understood in more detail. Such is the pattern in the development of many of the great concepts in the geosciences: the growth of continents, the nature of continental drift, the formation of mountain systems, the mechanisms of grantitization, the extinction of the dinosaurs, and the causes of glaciation, to name a few that apply to the Earth.

The chief contribution so far from the lunar and planetary exploration programs has been to define better the limits of the various physical and chemical properties of a planet. Thus, each genetic model can now be tested more rigorously, and eventually some one hypothesis (perhaps one not yet even conceived) will be raised to the status of a generally accepted theory. This approach as applied to the Moon is a classic illustration of the *scientific method*— initial definition of the problem, preliminary data gathering, development of multiple working hypotheses, execution of specific tests, prediction of further relationships, acquisition of critical data, interpretative analysis, rejection of unsatisfactory hypotheses, and formulation of a unifying theory by which observations and speculations are reconciled.

Origin of the Moon

We shall begin our final overview by surveying the principal hypotheses advanced for the formation of the Moon. A successful solution to that problem should apply to the other lithospheric planets if there has been a common process acting in their genesis. As modified by the results from space exploration, some important boundary conditions on the Moon's origin include the following (see also Tables 13-1, and 13-2).

(1) Moments of inertia very close to the theoretical value for a homogeneous sphere.
(2) A lower density than Earth, indicating a depletion in iron.
(3) Bulk composition of Moon not chondritic or solar (i.e., not primitive).
(4) Absence of water.
(5) Pronounced deficiencies in volatile and siderophile elements and enrichments in refractory elements relative to Earth and chondrites.
(6) Layered lunar interior (compositional shells).
(7) The great ages of rocks now at the surface.
(8) Evolution of the Moon mainly during the first billion years.
(9) Indications that the Moon was once much closer to the Earth.
(10) Rapid growth and development (in 10-20 million years) of the lunar body from the solar nebula at essentially the same time as the Earth and other planets.

No one hypothesis satisfies all these conditions, but among those still being seriously considered are the following:

I. Capture

Those unusual properties of the Moon that distinguish it from Earth are attributed by Gerstenkorn and by others to an origin in another part of the solar system (probably farther out) where metals and volatiles such as Sr, Rb, K, Pb, Zn, Hg, and In were already in short supply. The Moon, in a retrograde orbit, on occasions passed close enough to Earth to be diverted ultimately into its gravitational field as a captive. This should have happened about 2 billion years ago if the present rate of recession resulting from tidal interactions has remained constant. However, at that time melting would be induced in the Moon because of the strong forces exerted on it when close (near the Roche limit) to Earth; this is inconsistent with the ages now assigned to the lunar lava rocks.

Alfven and Arhennius (1972) believe that the Earth originally had a group of five to ten tiny satellites developed simultaneously with the parent planet. The present Moon, however, formed as an independent body from a stream of nebular material having a different overall composition from Earth, in a more elliptical orbit farther from the Sun. Repeated interactions of orbits whenever both planetary bodies approached close enough to exert mutual influence led to eventual capture of the Moon. Because of its large mass, compared with the other smaller satellites, the Moon experienced significant tidal interactions with Earth, causing its initial retrograde orbit to expand and convert to the prograde, locked state now observed. The Earth's legitimate satellites were all swept up by the adopted Moon. These infalls are responsible for the major lunar basins; mascon excesses in these basins balance closely with the estimates of 10^{21}-10^{22} g considered to be the total mass of the original satellites.

J.A. Wood and H.E. Mitler have proposed (at the Fifth Lunar Science Conference, March, 1974) a modified capture mechanism for the Moon's origin in which one or more passing small planets (protomoons) traveling at or below parabolic velocity are drawn into Earth orbit at or near the Roche limit of 2.5 Earth radii. As these bodies are disrupted by tidal stresses, those nearest the Earth are retained in closed orbit at less than parabolic velocities and those farthest from Earth escape in hyperbolic orbits. The relative proportions of fragments actually captured or rejected will vary depending on the approach speed of each protomoon with respect to Earth's escape velocity ($\sim$2 km/sec) and on the particular perigee distance during encounter. If, as likely, the protomoons had already differentiated into core and mantle, the bulk of an iron core would be lost in hyperbolic orbits for most combinations of approach velocity and perigee distance. The resulting Moon, built up by accretion of the remaining debris, would be enriched in mantle-crustal materials to give rise to the lower density Moon. A similar model was presented by J. V. Smith at the same conference.

II. Fission from the Earth

This concept was originally advanced by G. H. Darwin (son of Charles) at the beginning of this century and has received more modern treatments by Singer, Wise, Cameron, and O'Keefe. Material to make the Moon is supposed to

TABLE 13-1

Some Basic Observations on the Moon from Apollo Exploration

1. Moon may now be heating up (or is hotter than expected in its outer zones).
2. The heat flow is approximately equal at the two sites where measured.
3. Heat flow of 3×10^{-6} W/cm^2 is 3 to 4 times greater than expected; ½ that of Earth.
4. Temperatures at depth (higher pressures) less than needed to melt ultramafic rocks.
5. Presently, weak local magnetic fields but no global field.
6. Evidence of stronger ancient magnetic fields active at least from 3.9 to 3.2 AE ago.
7. Magnetic field strengths at times of cooling of crustal rocks in range of 10^2–10^3 γ (20 to 200 times greater than field in the solar wind).
8. Rigid crust and outer mantle, with low seismic and tectonic activities.
9. Most moonquakes at depths of 800–1000 km; occur periodically (tidal).
10. Accessible lunar rocks formed under anhydrous, strongly reducing conditions.
11. Notable depletions in alkali, siderophile, and chalcophile elements and enrichments in refractory elements relative to chondrites and to terrestrial rocks.
12. Radioactivity concentrated in the outer zones; local to regional pockets of high K, U, and Th; average U content of 0.06 ppm for entire Moon.
13. Less metallic iron in Moon than Earth; indigenous carbon very low (1-20 ppm).
14. Moon may now have partly fluid metallic core (but incapable of dynamo production of magnetism), with diameter ~1/5 that of entire Moon.
15. Upper mantle rocks pyroxene-rich, with some olivine and/or plagioclase.
16. All rocks returned from the Moon are products of magmatic differentiation (no primitive, undifferentiated rocks have been found); Moon must therefore be layered or thus contain materials in outer zones different from those that were in deep interior at time of accretion.
17. Outer shell(s) of Moon have low densities (~3.0 g/cm^3) and considerable strength, able to support mascons of higher density, more basic rocks; density rises to 3.5 between 60 and 200 km but reverses to 3.3 g/cm^3 in deeper mantle.
18. Thickness of crust may not be uniform (varying between 25 and 70 km).
19. Crustal rocks high in Al and Ca, lower in Mg, low in Fe; consist mainly of anorthositic gabbros, anorthosites, and KREEP basalts.
20. Moon formed ~4.6 AE ago, KREEP component of crust about 4.3 AE, and highlands basalts about 4.1 AE.
21. Most lunar basins formed about 0.5 to 0.7 billion years after formation of Moon; ejecta units (e.g., Fra Mauro) and light plains units related to basins.
22. Basins on nearside largely basalt-filled; those on farside contain few lavas.
23. Mare lavas extruded over 0.5-AE interval beginning ~3.7 AE ago.
24. Mare lavas rich in Fe and Mg; contain highly varying amounts of Ti; depleted in Eu.
25. Three types of basaltic rocks: (a) mare; (b) highlands; (c) KREEP-rich.

Source: Adapted *in part* from tables prepared by H. H. Schmitt and presented at American Geophysical Union spring meeting, April 1973, Washington, D.C.

have split off as liquids or solids from a rapidly spinning early Earth so that a part of the mantle (having a composition and density appropriate to the Moon) escaped to a distance beyond the Roche limit and there organized into a separate planet. One variant of this idea looks upon the Pacific Ocean basin as the ancient scar (about 150 to 200 km deep) left behind after a segment of the outer Earth was drawn off.

Cameron considers the lunar materials to have originated as solid fragments wrenched away mainly from a fast-spinning equatorial region along with some of the primitive atmosphere. As espoused by Wise and by O'Keefe, an essential condition for fission is a liquid Earth rotating at a period of 2 hours. A metal core had already formed. Wise selects thermal contraction, in which a body retains constant angular momentum but increases its angular velocity, as the cause of fission. O'Keefe, in describing a process he calls *spinoff,* also notes that any spherical body whose radius decreases while retaining its angular momentum must experience a gain in angular velocity. He assumes that the early, fluidized Earth had a thick atmosphere of hydrogen and helium. This atmosphere gradually was lost by flow of ionized gases toward the poles and into outer space. Simultaneously, these gases, moving slower than the

TABLE 13-2

Geological Constraints for the Moon from Apollo

Observational	*Implications*
• Crust melted soon after accretion Differentiated, igneous-textured old crystalline rocks Non-meteoritic or solar nature of outer crust	• Splashing of crustal melt Vacuum loss of volatiles Radiometric clocks not set until crust is largely solid • Fractional crystallization of crustal melt Early setting of Fe-S liquid Movement to core continues as mantle reaches ~1000°C Mafic minerals settle into lower crust Early geochemical isolation of basal zones and interstitial liquid Feldspathic minerals may float or be concentrated upwards Interstitial liquid provides gabbroic upper crust component Residual liquid rich in K, REE, P, U, Th, and volatiles
• Old crust saturated by impact craters > 50 km in diameter	• Outer crust pulverized and reheated to >10 km minimum depth Little primordial crystalline crust remains Most outer crust radiometric clocks partially reset Al- and Ca-rich powder abundant Early eruptions through outer crust will be contaminated • Pervasive fracturing to ~20 km depth Highly permeable outer crust
• Relatively: non-mascon basins are old, irregular, shallow; mascon basins young, circular, deep	• Relation between strength of crust and time of mascon basin formation
• Old, non-mare basins partially filled by pre-mare light plains deposits	• Lunar-wide, pre-mare eruptive event(s) with tuff-breccia characteristics
• Basins >100 km in diameter have flat original floors	• Lateral excavation dominant over vertical • Ejected material largely pulverized outer crust
• Molten or partially molten material was abundant inside the second ring of large circular basins	• Major surface unit prior to mare basalt extrusion Major exotic rock type on non-mare surfaces
• Mare basalts extruded from multiple centers	• If extrusion rate high, most basin fill is single cooling unit
• Last mare eruptive activity was from definable vents	• Fountaining will produce local glass bead deposits Dark or orange mantle distribution
• Regional systems of volcanic ridges, grabens and swirls are post-mare in age	• Requires regional stress systems to be imposed on the crust • May require regional eruptions of volatiles

Source: Adapted from table prepared by Astronaut H. H. Schmitt and presented at the American Geophysical Union's spring meeting, April, 1973, Washington, D.C.

rotating liquid interior, added to the body's instability by exerting a torque on the fluid outer layers. Eventually these factors, combined with a flattening and elongation of the liquid body, led to a separation of about a tenth of the total mass. Afterward, the Earth regained its spherical shape, leaving the liquid Moon orbiting at a distance somewhat greater than the Roche limit (~3 Earth radii). At this time the two planetary masses each rotated every 2 hours and revolved around one another in 4 hours, but tidal friction in time forced them apart. Both bodies then cooled, but the strong tides, still high temperatures, and possibly a T Tauri solar wind, drove away most of the gases. In this way, the Moon has lost most of its volatile elements, its water, and some of the chalcophile elements (Cu, Zn, S, Se, As, Bi).

O'Keefe has effectively used geochemical arguments to support his general hypothesis. He has compared the average chemical composition of the Earth's crust to the latest values for cosmic elemental abundances (inferred from analyses of the Sun's atmosphere and of meteorites) using the periodic table as a frame of reference (Figure 13-1a). The analyses of *Apollo 11* rocks are plotted in like manner (Figure 13-1b). It is immediately evident that most element variations with respect to cosmic abundances follow similar trends in both Earth and Moon, although the extent of excess or deficiency in the lunar materials will no doubt change somewhat as more samples are analyzed from new localities. Of particular note are the larger relative depletions in H, Na, K, Rb, Cs, Ca, Zn, F, Cl, and Br (volatile elements) and enrichments in Ti, Zr, Hf, Y, and

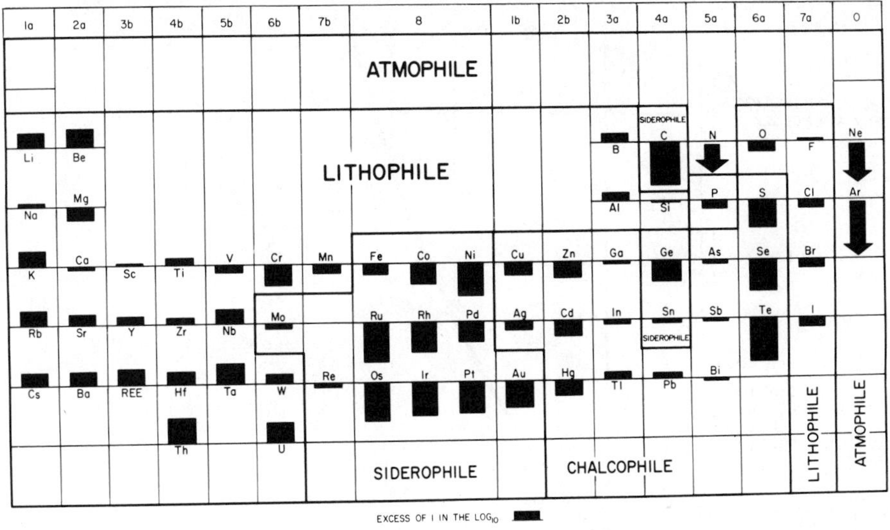

(a)

EXCESS APOLLO 11 OVER COSMIC ABUNDANCES

ATMOPHILE

LITHOPHILE

SIDEROPHILE

CHALCOPHILE

EXCESS OF 1 IN THE LOG_{10}

(b)

Figure 13-1 The relation between cosmic abundances of the elements and abundances in the Earth's crust (a, left) and *Apollo 11* crystalline rocks (b, right) as plotted on the Periodic Table of the Elements. Bars above the line indicate an excess, and below the line a deficiency, of an element in the terrestrial or lunar rocks with respect to cosmic abundances determined from the meteorites. The bars are scaled in increments of 1 in the $\log_{10}$ base (thus, a bar 4 times the length of the unit bar below the table represents an excess or deficiency of 10^4 or 10,000). (From O'Keefe, 1970.)

Sc (refractory elements) in the lunar rocks. Despite these differences, a strong kinship between the two planets is indicated. It will be interesting to compare analyses of Martian rocks, when obtained, with those shown here to see if Mars more closely resembles the meteorites, the Earth, or the Moon.

III. Sediment-Ring Accretion

As originally proposed by Öpik, the Moon resulted from accretion of Earth-orbiting planetesimals extending in ring(s) about the Earth that may have resembled those surrounding Saturn. The planetesimals within the rings—presumably derived from Earth (as in Cameron's hypothesis)—remained apart as though fragmented until they had receded to distances beyond the Roche limit, where they could begin to coagulate. Supposed evidence for this accretionary model is presented by traces of elliptical craters on the ancient lunar highlands—these mark impacts at low angles as would be expected from planetesimals moving in orbits more or less concentric about the accreting Moon.

Ringwood has modified this model to include some aspects of the fission hypothesis and to incorporate data from meteorite and lunar rock analyses. He builds an Earth in about a million years from accreting carbonaceous chondrites containing volatiles and oxidized iron. Very early in its history, the Earth began to melt from gravitational and impact energy (Figure 13-2). Carbon reduced the iron, and an atmosphere of H and CO accumulated. As temperatures rose, many volatile elements also became gaseous. When temperatures reached 1500°C, silicate minerals were reduced, some Mg and SiO was lost, and the iron core formed. In the outer mantle, temperatures then rose above 2000°C, causing vaporization of most silicates, which were added to the primitive atmosphere. Then, the Sun entered its T Tauri stage and most of these alkali-enriched gases were blown off; alternatively, this loss may

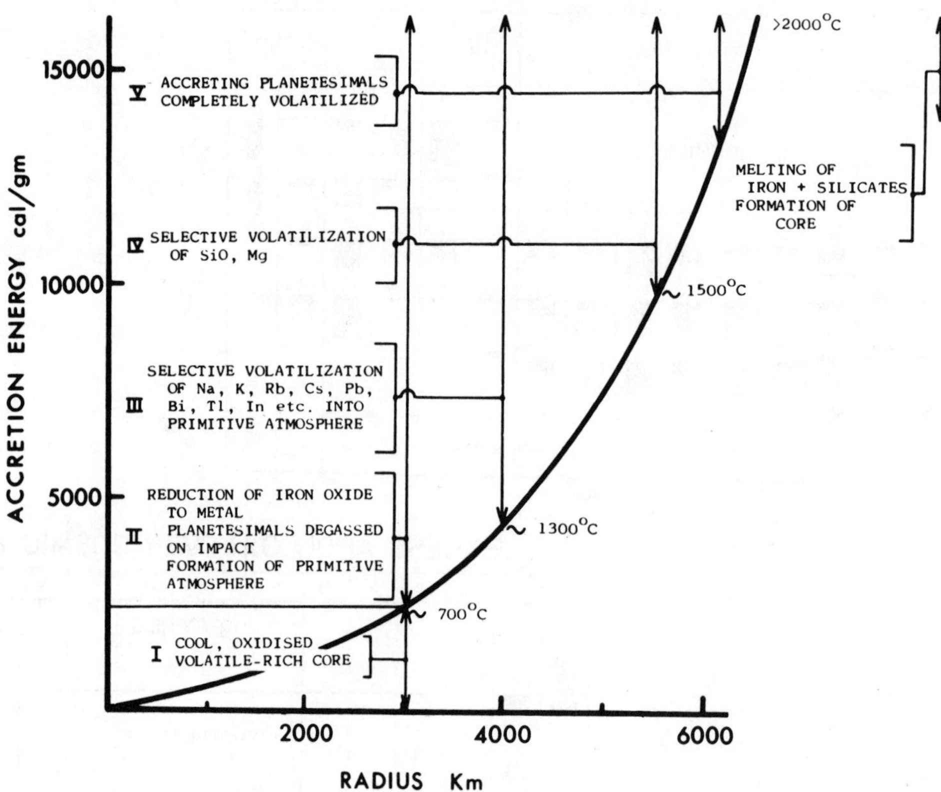

Figure 13-2 Stages in development of an Earth growing by accretion, in terms of increases in accretion energy and corresponding rises in near-surface temperatures. (A. E. Ringwood, *Journal of Geophysical Research*, Vol. 75, pp. 6453-6479, 1970.)

have resulted from rotational instability or magnetic effects. The atmosphere collected in a suitable orbit and then evolved into a sediment ring as silicates condensed on cooling and light-element gases mixed with part of the volatile lithophile elements flowed away.

Thus, the major chemical fractionation by which Earth-derived materials are differentiated into lunar materials occurred first in the "boiling off" from the lithosphere and then in the primitive atmosphere. In contrast to Earth, the Moon's outer zones are now depleted in volatile elements and show also a very low degree of oxidation. This is explained by assuming that the volatile elements initially were concentrated in the outer part of the sediment ring. The gaseous ring then condensed and accreted inward so that the last materials added to the growing lunar body were dominated by the nonvolatile, reduced fractions. Further changes, including some melting, took place during final accretion of the Moon. The total time involved in forming the Moon in this manner may have been as short as a thousand years.

Both Ruskol and MacDonald propose variants of the orbiting-ring hypothesis. They each envision formation of several small moons by an accretion of fission products. These then accreted to form the present Moon. Some of the large impact basins are scars from the last collisions. Another variant—the binary planet hypothesis (Latimer, Orowan, and others)—assumes that both Earth and Moon accreted independently of one another within the same region of the solar system. Ganapathy asserts that, for some reason, iron concentrated preferentially within the Earth; the Moon's smaller size prevented it from attracting as much of the heavier elements or holding a higher proportion of volatiles.

Anderson (1973) has devised a model for the Moon's formation, based on *Apollo* results, that belongs to the sediment-ring category. He recognizes simultaneous growth of bodies of varying sizes by condensation-evaporation processes throughout the gas-dust disc that developed during the evolution of the Sun. The earliest condensates were refractory-rich, making up assemblages of Ca, Al, and Ti compounds such as spinel, corundum, perovskite, merwinite, grossularite, and diopside. The Allende carbonaceous chondrite (p. 40) may well be an unaltered example of these primitive high-temperature condensates. An Earth-like body would form as a primary somewhat before any associated secondary satellites or moons are produced. The larger body, whose initial core would consist of the refractory phases, then grew by addition of more volatile components. A smaller satellite, formed later, built up preferentially from delayed condensation of

Allende-like refractories that accumulate as dust in a ring about the primary. The Moon, as an example, thus accreted mainly from these refractory phases but also received some of the Fe and Mg phases in late stages. The deep lunar interior still consists of spinels, perovskite, and melilite. The mantle contains Ca-garnet, merwinite, diopside, and spinel—an assemblage with an average density of 3.4 and compressional velocities up to 9.2 km/sec. Because of large drops in temperature by the final stages of accretion, phases with lower crystallization temperatures such as anorthite, olivine, and orthopyroxene developed in the outer shell. The original distribution of elements and their resulting mineral phases were further modified by early, extensive melting and differentiation, in which fractional crystallization aided in further separation of anorthite and FeO-bearing phases into one or more outer layers of ~250-350-km total thickness having a higher electrical conductivity than the refractory interior. This interior remained nearly devoid of iron and, despite being hot, never developed a metallic core.

The results from Project *Apollo* have not yet forced universal acceptance of any of these hypotheses as correct or most probable. It can now be said that samples of undifferentiated lunar materials will afford the best evidence of the Moon's origin. If, for example, the lunar interior proves similar to the Earth's mantle, then the fission hypothesis would acquire strong support. If, instead, that interior were shown to be compositionally like the carbonaceous chondrites, then the capture hypothesis would be favored. But, if the composition cannot be readily related to either the terrestrial mantle or the chondrites, the sediment-ring hypothesis would most likely explain this difference. Evidence from lunar crustal rocks, from which inferences on the interior can be constructed, now points to an overall lunar composition that is neither Earth-like nor primitive. At the moment, therefore, the concept of derivation of the Moon by a sedimentation process seems to have gained the most from the lunar sampling missions.

Implications of the Moon's Thermal History

Running as common themes through all the above hypotheses for the Moon's origin are the ideas of accretion as the dominant formative process and the closer proximity of the Moon to Earth in the distant past. The accreted materials had formed earlier by condensation over a wide range of temperatures. At high temperatures volatiles were driven off, leaving an excess of refractories. The low oxidation state of iron in the basalts results from equilibration temperatures of 200-300°C during the last stages of condensation. Both the Earth and the Moon accumulated from condensates that were strongly fractionated relative to solar abundances.

Following accretion, the Moon underwent some melting of its outer layers and perhaps much of its interior. Thus,

Figure 13-3 A simplified sketch map of the front (left) and far (right) sides of the Moon showing the distribution of dark (low albedo) mare lavas (stippled) in relation to highlands or cratered crust (white). The diagrams demonstrate that over 80% of the mare surfaces occur on the frontside. This selective distribution of maria on the present Earthside face may have resulted from tidal attractions which moved the melted rocks closer to the surface nearest Earth. Another explanation considers the crust on the nearside to be thinner than the crust on the opposite face, so that large impact basins would extend deeper towards the melt zone (with accompanying overburden reduction) in which the lavas were generated. The black arcuate lines represent the approximate positions of well-defined or discernible upraised rings or scarps (e.g., Apennine Mountains near Mare Imbrium) associated with large basins. (From Stuart-Alexander et al., 1970.)

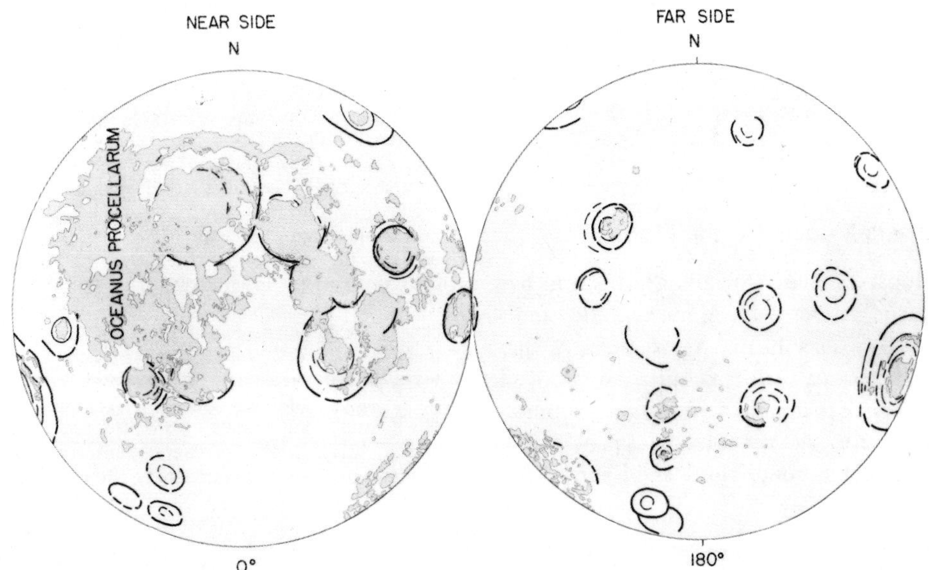

initially the accreted Moon became *hot*, so that advocates of this view can claim a victory of sorts. The proponents of a *cold* Moon can draw some comfort from its present status as an inert body with little or no internal magmatic activity. It should be emphasized that *hot* and *cold* do not refer to temperature extremes—such as those associated with molten lavas versus surface outcrop rocks on Earth—but describe instead conditions within the Moon in which temperatures are generally *above* or *below* the pressure-adjusted melting points of rocks throughout much of the lunar interior. A rigid crust is evidenced by the mascons and the presence of an Earthward bulge. A solid or non-mobile mantle is indicated by the low number of weak, internally originated moonquakes recorded by the seismometers. The absence of a general magnetic field points to a nonfluid center (including at most only a small solid iron core). These all attest to the conclusion that temperatures in the lunar interior today are generally below the melting points of most rock types inferred to exist below the observable exterior.

One sure conclusion from *Apollo* sample analyses is that parts of the Moon experienced some remelting that began a few hundred million years after the initial crust solidified and lasted intermittently for the next 600 million years. Differences of opinion exist as to the cause and significance of this later, basin-filling period of igneous activity. Questions still being asked are:

(1) Was this melting a general effect throughout the Moon resulting from a gradual rise in temperature, or do the lavas come mainly from rapid heating in response to local or regional conditions?
(2) Do the lavas originate from near-surface or deep-seated melting?
(3) Could the lavas represent residues of unsolidified liquid formed during a first melting of the Moon shortly after its formation?
(4) Is the melting initiated by heat build-up from radioactive decay, from impacts, from tidal friction, or from other sources of thermal energy?
(5) Why are most mare lavas distributed on the side of the Moon facing Earth (Figure 13-3) even though the number of multiringed basins is about equal on both front- and farside?

Genetic Models for the Moon

Drawing upon lunar samples studies, investigators have sought to define the nature of the Moon's interior and to reconstruct its history since birth. As examples of these speculative hypotheses, we shall next survey some of the major models that seek to define the origin, development, and internal structure of the Moon (see also papers in the proceedings of the First, Second, Third, and Fourth Lunar Science Conferences*). Physical, chemical, age, and geological constraints on the validity of any model have been provided by these consensus opinions (as of mid-1973), outlined in Tables 13-1 and 13-2, on the basic observations about the Moon that stem largely from *Apollo*.†

Petrologic Models

Smith et al. (1970) at the First Lunar Science Conference proposed early total melting of a Moon with modified "chondritic" composition (high in Mg and Si, low in Fe and alkalis) followed by general differentiation to produce a layered body as depicted in Figure 13-4. These layers represent extensive fractional crystallization that yields a small metallic core and an olivine-pyroxene mantle. It was also suggested that the crust was dominated by plagioclase-rich rocks and basalts; this is consistent with the anorthosites, norites, troctolites, various basalts, and felsitic differentiates found in soils and rocks from the *Apollo* sites. The residual liquid occurred between the crust and mantle and was a ferrobasalt. A thin layer of rock rich in ilmenite and sulfides developed near the interface between mantle and residual liquid.

The crystal-liquid differentiation occurred under the influence of a gravitational field directed from Earth. It was proposed that the Moon rotated synchronously and that, under the influence of the inhomogeneous gravitational potential, the floating plagioclase tended to concentrate on the farside, giving a thick crust. The residual liquid, being more dense, tended to concentrate on the nearside. Impacts on the farside created deep craters with little or no lava. On the frontside the crust was thin and temperatures remained high, thereby allowing impacts to produce large lava lakes at a late stage. Smith and his co-workers (especially I. M. Steele) have modified their views somewhat as the result of new data (J. V. Smith; personal communication). They still subscribe to total melting of the Moon, but they now believe that mare basalts stem from remelting (not necessarily at great depth). Small

*The reader's attention is called especially to these conferences. Beginning in 1970, each Lunar Science Conference has been held early each year at or near the Johnson Spacecraft Center in Houston, Texas. The meetings provide lunar samples investigators with the opportunity to present formal reports on their studies of materials returned from the Moon, particularly from Apollo missions since the preceding conference. These papers are then published in a 3-volume set of Proceedings (see Bibliography). Also, within four to six months of a conference, an "official" review of the major findings is presented in an issue of ***Science***, the weekly periodical published by the American Association for the Advancement of Science.

†An excellent general review of the major new findings from the Apollo program through the last mission appears in the article by J. Wood and others of the Lunar Sample Analysis Planning team. Entitled, "Fourth Lunar Science Conference," it can be found in the August 17, 1973 issue of ***Science, 181***, pp. 615-621. As far as possible, new ideas and data from the article have been taken into account in this book, but the reader is encouraged to consult the article for a comprehensive survey of some principal results from Apollo as an update to parts of Chapters 10, 11, and 13.

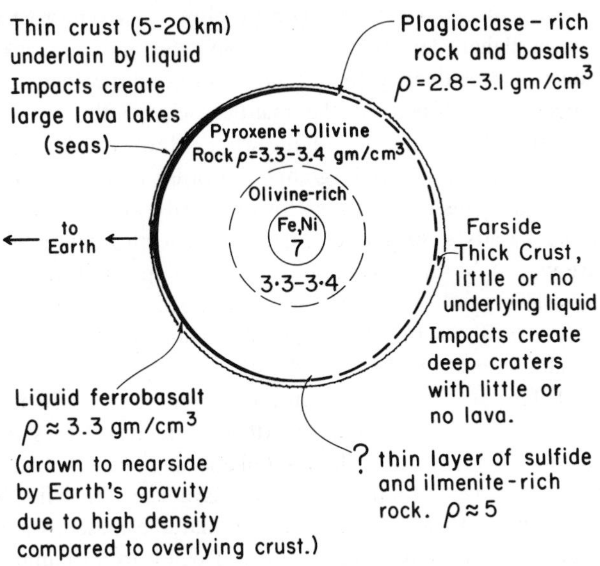

Figure 13-4 A model for the distribution of rock types developed at a late stage of crystallization of a completely melted Moon. (Smith et al., 1970.)

rock fragments containing Mg-rich olivine and pyroxene may represent samples of the mantle or of rare ultrabasic lavas which penetrate the crust from below. Basalts rich in the KREEP elements are explained as the products of primary differentiation and remobilization. The possibility that the Moon is composed of a high-temperature condensate (as proposed by D. L. Anderson) is taken seriously, but a composition dominated by Mg-rich silicates is still regarded as the best source for the basaltic rocks. A substantial amount of sulphur in the core (as proposed by V. Rama Murthy and by R. Brett) is thought to be likely. Tidal heating from the Earth (as proposed by H. R. Shaw) is regarded as a substantial contributor to the heating of the Moon.

Ringwood and Essene (1970) claim that, if the Moon melted shortly after formation, it would cool rapidly and solidify in much less than a billion years. This would rule out a partly liquid upper mantle. Their original Moon was not chondritic in composition, an assumption supported by the low initial Sr^{87}/Sr^{86} values in the basalts, indicative of Rb/Sr ratios incompatible with chondrites. The Moon derived instead as accreted condensates from vaporization of part of the Earth's mantle. In its first stage as a planet, the Moon developed a mantle composed of Fe-Mg pyroxenes and some olivine (but lower in Al and Ca than the terrestrial mantle) and an outer layer of eucritic magma. This layer further differentiated into calcium-free and calcium-bearing pyroxene overlain by a thick crust of anorthosites and anorthositic gabbros and a thin, now-dispersed layer of silicic rock (probably a vesicular, fine-grained rhyolite). Ringwood's model for the outer regions of the lunar sphere is shown in Figure 13-5. The Moon, in this view, was essentially solid before the subsequent period of mare lava flooding. At depths between 200 and 600 km, the pyroxenites contained enough U, Th, and radioactive K in the mesostasis to provide sufficient heat for

Figure 13-5 A section through the outer 500 km of the Moon indicating the two stages of magmatic history, in which the original (4.7 b.y.) pyroxenite deep in the interior formed from early fusion of the primitive Moon along with overlying differentiates of orthopyroxene, pyroxene cumulates, anorthosite, and a thin quartzo-feldspathic outer crust, remelted (3.7 b.y. ago) and moved into lower regions near the surface to form mare lavas of basalt underlain by gabbro. Mascons result in this model from conversion under pressure of these lavas to the denser rocks of eclogitic composition. (A. E. Ringwood, *Journal of Geophysical Research*, Vol. 75, pp. 5453-6479, 1970.)

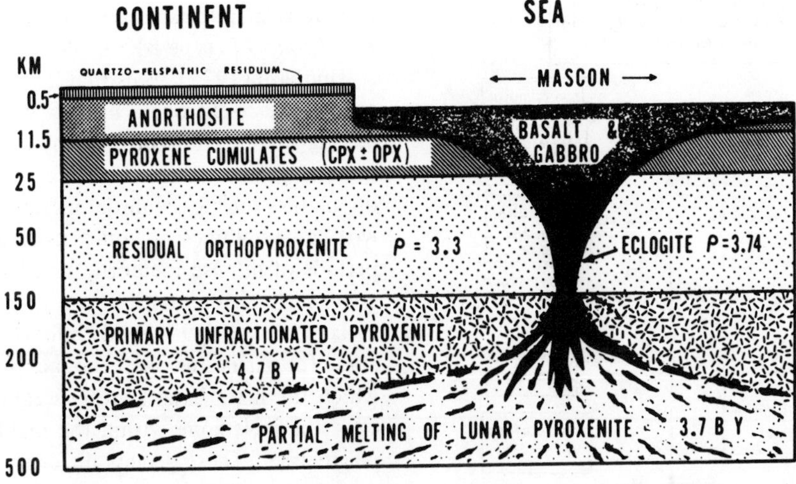

extensive *partial fusion* in various zones about a billion years later. Minerals with lower melting points, such as those in the mesostasis, were selectively fused so that some of the volatiles (remaining in a Moon initially with lower amounts of those elements) and most of the rare earth elements were mobilized. As postulated by Ringwood and Essene, europium stayed behind in the source rocks because its divalent charge (Eu^{+2}, the reduced form) allows it to associate with Ca-pyroxenes even as the other rare earths with their trivalent charges are released from the mesostasis. The melted fractions then extruded along fissures into previously formed basins and lowlands. The melt that passed into these depressions was basaltic in composition and therefore contained elements capable of crystallizing as feldspars. The titanium enrichment is explained as the result of selective fusion of the lower-melting titanium mineral *geikeite*. The portion of the basalts that

Figure 13-6 Flow chart defining the model of Ford et al. (1972) for a possible evolution of the Moon, starting with a water-rich aggregate of uncertain nature, and leading to the various rock types shown. (KREZP is a variant of KREEP used to indicate a high zirconium content.)

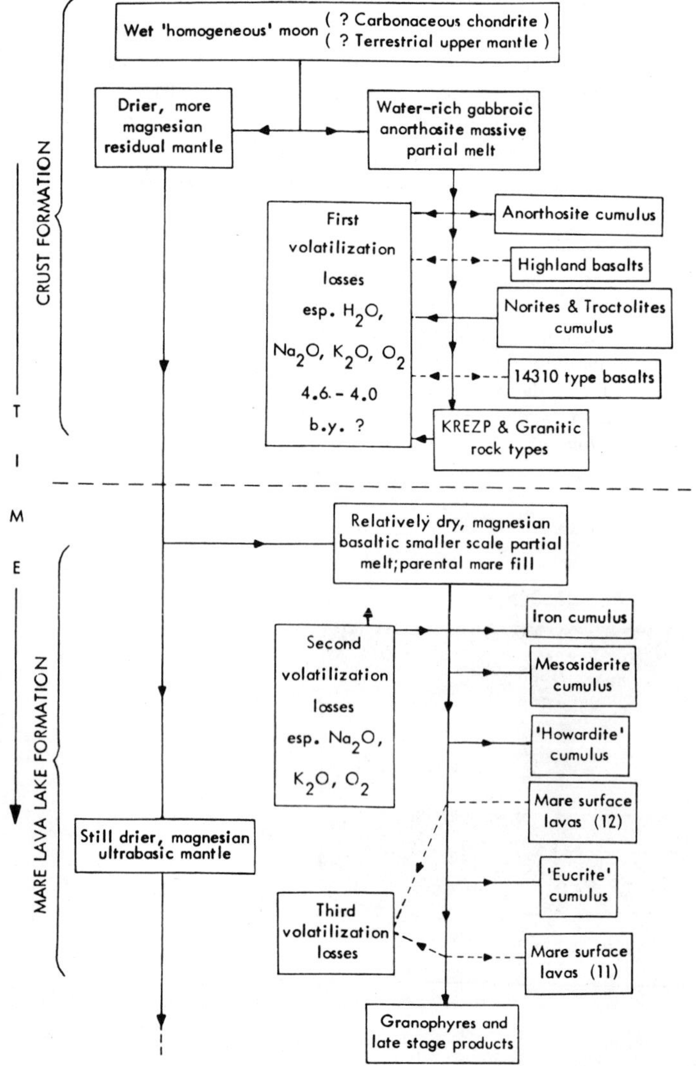

cooled in feeders well below the surface was under enough pressure to convert to *eclogite* (a garnet-pyroxene rock) whose higher density would account for the mascons. However, this density (~3.75 g/cm^3) rules out eclogite as a major constituent of the mantle, because this would raise the average density of the whole Moon to values in excess of the actual 3.35 g/cm^3. In order to account for this mean density, rocks of lower density at the surface (mare basalts, feldspathic basalts, and anorthosites) must give way at depth to rocks of different composition whose densities are not much different from the Moon as a whole. Only a small fraction of the Moon's volume is likely to be metallic iron of much higher density.

O'Hara et al. (1970) assign most of the characteristics of the lunar basaltic rocks to extensive fractional crystallization and loss of volatiles (alkalis and water) while these lavas were still liquid in great "lava lakes" that filled the basins. These workers suggest that the mascons are indications of basal accumulations of dense iron-titanium oxides that crystallized early and settled out of the lava lakes. However, observations of possible multiple layers of lava rock in the Hadley Rille near the *Apollo 15* site weakens the argument for a single deep lake capable of undergoing extensive differentiation by convection and settling. In a later paper (Ford et al., 1972), this same group extended their model to include formation of the lunar crust (Figure 13-6). They start with a primitive water-bearing material (possibly carbonaceous chondrites) that melts and devolatilizes into the series of rock types shown in their flow diagram.

Brown et al. (1972) also enlist near-surface crystal fractionation to explain differences between feldspathic basalts and ferromagnesian basalts. They envision deep lava pools in which norites and anorthosites concentrate as cumulates by flotation leaving the silica-poor, iron-rich basalts to develop at greater depths before extrusion, KREEP-rich liquids are final differentiates that intrude locally or regionally into older crustal rocks. Brown et al. have arranged the basaltic rocks at the *Apollo* sites in the following fractional crystallization sequence, based on variations in pyroxene composition: *Apollo 14*→*Apollo 12*→*Apollo 11* (type B)→*Apollo 15*.

Hubbard (1972) notes that large-scale losses of volatiles from surface-extruded basalts would require extensive mixing and exposure of melts by convection, gas bubbling, spattering, and lava fountaining. He prefers an early loss of volatiles during the initial melting of much of the Moon, with large volumes of liquids constantly stirred up during periods of intense meteorite bombardments, followed by grand-scale differentiation.

Haskin et al. (1970) and Philpotts and Schnetzler (1970) argue for partial fusion of the mantle as one mechanism by which the mare basalts are generated. In each view, however, the melt zone within the mantle must contain some plagioclase. These groups differ somewhat in the import-

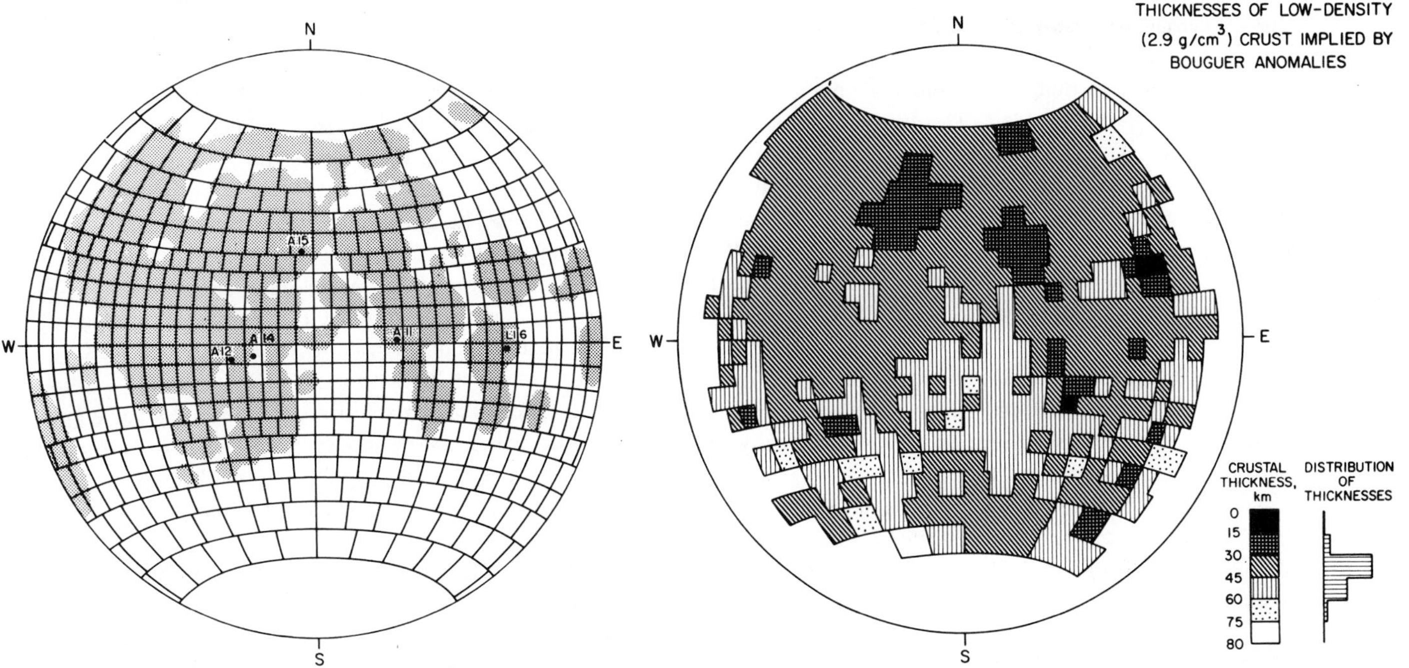

Figure 13-7 Variations in thickness of the lunar crust assuming it to consist of a single rock type–anorthositic gabbro–with a density of 2.9 g/cm^3. This crustal model was developed by Taylor et al. (1972) using gravity measurements (Wong et al., 1971) obtained from the S-band transponder experiments that involve tracking of spacecraft in orbit around the Moon. These measurements determine the distribution of mass deficiencies or excesses within a series of square computation cells. When a correction for topographic effects is applied, the residual deficiencies or excesses (Bouguer anomalies) can be converted to different thicknesses whose values depend on the rock densities selected.

ance they attach to subsequent fractional crystallization, but they agree that the europium anomaly and rare earths distributions are related to the behavior of feldspars during fusion and/or later crystallization. Schnetzler and Philpotts (1971) suggested later that the mare lavas were originally emplaced at the present surface almost at the start of lunar history in order to account for the 4.6-AE model ages of these rocks; the lava rocks subsequently remelted to produce the younger (3.2-3.7-AE isochron ages) crystalline rocks sampled at the *Apollo* sites.

Wakita and Schmitt (1970), upon finding europium enrichment in the anorthosite fragments from the *Apollo 11* soil, devised a method for determining the total depths to which the Moon has melted. They assume an inner core of olivine, overlain surfaceward by orthopyroxene, basalt, and anorthosite layers. By redistributing the rare earths according to an initial concentration similar to chondrites, they calculate the thickness of each layer for different extents of interior melting. Thus, if only the upper 200 km had ever been melted, the anorthositic highlands would be ~10 km thick, but total melting produces a highlands thickness as great as 30 km or more.

Wood et al. (1970) and Marvin et al. (1971) propose a three-layered Moon–an inner core composed of olivine and some pyroxene, a thick mantle of gabbro, and a thin outer crust of anorthositic gabbro, with some norite and anorthosite–developed by general melting and differentiation soon after the planet's formation. In the region now occupied by Mare Imbrium, the crust (as sampled in *Apollo 14* ejecta) was dominantly noritic, with smaller local segregations or cumulates of anorthosite and anorthositic gabbros. This probably serves as a Moonwide description of the lithologic character of the original terra. The crust typically is from 30 to 70 km thick under the terrae, about 30 to 45 km thick beneath most maria, and 0 to 30 km thick in the mascon regions (Figure 13-7). Penetration and removal of the crust in large impact events leads to melting of the gabbro. On solidification after eruption, the gabbroic magma crystallizes as basalt without much further differentiation. When cooled, the basaltic lavas (density ~3.0 g/cm^3) contract to denser (~3.3 g/cm^3) rocks. Further extrusion to fill the vacated volume, plus foundering of surface crust, brings about an excess of mass which accounts for the mascons.

Wood (1973) later suggested another mechanism to account for the variable thickness of the lunar crust and its noncoincidence between centers of figure and mass. His model involves selective removal of parts of the original crust by impact bombardment. After spin-orbit coupling locked the frontside of the Moon into its fixed state relative to Earth, the combined effects of gravitational focusing of incoming planetesimals by the Earth onto the frontside and of the greater exposure of the west limb or "leading edge" of the lunar sphere to these planetesimals led to a notable increase in numbers of impacts on the frontside compared with the farside. The largest number is expected to cluster on the western equatorial (Oceanus Procellarum) region. This preferential bombardment gradually reduced crustal thicknesses over much of the nearside, leaving the crust an average of 15 to 20 km thinner there than on the terra that dominate the farside. As a result, there was a reorientation of the Moon such that the principal axes of inertia shift to allow the minimum moment to point toward the Earth as now observed. Later, mobilized basaltic lavas invaded the thinner, Earth-facing crustal regions, in part because the Moon's center of gravity had been displaced toward these regions.

Chao et al. (1972) have analyzed impact glasses from Fra Mauro units at the *Apollo 14* site so as to reconstruct the bulk composition of their parent rock types. The following principal varieties are recognized:

(1) Granitic or sialic composition
(2) Anorthositic composition
(3) Anorthositic gabbro composition
(4) Basaltic composition
(5) Troctolitic composition
(6) Peridotite composition
(7) Feldspathic peridotite composition
(8) Mare basalt composition

Their group 4 is the most abundant present in the breccias. These heterogeneous glasses can be subdivided into several types, among which are the KREEP basalts and norites. Chao et al. conclude that noritic rocks prevail in the highlands followed by anorthositic gabbros and anorthosites in order of abundance. They find essentially these same highlands rock groups at the *Apollo 15* site (Best et al., 1972) but note also ilmenite and ilmentite-olivine basalts representing mare lavas.

Reid et al. (1972), in noting a distinction between Fra Mauro basalts (KREEP-rich) and highlands basalts (feldspathic, including also the anorthositic gabbros), propose a model for development of the Moon's outer shell as shown in Figure 13-8. The model recognizes a single parent basalt which separates into three fractions: a feldspar-rich cumulate that floats to form the outer crust, an iron-magnesium-rich deep zone produced by sinking of the higher-density minerals, and a residual liquid enriched in K, U, and Th that crystallizes to rocks at intermediate depths. These fractions are characterized chemically by a decrease in CaO and an increase in FeO with depth relative to Al_2O_3. It was the intermediate-depth fraction that was tapped during the Imbrium Basin event and transported to the Fra Mauro site.

Meyer (1972) considers the nonmare basalts to be the prime clues to the early history of the Moon. He suggests that the original crust, now surviving in the highlands, consisted of massive bodies of Al-rich rocks rather than a series of stratiform sheets such as may explain the layers in Mount Hadley at the *Apollo 15* site. His model follows this sequence of development: (1) initial accretion accompanied by melting to produce magma that covered the primeval Moon; (2) separation of crystallized plagioclase from ferromagnesian-rich liquids through convection, resulting in accumulation in surficial zones; (3) late-stage crystallization of a more silicic residue at depth; (4) a second melting of this residue through radiogenic heating to form the KREEP-rich basalts; (5) large impacts that produced the Imbrium and other basins, as evidenced by KREEP-bearing breccias; and (6) a third planetary heating by such possible mechanisms as Moon capture by Earth, tidal friction, and/or radioactivity which gives rise to the mare basalts.

Gast and McConnell (1972) and Gast (1972) point to different immediate sources for the KREEP-rich and mare

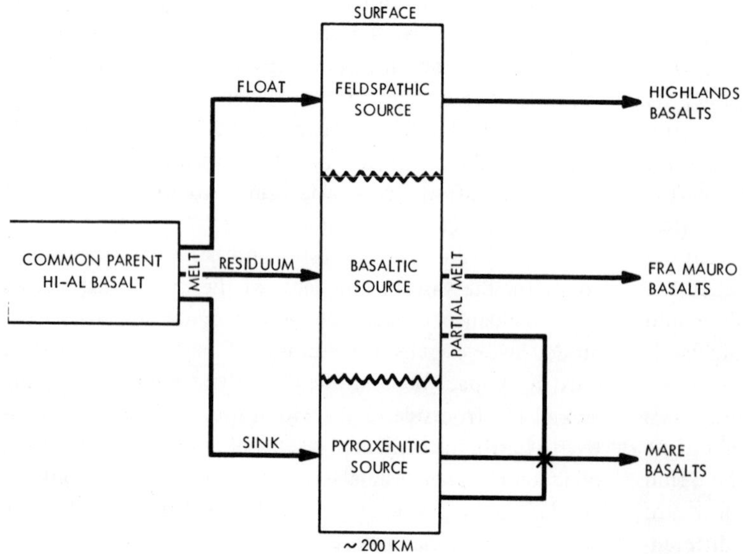

Figure 13-8 A schematic flow chart that traces the derivation of the three principal types of basalts found in surface rocks and soils sampled in the *Apollo* program. In this genetic sequence, a deep-seated, alumina-rich basaltic parent is melted into three differentiated fractions that form layers extending from the surface to ~ 200 km. The mare basalts, and possibly some Fra Mauro basalts, are emplaced later after partial remelting. (Adapted from Reid et al., 1972; courtesy J. Warner.)

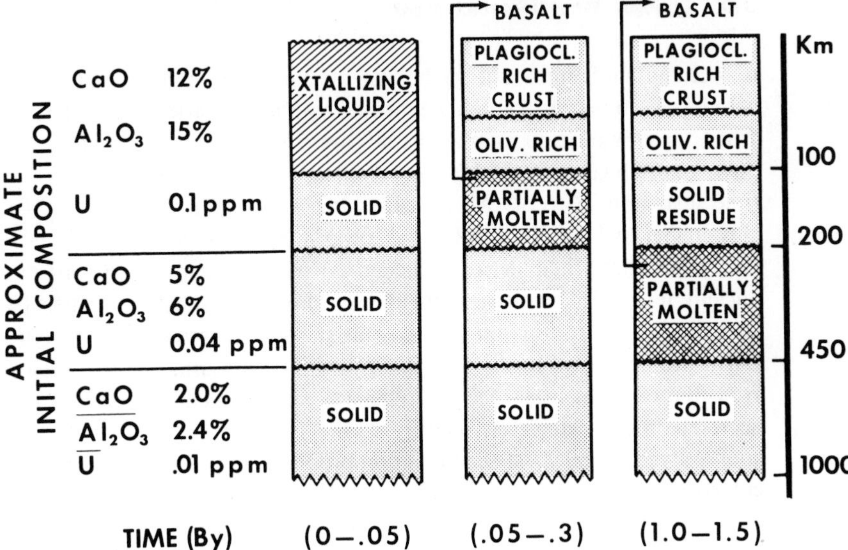

Figure 13-9 Diagram showing three stages of development of a layered crust and mantle in the outer 1000 km of the Moon, according to the model of Gast and McConnel (1972); see text for details. Time is expressed in billions of years (b.y.). (From "Lunar Sample Analysis Planning Team," *Science*, 1972, copyright 1972 by the American Association for the Advancement of Science.)

basalts. These source materials were, however, themselves derived by differentiations of a more primitive interior having many of the compositional characteristics of chondrites including Allende types. KREEP-rich basalts are enriched in LIL (*l*arge *i*on *l*ithophile) elements, such as K, Ba, Rb, and U, relative to LIL-poor mare basalts. The LIL elements Eu and Sr became especially enriched within anorthositic rocks as those rocks segregated by selective fusion. The KREEP-rich and mare basalts derive from plagioclase-rich and pyroxene-rich sources, respectively. Both basalt types were thus generated over separated time intervals within different depths in a chemically layered Moon (Figure 13-9) in which Ca, Al, and U contents decrease toward the interior while Fe and Mg increase. This layering began to develop even as the Moon was accreting into a heterogeneous body in which refractory elements (Ca, Al, Fe, Mg, REE) concentrated in the outer lunar shell. Separation of volatile and refractory elements into two distinct rock phases had previously taken place during condensation from a hot solar nebula but supplemental fractionation may also have resulted from volatile losses in the outer parts of the accreting Moon. Because heat-generating uranium is inferred to decrease with depth (from ~0.1 ppm in surficial rocks where U has been concentrated by differentiation and convection to ~0.01 ppm at a 1000-km depth), the calculated maximum temperatures during the early evolution of the Moon ranged from 1200°K to about 273°K downward between the outer shell and the Moon's center. Thus, this negative (decreasing) temperature gradient prevented the Moon from becoming largely molten at its outset. Instead, melting and the first differentiation were confined to the outer 150 km of the primitive Moon, perhaps in response to an external heat source from the Sun. This stage of melting produced KREEP-rich basalts (including the Cayley Formation debris sampled at *Apollo 16*) from the original feldspar-rich crust whose model ages of 4.4 to 4.6 AE date the time of its first crystallization rather than the accretion of the outer shell. The anorthosites were generated soon thereafter by partial melting of the plagioclase-rich source at shallow depths. Over the next 1 to 1.5 AE the base of the melt zone migrated inward to depths of 200 to 400 km, and the compositional trend of the volcanic products changed from Fe-poor-Al-rich to Fe-rich-Al-poor rocks. Partial fusion, affecting from 3 to 30% of the pyroxene-rich source rocks, generated the magmas that now constitute the mare basalts. In the Gast model, lunar rocks at even greater depths never melted generally and ever since have remained in essentially their original state as chondritic materials. By analogy to the stony meteorites, the lunar layers from surface to interior would follow the sequence:

eucrites→howardites→Type II carbonaceous chondrites

Based on phase equilibria studies, D. Walker et al. (*Earth and Planet. Sci. Letters*, vol. 20, pp. 325-339, 1973) suggest that (1) highlands basalts can be produced by almost total partial melting of types of feldspathic highlands soils such as found around *Apollo 16*, (2) very high alumina (VHA) basalts derive from ~60% partial melting of such soils or of feldspathic spinel troctolites, and (3) KREEP rocks are generated through limited (10-35%) partial melting of any highlands soil type. J. L. Warner et al., at the Fifth Lunar Science Conference in March, 1974, present arguments for development of most of the crystalline materials associated with highlands breccias by impact-induced fractionation processes involving varying degrees of partial melting of ejecta blanket material. Intense bombardment during the first 0.6-0.8 billion years of lunar history served to continuously crush and mix the outer 10 km or so of the primitive lunar crust, yielding a diversity of products from light matrix breccias, KREEP rock, and metamorphosed breccia, through glassy breccias and regolith.

Engle et al. (1972) draw attention both to similarities and differences between terrestrial and lunar basalts (Figure 13-10). The highlands basalts are roughly equivalent to oceanic tholeiites or ridge basalts, whereas the mare basalts are similar to certain nepheline basalts on Earth. When the compositions of mare basalts are compared with those of Archean terrestrial basalts emplaced in the same time span some 3 to 3.5 AE ago, the lunar rocks are notably higher in FeO, Zr, and Ba and lower in SiO_2, Na_2O, and H_2O than their terrestrial counterparts. Yet the ancient terrestrial basalts are chemically almost identical to basalts now erupting on Earth. This suggests that the Earth has been continually defluidizing its excess volatiles from the outer mantle since it formed as a relatively cold, accreted body. However, melting of the interior in response to radiogenic heat accumulation produced the present core and mantle. In contrast, the Moon in its early history underwent nearly complete devolatilization as it was heated through accretionary and impact processes and radioactive decay.

Figure 13-10 Variations of $K_2O + Na_2O$ with respect to SiO_2 in *Apollo 11* and *12* crystalline lavas compared with several major igneous rock series on Earth. Circles represent whole rock analysis of the lavas; the squares include analyses of low silica-iron-rich and high silica-potash-rich mesostasis glass (last differentiates to solidify) in the *Apollo* rocks. The compositional field of the *Apollo* rocks is defined by fine stippling. Heavy dot stippling indicates the field for Hawaiian volcanic rocks, and cross-hatching refers to the compositional range of granodiorite–granite series rocks in the Adirondack plutonic intrusions. The differentiation trend followed by the Skaergaard layered igneous sequence in Greenland is shown by a solid line. Lunar basalts on the whole are distinguished from terrestrial counterparts by the low $K_2O + Na_2O$ values, but some lunar mesostasis glasses overlap into terrestrial composition fields. When the ratio of K_2O/Na_2O versus SiO_2 is similarly plotted, differences between terrestrial and lunar rocks (and between *Apollo 11* and *12* basalts as well) are even more evident. (From Engel et al., 1971.)

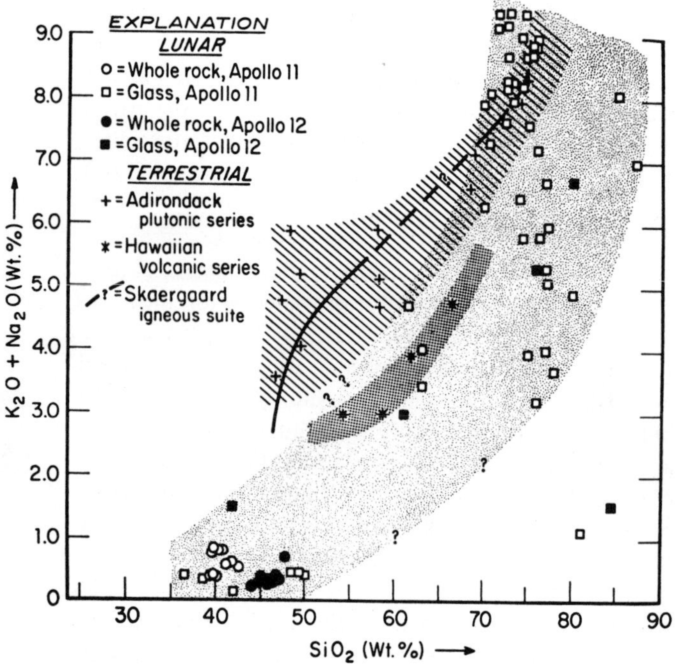

Geophysical Models

Sonett et al. (1971) have conceived a model for a radially stratified Moon (Figure 13-11) in which a primitive core of ~1400 km radius has remained unmelted since the time of initial accretion of the planet. They further recognize a two-layer mantle and a thin surficial crust which resulted from melting of the outer part of the Moon soon after its formation. Their conclusions depend on interpretation of *Apollo 12* magnetometer data (p. 180), from which they have derived an average electrical conductivity profile. Laboratory measurements of conductivities in lunar and terrestrial rocks indicate that olivine-rich rocks have very low conductivities, whereas the ferrobasalts from the *Apollo* sites are among the most highly conducting rock materials known. Conductivities vary according to the temperatures attained at different depths. By starting with a temperature range of 750 to 800°C for a substrate in the core 800 km from the Moon's center, they arrive at a gradual drop in temperature to 550°C at 1400 km from the center, then a sharp temperature reversal or increase of at least 100°C in the interval 250-350 km from the surface (corresponding to the conductivity peak at 1500 km), followed by another decrease in temperature to low values near the surface. These assumed temperature variations with depth give rise to thermal gradients in the outer mantle estimated to be 2-4°C/km. These investigators deduce a compositional distribution in which the core consists of olivine or olivine-pyroxene; the lower mantle contains basaltlike substances enriched in refractory minerals and is separated from an upper mantle composed of basalts by a zone dominated by olivine (corresponding to the strong conductivity peak near 1500 km).

Sonett and his colleagues' estimate of heat flux at the surface, based on this conductivity-temperature model, leads to a lower value of $\sim 1.6 \times 10^{-7}$ cal-cm^2-sec (raised somewhat if olivine dominates the lunar interior). This is considerably lower than values associated with Precambrian crustal rocks on Earth and is about 5 times lower than the average heat flow values reported from in-hole measurements at the *Apollo 15* site. The lower heat flow values, if valid, would preclude general melting of the entire Moon in the past.

Using the same *Apollo 12* magnetometer data, Dyal and Parkin have derived somewhat different electrical conductivity variations with depth. They have subdivided the lunar interior into three layers as shown in Figure 13-12. Each layer is assumed to consist of one of three rock types—dunite (olivine), peridotite (olivine + pyroxene), or mare basalt—whose conductivity (σ)—temperature (T) relationships have been measured. The *Apollo*

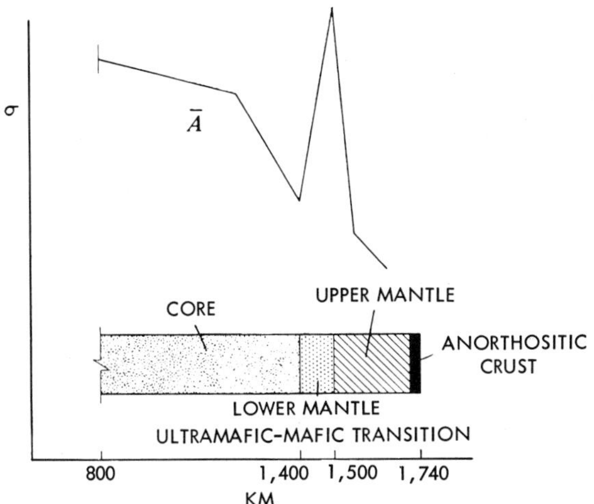

Figure 13-11 A simplified model for the nature and thickness of layers making up the lunar core, mantle, and crust based on the conductivity-temperature distribution model of Sonett et al. (1971). Conductivity (in mhos/meter) increases upward. The curve labeled A represents the profile of change of electrical conductivity with depth; A is an empirical transfer function calculated from time-dependent values of the magnetic field induced in the Moon and the interplanetary magnetic field.

12 electrical conductivity values can then be applied to calculation of the internal temperature distribution within each layer as a function of assumed rock type.

The remanent magnetization found in samples from the lunar surface and shallow crust has fostered several models which relate the thermal history of the Moon to conditions favorable to acquisition of the present magnetic states. A magnetizing field of at least 10^2-$10^3\,\gamma$ is needed to impress the magnetization observed on the rocks at the *Apollo* sites as measured by the stationary and portable magnetometers. A strong dipolar field could be induced by the dynamo effect in a small fluid metallic core. Formation of such a core requires the deep interior to be hot enough to melt and differentiate by the time crustal and mare rocks were being formed. However, if such melting did not occur during the first billion years—a view supported by many investigators—then some other field source must be found. Runcorn, Urey, Alfven, and others have resorted to the action of a strong external field from the early Sun or from closer proximity of the Moon to Earth. Strangway et. (1973) expand upon this idea in a model which ties the several types of remanent magnetism now recognized in lunar samples with a plausible heating history for the Moon (Figure 13-13).

Toksöz et al. (1972) have utilized heat flow, seismic, and magnetic data in setting the boundary conditions for a computer-generated model of the thermal evolution of the Moon. Different combinations of input parameters were selected for each computer run. Initial base temperatures shortly after accretion were fixed at 0°, 500°, 800^oC, and values were set leading to complete melting to simulate (1) a cold Moon, (2) a Moon with cold interior and hot outer shell, and (3) a fully molten Moon. This heat is supplied at first from gravitational accretion energy but thereafter derives mainly from radioactive decay. Average uranium contents in the interior were varied between 0.01 and 0.023 ppm. Values for heat flux at the surface ranged between 21 and 29 ergs/cm^2-sec. Transfer of heat is aided by mass convection wherever the viscosity of interior materials is reduced below $\sim 10^{24}$ poises. Results of the calculations include a series of curves relating temperatures to depths at various times in the Moon's history (Figure 13-14). Each curve may intersect the depth-dependent solidus (a boundary curve that defines the temperature above which melting must occur) over some time interval. From this the region of the lunar interior that experiences complete or partial fusion at any period in the Moon's history can be deduced. For the parameter chosen, the initially cold or initial molten *in toto* models are inconsistent with present conditions and are rejected as unlikely. The favored model (Figure 13-15) leads to a Moon that accreted rapidly so that its outer half melted early and underwent extensive differentiation to depths of

Figure 13-12 A cutaway section showing the interior of the Moon in terms of a three-layer conductivity model. For the electrical conductivity assigned to each region (bounded by shells whose radii are fractions of that of the Moon, R_m), corresponding temperatures are calculated for three rock types—dunite, peridotite, and mare basalt—as possible constituents of each layer. (From Dyal and Parkin, 1971.)

THREE-LAYER LUNAR MODEL

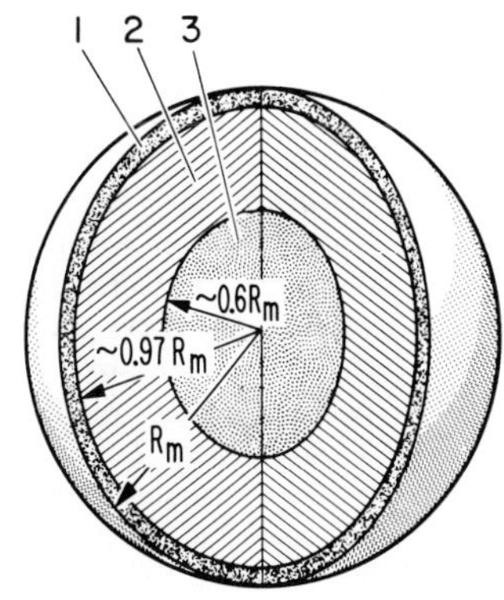

REGION	ELECTRICAL CONDUCTIVITY, σ, mhos/meter	TEMPERATURE, °K		
		OLIVINE	PERIDOTITE	APOLLO II SURFACE SAMPLE
1	$<10^{-9}$	<440	<430	<300
2	$\sim 3.5\times10^{-4}$	890	1000	580
3	$\sim 10^{-2}$	1240	1270	740

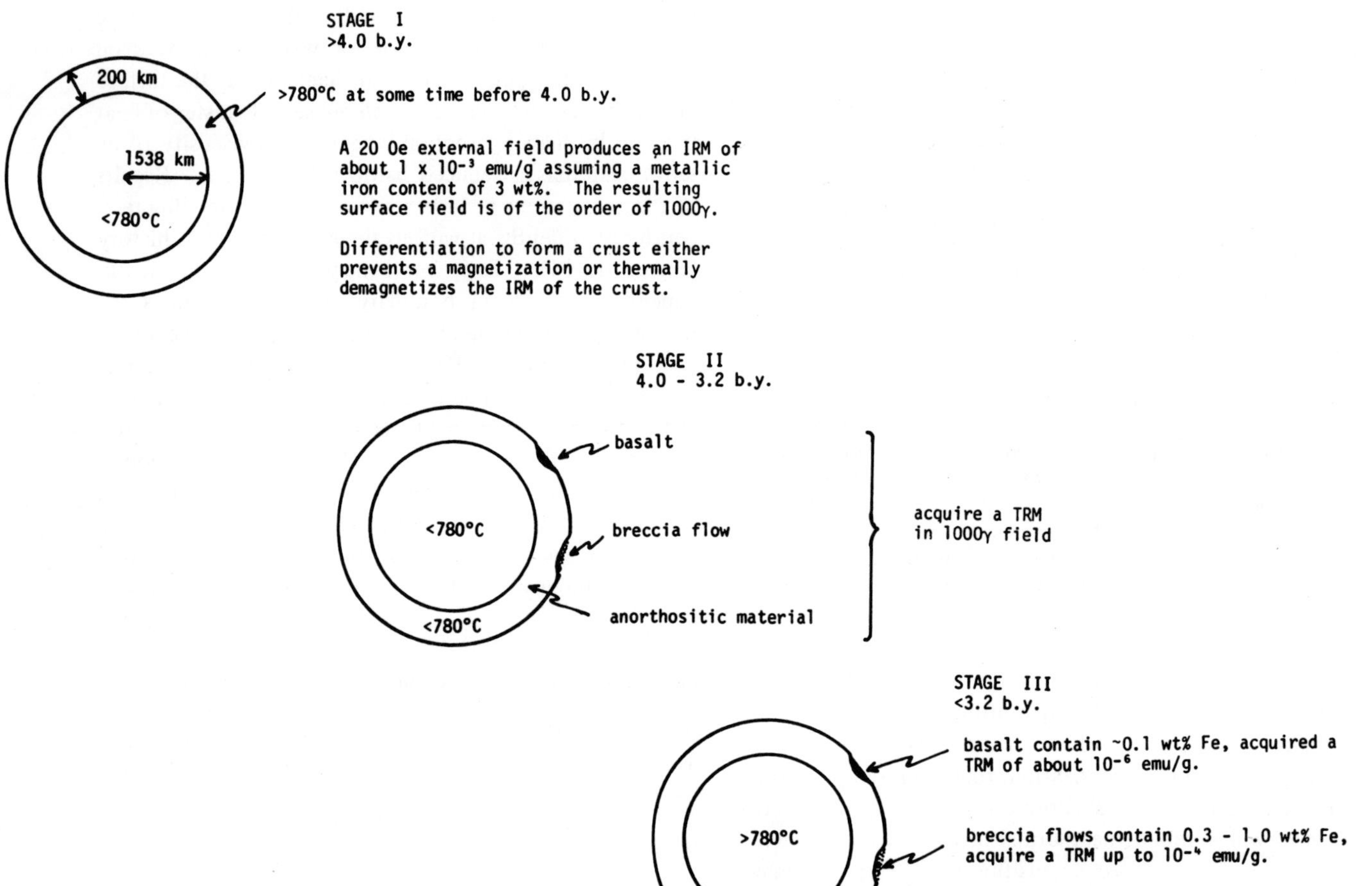

Figure 13-13 Three-stage model for the development of observed magnetic effects in lunar samples as a function of temperatures within the Moon. In the first stage the outermost 200 km of the Moon have probably melted more or less completely as the initial crust is produced. During the time when lunar materials were above the Curie point of iron (780°C), a strong external field (selected to be 20 gauss) produced an isothermal remanent magnetization (IRM) within cooler rocks in the interior that gave rise to a dipole moment of 5 to 10×10^{22} cgs units. No IRM is impressed or retained in the crustal rocks during this stage. Even after the external field had abated, the IRM of interior rocks was sufficient to maintain surface fields of 1000 to 2000 γ. Stage 2 began as the crust cooled below the Curie temperature, so that its rocks acquired a thermoremanent magnetism (TRM) in the presence of the residual IRM field from the interior. The resulting TRM is less than 10^{-5} emu/gm in rocks containing less than 1% metallic iron. In the final stage, the interior heated up to temperatures in excess of 780°C, thereby erasing any IRM and eliminating these deep rocks as a field source. The exterior rocks continue to preserve the TRM inherited from the second stage. The weak, variable magnetization remaining at the surface, in the absence of any strong internal field, is consistent with present-day observations. (From Strangway et al., 1973.)

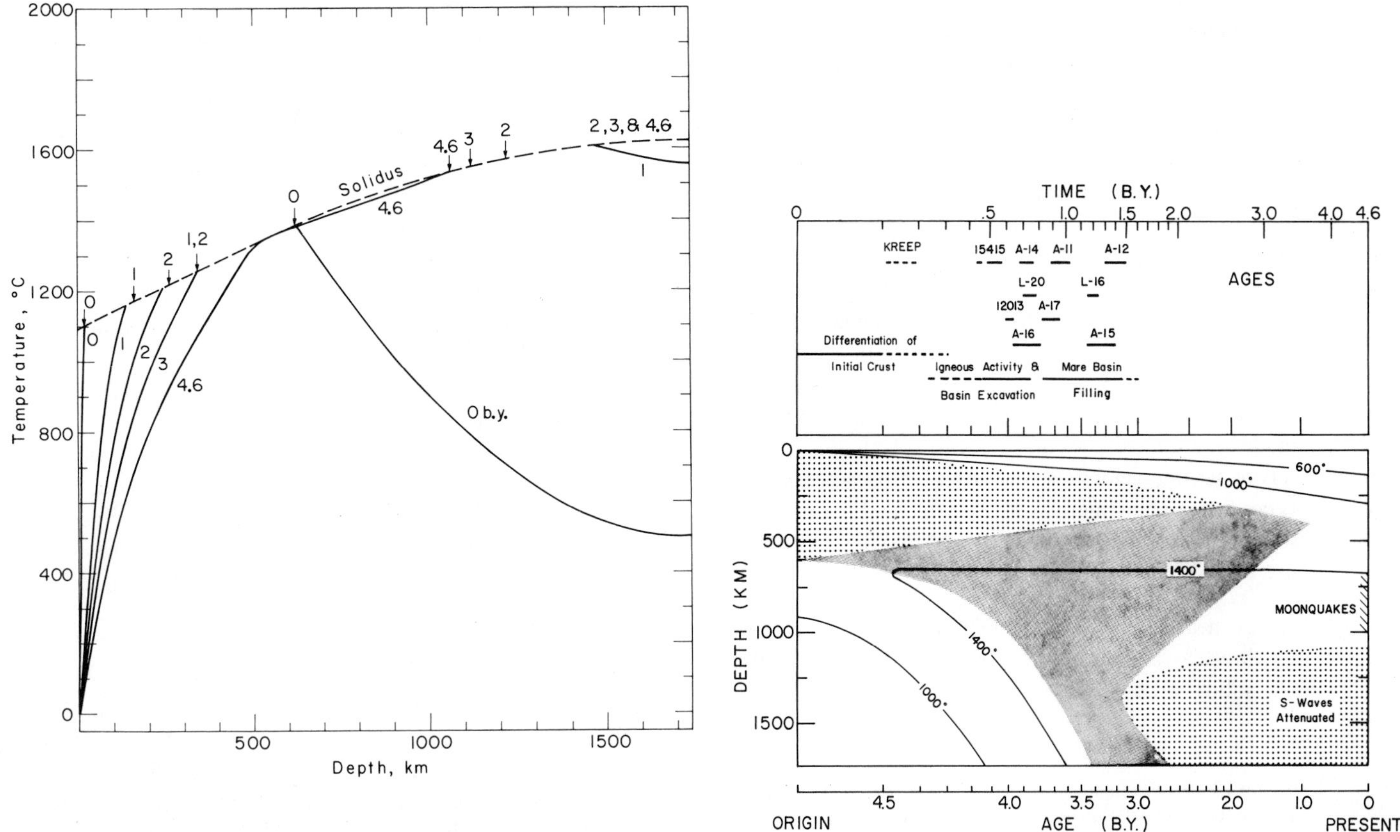

Figure 13-14 (left) One of the thermal evolution models of Toksöz et al. (1973) for the Moon. In this version, the Moon accreted rapidly over ~ 100 years and heated up to a starting temperature of 500°C at the base (center). Each temperature curve shown represents the distribution of temperature with depth at the time (0, 1, 2, etc.) in billions of years after initial formation. Where a curve intersects the solidus, melting at that time is assumed to be occurring within the indicated depth intervals (two intersection points). Thus, initially at time 0, the Moon was molten between a few tens of kilometers and about 650 km. After 1 b.y. the crust had differentiated and solidified to a depth of 120 km, but the lower limit of melting by then extended to more than 1400 km. With age the outer limit of solidification moved downward into the Moon at a rate of crustal thickening of ~ 120 km/b.y., while the entire interior became molten. At the present time the outer 1000 km are now largely solid (with possible localized melt zones), and the deeper regions are still hot enough to be more or less melted. The small arrows indicate boundaries of complete melting. This model is based on a present-day heat flow value of 30 ergs/cm^2/sec, consistent with observations at *Apollo 15* and *17* sites. (Courtesy N. Toksöz, M.I.T.)

Figure 13-15 (right) A best-case thermal evolution model for the Moon, proposed by Toksöz et al. (1973), coordinated with the ages of recognized events as determined radiometrically from *Apollo* and *Luna* rock materials. The location of isotherms (°C) as functions of depth and time assists in calculating melt states within the Moon. At its outset the Moon's outer surface was partially molten (upper dotted pattern) to about 600 km. With time complete melting extended both upward and to greater depths. A thin surficial crust formed almost at once and gradually thickened above a diminishing zone of incomplete melting. Most of the deeper Moon had melted totally (shaded pattern) after about 1.0 b.y. In the next 1.5 b.y., the zone of total melting narrowed at intermediate depths while the core region partially solidified. From the last billion years to the present, the lithosphere of the outer 1100 km of the Moon completely solidified and developed a firm rigidity except in the 800-1000 km interval in which relief of small stress differences gives rise to weak moonquakes. The asthenospherelike deep interior (lower dotted pattern) is the only remaining part of the Moon where some partial melting has persisted. (Courtesy N. Toksöz, M.I.T.)

600-800 km. Cooling proceeded inward during the first two billion years, but a zone of partially molten rock persisted in the outermost 300 km in that period. Partial melting and convection continues at present below about 1100 km (a transition region, overlain by rigid crust, that marks a zone of seismic activities).

The seismological and thermal data from the Moon, as interpreted by Latham et al. (1972), Toksöz et al. (1973), and other workers, places some definite contraints on its internal structure and extent of tectonic adjustments. The velocity-depth relations deduced from signals received at the Apollo seismic stations, together with seismic propagation properties measured on lunar and terrestrial rocks (see Figure 10-43), allow some specific inferences to be made regarding the Moon's interior. Thus, by fitting observed seismic velocities with those of possible lunar rock types, after adjustments for the effects of increasing pressures with depth, this model of the Moon's interior (Figure 13-16) has been proposed:

(1) A near surface layer (1-2 km thick) consisting of breccias and broken rocks in varying degrees of consolidation.

Figure 13-16 The model of lunar internal structure proposed by Toksöz et al. (1973) on the basis of seismic data. The solid black represents mare basalts which are concentrated primarily on the near side of the Moon. The crust is mainly anorthositic gabbros overlying a mantle of pyroxenite. The high velocity zone, detected so far only on the front side, is possibly a garnet- or spinel-rich assemblage of higher density basic rocks. The shaded "core" is considered to be an asthenosphere of more or less melted rocks extending out to a depth of 1000 km. Most moonquakes emanate from a region just above the outer asthenosphere. More recently, G. Latham (personal communication, April, 1974) has proposed a model for the lunar interior that includes a small iron (or FeS) core that is likely to be at least partly molten (needed to explain a drop in P wave velocity to ~5.5 km/sec). (Courtesy N. Toksöz, M.I.T.)

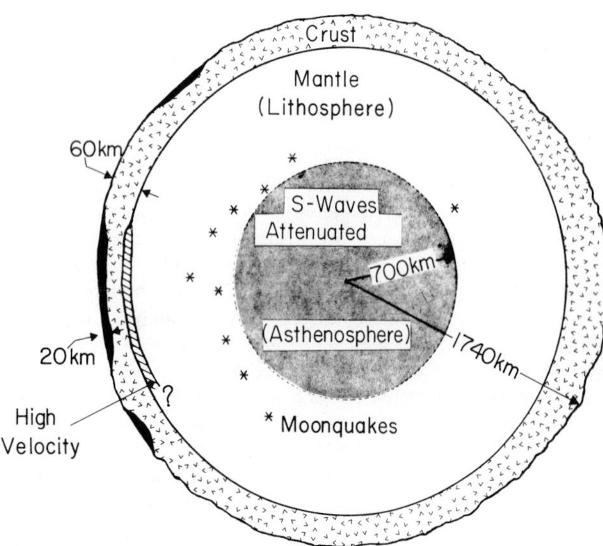

(2) A basaltic layer down to 20 km.
(3) A second solid layer between 20 and 60-70 km whose seismic properties can be matched by anorthosite, gabbro, norite, and possibly pyroxenite (any or all of these).
(4) A transition zone consisting of an intermittent high velocity layer rich in garnet (eclogite) or spinel that may occupy the top 40 km of the mantle over parts of the Moon; disappearance of this high density zone with depth is attributed to a phase reversal to a lower density form (rich in plagioclase) brought about by increased temperatures.
(5) A thick, rigid mantle of magnesium-rich pyroxenite and/or dunite.
(6) An inner zone or "core" of partly melted rock which fails to transmit S waves. The only sharp seismic discontinuities, defined by abrupt rises in velocities, occur at 20 and 60 km depths and apparently mark the boundaries between the basaltic and gabbroic rocks and between the crust and mantle.

The very low near-surface velocities and their steady increase down to depths of 15 to 20 km are consistent with a coherent shell which is progressively more jointed and fractured as the surface is approached. This could mean that the crust and upper mantle have been severely disrupted by continual tidal flexing. Alternatively, this may imply that the outer shell of the Moon—and perhaps much of its interior—still consists of great blocks and pieces of aggregated materials, diverse in composition and size, left over from the original infall of planetesimals that built up the Moon. Melting, in this view, was not general, and most of the Moon remains a consolidated rubble composed of debris derived from some older parent source (asteroids?; the Earth?). The low-energy release and infrequency of moonquakes—the characteristics of a quiet Moon—are indicative of a rigid, cool crust. Whatever volcanism has occurred in the past acted mainly to fill the mare basins and has affected only a small fraction of the total volume of the lunar interior. Even today, though, localized pockets of melt may ebb and flow within deep fractures in the Moon's subsurface regions as these open and close during monthly tidal flexing, registered as a set of regularly recurring, weak moonquakes that culminate near perigee.

Runcorn (1972, 1973) concurs with this concept of a generally static, rigid Moon enclosed within a cold crust now having finite strength. According to Runcorn's model, the Moon is assumed to have experienced total melting early in its history, and its fluid body adopted a hydrostatic shape (essentially spherical) in response to equipotential forces set up by the lunar gravitational field. Most of the lunar interior cooled in 100 million to 200 million years by heat transfer through convection. Then as a crust thickened and developed strength it gradually became dis-

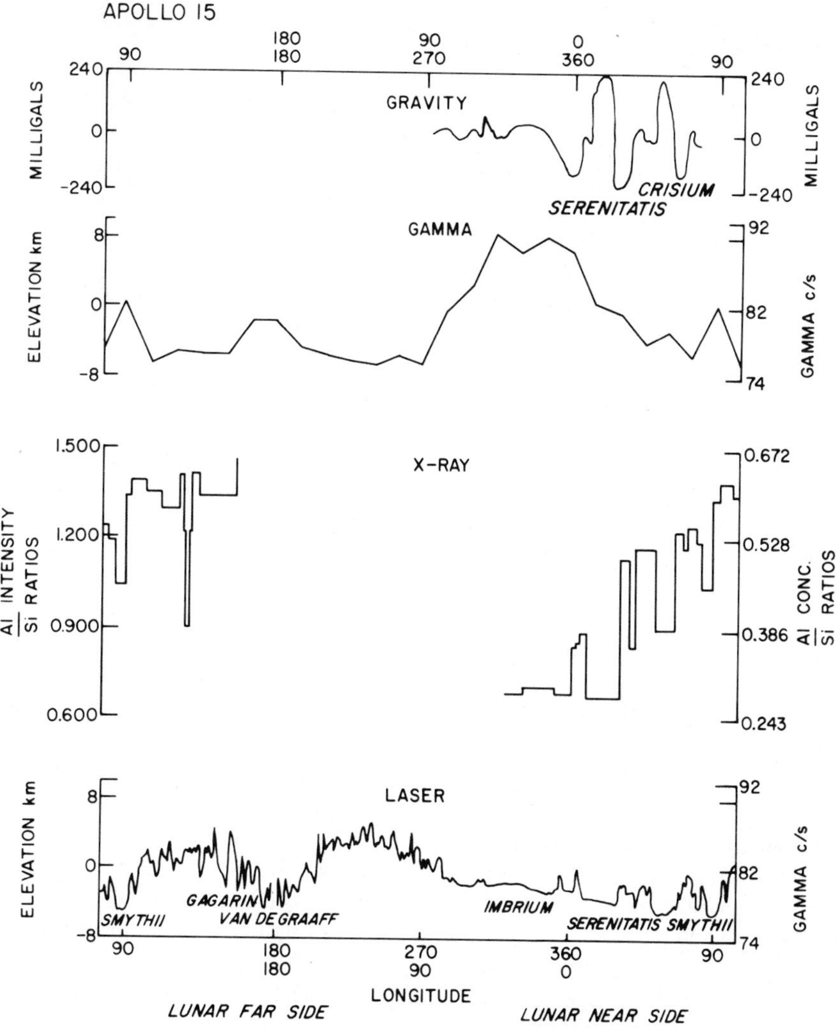

Figure 13-17 Variations of lunar gravity, gamma ray intensities (in counts/sec, or c/s), Al/Si intensity or concentration ratios generally correspond to passes over the lunar highlands. By comparing these ratios with those determined from lunar samples, it is evident that the high-high gamma-ray intensities occur over the Imbrium Basin and Oceanus Procellarum. The increases represent higher K and/or U concentrations in the surface materials. The higher Al/Si intensity or concentration ratios generally correspond to passes over the lunar highlands. By comparing these ratios with those determined from lunar samples, it is evident that the highlands could consist of gabbros and gabbroic anorthosites but probably do not contain large areas of pure anorthosites. Maximum elevations on the Moon are found mainly on the lunar far side; however, a large depression occurring in the vicinity of Van de Graaff is characteristic of basins observed on the near side. (Courtesy H. Masursky, U.S. Geological Survey.)

torted (mainly after emplacement of mare lavas), owing to the dragging action of thermal convection currents in the mantle, into the presently observed nonhydrostatic shape especially evident in the 2-km frontside bulge. The great thickness of the lunar crust has prevented it from breaking up into mobile plates despite interior convection. But even now the distortion of the Moon's figure may still be maintained by residual convective currents deep within a hot lunar interior. Mascons developed where the crust was breached through impacts whenever the resulting basins were not extensively backfilled with thick rubble deposits. Thus, those circular basins that were largely empty at times of extrusion would fill with deep lava flows which, on cooling, shrank in volume to produce the observed mass excesses. These basins have resisted destruction by isostatic adjustments for more than 3 billion years because of the strength of the crust, but their continuing settling may be responsible for some of the moonquakes beneath them.

Geochemistry of the Crust

The key to unlocking the mysteries of the lunar interior—and hence its long but seemingly rather uncomplicated history—may ultimately be found in the rocks making up the ancient highlands. The idea, put forth at the First

Lunar Science Conference, of a simple continuous primordial crust of anorthosite has since undergone critical reevaluation. Flotation of feldspars to produce this anorthosite would have required complete melting of much of the outer Moon to great depths very shortly after aggregation. The radiometric ages of anorthosite fragments between 4.1 and 4.25 AE appear to rule out such a general process at an early stage. More likely, the highlands were built up by a series of intrusions and outpourings of moderately differentiated primitive materials. The variety of rock types found in *Apollo 14* breccias could come from either tapping of several different sequential layers during a single event (Imbrium Basin) or from additions from several source areas containing diverse representatives of a crust differentiated into norites, anorthosites, anorthositic gabbros, troctolites, and peridotites, or from both factors.

If the crust is layered, the task at hand will be to work out the correct emplacement sequence and to determine which layer, if any, ever extended over the entire Moon as the principal surficial crust. The answer to this may never be available from direct observation and sampling; the intensely cratered highlands appear to be covered everywhere by a thick debris layer comprised of multiple ejecta blankets, and the lowlands are still largely topped by lava units. Statistical analyses of rock type distributions at individual sites in relation to possible sources, geophysical methods for "seeing" into the interior, and refinements in models for development of a differentiated Moon together provide the best hopes of determining the nature of the Moon below its visible surface short of actually drilling deep holes from some future lunar base.

The SIM bay experiments from orbit during the *Apollo 15* and *16* missions are a major advance in the geochemical approach. The data sensed by the spectrometers integrate variations and differences over a wide area along the ground swath, so that the results are representative of large parts of the lunar surface. Without doubt one of the most significant accomplishments in the *Apollo* program has been the identification by the X-ray and gamma-ray spectrometers of Al-rich, K-poor rocks in the highlands of the front and farside of the Moon (Figure 13-17). The Al/Si and Mg/Si ratios, in particular, strongly suggest that one or more outer layers in these regions are made up of anorthositic gabbro. However, presence of considerable pyroxenes and/or olivine in rocks on the surface of the central highlands, as indicated by infrared spectroscopy (Logan et al. 1972) favors the view that gabbroic or noritic rocks (feldspathic basalt group) dominate most of the outer crust.

If these various models of the lunar interior and its petrologic history, reviewed on the previous pages, are looked at collectively, one must conclude that each explains some observations but encounters difficulties with others. Further data from the Apollo program will certainly help to decide which is the most plausible model. Nevertheless, despite the remaining uncertainties, the Moon's surface itself is proving to be a classic area in which to study the petrology of basic igneous eruptions and the differentiation of primitive planetary bodies.

Lunar Time Scales

While the chemistry and mineralogy of the lunar rocks have proved of utmost value in unraveling the development of a differentiated Moon, the age-dating of these rocks now provides a remarkable picture in detail of the sequence of events in this development. The methods of geoscientists have often been likened to those of detectives who must piece together usually limited, incomplete, and often contradictory evidence to reconstruct a series of events that focus on some specific act. Whenever the times of these events can be sorted out in proper order, the critical factors, causes, interrelationships, and consequences associated with that act can normally be determined. In deciphering lunar history, some exceptional examples of "detective work" result from setting the precise time when some observed surface feature or rock sample assumed a characteristic state (Figure 13-18).

Thus, the cooling and crystallization times when the last lavas flowed into several mare regions have now been reliably established. Mare lavas represent a spasm of outpourings that lasted no longer than 600 million years. So far, no extrusive event younger than 3.1 AE has been reported. Other lavas—mainly highlands types—can be traced back to a period between 3.8 and 4.15 AE ago. The anorthositic "Genesis Rock" from the *Apollo 15* site, if actually a part of an outer crust, fixes the time at which these layers were formed at approximately 4.15 AE. No fragmental inclusions with crystallization ages older than 4.3 AE have yet been recovered from any breccias—either those at the highlands sites or those extracted from mare-covering regolith.* This value presently comprises the

*A.L. Albee et al. report (at the Fifth Lunar Science Conference in March, 1974) a whole-rock isochron Rb-Sr age of 4.6 AE for an annealed dunite fragment (sample 72415) from an *Apollo 17* breccia. They claim this to be a crystallization age for a rock from a cumulate layer formed well below the surface as an early differentiate from primitive upper mantle material. This material would then be considerably older than the 4.0-4.3 AE ages obtained on other crystalline-appearing breccia clasts from *Apollo* sites. Strong shock damage in this specimen, along with other factors, raises some doubt as to the validity of the reported age.

Another sample, 76535, consists of an annealed (but unshocked) troctolitic granulite (plagioclase, olivine, some bronzite). The coarse-grained texture, together with other considerations, points to this rock as a fractionated cumulate that crystallized some 4.3 AE (Ar-Ar age) to 4.4 AE (Rb-Sr age) (D.D. Bogard et al, Fifth Lunar Science Conference) at a depth between 10 and 30 km.

In general, most KREEP-bearing highlands crystalline rocks, found mainly as breccia clasts, give the whole-rock isochron ages around 4.3 AE. Anorthosites, by the same methods, show ages of 4.6 AE. Internal (mineral) isochron ages for these rocks cluster around 3.8-4.0 AE.

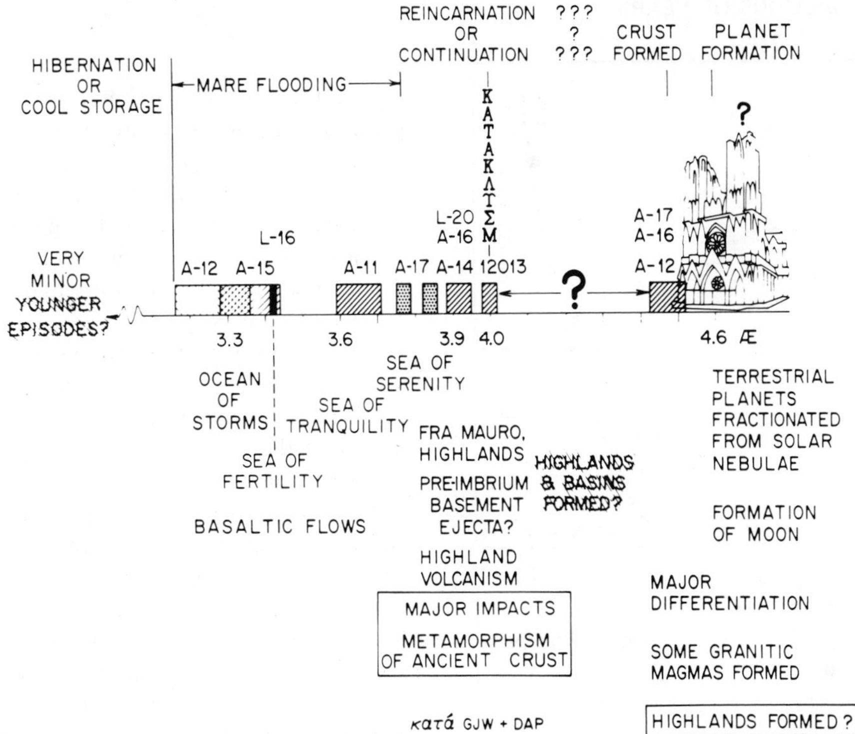

Figure 13-18 The "Cathedral" diagram of Wasserburg and Papanastassiou illustrating their interpretation of the history of the Moon during its first 1.5 billion years based on age dating of lunar samples. Back to about 3.9 AE, the bars represent mainly the Rb-Sr crystallization ages of mare lavas; most ages between 3.9 and 4.1 AE are associated with breccias. Those ages near 4.5 AE are typical model ages of both mare lavas and regolith. The event recorded by 12013 is identified by the Greek letters for "Cataclysm" that describe the formation of the Imbrium Basin. Excavation of other basins and the development of the highlands took place largely in the interval between 4.1 and 4.5 AE, for which direct age-dating evidence is poorly represented in rocks returned from the *Apollo* sites. (Courtesy G. J. Wasserburg, California Institute of Technology.)

upper limit for the oldest recorded ages of solidification of any lunar differentiates. If this holds up, then most of the crustal terra rocks must have crystallized 400-600 million years after the Moon aggregated from very primitive solar materials.

The ages of the terra units making up the lunar highlands have proved more difficult to establish. Rb-Sr and Ar-Ar ages of breccias at *Apollo 16* and *17* range between 3.90 and 4.25 AE. These ages reflect mainly the events which produced the ejecta blankets from which the outer highlands layers are constructed. The Imbrium event produced much of the material now at or near the surface. In general, according to Fredriksson et al. (1973), all lunar rocks older than about 3.9 AE have been strongly modified by impact processes; most, therefore, will be brecciated, shattered, or recrystallized from impact-generated melts. By selective narrowing, the time for formation of the Imbrium Basin can be pinpointed to an interval between 3.88 and 4.05 AE. Various investigators arrive at ages of 3.9 to 4.0 AE for KREEP-basalt and norite fragments in the Fra Mauro breccias that, by general agreement, are part of the Imbrium ejecta blanket. Unless shock processes "reset the clock" that dates these samples, they mark the minimum (youngest) time when pre-Imbrian layers with these lithologies could have consolidated. The "Genesis Rock" places an upper limit (given as 4.05 AE by Wasserburg) on the Imbrium impact, provided the Hadley Delta front is part of the crust that was uptilted as a ring scarp when the Imbrium Basin formed. The lava sample 14310, dated at 3.88 AE by Wasserburg, may set a lower limit if it belongs to a post-Imbrium lava unit. Otherwise, the Imbrium event could be as young as 3.3 to 3.2 AE, judging from *Apollo 15* lava rock ages, although these mare rocks could have come into contact with the pre-Imbrian ring rocks long after basin formation. Age dates clustering near 3.9 AE for many fragments in *Apollo 16* breccias—held to have been derived from the Orientale Basin-forming event—suggest that both the Orientale and Imbrium Basins were excavated during a narrow time interval in lunar history (see p. 321). From assumptions regarding the source(s) of age-dated ejecta at several *Apollo* landing sites, coupled with crater counts within filling lavas, O. E. Schaeffer et al. (Fifth Lunar Science Conference, March, 1974) propose the following ages for these multi-ring basin events: Nectaris, 4.20 ± 0.05 AE; Humorum, between 4.13 and 4.20 AE; Crisium, 4.13 ± 0.05 AE; Imbrium, 4.10 ± 0.05 AE; and Orientale, 3.85 ± 0.05 AE.

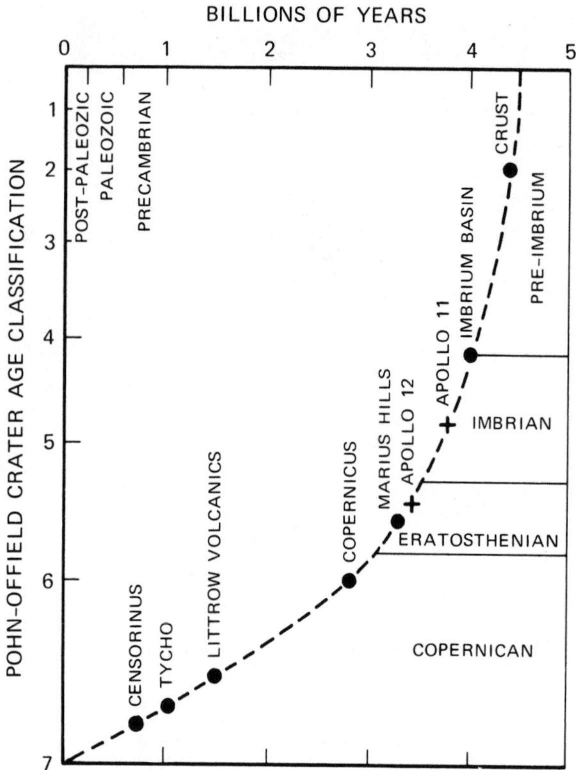

Figure 13-19 A general time scale for major lunar features in relation to crater age (Pohn-Offield classification) of their surrounding surfaces. The radiogenic ages of the *Apollo 11* and *12* crystalline rocks were used to calibrate the curve. The age boundary between the Imbrian and Eratosthenian Periods is approximate and that between the Eratosthenian and Copernican is probably younger than indicated here (see Table 9-1). (Courtesy H. Pohn.)

The Copernicus Crater has been assigned a relatively young genetic age by many workers. A ray from Copernicus crosses the *Apollo 12* site. Assuming that KREEP fragments at that site came from Copernicus, these show exposure ages which, after being corrected for times of burial when the ejecta were shielded from irradiation, indicate that the Copernicus event occurred some 800 to 1000 million years ago.

After the early *Apollo* missions, Pohn and Offield added a time scale (Figure 13-19) for the major events and stratigraphic subdivisions marking the lunar surface as defined in their earlier work (1969). Using their own classification (see Chapter 9), they have plotted the relative ages, from crater studies of the regions around the *Apollo 11* and *12* sites, against the radiogenic ages of the mare rocks underlying the cratered surfaces. A curve is then drawn through those plotted points and the age assigned to the Moon's origin. The boundaries between lunar stratigraphic systems are located according to the same crater age criterion. In this way, a time estimate of the beginning and end of each division is obtained and its relation to terrestrial geologic eras becomes evident. The absolute ages of some major lunar events that fall on this curve are then estimated. The age of the Imbrium Basin-forming event obtained from the curve is almost 4.0 AE—a value very near those reported by various other investigators. The age of the lunar crust would be about 4.3 to 4.4 AE—again,

Figure 13-20 A curve drawn through the plotted areas defined by the geomorphic index (see text) of the region around lunar sampling sites and the spread of ages for rocks from these sites. Extrapolation of the curve to older and younger times is tenuous because of the absence of samples with meaningful ages for these periods. (From Ronca, 1973.)

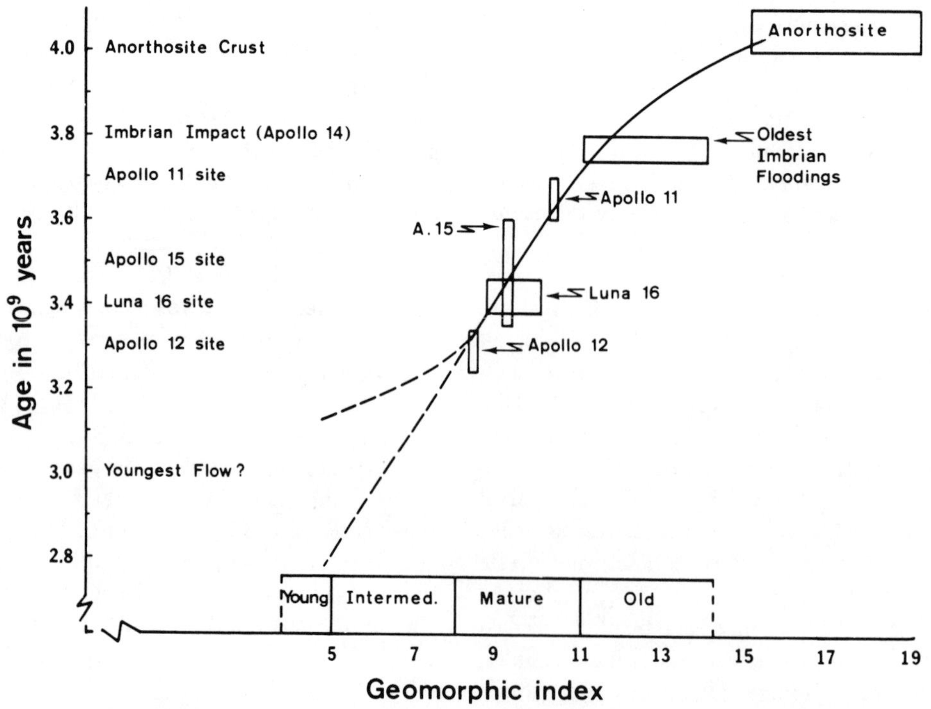

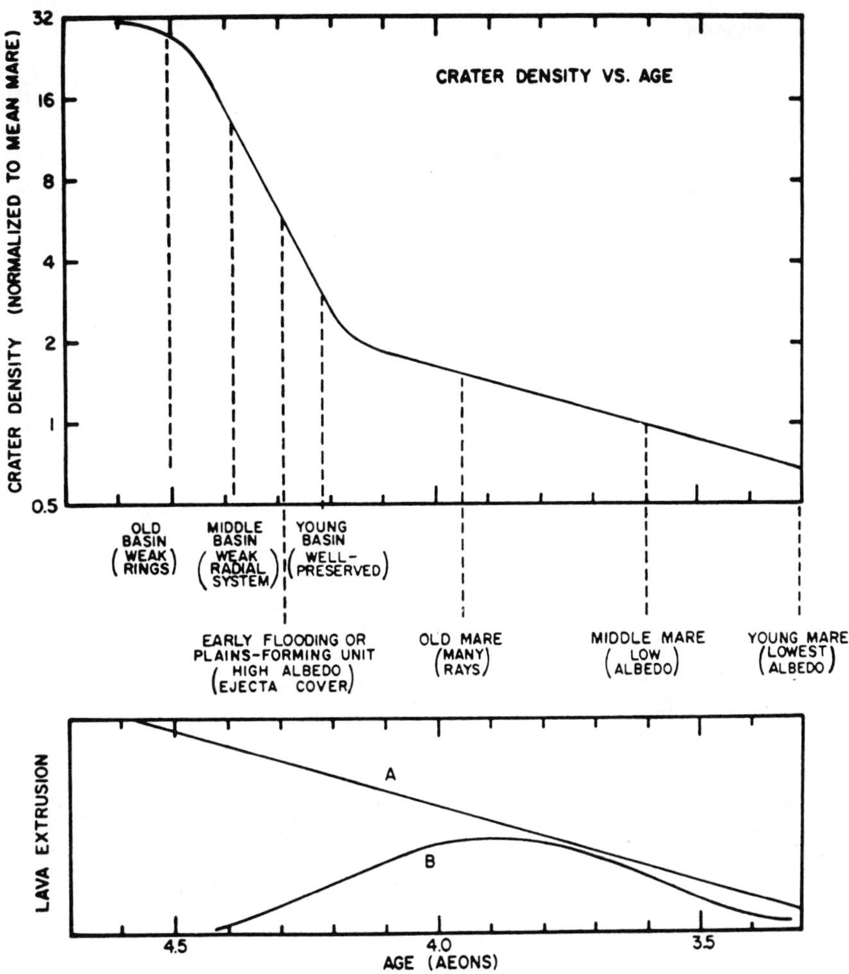

Figure 13-21 Summary of lunar history from its origin (4.6 AE) to the apparent end of mare lava flooding (about 3.2 AE) based on progressive changes in crater densities in open maria and mare-filled basins and on the highlands. Age values (shown on the abscissa of the lower diagram) result from *Apollo* sample dating of mare rocks assumed to be representative of the surrounding regions in which crater densities have been measured. The lower diagram shows two hypothetical sequences of mare flooding (in terms of volume of extruded lava) over time; curve B appears to best represent basin filling with *Apollo 11, 12,* and *15* types of basalt. (Reprinted from W. K. Hartmann, and C. A. Wood, "Moon: Origin and Evolution of Multi-Ring Basins," The Moon, Vol. 3, No. 1, pp. 3-78. Copyright 1971 by D. Reidel Publishing Company.)

a value similar to those suggested for the crust from studies of highlands rocks at *Apollo 15, 16* and *17* sites. Note, however, that if exposure ages for Copernicus ejecta correctly define its time of formation as well, Pohn and Offield's age estimate for that structure—and hence for the beginning of the Copernican period—is considerably in error.

Ronca (1973) has followed a comparable approach with similar results. He plots the geomorphic index applicable to the region around each site against the radiometric age of typical rock units for that region as sampled at the site (Figure 13-20). The geomorphic index (see the caption to Figure 13-20 and Ronca and Green, 1970) is a statistically sensitive measure of crater classes (defined primarily by the relative freshness or state of erosion of crater rims) as a function of their frequency of distribution per unit area (crater density). Mare regions contain craters whose geomorphic index ranges from less than 5 to about 14; highlands have characteristic indices between 15 and 19.

Hartmann and Wood (1971) have reconstructed the sequence in which the major multi-ring basins (e.g., Imbrium, Crisium) were formed and have used the distribution of crater diameters within and around these basins as indicators of the time of flooding. Their studies are summarized in two charts (Figure 13-21) showing lunar history as now calibrated by *Apollo* age-dating. Apparently, all of the larger basins were excavated during a narrow span of lunar time—perhaps by infall of protomoons or excess material left over from fission, depending on the model adopted for the Moon's origin. Considerable time elapsed before these basins were filled by mare basalts, as evidenced by the many large craters (Plato, Gassendi, Fracastorius, Archimedes) that were impressed on the basins and then partially covered by the lavas. This basin-filling took place through multiple outpourings that piled up a series of flows rather than through a single, continuous, and rapid flooding. Their assumed ages for the Imbrium and Orientale Basins (young in the sequence) are older than those indicated by *Apollo 14* results. The Orientale Basin must be older than the Oceanus Procellarum lava rocks because of absence of ejecta rays from that structure on the mare surfaces to the northeast. Hartmann and Wood have adopted two models for flooding of the surface by crust-forming lavas. The first (a) assumes a progressive decrease in the rate of extrusion, in response to gradual cooling of the lunar interior, such that the mare lavas (3.7-3.2-AE interval) are volumetrically only a fraction of earlier amounts. The second (b) describes an increase in flooding, in response to a thermal rise from radiogenic decay, to a peak at 3.9 AE (corresponding to some KREEP-basalt and nonmare basalt ages) followed by a diminishing in surface activity.

Schaber (1973) estimates that the filling of the Imbrium Basin with mare lavas may have taken up to 500 million years. He has identified three distinct episodes of filling over this time span. Some 10-20 X 10^3 km^3 of lava now occupy the basin, extending over an area of more than 2 X 10^5 km^2, and representing an accumulated thickness of flows of ~0.1 km. The low-viscosity lavas appear to have spread out from a single source area (18-23°N, 28-32°W) near the crater Euler.

EVOLUTION OF CRATERING RATES

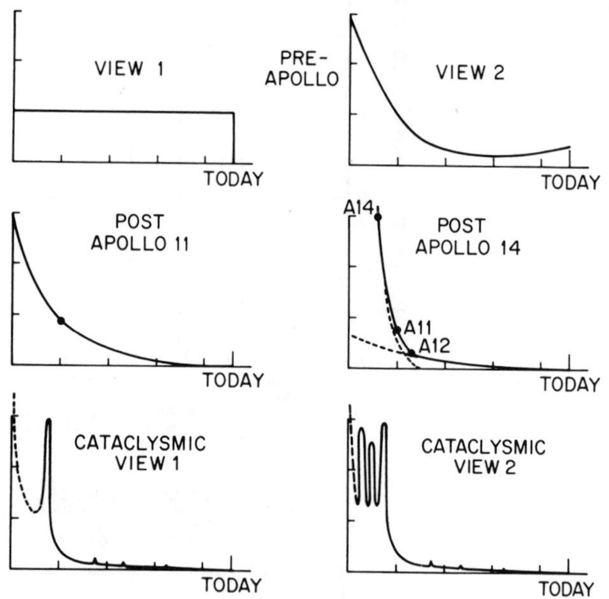

Figure 13-22 Comparison of several models for the variations in cratering rates from the time of formation of the outer lunar crust to the present. One early view held cratering to have been nearly constant since the Moon was formed. A more popular opinion considered the rates of bombardment of the lunar surface to have decreased exponentially over most of lunar history until the last 1½ billion years, when cometary impacts increased in frequency. After *Apollo 11* resultant ideas favored a continual drop-off in cratering to a very low rate at present. Following *Apollo 14* a new view recognized two distinct and nearly independent cratering histories: one (*Apollo 11* and *12* curve) resulted from the rapid and almost total insweep of meteoritic materials, including planetesimals, during the first million or so years; the second reflects the continuing presence of cometary bodies in the solar system, giving rise to moderately increasing numbers of impacts on the Moon over time because of greater probabilities of collisions with comets in the earlier periods. The cataclysmic views, espoused by Tera and Wasserburg (1973), are discussed in the text. (Courtesy G. J. Wasserburg, California Institute of Technology.)

Calibration of Crater Ages

Shoemaker (1972) has also interrelated cratering rates and saturation of surfaces with craters to the radiogenic ages of rocks involved in basin formation and lava extrusion. He begins by assuming that the Moon could have broken from the Earth by fission after the Earth had grown from accreting matter to about a third its present size. Thereafter, the Earth continued to grow by infall of planetesimals, while the smaller Moon lost some of its mass (hence changing its angular momentum relative to Earth) through ejection as it also was bombarded by planetesimals. The Moon melted early and developed a crust over which feldspathic basalt flows poured long before the present mare lavas were emplaced. This crust solidified near its surface before the majority of highlands craters were impressed on it. Volcanism from the early days of the Moon gradually diminished in extent and volume of extrusion as the lunar crust continued to thicken; most volcanic activities practically ceased some 3.2 AE ago.

The rate of planetesimal impact on the lunar surface in the first few hundred million years was extremely high. The "half-life" of these early planetesimals (time required for one-half to be destroyed by interplanetary collisions or infalls during accretion) is estimated to be 2 X 10^8 yr. The corresponding rate of planetesimal impact on the lunar surface in the first few hundred million years was extremely high—more than 300 times the present rate. Nevertheless, the total volume of planetesimals infalling during the terminal stage of accretion, when many of the craters still visible in the highlands were formed, was sufficient to build up a thick layer of added material. Depending on exact influx rates, all original crust and older flow surfaces would be completely saturated with craters larger than 1 km, and thus should be consumed by fragmentation into a continuous rubble deposit, sometime between 4.0 and 4.3 AE ago. The first lavas to survive without total destruction were those in the Mare Fecunditatis area, where some surfaces are estimated by Shoemaker to be as old as 3.85 billion years. During the interval from 3.85 to 3.20 AE age, cratering persisted at reduced (but still high) rates as remaining planetesimals were swept in, but none of the mare surfaces formed in that interval were ever saturated with craters. From 3.20 AE to the present, cratering rates were as much as 100 times lower than during the preceding episodes of mare formation. This reflects the scarcity of planetesimals in this part of the solar system. Most impacts since 3.20 AE result from collisions with comets and cometary fragments. Tycho, one of the freshest of large craters, may have formed by cometary impact only 200 to 300 million years ago, judging from the relative crater frequencies in its interior.

Shoemaker, Hartmann, Gault, and others have called attention to this change in crater frequency with time. Both the numbers of craters and crater-forming bodies in space appear to decrease systematically with time according to some exponential function (see p. 95). If, instead, these changes were linear, meteorite flux rates based on terrestrial crater counts and interplanetary measurements would lead to estimates of 10 to 100 AE needed to develop the crater densities observed on the lunar highlands. The age span from 4.0 to 4.6 AE assumed for the highlands requires an exponential increase in impact rates to account for the differences between terrae and maria.

Tera and Wasserburg (1973) have reviewed the various models of change of cratering rates with time in light of the age of some of the significant events recorded in the lunar samples (Figure 13-22). They have modified earlier views of the cratering history of the Moon by considering the cluster of age dates between 3.9 and 4.2 AE in ejecta breccias as evidence of a brief period of intensive bombardment by large planetesimals during which most of the major basins were formed. Thus, the Imbrium Basin resulted from one such cataclysmic event, and at least several other events of similar magnitude, either somewhat older (e.g., the Serenitatis Basin) or slightly younger (e.g., the Orientale Basin), are inferred although samples from the source region have not been uniquely identified. This doctrine of *cataclysm*, tied to a single brief interval of large-scale impacts that tore apart the outer crust over the entire Moon, has now become widely accepted by most investigators. Nearly all the terra breccias were derived as ejecta deposits from the bombardments and still retain evidence of the changes imposed on them by these dramatic events.

Short and Forman (1972) have calculated the thickness of ejecta produced entirely from visible craters and the larger circular basins. If uniformly distributed over the entire frontside, this ejecta would reach an average thickness of 1.55 to 1.90 km (depending on assumptions used) in the terrae. Near the edges of large basins, however, they estimate local thicknesses up to 3-4 km (Figure 13-23). Their model for the outermost layers of the Moon is supported by observations of breccias associated with the high relief hills at the *Apollo 15, 16,* and *17* sites and by seismic

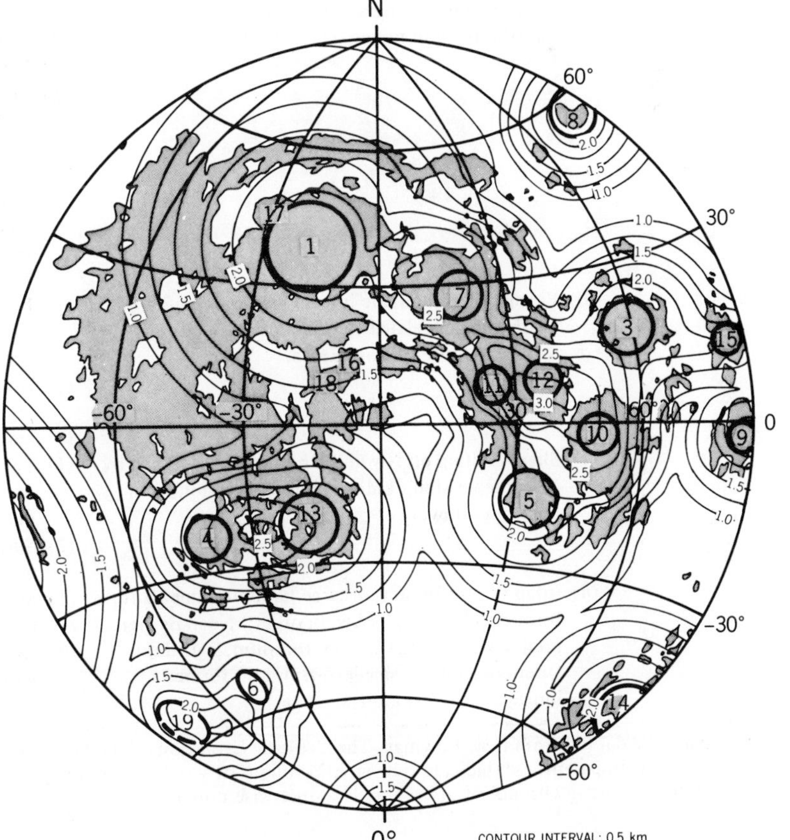

Figure 13-23 Generalized isopach map showing distribution of ejecta blanket thicknesses on the front face of the Moon. A minimum average thickness of 1 km is chosen for those areas, such as west (to the left) of Oceanus Procellarum and in the southern Highlands, which have not received extensive deposits from the indicated basins of excavation. These basins are: (1) Imbrium; (2) Orientale; (3) Crisium; (4) Humorum; (5) Nectaris; (6) near Schiller; (7) Serenitatis; (8) Humboldtianum; (9) Smythi; (10) Fecunditatis; (11) W. Tranquillitatis; (12) E. Tranquillitatis; (13) Nubium; (14) Australe; (15) Marginis; (16) Vaporum; (17) Sinus Iridum; (18) Sinus Aestium; (19) SE Limb. (From Short and Forman, 1972.)

TABLE 13-3

Geologic Evolution of the Moon

*Stage**	*Events†*	*Evidence*
I (4.7-4.6). . . .	(a) Formation of Moon, near the Earth or from it, by precipitation or fission (?). Preceded or accompanied by depletion of protomoon material in volatile elements (e.g., K, Pb, Rb). Process of formation took about 10 m.y. (million years), possibly less than 1000 years.	(1) Age of oldest lunar soil component, 4.6 b.y. (billion years), sets lower limit to time of Moon's formation. (2) Similar Sr^{87}/Sr^{86} ratios in ***Apollo 11*** and ***12*** samples and achondritic meteorites indicates separation of Moon from solar nebula in less than 10 m.y., 4.6 b.y. ago. (3) Radiogenic gas (Ar, He, Xe) retention in meteorites began 4.6 b.y. ago. (4) 176-179 m.y. interval between last stages of nucleosynthesis and radiogenic xenon retention (from iodine-129) suggests rapid formation of planetesimals. (5) U-Pb concordia diagrams for ***Apollo 11*** samples indicate last major U-Pb fractionation in Moon 4.65 b.y. ago. (6) ***Apollo 11, 12,*** and ***14*** samples systematically low in volatile elements compared with similar terrestrial rocks.
	(b) Strong heating, to temperatures over $1000^{\circ}C$, of outer parts of Moon in later stages of formation, from energy of accretion, fission, and tidal interaction with Earth. Short-lived isotopes (e.g., aluminum-26) possibly important.	(1) Evidence for early differentiation and magma generation discussed below implies early heating. (2) Present electrical conductivity and temperature profiles implied by magnetometer data suggest melting to depths of not more than 1000 km. (3) Calculations on energy of fission and accretion indicate high temperatures attained. (4) Remnant magnetism of mare rocks suggests early lunar magnetic field, possibly produced by internal dynamo mechanism.
II (4.6-3.7). . . .	(a) First differentiation of Moon by partial melting, secondary magmatic differentiation, and volcanism, to form global crust (now lunar highlands) of aluminum-rich basalt, gabbroic anorthosite, anorthosite, and minor acidic rocks.	(1) Relative age of highlands inferred from great density of craters and superposition of maria. (2) Absolute age of highlands (~4-4.5 b.y.) inferred from Rb-Sr whole-rock ages of 12013, LR-1, anorthosite 15415, and other allocthonous samples (e.g., KREEP). (3) Composition of highlands inferred from ***Apollo 14*** Fra Mauro samples, allocthonous ***Apollo 11*** and ***12*** mare samples, Al/Si ratios from ***Apollo 15*** X-ray fluorescence experiment, and ***Surveyor*** *7* α-scattering experiment. (4) Existence of highland volcanic flows indicated by layering in Mount Hadley and Hadley Delta, by level intercrater areas in highlands, and by old highland mare areas (e.g., Schiller Basin).
	(b) Formation of premare impact craters by infall of nearby bodies related to origin of Moon, during recession of Moon from Earth.	(1) Post-highland-crust early intense bombardment inferred from high crater density, coupled with survival of highland layering and intercrater smooth areas. (2) Impact origin of highland craters indicated by morphologic similarity to postmare Copernican and Eratosthenian craters.
	(c) Shear faulting on NW and NE directions, caused by N-S compression of Moon during recession from Earth and slowing of rotation.	(1) Telescopic study and ***Lunar Orbiter*** photographs show existence of "lunar grid" consisting of NW- and NE-trending shear (?) faults with slight horizontal displacement. (2) ***Surveyor*** and ***Apollo*** surface photos show local NW and NE lineaments concordant with regional pattern.
	(d) Infall of protomoons or large fission fragments to form circular mare basins, in order: Nectaris, Serenitatis, Humorum, Crisium, Imbrium, S. Iridum, Orientale. (Other smaller basins also formed by impact, exclusion arbitrary.)	(1) Morphologic continuity between Copernican craters and circular mare basins implies impact origin for latter. (2) Impact origin of Imbrium Basin suggested by fragmental nature of ***Apollo 14*** Fra Mauro formation samples.

Source: Adapted from P. D. Lowman, The Geologic Evolution of the Moon, J. Geol., 80, pp. 125-166, 1972.
*Numbers in parentheses indicate time in AE (billions of years) before the present.
†Subevents are in chronological order unless otherwise noted.

TABLE 13-3 (continued)
Geologic Evolution of the Moon

*Stage**	*Events†*	*Evidence*
II Continued		(3) Relatively low crater density of Fra Mauro Formation and similar units indicates post-highland-crust age. (4) Ejecta from youngest mare basin (Orientale), Helvelius Formation, not deposited on O. Procellarum surface, indicating premare age of circular mare basins.
	(e) Formation of Archimedian-generation craters (after mare basins but before mare filling).	(1) Archimedes and similar craters are in or on rims of mare basins, so must be postbasin. (2) Mare lavas fill and partly surround Archimedes, so crater is older than mare lavas.
	(f) Emplacement of highland units such as Cayley Formation; roughly contemporaneous with formation of Archimedian craters.	(1) Cayley Formation is not overlain by Fra Mauro Formation but is more heavily cratered and lighter toned than Imbrium mare material, showing postbasin, premare age. (2) Crater density in Ptolemaeus (Cayley Formation) intermediate between that of terrae and maria.
III (3.7-3.2). . . .	Second differentiation of Moon, by repeated localized partial melting of outer 500 km of Moon to produce iron-rich basaltic magma, erupted in multiple flows to form maria. Main eruptions occurred in interval of a few hundred million years, with minor mare eruptions considerably later. Tsiolkovsky mare lavas probably much younger, possibly Eratosthenian.	(1) Igneous nature and internal derivation of mare material shown by *Apollo 11, 12,* and *15* samples. (2) K-Ar and Rb-Sr data give concordant ages of crystallization and magma generation (3.2-3.7 b.y.) for basalts from M. Tranquillitatis, O. Procellarum, and P. Putredinus. (3) Variance in initial Sr^{87}/Sr^{86} ratios among mare basalts suggests several different sites of magma generation rather than one for each location. (4) Layering in maria (e.g., Hadley Rille) shows multiple flows; also inferred from crater morphology and sample analysis.
IV (3.2 to present) . . .	Concurrent events: (a) Local mare and highland vulcanism, forming Marius Hills, Flamsteed Ring, chain craters, sinuous rilles, Gambart-type craters, Sulpicius Gallus Formation, and other features. Minor gas eruptions from deep sources.	(1) Evidence for volcanic origin from photogeology and similarity to terrestrial volcanic landforms. (2) ALSEP instruments indicate low level of deep seismic activity and possible gas eruptions. (3) Evidence for relative age given on U.S. Geological Survey maps.
Present condition of Moon	Cold and rigid to depths of about 1000 km, but with pressure-temperature conditions for magma generation at greater depths. Occasional minor volcanism, largely gaseous, and accompanying seismic activity, indicated by lunar transient phenomena and ALSEP instruments. Periodic mild moonquakes triggered by Earth's attraction at perigee of Moon's orbit.	
	(b) Minor tension faulting, in and around maria and in highlands.	Photogeology: Straight Wall, "normal" rilles, Apennine Front.
	(c) Sporadic impact of meteoroids from asteroid belt and of cometary fragments, forming ray craters (including Eratosthenian) such as Copernicus; possibly followed by short-lived volcanic activity.	(1) Impact origin indicated by morphology, terrestrial analogues, and age of maria predicted from crater density and meteoroid flux. (2) Relative age indicated by superposition and albedo. (3) Postimpact volcanism indicated by variance in crater counts among flow units.
	(d) Continual slow landscape degradation and regolith formation by meteoritic and secondary fragment impact and ejecta deposition, thermal shock, and radiation. Seismic waves from impacts possibly effective.	(1) Impact origin of regolith indicated by petrography of soil and breccia samples. (2) Mass-wasting effects visible at all *Surveyor* and *Apollo* landing sites. (3) Cosmogenic isotopes indicate exposure times and hence erosion and turnover rates.

*Numbers in parentheses indicate time in AE (billions of years) before the present.
†Subevents are in chronological order unless otherwise noted.

data showing low near-surface velocities. The massifs around the *Apollo 17* sites appear to be entirely composed of ejecta breccias whose thicknesses are at least the height of the highest surrounding hills—more than 2.5 km for South Massif. Contributions from buried craters and from planetesimals could increase average thickness values for the entire highlands by several more kilometers.

Chronology of Lunar Events

Drawing upon many of the models just described for the origin and early history of the Moon, Lowman (1972) has combined the main results from *Apollo* with studies made by the U.S. Geological Survey's Astrogeology Branch and by others into a four-stage evolutionary sequence, from the Moon's birth to the present day, reproduced here as Table 13-3. Some of the events recorded in his summary are still speculative, but the broad pattern of growth and modification is consistent with the evidence cited. McCauley and Wilhelms, aided by an artist, have pictured the lunar surface at times equivalent to the end of stage II(c) (Figure 13-24a) and stage III (Figure 13-24b) in Lowman's chronology. Further modification of the Moon's front face since the beginning of stage IV can be deduced by comparison of the second figure with any full-moon photograph or map.

Ages of Other Planetary Surfaces

The surface of Mars can be age-dated in much the same way as the lunar surface. Unlike the Earth's present surface, much of the Martian surface remains pockmarked with craters formed early in its history. Variations in crater density on Mars are a direct result of a changing flux of infalling planetesimals over time. If the distribution of planetesimals around Mars were the same as around the Earth-Moon system since the formation of the planets, it would be possible to use any of the lunar radiometric age-calibrated flux curves discussed in the preceding two sections to make a fairly accurate estimate of absolute age of any cratered region of Mars. However, because Mars is closer to the asteroidal belt—a main source of many planetesimals—it is probable that the flux of large (meteoritic) particles was somewhat higher around that planet over most of geologic time. By making several plausible assumptions regarding the history of the flux in the vicinity of Mars, workers in the U.S. Geological Survey have derived a Martian flux curve (Figure 13-25) and have plotted it in relation to the age-calibrated lunar flux curve defined for a crater diameter interval between 4 and 10 km. On this basis, the oldest cratered surfaces on Mars were produced 3 to 3.5 billion years ago and are thus at least 500 to 700 million

Figure 13-24 (a) An artist's conception of the surface of the Moon just after the formation of the Imbrium Basin (upper left in drawing) and prior to filling of that and earlier-formed basins by mare lavas. (b) The lunar surface immediately after the major episodes of mare lava extrusion into the basins and lowlands on the frontside (producing the Procellarum Group). None of the Eratosthenian or Copernican craters have as yet been impressed on these lava units or the highlands (compare this view with the full moon photo of Figure 1-3). (Illustrations courtesy of J. McCauley and D. E. Wilhelms, U. S. Geological Survey.)

(a)

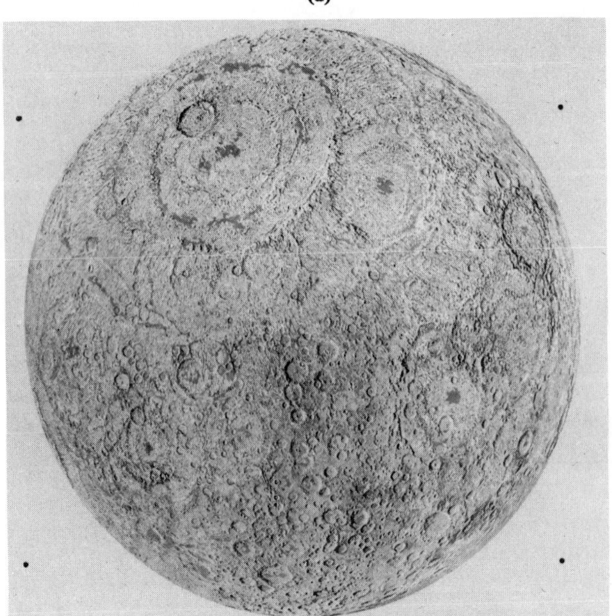

(b)

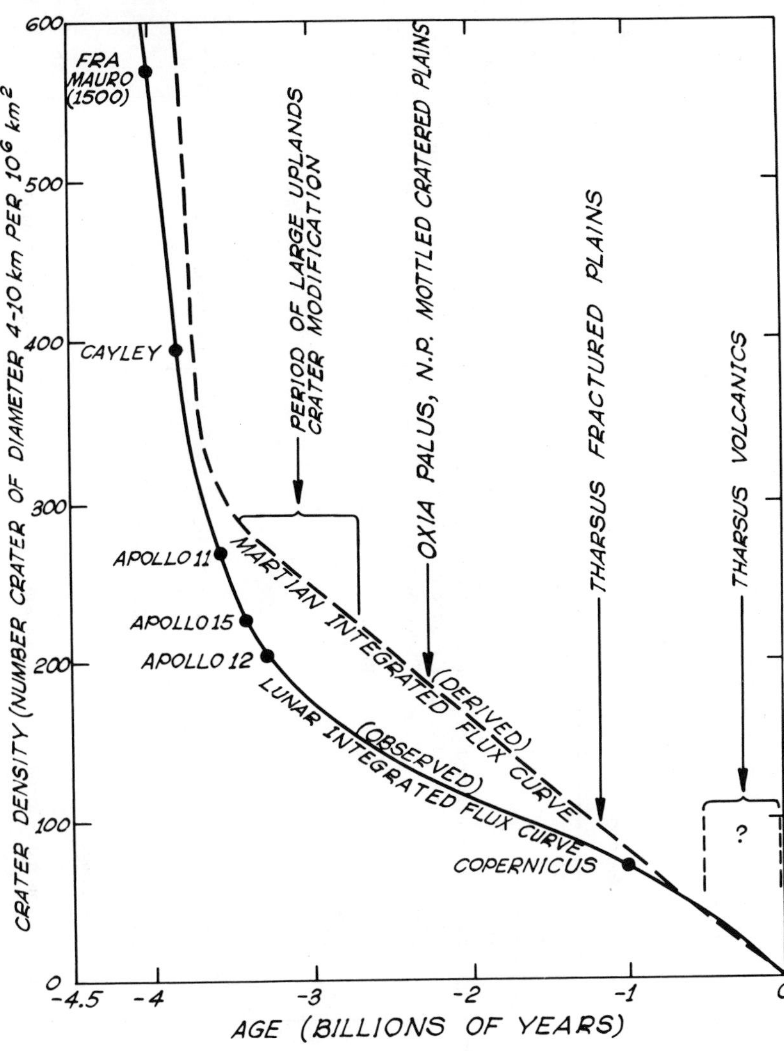

Figure 13-25 Plot of the observed lunar flux curve and derived Martian flux curve, based on crater counts for the size interval 4-10 km, as a function of the age of the solar system. Crater counts in the 10^6 km^2 regions that include the *Apollo 11, 12, 14, 15,* and *16* sites are plotted in terms of the characteristic age of representative rocks at each site to calibrate the steeper segment of the lunar curve. The Copernicus event is dated by exposure ages of *Apollo 12* rocks assumed to be ejecta from that crater. (Courtesy H. Masursky, U. S. Geological Survey.)

years younger than the lunar highlands. On the same time scale, fracturing of the Martian crust, giving rise to Coprates Canyon and other major fissures, would have begun about 1.2 billion years ago. This activity culminated in development of the great shield volcanoes from the last 500 million years to the present.

Estimates of ages of surfaces on Venus and Mercury must await determination of the frequency distributions of craters on each planet.

Volcanological Model of the Planets

Most models for the evolution of the Moon and the other inner planets such as Mars and Mercury assign an important to dominant role to impact during the early stages of present surface development. Without doubt volcanism has also greatly influenced the growth and modification of the surface of these planets as well as that of Venus. Evidence that volcanic activities persisted long after the major surface features were formed is clearcut for Earth and strong for Mars. Because of the still-unexplained lunar transients, the possibility of now-active volcanism on the Moon cannot be discounted, but the hope that evidence of this would be found in the orange soil at *Apollo 17* died with the dating of this soil at 3.71 AE.

There remains even today a small but vocal and convinced school of experts who contend that impact cratering is a rare and trivial process on planets in general. This group turns to volcanism as the prime factor in the early history of each inner planet. In their opinion all circular depressions—from large basins to small craterlets—have been formed exclusively by one or more volcanic processes. On the Moon, ringed basins like Imbrium and Orientale are huge volcano-tectonic complexes, and craters such as Copernicus, Tycho, and Alphonsus are calderas (Green, 1971); smaller craters are maars and gas vents. Shock damage in regolith materials is admittedly an indication of some impacting by interplanetary debris, but, it is argued, this is confined to micrometeorites and occasional larger particles.

The volcanological viewpoint has recently been effectively summarized by J. Green for the terrestrial planets

TABLE 13-4

Inner-Planet Evolution: A Volcanological Overview

Aeons Ago	Interval	Earth	Moon	Mercury	Venus	Mars	Aeons Ago
0.65 1.8	Phanerozoic Upper Proterozoic	▷ Tertiary volcanism ▷ Formation of modern ocean basins ▷ Carbonate buffer in sea METAZOA Change in S^{32}/S^{34} ratio ▷ End of "greenhouse" period Ozone screen; ***cold traps*** in atmosphere; water retained EUCARYOTA ▷ Red beds; oxidized sediments ▷ Storage of CO_2 as carbonates Oxygenous atmosphere; ozone present First drift and subduction (Grenville); geosynclinal activity; end of cratonization Ring structures	↑ Apogee-perigee-pulsed moonquakes; transients and volatiles, including water Postmare highland calderas (Aristarchus, Copernicus, Tycho, etc.) Continued buildup of radioactivity	Circular dark areas ↑ Carbon-enriched, silica-rich surface dust Extreme solar wind bombardments Quiescence	Observed ring structures Hydrated iron chlorides and other other halogens in atmosphere Extreme greenhouse effect Hydrothermal silica-clay alteration Photodissociation of H_2O No cold traps in atmosphere Buildup of CO_2 and loss of H_2O Evolution of calderas?	↑ Observated ring structures Water at poles Wind erosion Water erosion enhancing relief Lesser gravity and absence of cold traps in atmosphere promoting loss of H_2O and CO_2 ↑	0.65 1.8
2.6	Lower Proterozoic	▷ Local resetting of radiometric clock ▷ (Alberta) Oxygen-mediating enzymes First carbonates; no red beds Iron-dependent photosynthesis Global platforms; geosynclines Banded iron-ore formations	Massive loss of volatiles relative to Earth Continued crustal fracturing but no crustal plate spreading; crust thickens	No crustal plate spreading Loss of volatiles Evolution of calderas?	No storage of CO_2 as carbonates Gradual release of volatiles with heat-up ↑ Acid oceans	Evolution of calderas Volcanism and rifting	2.6
3.5	Archean	▷ Ring structures (Canada, Rhodesia, ▷ French Guiana) PROCARYOTA ▷ Anorthosites; hypersthene granulites; gray granodiorite; gneiss with graphite; first life in volcanic clays (?) Little free oxygen; no ozone	Local atmospheres Eruption of low-alkali basalts in maria; widespread volcanism Anorthosites by magmatic fractionation	↑ Volcanism, including lopoliths Maria and highlands Possible low-density atmosphere on dark side enriched in Ar, Rn, CO_2	Volcanism Radioactive heat buildup Gradual loss of angular momentum	Evolution of maria Capture of moons?	3.5
4.6	Selenian	▷ Resetting of radiometric clock Ring structures Nucleation of continents Mantle flexing; release of volatiles Widespread volcanism; calderas Metavolcanics; schists; gneisses Mantle purging of volatiles by close-by Moon of high eccentricity	Thermal events resetting radiometric clocks Volcano-tectonic circular mare basins Tidal braking of rotation Highlands calderas Crustal differentiation	Gradual solar braking to present 2/3 sidereal rotation	No major flexing or degassing or stripping away of volatiles	Formation of planetary rift pattern Early volcanism Enrichment of S in core and Fe in mantle	4.6

Source: Modified slightly from two (combined) versions prepared by J. Green, California State University at Long Beach.
Terrestrial tectonic periods of activity indicated by ▷.
Planetary volcanological periods indicated by ↑.

(Table 13-4). Note that the amount of information recorded for each planet is broadly in direct proportion to the extent to which it has been explored. Most of the categories of information entered into this tabular outline, including chronological sequence, are generally accepted by all planetologists, but the absence of effects related to impact cratering remains a point of considerable dispute.

Similarities and Differences in Planetary Evolution

When we compare the Moon to the Earth, we feel almost intuitively that these two planets were born from "common stock." Both have experienced general melting and extensive differentiation. The Moon and Mars also share certain major characteristics—mainly, by inference, primitive cratered surfaces. We are tempted to conclude that the Earth also underwent a similar cratering of its surface in its early life but that erosion, deposition, and numerous crustal meltings have erased the proof needed to sustain this assumption. Additional insight into the early history of planets may be revealed when the surfaces of Venus, Mercury, and Mars are examined first hand.

As a general hypothesis, then, we can postulate that all the inner or terrestrial planets originated in much the same way, and that thereafter they followed roughly the same plan of development for the next 1-2 billion years. But today there are sharp differences among these planets. Why?

Part of the answer depends on the extent to which a planetary interior melts and reorganizes into a layered body. However, its outer appearance results chiefly from the nature and history of the atmospheres developed from the outgasing of volatiles and carbonaceous compounds during the more intense igneous activities. The first products were probably H_2, CH_4, and NH_3. On the larger planets any free oxygen, if expelled in large enough quantities, would gradually oxidize the initially reducing atmosphere so that compounds such as H_2O, CO, CO_2, N_2O, and NO_2 have slowly replaced the primitive constituents. The small sizes of Mercury and the Moon allowed almost all the lighter gases to escape soon after release. On the Moon, most heavy gases now present at its surface are continually being introduced externally from the solar wind. Mars, however, is just large enough for its gravitational pull to retain some CO_2 and heavier gases, but lighter molecules such as H_2O, CH_4, NH_3, N_2, and NO have apparently diffused into outer space since the time of first melting (probably over 4 billion years ago). Some of these substances may now be locked up in combination with silicates or oxides (e.g., limonite) on the Martian surface. Sufficient atmosphere survives on Mars to have modified the surface by aeolian processes.

Venus, on the other hand, is large enough to retain all of these gaseous compounds. Presently, however, only CO_2 occurs in abundance—in quantities far greater than on any other planet. The apparent loss of nitrogen and the scarcity of H_2O, which occurs almost entirely in the upper atmospheric Venusian clouds, remain something of a mystery. The nearness of Venus to the Sun could have caused some loss of water and nitrogen compounds, leading thus to a relative buildup of carbon dioxide. When CO_2 reached a certain concentration, the runaway greenhouse effect led to notable temperature increases. These higher temperatures then drove off the other gases at greater rates, while the CO_2 atmosphere evolved to its present state. The elevated temperatures and pressures that result give rise to a hot surface in which rocks are approaching their melting points and topographic variations are minimized.

Temperature conditions on the early Earth must have been low enough to lead to a different sequence of changes. Essential to its evolutionary pattern was the retention of water vapor, which then condensed to form oceans. These oceans constitute a reservoir which can store CO_2 (dissolved or as precipitated carbonates) and soluble nitrogen compounds. Solar radiation returns a steady supply of water vapor to the atmosphere in a continuing cycle. At a critical stage, reactions between inorganic molecules of carbon and nitrogen compounds, water, etc., produced the first primitive single-celled life forms (anaerobic—capable of existing without oxygen). In time, the fundamental process of photosynthesis, in which CO_2 and H_2O react to make starches and oxygen, became possible, and life evolved to more complex aerobic (oxygen-using) forms. Molecular oxygen thus entered the atmosphere, and at high altitudes some of this gas (along with some water vapor) was converted to ozone (O_3^+) by ultraviolet radiation. The development of an ozone layer in the stratosphere provided a shield that reduced the amount of tissue-damaging UV radiation reaching the surface. Under these circumstances, plant life began to advance rapidly and oxygen gradually became a major constituent of air along with molecular nitrogen. The net effect of this trend in atmospheric and hydrospheric development is the complex, varied Earth on which we live.

Cloud (1968) has proposed a widely cited model (Figure 13-26) for the early history of the Earth and the evolution of its primitive crust (see caption for details). This model interrelates the rock types characteristic of each eon with the oxidation state of the atmosphere and hydrosphere. Of great significance are the major forms of living organisms found in the three aeons beginning about 3.5 billion years ago. As these evolved into more complex types, they gradually were able to utilize oxygen in their metabolic processes. Eventually, higher forms of plant life—and, later, animals—became the common suppliers of released oxygen, which led to buildup of that element in the changing atmosphere.

When we compare the Earth to the Moon, Mars, Venus, and Mercury, it is obvious that our planet is significantly

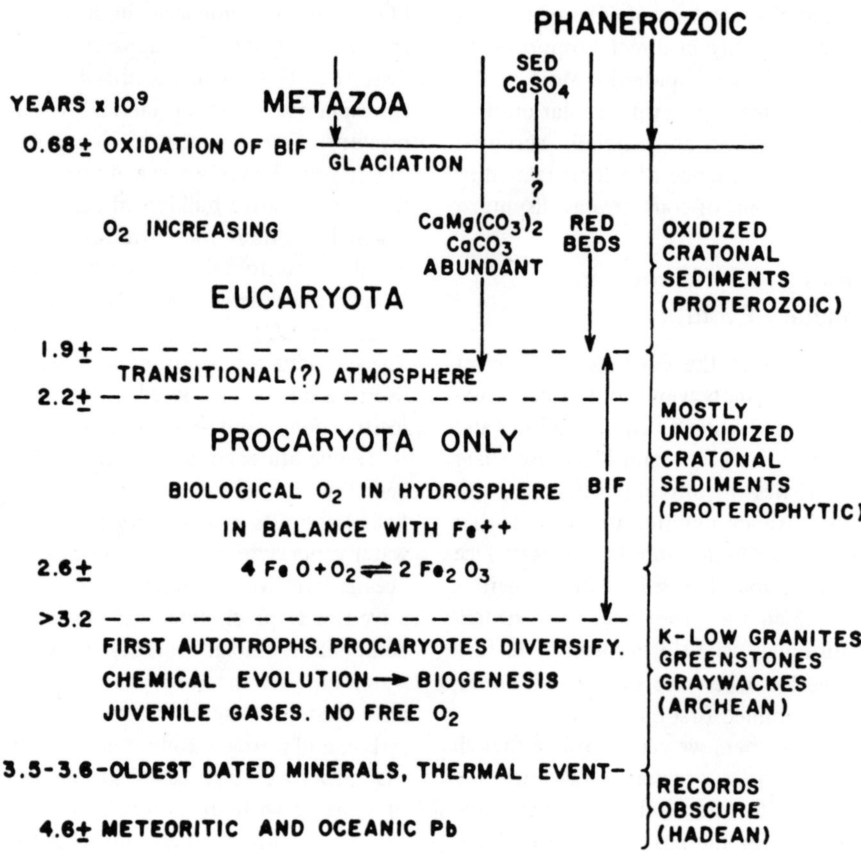

Figure 13-26 Summary diagram of the geochemical and biological evolution of the Earth according to the model of Cloud (1968, 1972). Times of major events appear on the left; characteristic types of sedimentary rocks are shown on the right (terms for the historical intervals, or "aeons" are given in parentheses). Biological terms are defined as follows: *autotrophs*, a simple organism capable of manufacturing its own food products by using sunlight or chemical energy to activate the process; *procaryotes*, unicellular microorganisms that build protonucleating and non-mitosing (non-splitting) cells; *eucaryotes*, unicellular microorganisms with mitosing (dividing) cells that utilize oxygen in their life processes; *metazoa*, multicellular organisms. The history developed in this diagram is one of increasing oxygen in the geochemical environment (reducing-to-oxidizing conditions) brought about, in part, by the metabolic processes of organisms that incorporated and released free oxygen to the atmosphere and hydrosphere. The first indication of the greater role of oxygen in the near-surface environment is oxidation of iron by organisms, giving rise to Banded Iron Formation (BIF). Red bed (hematitic) and oxygen-dependent single-celled organisms mark the period in which the atmosphere had become increasingly oxidizing. Build-up of an ozone layer in the upper atmosphere was instrumental in the introduction of the more complex metazoa at the onset of the Phanerozoic (just before the Cambrian).

different in two essential respects from all others composed mainly of rock materials. First, both today and seemingly throughout most of its past, the Earth has displayed a higher degree of internal energy being converted to active use than appears characteristic of the other inner planets. Second, the presence of water in and on the outer shell of Earth has played a dominant role in the inorganic evolution of our planet independent of its importance in promoting living organisms. Let us examine these two points in light of our new understanding of the similarities and differences among the several planets.

Role of Internal Energy in Planetary Development

All of the terrestrial planets can be considered to behave like heat engines. These drive both interior and surface processes, but each planet differs in its relative efficiency.

The Moon today shows few definite signs of any pronounced surface activities having operated extensively over intervals equivalent to the last 1 billion to 2 billion years of geologic time on Earth. The *Apollo* results suggest that the Moon heated up early in solar system time and then cooled over a period of 1.5 billion years to a state in which its surface has since remained passive to all but external influences. As yet it is not even clear whether this heating was general, leading to total planet melting and differentiation followed by overall solidification, or was confined mainly to the outer regions with formation of several shells, or was even more localized in that melting was restricted to pockets of magma near the surface which eventually extruded to form the maria. Mars seems to have advanced somewhat further. Differentiation of primitive basic materials may have proceeded far enough to yield a crust containing 50-60% silica, as evidenced by infrared spectra data from *Mariner 9*. The present atmosphere of Mars still affects its surface, but lack of evidence for any on-going volcanism (despite the "young" shield volcanoes) and absence of ocean-sized fluid bodies indicates that its crust is relatively simple and inactive compared with Earth. Even so, the extensive equatorial fracture systems girdling Mars suggest at least a stage of incipient crustal fragmentation. Venus, with its high surface temperatures and turbulent atmosphere, probably is expending its internal energy in such dynamic processes as magmatism and volcanism and, possibly, is forming a differentiated metallic core. However, until more is known about its surface and interior, a comparison with the Earth's energy system cannot be made with any accuracy. For all of these planets, including Earth, the sources of the internal energy that has supplied the driving forces for any observed activities are primarily gravitational and radiogenic, although heat derived from the work of impact during accretion may have made a major contribution. In addition, on Earth, solar radiation exerts a controlling influence on atmospheric and some surficial processes; this influence on the Venusian and Martian atmospheres appears much less.

The great diversity of surface features and conditions on the Earth attest to long-continuing activities that depend on a steady supply of kinetic energy from both an efficient internal heat engine and an external source (the Sun). The internal structure of the Earth itself indicates that much or perhaps all of the planet was once molten long enough for a metallic core to segregate from overlying mantle silicates. Zone melting in the outer mantle led to extensive differentiation, culminating in a primitive crust of more silicic composition. Occurrence on all continents of numerous plutonic bodies with intrusion ages from early Precambrian to late Cenozoic confirms that the outer crust has been subjected to repeated cycles of heat accumulation sufficient to require regional melting. Evidence for mobilization of mantle material over much of geologic time is being obtained directly from radiometric dating of rocks exposed in the roots of geosynclines or brought up from great depths in certain volcanic structures. Both the continental volcanism that has persisted to the present day and the recent discoveries that vast amounts of lava have been added to the ocean basins in the last few hundred million years demonstrate the extent to which considerable internal energy has been released since the end of the Paleozoic.

Influence of Water on the Earth's History

The expulsion of water and other volatile constituents from the Earth's interior and their accumulation in the outer shell—giving rise, as we have seen, to spatial-physical entities conceptualized for convenience as an oxidized-hydrated upper lithosphere, a hydrosphere, an atmosphere, and a biosphere—are the result of gradual degasing of rocks undergoing continual heating since the beginnings of the planet. Retention of these gases, whose initial mobilization was fostered by the general melting of the interior, depended on the critical size of Earth. Thus, its gravitational attraction has selectively held those vapors whose molecular weights prevented their escape by thermal agitation from solar or internal heating. In the first billion or so years of Earth history, it is likely that much of the gases released to a primitive, dense atmosphere were eventually carried off to outer space by convection and diffusion activated by high surface temperatures and by "blowing off" in response to strong solar winds. In time, with cooling in the Earth and decreasing action from the Sun, a permanent buildup of oceans and a more stable atmosphere led to some kind of marine environment as far back as 3.2 billion years ago (age of the oldest dated sediments). Accumulation of these materials in protocontinents was initiated as a result of such processes as chemical and physical weathering, erosion, and redistribution of the weathered products by means of fluid movements (water and atmosphere). This has been followed by a steady conversion of the differentiated sediments through metamorphism, granitization, and anatexis (fusion) into the sialic (granitic) masses making up most of the continents between the surface veneer of sedimentary rocks and soils and the deep-seated basic (mostly gabbros, basalts, and peridotites) crustal and mantle layers.

Most of the Earth's mantle rocks are presumed to be anhydrous owing to initially low water contents coupled with losses incurred during degasing. These rocks are probably also deficient in alkalis (sodium, potassium) which would also be transferred outward in the vapor state or within hot aqueous solutions. However, much of the water and alkalis that migrated from the interior have been locked up in near-surface rocks as essential constituents. The presence of water in the crust is the principal reason for development of such diverse rock types as granites and rhyolites, schists and gneisses, serpentines, shales and carbonates, and salt beds, along with the superficial soils comprising a vital link between lithosphere and biosphere. Water serves three functions in this context; (1) aiding in the fusion of deeply buried rocks (by lowering the melting points and helping to transfer ions among mineral phases); (2) entering into the composition of many minerals (micas, clay minerals, hydrous oxides); and (3) acting as the key agent in physically and chemically separating the elements released from primitive mantle rocks and reworked crust in almost every dynamic process in geology from magmatic emplacement, formation of ore deposits, development of soil profiles, and weathering-transportation-deposition of sediments to sculpturing of highly diversified landscapes and sea floor topographies.

The Earth's Dynamic Character

The many geosynclines found within the continental masses as well as along their margins offer further proof that the Earth has experienced a progression of extended events of such dynamic magnitudes as to continually reshape its exterior. Remnants of geosynclinal deposits are widely distributed on the continents; their times of development date from the early Precambrian to the Recent. (Australia, for example, contains exposures of deposits correlated with more than twenty distinct geosynclines developed prior to the Mesozoic.) These geosynclines owe their very existence to the close interaction between various processes by which internal energy is exchanged from the interior to the Earth's surface and to the presence of huge volumes of water on the crust and in rocks beneath.

The net effect of the influence of a thermally active Earth moderated by the concentration of water and free oxygen in the outer spheres is an *imbalance* in the distribution of mass and energy in the upper mantle and crust. This has brought about a segregation of less dense alkali- and silica-rich materials into continental blocks whose past and present locations were apparently never symmetric on the global sphere. The imbalance is reflected in other subtle or overt ways, all of which describe a heterogeneity of composition, density, spatial distribution, and age of materials in the complex region of Earth accessible to direct observation. A geologic map of any continent is marked by surface exposures of a wide variety of rock types representing many stratigraphic ages and arranged in attitudes from near-horizontal (undisturbed) to intricately folded, thoroughly fractured, interleaved by faulting, or intimately mixed through multiple intrusions and plastic deformations. Overall, these surface exposures tend to assume linear trends on a regional scale. Geosynclines, mountain systems, volcanic chains, mineralized belts, major fracture-fault zones, and even outcrop tracings of flat-lying sedimentary rocks on stable cratons are all characterized by elliptical-to-much-elongated (straight-to-curvilinear) patterns of spatial distribution. This complexity persists beneath the surface and only gradually with depth gives way to a more uniform compositional and geometric sequence of thick layers in the lower crust and mantle. That this imbalance is continuing and, in fact, is increasing the heterogeneity of the outer shells is clearly indicated since recorded historical times by the worldwide recurrence of earthquakes, volcanic eruptions, differential erosion of the land, nonuniform submergence or emergence of the continental shelves, and periodic modification of climates (some with drastic effects such as glacial advances or conversion of once-vegetated regions to deserts). We are only now beginning to connect these long-term changes in the Earth's natural imbalance with factors that affect ecological conditions within the biospheric environment.

Early History of the Earth

For decades some scientists speculating on the Earth's early history had generally assumed that an original crust of sialic or granitic rocks, differentiated directly by crystal flotation and convection from basic mantle rocks, existed as a single continuous layer around the globe. Others considered the differentiated sialic crust to be discontinuous in extent and irregular in location because of a nonsymmetrical distribution of convection megacells within a mantle once more or less completely molten. Still others believed that the sialic continents owe their origin and positioning to the chance redistribution of the first sediments, enriched in silica and alkalis, in lower parts of a primitive crustal surface and their subsequent metamorphism into sialic protocontinents. The first hypothesis requires loss of much of a primordial crust, while the second and third presume that continental masses have grown over time from primitive nuclei (represented, for example, by shield areas containing very old rocks).

Inasmuch as rocks older than about 3.9 billion years have not yet been found on Earth, the question of whether there ever existed a primordial sialic crust formed during the first billion years of Earth history still remains unanswered. Such a crust may have been completely destroyed by the time that oceanic waters began to cover the Earth, or it may have been completely changed by erosion and by melting and granitization of the resulting sediments, or its remnants may still be preserved deep within the crust. However, with the general acceptance in recent years of continental drift, plate tectonics, and formation of new oceanic crust as real processes in the evolution of the lithosphere, some further insight into the nature and history of the continents is at last in focus.

One broad view of this is as follows: The continents began to form sometime in the first billion years following accretion of the Earth and its subsequent near-complete melting. Most likely, only two or three great masses of sialic differentiates reached the surface by convection. These protocontinents continued to grow as sediments accumulating along continental margins and within interior geosynclines were converted by granitization into igneous-metamorphic complexes. Additional magmatic differentiates from the mantle were also injected into the continental bases. Some invading magmas penetrated upward to regions near the surface, and a fraction of these actually poured out as lava flows. Ultimately, these supercontinents grew to critical sizes, began to break up, and have been drifting across the globe as a set of great plates (approximately six major plates) of crust and upper mantle that move over a plastic region (asthenosphere) within the upper mantle. The violent ruptures involved have allowed large volumes of basic lavas (mostly basalts) to pour onto the surfaces along thermal rises (e.g., the Mid-Atlantic Ridge). These outflows produce new oceanic layers and the resulting sea floor spreading helps to propel the plates along their drift courses. Everywhere on Earth, the basaltic crust beyond continental margins consists of layers of post-Paleozoic lavas veneered by sediments whose ages are not older than Cretaceous. The present ocean basins are thus relatively recent and are continually changing in dimensions. The continents, on the other hand, are being consumed in some place along subduction zones (great fault zones and trenches into which crustal material is dragged down) while being added to elsewhere along active geosynclines.

In the above concept, we must envision that some kind of primitive crust older than the sialic materials of the continents has been progressively encroached upon by growing and drifting continents. Such a crust apparently now is completely covered by the younger lavas and sediments. This old crust, if composed of denser basic rocks, probably occupied the lower parts of the Earth's exterior and was covered by ancient oceans. After the great spasms that accompanied the inception of drifting, some parts of this now-lost noncontinental crust have foundered or were ingested by lavas, while other parts remain below the upper lava outpourings.

A tantalizing variation of this general view postulates the entire Earth to have once been surrounded by a thin sialic crust. Over time this crust broke up as heatings of the mantle produced lavas that reached the surface in copious amounts. Great segments of the crust were periodically inundated and were carried downward by foundering and dragging into subduction zones. These crustal materials were then remixed to form basalts and andesites that now comprise a continuous zone around the Earth. The last survivors of the crustal destruction were Pangea, Gondwanaland, and Laurasia—the supercontinents that are now fragmenting and moving into subduction zones. In this view, today's continents are in the process of gradually being destroyed by the process of *oceanization*. A crustal evolution model for the Earth combining aspects of this increasingly popular concept with new knowledge of the Moon, Mars, and Venus has recently been developed by P. D. Lowman (Table 13-5).

Earth and Moon Compared

This résumé of the Earth's present physical state and earlier history contrasts sharply with the picture of the Moon now available from space exploration. The Moon underwent periods of general melting in its first billion years. The Earth probably also was subjected to at least one, perhaps several, periods of melting on the same time interval, but evidence for this in terrestrial rocks is obscure. It is interesting—although perhaps coincidental—that the youngest Moon rocks dated so far are about the same ages as the oldest igneous (3.6-AE) and sedimentary (3.2-AE) rocks exposed on Earth. The Moon's span of surface activities seems to have ended just when the Earth started to experience a great diversification at its exterior. From the surface evidence, in a sense the Moon's heat engine was throttled back almost contemporaneous with the acceleration of the engine that drove the Earth's dynamic growth.

The Moon almost surely formed a continuous crust at its surface soon after birth. Most of this crust seemingly survives under the lunar highlands but is now covered by impact rubble. The lowlands containing the mare lavas probably owe their origin largely to impacts that carved deep excavations. In turn, vast quantities of lava were released to replace the crust in those areas. As far as we know today, these lava-emplacement episodes and the activities responsible for producing the original crust are the only two definite magmatic events recorded on the Moon. It is tempting to speculate on an alternative explanation for the maria; in this idea, the basins occupied by these lavas are the first stages of crustal foundering and responsive outpouring analogous to the aforementioned variant for the origin of terrestrial ocean basins.

Compared to rocks on Earth, lunar rocks are relatively simple in lithologic variety. The suites of samples returned from *Apollo* missions all fall within the range of differentiates associated with melting of rocks similar to ultrabasic mantle materials in the Earth. Although lunar peridotites, norites, gabbros, and ferrobasalts are distinguishable petrographically, they still represent a narrow group of interrelated rock types when contrasted with the many different lithologies in the Earth's outer regions. Had water remained in lunar rocks (if ever there) and had the alkalis been present in higher abundance, a much greater diversity of rock types would have developed in the Moon's outer crust.

This very absence of water in the observable regions of the Moon, together with its lower energy state over most of lunar time, account for nearly all the striking differ-

TABLE 13-5

Crustal Evolution of the Earth

Stage	*Event*	*Mechanism/Cause/Source*
I (4.6 b.y. ago)	(a) Formation of earth	(a) Condensation and accretion from solar nebula, possibly leading to initial chemical stratification by zoned accumulation.
	(b) Rapid heating and initial reduction of silicates to metallic iron	(b) Heat sources include energy of accretion, solar radiation (incl. particulate), short-lived isotopes, adiabatic compression, possibly electromagnetic resistive heating.
	(c) Formation of Moon	(c) Mechanism unknown; possibilities include precipitation, fission, and capture. Process may have raised temperature of earth still further.
II (4.5 to 3.5 b.y. before present)	(a) First differentiation, forming global crust, complementary to segregation of iron core; initial degasing to form atmosphere of CO_2, N_2, etc.	(a) Crust formed by repeated eruption and intrusion of intermediate composition magmas, high in Al. Core formation raised temperature of earth by several hundred degrees or more. Process probably overlapped by later stages of accretion, i.e., intense but declining infall of planetesimals.
	(b) Infall of several large proto-planetary bodies or fission fragments, disrupting global crust	(b) Bodies formed simultaneously near earth from solar nebula, or possibly by fission of earth to form moon.
	(c) Initial oceanization and beginning of second differentiation by generation of basaltic magma	(c) Large impact basins localized eruption of basaltic magmas, by deep fracturing, possibly addition of heat to earth, and relief of pressure under impact basins. Basaltic eruptions possibly assymetric concentrated in Pacific hemisphere.
III (3.5 b.y. to 600 m.y. before present)	(a) Initial tectonic oceanization, by fragmentation and subsidence of crust along global fracture systems	(a) Tension faulting, caused by mantle processes similar to those operating at present to form rift valleys and mid-ocean ridge system.
	(b) Sea-floor spreading and subduction of oceanic and some continental crust, leading to decrease in area but increase in thickness of continental crust.	(b) Sea-floor spreading probably result of some form of mantle convection and diapiric intrusion along oceanic ridges. Decrease in area of continental crust due to direct or indirect erosion and underthrusting of sialic crust at continental margins.
IV (600 m.y. ago to present; Phanerozoic Era)	Continuation of processes of Stage III, but at a slower rate, perhaps decreasing. Some secular addition of material to continents. Net result is enlargement of ocean basins.	Secular addition of material to continents by partial melting of subducted oceanic crust, continental vulcanism, minor emplacement of mantle-derived acidic magmas, and subordinate local accretion of eugeosynclinal sediments and volcanics.

From P. D. Lowman, Jr., 1974, "Evolution of the Earth's Crust: Evidence from Comparative Planetology," *J. Geol.*, in press..

ences between the Earth and its satellite. Thus, we see no geosynclinal sedimentary basins or folded mountains on the lunar surface. Linear patterns associated with subduction zones, rifts, thrusts, fault blocks, metamorphic terrains, and stream valleys are lacking on the Moon. "Rings of Fire" defined by alignments of volcanoes are almost surely absent on the Moon—although some lunar transient events seem to concentrate along the edges of mare basalts. Instead, the linear trends so characteristic of Earth are replaced by circular patterns of all sizes without much evident orientation.

Even though volcanism once played a part in shaping the Moon's surface, most of its current state is attributed to the effects of bombardment by extralunar materials. For all practical purposes, this shaping process was completed long before the Earth's present surface began to take form. The Moon can be considered as "essentially dead"—a conclusion confirmed by its very low seismicity and by absence of rocks younger than about 3 billion years. Its size, its lower energy content, and its deficiency in water (or inability to hold on to it) have led to the Moon's demise as a dynamic member of our solar system.

Figure 13-27 View of the nearly fully illuminated Earth as seen from *Apollo 17* enroute to the Earth after its trip to the Moon. The scene centers on much of the southern hemisphere and shows part of the Antarctic, nearly all of Africa and Asia Minor, and the Indian Ocean and western South Pacific. (NASA photo.)

Concluding Observations

Why do we study the Moon or Mars, or the other planets in our system? George Leigh-Mallory's famed reply to the question of why he tried to climb Mount Everest—"Because it is there"—is almost self-justifying when applied to man's climb to the Moon. But more practical reasons can be cited. One overrides all others: by reconstructing the composition, structure, and sequence of changes for the early Moon, we can perhaps also decipher the general nature and pattern of events of our early Earth provided both planets had a similar "childhood." The useful consequences of this reconstruction are many For example, the various ways in which ore deposits have been generated and dispersed in the Earth are dependent in large part on the initial states of the crust and mantle before widespread differentiation. Again, the beginnings of life on our planet were certainly influenced by events in the first 1 billion to 2 billion years, when we may assume that meteorite impacts were more common, terrestrial lithologies were less varied, igneous processes dominated, and the atmosphere, if any, was reducing. The sterile Moon of today reproduces most of these conditions.

In retrospect, as we contemplate the panorama of planets in our solar system, we can marvel at the almost providential fact that each planet seems to provide another set of essential clues for unraveling the mystery of creation. Every planet has proceeded along some peculiar path of evolution all its own—yet each still defines a certain stage in a general process. Thus, Jupiter represents a relic of what may be the first stage: accumulation of a dense, reducing atmosphere of hydrogen and helium around a small (silicate?) nucleus. Uranus and Neptune describe a more advanced state in which ammonia and methane are mixed with the hydrogen. Saturn displays orbiting rings similar to those from which the Moon and other satellites could have condensed by accretion. Venus and Earth are surrounded by highly evolved atmospheres that severely affect their rocky surfaces. Mars reveals the condition attained by a planet after it has lost most of its gaseous envelope. The Moon and Mercury disclose the fate of planetary surfaces left unprotected by any atmosphere. Finally, a group of small planets have conveniently come apart to expose (as meteorites) the kinds of interiors that can develop under varying conditions of heating. To be sure, some of these observed differences among the planets depend on their ultimate positions relative to the Sun, or on their final dimensions, or on physicochemical fluctuations in the nebula at the times of their formation. However, taken together, the planets make up an elegant picture of what we now believe is a typical system of condensed bodies around a central star.

As we move on to new adventures in planetary exploration, we can rightfully ask ourselves what is being accomplished for mankind. The *Apollo 17* view of the Earth (Figure 13-27) suggests a cogent answer. By releasing ourselves from the confines of our planet, we are led to a reawakened appreciation of *"Spaceship Earth"* and its unique sets of ecological conditions that favor our very existence. Thus, the studies of the planets and stars from outer space itself will inevitably turn back this new knowledge of the universe to us and our own world. And, at the same time, by extending our insatiable desire to fathom the unknown to the surrounding solar system, someday all men may be brought together through this common awareness of their inseparable bond to the Cosmos.

REVIEW QUESTIONS

1. State the principal steps in reaching truth by the scientific method; discuss, by giving an example, how it has been applied to lunar studies.

2. Develop, in broad terms, a summary of prevailing ideas on the origin of the solar system.

3. What is a T Tauri event? Mention several problems in planetary development which might be explained by this process?

4. How might chondritic materials be formed and subsequently melted?

5. List the boundary conditions appropriate to the solution of the Moon's origin.

6. Discuss the main aspects of either the fission or the sediment-ring hypothesis for the formation of the moon.

7. Offer your own answers to the five questions raised on p. 304.

8. Compare—or contrast, if appropriate—the models espoused by Ringwood and Essene and by Gast and McConnell for the Moon's evolution.

9. Attempt to extract certain common themes (ideas or conclusions for which there is some general agreement) from the various petrologic models of the Moon.

10. Review the information about the composition and state of the lunar interior provided by (a) seismic, (b) electrical conductivity, (c) gravitational, and (d) radiometric studies.

11. How have Hartmann and Wood and also Shoemaker utilized cratering-rate data to reconstruct a history of the lunar surface?

12. How might differences in atmospheric evolution have operated to bring about major differences among the terrestrial planets?

13. Summarize, in a general way, the roles and effects of heat and of water in developing the Earth's crust.

14. Compare the present-day state of the surfaces on Earth, Mars, and the Moon.

15. List at least one unique feature associated with each planetary body in the solar system that has contributed to our understanding of the origin of planets.

References

Note: NASA publications can be obtained from the Superintendent of Documents, U.S. Government Printing Office, Washington, D.C. 20402.

*Entries preceded by a star refer to text and source books, summary reports (NASA and others), and special issues of periodicals, which together serve as a general reading list.

ADAMS. J. B. and T. B. McCord, 1971*a*. Alteration of Lunar Optical Properties: Age and Composition Effects. *Science,* **171**, 567-571.

---, 1971*b*, Optical Properties of Mineral Separates, Glass, and Anorthositic Fragments from Apollo Mare Samples. In Levinson (1971), v. 3, pp. 2183-2195.

*ADLER, I. and J. I. Trombka, 1970, *Geochemical Exploration of the Moon and Planets,* Springer-Verlag, New York, 243 pp.

ADLER, I., J. I. Tromka, P. D. Lowman, Jr., R. Schmadebeck, H. Blodget, E. Eller, L. Yin, R. Lamothe, G. Osswald, J. Gerard, P. Gorenstein, B. Bjorkholm, H. Gursky, B. Harris, J. Arnold, A. Metzger, and R. Reidy, 1973, Apollo 15 and 16 Results of the Integrated Geochemical Experiment. *The Moon,* **7**, nos. 3/4, 487-504.

ALFVEN, H., 1963, The Early History of the Moon and the Earth. *Icarus,* **4**, 357-363.

ALFVEN, H. and G. Arrhenius, 1972, Origin and Evolution of the Earth-Moon System. *The Moon,* **5**, 210-230.

ALTER, D. (ed), 1968, *Lunar Atlas,* Dover, New York, 351 pp.

ANDERSON, D. L., 1973, The Formation of the Moon. In Chamberlin and Watkins (1973), pp. 40-43.

ANDERS, E., 1963, Meteorites and the Early History of the Solar System. In Jastrow and Cameron (1963), pp. 95-142.

---, 1970, Water on the Moon. *Science,* **169**, 1309-1310.

BAEDECKER, P. A., C. L. Chou, E. B. Grudewicz, and J. L. Wasson, 1973, The Flux of Extralunar Materials onto the Lunar Surface as a Function of Time. In Chamberlin and Watkins (1973), 45-47.

*BALDWIN, R. B., 1963, *The Measure of the Moon,* University of Chicago Press, Chicago, 488 pp.

---, 1964, Lunar Crater Counts. *Astronom. Jour.,* **69**, 377-392.

*---, 1965, *A Fundamental Survey of the Moon,* McGraw-Hill, New York, 149 pp.

---, 1969, Absolute Ages of the Lunar Maria and Large Craters. *Icarus,* **11**, 320-331.

---, 1970, Summary of Arguments for a Hot Moon. *Science,* **170**, 1297-1300

——, 1971*a*, The Question of Isostasy on the Moon. *Phys. Earth Planet. Interiors,* **4**, 167–179.

——, 1971*b*, On the History of Lunar Impact Cratering: The Absolute Time Scale and the Origin of Planetesimals. *Icarus,* **14**, 36–52.

*BARBOUR, J. (ed.), 1969, *Footprints on the Moon,* by the writers and editors of the Associated Press; published by the Associated Press, New York.

BARNES, V. E., 1961, Tektites, *Scientific Amer.,* **205**, no. 5, 58–65.

BEALS, C. S., M. J. S. Innes, and J. A. Rothenberg, 1963, Fossil Meteorite Craters. In Middlehurst and Kuiper (1963), v. 4, 235–284.

*BEDINI, S. M., W. Von Braun, and F. L. Whipple (edited by D. Thomas), 1970, *The Moon: Man's Greatest Adventure,* Harry N. Abrams, New York, 267 pp.

BEST, J. B., 1972, Apollo 15 Lunar Samples. In Watkins (1972), pp. 33–39.

BINDER, A. B., 1969, Internal Structure of Mars. *J. Geophys. Res.,* **74**, 3110–3118.

BINDER, A. B., D. P. Cruickshank, and W. K. Hartmann, 1965, Observations of the Moon and of Terrestrial Rocks in the Infrared. *Icarus,* **4**, 415–420.

BIRKEBAK, R. and J. Dawson, 1973, Apollo 15 and 16 Directional Spectral Reflectance. In Chamberlin and Watkins (1973), 75–77.

BLANDER, M. and J. L. Katz, 1967, Condensation of Primordial Dust. *Geochim. Cosmochim. Acta.,* **31**, 1025–1034.

*BOWKER, D. E. and J. K. Hughes, 1972, *Lunar Orbiter Photographic Atlas of the Moon,* NASA Special Publ. SP-206, 724 pp.

BROCK, B. B., 1965, Lunar and Terrestrial Structural Models on a Common Sphere. In H. E. Whipple (1965), pp. 617–630.

BROWN, G. M., 1970, Petrology, Mineralogy, and Genesis of Lunar Crystalline Rocks. *J. Geophys. Res.,* **75**, 6480–6496.

BROWN, G. M., C. H. Emelius, J. G. Holland, A. Peckett, and R. Philips, 1972, Mineral–Chemical Variations in Apollo 14 and 15 Basalts and Granitic Fractions. In King et al. (1973), v. 1, pp. 141–157.

BROWN, H., 1964, Planetary Systems Associated with Main Sequence Stars. *Science,* **145**, 1177–1181.

BURNETT, D. S., W. A. Fowler, and F. Hoyle, 1965, Nucleosynthesis in the Early History of the Solar System. *Geochim. Cosmochim. Acta.,* **29**, 1209–1241.

CAMERON, A. G. W., 1962, The Formation of the Sun and Planets. *Icarus,* **1**, 13–69.

——, 1970, Formation of the Earth-Moon System. *EOS, Amer. Geophys. Union Trans.,* **51**, 628–633.

——, 1973, Properties of the Solar Nebula and the Origin of the Moon. *The Moon,* **7**, no. 3/4, 377–383.

CAVARETTA, G., R. Funiciello, H. Giles, G. D. Nicholls, A. Taddeucci, and J. Zussman, 1972, Geochemistry of Green Glass Spheres from Apollo 15 Samples. In Watkins (1972), pp. 202–205.

*CHAMBERLAIN, J. W. (ed.), 1972, *Post-Apollo Lunar Science,* Lunar Science Institute, Houston, 104 pp.

*CHAMBERLAIN, J. W. and C. Watkins (ed.), 1973, *Fourth Lunar Science Conference* (abstracts), Lunar Science Institute, Houston, 848 pp.

CHAO, E. C. T., 1967, Shock Effects in Certain Rock-forming Minerals. *Science,* **156**, 192–202.

CHAO, E. C. T., J. B. Best, and J. A. Minkin, 1972, Apollo 14 Glasses of Impact Origin and their Parent Rock Types. In King et al. (1972), v. 1, pp. 907–926.

CHAO, E. C. T., J. A. Boreman, J. A. Minkin, O. B. James, and G. A. Desborough, 1970, Lunar Glasses of Impact Origin: Physical and Chemical Characteristics and Geologic Implications. *J. Geophys. Res.,* **75**, 7445–7479.

CHAO, E. C. T., L. A. Soderblom, J. M. Boyce, D. E. Wilhelms, and C. A. Hodges, 1973, Lunar Light Plains Deposits (Cayley Formation)–a Reinterpretation of Origin. In Chamberlin and Watkins (1973), pp. 127–128.

CHAPMAN, C. R. and R. Haefner, 1967, A Critique of Methods for the Analysis of the Diameter–Frequency Relation for Craters with Special Applications to the Moon. *J. Geophys. Res.,* **72**, 549–558.

CHENOWITH, P. A., 1962, Comparison of the Ocean Floor with the Lunar Surface. *Bull. Geol. Soc. Amer.,* **73**, 199–210.

*CHERRINGTON, E. H., 1969, *Exploring the Moon with Binoculars,* McGraw–Hill, New York, 211 pp.

CHRISTIAN, R. P., C. S. Annell, M. K. Carron, F. Cuttitta, E. J. Dwornik, D. T. Ligon, Jr., and H. J. Rose, 1972, Chemical Composition of some Apollo 15 Igneous Rocks. In Watkins (1972), pp. 206–209.

CLIFF, R. A., C. Lee-Hu, and G. W. Wetherill, 1972, K, Rb, and Sr Measurements in Apollo 14 and 15 Material. In Watkins (1972), pp. 131–133.

CLOUD, P. E., JR., 1968, Atmospheric and Hydrospheric Evolution of the Primitive North. *Science,* **160**, 729–735.

——, 1972, A Working Model of the Primitive Earth. *Amer. J. Science,* **272**, 537–548.

COLEMAN, P. J., JR., B. R. Lichtenstein, C. T. Russell, L. R. Sharp, and G. Schubert, 1972, Magnetic Fields near the Moon. In King et al. (1972), v. 3, pp. 2271–2286.

COMPSTON, W., B. W. Chappell, P. A. Arriens, and M. J. Vernon, 1970, The Chemistry and Age of Apollo 11 Lunar Material. In Levinson (1970), v. 2, pp. 1007–1027.

*COOK, J. J., 1967, A Survey of Lunar Geology, *NASA Earth Resources Survey Program Tech. Letter,* NASA-95. 43 pp.

*COOPER, H. S. F., 1970*a*, *Apollo on the Moon,* Dial Press, New York, 144 pp.

*——, 1970*b*, *Moon Rocks,* Dial Press, New York, 197 pp.

*CORTWRIGHT, E. M., 1968, *Exploring Space with a Camera,* NASA Special Publication SP-168, 214 pp.

CROZAZ, G., R. Drozd, C. M. Hohenberg, H. P. Hoyt, Jr., D. Ragan, R. M. Walker, and D. Yuhas, 1972, Solar Flare and Galactic Cosmic Ray Studies of Apollo 14 and 15 Samples. In King et al. (1972), v. 3, pp. 2917–2931.

DALY, R. A., 1946, Origin of the Moon and Its Topography. *Proc. Amer. Philosophical Society,* **90**, no. 2, 104–119.

*DAVIES, M. E. and B. C. Murray, 1971, *The View from Space: Photographic Exploration of the Planets,* Columbia University Press, New York, 163 pp.

DENCE, M. R., 1968, Shock Zoning at Canadian Craters: Petrography and Structural Implications. In French and Short (1968), pp. 169–184.

DENCE, M. R., J. A. V. Douglas, A. G. Plant, and R. J. Traill, 1970, Petrology, Mineralogy, and Deformation of Apollo 11 Samples. In Levinson (1970), pp. 315–340.

DESSLER, A. J., 1967, Solar Wind and Interplanetary Magnetic Field. *Rev. Geophys.,* **5**, 1–47.

DEWEY, J., 1972, Plate Tectonics. *Scientific Amer.,* **226**, no. 5, 56–68.

DIETZ, R. S., 1949, The Meteoritic Impact Origin of the Moon's Surface Features. *J. Geol.* **54**, 359–375.

——, 1961, Astroblemes. *Scientific Amer.,* **205**, no. 2, 50–58.

DIETZ, R. S. and J. C. Holden, 1965, Earth and Moon: Tectonically Contrasting Realms. In H. E. Whipple (1965), pp. 631–640.

DOBAR, W. L., 1965, Behaviour of Lava on the Lunar Surface. In H. E. Whipple (1965), pp. 495–515.

DODD, R. S., JR., J. W. Salisbury, and V. G. Smalley, 1963, Crater Frequency and the Interpretation of Lunar History. *Icarus,* **2**, 466–480.

DOLLFUS, A., 1962, The Polarization of Moonlight. In Kopal (1962), pp. 131–160.

DONN, W. L., B. D. Donn, and W. G. Valentine, 1965, On the Early History of the Earth. *Bull. Geol. Soc. Amer.,* **76**, 287–306.

DUBA, A. and A. E. Ringwood, 1973, Electrical Conductivity, Internal Temperatures, and Thermal Evolution of the Moon. *The Moon,* **7**, nos. 3/4, 357–376.

DYAL, P. and C. W. Parkin, 1971, The Magnetism of the Moon. *Scientific Amer.,* **225**, no. 2, 62–73.

——, 1972, Lunar Properties from Transient and Steady Magnetic Field Measurements. *The Moon,* **11**, 173–197.

EL-BAZ, F., 1973, The Lunar Dark Mantle: Its Distribution and Geologic Significance. In Chamberlain and Watkins (1973), pp. 217–218.

ELSTON, W. E., 1971, Evidence for Lunar Volcano-Tectonic Features. *J. Geophys. Res.,* **76**, 5690–5702.

ENGEL, A. E. J. and C. G. Engel, 1970, Lunar Rock Compositions and some Interpretations. *Science,* **167**, 527–528.

ENGEL, A. E. J., C. G. Engel, A. L. Sutton, and A. T. Myers, 1971, Composition of Five Apollo 11 and Apollo 12 Rocks and One Apollo 11 Soil and some Petrogenic Considerations. In Levinson (1971), v. 1, pp. 439–448.

ENGELHARDT, W. V., J. Arndt, W. F. Müller, and D. Stöffler, 1970, Shock Metamorphism of Lunar Samples and Origin of the Regolith at the Apollo 11 Landing Site. In Levinson (1971), v. 1, pp. 363–384.

ESCHELMAN, V. R., 1969, The Atmospheres of Mars and Venus. *Scientific Amer.,* **220**, no. 4, 78–88.

EVANS, J. and T. Hagfors, 1964, On the Interpretation of Radar Reflections from the Moon. *Icarus,* **3**, 151–160.

*FAIRBRIDGE, R. W. (ed.), 1967, *The Encyclopedia of Atmospheric Sciences and Astrogeology,* Reinhold, New York, 1200 pp.

*FAUL, H., 1966, *Ages of Rocks, Planets, and Stars,* McGraw-Hill, New York, 109 pp.

FESSENKOV, V. G., 1962, Photometry of the Moon. In Kopal (1962), pp. 99–130.

*FIELDER, G., 1961, *Structure of the Moon's Surface,* Pergamon Press, New York, 266 pp.

——, 1963*a*, Erosion and Deposition on the Moon. *Planetary and Space Sci.,* **11**, 1335–1340.

——, 1963*b*, Nature of the Lunar Maria, *Nature,* **198**, 1256.

*——, 1965, *Lunar Geology,* Butterworth, London, 184 pp.

——, 1967*a*, Volcanic Rings on the Moon. *Nature,* 213, 333–336.

——, 1967*b*, Evidence for Volcanism and Faulting on the Moon. In Runcorn (1967), pp. 461–472.

*—— (ed), 1971, *Geology and Physics of the Moon: A Study of Some Essential Problems,* American Elsevier, New York.

*FIRSOFF, V. A., 1961, *Surface of the Moon,* Hutchinson, London, 128 pp.

*——, 1962, *Strange World of the Moon,* Science Editions, New York, 226 pp.

FLEISCHER, R. L. and H. R. Hart, Jr., 1973, Surface History of Lunar Soil and Soil Columns. In Chamberlain and Watkins (1973), pp. 251–253.

FORD, C. E., G. M. Biggar, D. J. Humphries, G. Wilson, D. Dixon, and M. J. O'Hara, 1972, Role of Water in the Evolution of the Lunar Crust; an Experimental Study of Sample 14310; an Indication of Lunar Calc-Alkaline Volcanism. In King et al. (1972), v. 1, pp. 207–229.

FRENCH, B. M., 1968, Shock Metamorphism as a Geologic Process. In French and Short (1968), pp. 1–18.

——, Possible Relations between Meteorite Impact and Igneous Petrogenesis, as Indicated by the Sudbury Structure, Ontario, Canada, *Bull. Volcanologique,* **34**, 466–517.

*FRENCH, B. M. and N. M. Short (eds.), 1968, *Shock Metamorphism of Natural Materials,* Mono Press, Baltimore, 644 pp.

FRICKER, P., R. Reynolds, and A. Summers, 1967, On the Thermal History of the Moon. *J. Geophys. Res.,* **72**, 2649–2662.

FRONDEL, C., C. Klein, Jr., J. Ito, and J. C. Drake, 1970, Mineralogical and Chemical Studies of Apollo 11 Lunar Fines and Selected Rocks. In Levinson (1970), v. 1, pp. 445–474.

FULMER, C. V. and W. A. Roberts, 1967, Surface Lineaments Displayed on Lunar Orbiter Pictures. *Icarus,* 7, 394–406.

GAST, P. W., 1972, The Chemical Composition and Structure of the Moon. *The Moon,* **4**, 121–147.

——, 1973, Lunar Magmatism in Time and Space. In Chamberlain and Watkins (1973), pp. 275–277.

GAST, P. W., N. J. Hubbard, and H. Wiesmann, 1970, Chemical Composition and Petrogenesis of Basalts from Tranquillity Base. In Levinson (1970), v. 1. pp. 1143–1164.

GAST, P. W. and R. K. McConnell, Jr., 1972, Evidence for Initial Chemical Layering of the Moon. In Watkins (1972), pp. 257-258.

GAULT, D. E., 1970, Saturation and Equilibrium Conditions for Impact Cratering on the Lunar Surface, Criteria and Implications. *Radio Science,* **5**, 273-291.

GAULT, D. E., W. L. Quaide, and V. R. Oberbeck, 1968, Impact Cratering Mechanics and Structures. In French and Short (1968), pp. 87-100.

GEHRELS, T., 1964, A Model of the Lunar Surface. *Icarus,* **3**, 491-496.

GEHRING, J., A. C. Charters, and R. L. Warnica, 1964, Meteoroid Impact on the Lunar Surface. In Salisbury and Glaser (1964), pp. 215-264.

GIBSON, E. K., JR. and G. W. Moore, 1973, Volatile-rich Lunar Soil: Evidence of Possible Cometary Impact. *Science,* **179**, 74-76.

GILBERT, G. K., 1893, The Moon's Face: A Study of the Origin of its Features. *Bull. Philosophical Soc. of Washington,* **12**, 241-292.

GILVARRY, J. J., 1960, Origin and Nature of the Lunar Surface Features. *Nature,* **188**, 886-890.

——, 1969, What are the Mascons? *Saturday Rev. of Lit.,* **52**, no. 23, 54-57.

*GLASSTONE, S., 1965, *Sourcebook on the Space Sciences,* D. Van Nostrand, Princeton, N.J., 937 pp.

GOLD, T., 1955, The Lunar Surface, *Month. Not. Royal Acad. Sci.,* **115**, 585-603.

——, 1966, The Moon's Surface. In Hess et al. (1966), pp. 107-124.

GOLDREICH, P., 1972, Tides and the Earth-Moon System. *Scientific Amer.,* **226**, no. 4, 42-52.

GOLDSTEIN, J. and A. S. Doan, 1972, The Effects of Phosphorus on the Formation of the Widmanstätten Pattern in Iron Meteorites. *Geochim. et Cosmochim. Acta,* **36**, 51-69.

GREELEY, R., 1971, Lunar Hadley Rille: Considerations of its Origin. *Science,* **172**, 722-725.

GREEN, D. H. and A. E. Ringwood, 1972, Significance of Apollo 15 Mare Basalts and 'Primitive' Green Glasses in Lunar Petrogenesis. *Earth Planet. Sci. Newsletters.*

GREEN, J., 1962, The Geosciences Applied to Lunar Exploration. In Kopal and Milhailov (1962), pp. 169-257.

——, 1965*a*, Hookes and Spurrs in Selenology. In H. E. Whipple (1965), pp. 373-402.

——, 1965*b*, Tidal and Gravity Effects Intensifying Lunar Defluidization and Volcanism. In H. E. Whipple (1965), pp. 403-469.

——, 1971*a*, Origin of Glass Deposits in Lunar Craters, *Science,* **168**, 608-609.

——, 1971*b*, Copernicus as a Lunar Caldera. *J. Geophys. Res.,* **76**, 5719-5731.

*GREEN, J. and N. M. Short (eds.) 1971, *Volcanic Landforms and Surface Features: A Photographic Atlas and Glossary,* Springer-Verlag, New York, 519 pp.

GREENSPAN, J. and T. Owen, 1967, Jupiter's Atmosphere: Its Structure and Composition. *Science,* **156**, 1489-1493.

GUEST, J. E. and G. Fielder, 1968, Lunar Ring Structure and the Nature of the Maria. *Planetary and Space Sci.,* **16**, 665-673.

GUEST, J. E. and J. B. Murray, 1969, Nature and Origin of Tsiolkovsky Crater, Lunar Farside. *Planetary and Space Sci.,* **17**, 121.

*GUTSCHEWSKI, G. L., D. C. Kinsler, and E. Whitaker, 1971, *Atlas and Gazetteer of the Near Side of the Moon,* NASA Special Publication SP-241.

HACKMANN, R. J., 1961, Photointerpretation of the Lunar Surface. *Photogrammetric Engineering,* **27**, 377-386.

HAGFORS, T., 1966, Review of Radar Observations of the Moon. In Hess et al. (1966), pp. 229-240.

HAMMOND, A. L., 1972, Lunar Research: No Agreement on Evolutionary Models. *Science,* **175**, 868-870.

HANEL, R., B. Conrath, W. Hovis, V. Konde, P. Lowman, W. Maguire, J. Pearl, J. Pirraglia, C. Prabhakara, B. Schlachman, G. Lovin, P. Straat, and T. Burke, 1972, Investigation of the Martian Environment by Infrared Spectroscopy on Mariner 9. *Icarus,* **17**, 423-442.

HAPKE, B. W., 1966, Optical Properties of the Moon's Surface. In Hess et al. (1966), pp. 141-154.

HAPKE, B. W., 1968, Lunar Surface: Composition inferred from Optical Properties. *Science,* **159**, 76-79.

HAPKE, B. W., 1970, Inferences from the Optical Properties of the Moon Concerning the Nature and Evolution of the Lunar Surface. *Radio Science,* **5**, 293-299.

HAPKE, B. W. and H. Van Horn, 1963, Photometric Studies of Complex Surfaces, with Applications to the Moon. *J. Geophys. Res.,* **68**, 4545-4570.

HARRISON, J. C., 1963, An Analysis of Lunar Tides. *J. Geophys. Res.,* **68**, 4269-4280.

HARTMANN, W. K., 1965, Terrestrial and Lunar Flux of Large Meteorites in the last Two Billion Years. *Icarus,* **4**, 157-165.

——, 1966*a*, Early Lunar Cratering. *Icarus,* **5**, 406-418.

——, 1966*b*, Martian Cratering. *Icarus,* **5**, 565-576.

——, 1970*a*, Lunar Cratering Chronology. *Icarus,* **13**, 299-301.

——, 1970*b*, *Moons and Planets: An Introduction to Planetary Science,* Bogden and Quigley, New York, 404 pp.

——, 1973, Martian Cratering, 4 Mariner 9 Initial Analysis of Cratering Chronology. *J. Geophys. Res.,* **78**, 4096-4116.

HARTMANN, W. K. and C. A. Wood, 1971, Moon: Origin and Evolution of Multiring Basins. *The Moon,* **3**, 3-78.

HASKIN, L. A., R. O. Allen, P. A. Helmke, T. P. Paster, M. R. Anderson, R. L. Korotev, and K. A. Zweifel, 1970, Rare-earths and other Trace Elements in Apollo 11 Lunar Samples. In Levinson (1970), v. 2, pp. 1213-1232.

HAWKINS, G. S., 1964, Interplanetary Debris near the Earth. *Ann. Review Astronomy and Astrophysics,* **2**, 140-164.

*——, 1965, *Meteors, Comets, and Meteorites,* McGraw-Hill, New York, 128 pp.

*HEIDE, F., 1964, *Meteorites,* University of Chicago Press, Chicago, 144 pp.

*HESS, W. N. (ed.), 1965, *Introduction to Space Science*, Gordon and Breach, New York, 919 pp.

HESS, W. N., R. Kovach, P. W. Gast, and G. Simmons, 1969, The Exploration of the Moon. *Scientific Amer.*, **221**, no. 4, 54–75.

*HESS, W. N., D. H. Menzel, and J. A. O'Keefe (eds.), 1966, *The Nature of the Lunar Surface, The Johns Hopkins Press,* 320 pp.

HIBBS, A. R., 1967, The Surface of the Moon. *Scientific Amer.,* **216**, no. 3, 60–77.

HINNERS, N. W., 1971, The New Moon: A View. *Rev. of Geophys. & Space Phys.,* **9**, 447–522.

HINZE, W. J., 1967, Use of an Earth Analog in Lunar Mission Planning. *Icarus,* **6**, 444–452.

HOHENBERG, C. M., 1969, Radioisotopes and the History of Nucleosynthesis in the Galaxy. *Science,* **166**, 211–215.

*HÖRZ, F. (ed.), 12971, *Meteorite Impact and Volcanism,* entire issue of *J. Geophys. Res.,* **76**, no. 23, pp. 5381–5798.

HOWARD, K. A., J. W. Head, and G. A. Swann, 1972, Geology of Hadley Rille. In King et al. (1972), v. 1, pp. 1–14.

HUBBARD, N. J., P. W. Gast, J. M. Rhodes, B. M. Bansal, H. Wiesmann, and S. E. Church, Non-Mare Basalts, Part II. In King et al. (1972), v. 2, pp. 1161–1179.

HUNT, G. R., J. W. Salisbury, and R. K. Vincent, 1968, Infrared Images of the Eclipsed Moon. *Sky and Telescope,* **36**, 1–4.

HUSAIN, L. and O. A. Schaeffer, 1973, ^{40}Ar–^{39}Ar Crystallization Ages and ^{38}Ar–^{37}Ar Ages of Samples from the Vicinity of the Apollo 16 Landing Site. In Chamberlain and Watkins (1973), pp. 406–408.

JACKSON, E. D. and H. C. Wilshire, 1968, Chemical Composition of the Lunar Surface at the Surveyor Landing Sites. *J. Geophys. Res.,* **73**, 7621–7629.

JAFFE, L. D., 1969, The Surveyor Lunar Landings, *Science,* **164**, 775–788.

JAFFE, L. D., R. Choate, and R. B. Coryell, 1972, Spacecraft Techniques for Lunar Research. *The Moon,* **5**, 348–367.

JAMES, O. B. and E. D. Jackson, 1970, Petrology of the Apollo 11 Ilmenite Basalts. *J. Geophys. Res.,* **75**, 5793–5824.

JASTROW, R., 1960, Exploration of the Moon. *Scientific Amer.,* **202**, no. 5, 61–69.

——, 1968, The Planet Venus, *Science,* **160**, 1403–1410.

*JASTROW, R. and A. G. W. Cameron (eds.), 1963, *Origin of the Solar System,* Academic Press, New York, 176 pp.

KATTERFELD, G. N., 1967, Types, Ages, and Origins of Lunar Ring Structures—Statistical and Comparative Geological Approach. *Icarus,* **6**, 360–380.

*KAULA, W. M., 1968, *An Introduction to Planetary Physics,* John Wiley and Sons, New York, 490 pp.

——, 1969*a*, The Gravitational Field of the Moon. *Science,* **166**, 1581–1588.

——, 1969*b*, Interpretation of Lunar Mass Concentrations. *Phys. Earth and Planet. Interiors.* **2**, 123–137.

KAULA, W. M., G. Schubert, R. E. Lingenfelter, W. L. Sjogren, and W. R. Wollenhaupt, 1972, Analysis and Interpretation of Lunar Laser Altimetry. In King et al (1972), v. 3, pp. 2189–2204.

KEIFFER, H., 1967, Calculated Physical Properties of Planets in Relation to Composition and Gravitational Layering. *J. Geophys. Res.,* **72**, 3179–3198.

KEIL, K., M. Prinz, and T. E. Bunch, 1971, Mineralogy, Petrology, and Chemistry of Some Apollo 12 Samples. In Levinson (1971), v. 1, pp. 319–342.

*KING, E. (v. 1), D. Heymann (v. 2), and D. R. Criswell (v. 3) (eds.), 1972, *Proceedings of the Third Lunar Science Conference,* Suppl. 3, *Geochim. Cosmochim. Acta,* The MIT Press, Cambridge, Mass., 3263 pp.

KIRSTEN, T., J. Deubner, P. Horn, I. Kaneoka, J. Kiko, O. A. Schaeffer, and S. K. Thio, 1972, The Rare Gas Record of Apollo 14 and 15 Samples. In King et al. (1972), v. 2, pp. 1865–1889.

KLEIN, C., JR., 1972, Lunar Materials: Their Mineralogy, Petrology, and Chemistry. *Earth Sci. Rev.,* **8**, 169–204.

KLEIN, C., JR., J. C. Drake, and C. Frondel, 1971, Mineralogical, Petrological, and Chemical Features of Four Apollo 12 Lunar Microgabbros. In Levinson (1971), v. 1, pp. 265–284.

KOPAL, Z., 1961, The Internal Constitution of the Moon. *Planetary and Space Sci.,* **2**, 249–255.

*——, 1962, *Physics and Astronomy of the Moon,* Academic Press, New York, 538 pp.

——, 1963, Stress History of the Moon and Terrestrial Planets. *Icarus,* **2**, 376–395.

——, 1965*a*, The Luminescence of the Moon. *Scientific Amer.,* **212**, no. 5, 28–37.

——, 1965*b*, Topography of the Moon. *Space Science Reviews,* **4**, 737–855.

*——, 1966, *An Introduction to the Study of the Moon,* D. Reidel, Dordrecht, Holland, 464 pp.

——, 1969, The Earliest Maps of the Moon. *The Moon,* **1**, 59–66.

——, 1970, On the Depth of the Lunar Regolith. *The Moon,* **1**, 451–461.

*KOPAL, Z., J. Klepesta, and T. W. Rackham (eds.), 1965, *Photographic Atlas of the Moon,* Academic Press, New York, 277 pp.

*KOPAL, Z. and Z. K. Mikhailov (eds.), 1962, *The Moon: Proceedings of the International Astronomical Union Symposium 14, Leningrad, 1960,* Academic Press, London, 571 pp.

*KOSOFSKY, F. J. and F. El-Baz, 1970, *The Moon as Viewed by Lunar Orbiter,* NASA Special Publication SP-200, 152 pp.

KOVACH, R. and D. Anderson, 1965, The Interiors of the Terrestrial Planets. *J. Geophys. Res.,* **70**, 2873–2882.

KOVACH, R., J. S. Watkins, A. Nur, and M. Talwani, 1973, The Properties of the Shallow Crust: An Overview from Apollo 14, 16, and 17. In Chamberlain and Watkins (1973), pp. 444–445.

KOZYREV, N., 1962, Physical Observations of the Lunar Surface. In Kopal (1962), pp. 361–384.

*KRINOV, E. L., 1960, *Principles of Meteoritics,* Pergamon Press, New York, 535 pp.

KUIPER, G. P., 1954, On the Origin of the Lunar Surface Features. *Proc. Nat. Acad. Sciences,* **40**, 1096–1112.

*——, (ed.), 1960, *Photographic Lunar Atlas,* University of Chicago Press, Chicago, 23 pp. and 230 photos.

——, 1964, The Moon and the Planet Mars. *In Advances in Earth Sciences* (P. M. Hurley, ed.), M.I.T. Press, Cambridge, Mass., pp. 21–68.

——, 1965, Lunar Results from Rangers 7 to 9. *Sky & Telescope,* **29**, 293–305.

*KUIPER, G. P. and B. M. Middlehurst (eds.), 1961, *The Solar System, Part 3: Planets and Satellites,* University of Chicago Press, Chicago, 601 pp.

LANGSETH, M. G., JR., S. P. Clark, Jr., J. L. Chute, Jr., S. J. Leihm, and A. E. Wechsler, 1972, The Apollo 15 Lunar Heat Flow Measurement. *The Moon,* **4**, 390–410.

LANPHERE, M. A., G. J. Wasserburg, and A. L. Albee, 1964, Redistribution of Sr and Rb Isotopes during Metamorphism, Panamint Range, California. In *Isotopic and Cosmic Chemistry* (Craig, Miller, and Wasserburg, eds.), North-Holland Publ.Co., Amsterdam, pp. 269–320.

LARIMER, J. W. and E. Anders, 1967, Chemical Fractionations in Meteorites: II Abundance Patterns and their Interpretation. *Geochim. Cosmochim. Acta,* **31**, 1239–1270.

LATHAM, G., M. Ewing, F. Press, G. Sutton, J. Dorman, Y. Nakamura, N. Toksöz, R. Wiggins, J. Derr, and and F. Duennebier, 1970, Apollo 11 Passive Seismic Experiment. *Science,* **167**, 455–467.

LATHAM, G., M. Ewing, F. Press, G. Sutton, J. Dorman, Y. Nakamura, N. Toksöz, R. Wiggins, and R. Kovach, 1970*b*, Seismic Data from Man-made Impacts on the Moon. *Science,* **170**, 620–626.

LATHAM, G., M. Ewing, F. Press, G. Sutton, J. Dorman, Y. Nakmura, N. Toksöz, D. Lammlein, and F. Duennebier, 1971, Moonquakes. *Science,* **174**, 687–692.

LATHAM, G., M. Ewing, J. Dorman, Y. Nakamura, F. Press, N. Toksöz, G. Sutton, F. Duennebier, and D. Lammlein, 1973, Lunar Structure and Dynamics–Results from the Apollo Passive Seismic Experiments. *The Moon,* **7**, no. 3/4, 341–356.

LATHAM, G., M. Ewing, J. Dorman, F. Press, N. Toksöz, G. Sutton, F. Duennebier, and Y. Nakamura, 1972, Moonquakes and Lunar Tectonism. *The Moon,* **4**, 374–382.

LeROY, L., 1961, Lunar Features and Lunar Problems. *Bull. Geol. Soc. Amer.,* **72**, 591–604.

LEVIN, B. J., 1966, Thermal History of the Moon and Development of its Surface. In Hess et al. (1966), pp. 267–274.

LEVIN, E., D. D. Viele, and L. B. Eldrenkamp, 1968, The Lunar Orbiter Missions to the Moon. *Scientific Amer.,* **218**, no. 5, 58–78.

*LEVINSON, A. A. (ed.), 1970, *Proceedings of the Apollo 11 Lunar Science Conference,* Suppl. 1, *Geochim. Cosmochim. Acta,* Pergamon Press, New York, 2380 pp.

*——, 1971, *Proceedings of the Second Lunar Science Conference,* Suppl. 2, *Geochim. Cosmochim. Acta,* M.I.T. Press, Cambridge, Mass., 2818 pp.

*LEVINSON, A. A. and S. R. Taylor, 1971, *Moon Rocks and Minerals,* Pergamon Press, New York, 222 pp.

LINGENFELTER, R., S. Peale, and G. Schubert, 1968, Lunar Rivers. *Science,* **161**, 267–269.

LOGAN, L. M., G. R. Hunt, S. R. Balsamo, and J. W. Salisbury, 1972, Midinfrared Emission Spectra of Apollo 14 and 15 Soils and Remote Compositional Mapping of the Moon. In King et al. (1972), pp. 3069–3076.

*LOGSDON, J. M., 1970, *The Decision to Go to the Moon,* M.I.T. Press, Cambridge, Mass., 188 pp.

LOW, F. J. and A. W. Davidson, 1965, Lunar Observations at a Wave Length of 1 mm. *Astrophys. J.,* **142**, 1278.

LOWMAN, P. D., JR., 1966, The Scientific Value of Manned Lunar Exploration. *Annals of the New York Acad. Sciences,* **140**, art. 1., 623.

*——, 1967, *Space Panorama,* Weldflugbild, Zurich, 164 pp.

*——, 1969*a*, *Lunar Panorama: A Photographic Guide to the Geology of the Moon,* Weldflugbild, Zurich, 105 pp.

——, 1969*b*, Composition of the Lunar Highlands: Possible Implications for the Evolution of the Earth's Crust. *J. Geophys. Res.,* **74**, 495–504.

——, 1969*c*, The Moon's Resources. *Science Journal,* v. 5, no. 2, 90–95.

——, 1972*a*, The Geological Evolution of the Moon. *J. Geol.,* **80**, 125–166.

*——, 1972*b*, *The Third Planet,* Weldflugbild, Zurich, 170 pp.

LSAPT (Lunar Sample Analysis Planning Team), 1972, Third Lunar Science Conference. *Science,* **176**, 975–981.

LSAPT, 1973, Fourth Lunar Science Conference. *Science,* **181**, 615–621.

LSPET (Lunar Sample Preliminary Examination Team), 1969, Preliminary Examinations of Lunar Samples from Apollo 11. *Science,* **165**, 1211–1227.

LSPET, 1970, Preliminary Examinations of Lunar Samples from Apollo 12. *Science,* **167**, 1325–1339.

LSPET, 1971, Preliminary Examinations of Lunar Samples from Apollo 14. *Science,* **173**, 681.

LSPET, 1972, The Apollo 15 Lunar Samples: A Preliminary Description. *Science,* **175**, 363–375.

LSPET, 1973, The Apollo 16 Lunar Samples: Petrographic and Chemical Description. *Science,* **179**, 23–34.

*LUNA 16, 1972, Special issue. *Earth & Planet. Sci. Letters,* **13**, no. 2, pp. 223–466.

LYON, R. J. P. and P. C. Badgley, 1965, Lunar Exploration from Orbital Altitudes. In H. E. Whipple (1965), pp. 1198–1219.

LYTTLETON, R., 1965, On the Internal Structure of Mars. *Month. Notices Royal Astron. Soc.,* **129**, 21–39.

MacDONALD, G. J. F., 1959, Calculations on the Thermal History of the Earth. *J. Geophys. Res.,* **64**, 1967–2000.

——, 1960, Stress History of the Moon. *Planetary and Space Sci.,* **2**, 249–255.

——, 1963, The Internal Constitution of the Inner Planets and Moon. *Space Science Reviews,* **2**, 473.

——, 1964*a*, Tidal Friction. *Rev. of Geophys.,* **2**, 467–541.

——, 1964*b*, Earth and Moon: Past and Future. *Science,* **145**, 881–890.

——, 1966, What is a Planet? *Internat. Science & Technol.,* March, 48–59.

MACKIN, J. H., 1969, Origin of Lunar Maria. *Bull. Geol. Soc. Amer.*, **80**, 735–748.

*MANDELBAUM, L., 1969, Apollo: How the U.S. Decided to Go to the Moon. *Science*, **163**, 649–653.

*MARINER TO MARS, 1971, Collection of Papers in *Science*, **175**, 293–324.

*MARINER 9, 1973, Special Issue of 34 papers, *J. Geophys. Res.*, **78**, 4007–4440.

*MARKOV, A. V. (ed.), 1962, *The Moon: A Russian View*, University of Chicago Press, Chicago, 391 pp.

*MARSDEN, B. G. and A. G. W. Cameron (ed.), 1966, *The Earth–Moon System*, Plenum Press, New York, 288 pp.

MARVIN, U. B., 1973, The Moon after Apollo. *(M.I.T.) Technology Review*, **75**, no. 8, 12–23.

MARVIN, U. B., J. A. Wood, G. J. Taylor, J. B. Reid, Jr., B. N. Powell, J. S. Dickey, Jr., and J. F. Bower, 1971, Relative Proportions and Probable Sources of Rock Fragments in the Apollo 12 Soil Samples. In Levinson (1971), pp. 679–700.

*MASON, B., 1962, *Meteorites*, John Wiley and Sons, New York, 274 pp.

——, 1967*a*, Meteorites. *Amer. Scientist*, **55**, 429–455.

——, 1967*b*, Extraterrestrial Mineralogy. *Amer. Mineralogist*, **52**, 307–325.

——, 1971, The Lunar Rocks. *Scientific Amer.*, **225**, no. 4, 49–58.

*MASON, B. and W. G. Melson, 1970*a*, *The Lunar Rocks*, Wiley Interscience, New York, 179 pp.

——, 1970*b*, Comparison of Lunar Rocks with Basalts and Stony Meteorites. In Levinson (1970), v. 1, 661–671.

MASURSKY, H., 1973, An Overview of Geological Results from Mariner 9. *J. Geophys. Res.*, **78**, 4009–4030.

MAXWELL, J. A., L. C. Peck, and H. B. Wiik, 1970, Chemical Composition of Apollo 11 Lunar Samples 10017, 10020, 10072, 10084. In Levinson (1970), v. 2, pp. 1369–1374.

MAYER, C. H., 1961, Temperatures of the Planets. *Scientific Amer.*, **204**, no. 3, 58–65.

McCAULEY, J. F., 1967, The Nature of the Lunar Surface as Determined by Systematic Geologic Mapping. In Runcorn (1967), pp. 431–460.

——, 1968, Geologic Results from the Lunar Precursor Probes. *Amer. Inst. of Aeronaut. and Astronaut. Jour.*, **6**, 1991–1996.

*——, 1969, *Moon Probes*, Silver-Burdett, Morristown, N.J., 64 pp.

McCAULEY, J. F., M. H. Carr, J. A. Cutts, W. K. Hartmann, H. Masursky, D. J. Milton, R. P. Sharp, and D. E. Wilhelms, 1972, Preliminary Mariner 9 Report on the Geology of Mars. *Icarus*, **17**, 289–327.

McCONNELL, R. K., JR., L. A. McClaine, D. W. Lee, J. R. Aronson, and D. U. Allen, 1967, A Model for Planetary Igneous Differentiation. *Rev. of Geophys.*, **5**, 121–172.

McCORD, T., 1967, Observational Study of Lunar Visible Emission. *J. Geophys. Res.*, **72**, 2087–2097.

——, 1969, Color Differences on the Lunar Surface. *J. Geophys. Res.*, **74**, 3131–3142.

McCORD, T. and J. B. Adams, 1969, Spectral Reflectivity of Mars. *Science*, **163**, 1058–1060.

——, 1972, Mercury: Surface Composition from the Reflection Spectrum. *Science*, **178**, 745–746.

——, 1973, Progress in Remote Optical Analysis of Lunar Surface Composition. *The Moon*, **7**, no. 3/4, 251–278.

McCORD, T., T. V. Johnson, and H. H. Kieffer, 1969, Differences between Proposed Apollo Sites: 2. Visible and Infrared Reflectivity Evidence. *J. Geophys. Res.*, **74**, 4385–4388.

McGETCHIN, T. R. and G. W. Ulrich, 1973, Xenoliths in Maars and Diatremes with Inferences for the Moon, Mars, and Venus. *J. Geophys. Res.*, **78**, 1832–1852.

McKAY, D. S. and D. A. Morrison, 1971, Lunar Breccias. *J. Geophys. Res.*, **76**, 5658–5669.

McKAY, D. S., G. H. Heiken, R. M. Taylor, U. S. Clanton, D. A. Morrison, and G. H. Ladle, 1972, Apollo 14 Soils: Distribution and Particle Types. In King et al. (1972), v. 1, pp. 983–994.

MEGRUE, G. H. and F. Steinbrunn, 1972, Classification and Source of Lunar Soils, Clastic Rocks, and Individual Mineral, Rock and Glass Fragments from Apollo 12 and 14 Samples as Determined by the Concentration Gradients of the Helium, Neon, and Argon Isotopes. In King et al. (1972), v. 2, pp. 1899–1916.

MELSON, W. G. and B. Mason, 1971, Lunar "Basalts": Some Comparisons with Terrestrial and Meteoritic Analogs, and a Proposed Classification and Nomenclature. In Levinson (1971), v. 1, pp. 459–467.

METZGER, A. E., J. I. Tromka, L. E. Peterson, R. C. Reedy, and J. R. Arnold, 1973, Lunar Surface Radioactivity: Preliminary Results of the Apollo 15 and Apollo 16 Gamma-ray Spectrometer Experiments. *Science*, **179**, 800–803.

MEYER, C., Jr., R. Brett, N. J. Hubbard, D. A. Morrison, D. S. McKay, F. K. Aitken, H. Takeda, and E. Schonfeld, 1971, Mineralogy, Chemistry and Origin of the KREEP Component in Soil Samples from the Ocean of Storms. In Levinson (1971), v. 1, pp. 393–411.

MIDDLEHURST, B. M. 1967, An Analysis of Lunar Events. *Rev. of Geophys.*, **5**, 173–189.

*MIDDLEHURST, B. M. and G. P. Kuiper (ed.), 1963, *The Solar System, Part 4: The Moon, Meteorites, and Comets*, University of Chicago Press, Chicago, 810 pp.

MILLS, G. A., 1968, Absolute Coordinates of Lunar Features. *Icarus*, **8**, 90–116.

MILTON, D. J., 1973, Water and Processes of Degradation in the Martian Landscape. *J. Geophys. Res.*, **78**, 4037–4048.

*MOON ISSUE, 30 January 1970, *Science*, **167**, no. 3918, pp. 447–784.

MOORE, C. B., C. F. Lewis, E. K. Gibson, and W. Nichiporuk, 1970, Total Carbon and Nitrogen Abundances in Lunar Samples. *Science*, **167**, 496–497.

MOORE, H. J., 1971, Geological Interpretation of Lunar Data. *Earth–Science Reviews*, **7**, 5–33.

MORGAN, J. W., U. Krähenbühl, R. Ganapathy, and E. Anders, 1972, Trace Elements in Apollo 15 Samples: Implications for Meteorite Influx and Volatile Depletion in the Moon. In King et al. (1972), v. 2, pp. 1361–1376.

MUELLER, R. F., 1969, Planetary Probe: Origin of Atmosphere of Venus. *Science,* **163**, 1321–1324.

MULLER, P. M. and W. L. Sjogren, 1968, Mascons: Lunar Mass Concentrations. *Science,* **161**, 680–684.

*MURRAY, B. C. et al., 1971, The Surface of Mars: 1. Cratered Terrains (Murray); 2. Uncratered Terrains (Sharp); 3. Light and Dark Markings (Cutts); 4. South Polar Cap (Sharp). A series of Papers (Senior author indicated) in *Science,* **76**, 313–368.

MURRAY, B. C., L. A. Soderblom, J. A. Cutts, R. P. Sharp, D. J. Milton, and R. B. Leighton, 1972, Geological Framework of the South Polar Region of Mars. *Icarus,* **17**, 328–245.

MURRAY, J. B. and J. E. Guest, 1970, Circularities of Craters and Related Structures on Earth and Moon. *Modern Geology,* **1**, 149–159.

*MUTCH, T. A., 1970, *Geology of the Moon.* Princeton University Press, Princeton, N.J., 324 pp; 2nd ed., 1972.

*NATIONAL AERONAUTICS AND SPACE ADMINISTRATION, 1969, *Surveyor Program Results (Final Report),* NASA Special Publication SP-184, 423 pp.

——, 1969, *Mariner Mars 69: A Preliminary Science Report,* NASA Special Publication SP-225, 148 pp.

——, 1969, *Apollo 8: Analysis of Photography and Visual Observations,* NASA Special Publication SP-201, 321 pp.

*——, 1969, *Apollo 11: Preliminary Science Report,* NASA Special Publication SP-214, 204 pp.

*——, 1970, *Apollo 12: Preliminary Science Report,* NASA Special Publication SP-235, 227 pp.

*——, 1971, *Apollo 14: Preliminary Science Report,* NASA Special Publication SP-272, 309 pp.

*——, 1972, *Apollo 15: Preliminary Science Report,* NASA Special Publication SP-289, 496 pp.

*——, 1972, *Apollo 16: Preliminary Science Report,* NASA Special Publication SP-315, 622 pp.

*——, 1973, *Apollo 17: Preliminary Science Report,* NASA Special Publication SP-330, 710 pp.

*——, 1967, *Handbook of Physical Properties of the Planets Series:* (1) *Venus* (NASA SP-3029); (2) *Mars* (NASA SP-3030); (3) *Jupiter* (NASA SP-3031).

*NICKS, O. W. (ed), 1970, *This Island Earth,* NASA Special Publication SP-250, 182 pp.

NININGER, H. H., 1952, *Out of the Sky,* University of Denver Press, Denver, 336 pp.

NELEN, J. A., Noonan, and K. Fredriksson, 1972, Lunar Glasses, Breccias, and Chondrules. In King et al. (1972), v. 1, pp. 723–738.

OBERBECK, V. R., 1971, A Mechanism for the Production of Lunar Crater Rays. *The Moon,* **2**, 263–278.

OBERBECK, V. R. and W. L. Quaide, 1968, Genetic Implications of Lunar Regolith Thickness Variations. *Icarus,* **9**, 446–465.

OBERBECK, V. R., W. L. Quaide, and R. Greeley, 1969, On the Origin of Lunar Sinuous Rilles. *Modern Geology,* **1**, 75–80.

O'HARA, M. J., G. M. Biggar, S. W. Richardson, C. E. Ford, and B. G. Jamieson, 1970, The Nature of Seas, Mascons, and the Lunar Interior in the Light of Experimental Studies. In Levinson (1970), v. 1, pp. 695–710.

*O'KEEFE, J. A. (ed), 1973, *Tektites,* University of Chicago Press, Chicago, 228 pp.

——, 1966, Lunar Ash Flows. In Hess et al. (1966), pp. 259–266.

——, 1968, Isostasy on the Moon. *Science,* **162**, 1405–1406.

——, 1969, Origin of the Moon. *J. Geophys. Res.,* **74**, 2758–2767.

——, 1970*a*, Apollo 11: Implications for the Early History of the Solar System. *Trans. Amer. Geophys. Union,* **51**, 633–636.

——, 1970*b*, Origin of the Moon. *J. Geophys. Res.,* **75**, 6565–6574.

O'KEEFE, J. A. and W. S. Cameron, 1962, Evidence from the Moon's Surface Features for the Production of Lunar Granites. *Icarus,* **1**, 271–285.

O'KEEFE, J. A., P. D. Lowman, Jr., and W. S. Cameron, 1967, Lunar Ring Dikes from Lunar Orbiter I. *Science,* **155**, 77–79.

O'KELLEY, G. D., J. S. Eldridge, E. Schonfeld, and K. J. Northcutt, 1972, Primordial Radioelements and Cosmogenic Radionuclides in Lunar Samples from Apollo 15, *Science,* **175**, 440–442.

O'LEARY, B. T., 1966, On the Occurrence and Nature of Planets Outside the Solar System. *Icarus,* **5**, 419–436.

ÖPIK, E. J., 1960, The Lunar Surface as an Impact Counter. *Month. Notices. Royal Acad. Sci.,* **120**, 404–411.

——, 1962, The Lunar Atmosphere. *Planetary and Space Sci.,* **9**, 211–244.

——, 1962, Jupiter: Chemical Composition, Structure, and Origin of a Giant Planet. *Icarus,* **1**, 200–257.

——, 1966, The Martian Surface. *Science,* **153**, 255–265.

——, 1967, Evolution of the Moon's Surface, I. *Irish Astronom. Jour.,* **8**, 38–52.

PAPANASTASSIOU, D. A. and G. J. Wasserburg, 1970, Rb-Sr Ages from the Ocean of Storms. *Earth and Planetary Sci. Letters,* **8**, 269–278.

PARKIN, D. W. and D. Tilles, 1968, Influx Measurements of Extraterrestrial Materials. *Science,* **159**, 936–946.

PHILPOTTS, J. A. and C. C. Schnetzler, 1970, Apollo 11 Lunar Samples: K, Rb, Sr, Ba and Rare Earth Concentrations in Some Rocks and Separated Phases. In Levinson (1970), v. 5, pp. 1471–1486.

POHN, H. A. and T. W. Offield, 1969, Lunar Crater Morphology and Relative Age Determination of Lunar Geologic Units. U. S. Geological Survey Interagency Report, *Astrogeology,* **13**, 35 pp.

POHN, H. A., R. L. Wiley, and G. E. Sutton, 1970, A Photoelectric-Photographic Study of the Normal Albedo of the Moon, *U. S. Geological Survey Prof. Paper 599-E.*

PORTER, J., 1960, The Satellites of the Planets. *Jour. of the British Astronom. Assoc.,* **70**, 35–59.

PUGH, M. J. and J. A. Bastin, 1972, Infrared Observations of the Moon and Their Interpretation. *The Moon,* **4**, 17–3.

QUAIDE, W. L., 1965, Rilles, Ridges, and Domes—Clues to Maria History, *Icarus,* **4**, 37–389.

QUAIDE, W. L. and V. R. Oberbeck, 1968, Thickness Determinations of the Lunar Surface Layer from Lunar Impact Craters. *J. Geophys. Res.*, **73**, 5247–5270.

REA, D. G. and B. T. O'Leary, 1968, On the Composition of the Venus Clouds. *J. Geophys. Res.*, **73**, 665–676.

REID, A. M., J. Warner, W. I. Ridley, D. A. Johnston, R. S. Harmon, P. Jakeš, and R. W. Brown, 1972, The Major Element Compositions of Lunar Rocks as Inferred from Glass Compositions in the Lunar Soils. In King et al. (1972), pp. 363–378.

REYNOLDS, J. H., 1960, The Age of the Elements in the Solar System. *Scientific Amer.*, **202**, 171–185.

RINGWOOD, A. E., 1961, Chemical and Genetic Relationships Among Meteorites, *Geochim. et Cosmochim. Acta.*, **24**, 159–197.

——, 1966, Chemical Evolution of the Terrestrial Planets. *Geochim. Cosmochim. Acta*, **30**, 41–104.

——, 1970*a*, Origin of the Moon: The Precipitation Hypothesis. *Earth and Planetary Sci. Letters*, **8**, 131–140.

——, 1970*b*, Petrogenesis of Apollo 11 Basalts and Implications for Lunar Origin. *J. Geophys. Res.*, **75**, 6453–6479.

RINGWOOD, A. E. and E. Essene, 1970, Petrogenesis of Apollo 11 Basalts and the Internal Constitution and Origin of the Moon. In Levinson (1971), v. 1, pp. 769–799.

ROEDDER, E. and P. W. Weiblen, 1971, Petrology of Silicate Melt Inclusions, Apollo 11 and 12 and Terrestrial Equivalents, In Levinson (1971), v. 1, pp. 507–528.

ROGERS, A. E. E. and R. P. Ingalls, 1969, Venus: Mapping the Surface Reflectivity by Radar Interferometry. *Science*, **165**, 797–799.

RONCA, L. B., 1966, Introduction to the Geology of the Moon. *Proc. Geol. Assoc. London*, **77**, 101–126.

——, 1967, Meteorite Impact and Volcanism. *Icarus*, **5**, 515–520.

——, 1971, Ages of Lunar Mare Surfaces. *Bull. Geol. Soc. Amer.*, **82**, 1743–1748.

RONCA, L. B. and R. R. Green, 1970, Statistical Geomorphology of the Lunar Surface. *Bull. Geol. Soc. Amer.*, **81**, 337–352.

RONCA, L. B. and J. W. Salisbury, 1966, Lunar History as Suggested by the Circularity Index of Lunar Craters. *Icarus*, **5**, 130–138.

*RUNCORN, S. K. (ed.), 1967*a*, *Mantles of the Earth and Terrestrial Planets*, Wiley Interscience, London, 584 pp.

——, 1967*b*, Convection in the Planets. In Runcorn (1967), pp. 513–524.

——, 1972, Implications of the Magmatism and Figure of the Moon. In Watkins (1972), pp. 590–592.

RUSS, G. P. III, D. S. Burnett, and G. J. Wasserburg, 1972, Lunar Neutron Stratigraphy. *Earth Planet. Sci. Letters*, **15**, 172–181.

SAARI, J. M. and R. W. Shorthill, 1963, Isotherms of Crater Regions on the Illuminated and Eclipsed Moon. *Icarus*, **2**, 115–136.

——, 1966, Infrared and Visual Images of the Eclipsed Moon of December 19, 1964. *Icarus*, **5**, 635–659.

*SAGAN, C. (ed.), 1972, *The Viking Mission to Mars.* Special Issue Icarus, **16**, 1–227.

——, 1973, Sandstorms and Eolian Erosion on Mars. *J. Geophys. Res.*, **78**, 4155–4162.

*SAGAN, C., J. N. Leonard, and the editors of *Life*, 1966, *Planets*, Time Inc., New York.

SAGAN, C., J. Veverka, P. Fox, R. Dubisch, R. French, P. Gierasch, L. Quam, J. Lederberg, E. Leninthal, R. Tucker, B. Gross, and J. B. Pollack, 1973, Variable Features on Mars, 2, Mariner 9 Global Results. *J. Geophys. Res.*, **78**, 4163–4196.

SALISBURY, J. W., 1966, The Light and Dark Areas of Mars. *Icarus*, **5**, 291–298.

*SALISBURY, J. W. and P. E. Glaser (eds.), 1964, *The Lunar Surface Layer: Materials and Characteristics*, Academic Press, New York, 532 pp.

SALISBURY, J. W. and G. R. Hunt, 1967, Infrared Images: Implications for the Lunar Surface. *Icarus*, **7**, 47–58.

SALISBURY, J. W. and J. B. Pollack, 1966, On the Nature of the Canals of Mars. *Nature*, **212**, 117–121.

SCHNETZLER, C. C. and J. A. Philpotts, 1971, Alkali, Akaline Earth, and Rare-earth Element Concentrations in some Apollo 12 Soils, Rocks, and Separated Phases. In Levinson (1971), v. 2, 1101–1122.

SCHNETZLER, C. C., J. A. Philpotts, D. F. Nava, S. Schuhmann, and H. H. Thomas, 1972, Geochemistry of Apollo 15 Basalt 15555 and Soil 15531. *Science*, **175**, 426–428.

SCHONFELD, E. and C. Meyer, Jr., 1972, The Abundances of Components of the Lunar Soils by at Least Squares Mixing Model and the Formation Age of KREEP. In King et al. (1972), v. 2, pp. 1397–1420.

SCHUBERT, G., R. E. Lingenfelter, and S. J. Peale, 1970, The Morphology, Distribution, and Origin of Lunar Sinuous Rilles. *Reviews of Geophys. and Space Phys.*, **8**, 199–224.

SCHUMM, S. A., 1970, Experimental Studies on the Formation of Lunar Surface Features by Fluidization. *Bull. Geol. Soc. Amer.*, **81**, 2539–2552.

SCHURMEIER, H. M., R. J. Heacock, and A. E. Wolfe, 1966, The Ranger Missions to the Moon. *Scientific Amer.*, **214**, 52–67.

SCOTT, R. F., 1967, The Feel of the Moon. *Scientific Amer.*, **217**, 34–43.

SCOTT, R. F., 1967, Viscous Flow of Craters. *Icarus*, **7**, 139–148.

SELLERS, G. A., C. C. Woo, M. L. Bird, and M. B. Duke, 1971, Composition and Grain-size Characteristics of Fines from the Apollo 12 Double-core Tube, In Levinson (1971), 665–678.

SHALER, N. S., 1903, A Comparison of the Features of the Earth and Moon. *Smithsonian Contributions to Knowledge*, **34**, 1–130.

SHARP, R. P., 1973, 1. Mars: Troughed Terrain; 2. Mars: Fretted and Chaotic Terrains. *J. Geophys. Res.*, **78**, 4063–4083.

*SHELTON, W. R., 1968, Man's Conquest of Space, *National Geographic Society*, Washington, D. C., 200 pp.

SHOEMAKER, E. M., 1962, Exploration of the Moon's Surface. *American Scientist,* **50**, 99–13.

——, 1962, Interpretation of Lunar Craters. *In Physics and Astronomy of the Moon,* (Kopal, ed.), 283–360.

——, 1963, Impact Mechanics at Meteor Crater, Arizona. In Middlehurst and Kuiper (1963), pp. 301–336.

——, 1964, Geology of the Moon. *Scientific Amer.,* **211**, 38–47.

——, 1971, Origin of Fragmental Debris on the Lunar Surface and the History of Bombardment of the Moon. In *I Seminario de Geologia Lunar, Instituto de Investigaciones Geologicas,* Universidad de Barcelona, Numero Especial, vol. XXV, 27–56.

——, 1972, Cratering History and Early Evolution of the Moon. In Watkins, pp. 612–614.

SHOEMAKER, E. M. and R. J. Hackman, 1962, Stratigraphic Basis for a Lunar Time Scale. In Kopal and Mikhailov (1962), pp. 289–300.

SHOEMAKER, E. M., R. J. Hackman, and R. E. Eggleston, 1962, Interplanetary Correlation of Geologic Time. In Advances in the Astronautical Sciences, 8, *Plenum Press,* New York, 70–89.

SHOEMAKER, E. M., M. H. Hait, G. A. Swann, D. L. Schleicher, G. G. Schaber, R. L. Sutton, D. H. Dahlem, E. N. Goddard, and A. C. Waters, 1970, Origin of the Lunar Regolith at Tranquility Base. In Levinson (1970), v. 3, pp. 2399–2412.

SHORT, N. M., 1966, Shock Processes in Geology. *J. Geol. Education,* **14**, 149–166.

——, 1967, A Review of Shock Processes Pertinent to Fragmentation and Lithification of the Lunar Terrane. In Interpretation of Lunar Probe Data (Green, ed.), *Amer. Astronautical Soc. Science and Technol. Series,* **14**, 17–60.

——, 1971, The Nature of the Moon's Surface: Evidence from Shock Metamorphism in Apollo 11 and 12 Samples. *Icarus,* **13**, 383–413.

SHORT, N. M. and T. E. Bunch, 1968, A Worldwide Inventory of Features Characteristic of Rocks Associated with Presumed Meteorite Impact Craters. In French and Short (1968), pp. 255–266.

SHORT, N. M. and M. L. Forman, 1972, Thickness of Impact Crater Ejecta on the Lunar Surface. *Modern Geology,* **3**, 69–91.

SHORTHILL, R. W., 1970, Infrared Moon: A Review. *J. of Spacecraft and Rockets,* **7**, 385–397.

SILVER, L. T., 1972, Lead Volatilization and Volatile Transfer Processes on the Moon. In Watkins (1972), pp. 617–619.

——, 1973, Uranium–Thorium–Lead Isotope Characteristics in some Regolith Materials from the Descartes Region. In Chamberlain and Watkins (1973), pp. 672–674.

SINTON, W., 1967, Temperatures on the Lunar Surface. In Kopal (1962), pp. 407–428.

*SIMMONS, G., 1971, On the Moon with Apollo 15; *A Guidebook to Hadley Rille and the Apennine Mountains,* NASA Education Publication EP–94, 32 pp.

*——, 1972*a, On the Moon with Apollo 16: A Guidebook to the Descartes Region,* NASA Education Publication EP–95, 90 pp.

*——, 1972*b, On the Moon with Apollo 17: A Guidebook to Taurus-Littrow,* NASA Education Publication EP–101, 111 pp.

SMITH, E. I., 1973, Identification, Distribution, and Significance of Lunar Volcanic Domes. *The Moon,* **6**, 3–31.

SMITH, J. V., A. T. Anderson, R. C. Newton, E. J. Olsen, and P. J. Wyllie, 1970, A Petrologic Model for the Moon based on Petrogenesis, Experimental Petrology, and Physical Properties. *J. Geol.,* **78**, 381–405.

SMITH, R. L., 1966, Terrestrial Calderas, Associated Pyroclastic Deposits, and Possible Lunar Applications. In Hess et al. (1966), pp. 241–258.

SONETT, C. P., D. S. Colburn, P. Dyal, C. W. Parkin, B. F. Smith, G. Schubert, and K. Schwartz, 1971, Lunar Electrical Conductivity Profile. *Nature,* **230**, 359–362.

SONETT, C. P., B. F. Smith, D. S. Colburn, G. Schubert, and K. Schwartz, 1972, The Induced Magnetic Field of the Moon: Conductivity Profiles and Inferred Temperature. In King et al. (1972), v. 3, pp. 2309–2336.

*SPURR, J. E., 1944, 1945, 1948, 1949, *Geology Applied to Selenology,* v. 1–4, Lancaster (Pa.) Science Press and Rumford Press, Concord (N.H.).

STEINBACHER, R. H. and S. Gunter, 1970, The Mariner Mars 1971 Experiments: Introduction. *Icarus,* **12**, 3–9.

STRANGWAY, D. W., H. N. Sharpe, W. A. Gose, and G. W. Pearce, 1973, Magnetism and the Early History of the Moon. In Chamberlain and Watkins (1972), pp. 8–10.

STROM, R. G., 1971, Lunar Mare Ridges, Rings, and Volcanic Ring Complexes. *Modern Geol.,* **2**, 133–158.

STUART-ALEXANDER, D. E. and K. A. Howard, 1970, Lunar Maria and Circular Basins–A Review. *Icarus,* **12**, 440–446.

TATSUMOTO, M., 1970, Age of the Moon: An Isotopic Study of U–Th–Pb Systematics of Apollo 11 Lunar Samples. In Levinson (1970), v. 2, pp. 1595–1612.

TATSUMOTO, M., C. E. Hodge, B. R. Doe, and D. M. Unruh, 1972, U–Th–Pb and Rb–Sr Measurements on Some Apollo 14 Lunar Samples. In King et al. (1972), v. 2, pp. 1531–1555.

TAYLOR, G. J., M. J. Drake, J. A. Wood, and U. B. Marvin, 1973, Petrogenesis of KREEP-rich and KREEP-poor Nonmare Rocks. In Chamberlain and Watkins (1972), pp. 708–710.

TAYLOR, G. J., U. B. Marvin, J. B. Reid, Jr., and J. A. Wood, 1972, Noritic Fragments in the Apollo 12 and 14 Soils and the Origin of Oceanus Procellarum. In King et al. (1972), v. 1, pp. 995–1014.

TAYLOR, S. R., P. H. Johnson, R. Martin, D. Bennett, J. Allen, and W. Nance, 1970, Preliminary Chemical Analyses of Apollo 11 Lunar Samples. In Levinson (1970), v. 2, pp. 1627–1636.

TERA, F., D. A. Papanastassiou, and G. J. Wasserburg, 1973, A Lunar Cataclysm at ~ 3.95 AE and the Structure of the Lunar Crust. In Chamberlain and Watkins (1972), pp. 723–725.

TER HAAR, D. and A. G. W. Cameron, 1963, Historical Review of the Origin of the Solar System. In Jastrow and Cameron (1963), pp. 1–38.

TOKSÖZ, M. N., F. Press, K. Anderson, A. Dainty, G. Latham, M. Ewing, J. Dorman, D. Lammlein, G. Sutton, F. Duennebier, and Y. Nakamura, 1972*a*, Lunar Crust: Structure and Composition. *Science,* **176,** 1012–1016.

TOKSÖZ, M. N., F. Press, A. Dainty, K. Anderson, G. Latham, M. Ewing, J. Dorman, D. Lammlein, G. Sutton, and F. Duennebier, 1972*b*, Structure, Composition, and Properties of Lunar Crust. In King et al. (1972), v. 3, pp. 2527–2544.

TOKSÖZ, M. N., A. M. Dainty, S. C. Solomon, and K. Anderson, 1973*a*, Velocity Structure and Evolution of the Moon. In Chamberlain and Watkins (1973).

TOKSÖZ, M. N. and S. C. Solomon, 1973*b*, Thermal History and Evolution of the Moon. *The Moon.* **7,** 251–278.

TOZER, D. C., 1972, The Moon's Thermal State and an Interpretation of the Lunar Electrical Conductivity Distribution. *The Moon,* **4,** 90–105.

TRASK, N. J. and L. C. Rowan, 1967, Lunar Orbiter Photographs: Some Fundamental Observations. *Science,* **158,** 1529–1535.

*TSCHERMAK, G., 1883, *The Microscopic Properties of Meteorites* (translated by J. A. Wood), *Smithsonian Contrib. Astrophys.,* **4,** no. 6 (1964).

TURKEVICH, A. L., 1972, Comparison of the Analytical Results from the Surveyor, Apollo, and Luna Missions. *The Moon,* **5,** 411–421.

TURKEVICH, A. L., E. J. Patterson, and J. H. Franzgrote, 1968, The Chemical Analysis of the Lunar Surface. *Amer. Scientist,* **56,** 312–343.

TURNER, G., J. C. Huneke, F. A. Podesek, and G. J. Wasserburg, 1972, Ar^{40}–Ar^{39} Systematics in Rocks and Separated Minerals from Apollo 14. In King et al. (1972), v. 2, pp. 1589–1612.

TYLER, G. L., 1966, The Bistatic, Continuous-Wave Radar Method for the Study on Planetary Surfaces. *J. Geophys. Res.,* **71,** 1559–1567.

UREY, H. C., 1951, The Origin and Development of the Earth and Other Terrestrial Planets. *Geochim. Cosmochim. Acta,* **1,** 209–277.

*——, 1952*a*, *The Planets,* Yale University Press, New Haven, 245 pp.

——, 1952*b*, Boundary Conditions for Theories of the Origin of the Solar System. In *Physics and Chemistry of the Earth,* v. 2, Pergamon Press, London, 46–75.

——, 1962, Origin and History of the Moon. In Kopal (1962), pp. 471–525.

——, 1969, The Contending Moons. *Astronautics & Aeronautics,* Jan., 37–41.

——, 1971, Was the Moon Originally Cold? *Science,* **172,** 403–404.

*VALLEY, S. L. (ed.), 1965, *Handbook of Geophysics and Space Environments,* McGraw-Hill, New York, 696 pp.

VAN DORN, W. G., 1969, Lunar Maria: Structure and Evolution. *Science,* **165,** 693.

VAN SCHMUS, W. R. and J. A. Wood, 1967, A Chemical–Petrology Classification for the Chondritic Meteorites. *Geochim. et Cosmochim. Acta,* **31,** 747–765.

VAN TASSEL, R. A. and J. W. Salisbury, The Composition of the Martian Surface. *Icarus,* **3,** 264–269.

WAKITA, H., R. A. Schmitt, and P. Rey, 1970, Elemental Abundances of Major, Minor and Trace elements in Apollo 11 Lunar Rocks, Soil and Core Samples. In Levinson (1970), v. 2, pp. 1685–1719.

WAKITA, H. and R. A. Schmitt, 1970, Lunar Anorthosites: Rare-Earth and Other Elemental Abundances. *Science,* **170,** 969–974.

WALTER, L. S., 1965, Lunar Differentiation Processes. In H. E. Whipple (1965), pp. 470–480.

WARNER, J. L., 1972, Metamorphism of Apollo 14 Breccias, *Proc. Third Lunar Science Conf.,* pp. 623–643.

WASSERBURG, G. J., G. Turner, F. Tera, F. A. Podesek, D. A. Papanastassiou and J. C. Huneke, 1972, Comparison of Rb–Sr, K–Ar, and U–Th–Pb Ages; Lunar Chronology and Evolution. In Watkins (1972), pp. 695–697.

*WATKINS, C., 1972, *Third Lunar Science Conference (revised abstracts),* Lunar Science Institute, Houston, Contr. No. 88, 813 pp.

WEAVER, K. F., 1973, Have We Solved the Mysteries of the Moon? *National Geographic* **144,** no. 3, 309–325.

WELLS, J. W., 1966, Paleontological Evidence of the Rate of the Earth's Rotation. In Marsden and Cameron (1966), pp. 70–81.

WESTON, C. R., 1965, A Strategy for Mars (Search for Life). *Amer. Scientist,* **53,** 495–507.

*WHIPPLE, F. L., 1968, *Earth, Moon, and Planets* (3rd Edition), Harvard University Press, Cambridge, Mass., 278 pp.

*WHIPPLE, H. E. (ed.), 1965, *Geological Problems in Lunar Research, Annals of the New York Academy of Sciences,* v. 123, art. 2, pp. 367–1257.

WHITAKER, E. A., 1966, The Surface of the Moon. In Hess et al. (1966) pp. 79–98.

——, 1972, Lunar Color Boundaries and their Relationship to Topographic Features: A Preliminary Survey. *The Moon,* **4,** 348–355.

*WILFORD, J. N., 1969, *We Reach the Moon: The New York Times Story of Man's Greatest Adventure,* Bantam Books, New York, 331 pp.

*WILHELMS, D. E., 1970, *Summary of Lunar Stratigraphy-Telescopic Observations,* U.S. Geological Survey Professional Paper 599–F, 47 pp.

*WILHELMS, D. E. and J. F. McCauley, 1971, *Geological Map of the Near Side of the Moon,* U.S. Geological Survey, Miscell. Invest. Map I-703.

WILHELMS, D. E. and N. J. Trask, 1965, Polarization Properties of Some Lunar Geologic Units, *In Astrogeol. Studies Ann. Progress Report,* July 1964–July 1965. pt. A., *U.S. Geological Survey Open File Report,* 63–80.

WILSHIRE, H. G., D. E. Stuart-Alexander, and E. D.

Jackson, 1973, Apollo 16 Rocks: Petrology and Classification. *J. Geophys. Res.,* **78**, 2379-2392.

WISE, D. U., 1969, Origin of the Moon from the Earth; Some New Mechanisms and Comparisons. *J. Geophys. Res.,* **74**, 6034-6045.

WISE, D. U. and M. T. Yates, 1970, Mascons as Structural Relief on a Lunar "Moho". *J. Geophys. Res.,* **75**, 261-268.

WOOD, J. A., 1963, Physics and Chemistry of Meteorites. In Middlehurst and Kuiper (1963), pp. 337-401.

——, 1967, The Early Thermal History of Planets: Evidence from Meteorites. In Runcorn (1967), pp. 1-14.

*——, 1968, *Meteorites and the Origin of Planets,* McGraw-Hill, New York, 117 pp.

——, 1970*a,* Petrology of the Lunar Soil and Geophysical Implications. *J. Geophys. Res.,* **75**, 6497-6513.

——, 1970*b,* The Lunar Soil. *Scientific Amer.,* **223**, no. 2, 14-23.

——, 1972*a,* Thermal History and Early Magmatism on the Moon. *Icarus,* **16**, 229-240.

——, 1972*b,* Fragments of Terra Rock in the Apollo 12 Soil Samples and a Structural Model of the Moon. *Icarus,* **16**, 462-501.

——, 1973, Asymmetry of the Moon. In Chamberlain and Watkins (1973), 790-792.

WOOD, J. A., J. S. Dickey, Jr., U. B. Marvin, and B. N. Powell, 1970, Lunar Anorthosites and a Geophysical Model of the Moon. In Levinson (1970), v. 1, 965-988.

WOOLUM, D. S., D. S. Burnett, and C. A. Bauman, 1973, The Apollo 17 Lunar Neutron Probe Experiment. In Chamberlain and Watkins (1973), 793-795.

ZELLER, E. J. and L. B. Ronca, 1967, Space Weathering of Lunar and Asteroidal Surfaces. *Icarus,* **7**, 372-379.

Sources of Materials

Many of the illustrations appearing in this book are in the public domain and are available on request or by purchase from both government and private sources.

Most of the geologic maps of the Moon are available in color at scales from 1:100,000 to 1:1,000,000 by ordering from:

Distribution Section
U.S. Geological Survey

1200 South Eads St.	or	Federal Center
Arlington, Va. 22202		Denver, Colo. 80225

In addition, maps and charts are sold over the counter at most of the larger offices of the Geological Survey. Prices are quoted on request; a typical cost is $1.00 per map. The maps are identified by an I number (e.g., I-491) and a descriptive title (e.g., Geologic Map of the Hevelius Region of the Moon). A second identifier (not given here) is the LAC number (e.g., LAC-56).

Special-purpose lunar maps (e.g., Albedo Map) are also available from the U.S. Geological Survey. Also, maps of Mars (entire planet and selected quadrangles) are now largely completed using Mariner 9 and other data, and those of Mercury are in preparation; many will be released by the Geological Survey by the time this book is published.

Based largely on telescope observations, a series of 1:1,000,000 lunar charts (identified by LAC number) on which major craters and other features are sketched on Mercator or Lambert Conformal projections have been prepared by the U.S. Air Force's Aeronautical Chart and Information Center (ACIC), St. Louis, Mo. These serve as base maps for quadrangles in the geological series produced by the U.S. Geological Survey. Other similar charts at 1:500,000 (AIC series) result from subdividing the LAC charts into 4 smaller maps. Most of these maps now also have elevation data, and many are contoured, usually at a 300-meter interval. The charts can be obtained from:

Superintendent of Documents
U.S. Government Printing Office
Washington, D.C. 20402

along with an index map of the entire LAC and AIC series.

The ACIC has also produced maps of the entire near

and far sides and polar regions of the Moon using different projections. The National Geographic Society of Washington, D.C., sells an outstanding version of full moon maps of both front and far sides. Rand McNally and Co. and other suppliers market lunar globes.

The photographs and other supporting data from the Ranger, Surveyor, Lunar Orbiter, Apollo, and Mariner Mars missions can be obtained by qualified individuals (e.g., scientists, advanced students) and organizations from:

National Space Science Data Center Code 601.4, NASA Goddard Space Flight Center Greenbelt, Maryland 20771, USA	or	World Data Center A, Rockets and Satellites, Code 601 Goddard Space Flight Center Greenbelt, Maryland 20771, USA
	for those:	
within the U.S.		outside the U.S.

It is ultimately necessary to give specific identification numbers in requesting any item from these missions. In making inquiry about availability of such materials, one should ask for the Data Users Note and Index Maps for the mission(s) required. An order blank and price list is included.

Most NASA scientific publications dealing with planetary missions and studies are sold through the Government Printing Office. Science reports and program results for Surveyor, Orbiter, Apollo, and Mariner are included in the NASA SP series. Additional information about these publications can be requested from:

Scientific & Technical Information Office
National Aeronautics & Space Administration (Hq)
Washington, D.C. 20546

Some general materials concerning the exploration of space are commonly available from the Education Office of NASA Headquarters or the major NASA centers across the country.

Space photographs, maps, etc., are also sold commercially. Advertisments listing such materials and their suppliers appear frequently in the magazine *Sky & Telescope.* This magazine is also an excellent source of up-to-date semitechnical reviews of major achievements in the exploration of space. The weekly periodical *Aviation Week and Space Technology* includes new results of lunar and planetary missions usually within one to two weeks after their occurrence or release. Other magazines devoted largely to reporting scientific studies of interest to students of planetary geology include *Icarus, The Moon, Earth and Planetary Letters, Modern Geology,* and *Journal of Geophysical Research.* Many issues of *Nature* and *Science,* published weekly, contain one or more articles related to the exploration of space.

Name Index

Individuals cited as references in the captions are not included.

Subject Index

All entries referring directly or primarily to the Moon (or Lunar) are set in roman type; all other entries are set in italic type.

NATIONAL AERONAUTICS AND SPACE ADMINISTRATION

LUNAR CHART

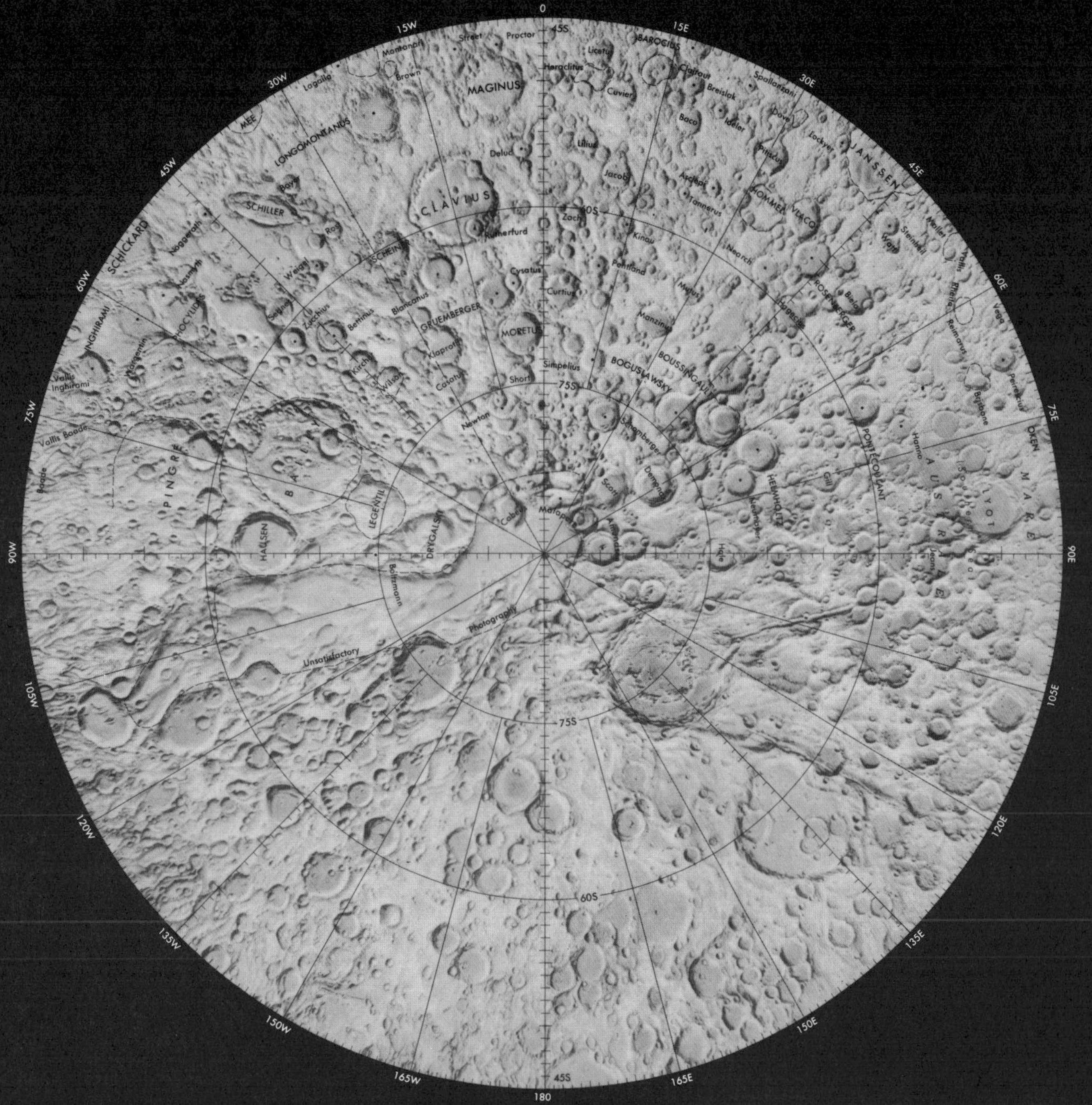

SOUTH POLAR REGION

NOTES

The Lunar Surface Features shown on this chart are interpreted from the photographic records of Lunar Orbiter Missions I, II, III, IV, and V. Ray patterns and Albedo differences on the earthside limb and farside hemisphere are incomplete due to limitations of the source photographs. Horizontal position of the features is based on the ACIC Positional Reference System, 1969. Feature names are adopted from the 1935 International Astronomical Union Nomenclature System, as amended by I.A.U. in 1961 and 1964. The application of names is restricted to only those associated features that have been positively identified.

LUNAR DATA

Earth/Moon Mass Ratio	M_e/M_m 81.3015
Density (mean)	3.34 g/(cm)^3
Synodic Month (new Moon to new Moon)	29.530 588d
Sidereal Month (fixed star to fixed star)	27.321 661d
Inclination of Lunar orbit to ecliptic	5°8′43″
Inclination of equator to ecliptic	1°32′40″
Inclination of Lunar orbit to Earth's equator	18°.5 to 28°.5
Distance of Moon from Earth (mean)	238,328M (384,400km)
Optical libration in longitude	±7°.6
Optical libration in latitude	±6°.7
Albedo (average)	0.07
Magnitude (mean of full Moon)	−12.7
Temperature	−244°F to +273°F (120°K to 407°K)
Escape velocity	1.48 mi/sec (2.38km/sec)
Diameter of Moon	2160 mi (3476km)
Surface gravity	162.2 cm/sec^2
Orbital velocity	0.64 mi/sec (Moon) 18.5 mi/sec (Earth)

NATIONAL AERONAUTICS AND SPACE ADMINISTRATION

LUNAR FARSIDE CHART

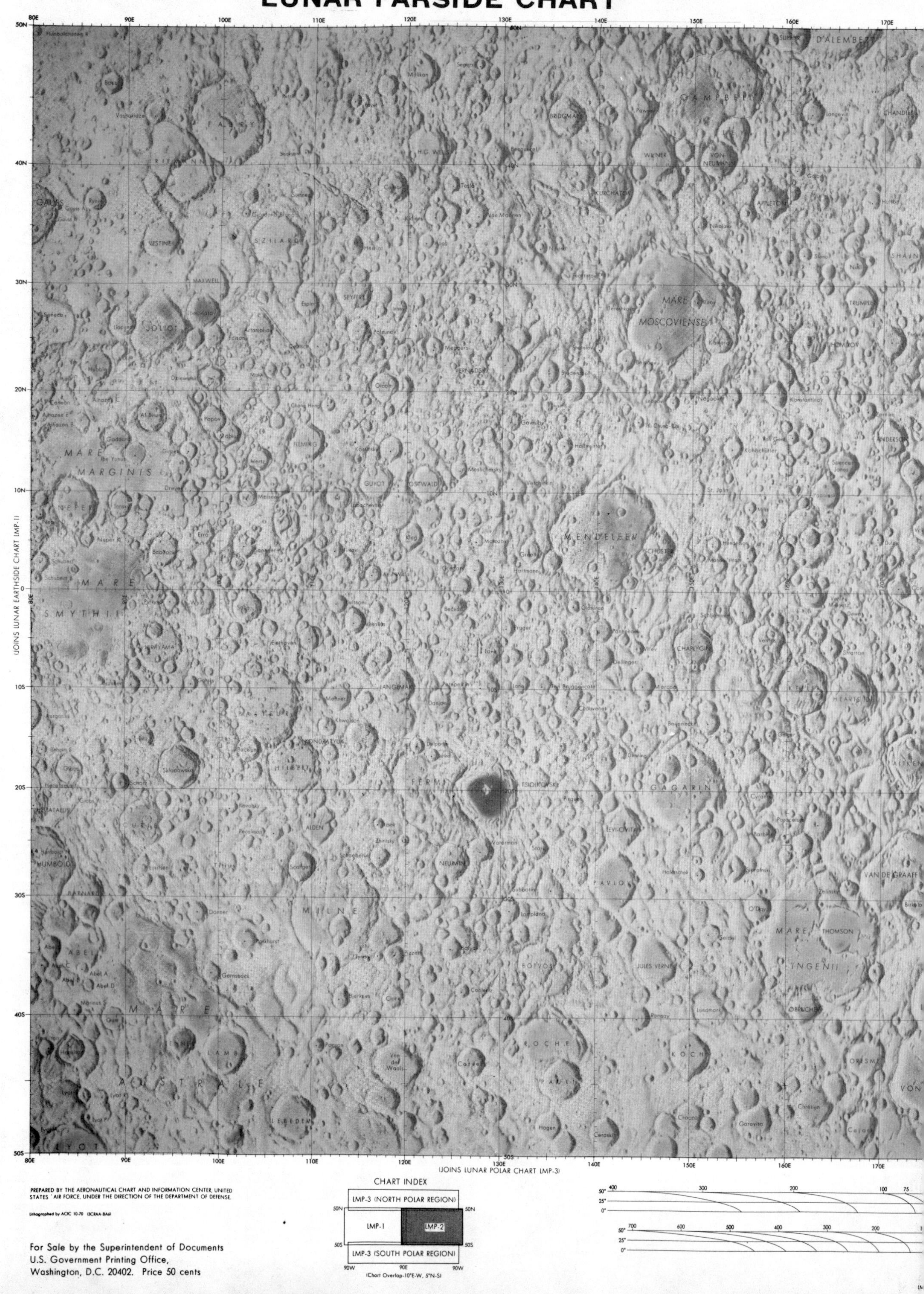

PREPARED BY THE AERONAUTICAL CHART AND INFORMATION CENTER, UNITED STATES AIR FORCE, UNDER THE DIRECTION OF THE DEPARTMENT OF DEFENSE.

Lithographed by ACIC 10-70 (3CBAA-BA6)

For Sale by the Superintendent of Documents
U.S. Government Printing Office,
Washington, D.C. 20402. Price 50 cents